Decarbonization Technology

ICDT2024

International Conference on Decarbonization Technology 2024 (ICDT2024), Sabah International Convention Centre, Sabah, Malaysia, 10-11 September 2024

Editors

Nik Abdul Hadi Sapiaa[1], Lam Man Kee[2], Khairulazhar Jumbri[1], Bamidele Victor Ayodele[3], Syaza Izyanni Ahmad[4]

[1] Centre of Research in Ionic Liquid, Institute of Sustainable Energy & Resources, Chemical Engineering Department, Universiti Teknologi PETRONAS

[2] Institute of Sustainable Energy & Resources, Chemical Engineering Department, Universiti Teknologi PETRONAS

[3] Centre of Carbon Capture, Utilisation, & Storage, Institute of Sustainable Energy & Resources, Chemical Engineering Department, Universiti Teknologi PETRONAS

[4] Centre of Advanced Process Safety, Institute of Smart & Sustainable Living, Chemical Engineering Department, Universiti Teknologi PETRONAS

Peer review statement

All papers published in this volume of "Materials Research Proceedings" have been peer reviewed. The process of peer review was initiated and overseen by the above proceedings editors. All reviews were conducted by expert referees in accordance to Materials Research Forum LLC high standards.

Published under License by **Materials Research Forum LLC**
Millersville, PA 17551, USA

Published as part of the proceedings series
Materials Research Proceedings
Volume 53 (2025)

ISSN 2474-3941 (Print)
ISSN 2474-395X (Online)

ISBN 978-1-64490-356-8 (Print)
ISBN 978-1-64490-357-5 (eBook)

Distributed worldwide by

Materials Research Forum LLC
105 Springdale Lane
Millersville, PA 17551
USA
https://mrforum.com

Manufactured in the United State of America
10 9 8 7 6 5 4 3 2 1

Table of Contents

Preface

The International Conference on Decarbonization Technology (ICDT2024) was successfully held physically in Sabah, Malaysia from 10 – 11 September 2024. This marked the 8th conference series organized under the World Engineering, Science and Technology Congress (ESTCON) umbrella by Universiti Teknologi PETRONAS (UTP) since 2010. ICDT2024 served as a vital global platform where academics, scientists, and industry leaders gathered to share cutting-edge knowledge, showcase innovative research, and inspire advancements in decarbonization technology. The conference highlighted its crucial role in promoting a sustainable future by covering a wide range of topics, including Hydrogen, Solar and Thermal Energy, Biomass and Biofuel, Carbon Capture and Utilization, Green Processes and Materials, and Carbon Offsets and Accounting. All accepted papers underwent a rigorous peer-review process conducted by experts to ensure alignment with the highest international publication standards.

Over 100 papers were submitted, and 65 were selected for inclusion in the conference proceedings, representing a collection of transformative insights and innovative outcomes. The technical sessions featured numerous presenters from both national and international backgrounds, leading to engaging discussions and stimulating debates among participants. Feedback was overwhelmingly positive, with attendees appreciating the valuable insights shared throughout the conference.

The organizing committee extends its deepest gratitude to all individuals and institutions whose invaluable support contributed to the success of ICDT2024. The committee hopes that ICDT2024 will inspire collaborative research, ignite new ideas, and drive sustainable future through decarbonization. Ultimately, the proceedings from ICDT2024 will serve as an essential resource for the scientific and engineering communities to discover groundbreaking findings and develop innovative decarbonization technologies that drive sustainable progress.

Organizing Committee of ICDT2024

ORGANISING COMMITTEE

Chairman
Assoc. Prof. Ir Dr Abdul Halim Shah Maulud

Co Chairman
ChM Assoc. Prof. Dr Maizatul Shima Shaharun

Secretary
Ts Dr Khairiraihanna Johari
Ts Dr Nooraini Zainuddin

Promotion/Marketing/Website Master
Ts Dr Mohd Hizami Mohd Yusoff
Dr Nabila Farhana Jamaludin
Ts Dr Muhammad Fadhlullah Abd Shukur
Dr Madiah Omar

Technical
Ts Dr Nik Abdul Hadi Sapiaa @ Nordin
Assoc. Prof. Ts Dr Lam Man Kee
ChM Assoc. Prof. Ts Dr Khairulazhar Jumbri
Ts Dr Bamidele Victor Ayodele
Dr Syaza Izyanni Ahmad

Publication
ChM Ts Dr Lim Jun Wei
Dr Nurul Aini Amran

Industrial Track
Assoc. Prof. Ir Dr Haslinda Zabiri

Preface

The International Conference on Decarbonization Technology (ICDT2024) was successfully held physically in Sabah, Malaysia from 10 – 11 September 2024. This marked the 8th conference series organized under the World Engineering, Science and Technology Congress (ESTCON) umbrella by Universiti Teknologi PETRONAS (UTP) since 2010. ICDT2024 served as a vital global platform where academics, scientists, and industry leaders gathered to share cutting-edge knowledge, showcase innovative research, and inspire advancements in decarbonization technology. The conference highlighted its crucial role in promoting a sustainable future by covering a wide range of topics, including Hydrogen, Solar and Thermal Energy, Biomass and Biofuel, Carbon Capture and Utilization, Green Processes and Materials, and Carbon Offsets and Accounting. All accepted papers underwent a rigorous peer-review process conducted by experts to ensure alignment with the highest international publication standards.

Over 100 papers were submitted, and 65 were selected for inclusion in the conference proceedings, representing a collection of transformative insights and innovative outcomes. The technical sessions featured numerous presenters from both national and international backgrounds, leading to engaging discussions and stimulating debates among participants. Feedback was overwhelmingly positive, with attendees appreciating the valuable insights shared throughout the conference.

The organizing committee extends its deepest gratitude to all individuals and institutions whose invaluable support contributed to the success of ICDT2024. The committee hopes that ICDT2024 will inspire collaborative research, ignite new ideas, and drive sustainable future through decarbonization. Ultimately, the proceedings from ICDT2024 will serve as an essential resource for the scientific and engineering communities to discover groundbreaking findings and develop innovative decarbonization technologies that drive sustainable progress.

Organizing Committee of ICDT2024

ORGANISING COMMITTEE

Chairman
Assoc. Prof. Ir Dr Abdul Halim Shah Maulud

Co Chairman
ChM Assoc. Prof. Dr Maizatul Shima Shaharun

Secretary
Ts Dr Khairiraihanna Johari
Ts Dr Nooraini Zainuddin

Promotion/Marketing/Website Master
Ts Dr Mohd Hizami Mohd Yusoff
Dr Nabila Farhana Jamaludin
Ts Dr Muhammad Fadhlullah Abd Shukur
Dr Madiah Omar

Technical
Ts Dr Nik Abdul Hadi Sapiaa @ Nordin
Assoc. Prof. Ts Dr Lam Man Kee
ChM Assoc. Prof. Ts Dr Khairulazhar Jumbri
Ts Dr Bamidele Victor Ayodele
Dr Syaza Izyanni Ahmad

Publication
ChM Ts Dr Lim Jun Wei
Dr Nurul Aini Amran

Industrial Track
Assoc. Prof. Ir Dr Haslinda Zabiri

Monocationic and dicationic phosphonium based ionic liquids: Synthesis and properties

AMINUDDIN Muhammad Syahir[1,3], BUSTAM Mohamad Azmi[1,3] and JOHARI Khairiraihanna[2,3*]

[1]Centre of Research in Ionic Liquids (CORIL),Universiti Teknologi PETRONAS, 32610 Bandar Seri Iskandar, Perak, Malaysia

[2]Centre for Carbon Capture, Utilization and Storage (CCCUS), Universiti Teknologi PETRONAS, 32610 Bandar Seri Iskandar, Perak, Malaysia

[3]Chemical Engineering Department, Universiti Teknologi PETRONAS, 32610 Bandar Seri Iskandar, Perak, Malaysia

[a]mr.m.syahir.aminuddin@gmail.com, [b]azmibustam@utp.edu.my, [c]khairiraihana.j@utp.edu.my

Keywords: Monocationic Ionic Liquids, Dicationic Ionic Liquids, Refractive Index, Thermal Behavior, Derived Properties

Abstract. In the current work, two monocationic ionic liquids (MILs) and two dicationic ionic liquids (DILs) with similar trihexyl(tetradecyl)phosphonium cation, $[P_{66614}]^+$ and different metal chloride anions such as trichlorostannate anion, $[SnCl_3]^-$, pentachlorozirconate anion, $[ZrCl_5]^-$, tetrachloronickelate(II) anion, $[NiCl_4]^{2-}$ and tetrachlorocuprate(II) anion, $[CuCl_4]^{2-}$ were successfully synthesized and characterized, respectively. The structures of ILs were confirmed using FT-IR spectroscopy and CHNS elemental analysis. The effect of temperature and anion type on thermophysical properties such as viscosity, density and refractive index were studied. Viscosity and density were measured within temperature range of 25 °C to 100 °C. Moreover, thermal expansion coefficient, molecular volume, molar volume and free volume were derived from viscosity and density values. Refractive index values were measured in the temperature range of 10 °C to 60 °C and were used further to estimate the electronic polarizability using Lorenz-Lorentz equation. Thermal properties of the synthesized MILs and DILs were determined using thermogravimetric analyzer (TGA) and differential scanning calorimetry (DSC).

Introduction

Ionic liquids (ILs) are well-known as green solvents due to numerous reasons such as highly stable thermally and chemically, low vapor pressure, non-flammable, wide temperature range and diverse structural potentials making them a viable alternatives in various reactions [1]. ILs are used in chemical synthesis, catalysis, and extraction processes as eco-friendly alternatives to traditional organic solvents which can contribute to environmental pollution. Moreover, ILs can be tailored in limitless ways by modifying their anions and cations to achieve specific functionalities [2]. Due to their ease of tunability, ILs are widely harnessed in different applications such as separation processes, pharmaceutical and biochemical synthesis, biomass pretreatment and engineering materials [3, 4].

Dicationic ILs are a specific type of ILs composed of molecules carrying two positively charged sites, or cations, within their structure. Unlike traditional ILs consisting of a single cation, dicationic ILs exhibit distinctive properties due to their dual charged nature. Dicationic ILs are more tunable as compared to the monocationic ILs due to large number of combination possibilities between the cations, anions and linker chains. The physical and chemical properties of dicationic ILs, such as density, viscosity, melting points and thermal behavior can be finely adjusted to a greater degree than those of conventional monocationic ILs by altering the cations, anions, and linker chain lengths

[5, 6]. Jadhav and Kim [7] have prepared dicationic ILs consisting of dual imidazolium cations with paratoluene sulfonate and methanesulfonate anions with ethylene glycol linker chains, which were found to be effective for azidation of alkyl halides. Similar ILs was also reported to achieve 92.3% yield in 40 min at 70°C for catalytic conversion of fructose to 5-hydroxymethylfurfural applications [8].

Over the years, more and more exclusively tailored ILs being designed especially dicationic ILs and are gaining worldwide recognition as green solvents for various applications including surfactants, enzymatic treatments, organic reactions and chemical analysis. Furthermore, dicationic ILs are also reported to dissolve carbohydrates and cellulose to a higher extent. In addition, tris-imidazolium ILs containing hexadecyl chains and mesitylene linkers were studied for the stabilization of palladium nanoparticles [9]. Besides, two different classes of trigonal tricationic ILs with 1,3,5-trimethylbenzene and triethylamine cations were studied as surfactants in aqueous solution [10]. The design of industrial processes and the production yield of products based on ILs rely significantly on their thermophysical properties, such as refractive index, viscosity, density and thermal behavior. Hence, precise and methodical measurements of these thermophysical properties in ILs are crucial to identify the most suitable ILs for specific applications.

To investigate the impact of anions, number of cations and temperature on diverse thermophysical properties of ILs, a range of monocationic and dicationic ILs were synthesized. These ILs featured the trihexyl(tetradecyl)phosphonium cation paired with four anions derived from different metal chlorides such as tin(II) chloride, zirconium(IV) chloride, nickel(II) chloride, and copper(II) chloride, using the documented method. [11]. To understand the properties of the synthesized monocationic and dicationic ILs, their physicochemical properties such as refractive index, density and viscosity were investigated as a function of temperature. The density experimental data served as a basis for deriving additional properties, including the thermal expansion coefficient, standard molar volume, and molecular volume of the ILs. Likewise, the refractive index values were utilized in determining the electronic polarizability of the synthesized ILs. Thermogravimetric analysis (TGA) and differential scanning calorimetry (DSC) were employed to examine the thermal characteristics of these ILs.

Experimental procedures

Materials. All chemicals utilized in this research were of analytical grade and employed without additional purification. Trihexyl(tetradecyl)phosphonium chloride (95%), tin(II) chloride (98%), zirconium(IV) chloride (99.99%), nickel(II) chloride (98%) and copper(II) chloride (97%) were supplied by Sigma Aldrich.

Synthesis of Ionic Liquids. Two monocationic ionic liquids (MILs) and two dicationic ionic liquids (DILs) composed of trihexyl(tetradecyl)phosphonium, $[P_{66614}]^+$ cation and four different metal chloride anions such as $[SnCl_3]^-$, $[ZrCl_5]^-$, $[NiCl_4]^{2-}$ and $[CuCl_4]^{2-}$. MILs that were synthesized in this work involves $[P_{66614}][SnCl_3]$ and $[P_{66614}][ZrCl_5]$, whereas the synthesized DILs consist of $[P_{66614}]_2[NiCl_4]$ and $[P_{66614}]_2[CuCl_4]$. MILs were synthesized by metathesis reaction between a cation and an anion, whereas DILs were synthesized simlarly between two cations and single anion using the reported approach [11].

Initially, trihexyl(tetradecyl)phosphonium chloride (0.1 mol) and tin(II) chloride (0.1 mol) were charged into round bottom flask and stirred for 24 h at 80 °C under N_2 atmosphere. The synthesized MILs, trihexyl(tetradecyl)phosphonium trichlorostannate, $[P_{66614}][SnCl_3]$ was washed with water and ethyl acetate to remove unreacted materials and dried in a vacuum oven at 70 °C for 24 h. The synthesis scheme for another MILs, trihexyl(tetradecyl)phosphonium pentachlorozirconate, $[P_{66614}][ZrCl_5]$ also followed similar procedure.

To synthesize DILs such as trihexyl(tetradecyl)phosphonium tetrachloronickelate(II), $[P_{66614}]_2[NiCl_4]$ and trihexyl(tetradecyl)phosphonium tetrachlorocuprate(II), $[P_{66614}]_2[CuCl_4]$, 0.05 mol of trihexyl(tetradecyl)phosphonium chloride was reacted with 0.1 mol of the respective metal chlorides in the round bottom flask and stirred for 24 h at 80 °C. The products were then washed with water and ethyl acetate following the procedures mentioned earlier. The formal names, abbreviations, molecular weights and structures of cation and anions used for the synthesis of both MILs and DILs are shown in Table 1.

Table 1. *List of cation and anions used for synthesis.*

Formal name	*Abbreviation*	*MW (g/mol)*	*Structure*
Cation			
Trihexyl(tetradecyl) phosphonium	$[P_{66614}]^+$	483.86	
Anions			
Trichlorostannate	$[SnCl_3]^-$	225.07	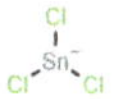
Pentachlorozirconate	$[ZrCl_5]^-$	268.49	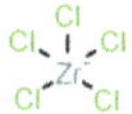
Tetrachloronickelate	$[NiCl_4]^{2-}$	200.51	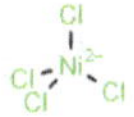
Tetrachlorocuprate	$[CuCl_4]^{2-}$	205.36	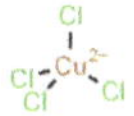

Characterization of Ionic Liquids. CHNS Analysis. Physicochemical properties measurement The elemental composition of the synthesized ILs was verified using CHNS-932 (LECO instruments) for elemental analysis. CHNS analysis (carbon, hydrogen, nitrogen, sulfur analysis) is an established method widely used by researchers for the characterization of ILs. This critical and reliable analytical technique helps to determine the elemental composition of ILs, ensuring their accuracy and successful formation by comparing synthesized and experimental values. CHNS analysis is frequently employed in the characterization of ILs due to its reliability and precision. The use of CHNS analysis in determining the elemental composition of ILs is well-documented in various research studies [12-14]. This method allows for the accurate quantification of the primary elements within the synthesized compounds, providing a basis for assessing their purity and composition [15]. The effectiveness of CHNS analysis is evident when comparing calculated values of synthesized ILs to their experimental counterparts. Minimal deviation between these values is a strong indication of the successful synthesis and accurate formation of ILs. This is crucial in validating the synthesis process, ensuring that the intended chemical structures are achieved without significant impurities [16-18].

Physicochemical Properties Measurement. The thermophysical properties such as density, viscosity, refractive index and thermal behavior of the synthesized monocationic and dicationic ILs are further elaborated in the upcoming sections. The refractive index values of the synthesized ILs were examined using an automated ATAGO digital refractometer (RX-5000α) at temperature intervals from 10 °C to 60 °C, with 5 °C increments. The viscosity measurement of the synthesized ILs was conducted using a CAP 2000+ Cone & Plate viscometer (AMETEK Brookfield) in the temperature ranging from 25 °C to 100 °C, with 25 °C intervals. The density was measured by a DA-645 Kyoto density meter from 20°C to 90°C at 10°C interval each. All instruments were callibrated before measurements were taken. All measurements were conducted three times, and the recorded values represent their respective averages.

TGA Analysis. The thermal decomposition temperature of the synthesized monocationic and dicationic ILs was analyzed using a Thermogravimetric Analyzer (Pyris V-3.81, Perkin Elmer). The thermal behavior of these ILs was measured in the temperature range from 50°C to 700°C at a heating rate of 10 °C/min under a nitrogen-rich environment.

DSC Analysis. DSC analysis is a is an advanced thermal analysis technique that drives understanding of material properties. The glass transition temperature (T_g) and melting point temperature (T_m) of the ILs were assessed utilizing a differential scanning calorimeter (Pyris v1, Perkin Elmer). The samples underwent heating in aluminium pans ranging from -80 °C to 200 °C, at a rate of 10 °C/min under a nitrogen-rich atmosphere. (1)

Results and discussion

CHNS Analysis.

Trihexyl(tetradecyl)phosphonium trichlorostannate, $[P_{66614}][SnCl_3]$.

Molecular weight: 708.91 g/mol

Calculated (%): C: 54.22, H: 9.67, N: 0, S: 0.

Found (%): C: 54.19, H: 9.71, N: 0, S: 0.

Trihexyl(tetradecyl)phosphonium pentachlorozirconate, $[P_{66614}][ZrCl_5]$

Molecular weight: 752.35 g/mol

Calculated (%): C: 51.09, H: 9.11, N: 0, S: 0.

Found (%): C: 51.12, H: 9.08, N: 0, S: 0.

Trihexyl(tetradecyl)phosphonium tetrachloronickelate(II), $[P_{66614}]_2[NiCl_4]$

Molecular weight: 1168.22 g/mol

Calculated (%): C: 65.80, H: 11.73, N: 0, S: 0.

Found (%): C: 65.82, H: 11.71, N: 0, S: 0.

Trihexyl(tetradecyl)phosphonium tetrachlorocuprate(II), $[P_{66614}]_2[CuCl_4]$

Molecular weight: 1173.07 g/mol

Calculated (%): C: 65.53, H: 11.69, N: 0, S: 0.

Found (%): C: 65.51, H: 11.71, N: 0, S: 0.

Physical properties correlation of ILs. The density, viscosity, and refractive index values of the synthesized MILs and DILs were modeled using the Least Squares Method (LSM) through the linear expressions.

$$n_D = A_0 + A_1T \quad (1)$$

$$\rho = A_2 + A_3T \quad (2)$$

where T represents absolute temperature (K) and A_0, A_1, A_2 and A_3, as adjustable parameters. The standard deviation (σ) of the physical properties were computed using the following expression:

$$\sigma = \sqrt{\frac{\sum_i^N (z_{exp} - z_{calc})^2}{N}} \tag{3}$$

The values obtained for estimated fitting parameters of refractive index, A_0 and A_1 using Equation 1 is tabulated in Table 2, whereas the density fitting parameters, A_2 and A_3 were calculated using Equation 2 is tabulated in Table 3.

Table 2. The estimated fitting parameters, A_0 and A_1

ILs	σ (×10^{-3})	R^2	A_0	A_1
$[P_{66614}][SnCl_3]$	5.340	0.999	1.596	-3.377×10^{-4}
$[P_{66614}][ZrCl_5]$	6.103	0.999	1.623	-3.860×10^{-4}
$[P_{66614}]_2[NiCl_4]$	5.313	0.999	1.577	-3.360×10^{-4}
$[P_{66614}]_2[CuCl_4]$	5.112	0.999	1.588	-3.233×10^{-4}

Table 3. The estimated fitting parameters, A_2 and A_3

ILs	σ (×10^{-2})	R^2	A_2	A_3
$[P_{66614}][SnCl_3]$	1.063	0.999	1.225	-6.723×10^{-4}
$[P_{66614}][ZrCl_5]$	0.999	0.999	1.306	-6.316×10^{-4}
$[P_{66614}]_2[NiCl_4]$	1.068	0.999	1.130	-6.756×10^{-4}
$[P_{66614}]_2[CuCl_4]$	0.996	0.999	1.159	-6.302×10^{-4}

Refractive Index (*RI*) represents the ratio of the speed of light in a vacuum to its speed within the measured medium. *RI* is an important physical property which can be used to determine the purity, solute concentration and electronic polarizability (R_m). The refractive indices of the synthesized ILs at different temperatures were displayed in Figure 1. The findings suggested a linear decrease in the refractive indices of ILs with rising temperatures, consistent with previous reports. The *RI* values of the synthesized ILs were found to be in the decreasing order of $[P_{66614}][ZrCl_5]$ > $[P_{66614}][SnCl_3]$ > $[P_{66614}]_2[CuCl_4]$ > $[P_{66614}]_2[NiCl_4]$.

At 25 °C, the ILs with the highest density, $[P_{66614}][ZrCl_5]$ demonstrated the highest refractive index value of 1.5083, whereas the ILs with the lowest density, $[P_{66614}]_2[NiCl_4]$ displayed the lowest refractive index value of 1.4765. Due to limited availability of literature on thermophysical properties of monocationic and dicationic ILs, comparisons were made among the synthesized ILs.

It was observed that refractive indices of dicationic ILs are slightly higher as compared to the monocationic ILs. At 60 °C, the *RI* values for MILs such as $[P_{66614}][ZrCl_5]$ and $[P_{66614}][SnCl_3]$ are 1.4948 and 1.4839, respectively which are higher than the recorded values for DILs such as $[P_{66614}]_2[NiCl_4]$ and $[P_{66614}]_2[CuCl_4]$ at 1.4647 and 1.4801, respectively. This probably occurs due to the presence of two charged moieties in dicationic ILs, which leads to a larger molecular volume as compared to monocationic ILs with only single charged moiety.

Materials Research Forum LLC
https://doi.org/10.21741/9781644903575-1

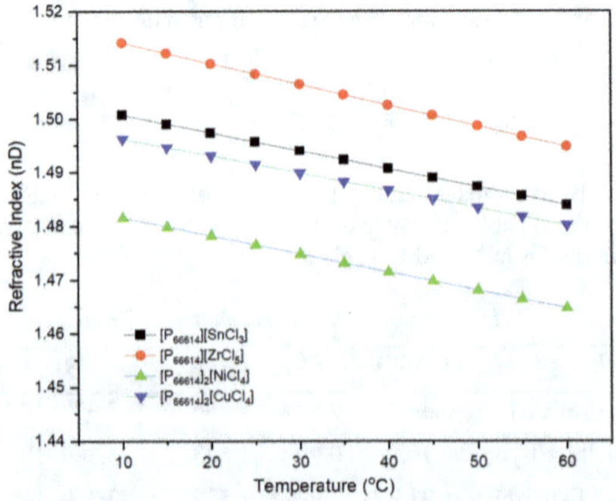

Fig. 1. Refractive index of the synthesized MILs and DILs

The increase of molecular volume in dicationic ILs results in lower packing efficiency, leading to lower density. In addition, the presence of two cations in dicationic ILs might affect the intermolecular interactions. The repulsion among the charged sites can cause a larger intermolecular distance, reducing the overall density and refractive index as compared to monocationic ILs where such repulsion might not be as prominent. The obtained values of fitting parameters A_0 and A_1, along with the standard deviation of *RI* values for the synthesized ILs using Equation 1 are summarized in Table 2 of the previous section.

Density analysis. Based on the overall results obtained, the density of metal chloride functionalized ILs varied from 0.883 g/cm^3 to 1.120 g/cm^3. These results agreed with Valderrama and Zarricueta [19] who predicted accurately that the density of ILs are within range of 0.9 g/cm^3 to 1.7g/cm^3. The density of the synthesized ILs from 20 °C to 90 °C, with a 10 °C interval was displayed in Figure 2. The presence of long and bulky alkyl chain of phosphonium cation in the synthesized ILs decreases the density of ILs and is observable systematically. As the length and size of alkyl chain in the cation of ILs increases, the density of ILs also decreases. These results agreed with the recent studies conducted by several researchers [20-23]. The interactions between long alkyl chains of phosphonium cations in ILs allow the ions to order among themselves and slide each other easily, thus making the ILs less dense.

The density of metal chloride anion-based ILs at 90 °C were found to be in the increasing order of $[P_{66614}]_2[NiCl_4]$ > $[P_{66614}]_2[MnCl_4]$ > $[P_{66614}]_2[CuCl_4]$ > $[P_{66614}]_2[ZnCl_4]$ > $[P_{66614}][AlCl_4]$ > $[P_{66614}][SnCl_3]$ > $[P_{66614}][GaCl_4]$ > $[P_{66614}][ZrCl_5]$, respectively as shown in Table 4.9. Furthermore, it was observed that the density of dicationic ILs were much lower as compared to the monocationic ILs. At 90 °C, the density values of $[P_{66614}][SnCl_3]$ and $[P_{66614}][ZrCl_5]$ are 0.981 g/cm^3 and 1.075 g/cm^3, respectively. These values are higher than those of $[P_{66614}]_2[NiCl_4]$ and $[P_{66614}]_2[CuCl_4]$ at 0.883 g/cm^3 and 0.930 g/cm^3, respectively. The presence of dual cations in the matrix of dicationic ILs often leads to a larger molecular volume and stronger electrostatic repulsion between molecules as compared to monocationic ILs with only single cation. The elevated molecular volume and electrostatic repulsion resulted in lower packing efficiency, thus leading to lower density.

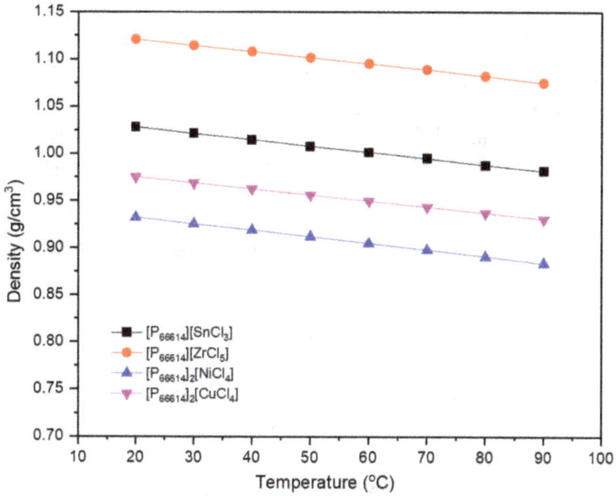

Fig. 2. Density (ρ) as a function of temperature (T) for the synthesized ILs

Table 4. Density of MILs and DILs at 90 °C

ILs	Density (g/cm^3)
$[P_{66614}][SnCl_3]$	0.981
$[P_{66614}][ZrCl_5]$	1.075
$[P_{66614}]_2[NiCl_4]$	0.883
$[P_{66614}]_2[CuCl_4]$	0.930

The decrease in density as temperature rises is a well-established phenomenon. This occurs because, at elevated temperatures, the molecules within ILs gain higher kinetic energy, allowing them to move more freely and slide past each other more readily. Consequently, this enhanced mobility among molecules leads to a reduction in the overall density of the ILs. Similar observation was recorded by Valderrama and Zarricueta [19] in their study on the synthesis and properties of several ILs.

Thermal expansion coefficient (α). Density values obtained were used further to compute the thermal expansion coefficient value (α) via following expression:

$$\alpha = -\frac{1}{\rho}\left(\frac{\partial \rho}{\partial T}\right) = -\frac{A_3}{A_2 + A_3 T} \tag{4}$$

where ρ is density of ILs, T is temperature and A_2 and A_3 as density fitting parameters calculated from Equation 2. The thermal expansion coefficients values of both MILs and DILs are presented in Table 5. Deviations of the α values with temperatures are significant among the synthesized MILs and DILs. The thermal expansion coefficients values vary in the range of 5.635×10^{-4}/K and 7.466×10^{-4}/K which are much lower as compared to the reported values of conventional organic solvents [24, 25]. Among these ILs, $[P_{66614}]_2[NiCl_4]$ has the highest α value at 7.249×10^{-4}/K while

$[P_{66614}][ZrCl_5]$ recorded the lowest value at 5.635×10^{-4}/K. The values measured were within range of the reported dicationic ILs [26].The thermal expansion coefficient values of the synthesized MILs and DILs are relatively lower than the organic solvents (1×10^{-3}/K) but far higher than the reported conventional ILs ($1\text{-}2\times10^{-4}$/K) [24, 25]. It was observed that the α values increases with the increase in temperature. The slight increase in α values with the rise of temperatures can be attributed to the reduced ordering of ILs at elevated temperatures [25].

Standard molar volume (V_m). Standard molar volume (V_m) is the volume taken up by one mole of a substance under standard temperature and pressure (STP) conditions. The standard molar volume is typically measured in units of cubic meters per mole (m^3/mol). V_m of ILs at different temperatures and atmospheric pressure can be computed using the molecular mass (M) by the following expression:

$$V_m = M/\rho \tag{5}$$

where V_m is molar volume (cm^3/mol), M is molecular weight (g/mol) and ρ is density (g/cm^3). The calculated V_m values of the synthesized ILs within temperature range of 293.15 to 333.15 K are displayed in Table 3. At 293.15 K, the molar volumes values of $[P_{66614}][SnCl_3]$, $[P_{66614}][ZrCl_5]$, $[P_{66614}]_2[NiCl_4]$ and $[P_{66614}]_2[CuCl_4]$ are 689.51, 671.48, 1253.72 and 1203.62 cm^3/mol, respectively. It was observed that the standard molar volumes of DILs are relatively higher than those of MILs due to the presence of dual cations in the matrix of DILs [27].

Molecular volume (V). The molecular volumes (V) of ILs were calculated from V_m and Avogadro's constant (N_A) using the following expression:

$$V = \frac{V_m}{N_A} \tag{6}$$

where N_A is 6.02214076×10^{23}/mol. The calculated molecular volumes of the synthesized ILs at various temperatures are tabulated in Table 5. The molar volumes values of $[P_{66614}][SnCl_3]$, $[P_{66614}][ZrCl_5]$, $[P_{66614}]_2[NiCl_4]$ and $[P_{66614}]_2[CuCl_4]$ are 1.145×10^{-21}, 1.115×10^{-21}, 2.082×10^{-21} and 1.999×10^{-21} cm^3, respectively. The obtained values of V clearly indicate that as the temperature rises, the molar volume of ILs also increases [28].

Electronic polarizability (R_m). Electronic polarizability (R_m) or also known as molar refraction refers to the ability of an atom or a molecule to polarize its electron cloud in the presence of an electric field and can be derived from their RI values. The electronic polarizability of a substance is a measure of how easily its electron cloud can be distorted by an external electric field and shows their molecular behaviour in solution. R_m values can be calculated from RI values as displayed in the Lorenz-Lorentz equation:

$$R_m = \left[\frac{\eta_D^2-1}{\eta_D^2+2}\right]\frac{M}{\rho} = \left[\frac{\eta_D^2-1}{\eta_D^2+2}\right]V_m \tag{7}$$

where η_D is refractive index, ρ is density, M is molar mass and V_m is molar volume. The calculated R_m values at different temperatures are displayed in Table 5. At 293.15 K, the electronic polarizability values of $[P_{66614}][SnCl_3]$, $[P_{66614}][ZrCl_5]$, $[P_{66614}]_2[NiCl_4]$ and $[P_{66614}]_2[CuCl_4]$ are 201.91, 200.91, 354.99 and 349.83 cm^3/mol, respectively. The R_m values increases slighlty with the increase in temperatures. It was also observed that the R_m values of DILs are much higher as compared to MILs due to the presence of two charged centers in DILs which enhances the electrostatic interactions within the molecule. This can lead to a higher electronic polarization as the electron cloud distribution becomes more easily distorted [21, 25, 29].

Free volume (V_f). Free volume (V_f) represents the space between the molecules or particles where movement or diffusion can occur. Free volume is crucial in understanding the mobility of molecules, especially in liquids and helps better understanding of the transport phenomenon in ILs. (V_f) values are calculated using the following equation:

$$V_f = V_m - R_m \tag{7}$$

where V_m is molar volume and R_m is electronic polarizability or molar refraction.

The V_f values of the synthesized ILs at various temperatures were computed and depicted in Table 5. At 293.15 K, the order of free volumes were increasing in the order of $[P_{66614}][ZrCl_5]$, $[P_{66614}][SnCl_3]$, $[P_{66614}]_2[CuCl_4]$ and $[P_{66614}]_2[NiCl_4]$ at 470.57, 487.60, 853.79 and 898.73 cm^3/mol, respectively. $[P_{66614}]_2[NiCl_4]$ recorded the highest free volume whereas $[P_{66614}][ZrCl_5]$ has the lowest free volume.

Table 5. Estimated values of thermal expansion coefficient (α), molecular volume (V), molar volume (V_m), electronic polarizability (R_m) and free volume (V_f) at different temperatures.

ILs	α (1/K)	V (cm^3)	V_m (cm^3/mol)	R_m (cm^3/mol)	V_f (cm^3/mol)
$[P_{66614}][SnCl_3]$					
293.15	6.540×10^{-4}	1.145×10^{-21}	689.51	201.91	487.60
303.15	6.583×10^{-4}	1.152×10^{-21}	693.98	202.05	491.93
313.15	6.627×10^{-4}	1.160×10^{-21}	698.50	202.18	496.32
323.15	6.671×10^{-4}	1.168×10^{-21}	703.38	202.41	500.97
333.15	6.716×10^{-4}	1.176×10^{-21}	707.95	202.50	505.45
$[P_{66614}][ZrCl_5]$					
293.15	5.635×10^{-4}	1.115×10^{-21}	671.48	200.91	470.57
303.15	5.667×10^{-4}	1.121×10^{-21}	675.26	200.75	474.51
313.15	5.699×10^{-4}	1.128×10^{-21}	679.08	200.58	478.50
323.15	5.732×10^{-4}	1.134×10^{-21}	682.94	200.41	482.53
333.15	5.765×10^{-4}	1.141×10^{-21}	687.00	200.28	486.72
$[P_{66614}]_2[NiCl_4]$					
293.15	7.249×10^{-4}	2.082×10^{-21}	1253.72	354.99	898.73
303.15	7.302×10^{-4}	2.097×10^{-21}	1262.76	355.40	907.36
313.15	7.356×10^{-4}	2.112×10^{-21}	1271.93	355.80	916.13
323.15	7.410×10^{-4}	2.127×10^{-21}	1281.24	356.21	925.03
333.15	7.466×10^{-4}	2.144×10^{-21}	1291.29	356.79	934.50
$[P_{66614}]_2[CuCl_4]$					
293.15	6.469×10^{-4}	1.999×10^{-21}	1203.62	349.83	853.79
303.15	6.511×10^{-4}	2.012×10^{-21}	1211.44	350.17	861.27
313.15	6.553×10^{-4}	2.025×10^{-21}	1219.36	350.48	868.88
323.15	6.597×10^{-4}	2.038×10^{-21}	1227.38	350.79	876.59
333.15	6.640×10^{-4}	2.052×10^{-21}	1235.59	351.09	884.50

The highest free volume in the matrix of $[P_{66614}]_2[NiCl_4]$ can be attributed to the tetrachloronickelate anion, $[NiCl_4]^{2-}$ which possess the lowest molecular weight as compared to other metal chloride anions among the synthesized ILs. ILs with higher free volume can potentially accommodate larger quantities of gases, which may contribute to increased gas solubility within the ILs [30]. Table 5 shows the estimated values of thermal expansion coefficient (α), molecular volume (V), molar volume (V_m), electronic polarizability (R_m) and free volume (V_f) at different temperatures for the synthesized ILs.

Viscosity analysis. Viscosity refers to the measure of a fluid's resistance to flow and represents the internal friction within a fluid as its molecules move past each other. A high viscosity fluid is thick and resists flowing easily, while a low viscosity fluid flows more readily. Viscosity is one of the most critical physical properties of Ils as it affects pumping, mixing, and transportation processes. Lower viscosity fluids will facilitate easier flow thus, making certain processes more efficient. Figure 3 shows the effects of temperature on the viscosity of the synthesized MILs and DILs.

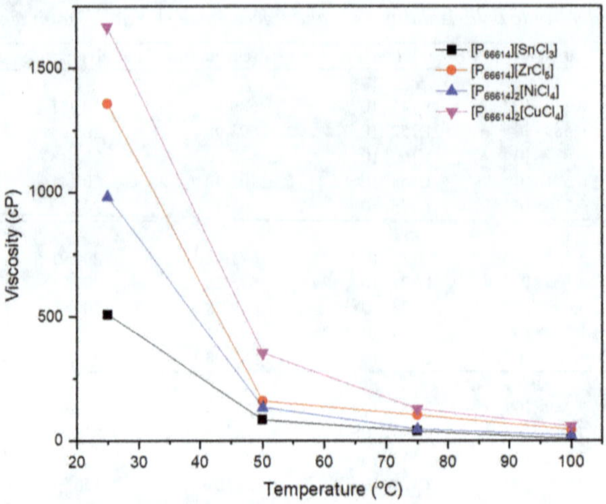

Fig. 3. Viscosity (η) as a function of temperature (T) for the synthesized ILs

At higher temperatures, the viscosity of ionic ILs tends to decrease significantly. This phenomenon occurs due to the increased thermal energy present in the system. At elevated temperatures, the kinetic energy of the ILs molecules intensifies, causing them to move more vigorously and disrupt the ordered structure within the liquid. The enhanced molecular motion leads to reduced intermolecular interactions and weaker cohesive forces between the ions in ILs matrix. As a result, the resistance to flow decreases, allowing the ILs to exhibit lower viscosity as reported in the previous studies [2, 11, 31].

At 25 °C, the synthesized ILs exhibited a viscosity range between 508 cP to 1666 cP. However, the viscosity of decreases significantly when the temperature reached 100 °C as depicted in Table 4.2. All values obtained are within the reported viscosity range of ILs by other researchers, which varied between 5 cP to 123500 cP [32]. It was observed that $[P_{66614}][SnCl_3]$ demonstrated the lowest viscosity at 25 °C, followed by $[P_{66614}]_2[NiCl_4]$, $[P_{66614}][ZrCl_5]$ and $[P_{66614}]_2[CuCl_4]$. Furthermore, it was also observed that the DILs possess relatively higher viscosities as compared to the MILs due to the presence of extra cation in the matrix of DILs. These additional cations occupy larger space in ILs, leading to stearic hindrance and increased resistance to flow, resulting in higher viscosity.

TGA analysis. The thermal stability of the synthesized ILs was evaluated in a rich N_2 atmosphere and displayed display a high thermal stability. The starting of the weight loss and the change in slope of the TGA curve was found and calculated by using TGA instrumental software Pyris V-3.81. The thermal decomposition values of the synthesized MILs and DILs were reported in terms of thermal onset temperature, T_{onset} and thermal decomposition temperatures, $T_{decomposition}$ as summarized in Table 6.

Table 6. T_{onset} and $T_{decomposition}$ for each ILs

ILs	T_{onset} (°C)	$T_{decomposition}$ (°C)
$[P_{66614}][SnCl_3]$	448.05	475.25
$[P_{66614}][ZrCl_5]$	367.49	395.14
$[P_{66614}]_2[NiCl_4]$	366.52	394.00
$[P_{66614}]_2[CuCl_4]$	372.89	434.20

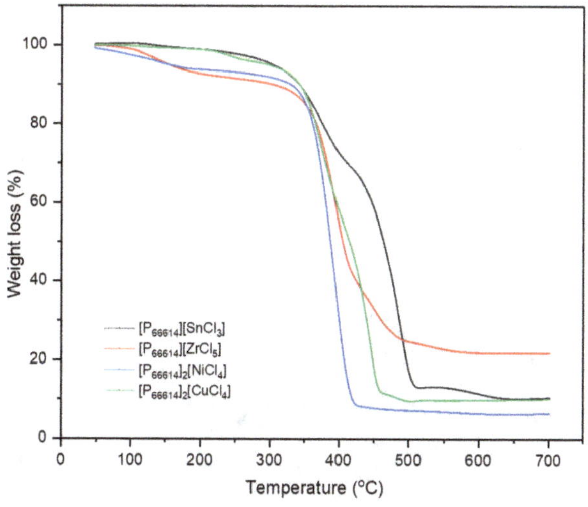

Fig. 4. TGA profile for the synthesized MILs and DILs

It was reported that the thermal decomposition temperatures of ILs are highly dependent on the length of alkyl chains of the cation and the type of anions [33]. Based on TGA results as shown in Figure 4, the thermal onset temperature of the synthesized ILs has appeared within the range of 366.52 °C to 448.05 °C. It was also clearly indicated that $[P_{66614}][SnCl_3]$ possessed a higher thermal

Materials Research Forum LLC
https://doi.org/10.21741/9781644903575-1

stability comparatively among the synthesized ILs. The thermal stability of the ILs displayed a decreasing trend of $[P_{66614}][SnCl_3]$ > $[P_{66614}]_2[CuCl_4]$ > $[P_{66614}][ZrCl_5]$ > $[P_{66614}]_2[NiCl_4]$. Since each of the synthesized ILs has T_{onset} of higher than 350 °C [34], it can be concluded that each of them is a very stable ILs thermally and can be used effectively at high temperature reaction conditions.

DSC analysis. DSC analysis is a thermal analysis technique used to measure the heat flow associated with physicochemical changes and transitions in materials including ILs. The glass transition temperature (T_g) and melting point temperature (T_m) are key thermal properties that characterize the behavior of ILs. Low T_g values among ILs is a good indication of desirable physicochemical properties such as low viscosity and high ionic conductivity. The T_g and T_m values of the respective ILs were determined under a rich N_2 atmosphere and are clearly represented in Table 7.

Table 7. T_g and T_m for each ILs

ILs	T_g (°C)	T_m (°C)
$[P_{66614}][SnCl_3]$	-70.07	90.62
$[P_{66614}][ZrCl_5]$	-72.15	94.31
$[P_{66614}]_2[NiCl_4]$	-66.64	91.53
$[P_{66614}]_2[CuCl_4]$	-65.48	96.98

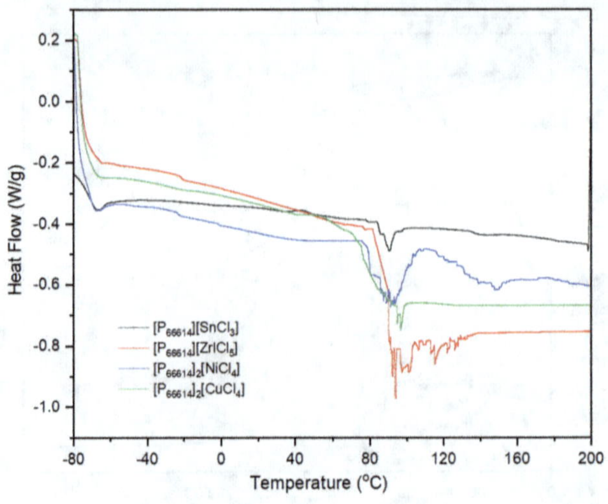

Fig. 5. DSC thermogram for the synthesized MILs and DILs

In general, the synthesized ILs characterized by large and bulky phosphonium cations with asymmetrical shape exhibit a rather high melting points due to strong intermolecular reactions among the cations side chains caused by the van der Waals forces [35]. Since the synthesized ILs share a common cation type, the variation of T_g and T_m values among them are relatively close. The DSC thermogram of the synthesized ILs is displayed in Figure 5. When comparisons were made among the synthesized ILs, it was observed that the DILs displayed slightly higher T_g and T_m values as compared to the MILs due to their structural differences and increased molecular complexity. DILs consist of larger and heavier molecular structures as compared to MILs. The presence of two cationic species within the molecule leads to an increased molecular mass, resulting in stronger intermolecular forces such as van der Waals forces and hydrogen bonding. These stronger interactions require more energy to break, hence requiring higher temperatures to reach the glass transition or melting point, which contribute to higher T_g and T_m values. The synthesized monocationic and dicationic ILs were found to be thermally stable over a wide range of temperatures making them highly attractive for various application purposes.

Conclusion

In the present work, two monocationic ILs such as $[P_{66614}][SnCl_3]$ and $[P_{66614}][ZrCl_5]$, and two dicationic ILs such as $[P_{66614}]_2[NiCl_4]$ and $[P_{66614}]_2[CuCl_4]$ were successfully synthesized and characterized. The effects of anions and temperature on physicochemical properties such as refractive index, density and viscosity were thoroughly studied and analyzed. The present study observed a decrease in both density and viscosity as the temperature increased within the investigated range. Furthermore, it was observed that $[P_{66614}]_2[CuCl_4]$ recorded the lowest density value at 50 °C, while $[P_{66614}][ZrCl_5]$ recorded the highest value. The viscosity values of $[P_{66614}][SnCl_3]$, $[P_{66614}][ZrCl_5]$, $[P_{66614}]_2[NiCl_4]$ and $[P_{66614}]_2[CuCl_4]$ are 83, 156, 131 and 350 cP, respectively at similar temperature. The thermal expansion coefficient, α and molar volumes, V_m were found to increase linearly with the increase of temperature. Additionally, R_m values of dicationic ILs were found to be higher than those of monocationic ILs.

The authors gratefully acknowledge Murata Manufacturing Co., Ltd. for funding the research work under the Murata Science and Education Foundation Research Grant Scheme (015ME0-392), Universiti Teknologi PETRONAS (UTP), Centre of Research in Ionic Liquids (CORIL), the research officers and postgraduate students for their assistance in each aspect of research.

References

[1] P. Wasserscheid and T. Welton, Ionic liquids in synthesis. John Wiley & Sons, 2008. https://doi.org/10.1002/9783527621194

[2] M. S. Aminuddin, Z. Man, M. A. Bustam Khalil, and B. Abdullah, "Screening of Metal Chloride Anion-based Ionic Liquids for Direct Conversion of Hydrogen Sulfide by COSMO-RS," E3S Web Conf., vol. 287, p. 02003, 2021. https://doi.org/10.1051/e3sconf/202128702003

[3] H. Zhang, L. Chen, Y. Chen, and Z. Wang, "Removal of sulfide from fuels by ionic liquids: prospects for the future," Brazilian Journal of Chemical Engineering, pp. 1-35, 2023.

[4] L. Peng et al., "Facilitated transport separation of CO2 and H2S by supported liquid membrane based on task-specific protic ionic liquids," Green Chemical Engineering, 2022. https://doi.org/10.1016/j.gce.2021.12.005

[5] T. Payagala, J. Huang, Z. S. Breitbach, P. S. Sharma, and D. W. Armstrong, "Unsymmetrical dicationic ionic liquids: manipulation of physicochemical properties using specific structural architectures," Chemistry of Materials, vol. 19, no. 24, pp. 5848-5850, 2007. https://doi.org/10.1021/cm702325a

[6] S. Werner, M. Haumann, and P. Wasserscheid, "Ionic liquids in chemical engineering," Annual review of chemical and biomolecular engineering, vol. 1, pp. 203-230, 2010. https://doi.org/10.1146/annurev-chembioeng-073009-100915

[7] A. H. Jadhav and H. Kim, "Short oligo (ethylene glycol) functionalized imidazolium dicationic room temperature ionic liquids: Synthesis, properties, and catalytic activity in

azidation," Chemical engineering journal, vol. 200, pp. 264-274, 2012. https://doi.org/10.1016/j.cej.2012.06.045

[8] A. H. Jadhav, A. Chinnappan, R. H. Patil, S. V. Kostjuk, and H. Kim, "Green chemical conversion of fructose into 5-hydroxymethylfurfural (HMF) using unsymmetrical dicationic ionic liquids under mild reaction condition," Chemical Engineering Journal, vol. 243, pp. 92-98, 2014. https://doi.org/10.1016/j.cej.2013.12.054

[9] M. Planellas, R. Pleixats, and A. Shafir, "Palladium Nanoparticles in Suzuki Cross-Couplings: Tapping into the Potential of Tris-Imidazolium Salts for Nanoparticle Stabilization," Advanced Synthesis & Catalysis, vol. 354, no. 4, pp. 651-662, 2012. https://doi.org/10.1002/adsc.201100574

[10] O. Nacham et al., "Interfacial and aggregation behavior of dicationic and tricationic ionic liquid-based surfactants in aqueous solution," Colloids and Surfaces A: Physicochemical and Engineering Aspects, vol. 469, pp. 224-234, 2015. https://doi.org/10.1016/j.colsurfa.2015.01.026

[11] M. S. Aminuddin, M. A. B. Khalil, and B. Abdullah, "Metal chloride anion based ionic liquids: synthesis, characterization and evaluation of performance in hydrogen sulfide oxidative absorption," RSC advances, vol. 12, no. 19, pp. 11906-11912, 2022. https://doi.org/10.1039/D2RA01494G

[12] A. Osman, A. K. Ziyada, A. M. Khan, and F. Rajab, "Alkylimidazolium-based ionic liquids with tailored anions and cations for CO 2 capture," RSC advances, vol. 14, no. 6, pp. 3985-3995, 2024. https://doi.org/10.1039/D3RA08335G

[13] A. K. Ziyada et al., "Impact of nitrile incorporated with hydroxyl, benzyl, and allyl on physicochemical properties of propanenitrile imidazolium-based dual functionalized ionic liquids," Ionics, vol. 29, no. 10, pp. 4067-4076, 2023. https://doi.org/10.1007/s11581-023-05106-8

[14] M. S. Aminuddin, M. A. Bustam, and K. Johari, "Evaluation of amine-based ionic liquids as potential solvents for hydrogen sulfide absorption using COSMO-RS: Computational and experimental validation," Chemical Engineering Research and Design, vol. 203, pp. 721-730, 2024. https://doi.org/10.1016/j.cherd.2024.01.054

[15] M. R. Chowdhury, R. M. Moshikur, R. Wakabayashi, M. Moniruzzaman, and M. Goto, "Biocompatible ionic liquids assisted transdermal co-delivery of antigenic protein and adjuvant for cancer immunotherapy," International Journal of Pharmaceutics, vol. 601, p. 120582, 2021. https://doi.org/10.1016/j.ijpharm.2021.120582

[16] J. C. Almeida, C. E. Cardoso, M. Neves, T. Trindade, M. G. Freire, and E. Pereira, "Ionic-liquid-supported magnetic nanoparticles: synthesis, characterization, and their use for the separation of Co and Ni from Li in aqueous solutions," European Journal of Inorganic Chemistry, p. e202400237.

[17] E. A. Hays, G. Eicher, A. Morales, D. S.-d. la Cruz, and K. M. Miller, "Dual Ionic Liquid-Functionalized Cellulosic Materials: Thermal, Conductive, and Morphological Properties," ACS Applied Polymer Materials, vol. 6, no. 5, pp. 2616-2625, 2024. https://doi.org/10.1021/acsapm.3c02815

[18] R. González-Martín et al., "Exploring a new generation of bimetallic magnetic ionic liquids with ultra-low viscosity in microextraction that enable direct coupling with high-performance liquid-chromatography," Analytica Chimica Acta, vol. 1301, p. 342448, 2024. https://doi.org/10.1016/j.aca.2024.342448

[19] J. O. Valderrama and K. Zarricueta, "A simple and generalized model for predicting the density of ionic liquids," Fluid Phase Equilibria, vol. 275, no. 2, pp. 145-151, 2009. https://doi.org/10.1016/j.fluid.2008.10.002

[20] F. J. Deive, M. A. Rivas, and A. Rodríguez, "Study of thermodynamic and transport properties of phosphonium-based ionic liquids," The Journal of Chemical Thermodynamics, vol. 62, pp. 98-103, 2013. https://doi.org/10.1016/j.jct.2013.02.012

[21] M. Tariq, P. Forte, M. C. Gomes, J. C. Lopes, and L. Rebelo, "Densities and refractive indices of imidazolium-and phosphonium-based ionic liquids: Effect of temperature, alkyl chain length, and anion," The Journal of Chemical Thermodynamics, vol. 41, no. 6, pp. 790-798, 2009. https://doi.org/10.1016/j.jct.2009.01.012

[22] C. M. Neves, P. J. Carvalho, M. G. Freire, and J. A. Coutinho, "Thermophysical properties of pure and water-saturated tetradecyltrihexylphosphonium-based ionic liquids," The Journal of Chemical Thermodynamics, vol. 43, no. 6, pp. 948-957, 2011. https://doi.org/10.1016/j.jct.2011.01.016

[23] L. I. Tomé, R. L. Gardas, P. J. Carvalho, M. J. Pastoriza-Gallego, M. M. Pineiro, and J. A. Coutinho, "Measurements and correlation of high-pressure densities of phosphonium based ionic liquids," Journal of Chemical & Engineering Data, vol. 56, no. 5, pp. 2205-2217, 2011. https://doi.org/10.1021/je101232g

[24] Q. Zhang et al., "Physicochemical properties of nitrile-functionalized ionic liquids," The Journal of Physical Chemistry B, vol. 111, no. 11, pp. 2864-2872, 2007. https://doi.org/10.1021/jp067327s

[25] A. K. Ziyada and C. D. Wilfred, "Effect of temperature and anion on densities, viscosities, and refractive indices of 1-octyl-3-propanenitrile imidazolium-based ionic liquids," Journal of Chemical & Engineering Data, vol. 59, no. 5, pp. 1385-1390, 2014. https://doi.org/10.1021/je400824z

[26] T. Peppel et al., "Low-Viscosity Paramagnetic Ionic Liquids with Doubly Charged [Co (NCS) 4] 2− Ions," Angewandte Chemie International Edition, vol. 49, no. 39, pp. 7116-7119, 2010. https://doi.org/10.1002/anie.201000709

[27] Y. Hu and X. Peng, "Effect of the structures of ionic liquids on their physical chemical properties," Structures and interactions of ionic liquids, pp. 141-174, 2014. https://doi.org/10.1007/978-3-642-38619-0_5

[28] B. Gonzalez, E. Gomez, A. Dominguez, M. Vilas, and E. Tojo, "Physicochemical characterization of new sulfate ionic liquids," Journal of Chemical & Engineering Data, vol. 56, no. 1, pp. 14-20, 2011. https://doi.org/10.1021/je1006357

[29] J. G. McDaniel and A. Yethiraj, "Influence of electronic polarization on the structure of ionic liquids," The journal of physical chemistry letters, vol. 9, no. 16, pp. 4765-4770, 2018. https://doi.org/10.1021/acs.jpclett.8b02120

[30] P. K. Chhotaray and R. L. Gardas, "Structural dependence of protic ionic liquids on surface, optical, and transport properties," Journal of Chemical & Engineering Data, vol. 60, no. 6, pp. 1868-1877, 2015. https://doi.org/10.1021/acs.jced.5b00077

[31] S. De Santis et al., "Cholinium-amino acid based ionic liquids: a new method of synthesis and physico-chemical characterization," Physical Chemistry Chemical Physics, vol. 17, no. 32, pp. 20687-20698, 2015. https://doi.org/10.1039/C5CP01612F

[32] E. Santos, J. Albo, A. Rosatella, C. A. Afonso, and A. Irabien, "Synthesis and characterization of magnetic ionic liquids (MILs) for CO2 separation," Journal of Chemical Technology & Biotechnology, vol. 89, no. 6, pp. 866-871, 2014. https://doi.org/10.1002/jctb.4323

[33] N. M. Yunus, M. A. Mutalib, Z. Man, M. A. Bustam, and T. Murugesan, "Thermophysical properties of 1-alkylpyridinum bis (trifluoromethylsulfonyl) imide ionic liquids," The Journal of Chemical Thermodynamics, vol. 42, no. 4, pp. 491-495, 2010. https://doi.org/10.1016/j.jct.2009.11.004

[34] Y. Cao and T. Mu, "Comprehensive investigation on the thermal stability of 66 ionic liquids by thermogravimetric analysis," Industrial & engineering chemistry research, vol. 53, no. 20, pp. 8651-8664, 2014. https://doi.org/10.1021/ie5009597

[35] M. H. Rahman, A. Khajeh, P. Panwar, M. Patel, A. Martini, and P. L. Menezes, "Recent progress on phosphonium-based room temperature ionic liquids: Synthesis, properties, tribological performances and applications," Tribology International, vol. 167, p. 107331, 2022. https://doi.org/10.1016/j.triboint.2021.107331

Decarbonization Technology: ICDT2024 Materials Research Forum LLC
Materials Research Proceedings 53 (2025) 17-42 https://doi.org/10.21741/9781644903575-2

Plasma-based technologies for sustainable hydrogen production: A data-driven bibliometric

Habeeb Bolaji Adedayo[1,a *], Thar Mohammed Albarody[1,b *], Kamaruzzaman Sopian[1,c], Ali Samer Muhsan[1,d], Aishat Adebonike Adedayo-Ojo[2,e], and Nabil Majd Alawi[3,f]

[1]Department of Mechanical Engineering, Universiti Teknologi Petronas, Seri Iskander, 32610, Perak, Malaysia

[2]Department of Management and Humanities, Universiti Teknologi Petronas, Seri Iskander, 32610, Perak, Malaysia

[3]Chemical Engineering Department, University of Technology-Iraq, Baghdad, Iraq

[a]habeeb_21002757@utp.edu.my, [b]dher.albarody@utp.edu.my, [c]ksopian@utp.edu.my, [d]ali.samer@utp.edu.my, [e]aishat_24001193@utp.edu.my, [f]nabil.m.alawi@uotechnology.edu.iq

Keywords: Plasma Technologies, Hydrogen Production, Sustainable Energy, Merged Dataset, Bibliometric Analysis

Abstract. The quest for sustainable hydrogen production has intensified in the past decade, driven by the global energy transition from fossil fuels to hydrogen-based alternatives. This urgency necessitates the exploration of innovative technologies beyond traditional methods. Therefore, this study pioneers a bibliometric analysis to evaluate the potential of plasma-based technologies for sustainable hydrogen production. Methodologically, the systematic data adheres strictly to the PRISMA 2020 framework. The Shiny App of the Bibliometrix R package is used to uncover the performance analysis and science mapping visualization of the bibliometric dataset. The findings revealed 581 published documents following database merging, with 152 records excluded by automated tools and another 152 through screening with PRISMA. This study is of great interest to academic and industrial relevance, as evidenced by 277 documents with annual growth rate (22.57%), authors' contributions (n = 1,185), co-authorship (30.32%), and average citation rate (n = 23). Notably, microwave plasma emerges as the most dominant technology for hydrogen production from natural gas reforming, achieving a 65% volumetric yield among non-thermal plasma variants, alongside warm plasma (48%) and thermal plasma (67%). Collaborative efforts between China and Australia in energy policy research emphasize the importance of international partnerships in advancing sustainable energy solutions. The emerging research trends in plasma gasification, water splitting, and hydrogen evolution reactions, signal a paradigm shift towards more sustainable, efficient hydrogen production techniques. Ongoing research is poised to enhance efficiency and environmental sustainability, contributing vitally to a net-zero emission future.

Introduction

Globally, the pursuit of sustainable energy sources has gained momentum in the last decade due to the pressing need to address environmental, health, economic, and climate impacts [1], [2]. The rapid transition from fossil-based fuels to cleaner alternatives is necessitated for a sustainable energy future. Hydrogen, as a clean, sustainable, and versatile energy carrier, has garnered considerable interest as a potential fuel for the future [3]. However, the traditional/conventional methods of hydrogen production (methane steam reforming, autothermal reforming, natural gas/hydrocarbon reforming, and coal gasification reforming) are associated with carbon emissions and other environmental concerns [4]. This has led researchers [5]–[10] to a growing interest in exploring sustainable methods for hydrogen production.

Plasma-based technologies (PbT's) offer a promising avenue for sustainable hydrogen production due to their unique opportunities as an efficient and eco-friendly fuel in the conversion of various feedstocks (natural gas, biomass, and wastewater) into hydrogen [11]–[14], In its simplest form, plasma is indeed considered the 4th state of matter, created by the application of electric energy or intense heat to gas. Moreover, PbT's facilitate reaction rates, high efficiency, versatility, scalability, and low environmental impact, making it an attractive area of research in the production of sustainable hydrogen [15], [16].

In this regard, numerous studies have investigated plasma for hydrogen production from vaporized fuel (liquid or gaseous) [11], [12], [17]–[20], renewables [21], [22], coal and plasma gasification [23]–[28], towards reducing environmental pollution. Some of these studies have considered different plasma reactors for hydrogen production such as dielectric barrier discharge (DBD) [29]–[33]; gliding arc discharge (GAD) [34], [35], corona discharge [36], microwave (MW) discharge [17], [18], [37], and radio frequency (RF) plasma [38], [39], at various operational conditions. These plasma reactors operate at precise temperature, pressure, flow rate, and conversion rates, based on their desired feedstocks, as reported [40]. It is noteworthy that the highest conversion rate from 5 to above 90% is achievable through MW plasma reactor, making it distinct from other reactors in terms of the methane-to-hydrogen conversion process [40]. Other research advances in the development of hybrid plasma reactors and catalytic enhancement were observed in the last decade for performance efficiency and productivity [21], [41]–[45].

The world continues to revolve around data due to its unleashing insights in the research domain towards evidence-based decision-making, accuracy and reliability, efficiency, identification of emerging trends and patterns, and predictive analytics, as well as embracing digital transformation [46]. Therefore, a data-driven bibliometric analysis becomes crucial for research on plasma-based technologies for sustainable hydrogen production, as implemented in other energy research domains [2], [47]–[49], providing quantitative insights into the trends, dynamics, and impact of scientific literature. Contrary to traditional literature reporting, which often conducts a qualitative investigation, a bibliometric analysis provides a systematic method to evaluate the scholarly publications' output. Through a comprehensive assessment of the performance analysis (journal articles, authors, institutions/affiliations, and countries), and science mapping (bibliographic coupling, co-word analysis, co-authorship analysis, citation analysis and co-citation analysis), researchers can identify key research themes, influential publications, and emerging area of interest [50]. The outcome of the bibliometric findings will not only help in understanding the current state of research but also serve as a guide for future research directions and policy decision-makers who aim at advancing sustainable hydrogen production technologies by the application of plasma.

Despite the numerous contributions of studies in plasma-based technologies for hydrogen production, there exists a notable research gap in comprehensively identifying the most predominant sustainable reactors for efficient hydrogen yield. This study aims to bridge the gap by providing a detailed analysis of the performance and science mapping workflow, based on the data retrieved from the merged databases (Web of Science and Scopus) in answering the following research questions, **RQ1***:* what are the pattern of publications and citations on PbT's for sustainable hydrogen production from 2001 to 2023?; **RQ2:** what are the most predominant reactors for efficient hydrogen production in the past two decades of intensive research?; **RQ3:** which countries are the leading scholarly publications on PbT's for sustainable hydrogen production in the multidisciplinary databases between 2001 and 2023?; **RQ4:** what are the most researched and understudied scholarly themes/keywords, and how do they relate within the PbT's for sustainable hydrogen production?; **RQ5:** what is the social structure of knowledge on PbT's for sustainable hydrogen production based on the countries' collaboration mapping?; and **RQ6:** what are the emerging trend topics for future research on PbT's for sustainable hydrogen production? Thus, uncovering the potential of plasma-based technologies for sustainable hydrogen production through a data-driven bibliometric approach.

Research method

A Search query strategy

Fig. 1 (a) and **(b)** show the data obtained from the search strings in the Web of Science (WoS) and Elsevier Scopus repository, respectively, between the time frame 2001–2023, dated 22 January 2024. The search string consists of search terms and Boolean operators that were selected for the investigation of plasm-based technologies for sustainable hydrogen production. Noteworthy among these terminologies, as identified by the panel of experts in the research field and searched in the WoS database include, TS = ("plasma*") AND ("technolog*" OR "react*" OR "system*" OR "method*" OR "source*") AND ("sustainab*") AND ("hydrogen" OR "H2") AND ("produc*" OR "generat*") (Topic) and 2024 (Exclude – Publication Years). While the Scopus database search employs the notation TITLE-ABS-KEY (("plasma*") AND ("technolog*" OR "react*" OR "system*" OR "method*" OR "source*") AND ("sustainab*") AND ("hydrogen" OR "H2") AND ("produc*" OR "generat*")) AND PUBYEAR < 2024. The use of TITLE-ABS-KEY (Article title, Abstract, Keywords) in the Scopus database advanced query, represents the same as the TS (Topic: Title, Abstract, Author keywords) notation in the WoS database. Also, the PUBYEAR corresponds to PY, both denoting the publication year. The differences in notations resulted from database criteria in ensuring distinctiveness across their respective platforms.

All the records (title and abstract) were processed and retrieved on the same day to avoid any potential bias from the daily updates of the databases. Moreover, the selected databases are recognized as veteran platforms, esteemed for their productivity, credibility, and reliability, standing as the longest-running repositories of bibliographic documents covering various disciplines [51], [52].

Materials Research Forum LLC
https://doi.org/10.21741/9781644903575-2

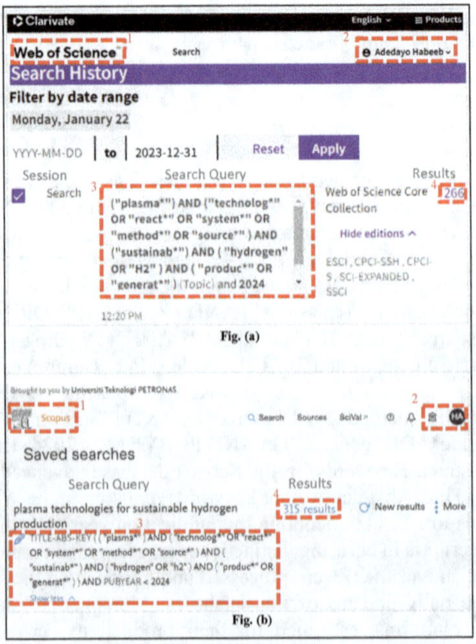

*Fig. 1: Search queries on PbT' for sustainable hydrogen production: **(a).** WoS database **(b).** Scopus database*

***Note(s):** 1: Database identification; 2: Author profile name; 3: Search string 4: Results*

B Merged databases

When conducting a bibliometric analysis, researchers are encouraged to merge the WoS with Scopus databases because of their unique characteristics, as stated by [53], [54], in achieving the desired results. WoS offers the most high-impact scientific studies and reliability, while Scopus offers greater publications coverage. Therefore, **Fig. 2** is introduced as a procedural step for merging both the WoS and Scopus databases for the search string on PbT's for sustainable hydrogen production towards a visualization word cloud. The data retrieved from this search string on each database were exported in "BibTex" file format. These files were automatically combined, and duplicates were removed using Rstudio software to ensure accuracy and timeliness, following the steps presented by [53]. The combined file in Excel format (.xlsx) is then screened for eligibility with PRISMA, and uploaded to Biblioshiny for the visualisation network analysis.

C Systematic analysis: PRISMA

The methodological approach adapted for reporting the scientific documents on plasma-based technologies for sustainable hydrogen production adheres strictly to the criteria outlined in the "Preferred Reporting Items for Systematic Reviews and Meta-Analyses" (PRISMA) 2020 guidelines [55]. This is to ensure that the data is transparent, valid, reliable, and replicable towards presenting credible and unbiased findings in the reporting document. A Systematic literature review (SLR) adapting PRISMA is crucial for consolidating existing/traditional research and providing a comprehensive understanding of the qualitative aspect of PbT's for sustainable hydrogen production. This qualitative method ensures a transparent and rigorous systematic review process, facilitating the records identification, selection, screening, eligibility, and synthesis of relevant studies [55].

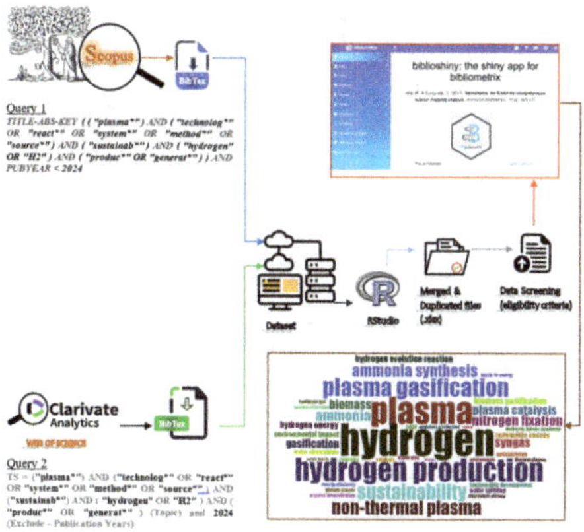

Fig. 2: Procedural steps for the merged databases on the plasma-based technologies for hydrogen production

Source: *Authors (2024)*

Fig. 3 presents the PRISMA flow diagram illustrating the search process for plasma-based technologies for sustainable hydrogen production. This consists of four distinct phases namely identification, screening, eligibility, and inclusion, as previously used in other systematic studies [56], [57]. In the identification phase, 581 published documents were retrieved from the WoS and Elsevier Scopus databases using the search string ("plasma*") AND ("technolog*" OR "react*" OR "system*" OR "method*" OR "source*") AND ("sustainab*") AND ("hydrogen" OR "H2") AND ("produc*" OR "generat*"), these documents were estimated to be 266 (45.8%) and 315 (54.2%) from their respective databases. Subsequently, 152 (26.2%) records were removed before screening due to duplication and ineligibility by the automation tool. During the screening process, 429 (73.8%) records were assessed based on their titles and abstracts, resulting in the exclusion of 96 records deemed irrelevant to the analysis of PbT's for sustainable hydrogen production. The next step of the screening process is the assessment of 333 reports sought for retrieval, 9 were considered unretrievable due to unidentified/anonymous authors. In the eligibility phase, 324 reports underwent further scrutiny, with records excluding languages other than English (n = 9), low quality (n = 32), and Year 2024 (n = 6). Ultimately, 277 reports were considered in the newly included studies.

D Bibliometric analysis: Biblioshiny

The use of bibliometric analysis is to apply mathematical, statistical, and quantitative methods to publications on PbT's for hydrogen production. This is done to present the conceptual and intellectual structure, as well as emerging trends in the research domain during the science mapping analysis. Furthermore, the Shiny App for Bibliometrix was introduced with the specific

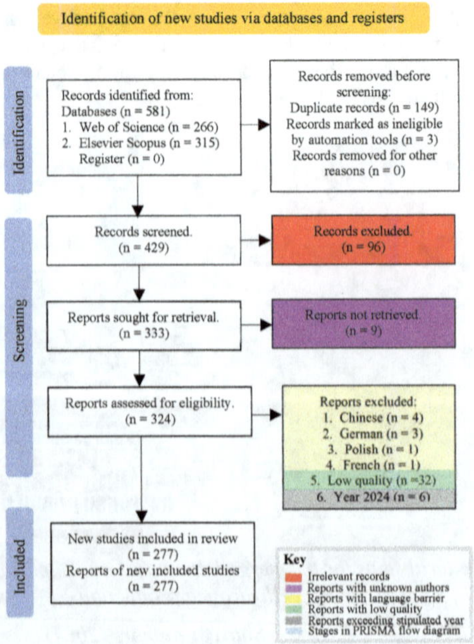

Fig. 3: *PRISMA flow diagram for the search string*

Source: *Adapted from Haddaway et al [58].*

purpose of uncovering and visualizing Bibliometric datasets. The Biblioshiny employs an R package for conducting the bibliometric research process, as it supports the features of Bibliometrix such as the analysis of authors, documents, clustering by coupling, intellectual structure, and social structure. Bibliometrix is an open-source tool for "quantitative research in bibliometrics and scientometrics" [59]. Performance analysis and science mapping are the procedures adopted for the bibliometric analysis techniques as encouraged by [50]. The former denotes the "*contributions* of research constituents" and the latter presents the "*relationships* between the research constituents.

Results and discussion

In this study, two techniques for categorizing bibliometric analysis on plasma-based technologies for sustainable distributed hydrogen production are discussed: performance analysis and science mapping. The former explains the "*contributions* of research constituents" and the latter presents the "*relationships* between the research constituents" [50], based on the available dataset. Moreover, the performance analysis summarizes the most prolific research constituents in terms of the journal articles, authors, institutions/affiliations, and countries, while the science mapping indicates the relationship of how different scientific data can be visualized as an entity in presenting the bibliographic coupling, co-word analysis, co-authorship analysis, citation analysis and co-citation analysis. Thus, addressing various research questions (**RQ 1–4**).

A Bibliometric dataset

Fig. 4 presents the main information on the search string (("plasma*") AND ("technolog*" OR "react*" OR "system*" OR "method*" OR "source*") AND ("sustainab*") AND ("hydrogen" OR "H2") AND ("produc*" OR "generat*")). These findings are obtained from the comprehensive analysis of the merged databases, as presented in the descriptive statistics.

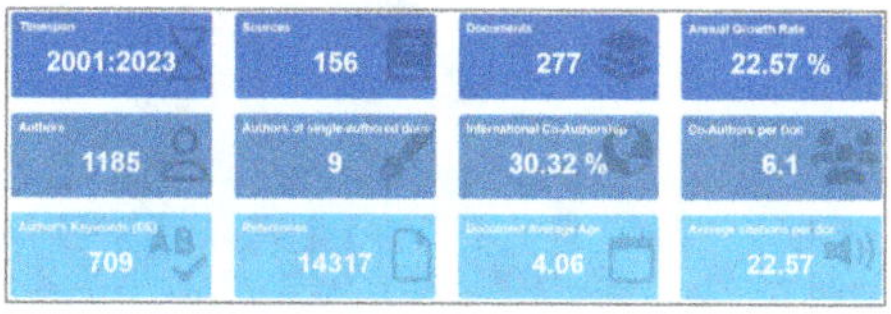

Fig. 4: *Main information on PbTs for hydrogen production*

Source: *Biblioshiny using the Authors' dataset [59]*

Considering the available results, this research domain is of great interest in academia. This is evidenced by the 277 published documents, 22.57% annual growth rate, 1,185 contributed authors, 30.32% international co-authorship, 14,317 references, and an average of approximately 23 citations per document. The reports of the included studies in the databases were accessed from 2001 to 2023, excluding the information from the incomplete year, 2024. Meanwhile, the Scopus database began reporting on topics related to plasma-based technologies for sustainable distributed hydrogen production in 1993, while WoS started reporting in 2006, despite being the oldest scientific repository database. The merging of the databases, removal of duplicate records by the automation tool, and further screening of the eligible documents through PRISMA (**Fig. 3**), resulted in 277 studies included in the review that span 2001–2023. Furthermore, the main information of the dataset is used for performance analysis and science mapping. The results are intertwined with insightful discussions that aimed at contributing to the ongoing discourse within the scope of the author's search string, fostering a deeper comprehension of the subject, and delineating avenues for future investigations.

B Performance analysis

In the process of performance analysis, 277 reports of included studies have been synthesized to provide the publication-related metrics, citation-related metrics, and citation and publication-related metrics. These metrics have also been considered in the work conducted by [50], [60]–[63] as the basis for their performance analysis.

1) Annual publications and citations: **Fig. 5**, addresses the first research question (**RQ 1**), which investigates the evolution of publication and citation patterns on plasma-based technologies for sustainable distributed hydrogen production. The primary objective was to determine the annual scientific production and document citation trends over 22 years of intensive research, based on four distinct phases. Phase A (2001-2006) shows the minimum activity in publications, with only 3 (1.08%) articles experiencing a sporadic trend. This resulted in a low mean total citation per year (MeanTCperYear), which peaked at 0.36 in 2003. This phase is characterized by a nascent interest in plasma-based technologies for sustainable distributed hydrogen production. This timeframe does not denote the commencement of research activities on the topic but rather the beginning of scientific repositories in the WoS and Scopus databases.

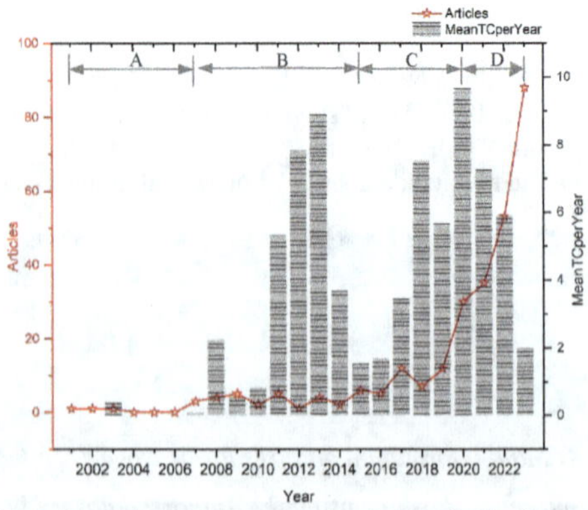

***Fig. 5:** Evolution of articles on plasma technologies for hydrogen sustainability*

***Source:** Author (2024)*

***Note(s):** A: Early development and exploration (2001-2006); B: Initial growth and research expansion (2007-2014); C: Significant growth and maturation (2015-2019); D:Accelerated expansion and advancement(2020-2023)*

Table 1 presents the published documents on plasma-based reactors for distributed hydrogen production towards the future generation of gas-cleaning systems/zero-emission systems. The Phase A category has critically investigated the techniques for reactor pollution control and sustainable development in large-scale industrial applications, such as pilot plasma and cyclone reactors for chemical waste disinfection applications [64], as well as pulsed corona plasma systems in methane reforming processes for hydrogen production from fossil fuels, synthesis gas cleaning, air cleaning, soil cleaning, exhaust gas cleaning, surface treatment, and production of nanoparticles, aiming at reducing environmental pollution [65], [66]. The rate of hydrogen yield has not been reported, as the publications in this phase focus on waste pollution control towards sustainable development.

Table 1, Phase B (2007-2014), witnessed a notable increase in the number of publications, with only 26 (9.39%) articles, reaching its peak (5 articles) in 2009 and 2011. This resulted in fluctuations in MeanTCperYear, which generally increased from 0.09 in 2007 to 8.90 in 2013, indicating a period of active research and publication in the PbT's for sustainable hydrogen production. Research in this area gained momentum, leading to a significant rise in both articles and citations compared to Phase A.

Subsequently, Phases B, C and D have been streamlined to focus more on the rate of sustainable production of hydrogen in published articles, among others. To identify the most efficient plasma-based reactors for sustainable hydrogen production.

In Phase B, non-thermal plasma reactors [67], natural gas catalytic decomposition reactors [67], rotary-type solar reactors [68], and dielectric barrier discharge reactors [69] have been considered for efficient hydrogen production towards environmental sustainability. The efficiency of these reactors may vary depending on various factors such as operating conditions, scale, and specific requirements.

***Table 1:** Plasma-based technologies towards sustainable distributed hydrogen production*

Phase A: 2001–2006 (Early development and exploration)

Reactors: Pilot plasma	Cyclone plasma	Pulsed corona	H_2 production	Applications	Ref(s)
✓	✓		Not reported	Industrial waste treatment with low environmental pollution and energy consumption	[64]
		✓	Not reported	Hydrogen production, nanoparticle generation, surface treatment, wastewater, and synthesis gas cleaning	[65], [66]

Phase B: 2007–2014 (Initial growth and research expansion)

Reactors: Non-thermal plasma	Catalytic reactor	Rotary-type solar	Dielectric barrier	H_2 production rate	Applications	Ref(s)
✓	✓			0.8-2.1 mmole H_2/min-g(carbon), 875 °C	In the H_2-rich gas and carbon with high energy efficiency of 83.9%.	[67]
		✓		3.17 H_2 (cm^3/g) at 1,473 K maximum	The study is relevant to hydrogen production via a solar concentration system.	[68]
			✓	Selectivity: 99% H_2, 773 K	In industrial applications for sustainable energy production	[69]

Phase C: 2015–2019 (Significant growth and maturation)

Reactors: Microwave plasma	Gliding arc discharge	Warm plasma reactor	H_2 production rate	Applications	Ref(s)
		✓	H_2 yield: 48%; CO yield: 51% (vol%)	Syngas (H_2 and CO) production, and reactant treatment	[70]
✓			H_2 yield: 12.84-13.96% (vol%)	In biogas reform into synthetic gas or high-quality product	[71]
✓			H_2 yield: 18.4%; CO yield: 37.2% (vol%)	In large-scale industrial syngas production to obtain sustainable and clean energy	[72]
✓			Syngas: H_2 to CO ratio: 1.6	In developing sustainable methods for tar destruction and biomass gasification	[73]
	✓		Syngas: 27.5% H_2; 30.8% CO (vol%)	In developing sustainable methods for tar destruction and biomass gasification	[74]

Phase D: 2020–2023 (Accelerated expansion and advancement)

Reactors: Microwave plasma	Dielectric discharge	Gliding arc discharge	Thermal plasma	Corona-DBD Hybrid	H_2 production yield	Applications	Ref(s)
✓					30.80-37.58 H_2 (vol%)	Sustainable production of hydrogen through biological processes (bio-hydrogen).	[37], [75]
✓					45-65 H_2; 25-30 CO (vol%)	In the generation of H_2-rich syngas for biomass energy conversion	[76]
	✓				7-10 H_2 (vol%)	In the reform of CO2 and CH4 into H2; the treatment of greenhouse gas emissions	[77]
✓					22-63 H_2; 17-45 CO (vol%)	A potential option in the energy sector for H_2 fuel from coal gasification	[78], [79]
	✓				18-57 H_2; 34-42 CO (vol%)	A potential option in the energy sector for H_2 fuel from coal gasification	[25], [78]
		✓			35-45 H_2; 32-37 CO (vol%)	A potential option in the energy sector for H_2 fuel from coal gasification	[78], [80]
	✓				2.66 H_2 (vol%)	Suitable in ammonia decomposition systems for hydrogen production	[81]
			✓		67 H_2; 43 CO (vol%)	For efficient waste utilization and management for producing H_2-rich syngas	[82]
✓					43.31 H_2; 11.88 CO (vol%)	To pyrolyze battery separator wastes without pre-processing and pre-sorting.	[83]
✓					37.07 H_2; 4.61 CO (vol%)	To pyrolyze plastic wastes without pre-processing and pre-sorting, for syngas	[83]
✓					40.68 H_2; 35.43 CO (vol%)	To pyrolyze silk wastes without pre-processing and pre-sorting, for syngas	[83]
				✓	23.9 g/kWh	H_2 production via water plasmolysis for decentralized power system in remote area	[84]

Source: Author (2024)

Phase C (2015-2019), continues the trend of increasing research activity and citations observed in Phase B. The highest number of articles (12) was published in both 2017 and 2019, indicating sustained interest and research output in the field, totaling about 42 (15.16%) articles. The MeanTCperYear remains relatively high, with fluctuations but generally maintaining levels above 1.5. This phase is characterized by sustained interest and research output in plasma-based technologies for sustainable distributed hydrogen production, with a focus on maintaining the momentum gained in Phase B. Studies within this period have predominantly considered MW plasma reactors due to their potential for achieving the highest H_2 yield, feedstock flexibility, and productive, as well as for their use as catalysts in improving hydrogen selectivity [85]. Furthermore, MW plasma has been deployed in diverse environments, including oceans, rivers, and lakes [86] and also, in the conversion of municipal solid waste to synthetic gas and in enhancing oil recovery [87]. Other studies have reviewed non-thermal PbTs (gliding arc plasma reactor, dielectric barrier discharge plasma reactor, corona discharge reactor, and MW plasma rector) for hydrogen production, but precise accounts of their yield are not provided [88], [89].

Phase D (2020-2023) witnessed accelerated expansion and advancement in research output related to plasma-based technologies for sustainable hydrogen production. The number of documents more than doubled from 35 in 2021 to 88 in 2023 out of a total of 206 (74.37%). However, the mean total citation per year declined from 7.27 in 2021 to 2.00 in 2023. This suggests that while there was a rapid increase in publications, the citation impact per article decreased, especially in recent publications with fewer citations. It is anticipated that this trend is likely to increase in the future as years progress. Moreover, there are substantial advances in MW plasmas for sustainable hydrogen production through biological processes, such as the fermentation of organic materials by microorganisms (bio-hydrogen) [37], [75], and biomass [76]. Additionally, coal gasification for H_2 production through the thermo-chemical process, herein, plasma gasification, offers a more sustainable, efficient, and environmentally friendly approach compared to other gasification technologies such as fluidized bed, fixed bed, and entrained bed gasifiers [78]. The flexibility of the microwave plasma reactor enables the pyrolysis of plastic waste, silk waste, and battery separator waste without the need for pre-processing and pre-sorting, resulting in high H_2 yield at 37.07, 40.68, and 43.31 H_2 (vol%), respectively.

On the other hand, thermal plasma has been researched for processing waste streams (refuse-derived fuel/sewage sludge) [82] due to its potential to operate at extreme temperatures for hydrogen production compared to non-thermal plasma, which remains at or near room temperature [90]. Non-thermal plasma processes for hydrogen production are generally more energy-efficient compared to thermal plasma, as they operate at lower temperatures [91], [92]. Both types of plasma can be used for hydrogen production, but the specific processes and conditions vary based on the desired temperature, energy efficiency, and applications. Recent advances demonstrate the effectiveness of combining non-thermal plasma (Corona-DBD hybrid) for efficient hydrogen production through water plasmolysis for decentralized power systems in remote areas [84]. Other studies [22], [93]–[100] report on plasma-based catalytic enhancement for high-yield products.

Overall, the analysis indicates a growing trend in research activity and interest in PbT's for sustainable H_2 production, with the most substantial growth observed in the latest phase (2020-2023). However, the citation impact per article showed some fluctuations, indicating the need for further assessment of the quality and impact of the research in this field.

2) PbT's for H_2 yield (vol%) productivity: **Fig. 6**, addresses the second research question (**RQ 2**) by analyzing fifteen prominent reviews related to plasma-based technologies that report their hydrogen yield productivity (vol%) from 2001 to 2023. These reviews have identified MW plasma as the most prominent PbT's with the highest rate of hydrogen production (45-66 vol%) among the non-thermal plasma reactors category. According to Budhraja and Mishra [85], MW plasma yields the highest H_2 due to its efficiency, productivity, catalyst involvement, and flexibility in the conversion of methane, carbon dioxide, or other hydrocarbon feedstock into hydrogen. Other plasma reactors, such as warm plasma and thermal plasma, have focused on waste treatment, utilization, and management in the pursuit of sustainable and efficient energy solutions.

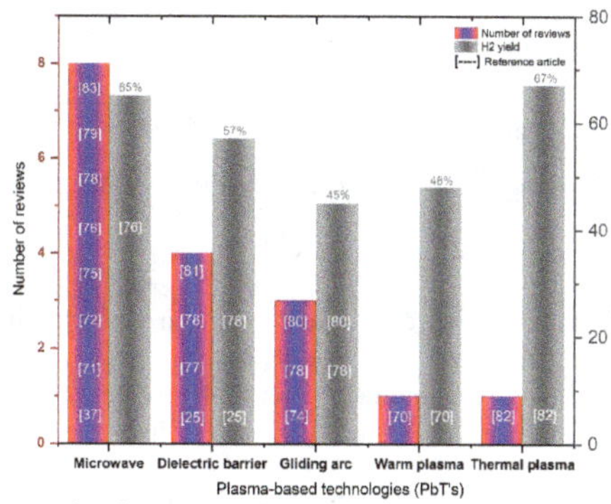

Fig. 6: *Plasma technologies articles for hydrogen yield in vol%, (2001-2023)*

3) Active countries' production over time: **Fig. 7** addresses the third research question (**RQ 3**), which identifies the top five most active research countries contributing to the published documents related to PbT's for sustainable hydrogen production between 2001 and 2023.

China shows a remarkable increase in scientific publications from 2017 onwards, with a major spike in 2023 (153 articles). This rapid increase may be influenced by its commitment to carbon neutrality by 2060, driving research into sustainable energy technologies as reflected in its energy and sustainable policies [101], [102].

Italy witnessed the first surge in documents (14 articles) in 2013 compared to other countries, after which the trend shows a gradual increase until 2023. This is likely due to the implementation of Italy's National Energy Strategy, and EU Strategic Energy Technology Plan (SET-Plan), which aimed to accelerate the development and deployment of low-carbon technologies, including hydrogen technologies, for more competitive and sustainable energy [103], [104].

The USA, Australia, and Germany follow a similar trend with a lagging effect in publications related to PbT's for sustainable hydrogen production. The low publications by these three countries can be attributed to more diverse methods of hydrogen production from fossil fuels through hydrocarbon reforming, hydrocarbon pyrolysis, and also from renewable sources through biomass process and water splitting [4], [105]. Countries globally are exploring these methods in the application to sustainable hydrogen production as guided by their respective energy policy initiatives towards net zero emissions.

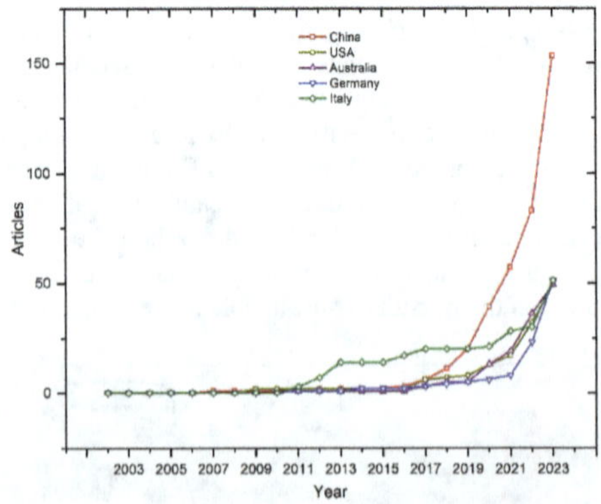

Fig. 7: *Top counties' production over time on plasma technologies for hydrogen yield*

C Science mapping

The relationships between the research constituents were investigated by performing a science mapping on the 277 published documents through co-word analysis, co-citation analysis, co-authorship analysis, and bibliographic coupling. This analysis aims to present the intellectual and bibliographic structure of the research domain [50], [61] on PbTs for hydrogen production.

1) Word cloud: **Fig. 8**, addresses the fourth research question (**RQ 4**) by analyzing the authors' scholarly keywords along with their highest frequency (freq) of occurrences in the dataset. In the word cloud, the size of each word denotes its frequency of occurrences in the publications' titles, abstracts, and/or keywords. The first three predominant words reflect the terms used in the search strings, followed by the hierarchy of occurrences of the words "plasma gasification", "sustainability", and "non-thermal plasma". This suggests a strong focus on plasma-related processes for sustainable H_2 production, including topics such as ammonia synthesis, biomass and nitrogen fixation. Meanwhile, the least researched author keywords, including waste-to-energy, biogas, and carbon dioxide, suggest less emphasis on topics related to these specific aspects in the context of PbT's for H_2 production. This could indicate areas where further research and attention may be needed to fully understand their potential impact on sustainable hydrogen production.

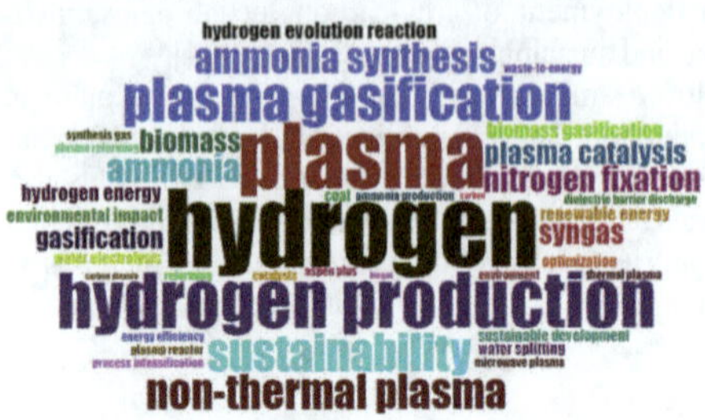

Terms	Freq
Hydrogen	33
Plasma	28
Hydrogen production	24
Plasma gasification	19
Sustainability	16
Non-thermal plasma	15
Ammonia synthesis	13
Ammonia	12
Nitrogen fixation	11
Biomass	10
Plasma catalysis	10
Syngas	10

Fig. 8: *Keywords cloud on plasma technologies for sustainable hydrogen yield*

2) Co-occurrence network mapping: **Fig. 9** went further to reveal the conceptual structure of the scholarly publications' co-occurrence keywords. This is sometimes known as a semantic network, as it examines the potential relationships between keywords that appear frequently in the study [106]. The analysis of the co-occurrence becomes crucial for researchers to recognize the research hotspots by providing insights into the key topic, trends, and relationships within a research field. Thus, helping researchers understand the structure and dynamics of knowledge within their cluster.

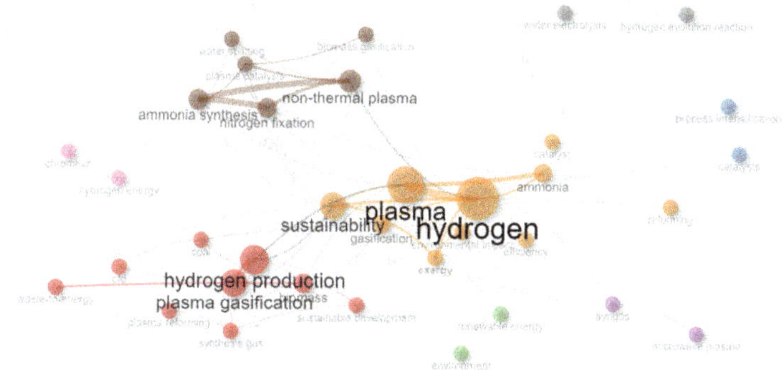

Fig. 9: *Co-citation network mapping analysis*

Note (s): *node: an entity (keywords); node size: frequency of keywords occurrence; the link between nodes: keywords co-occurrence; link thickness: occurrence of keywords co-occurrence; each colour: a thematic cluster. Also, the larger the node, the higher the keywords occurrence, and the thicker the link between the nodes, the higher the occurrence of keywords co-occurrence.*

Table 3: *Co-occurrence network of the top keywords in the research domain*

Node	Cluster	Betweenness	Closeness	PageRank
Hydrogen production	1	27.08316766	0.015384615	0.040048569
Plasma gasification	1	51.27609143	0.015873016	0.0540459
Biomass	1	28.04307166	0.015384615	0.037814986
Coal	1	5.503542062	0.012987013	0.019203128
Sustainable development	1	0	0.011904762	0.019936801
Plasma reforming	1	0	0.008928571	0.008896888
Synthesis gas	1	29	0.012048193	0.027124552
Waste-to-energy	1	0	0.013513514	0.015117288
Cfd	1	0.6	0.011494253	0.012624452
Catalysts	2	29	0.0125	0.017878896
Process intensification	2	0	0.009174312	0.011884245
Renewable energy	3	29	0.014492754	0.020092542
Environment	3	0	0.010204082	0.009978601
Syngas	4	29	0.01369863	0.020574233
Microwave plasma	4	0	0.009803922	0.01011508
Hydrogen	5	166.6304663	0.018867925	0.097222624
Plasma	5	116.1347153	0.018867925	0.073939197
Sustainability	5	100.8007451	0.019607843	0.065530942
Ammonia	5	0.842592593	0.014925373	0.025674618
Gasification	5	44.89560408	0.017857143	0.045259005
Environmental impact	5	0	0.014925373	0.02314632
Reforming	5	0	0.012195122	0.008129399
Catalyst	5	0	0.012195122	0.008129399
Efficiency	5	0	0.012195122	0.008129399
Exergy	5	0	0.014925373	0.021314463
Non-thermal plasma	6	17.86839827	0.013513514	0.042278921
Ammonia synthesis	6	12.43858422	0.014705882	0.043262315
Nitrogen fixation	6	14.01967063	0.014925373	0.041758764
Plasma catalysis	6	5.780017374	0.013888889	0.029181053
Biomass gasification	6	1.083333333	0.012820513	0.01344452
Water splitting	6	0	0.011363636	0.013977185
Hydrogen energy	7	0	1	0.028571429
Chromium	7	0	1	0.028571429
Hydrogen evolution reaction	8	0	1	0.028571429
Water electrolysis	8	0	1	0.028571429

Source: Biblioshiny using the Author dataset (Author, 2024)

In the co-occurrence network, eight clusters [cluster 1: red colour (9 nodes); cluster 2: blue colour (2 nodes); cluster 3: green colour (2 nodes); cluster 4: purple colour (2 nodes); cluster 5: orange colour (10 nodes); cluster 6: brown colour (6 nodes); cluster 7: pink colour (2 nodes); and cluster: grey colour (2 nodes)] were mapped from the 35 most predominant keywords related to the plasma-based technologies for sustainable hydrogen production. The clusters are to understand how the research domain is manifesting and developing over time.

Table 3 is the co-occurrence network metric presenting the betweenness centrality, closeness centrality and PageRank. The betweenness centrality shows the node's ability to disseminate information across unrelated/unconnected groups of nodes in the network [50]. This metric measures the influence or centrality of a node within the co-occurrence network. Higher betweenness values indicate that the corresponding keyword holds a central position in connecting others within the clusters. This has shown the impact of the highest betweenness centrality with an approximate value of 167, 116, and 100, for hydrogen, plasma, and sustainability, respectively, in cluster 5, indicating their pivotal role in connecting keywords towards developing sustainable hydrogen production methods, which is driven by the global focus on reducing carbon emissions and transitioning to clean energy sources. While clusters 7 (hydrogen energy and chromium), and 8 (hydrogen evolution reaction and water electrolysis) have been isolated based on their zero betweenness, indicating different aspects of hydrogen energy and its production, such as electrocatalysis and materials science. The closeness centrality is an information carrier to measure how close or effective a node is to all other nodes in the network [50]. Higher closeness centrality indicates a node that can quickly interact or connect with other nodes within the research domain. "Hydrogen energy," "chromium," "hydrogen evolution reaction," and "water electrolysis" have the highest closeness centrality of 1, suggesting their proximity to other keywords in the cluster. The PageRank reflects the significance of a node based on its publication influence in the research domain. A published article with a high PageRank denoted high quality and must be cited among the highly-cited published articles [50]. Moreover, a higher PageRank indicates a node that is more influential in the network. "Hydrogen" has the highest PageRank score, followed by "plasma" and "sustainability with approximately 0.097, 0.074, and 0.066, respectively.

In the relationship of the keywords to the research domain, the centrality of hydrogen within the network emphasizes its dominant role towards sustainable energy production. The plasma-based processes such as gasification, and catalyst enhancement, are crucial for efficient hydrogen yield, as evidenced by their high betweenness and PageRank values. Whereas, the inclusion of nodes related to renewable energy, environmental impact, and efficiency signifies the interdisciplinary nature of the research domain and its focus on sustainability. Specific applications such as non-thermal plasma, ammonia synthesis, and nitrogen fixation are also integral to the network, indicating their relevance to the broader topic.

Therefore, the intellectual structure of the co-occurrence network reflects the multidimensional nature of research on plasma-based technologies for sustainable hydrogen production, encompassing diverse fields such as energy sources, environmental impact, and technological processes.

3) *Countries' collaboration network:* **Fig. 10**, addresses the fifth research question (**RQ 5**) by analyzing the top countries' scholarly collaboration within the research domain. China and Australia have the highest collaboration network in research related to plasma-based technologies for sustainable hydrogen production with about 13 publications. This collaboration might be driven by China's strong focus on renewable energy development and Australia's abundant natural resources, including solar and wind energy. Making it possible for the two countries to reach a consensus on the "Renewable Energy Target and promulgated legislation and regulations to enlarge the scale of renewable energy" [107]. Subsequently followed by the collaboration between

Materials Research Proceedings 53 (2025) 17-42 https://doi.org/10.21741/9781644903575-2

Australia–United Kingdom, and China–United Kingdom, with about 8 publications each. The United Kingdom's (UK) involvement reflects its interest in sustainable technologies and climate change mitigation towards zero emissions and clean energy transition [108], [109]. Therefore, China, Australia and the UK have engaged in collaboration and cooperation in the field of energy and its policy towards sustainable development.

C Emerging research trends

Fig. 11 answers the sixth research question (**RQ 6**) by investigating the yearly emerging topics based on the authors' keywords. In 2016, biomass became an emerging trend with 10 frequency of publication occurrence (n = 10). Its relationship to the research domain encompasses renewable energy sources that convert hydrogen into various processes, including thermochemical (gasification, pyrolysis, combustion, and liquefaction), or biological (bio-photolysis, dark and photo fermentation) [105]. This suggests a focus on biomass as a potential feedstock for sustainable hydrogen production, highlighting the interest in diversifying hydrogen sources beyond fossil fuels.

Surprisingly, in 2017, coal (n = 5) emerged as a trending topic. Coal is traditionally associated with conventional energy production methods, but the presence of publications at this period suggests a possible exploration of coal-related technologies in the context of hydrogen production. This could be related to research on coal gasification or other processes to extract hydrogen from coal in a more sustainable manner [25], [79], [80].

In 2021, nitrogen (N_2) fixation, ammonia (NH_3) synthesis, and hydrogen production emerged as the trending topic with about 11, 13, and 24 publications, respectively. To the research domain, N_2 fixation is essential for sustainable agriculture and environmental conservation, as it involves converting atmospheric N_2 into NH_3 or other N_2–compounds. While not directly related to hydrogen production, N_2 fixation research could be linked to efforts to develop sustainable and efficient processes for producing ammonia, which is a precursor for hydrogen production. Moreover, ammonia synthesis became relevant to sustainable hydrogen production as it involves the utilization of hydrogen to produce ammonia, which is a key chemical in various industries. The number of publications in 2021 indicates ongoing research into improving the efficiency and sustainability of ammonia synthesis processes, which could contribute to the broader goal of sustainable hydrogen production. This is driven by the increase in hydrogen demand as a clean fuel and the need to scale up the production process to cater for future energy requirements.

In 2022, the emerging trend narrowed down to plasma gasification-based hydrogen production. These trend topics are plasma gasification (n = 19), plasma (n = 28), and hydrogen (n = 33). Firstly, plasma gasification is a thermal process that uses plasma to convert organic materials/feedstocks into syngas, which can then be utilized for H_2 production. The noticeable publications during this period denotes a growing interest in this technology for converting waste into valuable resources, as aligned with the sustainability goals of H_2 production. Research findings in **Table 1** have shown that plasma-based technologies are at the core of sustainable hydrogen production methods, as plasma can be used to dissociate water or reform hydrocarbons

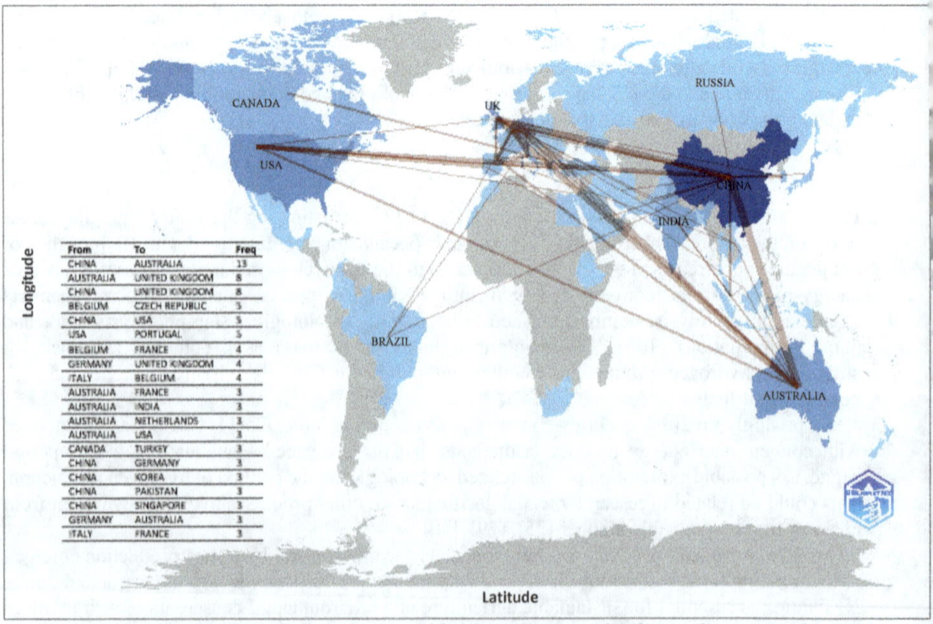

From	To	Freq
CHINA	AUSTRALIA	13
AUSTRALIA	UNITED KINGDOM	8
CHINA	UNITED KINGDOM	8
BELGIUM	CZECH REPUBLIC	5
CHINA	USA	5
USA	PORTUGAL	5
BELGIUM	FRANCE	4
GERMANY	UNITED KINGDOM	4
ITALY	BELGIUM	4
AUSTRALIA	FRANCE	3
AUSTRALIA	INDIA	3
AUSTRALIA	NETHERLANDS	3
AUSTRALIA	USA	3
CANADA	TURKEY	3
CHINA	GERMANY	3
CHINA	KOREA	3
CHINA	PAKISTAN	3
CHINA	SINGAPORE	3
GERMANY	AUSTRALIA	3
ITALY	FRANCE	3

***Fig. 10:** Top Countries' collaboration world map on plasma technologies for sustainable hydrogen production*

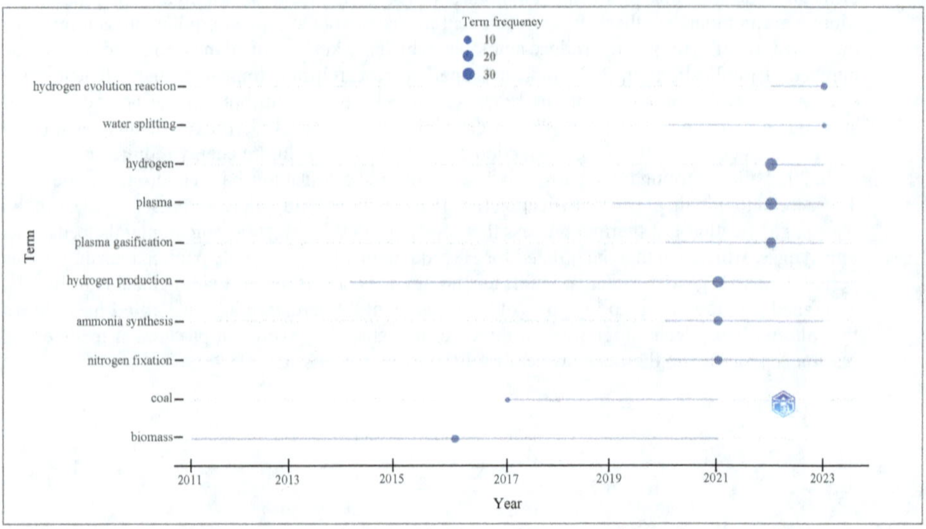

***Fig.11:** Emerging trend topics on plasma technologies for sustainable hydrogen production*
***Source:** Biblioshiny using the Authors' dataset [59]*

or other feedstocks into hydrogen. Lastly, the high frequency of publications in 2022 reflects a significant interest in hydrogen as an energy carrier and its role in sustainable energy systems.

In 2023, the emerging trends focused on water splitting (n = 5), and hydrogen evolution reaction (n = 7). The transition into these emerging trends in recent years suggests a deepening understanding towards a more specialized field of research in sustainable hydrogen production. Water splitting has been a fundamental process in H_2 production [110], [111], whereby water is split into hydrogen and oxygen using various methods such as thermolysis, photolysis, and electrolysis (alkaline, solid oxide, microbial, proton exchange membrane) [105]. These prompted the focus on hydrogen evolution research (HER) which involves the electrochemical process of generating hydrogen from water, using a catalyst to accelerate/speed the reaction. This is a crucial aspect of sustainable hydrogen production techniques that is driven by the need for cost-effective and efficient methods for hydrogen production.

Therefore, these trend topics reflect a multidisciplinary approach to advancing plasma-based technologies for sustainable hydrogen production, encompassing some specific areas including electrochemistry, environmental science, materials science, and renewable energy. The increasing number of publications in recent years depicts a growing research interest and a concerted effort to develop innovative and sustainable solutions for hydrogen production.

Conclusion

The analysis of plasma-based technologies for sustainable hydrogen production reveals a growing interest in academia, as evidenced by the increasing number of publications (n = 277), annual growth rate (n = 23%), authors' contributions (n = 1,185), references (n = 14,317), and international co-authorship (n = 30%). The field has evolved over four distinct phases from 2001 to 2023, with significant advancements in research output and citation impact. The key areas of plasma technologies include non-thermal (MW, GAD, DBD, Pulsed-Corona, and Corona-DBD hybrid), warm plasma, and thermal plasma, which show highly efficient rates of hydrogen production based on their specific areas of application. The study identifies the MW plasma reactor as the most prominent technology for hydrogen production, with the highest yield productivity (45-65 vol%), among the non-thermal plasma. In the analysis of countries, China has shown significant research output in this study, driven by its energy policies and commitments to sustainability. Also, China and Australia have engaged in strong collaboration and cooperation in the field of energy and its policy towards sustainable development. Moreover, the co-occurrence network analysis emphasizes the multidimensional nature of research in this domain, covering areas such as energy sources, environmental impact, and technological processes. The emerging research trends indicate a focus on biomass utilization, coal gasification, ammonia synthesis, plasma gasification, water splitting, and hydrogen evolution reaction, demonstrating a shift towards more sustainable and efficient methods for hydrogen production. Therefore, the analysis shows the importance of plasma-based technologies in the transition towards a more sustainable energy future, with ongoing research focusing on improving efficiency, sustainability, and environmental impact.

Based on the research findings, the following recommendations can be deduced:

- Technological advancements: Further research is needed to explore advancements in plasma-based technology such as hybrid plasma reactors to improve efficiency and sustainability.

- Environmental impact: Future research should focus on assessing the environmental impact of plasma-based technologies for hydrogen production, including life cycle assessments and carbon footprint analyses.

- Policy implications: It is crucial to consider the policy implications of adopting plasma-based technologies for hydrogen production, including regulatory frameworks and incentives for sustainable energy development.

- Collaborative research: Encouraging international collaboration can facilitate knowledge exchange and accelerate advancements in plasma-based technologies for sustainable hydrogen production.

- Interdisciplinary approach: Adopting an interdisciplinary approach that integrates fields such as energy, environmental science, and materials science can lead to innovative solutions for sustainable hydrogen production.

Conclusively, the findings suggest a growing interest and potential for plasma-based technologies to play a significant role in sustainable hydrogen production, with opportunities for further research and development in various aspects of the field.

Acknowledgement
The authors sincerely appreciate the financial assistance received from the Petroleum Technology Development Fund (PTDF), Nigeria. Towards an effective approach to industrial waste flare utilization in producing high-value products for sustainable energy generation. Under the sponsorship candidate number: PTDF/ED/OSS/PHD/HBA/1738/2020PHD046. Also, the prestigious citadel of learning – Universiti Teknologi PETRONAS is acknowledged.

References

[1] Q. Hassan, S. Algburi, A. Z. Sameen, H. M. Salman, and M. Jaszczur, "Green hydrogen: A pathway to a sustainable energy future," *Int. J. Hydrogen Energy*, vol. 50, pp. 310–333, 2024. https://doi.org/10.1016/j.ijhydene.2023.08.321

[2] H. B. Adedayo, S. A. Adio, and B. O. Oboirien, "Energy research in Nigeria: A bibliometric analysis," *Energy Strateg. Rev.*, vol. 34, no. 100629, pp. 1–18, 2021. https://doi.org/10.1016/j.esr.2021.100629

[3] L. Guo *et al.*, "Hydrogen safety: An obstacle that must be overcome on the road towards future hydrogen economy," *Int. J. Hydrogen Energy*, vol. 51, pp. 1055–1078, 2024. https://doi.org/10.1016/j.ijhydene.2023.08.248

[4] C. M. Kalamaras and A. M. Efstathiou, "Hydrogen Production Technologies: Current State and Future Developments," in *Conference Papers in Energy*, 2013, pp. 1–9. doi: 10.1155/2013/690627

[5] S. F. Ahmed *et al.*, "Sustainable hydrogen production: Technological advancements and economic analysis," *Int. J. Hydrogen Energy*, vol. 47, no. 88, pp. 37227–37255, 2022. https://doi.org/10.1016/j.ijhydene.2021.12.029

[6] I. Dincer and C. Acar, "Review and evaluation of hydrogen production methods for better sustainability," *Int. J. Hydrogen Energy*, vol. 40, no. 34, pp. 11094–11111, 2014. https://doi.org/10.1016/j.ijhydene.2014.12.035

[7] G. F. A. C. D. I. Naterer, "Review of photocatalytic water-splitting methods for sustainable hydrogen production," *Int. J. Energy Res.*, vol. 40, no. 11, pp. 1449–1473, 2016. https://doi.org/10.1002/er.3549

[8] D. W. M. W. G. S. Z. Z. Y. Xu, "Review of renewable energy-based hydrogen production processes for sustainable energy innovation," *Glob. Energy Interconnect.*, vol. 2, no. 5, pp. 436–443, 2019. https://doi.org/10.1016/j.gloei.2019.11.019

[9] M. S. M. N. S. G. A. O. E. E. I. A. A. A. H. B. P. S. A.-S. D. Kamal, "Hydrogen production, transportation, utilization, and storage: Recent advances towards sustainable energy," *J. Energy Storage*, vol. 73, p. 109207, 2023. https://doi.org/10.1016/j.est.2023.109207

[10] M. M. A. B. E. M. P. A. B. A. Fermeglia, "Sustainability analysis of hydrogen production processes," *Int. J. Hydrogen Energy*, vol. 54, pp. 540–553, 2024. https://doi.org/10.1016/j.ijhydene.2023.06.122

[11] J. Mizeraczyk and M. Jasiński, "Plasma processing methods for hydrogen production," *EPJ Appl. Phys.*, vol. 75, no. 2, pp. 1–7, 2016. https://doi.org/10.1051/epjap/2016150561

[12] J. Mizeraczyk, K. Urashima, M. Jasiński, and M. Dors, "Hydrogen Production from gaseos fuel by plasmas- A review," *Int. J. Plasma Environ. Sci. Technol.*, vol. 8, no. 2, pp. 89–97, 2014

[13] G. Chen, X. Tu, G. Homm, and A. Weidenkaff, "Plasma pyrolysis for a sustainable hydrogen economy," *Nat. Rev. Mater.*, vol. 7, no. 5, pp. 333–334, 2022. https://doi.org/10.1038/s41578-022-00439-8

[14] E. Tatarova, N. Bundaleska, J. P. Sarrette, and C. M. Ferreira, "Plasmas for environmental issues: from hydrogen production to 2D materials assembly," *Plasma Sources Sci. Technol.*, vol. 23, no. 6, p. 063002, 2014. https://doi.org/10.1088/0963-0252/23/6/063002

[15] K. W. K. K. J. F. H. M. U. D. Š. M. O. K. (Ken) H. S. C. U. Č. M. L. B. F. A. F. P. Becker, "The future for plasma science and technology," *Plasma Process. Polym.*, vol. 16, no. 1, p. 1800118, 2018. https://doi.org/10.1002/ppap.201800118

[16] A. S. R. Bogaerts, "Plasma technology – a novel solution for CO2 conversion?," *Chem. Soc. Rev.*, vol. 46, no. 19, pp. 5805–5863, 2017. https://doi.org/10.1039/C6CS00066E

[17] D. Czylkowski, B. Hrycak, R. Miotk, M. Jasiński, J. Mizeraczyk, and M. Dors, "Microwave plasma for hydrogen production from liquids," *Nukleonika*, vol. 61, no. 2, pp. 185–190, 2016. https://doi.org/10.1515/nuka-2016-0031

[18] B. Sun, X. Zhao, Y. Xin, and X. Zhu, "Large capacity hydrogen production by microwave discharge plasma in liquid fuels ethanol," *Int. J. Hydrogen Energy*, vol. 42, no. 38, pp. 24047–24054, 2017. https://doi.org/10.1016/j.ijhydene.2017.08.052

[19] J. Kierzkowska-Pawlak, H., Tyczkowski, A. Jarota, and H. Abramczyk, "Hydrogen production in liquid water by femtosecond laser-induced plasma," *Appl. Energy*, vol. 249, pp. 24–31, 2019. https://doi.org/10.1016/j.apenergy.2019.04.010

[20] K.-H. Chung, S. S. Lam, Y.-K. Park, and S.-C. Jung, "Enhanced hydrogen production from cracking of liquid toluene by applying liquid plasma and perovskite catalysts," *Int. J. Hydrogen Energy*, vol. 52, pp. 612–621, 2024. https://doi.org/10.1016/j.ijhydene.2023.05.092

[21] C. Du, J. Mo, and H. Li, "Renewable Hydrogen Production by Alcohols Reforming Using Plasma and Plasma-Catalytic Technologies: Challenges and Opportunities," *Chem. Rev.*, vol. 115, no. 3, pp. 1503–1542, 2014. https://doi.org/10.1021/cr5003744

[22] G. Akay, "Hydrogen, Ammonia and Symbiotic/Smart Fertilizer Production Using Renewable Feedstock and CO2 Utilization through Catalytic Processes and Nonthermal Plasma with Novel Catalysts and In Situ Reactive Separation: A Roadmap for Sustainable and

Innovation-Base," *Catalysts*, vol. 13, no. 9, pp. 1287–1287, 2023. https://doi.org/10.3390/catal13091287

[23] A. Midilli, H. Kucuk, M. E. Topal, U. Akbulut, and I. Dincer, "A comprehensive review on hydrogen production from coal gasification: Challenges and Opportunities," *Int. J. Hydrogen Energy*, vol. 46, no. 50, pp. 25385–25412, 2021. https://doi.org/10.1016/j.ijhydene.2021.05.088

[24] J. Favas, E. Monteiro, and A. Rouboa, "Hydrogen production using plasma gasification with steam injection," *Int. J. Hydrogen Energy*, vol. 42, no. 16, pp. 10997–11005, 2017. https://doi.org/10.1016/j.ijhydene.2017.03.109

[25] V. Galvita, V. E. Messerle, and A. B. Ustimenko, "Hydrogen production by coal plasma gasification for fuel cell technology," *Int. J. Hydrogen Energy*, vol. 32, no. 16, pp. 3899–3906, 2007. https://doi.org/10.1016/j.ijhydene.2007.05.039

[26] S. Yousef, A. Tamošiūnas, M. Aikas, R. Uscila, D. Gimžauskaitė, and K. Zakarauskas, "Plasma steam gasification of surgical mask waste for hydrogen-rich syngas production," *Int. J. Hydrogen Energy*, vol. 49, pp. 1375–1386, 2024. https://doi.org/10.1016/j.ijhydene.2023.09.288

[27] K. Yin *et al.*, "Thermodynamic analysis of a plasma co-gasification process for hydrogen production using sludge and food waste as mixed raw materials," *Renew. Energy*, vol. 222, pp. 119893–119893, 2024. https://doi.org/10.1016/j.renene.2023.119893

[28] P. V. Mallick, R., "Experimental studies on CO2-thermal plasma gasification of refused derived fuel feedstock for clean syngas production," *Energy*, vol. 288, pp. 129766–129766, 2024. https://doi.org/10.1016/j.energy.2023.129766

[29] Y. Wang *et al.*, "Hydrogen-rich gas production from tar model compound disintegrates over low-temperature plasma in dielectric barrier discharge reactor," *Int. J. Hydrogen Energy*, vol. 58, pp. 678–687, 2024. https://doi.org/10.1016/j.ijhydene.2024.01.250

[30] Z. Wang *et al.*, "H2 production from ammonia decomposition with Mo2N catalyst driven by dielectric barrier discharge plasma," *Int. J. Hydrogen Energy*, vol. 49, pp. 1375–1385, 2024. https://doi.org/10.1016/j.ijhydene.2023.06.173

[31] B. Sarmiento, J. J. Brey, I. G. Viera, A. R. González-Elipe, J. Cotrino, and V. J. Rico, "Hydrogen production by reforming of hydrocarbons and alcohols in a dielectric barrier discharge," *J. Power Sources*, vol. 169, no. 1, pp. 140–143, 2007. https://doi.org/10.1016/j.jpowsour.2007.01.059

[32] A. H. Khoja *et al.*, "Hydrogen Production from Methane Cracking in Dielectric Barrier Discharge Catalytic Plasma Reactor Using a Nanocatalyst," *Energies*, vol. 13, no. 22, p. 5921, 2020. https://doi.org/10.3390/en13225921

[33] Y. Nishida, H.-C. Chiang, T.-C. Chen, and C.-Z. Cheng, "Efficient Production of Hydrogen by DBD Type Plasma Discharges," *IEEE Trans. Plasma Sci.*, vol. 42, no. 12, pp. 3765–3771, 2014. https://doi.org/10.1109/tps.2014.2354695

[34] K. Khan, A., Rashid, M., Rehman, A., Saleem, F., Salman Razza Naqvi, Afzal, S., Qazi, U. Y., Ahmad, W., Iftikhar Nisar Butt, Mahmood, "A comprehensive review of the methane decomposition using a gliding arc discharge reactor for hydrogen generation," *J. Energy Inst.*, vol. 109, pp. 101309–10130, 2023. https://doi.org/10.1016/j.joei.2023.101309

[35] W. Baowei, W., Shize, L., Yeping, P., Chengyu, "Gliding arc plasma reforming of toluene for on-board hydrogen production," *Int. J. Hydrogen Energy*, vol. 45, no. 11, pp. 6138–6147, 2020. https://doi.org/10.1016/j.ijhydene.2019.12.184

[36] C. J. Zou, J. J., Zhang, Y. P., Liu, "Hydrogen production from partial oxidation of dimethyl ether using corona discharge plasma," *Int. J. Hydrogen Energy*, vol. 32, no. 8, pp. 958–964, 2007. https://doi.org/10.1016/j.ijhydene.2006.09.023

[37] Y.-H. Chang, K.-L., Lin, Y.-C., Shangdiar, S., Chen, S.-C., Hsiao, "Hydrogen production from dry spirulina algae with downstream feeding in microwave plasma reactor assisted under atmospheric pressure," *J. Energy Inst.*, vol. 93, no. 4, pp. 1597–1601, 2020. https://doi.org/10.1016/j.joei.2020.01.021

[38] H. Putra, A. E. E., Nomura, S., Mukasa, S., Toyota, "Hydrogen production by radio frequency plasma stimulation in methane hydrate at atmospheric pressure," *Int. J. Hydrogen Energy*, vol. 37, no. 21, pp. 16000–16005, 2012. https://doi.org/10.1016/j.ijhydene.2012.07.099

[39] N. Longmier, B. W., Gallimore, A. D., Hershkowitz, "Hydrogen production from methane using an RF plasma source in total nonambipolar flow," *Plasma Sources Sci. Technol.*, vol. 21, no. 1, p. 015007, 2012. https://doi.org/10.1088/0963-0252/21/1/015007

[40] S. Heijkers, M. Aghaei, and A. Bogaerts, "Plasma-Based CH4 Conversion into Higher Hydrocarbons and H2: Modeling to Reveal the Reaction Mechanisms of Different Plasma Sources," *J. Phys. Chem. C*, vol. 124, no. 13, pp. 7016–7030, 2020. https://doi.org/10.1021/acs.jpcc.0c00082

[41] W. Zhou, Z., Zhang, J., Ye, T., Zhao, P., Xia, "Hydrogen production by reforming methane in a corona inducing dielectric barrier discharge and catalyst hybrid reactor," *Chinese Sci. Bull.*, vol. 56, no. 20, pp. 2162–2166, 2011. https://doi.org/10.1007/s11434-011-4485-0

[42] E. Blanquet, M. A. Nahil, and P. T. Williams, "Enhanced hydrogen-rich gas production from waste biomass using pyrolysis with non-thermal plasma-catalysis," *Catal. Today*, vol. 337, pp. 216–224, 2019. https://doi.org/10.1016/j.cattod.2019.02.033

[43] W. Li *et al.*, "Non-thermal plasma assisted catalytic water splitting for clean hydrogen production at near ambient conditions," *J. Clean. Prod.*, vol. 387, pp. 135913–135913, 2023. https://doi.org/10.1016/j.jclepro.2023.135913

[44] O. Khalifeh, H. Taghvaei, A. Mosallanejad, M. R. Rahimpour, and A. Shariati, "Extra pure hydrogen production through methane decomposition using nanosecond pulsed plasma and Pt-Re catalyst," *Chem. Eng. J.*, vol. 294, pp. 132–145, 2016. https://doi.org/10.1016/j.cej.2016.02.077

[45] N. Zhu, Y. Hong, F. Qian, and J. Liang, "Research progress on plasma-catalytic hydrogen production from ammonia: Influencing factors and reaction mechanism," *Int. J. Hydrogen Energy*, vol. 59, pp. 791–807, 2024. https://doi.org/10.1016/j.ijhydene.2024.02.016

[46] M. C. R. Younas, "Emerging trends and technologies in big data processing," *Concurr. Comput. Pract. Exp.*, vol. 27, no. 8, pp. 2078–2091, 2014. https://doi.org/10.1002/cpe.3398

[47] J. G. Y. X. Q. G. J. Y. Y. Li, "Progress and framework of clean energy production: Bibliometric analysis from 2002 to 2022.," *Energy Strateg. Rev.*, pp. 101270–101270, 2024. https://doi.org/10.1016/j.esr.2023.101270

[48] L. J. M. P. W. Jaramillo, "Adaptive Forecasting in Energy Consumption: A Bibliometric Analysis and Review," *Data*, vol. 9, no. 1, p. 13, 2024. https://doi.org/10.3390/data9010013

[49] Y. Z. D. J. M. L. G. Tang, "An advanced bibliometric analysis and future research insights on safety of hydrogen energy," *J. Energy Storage*, vol. 77, pp. 109833–109833, 2024. https://doi.org/10.1016/j.est.2023.109833

[50] N. Donthu, S. Kumar, D. Mukherjee, N. Pandey, and W. M. Lim, "How to conduct a bibliometric analysis: An overview and guidelines," *J. Bus. Res.*, vol. 133, pp. 285–296, 2021. https://doi.org/10.1016/j.jbusres.2021.04.070

[51] A. Martín-Martín, M. Thelwall, E. Orduna-Malea, and E. Delgado López-Cózar, *Google Scholar, Microsoft Academic, Scopus, Dimensions, Web of Science, and OpenCitations' COCI: a multidisciplinary comparison of coverage via citations*, vol. 126, no. 1. Springer International Publishing, 2021. doi: 10.1007/s11192-020-03690-4

[52] R. Pranckutė, "Web of Science (Wos) and Scopus: The Titans of Bibliographic Information in Today's Academic World," *Publications*, vol. 9, no. 12, pp. 1–59, 2021. https://doi.org/10.3390/publications9010012

[53] S. Echchakoui, "Why and how to merge Scopus and Web of Science during bibliometric analysis: the case of sales force literature from 1912 to 2019," *J. Mark. Anal.*, vol. 8, no. 3, pp. 165–184, 2020. https://doi.org/10.1057/s41270-020-00081-9

[54] H. Kasaraneni and S. Rosaline, "Automatic merging of Scopus and Web of Science data for simplified and effective bibliometric analysis," *Ann. Data Sci.*, vol. 11, no. 3, pp. 785–802, 2024. https://doi.org/10.1007/s40745-022-00438-0

[55] M. J. Page *et al.*, "The PRISMA 2020 statement: An updated guideline for reporting systematic reviews," *Br. Med. J.*, vol. 372, no. 71, pp. 1–9, 2021. https://doi.org/10.1136/bmj.n71

[56] L. O. Soares, A. da C. Reis, P. S. Vieira, L. Hernández-Callejo, and R. A. M. Boloy, "Electric Vehicle Supply Chain Management: A Bibliometric and Systematic Review," *Energies*, vol. 16, no. 4, p. 1563, 2023. https://doi.org/10.3390/en16041563

[57] A. D. Samala *et al.*, "Metaverse Technologies in Education: A Systematic Literature Review Using PRISMA," *Int. J. Emerg. Technol. Learn.*, vol. 18, no. 5, pp. 231–252, 2023. https://doi.org/10.3991/IJET.V18I05.35501

[58] N. R. Haddaway, M. J. Page, C. C. Pritchard, and L. A. McGuinness, "PRISMA2020: An R package and Shiny app for producing PRISMA 2020-compliant flow diagrams, with interactivity for optimised digital transparency and Open Synthesis," *Campbell Syst. Rev.*, vol. 18, no. 2, p. e1230, 2022. https://doi.org/10.1002/cl2.1230

[59] M. Aria and C. Cuccurullo, "bibliometrix: An R-tool for comprehensive science mapping analysis," *J. Informetr.*, vol. 11, no. 4, pp. 959–975, 2017. https://doi.org/10.1016/j.joi.2017.08.007

[60] M. A. Rojas-Sánchez, P. R. Palos-Sánchez, and J. A. Folgado-Fernández, *Systematic literature review and bibliometric analysis on virtual reality and education*, vol. 28, no. 1. Springer US, 2023. doi: 10.1007/s10639-022-11167-5

[61] C. Bota-Avram, "Bibliometric analysis of sustainable business performance: where are we going? A science map of the field," *Econ. Res. Istraz.* , vol. 36, no. 1, pp. 2137–2176, 2023. https://doi.org/10.1080/1331677X.2022.2096094

[62] S. A. Samsul, N. Yahaya, and H. Abuhassna, "Education big data and learning analytics: a bibliometric analysis," *Humanit. Soc. Sci. Commun.*, vol. 10, no. 1, pp. 1–11, 2023. https://doi.org/10.1057/s41599-023-02176-x

[63] M. Afjal, "Bridging the financial divide: a bibliometric analysis on the role of digital financial services within FinTech in enhancing financial inclusion and economic development,"

Humanit. Soc. Sci. Commun., vol. 10, no. 1, pp. 1–27, 2023. https://doi.org/10.1057/s41599-023-02086-y

[64] P. G. Rutberg *et al.*, "Plasma technologies of solid and liquid toxic waste disinfection," in *IEEE Conference Record - Abstracts. PPPS-2001 Pulsed Power Plasma Science 2001. 28th IEEE International Conference on Plasma Science and 13th IEEE International Pulsed Power Conference (Cat. No.01CH37255)*, 2001, pp. 1178–1181. doi: 10.1109/ppps.2001.961002

[65] K. Yan, E. J. M. van Heesch, A. J. M. Pemen, and S. A. Nair, "Corona Plasmas: Fundamental Studies and Industrial Applications," *WIT Trans. Ecol. Environ.*, vol. 56, pp. 110–118, 2002. https://doi.org/10.2495/wm020121

[66] E. J. M. van Heesch *et al.*, "Repetitive pulsed power to serve nano technology, sustainability and hydrogen production," *Dig. Tech. Pap. PPC-2003. 14th IEEE Int. Pulsed Power Conf. (IEEE Cat. No. 03CH37472)*, vol. 1, pp. 441–444, 2003. https://doi.org/10.1109/ppc.2003.1277746

[67] N. Muradov, F. Smith, G. Bockerman, and K. Scammon, "Thermocatalytic decomposition of natural gas over plasma-generated carbon aerosols for sustainable production of hydrogen and carbon," *Appl. Catal. A Gen.*, vol. 365, no. 2, pp. 292–300, 2009, [Online]. Available: 10.1016/j.apcata.2009.06.031

[68] H. Kaneko, L. C., Y. Ishikawa, K. Hosogoe, and Y. Tamaura, "Solar H2 Production With Tokyo Tech Rotary-Type Solar Reactor to be Tested Using Solar Concentration System at CSIRO in Australia," in *In Energy Sustainability*, 2009, pp. 491–496. doi: 10.1115/es2009-90420

[69] Q. Wang, H. Shi, B. Yan, Y. Jin, and Y. Cheng, "Steam enhanced carbon dioxide reforming of methane in DBD plasma reactor," *Int. J. Hydrogen Energy*, vol. 36, no. 14, pp. 8301–8306, 2011. https://doi.org/10.1016/j.ijhydene.2011.04.084

[70] J. F. Pacheco, G. Soria, M. Pacheco, H. Ricardo Rueda Valdivia, Ramos, F., and M. Frías, M. Suarez Durán, Hidalgo, "Greenhouse gas treatment and H2 production, by warm plasma reforming," *Int. J. Hydrogen Energy*, vol. 40, p. 2015, 2015. https://doi.org/10.1016/j.ijhydene.2015.08.062

[71] N. Tippayawong, E. Chaiya, P. Thanompongchart, and P. Khongkrapan, "Sustainable Energy from Biogas Reforming in a Microwave Discharge Reactor," in *Procedia Engineering*, 2015, pp. 120–127. doi: 10.1016/j.proeng.2015.08.410

[72] B. Ibrahimoglu, A. Cucen, and M. Z. Yilmazoglu, "Numerical modeling of a downdraft plasma gasification reactor," *Int. J. Hydrogen Energy*, vol. 42, no. 4, pp. 2583–2591, 2017. https://doi.org/10.1016/j.ijhydene.2016.06.224

[73] J. Sun *et al.*, "Novel treatment of a biomass tar model compound via microwave-metal discharges," *Fuel*, pp. 121–125, 2017. https://doi.org/10.1016/j.fuel.2017.06.075

[74] S. Liu, D. Mei, Y. Wang, Y. Ma, and X. Tu, "Plasma reforming of toluene as a model tar compound from biomass gasification: effect of CO2 and steam," *Waste Dispos. Sustain. Energy*, vol. 1, no. 2, pp. 133–141, 2019. https://doi.org/10.1007/s42768-019-00011-1

[75] A. Inayat, A., Tariq, R., Khan, Z., Ghenai, C., Kamil, M., Jamil, F., Shanableh, "A comprehensive review on advanced thermochemical processes for bio-hydrogen production via microwave and plasma technologies," *Biomass Convers. Biorefinery*, pp. 1–10, 2020. https://doi.org/10.1007/s13399-020-01175-1

[76] S. Vecten, M. Wilkinson, N. Bimbo, R. Dawson, and B. M. Herbert, "Hydrogen-rich syngas production from biomass in a steam microwave-induced plasma gasification reactor.," *Bioresour. Technol.*, vol. 337, p. 125324, 2021. https://doi.org/10.1016/j.biortech.2021.125324

[77] H. Liu, D. Guo, X. Ma, Y. Wang, and J. Xie, "Effect of non-thermal plasma on carbon dioxide reforming of methane to hydrogen," *Proc. Inst. Civ. Eng.*, vol. 174, no. 2, pp. 67–78, 2021. https://doi.org/10.1680/jener.19.00030

[78] A. Midilli, H. Kucuk, M. E. Topal, U. Akbulut, and I. Dincer, "A comprehensive review on hydrogen production from coal gasification: Challenges and Opportunities," *Int. J. Hydrogen Energy*, vol. 46, no. 50, pp. 25385–25412, 2021. https://doi.org/10.1016/j.ijhydene.2021.05.088

[79] S. J. Yoon and J.-G. Lee, "Hydrogen-rich syngas production through coal and charcoal gasification using microwave steam and air plasma torch," *Int. J. Hydrogen Energy*, vol. 37, no. 22, pp. 17093–17100, 2012. https://doi.org/10.1016/j.ijhydene.2012.08.054

[80] J. Qiu *et al.*, "Coal gasification in steam and air medium under plasma conditions: a preliminary study," *Fuel Process. Technol.*, vol. 85, no. 8–10, pp. 969–982, 2004. https://doi.org/10.1016/j.fuproc.2003.11.035

[81] M. El-Shafie, S. Kambara, and Y. Hayakawa, "Energy and exergy analysis of hydrogen production from ammonia decomposition systems using non-thermal plasma," *Int. J. Hydrogen Energy*, vol. 46, no. 57, pp. 29361–29375, 2020. https://doi.org/10.1016/j.ijhydene.2020.08.249

[82] V. S. Sikarwar, A. Reichert, M. Pohorely, E. Meers, N. L. Ferreira, and M. Jeremias, "Equilibrium modeling of thermal plasma assisted co-valorization of difficult waste streams for syngas production," *Sustain. Energy Fuels*, vol. 5, no. 18, pp. 4650–4660, 2021. https://doi.org/10.1039/d1se00998b

[83] G. Xu *et al.*, "Self-perpetuating carbon foam microwave plasma conversion of hydrocarbon wastes into useful fuels and chemicals.," *Environ. Sci. Technol.*, vol. 55, no. 9, pp. 6239–6247, 2021. https://doi.org/10.1021/acs.est.0c06977

[84] M. Younas *et al.*, "Hydrogen Production through Water Vapors using Optimized Corona-DBD Hybrid Plasma Micro-Reactor," *Fuel*, vol. 331, p. 125838, 2023. https://doi.org/10.1016/j.fuel.2022.125838

[85] N. Budhraja, A. Pal, and R. S. Mishra, "Plasma reforming for hydrogen production: Pathways, reactors and storage," *Int. J. Hydrogen Energy*, vol. 48, no. 7, pp. 2467–2482, 2023. https://doi.org/10.1016/j.ijhydene.2022.10.143

[86] P. K. Panicker and A. Magid, "Microwave plasma gasification for the restoration of urban rivers and lakes, and the elimination of oceanic garbage patches," *Energy Sustain.*, vol. 50220, p. V001T02A011, 2016. https://doi.org/10.1115/ES2016-59632

[87] P. K. Panicker and A. Magid, "Microwave plasma gasification for enhanced oil recovery and sustainable waste management," in *ASME 2016 10th International Conference on Energy Sustainability, ES 2016, collocated with the ASME 2016 Power Conference and the ASME 2016 14th International Conference on Fuel Cell Science, Engineering and Technology*, 2016, pp. 1–13. doi: 10.1115/ES2016-59630

[88] G. Akay, "Sustainable ammonia and advanced symbiotic fertilizer production using catalytic multi-reaction-zone reactors with nonthermal plasma and simultaneous reactive separation," *ACS Sustain. Chem. Eng.*, vol. 5, no. 12, pp. 11588–11606, 2017. https://doi.org/10.1021/acssuschemeng.7b02962

[89] R. Peng, P., Chen, P., Schiappacasse, C., Zhou, N., Anderson, E., Chen, D., Liu, J., Cheng, Y., Hatzenbeller, R., Addy, M., Zhang, Y., Liu, Y., Ruan, "A review on the non-thermal plasma-assisted ammonia synthesis technologies," *J. Clean. Prod.*, vol. 177, pp. 597–609, 2018. https://doi.org/10.1016/j.jclepro.2017.12.229

[90] A. George *et al.*, "A Review of Non-Thermal Plasma Technology: A novel solution for CO2 conversion and utilization," *Renew. Sustain. Energy Rev.*, vol. 135, p. 109702, 2021. https://doi.org/10.1016/j.rser.2020.109702

[91] B. Tabu *et al.*, "Hydrogen from cellulose and low-density polyethylene via atmospheric pressure nonthermal plasma," *Int. J. Hydrogen Energy*, vol. 49, pp. 745–763, 2024. https://doi.org/10.1016/j.ijhydene.2023.07.173

[92] B. Tabu *et al.*, "Nonthermal atmospheric plasma reactors for hydrogen production from low-density polyethylene," *Int. J. Hydrogen Energy*, vol. 47, no. 94, pp. 39743–39757, 2022. https://doi.org/10.1016/j.ijhydene.2022.09.161

[93] M. Boscherini, A. Storione, M. Minelli, F. Miccio, and F. Doghieri, "New Perspectives on Catalytic Hydrogen Production by the Reforming, Partial Oxidation and Decomposition of Methane and Biogas," *Energies*, vol. 16, no. 17, p. 6375, 2023. https://doi.org/10.3390/en16176375

[94] D. A. Gonzalez-Casamachin, T. Qin, W. M. Huang, S. Rangarajan, L. Zhang, and J. Baltrusaitis, "Actively Learned Optimal Sustainable Operation of Plasma-Catalyzed Methane Bireforming on La0. 7Ce0. 3NiO3 Perovskite," *ACS Sustain. Chem. Eng.*, vol. 12, no. 1, pp. 610–622, 2023. https://doi.org/10.1021/acssuschemeng.3c06929

[95] H. M. Nguyen, A. Omidkar, W. Li, S. Meng, Z. Li, and H. Song, "Non-thermal plasma assisted catalytic nitrogen fixation with methane at ambient conditions," *Chem. Eng. J.*, vol. 471, pp. 144748–144748, 2023. https://doi.org/10.1016/j.cej.2023.144748

[96] H. M. Nguyen, A. Omidkar, and H. Song, "Technical Challenges and Prospects in Sustainable Plasma Catalytic Ammonia Production from Methane and Nitrogen," *Chempluschem*, vol. 88, no. 7, p. 2023, 2023. https://doi.org/10.1002/cplu.202300129

[97] E. Meloni, L. Cafiero, M. Martino, and V. Palma, "Structured Catalysts for Non-Thermal Plasma-Assisted Ammonia Synthesis," *Energies*, vol. 16, no. 7, pp. 3218–3218, 2023. https://doi.org/10.3390/en16073218

[98] K. P. Bhatt, S. Patel, D. S. Upadhyay, and R. N. Patel, "In-depth analysis of the effect of catalysts on plasma technologies for treatment of various wastes," *J. Environ. Manage.*, vol. 344, pp. 118335–118335, 2023. https://doi.org/10.1016/j.jenvman.2023.118335

[99] J. Ding *et al.*, "Sustainable ammonia synthesis from air by the integration of plasma and electrocatalysis techniques," *Inorg. Chem. Front.*, vol. 10, no. 19, pp. 5762–5771, 2023. https://doi.org/10.1039/d3qi00915g

[100] B. Xu *et al.*, "Plasma-enabled catalytic steam reforming of toluene as a biomass tar surrogate: Understanding the synergistic effect of plasma catalysis," *Chem. Eng. J.*, vol. 464, pp. 142696–142696, 2023. https://doi.org/10.1016/j.cej.2023.142696

[101] L. L. Sun, H. J. Cui, and Q. S. Ge, "Will China achieve its 2060 carbon neutral commitment from the provincial perspective?," *Adv. Clim. Chang. Res.*, vol. 13, no. 2, pp. 169–178, 2022. https://doi.org/10.1016/j.accre.2022.02.002

[102] J. Li, M. S. Ho, C. Xie, and N. Stern, "China's flexibility challenge in achieving carbon neutrality by 2060," *Renew. Sustain. Energy Rev.*, vol. 158, no. October 2021, p. 112112, 2022. https://doi.org/10.1016/j.rser.2022.112112

[103] Ministero dello Luilippe Economico, "Italy ' s National Energy Strategy: For a more competitive and sustainable energy," 2013. [Online]. Available: chrome-extension://efaidnbmnnnibpcajpcglclefindmkaj/https://www.mimit.gov.it/images/stories/documenti/SEN_EN_marzo2013.pdf

[104] C. A. Bollino, "The role of the National Energy Strategy in boosting italian economy," *Econ. Policy Energy Environ.*, vol. 2, no. 2013, pp. 19–35, 2013

[105] B. Haq, N. Salahu Muhammed, J. Liu, and H. Tong Chua, "Enhanced natural gas production using CO2 injection: Application to sustainable hydrogen production," *Fuel*, vol. 347, p. 128474, 2023. https://doi.org/10.1016/j.fuel.2023.128474

[106] J. K. Tamala, E. I. Maramag, K. A. Simeon, and J. J. Ignacio, "A bibliometric analysis of sustainable oil and gas production research using VOSviewer," *Clean. Eng. Technol.*, vol. 7, p. 100437, 2022. https://doi.org/10.1016/j.clet.2022.100437

[107] Y. Hua, M. Oliphant, and E. J. Hu, "Development of renewable energy in Australia and China: A comparison of policies and status," *Renew. Energy*, vol. 85, pp. 1044–1051, 2016. https://doi.org/10.1016/j.renene.2015.07.060

[108] M. A. Young and G. Clough, *Net Zero Emissions and Free Trade Agreements: Efforts At Integrating Climate Goals By the United Kingdom and Australia*, vol. 72, no. 2. 2023. doi: 10.1017/S002058932300012X

[109] B. Senior *et al.*, "Carbon capture and storage in China - Main findings from China- UK Near Zero Emissions Coal (NZEC) initiative," *Energy Procedia*, vol. 4, pp. 5956–5965, 2011. https://doi.org/10.1016/j.egypro.2011.02.598

[110] D. Jing *et al.*, "Efficient solar hydrogen production by photocatalytic water splitting: From fundamental study to pilot demonstration," *Int. J. Hydrogen Energy*, vol. 35, no. 13, pp. 7087–7097, 2010. https://doi.org/10.1016/j.ijhydene.2010.01.030

[111] L. Yuan, C. Han, M. Q. Yang, and Y. J. Xu, "Photocatalytic water splitting for solar hydrogen generation: fundamentals and recent advancements," *Int. Rev. Phys. Chem.*, vol. 35, no. 1, pp. 1–36, 2016. https://doi.org/10.1080/0144235X.2015.1127027

Influence of Tween 65 on Carbon dioxide Hydrate System

MANJUSHA Anipeddi[1,2,a], LAL Bhajan[1,2,b*], SOW MUN Serene Lock [1,c] and DESHANNAVAR Umesh [3,d]

[1]Chemical Engineering Department, Universiti Teknologi PETRONAS, 32610, Bandar Seri Iskandar, Perak Darul Ridzuan, Malaysia

[2]Center of Carbon Capture,Utilization and Storage(CCUS),Universiti Teknologi Petronas, 32610, Bandar Seri Iskandar, Perak Darul Ridzuan, Malaysia

[3]Chemical Engineering Department, Waranagar University, Waranagar, India

[a]a.manjusha21@gmail.com, [b]bhajan.lal@utp.edu.my, [c]sowmun.lock@utp.edu.my, [d]deshannavar@gmail.com

Keywords: CO2hydrate, Promoter, Concentration, Tween 65, Kinetics

Abstract. The effect of Tween 65 promoter on the kinetics for the hydrate of CO2 to form was being experimentally evaluated . The experiments were conducted using constant cooling method in RAMAN reactor at formation pressure of 4 MPa and 275.15 K. The experiments were performed at 500,1000,1500 and 2000 ppm concentration of Tween 65. The results showed that the induction time for 500 ppm was 3.23 which reduced to 3.18 with increase in concentration of promoter to 1000 ppm. Furthermore the CO2 to hydrate conversion was found to be 58.44 for 2000 ppm and average gas uptake was more for 1500 ppm concentration. The results in this work show that Tween 65 is an effective kinetic promoter.

Introduction

The warming of the Earth's atmosphere are undeniable realities, as evidenced by the observed increase in average air and ocean temperatures worldwide, the extensive melting of glaciers and polar ice at both the North and South Poles, and the subsequent rise in sea levels. These changes are primarily driven by human activities, which release greenhouse gas emissions and aerosols into the atmosphere. The concentration of carbon dioxide (CO2) in the atmosphere has surged by a staggering 51%, from 278 parts per million (ppm) in 1750 to 421 ppm in 2023. Over the past century and a half, global average temperatures have risen by 0.6°C, and sea levels have increased by 0.17 meters. To combat this global crisis, carbon dioxide capture has emerged as a critical strategy. containing CO2 emissions from various sources before they are released into the atmosphere. The aim is to mitigate the worsening of the greenhouse effect and the resulting global warming caused by excessive CO2 emissions [1,2,3]. Various methods are employed for carbon dioxide capture, including post-combustion capture from industrial emissions, pre-combustion capture from fossil fuel processes, and direct air capture from the ambient atmosphere. Consequently, there is a pressing need to develop efficient storage techniques that enable the permanent sequestration of CO2. Carbon capture and storage (CCS) technology plays a pivotal role in achieving this objective by integrating CO2 capture from diverse sources with effective carbon sequestration technologies[20,21].

Recently, gas hydrates have gained recognition as a promising medium for CO2 capture, particularly from industrial flue gases in post-combustion scenarios. Hydrate-based separation has garnered attention as a novel carbon capture technology due to its lower energy requirements compared to conventional methods. Notably, this technology relies primarily on water as the main chemical agent with a minimal amount of additives, making it an environmentally friendly process. Selectivity is a critical parameter in any capture system, and as a relatively new field, researchers

Materials Research Proceedings 53 (2025) 43-49 https://doi.org/10.21741/9781644903575-3

are actively conducting extensive studies to enhance understanding of the kinetic aspects of hydrate formation. CO2 gas hydrates exhibit remarkable thermodynamic stability, ensuring that once formed, they can persist as a secure and reliable medium for long-term CO2 storage. Research indicates that the CO2 concentration in the hydrate phase is at least four times higher than in the gas phase. Furthermore, CO2 prefers occupying cavities at lower pressures compared to gases like nitrogen (N2) and methane (CH4), offering possibilities for gas separation in complex gas mixtures. Hydrate-based carbon capture (HBCC) technology is gaining recognition as a secure, environmentally friendly, and economically efficient approach to addressing the urgent need for carbon dioxide capture and storage in efforts to combat climate change [4,5,6]. This study is significantly focused on delving deep into the intricate kinetic aspects of CO2 hydrate formation with Tween 65 as hydrate promoter.

All the formation experiments were performed at 40 bar and 2 0C with varying concentration of Tween 65. The parameters evaluated include induction time, rate. By dedicating efforts to a comprehensive examination of these fundamental processes, the aim is to make substantial contributions to the development and refinement of environmentally friendly and economically efficient carbon capture technologies.

Methodology

Apparatus and Materials. In the course of research, a specialized high-pressure stainless-steel cell that has been ingeniously crafted by Dixson FA Engineering Sdn. Bhd is used. This meticulously designed cell serves as a crucial component for measuring the formation and dissociation of gas hydrates, drawing inspiration from its utilization in a previous study conducted by Partoon et al. Figure 1 offers a comprehensive schematic overview of experimental setup. At its core, the apparatus features a high-pressure equilibrium cell with a volumetric capacity of 500 cm^3, enabling us to conduct experiments across a wide spectrum of temperatures, ranging from as low as 253 K to as high as 523 K. Significantly, this high- pressure cell boasts a maximum operating capacity of 20 MPa.

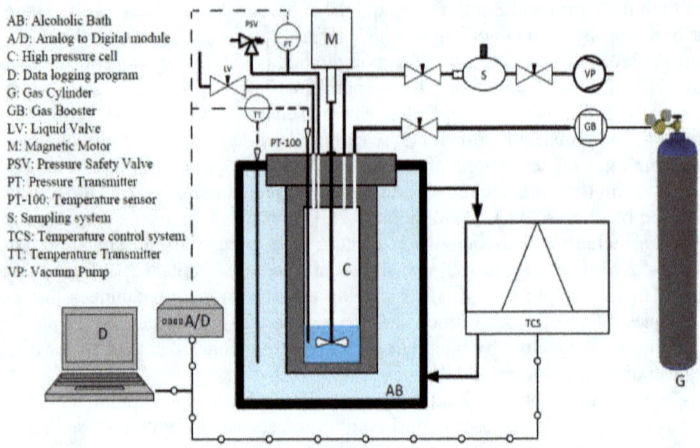

Fig. 1: Schematic Diagram of experimental setup [7,8]

Experimental procedures are characterized by meticulous precision, where critical parameters are diligently recorded. Temperature, volume, and pressure data are collected at one-second intervals, showcasing an impressive level of accuracy with ±0.1K for temperature, ±0.001 cm^3 for

Materials Research Proceedings 53 (2025) 43-49 https://doi.org/10.21741/9781644903575-3

volume, and ±0.01 MPa for pressure measurements. To facilitate optimal mixing of the test sample under the demanding conditions of experiments, the apparatus is thoughtfully equipped with a magnetic motor. This motor-driven mixing mechanism enhances the reliability and reproducibility of experimental outcomes, thereby establishing a robust foundation for this comprehensive investigations into gas hydrate formation and dissociation phenomena [7,8]. Through this diligent and well-equipped approach, the aim is to contribute valuable insights to the field of gas hydrate research, ultimately advancing understanding of this critical aspect of materials science and engineering.

Procedure. At the starting the cell is cleaned with distilled water and dried to start with experiments. Once the cell is washed it is filled with required concentration of the promoter by dissololving Distilled water 200 ml with calculated quantity of TWEEN-65 after which air is removed from the system by release of small quantity of pressure three times. At the starting stages of the experiment the temperature is being set to 285.15 K slight higher than the temperature of the equilibrium of the hydrates for the experimental pressure to avoid the formation of hydrate when the cell is being pressurized [20,21]. Then the stirrer is turned on 7 set to 50 rpm and parallelly reactor is pressurized to 4MPa and left to stabilize for 3 hrs. After the system gets stabilized the temperature is set to 275.15 K for formation of hydrate. Hydrate formation was observed with sudden drop in pressure and spike in temperature. The figure 2 below represents atypical temperature-pressure profile for CO2 hydrate. Completion of experiment is shown by no effect in temperature as well as pressure for complete day. The rise of temperature suddenly shows the start of induction time which is the time for initiation of experiment for first hydrate crystal formation[20,][21]. The induction time measurement is not so accurate for one time nature, hence it is must to perform atleast 3 times [22]. The parameters evaluated include induction time, water to hydrate conversion, gas uptake, gas to hydrate conversion.

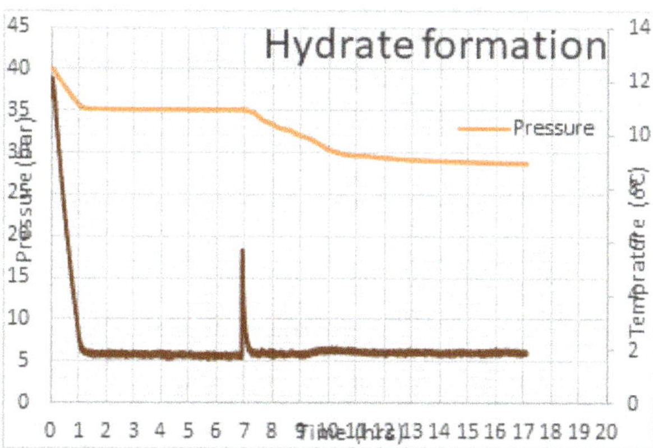

Fig. 2: Typical kinetic profile for CO2 hydrate formation

Kinetic Parameters. Induction Time. It indicates the time it takes to form first crystal of hydrate which can be calculated by the time for hydrate crystal formed subtracted by the time it takes to maintain the intial conditions required [20,21].

Water to Hydrate Conversion(%). It is given as Eq. 1 below:

$$C_{wh}(\%) = \frac{\Delta n_H \times \text{hydration number}}{n_{H_2O}} \times 100 \quad (1)$$

where ΔnH indicates the moles of gas consumed when the formation process ends that can be evaluated by moles consumed and nH2O is the number of moles of water present in the reactor [20,21,23,26,27]. Hydration number is defined as the number of water molecules needed to enclose hydrate former.

Gas Uptake. It can be obtained by the equation of gases for real system which is as follows

$$\Delta n_H = \left[\frac{PV}{zRT}\right]_0 - \left[\frac{PV}{zRT}\right]_t \quad (2)$$

In which volume is shown by Pressure by P, Compressibility factor by Z [20][21][24][26,27].

Gas to Hydrate Conversion(%). It can be obtained as follows [25]

$$C_{gh}(\%) = \frac{(\Delta n_H)}{|n_{start} - n_{equib}|} \times 100 \quad (3)$$

It represents the moles that have been utilized by system and n_{eqil} which is the moles when at the end , nstart implies before the start how many moles are present [20][21][26,27].

Results and Discussion

In this study experimental runs were performed for Tween 65 with concentration of 500,1000,1500 and 2000 ppm to determine their effect on the CO_2 hydrate formation. The effects of varying concentration on the kinetic parameters are shown in Fig. 3. It is observed that the induction time for 500 ppm is 3.23 whereas the time is reduced to 3.18 hrs for 1000 ppm concentration and for 2000 ppm is 3.42. The increase in concentration of tween 65 from 500 to 2000 ppm shows an increased inhibition effect on the kinetics of CO_2 hydrate formation. The inhibition impact of tween 65 is very strong at 2000 ppm exhibiting a much longer induction time and reduced CO_2 uptake.

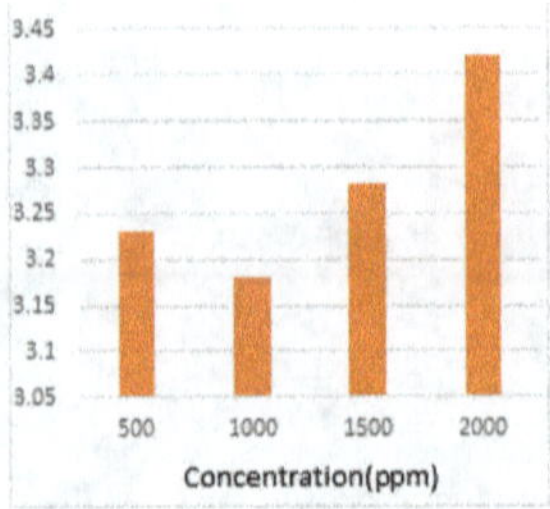

Fig. 3: Induction time v/s Varying Concentration of Tween 65

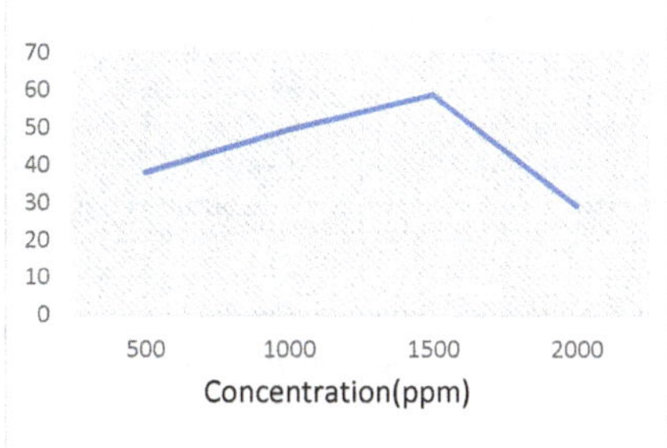

Fig. 4: Gas Uptake of CO_2 hydrate for Varying Concentration of Tween 65

The amount of CO_2 uptake is a very important parameter. The Figure 4 above shows the gas uptake for 4 different concentrations of the promoter. From the figure can be seen that the gas uptake increased by around 5% for concentration of 1500 ppm and with further increase in concentration gas uptake has decreased. The amount of CO_2 uptake is an important kinetic parameter to determine CO_2 hydrate storage capacity.With concentration of 1500 ppm CO_2 uptake is high indicating a critical concentration effect on moles of gas uptake.

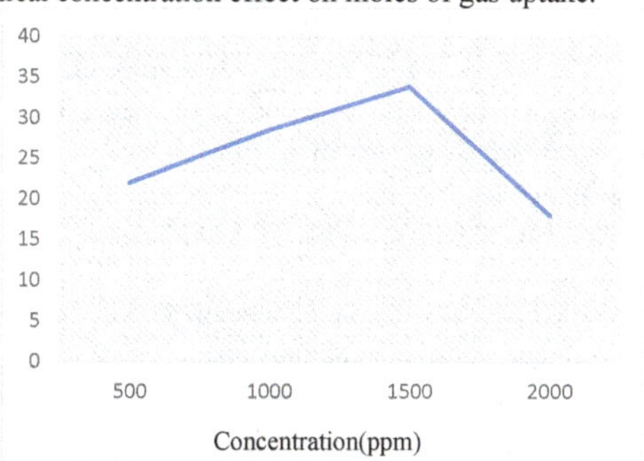

Fig. 5: Water to Hydrate conversion for varying Concentration of Tween-65

From Figure 5 it is visible that the Water to hydrate conversion is being reduced for higher concentration of the promoter which in turns gives the idea that that Tween 65 above 2000 ppm doesn't show further p promotion effect.

Conclusion

The hydrate formation kinetics is studied in presence of promoter Tween 65 using constant cooling method at 4 Mpa and 274.15 K in the high-pressure RAMAN reactor. From the results it can be seen that the higher concentration of TWEEN 65 will not have more impact on the parameters. In the pursuit of addressing the pressing issue of climate change, this study delved into the promising realm of hydrate-based carbon capture (HBCC) technology. This alignment not only underscores the robustness of experimental apparatus but also instills confidence in the readiness of HBCC for further research and application. The proposed method would be studied to determine the optimum operating conditions for separation [17,18,19]. The findings from this study would therefore be useful in handling the enormous emission of CO_2 from oil and gas fields. With a focus on mitigating the effects of greenhouse gas emissions, HBCC technology emerges as a secure and environmentally friendly approach to contribute to the global effort against climate change.

Through rigorous scientific inquiry, the way for more effective strategies in the battle against this critical global challenge.

Acknowledgement
This study was supported by the Department of Chemical Engineering, Universiti Teknologi PETRONAS, with funding from YUTP-FRG Research Grant (Grant no. 015LC0-387).

References

[1] Nesterov, A., Journal, A. R.-C. E. & 2019, undefined. New combination of thermodynamic and kinetic promoters to enhance carbon dioxide hydrate formation under static conditions. Elsevier. https://doi.org/10.1016/j.cej.2019.122165

[2] Nasir, Q., Suleman, H., and, Y. E.-J. of N. G. S. & 2020, undefined. A review on the role and impact of various additives as promoters/inhibitors for gas hydrate formation. Elsevier. https://doi.org/10.1016/j.jngse.2020.103211

[3] Prasad, P., Engineering, B. K.-J. of N. G. S. and & 2018, undefined. Are the amino acids thermodynamic inhibitors or kinetic promoters for carbon dioxide hydrates? Elsevier. https://doi.org/10.1016/j.jngse.2018.02.001

[4] Karaaslan, U. & Parlaktuna, M. Surfactants as hydrate promoters? Energy and Fuels 14, 1103-1107 (2000). https://doi.org/10.1021/ef000069s

[5] Li, Y., Gambelli, A., Rossi, F., Equilibria, S. M.-F. P. & 2021, undefined. Effect of promoters on CO2 hydrate formation: Thermodynamic assessment and microscale Raman spectroscopy/hydrate crystal morphology characterization. Elsevier. https://doi.org/10.1016/j.fluid.2021.113218

[6] Karaaslan, U., Fuels, M. P.-E. & & 2000, undefined. Surfactants as hydrate promoters? ACS Publications 34, 4397-4400 (2000).

[7] Arora, A., Cameotra, S., Kumar, R., reports, C. B.-S. & 2016, undefined. Biosurfactant as a promoter of methane hydrate formation: thermodynamic and kinetic studies. nature.com. https://doi.org/10.1038/srep20893

[8] Partoon, B., Engineering, J. J.-J. of C. & & 2013, undefined. Effect of mixed thermodynamic and kinetic hydrate promoters on methane hydrate phase boundary and formation kinetics. ACS Publications 58, 501-509 (2013). https://doi.org/10.1021/je301153t

[9] Kumar, A., Bhattacharjee, G., … V. B.-J. of environmental & 2016, undefined. Influence of kinetic and thermodynamic promoters on post-combustion carbon dioxide capture through gas hydrate crystallization. Elsevier. https://doi.org/10.1016/j.jece.2016.03.021

[10] Park, S. et al. Hydrate-based pre-combustion capture of carbon dioxide in the presence of a thermodynamic promoter and porous silica gels. Elsevier.

[11] Majid, A., Worley, J., Fuels, C. K.-E. & & 2021, undefined. Thermodynamic and kinetic promoters for gas hydrate technological applications. ACS Publications 35, 19288-19301 (2021). https://doi.org/10.1021/acs.energyfuels.1c02786

[12] Daraboina, N. & Von Solms, N. The combined effect of thermodynamic promoters tetrahydrofuran and cyclopentane on the kinetics of flue gas hydrate formation. J Chem Eng Data 60, 247-251 (2015). https://doi.org/10.1021/je500529w

[13] Bavoh, C. B., Lal, B. & Keong, L. K. Introduction to Gas Hydrates. Green Energy and Technology 1-25 (2020) doi:10.1007/978-3-030-30750-9_1. https://doi.org/10.1007/978-3-030-30750-9_1

[14] Wu, W. et al. Application of monocyclic compounds as natural gas hydrate promoters: A review. Elsevier.

[15] Herslund, P., Daraboina, N., Thomsen, K., J. A.-F. P. & 2014, undefined. Measuring and modelling of the combined thermodynamic promoting effect of tetrahydrofuran and cyclopentane on carbon dioxide hydrates. Elsevier. https://doi.org/10.1016/j.fluid.2014.08.015

[16] Lal, B. & Nashed, O. Chemical additives for gas hydrates. (2019). https://doi.org/10.1007/978-3-030-30750-9

[17] Mech, D., Gupta, P., and, J. S.-J. of N. G. S. & 2016, undefined. Kinetics of methane hydrate formation in an aqueous solution of thermodynamic promoters (THF and TBAB) with and without kinetic promoter (SDS). Elsevier. https://doi.org/10.1016/j.jngse.2016.06.013

[18] Nashed, O., Partoon, B., Lal, B., … K. S.-J. of N. G. & 2018, undefined. Review the impact of nanoparticles on the thermodynamics and kinetics of gas hydrate formation. Elsevier. https://doi.org/10.1016/j.jngse.2018.05.022

[19] Daraboina, N., Engineering, N. von S.-J. of C. & & 2015, undefined. The combined effect of thermodynamic promoters tetrahydrofuran and cyclopentane on the kinetics of flue gas hydrate formation. ACS Publications. https://doi.org/10.1021/je500529w

[20] Amirun Nissa Rehman, Rajashekhar Pendyala, Bhajan Lal. "Effect of brine on the kinetics of Carbon dioxide hydrate formation and dissociation in porous media", Materials Today: Proceedings, 2021 https://doi.org/10.1016/j.matpr.2021.04.024

[21] Amirun Nissa Rehman, Cornelius B. Bavoh, Bhajan Lal, Khalik M. Sabil, Jitendra S. Sangwai. " Insights on CO Hydrate Formation and Dissociation Kinetics of Amino Acids in a Brine Solution ", Industrial & Engineering Chemistry Research, 2022 [22] Lee, K.; Lee, S. H.; Lee, W. Stochastic Nature of Carbon Dioxide Hydrate Induction Times in Na-Montmorillonite and Marine Sediment Suspensions. Int. J. Greenhouse Gas Control 2013, 14, 15- 24 https://doi.org/10.1016/j.ijggc.2013.01.001

[23] D. Sun, P. Englezos, Storage of CO2 in a partially water saturated porous medium at gas hydrate formation conditions, Int. J. Greenh. Gas Control. 25 (2014) https://doi.org/10.1016/j.ijggc.2014.03.008

[24] Smith, J. M. Introduction to Chemical Engineering Thermodynamics. J. Chem. Educ. 1950, 27, 584, DOI: 10.1021/ed027p584.3 There is no corresponding record for this reference. https://doi.org/10.1021/ed027p584.3

[25] Bhattacharjee, Gaurav; Kumar, Asheesh; Sakpal, Tushar; Kumar, Rajnish ACS Sustainable Chemistry & Engineering (2015), 3 (6), 1205-1214CODEN: ASCECG; ISSN:2168-0485. (American Chemical Society) https://doi.org/10.1021/acssuschemeng.5b00171

[26] Sun, D.; Englezos, P. Storage of CO2 in a partially water saturated porous medium at gas hydrate formation conditions. Int. J. Greenhouse Gas Control 2014, 25, 1- 8 https://doi.org/10.1016/j.ijggc.2014.03.008

[27] Anderson, G. K. Enthalpy of Dissociation and Hydration Number of Carbon Dioxide Hydrate from the Clapeyron Equation. J. Chem. Thermodynamic. 2003, 35,1171-1183 https://doi.org/10.1016/S0021-9614(03)00093-4

Review on optimised process parameters of bioethanol production for environmental sustainability

MINAL P Deshmukh[1,a*], DHANASHRI S Pendse[2,b]

[1]Associate Professor, School of Petroleum Engineering, Dr. Vishwanath Karad MIT World Peace University, Pune -411038, India

[2]Research Scholar, School of Chemical Engineering, Dr Vishwanath Karad MIT World Peace University, Pune-411038, India

[a]minal.deshmukh@mitwpu.edu.in, [b]dhanashrijg@gmail.com

Keywords: Biofuel, Water Plants, Fermentation, Energy, Environmental Sustainability

Abstract. Water hyacinth is a commonly grown aquatic plant. It causes a menace in water bodies due to its large mat formation. It is removed from water bodies through physical, chemical, or biological methods. These methods are costly and ineffective as the plant grows again quickly. Since the plant is easily available, it has been explored as a potential source of lignocellulose biomass for biofuel production. This review paper reflects upon the important steps in processing water hyacinth plants for producing biofuel. A quick summary of the research concerning pre-treatment methods, enzyme hydrolysis, and fermentation studies shows that acid pre-treatment followed by cellulase enzyme hydrolysis and using acidic conditions for fermentation up to 3 days at room temperature gives optimum ethanol and sugar yields. The main challenge in the commercialization of bioethanol production from water hyacinth biomass is the low ethanol yield obtained as compared to other feedstock like sugarcane bagasse and corn stover. However, from an environmental point of view, by converting waste biomass into a valuable energy resource, bioethanol production from water hyacinth promotes a circular economy and sustainable resource management.

Introduction

Industrialization and population growth have led to increased energy consumption [1]. Countries around the world are dependent on imported petroleum fuel [2]. Combustion of fossil fuels poses a risk to human health and a potential risk to the environment[3]. To address the issues related to the supply of petroleum, escalating fuel prices and consequences of combustion of fossil fuels, research for renewable biofuels has risen sharply in the last few years [4][5]. The most desired method for making gasoline-ethanol is bioenergy, a renewable energy source created from biomass. Reduction in the utilization of non-renewable resources in the process of going green and sustainable is one of the major advantages of producing bioethanol [6][7]. Human edible feedstock leads to a food vs. fuel crisis as the primary foodstuff for humans and livestock cannot be diverted for energy production. In this regard, lignocellulosic biomass comes as an attractive alternative source for bioethanol production as it doesn't compete with the food supply. Apart from terrestrial plants, water plants can also serve as a biomass source [8][9]. Cellulose, hemicellulose, and lignin form the major constituents of lignocellulosic material. Bioethanol is obtained from cellulose and hemicellulose by saccharification and fermentation process [10][11]. Ethanol production can be cost-effective with proper utilization of lignocellulosic material and optimizing process parameters.

Eichhornia crassipes, commonly known as water hyacinth, is a free-floating plant species that lives in water. Due to its rapid growth, it has turned into weed. The growth of water hyacinth reaches 125 tonnes wet weight/ 1 ha area within 6 months [12].

The weed causes problems to a diverse group of organisms due to its nature of large mat formation. Underlying aquatic species are deprived of desired oxygen, mat acts like a habitat for disease-causing mosquitoes. Fishing, water sports, and activities are disturbed thus affecting local businesses [4]. Water quality is altered [13].
Some of the traditional approaches used to remove water hyacinth from water bodies pose several disadvantages as mentioned below

- The mechanical method of raking or skimming the pond surface consumes a lot of energy and it is an economically inefficient method
- The chemical method uses herbicides and toxic chemicals that negatively affect the aquatic life
- The biological method uses arthropods and pathogens that can consume water hyacinth plants, but such a method drastically affects other species of aquatic life in water bodies and water hyacinth resurfaces due to its rapid growth
- Management of water habitat methods by changing the water body's ecosystem also negatively impacts other plants and animals [14][15].

This noxious aquatic weed represents a promising lignocellulosic substrate for fuel production because of its high availability and high biomass yield [16] [17]. Utilization of water hyacinth biomass can be a part of water pollution control on one hand and a cheap substrate for fuel production on the other hand [18][19].

Review of Literature
The different steps involved in the production of bioethanol from water hyacinth powder as reported in the literature are described in Fig 1.

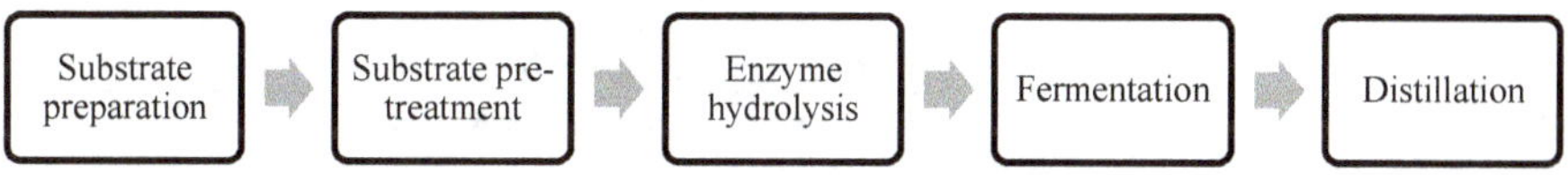

Figure 1: Schematic representation of steps involved in bioethanol production from water weed substrate

Substrate Preparation
Fresh water hyacinth plant collected from ponds or lakes is washed several times to remove dirt chopped into small pieces (1~2 cm) and then blended into small particles followed by drying at high temperature. The dried material is stored at room temperature until further use (Isarankura-na-Ayodhya and Tantimongcolwat 2007) [20]. Fig 2 shows the diagrammatic representation of the process.

Figure 2: Figure shows the procedure of substrate preparation

Substrate Pre-Treatment
Pretreatment of water hyacinth powder is crucial to break the bonds between lignin and hemicellulose which form a barrier for hydrolysis of lignocellulose into fermentable sugars [21].

The different pre-treatment methods used are mechanical-physical pre-treatment, thermal pre-treatment, chemical pre-treatment, and enzymatic and biological pre-treatment[22] [23]. Table 1 shows the details of pre-treatment methods and conditions reported by various authors.

Table 1: Different pre-treatment methods as reported in the available literature

Sl. No	Pre-treatment method	Pre-treatment conditions	Reference
1	Acid pre-treatment	1% or 10% H_2SO_4 to the dried powder, autoclaved at 121°C, 15 lbs, 15 min, filtered to get hydrolysate	[6]
2	Acid or alkali pre-treatment	10% biomass was treated with 0.88% H_2SO_4 or 2% NaOH for 1 hour at 95°C.	[20]
3	Acid or alkali pre-treatment	0.1 N H_2SO_4 and 1% NaOH	[4]
4	Acid pre-treatment	3 g sample with 100 ml of 3% H_2SO_4, autoclave at 121°C, 1 hour	[8]
5	Acid or alkali pre-treatment	3g sample treated with different acids HCl/H_2SO_4 /HCOOH(2% v/v) and alkali NaOH and KOH (3% v/v), autoclaved at 121°C, 15 lbs, 15 min, filtered to get hydrolysate.	[1]
6	Alkali pre-treatment	1 g sample + 10 ml of 5% NaOH, soaked for 1 hour and then heated in a furnace at 150°C for 10min, hydrolysate was collected by filtration.	[24]
7	Acid pre-treatment	Water hyacinth was treated with H_2SO_4 (0.1, 0.5, 1, 1.5, or 2 %) for a solid liquid ratio of 1:8 and the reactive mixture was kept at 121°C for 1 h.	[10]
8	Biophysical pre-treatment	White rot fungus, *G. boninense*, was used for pre-treatment for 7 days.	[25]
9	Combined pre-treatment	Acid- 10 g biomass with 2% dilute acid (H_2SO_4, HNO_3, HCl), 40°C, 60 min, Alkali- 10 g biomass with 2% NaOH, 40°C, 60 min, Steam treatment - 121°C, 60 min solid-liquid ratio- 2 to 15% w/w.	[26]
10	Acid or alkali pre-treatment	5 g sample treated with NaOH (1 to 10 g/L) or treated with H_2SO_4 (0.1 to 0.5%), solid-liquid ratio- 1:10 to 1:30, 60-135°C, 30-90 min.	[21]
11	Combined pre-treatment	Acid- 1% H_2SO_4, 100°C, 30 min, solid-liquid ratio- 1:30 Alkali- 0.5% NaOH, 40°C, 30 min, solid-liquid ratio- 1:16 Microwave alkali- 150 W microwave with 0.5% NaOH, 0.5 min, solid-liquid ratio-1:16	[3]
12	Combined pre-treatment	Microwave-assisted alkali – 20 g sample + 160 ml of 1% (w/v) NaOH solution, 900 W, 25 min Organosolvent pre-treatment – 1 g pre-treated sample was mixed with ethanol: water mixture (70:30 v/v) having 1% of concentrated sulphuric acid, acetic acid, hydrochloric acid, and phosphoric acid for 1 h at 70°C	[27]

13	Alkali pre-treatment	1g sample with 20 ml of 0.5% NaOH, room temperature, 24 hours or 121 °C, 30 min	[28]
14	Combined pre-treatment	Microbial - 4 loops of *Phanerochaete chrysosporium* were added to 9 gm biomass in an inorganic salt medium, 30°C, 60 hrs Microbial + acid- 1% H_2SO_4 to treated system, 100°C, 1hour Microbial + alkali- 4% NaOH to the treated system, 100°C, 1hour	[29]
15	Combined pre-treatment	Chemical – 4 g sample with 100 ml of 4% H_2SO_4, HCl, NaOH, or KOH, autoclaved at 121 °C, 20 min Fungal – 1 ml spore suspension to 4 g autoclaved sample, 30 °C, 10 days Chemical treatment and fungal hydrolysis Fungal treatment and chemical hydrolysis Fungal strains used - *Trichoderma viride*; *T. reesei*; *Aspergillus terreus*; *A. niger*	[30]
16	Acid pre-treatment	Oxalic acid and citric acid- organic acids used Hydrochloric acid- inorganic acid used Optimization studies were conducted to treat biomass at acid concentration – 0.5 – 12%, pre-treatment time- 30-90 min, pre-treatment temperature- 100-140°C	[31]
17	Steam explosion	Pretreatment of biomass was carried out using steam explosion at 190 °C with retention times of 1, 5, and 10 min	[32]
18	Wet oxidation method	Samples at 6g/L were processed at two different conditions to obtain cellulose liberation **1.** 125°C under 1.5 bar for 45 minutes **2.** 130°C under 1.8 bar for 75 minutes	[33]
19	Thermochemical pre-treatment	Dry biomass was treated with 3% H_2SO_4 at 121°C. Optimization studies were performed on solid load (10, 20 and 50 mg of biomass/mL H_2SO_4), reaction times (10, 20, 40, and 60 min), and particle sizes (0.15, 0.85, and 2.36 mm) on the release of reducing sugars.	[34]

According to [35] biomass characteristics, biomass availability, financial resources, and low negative environmental impacts can be considered for selecting the best pre-treatment method. For water hyacinth biomass, acid pre-treatment using sulphuric acid is a very commonly used method to obtain effective hydrolysis. Some authors have used alkaline solutions to disrupt the lignin structure and break the bonds between lignin and carbohydrates. Very few have worked on the biological pre-treatment method which uses the enzymes produced by fungi to degrade lignocellulose. Some authors have obtained good sugar yields using a combined pre-treatment method. The advantages and limitations of each method are illustrated in detail in Table 2 and Table 3.

Table 2: Advantages of different pre-treatment methods

Advantages	Acid pre-treatment	Alkali pre-treatment	Biological pre-treatment	Combined pre-treatment
Removal of lignin and hemicellulose	Yes	Partial	Yes	Yes
Hemicellulose solubility	Yes	-	-	Yes
Cellulose accessibility	Yes	-	Yes	Yes
Sugar recovery efficiency	Yes	-	-	Yes
Decrease in Polymerization	-	Yes	-	-
Decrease in crystallinity	-	Yes	-	-
Eco-friendly	-	-	Yes	-

Table 3: Limitations of different pre-treatment methods

Disadvantages	Acid pre-treatment	Alkali pre-treatment	Biological pre-treatment	Combined pre-treatment
Corrosive and hazardous	Yes	-	-	
Special Equipment	Yes	-	-	Yes
Formation of inhibitors	Yes, at low pH	-	-	-
Sugar loss/low yield	Yes	-	Yes	-
Neutralization and salt disposal/ toxic waste	Yes	-	-	Yes
Low digestibility	-	Yes	-	-
Long pre-treatment time	-	Yes	Yes	-
Huge water requirement	-	Yes	-	-
High chemical recovery cost	-	Yes	-	-
High energy demand	-	-	-	Yes

Enzyme Hydrolysis
Cellulose present in lignocellulosic feedstock is hydrolysed into fermentable sugars by a group of enzymes called Cellulases. Fig 3 shows the structural representation of cellulose breakdown. Hydrolysis product glucose is further used for producing many useful products such as ethanol, biofuel and other useful chemicals [36][37].

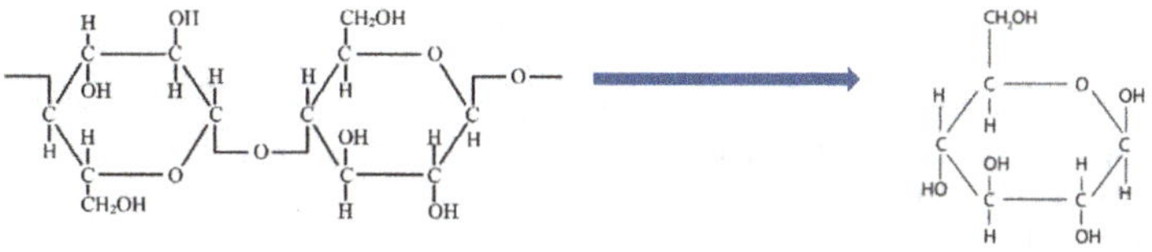

Cellulose (Cellulases) Glucose

Figure 3: Figure showing Cellulose hydrolysis using cellulase enzyme

Compared to chemical hydrolysis of cellulose, enzymatic hydrolysis is preferred as it is a green approach. However, the cost of ethanol production shoots up due to the usage of expensive enzymes for biomass hydrolysis [20][38]. Hence it is a common approach that researchers produce on-site enzymes by culturing various cellulase-producing micro-organisms. Table 4 shows the results of enzyme hydrolysis reported by various authors.

Table 4: Enzyme hydrolysis as reported in the available literature

Sl. No	Biomass loading	Enzymes used	Process of hydrolysis	Sugar yield	Reference
1	6.5% alkali-treated biomass	1. Cellulase (12 FPU/g) from *Trichoderma reesei* RUT C30 2. β-glucosidase (2400 FPU/g) from *Aspergillus niger* MTCC 7956	Biomass was treated with different concentrations of enzymes at 45°C in 50mM citrate buffer in a shaking water bath for 0-48 hours	4.4 % glucose yield	[20]
2	40g/L pre-sterilised water hyacinth biomass	1. Cellulase (12 FPU/g) from *Trichoderma reesei* ATCC 26921 2. β-glucosidase (2400 FPU/g) from *Aspergillus Phoenicis* ATCC 52007	Biomass was treated with enzyme blend (1:1.2) at 45°C, pH – 5.0 for 72 hours with agitation at 150 rpm	1.2 %glucose yield	[4]

3	1g acid-treated biomass hydrolysate	Cellulase GC 220 (0–1,000 IU)	The suspension was kept at 50 °C and stirred at 300 rpm, the reaction was monitored for saccharification (24–96 h).	0.2-0.3% glucose yield	[8]
4	Alkali-treated water hyacinth hydrolysate	Cellulase extracted from F13 fungal strain	1 ml of the enzyme was added to biomass and the reaction mixture was incubated on a rotary shaker at 50°C and 125 rpm for 12-60 h	2.1% reducing sugar yield	[24]
5	Biomass concentration (5-30% w/v)	Cellulase concentrations (10–30 FPU/g)	Enzymatic saccharification of pretreated water hyacinth was performed at pH (4.0–8.0), temperatures (40–55 °C) on a rotary shaker at 100 rpm for 48 h.	5% sugar yield	[26]
6	Pre-treated 1% w/v water hyacinth	GH5 cellulase (0.44 mg /mL) and GH43 hemicellulase (0.33 mg/ mL)	The flask was maintained at 30 °C and 120 rpm for 72 h with the collection of samples at intervals of every 6 h.	0.2% reducing sugar yield	[27]
7	(200-1000 mg) pre-treated biomass	Cellulase concentrations (20 FPU) from *Trichoderma reesei*	The flask was maintained at (20-60 °C), pH from 4.0-6.4, agitation of 150 rpm for 1-48 h incubation time	50.1% reducing sugar yield	[21]
8	1 g pre-treated water hyacinth sample	Cellulase from *Trichoderma viride*	To 1g pre-treated sample, 0.05 g cellulase and 0.05 g $CaCl_2$ was added at 120rpm, 45°C, 96 hours	40.2 % reducing sugar yield	[3]
9	100 mg Pre-treated sample	35 µl of CelluClast 1.5L, (produced by *Trichoderma*	Hydrolysis reaction was prepared in 20mL of 50mM	-	[31]

		reesei ATCC 26921) and 10 μl of β-glucosidase (produced by *Aspergillus niger*)	Sodium citrate buffer (pH 4.7) containing 200 μl of 2 M sodium azide with a 2.5% loading ratio of biomass. The mixture was incubated in a shaking incubator at 45°C, 200 rpm for 72 h		
10	20 g water hyacinth sample	Cellulase and Xylanase from *Trichoderma harzianum* PBLA	The enzymatic reaction was performed in 500-mL flasks with gentle agitation (125 rpm) in a water bath; the pH was adjusted to 5.2 with 0.1 M citrate buffer and maintained at 50 °C for 48 h.	51% reducing sugar yield	[32]

It can be inferred from Table 4 that cellulase enzymes are most commonly used for hydrolysis of cellulose after pre-treatment. Cellulase consists of three sets of enzymes i.e. a) Endoglucanase (cuts cellulose chains into fragments of glucose, cellobiose, and cellotriose) b) Exoglucanase (cleaves into cellobiose units) c) β glucosidase (breaks cellobiose units into glucose) [2]. Some of them are produced on-site and used for experiments while some authors have made use of commercial enzymes for hydrolysis. Along with cellulases, other enzymes are used depending on the requirement of hemicellulose or cellobiose degradation. Based on the composition of the source and type of pre-treatment method used, the concentration of enzyme varies and so accordingly its activity. In most of the studies reported, hydrolysis efficiency is achieved with the optimized temperature of (45-50°C), (pH 4.0-5.0), (100-200 rpm) shaking and hydrolysis time of 24-48 hours.

Fermentation

After the enzyme hydrolysis step, the supernatant containing various sugars like pentoses and hexoses is fermented into bioethanol and various micro-organisms including bacteria and fungi are used for the same [2][39]. Glucose fermenting yeast *Saccharomyces cerevisiae* is most commonly used. Apart from glucose, pentoses like xylose are present in high amounts in the supernatant and therefore xylose fermenting yeast is also used in most of the fermentation processes [8][40]. Table 5 shows the information regarding various studies reported by authors for the fermentation process along with the type of micro-organisms used and bioethanol yield obtained.

Table 5: Fermentation conditions and micro-organisms used as reported in the literature

Sl. No	Fermenting micro-organism	Fermentation conditions (optimum)	Bioethanol yield	Reference
1	Pichia stipitis	Inoculation cell density- 0.25 g/l , time-30h , aeration rate – 0.02 v/v/m, temperature-30°C, pH- 6.0	19.14 g/L	[41]
2	Candida shehatae	temperature-30°C, pH- 5.6 , time-3 weeks	1.01 g/L	[6]
3	Saccharomyces cerevisiae	Hydrolysate containing 10% reducing sugar, Inoculum- 5% v/v, temperature-28 ± 2°C	4.25 g/L	[20]
4	Saccharomyces Cerevisiae and Pachysolen tannophilus	1. Separate hydrolysis and fermentation (SHF) Cell suspension – 2% v/v (S.Cerevisiae), temperature- 30°C, time-48h , further with 3% v/v (P.tannophilus), temperature- 25°C, time-another 48h 2. Simultaneous accharification and fermentation (SSF) Crude culture supernatants of T.reseei and A. phoenicis along with 2% v/v (S.Cerevisiae), temperature-37°C, time-72h, further with 3% v/v (P.tannophilus), temperature- 25°C, time-another 48h 3. Single batch bioconversion (SBB) Heat inactivation of culture broth, 2% v/v (S.Cerevisiae), temperature-30°C, pH-5.0, time-48h , further with 3% v/v (P.tannophilus), temperature- 25°C, time-another 48h 4. Pre-fermentation hydrolysis Simultaneous saccharification and fermentation (PH-SSF) Hydrolysis for only 24 hours followed by SSF conditions	SHF- 4.5 g/L SSF- 5.2 g/L SBB- 5 g/L PH-SSF- 8.3g/L	[4]
5	Saccharomyces Cerevisiae and Pichia stipitis	Concentration solution- 10 ml, yeast cells-0.1 g, temperature -25°C, time-5 days, anaerobic conditions	9.61 g/L	[8]
6	Pichia stipitis	Time – 7.5 hours	3.9 g/L	[24]
7	Saccharomyces Cerevisiae Pachysolen tannophilus,	Inoculum – 10% v/v, pH- 4.5, temperature -28°C, 125 rpm, time-72 h.	43 g/L	[10]

	Candida intermedia and Pichia stipitis			
8	Saccharomyces Cecerevisiae and Zymomonas mobilis	The concentration of hydrolysate- 5%, time- 37.7 h, pH- 6.41, Saccharomyces to Zymomonas ratio of 1	13.6 g/L	[26]
9	Saccharomyces Cerevisiae and Candida shehatae	SSF in a 3L bioreactor, 5% w/v pre-treated water hyacinth, batch reaction at 35°C, 120 rpm for 3 days, pH – 5.4, aeration rate at 1 vvm, 40% DO	13.7 g/L	[27]
10	Saccharomyces Cerevisiae	SSF with fermentation temperature – 38.87°C, time- 81.87 h, inoculum- 6.11 ml	1.289 g/L	[3]
11	Kluveromyces marxianus and Saccharomyces Cerevisiae	SSF with fermentation time- 48 h at temperature - 42°C and 30°C, 8% solids loading (w/v)	8.0 g/L	[28]
12	Saccharomyces Cerevisiae	Yeast inoculum- 6g/L, fermentation temperature – 30°C, 120 rpm	1.4 g/L	[29]
13	Candida tropicalis Y-26	Batch fermentation in a 10 L bioreactor, pH – 5.5, 10 % (v/v) inoculum, time- 60 h, temperature – 30°C	14 g/L	[30]
14	Saccharomyces Cerevisiae	Fermentation temperature- 32°C, 100 rpm for fermentation time-48 h	35.5 g/L	[31]
15	Saccharomyces Cerevisiae	Fermentation temperature-38.87°C, time- 81.87 hours	1.289g/L	[42]
16	Saccharomyces Cerevisiae	pH-5.5, temperature- 30°C, oxygen flow rate- 50 ml/min	7.13 g/L	[32]

From Table 5, it can be inferred that most of the authors have used a combination of both glucose and xylose-fermenting yeasts to convert monosaccharides into bioethanol. It can also be inferred that optimum bioethanol yields are obtained at conditions like acidic pH, room temperature and fermentation time of 2-3 days.

Distillation

Distillation is used to separate ethanol from the reaction mixture. The boiling point of ethanol is 78.8°C [43][44]. At the end of fermentation, the reaction broth is subjected to fractional distillation at 78 to 79°C to collect ethanolic solution then the concentration of ethanol is determined. Water-free ethanol obtained is further used in vehicles as a blend with gasoline for transportation.

Characterization Results of the Bioethanol Produced from Water Hyacinth

By combining bioethanol made from water hyacinth, (Choudhary et al. 2016) investigated the emission and performance of a single-cylinder diesel engine under varied operating conditions. Using yeasts, the biomass was fermented after being pre-treated with 4% sulfuric acid to extract sugars. Different ratios of diesel were mixed with the produced biofuel (5%, 10%, 15%, 20%, and

25% v/v). The outcomes of their experiments demonstrated that the blends with 5% and 10% delivered the highest levels of brake power, brake thermal efficiency, and brake mean effective pressure along with the best exhaust emission profiles.

Similarly, the performance and emissions of a single-cylinder diesel engine with a 3.5kW rating have also been studied by [45]. Zymomonas mobilis CP4 was used to ferment the biomass of water hyacinth after it had been pre-treated with sulfuric acid. Different ratios of the produced biofuel (10%, 20%, and 30%) were mixed with diesel. According to their experimental findings, when compared to other mixes of bioethanol and diesel, 10% blends of bioethanol produced results for peak combustion pressure, brake power, and brake thermal efficiency that were much greater.

Table 6: Comparison of properties of biofuel blends

Property	Diesel	Bioethanol	5%	10%	15%	20%	25%	30%
Density at 20°C	840[a]	780 [a]	839 [a]	834 [a]	828 [a]	821 [a]	819 [a]	-
Density at 25°C (Kg/m^3)	837[b]	788 [b]	-	826 [b]	-	819 [b]	-	805 [b]
Viscosity (10^{-6} $v/m^2/s$) at 20°C	3.35 [a]	1.2 [a]	3.24 [a]	3.13 [a]	3.02 [a]	2.92 [a]	2.81 [a]	-
Viscosity (cSt) at 40°C	2.6 [b]	1.2 [b]	-	2.41 [b]	-	2.19 [b]	-	1.93 [b]
Lower heating value (MJ/Mol)	6261.3 [a]	1541 [a]	6073.7 [a]	5835.1 [a]	5596.6 [a]	5358 [a]	5119.4 [a]	-
Calorific value (MJ/Kg)	43-45 [b]	26.8 [b]	-	40 [b]	-	36 [b]	-	33 [b]

Value [a] – results from [46]
Value [b]- – results from [45]

Table 6 shows the comparison of three different fuel properties from two different authors. Both authors have compared the density, viscosity and calorific value of blend fuels along with pure diesel. It can be assumed that the physical characteristics of diesel fuel are altered when the percentage of bioethanol surpasses 20%. It lowers the blend's viscosity and calorific value. For most of the blends in both experiments, the emissions were found to be significantly lower than those from neat diesel. NOx (Oxides of Nitrogen) emissions fall with increasing loads. When comparing all studied bioethanol-diesel blends to pure diesel fuel, the levels of HC (hydrocarbons) and CO (carbon monoxide) were significantly higher. The outcome illustrates the viable conversion of water hyacinth plants into bioethanol, which possesses the necessary properties to function in a diesel engine. Not only does the 10% blend have good qualities, but it can also be utilized safely in diesel engines without requiring any changes, making it stand out from the rest.

Efficiency of Bioethanol Production Compared to Other Feedstock

The industrial production of biofuel from first-generation sources such as sugarcane, corn stover, sweet sorghum, and cassava is technically and economically feasible. It has several environmental benefits compared to pure gasoline in terms of carbon emissions. Since these sources compete with food security, second-generation sources are explored to achieve environmental benefits in terms of carbon sequestration, reduced soil loss and carbon emissions. However, the high production cost of producing bioethanol from second-generation sources makes it challenging for commercialization. [42] have reported that "Biofuels produced from switchgrass cost 17.8% more than corn and are 34.4% more costly than gasoline when measured on an energy equivalent basis". Since water hyacinth also falls in the category of lignocellulose biomass, it is less competitive in terms of bioethanol production at a commercial scale. From Table 7, we can infer that water hyacinth has lower cellulose content compared to other potential feedstock. However, it is looked upon as a potential feedstock due to its lower lignin content. The commercialization of bioethanol production from water hyacinth can be eco-friendly and profitable with the development of an efficient pre-treatment method to remove lignin for effective hydrolysis and by utilizing suitable microbes in the fermentation of sugars into bioethanol.

Table 7: Bioethanol yield comparison for different feedstock [42]

Feedstock	Cellulose	Hemicellulose	Lignin	Bioethanol yield (Kg/kg of feedstock)
Water hyacinth	18.07	28.21	7.03	0.1289-0.192
Corn stover	36.2	23.2	18.5	0.13-0.25
Cassava	46.7	32.6	16.9	0.418
Corn grain	18.7	28.7	7.5	033-0.39
Miscanthus	32.4	48.6	16.5	0.3
Switchgrass	32.0	25.2	18.1	0.3

Techno-Economic Analysis of Water Hyacinth Bioethanol Production

In Fig 4, the cost of water hyacinth is mainly associated with feedstock harvesting and transportation which accounts for 40.22 % of the unit cost of bioethanol. 15.32% of chemical cost is due to the use of acids and bases required for pre-treatment and neutralization. The use of commercial enzymes and yeast for hydrolysis and fermentation along with water consumption account for 15.65 % of the total cost. Capital costs (14.85%) are covered for one-time purchase of fixed assets including land, building, construction and purchase of equipment. Equipment repair and depreciation account for 3.93% and 8.3 % respectively. Insurance and taxes are covered under other costs (2.62%).

Labour cost includes salary for managers and skilled labour (9.21%) and the remaining 4.75% (office supplies, rent, utilities etc.) falls under overhead costs [42].

In India, currently, there are no commercial establishments that produce bioethanol from water hyacinth. Agricultural residues such as rice straw, wheat straw, and sugarcane bagasse are mostly commonly used feedstock for large-scale production of biofuel. [47] have reported in their study that in India, the total CAPEX (Capital expenditure) needed for building a 36.5 million litre cellulosic ethanol facility is INR(Indian rupees) 8.8 billion. Table 8 shows the cost of feedstock, ethanol yield and production cost per litre for three different feedstock as reported by [47].

From Table 8, it is inferred that the production cost is mainly dependent on the type of feedstock. Since sugarcane bagasse is cheaper than rice straw and wheat straw, the process of bioethanol production is economical even though the ethanol yield is less compared to rice straw.

Water hyacinth, in comparison to agricultural feedstock, is less expensive and can be explored at a commercial scale for bioethanol production as it is available in plenty.

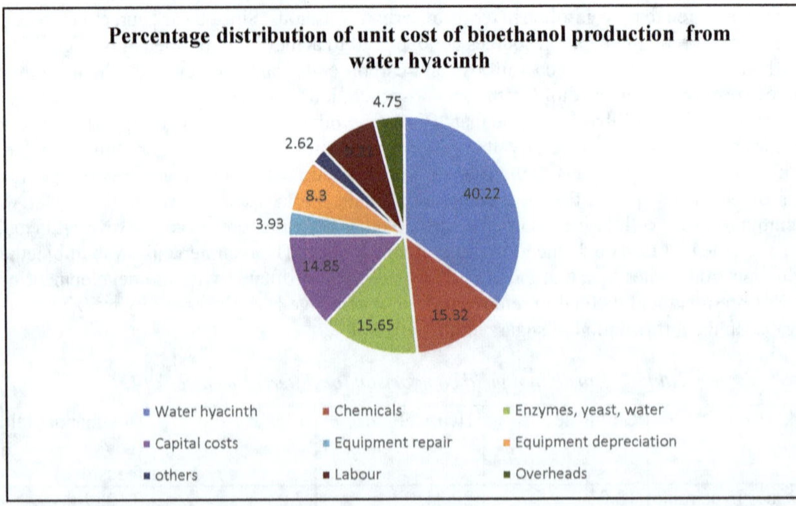

Figure 4: Percentage distribution of unit cost of bioethanol production from water hyacinth

Table 8: Average price, ethanol yield, and production cost for three types of feedstock

Feedstock	Average price/tonne (INR)	Ethanol yield (g/g) on a dry basis	Production cost per litre (INR)
Rice straw	3532	0.38	93.1
Wheat straw	3532	0.34	93.6
Sugarcane bagasse	2225	0.36	92.4

Conclusion

A potential weed-like water hyacinth plant, which otherwise causes health and environmental hazards when left untreated, could be used as an eco-friendly solution for bioethanol production. This review gives an overall picture of the basic steps involved in the bioconversion of water hyacinth biomass into bioethanol at the lab scale. Acid pre-treatment is most commonly used as a feasible pre-treatment method to break down the lignocellulose bond. Maximum sugar formation (~50g/L)is achieved using cellulase enzyme at a temperature range of 40-50°C at acidic pH. Glucose and pentose fermenting yeast yield up to 40g/L of bioethanol at a temperature range of 28-30°C at acidic pH for 3 days. From the basic understanding of process conditions required for processes like pre-treatment, hydrolysis and fermentation, one can further develop and commercialize the entire process with suitable modifications and scale-up. From a future perspective, optimization studies using biological agents in all three steps can make the entire process more sustainable and economical.

In a developing nation like India, fuel is one of the top three import products and 80% of the fuel demands are met by imports. In such a situation, in-house production of cellulosic ethanol has many advantages in terms of revenue generation and following a green approach for environmental protection. However, the economic challenges still prevail to scale up the technology in terms of high production costs. India's fixed ethanol price per litre is 65.6 INR. Ethanol industries need support from the government in terms of policies and subsidies to make commercial establishments profitable and sustainable.

Declarations

Credit authorship contribution statement
Dr Minal P Deshmukh – Conceptualization, Investigation, Supervision
Dhanashri S Pendse – Literature review, Writing draft and editing

Data availability
Data will be made available on request

Declaration of competing interest
The authors declare that they have no known competing financial interests or personal relationships that could have appeared to influence the work reported in this paper.

Funding statement
This declaration is "Not applicable" to our present study.

Ethical Approval
This declaration is "Not applicable" to our present study.

References

[1] E. Suarez, M. Awasthi, J. Kaur, and S. Rana, "Bioethanol Production Through Water Hyacinth, Eichhornia Crassipes Via Optimization of the Pretreatment Conditions," *ICERTSD2013 Int. J. Emerg. Technol. Adv. Eng.*, vol. 3, pp. 42–46, 2013, [Online]. Available: www.ijetae.com

[2] S. Premjet, "Potential of Weed Biomass for Bioethanol Production," *Fuel Ethanol Prod. from Sugarcane*, 2019. https://doi.org/10.5772/intechopen.77507

[3] Q. Zhang, C. Weng, H. Huang, V. Achal, and D. Wang, "Optimization of bioethanol production using whole plant of water hyacinth as substrate in simultaneous saccharification and fermentation process," *Front. Microbiol.*, vol. 6, no. JAN, pp. 1–9, 2016. https://doi.org/10.3389/fmicb.2015.01411

[4] S. Mukhopadhyay and N. C. Chatterjee, "Bioconversion of water hyacinth hydrolysate into ethanol," *BioResources*, vol. 5, no. 2, pp. 1301–1310, 2010.

[5] D. S. Pendse, M. Deshmukh, and A. Pande, "Different pre-treatments and kinetic models for bioethanol production from lignocellulosic biomass: A review," *Heliyon*, vol. 9, no. 6, p. e16604, 2023. https://doi.org/10.1016/j.heliyon.2023.e16604

[6] C. Isarankura-na-ayudhya and T. Tantimongcolwat, "Original article : Appropriate Technology for the Bioconversion of Water Hyacinth (Eichhornia crassipes) to Liquid Ethanol : Future Prospects for Community Strengthening and Sustainable Development," *EXCLI J.*, vol. 6, no. May, pp. 167–176, 2007.

[7] M. Deshmukh and A. Pande, "Comparative Study for Production of Biofuel from Potato Peel Waste as Feedstock by Different Enzymes," *Enzym. Eng.*, vol. 11, no. 1, pp. 1–6, 2022. https://doi.org/10.35841/2329-6674.22.11.1000175

[8] T. Takagi, M. Uchida, R. Matsushima, M. Ishida, and N. Urano, "Efficient bioethanol production from water hyacinth Eichhornia crassipes by both preparation of the saccharified solution and selection of fermenting yeasts," *Fish. Sci.*, vol. 78, no. 4, pp. 905–910, Jul. 2012. https://doi.org/10.1007/s12562-012-0516-2

[9] C. G. Sierra-Carmona, M. G. Hernández-Orduña, and R. Murrieta-Galindo, "Alternative Uses of Water Hyacinth (Pontederia crassipes) from a Sustainable Perspective: A Systematic Literature Review," *Sustain.*, vol. 14, no. 7, pp. 1–14, 2022. https://doi.org/10.3390/su14073931

[10] A. Manivannan and R. T. Narendhirakannan, "Bioethanol Production From Aquatic Weed Water Hyacinth (Eichhornia crassipes) by Yeast Fermentation," *Waste and Biomass Valorization*, vol. 6, no. 2, pp. 209–216, 2015. https://doi.org/10.1007/s12649-015-9347-6

[11] G. K. Gaurav, T. Mehmood, L. Cheng, J. J. Klemeš, and D. K. Shrivastava, "Water hyacinth as a biomass: A review," *J. Clean. Prod.*, vol. 277, p. 122214, 2020. https://doi.org/10.1016/j.jclepro.2020.122214

[12] T. Istirokhatun, N. Rokhati, R. Rachmawaty, M. Meriyani, S. Priyanto, and H. Susanto, "Cellulose Isolation from Tropical Water Hyacinth for Membrane Preparation," *Procedia Environ. Sci.*, vol. 23, no. Ictcred 2014, pp. 274–281, 2015. https://doi.org/10.1016/j.proenv.2015.01.041

[13] S. Rezania, M. Ponraj, M. F. M. Din, A. R. Songip, F. M. Sairan, and S. Chelliapan, "The diverse applications of water hyacinth with main focus on sustainable energy and production for new era: An overview," *Renew. Sustain. Energy Rev.*, vol. 41, pp. 943–954, 2015. https://doi.org/10.1016/j.rser.2014.09.006

[14] J. E. Santibañez-Aguilar, J. B. González-Campos, J. M. Ponce-Ortega, M. Serna-González, and M. M. El-Halwagi, "Sustainable processing of water hyacinth through a distributed biorefinery system," *AIChE Annu. Meet. Conf. Proc.*, no. April 2016, 2012.

[15] M. B. Djihouessi, M. Olokotum, L. C. Chabi, F. Mouftaou, and M. P. Aina, "Paradigm shifts for sustainable management of water hyacinth in tropical ecosystems: A review and overview of current challenges," *Environ. Challenges*, vol. 11, no. December 2022, p. 100705, 2023. https://doi.org/10.1016/j.envc.2023.100705

[16] S. M. M. Shanab, E. A. Hanafy, and E. A. Shalaby, "Water Hyacinth as Non-edible Source for Biofuel Production," *Waste and Biomass Valorization*, vol. 9, no. 2, pp. 255–264, 2018. https://doi.org/10.1007/s12649-016-9816-6

[17] M. A. Bote, V. R. Naik, and K. B. Jagadeeshgouda, "Review on water hyacinth weed as a potential bio fuel crop to meet collective energy needs," *Mater. Sci. Energy Technol.*, vol. 3, pp. 397–406, 2020. https://doi.org/10.1016/j.mset.2020.02.003

[18] S. Mukhopadhyay and B. Nandi, "Optimization of Cellulase Production by Trichoderma reesei ATCC 26921 Using a Simplified Medium on Water Hyacinth Biomass," *J. Sci. Ind. Res. (India).*, vol. 58, no. 2, pp. 107–111, 1999.

[19] D. S. Pendse and M. P. Deshmukh, "A comprehensive study on an integrated approach for water hyacinth management to conserve natural water resources in India," *Mater. Today Proc.*, no. December, 2024. https://doi.org/10.1016/j.matpr.2023.12.047

[20] U. S. Aswathy, R. K. Sukumaran, G. L. Devi, K. P. Rajasree, R. R. Singhania, and A. Pandey, "Bio-ethanol from water hyacinth biomass: An evaluation of enzymatic saccharification strategy," *Bioresour. Technol.*, vol. 101, no. 3, pp. 925–930, 2010. https://doi.org/10.1016/j.biortech.2009.08.019

[21] T. Ruan, R. Zeng, X. Y. Yin, S. X. Zhang, and Z. H. Yang, "Water Hyacinth (Eichhornia crassipes) Biomass as a Biofuel Feedstock by Enzymatic Hydrolysis," *BioResources*, vol. 11, no. 1, pp. 2372–2380, 2016. https://doi.org/10.15376/BIORES.11.1.2372-2380

[22] M. Deshmukh, A. Pande, and A. Marathe, "Heliyon Different particle size study of castor deoiled cake for biofuel production with an environmental sustainability perspective," *Heliyon*, vol. 8, no. January, p. e09710, 2022. https://doi.org/10.1016/j.heliyon.2022.e09710

[23] M. Carlini, S. Castellucci, and A. Mennuni, "Water hyacinth biomass: Chemical and thermal pre-treatment for energetic utilization in anaerobic digestion process," *Energy Procedia*, vol. 148, pp. 431–438, 2018. https://doi.org/10.1016/j.egypro.2018.08.106

[24] Kumari, Bhattacharya, Dey, Ganguly, and Chatterje, "Bioethanol Production From Water Hyacinth Biomass Using Isolated Fungal Strain From Local Environment," *J. Biol. Life Sci.*, vol. 2, no. 2, pp. 516–522, 2014.

[25] F. Amriani, F. A. Salim, I. Iskandinata, D. Khumsupan, and Z. Barta, "Physical and biophysical pretreatment of water hyacinth biomass for cellulase enzyme production," *Chem. Biochem. Eng. Q.*, vol. 30, no. 2, pp. 237–244, 2016. https://doi.org/10.15255/CABEQ.2015.2284

[26] A. Das, P. Ghosh, T. Paul, U. Ghosh, B. R. Pati, and K. C. Mondal, "Production of bioethanol as useful biofuel through the bioconversion of water hyacinth (Eichhornia crassipes)," *3 Biotech*, vol. 6, no. 1, pp. 1–9, 2016. https://doi.org/10.1007/s13205-016-0385-y

[27] S. P. Das, A. Gupta, D. Das, and A. Goyal, "Enhanced bioethanol production from water hyacinth (Eichhornia crassipes) by statistical optimization of fermentation process parameters using Taguchi orthogonal array design," *Int. Biodeterior. Biodegrad.*, vol. 109, pp. 174–184, 2016. https://doi.org/10.1016/j.ibiod.2016.01.008

[28] M. Narra *et al.*, "Cellulase production, simultaneous saccharification and fermentation in a single vessel: A new approach for production of bio-ethanol from mild alkali pre-treated water hyacinth," *J. Environ. Chem. Eng.*, vol. 5, no. 3, pp. 2176–2181, 2017. https://doi.org/10.1016/j.jece.2017.04.043

[29] Q. Zhang, Y. Wei, H. Han, and C. Weng, "Enhancing bioethanol production from water hyacinth by new combined pretreatment methods," *Bioresour. Technol.*, vol. 251, pp. 358–363, 2018. https://doi.org/10.1016/j.biortech.2017.12.085

[30] H. R. Madian, N. M. Sidkey, M. M. Abo Elsoud, H. I. Hamouda, and A. M. Elazzazy, "Bioethanol Production from Water Hyacinth Hydrolysate by Candida tropicalis Y-26," *Arab. J. Sci. Eng.*, vol. 44, no. 1, pp. 33–41, 2019. https://doi.org/10.1007/s13369-018-3247-6

[31] P. Tantayotai, P. Mutrakulchareon, A. Tawai, S. Roddecha, and M. Sriariyanun, "Effect of organic acid pretreatment of water hyacinth on enzymatic hydrolysis and biogas and bioethanol production," *IOP Conf. Ser. Earth Environ. Sci.*, vol. 346, no. 1, 2019. https://doi.org/10.1088/1755-1315/346/1/012004

[32] L. A. Figueroa-Torres *et al.*, "Saccharification of water hyacinth biomass by a combination of steam explosion with enzymatic technologies for bioethanol production," *3 Biotech*, vol. 10, no. 10, pp. 1–9, 2020. https://doi.org/10.1007/s13205-020-02426-8

[33] Y. S. Ahou *et al.*, "Wet oxidation pretreatment effect for enhancing bioethanol production from cassava peels, water hyacinth, and green algae (Ulva)," *AIP Conf. Proc.*, vol. 2255, no. September, 2020. https://doi.org/10.1063/5.0013560

[34] E. Romero-Borbón, A. E. Oropeza-González, Y. González-García, and J. Córdova, "Thermochemical and Enzymatic Saccharification of Water Hyacinth Biomass into Fermentable Sugars," *Processes*, vol. 10, no. 2, pp. 1–12, 2022. https://doi.org/10.3390/pr10020210

[35] S. Rezania *et al.*, "Review on pretreatment methods and ethanol production from cellulosic water hyacinth," *BioResources*, vol. 12, no. 1, pp. 2108–2124, 2017. https://doi.org/10.15376/biores.12.1.rezania.

[36] S. hao Zhao, X. hui Liang, D. liang Hua, T. suo Ma, and H. bing Zhang, "High-yield cellulase production in solid-state fermentation by Trichoderma reesei SEMCC-3.217 using water hyacinth (Eichhornia crassipes)," *African J. Biotechnol.*, vol. 10, no. 50, pp. 10178–10187, 2011. https://doi.org/10.5897/ajb10.748

[37] R. Sindhu *et al.*, "Water hyacinth a potential source for value addition: An overview," *Bioresour. Technol.*, vol. 230, pp. 152–162, 2017. https://doi.org/10.1016/j.biortech.2017.01.035.

[38] M. Deshmukh and A. Pathan, "Bambusa tulda: A potential feedstock for bioethanol and its blending effects on the performance of spark ignition engine," *Renew. Sustain. Energy Rev.*, vol. 192, no. December 2023, p. 114270, 2024. https://doi.org/10.1016/j.rser.2023.114270

[39] V. Randive, S. Belhekar, and S. Paigude, "Original Research Article Production of Bioethanol From Eichhornia crassipes (Water Hyacinth)," vol. 2, no. 2, pp. 399–406, 2015.

[40] M. Deshmukh and A. Pathan, "Transformations of bamboo into bioethanol through biorefinery," *Environ. Sci. Pollut. Res. Int.*, vol. 31, no. 3, pp. 3343–3360, 2024. https://doi.org/10.1007/s11356-023-31510-8

[41] J. N. Nigam, "Bioconversion of water-hyacinth (Eichhornia crassipes) hemicellulose acid hydrolysate to motor fuel ethanol by xylose-fermenting yeast," *J. Biotechnol.*, vol. 97, no. 2, pp. 107–116, 2002. https://doi.org/10.1016/S0168-1656(02)00013-5

[42] Z. Wang, F. Zheng, and S. Xue, "The economic feasibility of the valorization of water hyacinth for bioethanol production," *Sustain.*, vol. 11, no. 3, 2019. https://doi.org/10.3390/su11030905

[43] A. G. Bayrakci and G. Koçar, "Second-generation bioethanol production from water hyacinth and duckweed in Izmir: A case study," *Renew. Sustain. Energy Rev.*, vol. 30, pp. 306–316, 2014. https://doi.org/10.1016/j.rser.2013.10.011

[44] M. P. Deshmukh, A. Pande, V. Choudhari, and D. S. Pendse, "Investigation of bioethanol production from jatropha deoiled cake and its blending effects for environmental sustainability," *Environ. Sci. Pollut. Res.*, vol. 30, no. 47, pp. 103640–103651, Sep. 2023. https://doi.org/10.1007/s11356-023-29614-2

[45] O. Khan, A. K. Yadav, M. E. Khan, and M. Parvez, "Characterization of bioethanol obtained from Eichhornia Crassipes plant; its emission and performance analysis on CI engine," *Energy Sources, Part A Recover. Util. Environ. Eff.*, vol. 43, no. 14, pp. 1793–1803, 2021. https://doi.org/10.1080/15567036.2019.1648600

[46] A. K. Choudhary, H. Chelladurai, and C. Kannan, "Performance analysis of bioethanol (Water Hyacinth) on diesel engine," *Int. J. Green Energy*, vol. 13, no. 13, pp. 1369–1379, 2016. https://doi.org/10.1080/15435075.2016.1185724

[47] Y. Zhou, S. Searle, and S. Anup, "Techno-economic analysis of cellulosic ethanol in India using agricultural residues," *ICCT-White Pap.*, vol. June, no. 1, pp. 1–18, 2021.

Enhancing Co₂ adsorption in sandstone formation using fly ash nanoparticles: A geo-sequestration approach

Ahmed Abdulla Elryes[1,2,a], Berihun Mamo Negash[1,2,b] *, Nur Asyraf Md Akhir[1,2,c], Faizan Ali[1,2,d], Rubaya Tasnin Mim[1,2,e]

[1]Petroleum engineering department, Universiti Teknologi PETRONAS, 32610 Bandar Seri Iskandar, Perak Darul Ridzuan, Malaysia

[2]Institute of hydrocarbon recovery (IHR), Universiti Teknologi PETRONAS, Bandar Seri Iskandar, 32610, Perak, Malaysia

[a]ahmed_22009956@utp.edu.my, [b]bmamo.negash@utp.edu.my, [c]asyraf.akhir@utp.edu.my, [d]faizan_22005942@utp.edu.my, [e]rubaya_22005275@utp.edu.my

Keywords: CO_2 Adsorption, Fly Ash Nanoparticles, Nanofluid Technology, Sandstone Formation, Carbon Geo-Sequestration

Abstract. A primary environmental concern making global warming worse is the continuous release of CO_2 into the atmosphere. One feasible approach to this issue is the geo-sequestration of CO_2 involving CO_2 adsorption and sequestration in saline aquifer formations. This study explores the possibility of improving CO_2 adsorption in sandstone formations by using fly ash nanoparticles, a by-product. Pre- and post-nano-treated Berea sandstone samples are tested for CO_2 adsorption at pressures ranging from 0.1 to 10 MPa, temperature of 40°C, and salinity of 10 wt% NaCl. High-pressure-temperature volumetric CO_2 adsorption equipment was used to quantify CO_2 adsorption under reservoir conditions. The findings reveal that Xanthan Gum (XG) polymer concentration of 0.04 wt% stabilized and well-suspended the fly ash nanofluid, exhibiting a zeta potential below -32.2 mV. The findings revealed that the total CO_2 adsorption was 45.07 mmol/g for the nano-treated sample and 41.56 mmol/g for the untreated sample. Additionally, the fly ash nano-treated sample demonstrated superior CO_2 adsorption across all pressure points. Specifically, when CO_2 reached a supercritical state over 7.38 MPa, the nano-treated sample showed a maximum CO_2 adsorption of 39.17 mmol/g, while untreated samples reached 36.40 mmol/g. Adsorption isotherm analysis revealed that the Toth model provided the best fit for the experimental data, with an R^2 value of 0.9708, indicating heterogeneous and multilayer adsorption behavior. The isotherm analysis suggests that multilayer adsorption, driven by van der Waals forces, was predominant due to the energy favorability of forming multiple adsorption layers rather than a monolayer. The study outcome suggests the feasibility of fly ash nanoparticles as a promising nanofluid to improve CO_2 adsorption in CO_2 geo-sequestration projects.

Introduction

Using external and renewable energy sources is an innovative solution to reduce energy expenses related to CO_2 capture. Absorption is commonly utilized technique to decrease CO_2 levels in industrial applications and human activities. CO_2 absorption with nanofluids is becoming popular due to their adjustable properties and high surface area. Materials like nano-sized zeolites, metal oxides, metal-organic frameworks (MOFs), and covalent organic frameworks (COFs) enhance CO_2 absorption efficiency because of their porosity [1]. For example, CO_2 sequestration is now the sole action plan that may significantly cut CO_2 emissions. Anthropogenic CO_2 acquired from the surface environment has been extensively studied for storage and isolation in three primary geological formations: deep salt aquifers, un-mineable coal beds, and depleted oil and gas reservoirs [2]. The process of CO_2 adsorption during geo-sequestration in deep saline aquifers is a critical phenomenon when it comes to a secured absorbed CO_2 underground for the long term; the phenomena in which one or more elements are removed by adhering to a rock surface is called adsorption. The basis of adsorption is the intermolecular interactions between solid material surfaces, such as sandstone, and the adsorbed gas, such as CO_2. Gas partial pressure, temperature, adsorbent pore size, and surface forces are the variables that determine whether adsorption is single- or multilayer [3].

Nanofluids comprising nanosized components distributed in a base fluid are being studied for their ability to improve CO_2 absorption in gas-liquid systems [4]. Nanomaterials, with sizes ranging from 1 to 100 nanometers, have increased surface area compared to their volume, which is why they are classified as such [2]. Nanofluids are created using ceramics, metals, and inorganic oxides to enhance gas absorption rates and minimize mass transfer resistance. Nanoparticles, such as graphene oxide (GO) NPs, disperse in an absorption solvent and improves performance due to their high surface area, which induces Brownian motion, micro convection, and shuffle effects. The small size of nanoparticles increases the rate and capacity of CO_2 absorption compared to larger particles. However, exceeding the optimal concentration can cause nanoparticles to agglomerate, hindering gas absorption into the liquid phase [5]. Key factors affecting CO_2 adsorption are nanoparticle type, shape, concentration, base fluid properties, gas flow rate, and operating conditions, highlighting the complex nature of optimizing nanofluids for CO_2 collection applications [6].

Fly ash from coal-fired power plants is widely used in catalysis, construction, and zeolite synthesis. Many reports demonstrated that fly ash nanoparticles can stabilize gas flow and create foam in increased oil recovery. Phong et al., [7] used fly ash nanoparticles to make a robust CO_2 foam, showing that fly ash's carbonaceous material may stabilize emulsions and foams. Rezaee et al. conducted an experimental study demonstrating that a small amount of fly ash nanoparticles significantly enhances foam stability when used with a cationic surfactant (CTAB). The foam agents not only improved foam performance but also altered surface wettability from oil-wet to water-wet. Additionally, fly ash nanoparticles were found to reduce the interfacial tension (IFT) between the oil and water phases, which contributed to the overall oil recovery [8]. Another study by Wang et al., showed that even such anionic and nonionic surfactants working in tangent with fly ash nanoparticles had successfully produced stable emulsions that improved the oil recovery by 8.5% [9]. Fly ash nanoparticles have been broadly employed in enhanced oil recovery (EOR) and CO_2-EOR to improve sweep efficiency. Still, research has yet to study their CO_2 adsorption capability in rock/CO_2/saline aquifer systems. Thus, this experimental study evaluates fly ash's effect on CO_2 adsorption on crushed Berea sandstone before and after nano-treatment with 0.2 wt% fly ash nanoparticles under reservoir settings.

Experimental Preparation and Setup

Materials. A Berea sandstone sample was utilized to represent the reservoir rock. Ultrapure deionized (DI) water and Sodium Chloride ($\geq$ 99.5 mol%, from Sigma Aldrich) were used as base fluid to generate a brine solution (10 wt%, NaCl, 1.71 M). The dissolved air was extracted from the brine by vacuuming for 12 hours. Class C fly ash was obtained from Manjung coal-fired power station in Perak, Malaysia. The fly ash underwent fine milling using Mortar Grinder PULVERISETTE 2 and then was utilized to prepare the nanofluids. XG polymer ($\geq$ 98.5 mol%, from Sigma Aldrich) was used to enhance the stability of the fly ash nanoparticle suspension in brine [10], [11].

Fly Ash Nanofluid Preparation and Stability. This investigation used an ultrasonic homogenizer sonicator for 10 minutes to homogenize fly ash nanoparticles in the base fluid to create nanofluids after the hot plate magnetic stirrer could not disperse the nanoparticles in the brine to create a uniform suspension. Due to the nanofluid's high salinity, the nano-formulation was then supplemented with XG polymer to increase the repulsive force and supercharge the nanoparticles

[10], [11]. Utilizing a sonication power of 300 W, 20 mL batches of nanofluid were produced experimentally using a sonicator. Each batch was subjected to sonication for a total of 40 minutes. In this study, several concentrations of XG polymer (0, 04, 0.06, 0.10, and 0.14 wt%), brine salinity (10 wt% NaCl), and fly ash (2 wt%) were evaluated to determine the effect of nanofluid composition on treatment stability and efficiency.

Aging Procedure of Rock Sample in Fly Ash Nanofluid. To examine the efficiency of the fly ash nanofluid with respect to its CO_2 adsorption enhancement, the sandstone powder was submerged in the prepared nanofluid (0.2 wt% fly ash dispersed in XG-brine mixture). A fixed weight ratio of 5 g of fly ash nanofluid per 1 g of sandstone powder was utilized. Once 2 g of sandstone powder is submerged into 10 g of fly ash nanofluid, they are placed in a linear water shaking bath at 40 °C for 24 hours. The linear shaking water bath allows maximum nanofluid and sample interaction. After this process, the sandstone sample was centrifuged and then dried in an oven at 80 °C for 24 hours.

Adsorption Measurement. The adsorption enhancement of the fly ash nano-treated and untreated samples in the CO_2 system was measured using high-pressure-temperature volumetric gas adsorption equipment. The measuring equipment was built to operate under standard reservoir environments, including high pressure, temperature, and salinity. The untreated and fly ash nanoparticle-treated samples were placed in canister 1 and canister 2, respectively. Once both canisters were tightly secured, the temperature was raised to 40 °C, as the temperature stabilized, the test started by simultaneously pressurizing both canisters with CO_2. In the first stage, the canisters were pressurized up to 3 MPa and closed. The pressure across the canisters is recorded versus time by the computer. As the initial (3 MPa) pressure dropped and finally stabilized due to adsorption, the second stage starts, which is further pressurizing, this time up to 6 MPa. The same procedure was applied in the first stage for 6, 7, 8, and 10 MPa pressure points.

Results and Discussion

XG Polymer Impact on Fly Ash Nanofluid Stability. Fly ash nanofluid with 0 wt% XG concentration resulted in a zeta potential of 15 mV and pH of 6.5. However, adding XG to the fly ash nanofluid solution increased the negative zeta potential values, indicating better nanoparticle stability. Even though it is known that the electrolytes can significantly influence nanofluid properties, the results highlight that adding 0.04 wt% of XG polymer to the high salinity fly ash nanofluid (10 wt% NaCl) led to zeta potential as low as −32.2 mV. This finding is essential in nanofluid formulations since previous research has shown that nanofluids may remain stable provided the zeta potential of the nanosuspension is equal to or less than $-25\,mV$[12]. Thus, Creating a stable nanofluid with high salinity is feasible by adding the appropriate quantity of XG polymer.

Fly Ash Nanoparticles Effect on Total CO_2 Adsorption. Fig. 1 and Fig. 2 illustrate the total CO_2 adsorption on the untreated and fly ash-treated samples over time at 40 C. Among the two tested samples, the nano-treated sandstone sample seemed to experience more CO_2 absorption than the untreated sandstone sample. The fly ash nano-treated sample achieved a total CO_2 adsorption of 45.07 mmol/g, while the untreated sample resulted in CO_2 adsorption of 41.56 mmol/g. Fly ash nanoparticles' enhancement of the CO_2 adsorption capacity can be attributed to the amount of additional surface area the fly ash nanoparticle provides to the sandstone, improving its absorption capacity.

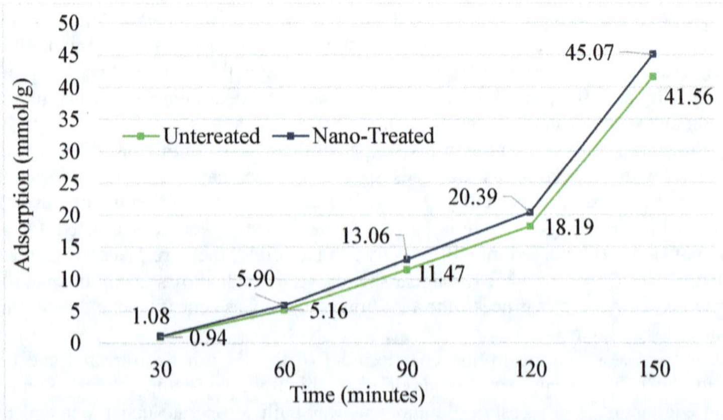

Fig. 1 Cumulative CO_2 adsorption of nano-treated and untreated sample at 40 °C

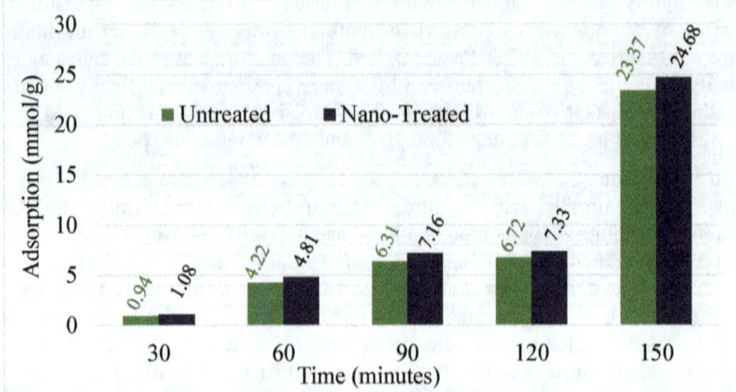

Fig. 2 Non-cumulative CO_2 adsorption of nano-treated and untreated sample at 40 °C

Pressure effect on Fly Ash CO_2 Adsorption. Fig. 3 shows that the fly ash nano-treated sample trend consistently exhibited higher CO_2 adsorption rates than the untreated sample at all tested pressures. Quantitatively, the major CO_2 adsorption capacities were determined to be 39.17 mmol/g and 36.40 mmol/g for the nano-treated and untreated samples, respectively, while the major adsorption occurring when pressure exceeded 7.38 MPa and CO_2 became in a supercritical phase. However, even in the subcritical zone (<7.38 MPa), the nano-treated sample demonstrated higher CO_2 adsorption capacities (5.90 mmol/g) compared to the untreated sample (5.16 mmol/g), indicating the continued effectiveness of the fly ash nano-treatment in enhancing adsorption performance in subcritical and supercritical zones.

Materials Research Proceedings 53 (2025) 67-73 https://doi.org/10.21741/9781644903575-5

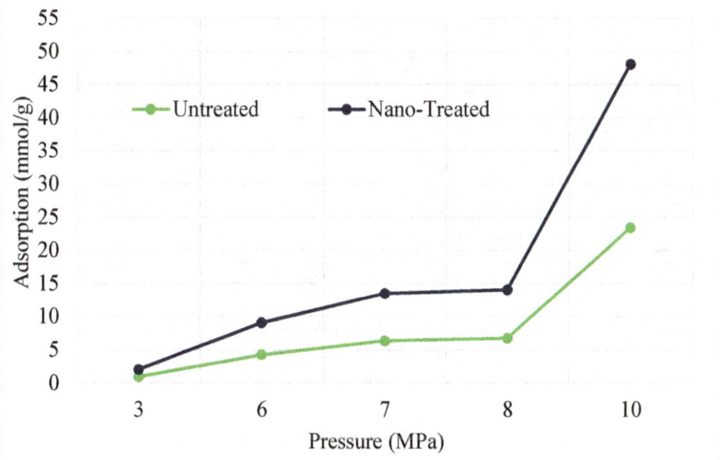

Fig. 3 Pressure Influence on CO_2 Adsorption of Fly Ash Treated and Untreated Sandstone Samples at °C.

Adsorption Isotherms Analysis. The adsorption of CO_2 on fly ash-treated Berea sandstone follows a Type III isotherm, as shown in Fig. 4, which indicates weak adsorbate-adsorbent interactions. Adsorption increases significantly at higher pressures due to cooperative adsorption, where adsorbed molecules make it easier for subsequent molecules to adsorb. Referring to Table 1, Toth isotherm provides the best fit for the experimental data among the other isotherms, with an R^2 value of 0.9708, indicating heterogeneous adsorption with multilayer characteristics. As the CO_2 molecules come into contact with the fly ash-treated sandstone, they have the option to either occupy available spaces on the surface or to directly adsorb onto previously adsorbed molecules, forming multiple layers. This behavior occurs because the heat of liquefaction of CO_2 is greater than the heat required to form a monolayer on the surface. As a result, instead of forming a distinct monolayer, CO_2 molecules tend to stack on top of each other, leading to direct multilayer formation. This process is stabilized by van der Waals forces between the adsorbed layers. The presence of fly ash enhances the surface area and increases the adsorption capacity of the sandstone, allowing for greater CO_2 uptake. The multilayer adsorption mechanism ensures that more CO_2 can be stored at higher pressures, which is crucial for efficient CO_2 sequestration in geological formations.

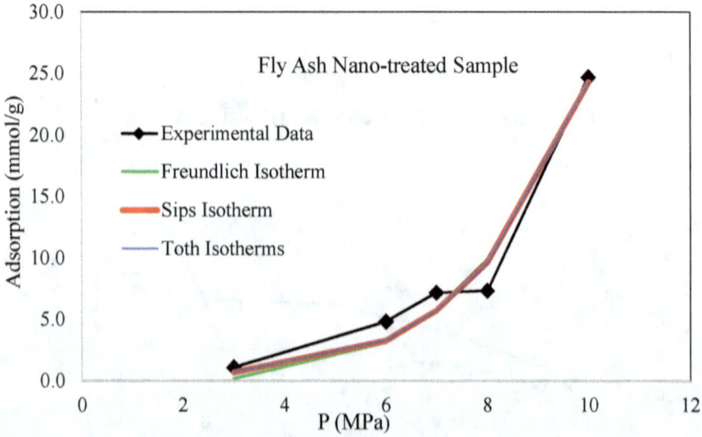

Fig 4. Fitting curves of experimental CO_2 adsorption data using Freundlich, Sips, and Toth isotherms

Table 1. Fitting constants and R^2 values for Freundlich, Sips, and Toth isotherms.

	Freundlich Isotherm			**Sips Isotherm**			**Toth Isotherm**		
Tem (C)	K_F	n_F	R^2	K_S	n_S	R^2	K_T	n_T	R^2
40	0.0024	0.2500	0.9630	0.2062	4.3645	0.9695	0.1968	4.5870	0.9708

Conclusion

This study demonstrates that incorporating fly ash nanoparticles significantly enhances the stability of nanofluids and improves CO_2 adsorption capacity in high-salinity environments. The addition of 0.04 wt% XG polymer to fly ash nanofluids resulted in a zeta potential of -32.2 mV, indicating superior nanoparticle stability. Experimental results showed that fly ash nano-treated sandstone achieved a higher total CO_2 adsorption of 45.07 mmol/g compared to untreated sandstone, which exhibited 41.56 mmol/g, attributable to the increased surface area provided by the nanoparticles. Additionally, the fly ash nano-treated samples consistently exhibited higher CO_2 adsorption rates across all tested pressures, including subcritical and supercritical zones. The adsorption isotherm analysis revealed that the Toth model provided the best fit, indicating heterogeneous and multilayer adsorption behavior with evidence of multilayer adsorption driven by van der Waals forces. These findings highlight the potential of fly ash nanoparticles to significantly enhance CO_2 geo-sequestration efficiency in saline aquifers.

Acknowledgment

The authors express their sincere gratitude to the YUTP grant (015LC0-484) for its crucial financial support, which played a vital role in both this work and the continuous progress of the project. Additionally, the researchers would like to extend their sincere gratitude to the Centre for Research in Enhanced Oil Recovery, Institute of Hydrocarbon Recovery (IHR), Universiti Teknologi PETRONAS, for their indispensable assistance and that which significantly contributed to the successful culmination of this research undertaking by providing a conducive working environment.

References

[1] W. Yu, T. Wang, A. H. A. Park, and M. Fang, "Review of liquid nano-absorbents for enhanced CO2 capture," Nanoscale, vol. 11, no. 37, pp. 17137–17156, Sep. 2019. https://doi.org/10.1039/C9NR05089B

[2] T. D. Rathnaweera and P. G. Ranjith, "Nano-modified CO2 for enhanced deep saline CO2 sequestration: A review and perspective study," Earth Sci Rev, vol. 200, p. 103035, Jan. 2020. https://doi.org/10.1016/J.EARSCIREV.2019.103035

[3] S. Zhang, Y. Shen, L. Wang, J. Chen, and Y. Lu, "Phase change solvents for post-combustion CO2 capture: Principle, advances, and challenges," Appl Energy, vol. 239, pp. 876–897, Apr. 2019. https://doi.org/10.1016/J.APENERGY.2019.01.242

[4] M. Mehdipour, P. Keshavarz, and M. R. Rahimpour, "Rotating liquid sheet contactor: A new gas-liquid contactor system in CO2 absorption by nanofluids," Chemical Engineering and Processing - Process Intensification, vol. 165, p. 108447, Aug. 2021. https://doi.org/10.1016/J.CEP.2021.108447

[5] H. Rashidi and S. Mamivand, "Experimental and numerical mass transfer study of carbon dioxide absorption using Al2O3/water nanofluid in wetted wall column," Energy, vol. 238, p. 121670, Jan. 2022. https://doi.org/10.1016/J.ENERGY.2021.121670

[6] Z. Zhang, J. Cai, F. Chen, H. Li, W. Zhang, and W. Qi, "Progress in enhancement of CO2 absorption by nanofluids: A mini review of mechanisms and current status," Renew Energy, vol. 118, pp. 527–535, Apr. 2018. https://doi.org/10.1016/J.RENENE.2017.11.031

[7] G. M. Phong, R. M. Pilus, A. Mustaffa, L. Thangavel, and N. M. Mohamed, "Relationship between fly ash nanoparticle-stabilized-foam and oil production in core displacement and simulation studies," Fuel, vol. 266, p. 117033, Apr. 2020. https://doi.org/10.1016/J.FUEL.2020.117033

[8] M. Rezaee, S. M. Hosseini-Nasab, J. Fahimpour, and M. Sharifi, "New Insight on improving foam stability and foam flooding using fly-ash in the presence of crude oil," J Pet Sci Eng, vol. 214, p. 110534, Jul. 2022. https://doi.org/10.1016/J.PETROL.2022.110534

[9] W. Wang et al., "Fly Ash Nanoparticle-Stabilized Emulsions for Improve Mobility Control Application," Society of Petroleum Engineers - SPE EuropEC - Europe Energy Conference featured at the 83rd EAGE Annual Conference and Exhibition, EURO 2022, Jun. 2022. https://doi.org/10.2118/209646-MS

[10] A. María Díez-Pascual et al., "Effect of Silica Nanoparticles in Xanthan Gum Solutions: Evolution of Viscosity over Time," 2022. https://doi.org/10.3390/nano12111906

[11] D. L. Buitrago-Rincon et al., "Silica Nanoparticles in Xanthan Gum Solutions: Oil Recovery Efficiency in Core Flooding Tests," Nanomaterials, vol. 13, no. 5, Mar. 2023. https://doi.org/10.3390/NANO13050925

[12] J. Wang, X. Yang, J. J. Klemeš, K. Tian, T. Ma, and B. Sunden, "A review on nanofluid stability: preparation and application," Renewable and Sustainable Energy Reviews, vol. 188, p. 113854, Dec. 2023. https://doi.org/10.1016/J.RSER.2023.113854

Decarbonization Technology: ICDT2024 Materials Research Forum LLC
Materials Research Proceedings 53 (2025) 74-80 https://doi.org/10.21741/9781644903575-6

Lithium recovery from aqueous solution through adsorption using batch study

AL-DHAWİ Baker Nasser Saleh[1,a*], MOHAMED KUTTY Shamsul Rahman[1,b], JAGABA Ahmad Husaini[2,c], ALMAHBASHİ Najib Mohammed Yahya[1,d], AL-YAARİ Abdullah[3,e]

[1]Department of Civil and Environmental Engineering, Universiti Teknologi PETRONAS, Bandar Seri Iskandar 32610, Perak Darul Ridzuan, Malaysia

[2]Interdisciplinary Research Centre for Membranes and Water Security, KingFahd University of Petroleum and Minerals, Dhahran 31261, Saudi Arabia

[3]Department of Fundamental and Applied Sciences, Universiti Teknologi PETRONAS, UTP, Seri Iskandar, 32610, Perak, Malaysia

[a]bakeraldhawi@gmail.com, [b]shamsulrahman@utp.edu.my, [c]ahmad.jagaba@kfupm.edu.sa, [d]najib_17005375@utp.edu.my, [e]abdullah_20001447@utp.edu.my

Keywords: Lithium, Batch Study, Adsorption, Kinetic, Isotherm

Abstract. Treating lithium in aqueous solutions through adsorption is crucial due to the rising demand for lithium in batteries and renewable energy. This method offers a cost-effective and efficient way to capture lithium ions. Effective treatment methods are essential to mitigating these environmental risks. To address this concern, separation technologies like adsorption have been utilized to remove lithium from solutions with minimal concentrations. This study specifically investigates the removal of residual lithium from a synthesized solution using aluminum hydroxide $Al(OH)_3$ as the adsorbent. The research examines the impact of adsorbent dosage, lithium concentration, and pH on residual lithium removal. Results underscore the substantial influence of pH on residual lithium efficiency, with optimal removal observed at pH 4.5, an adsorbent dosage of 1150 mg/L, and a lithium concentration of 1150 mg/L, achieving an 89% removal rate. The utilization of $Al(OH)_3$for lithium adsorption has been demonstrated as an efficient technique for for recovering lithium from synthesis solutions. The process followed a pseudo-first-order kinetic model ($R^2 = 0.917$) and fit the Langmuir isotherm model ($R^2 = 0.91$). These findings contribute to a better understanding of $Al(OH)_3$ adsorption behavior and provide perspectives on the most favorable operational parameters for effective lithium recovery.

Introduction

As the need, for energy continues to grow the extraction and processing of hydrocarbons from sources like shale formations have increased worldwide. This rise in production results in the creation of amounts of wastewater, known as produced water, which contains pollutants such as heavy metals organic compounds and salts. Of these pollutants lithium has garnered interest [1, 2]. The extraction of lithium from produced water poses an economic challenge leading to the development of effective and sustainable treatment methods. One promising solution involves using adsorption processes with adsorbents like aluminium $Al(OH)_3$ activated carbon, zeolites and materials based on graphene [3, 4]. Each of these adsorbents offers features and benefits for removing lithium from water solutions. Aluminum hydroxide stands out for its surface area and ability to attract lithium ions from produced water [5]. Activated carbon is renowned for its surface area and porous structure that enable it to adsorb contaminants efficiently including lithium ions. Zeolites with their crystalline structure and adjustable pore sizes provide sites, for capturing lithium during the removal process. Graphene derived substances, with their durability and extensive surface area show capabilities as absorbers of lithium ions, in water based solutions [6].

This paper aims to explore the efficacy of aluminum hydroxide as an adsorbent for lithium removal from produced water. The following sections will delve into the principles of adsorption, and the factors influencing the adsorption process. Furthermore, the paper will discuss recent advancements in the application of aluminum hydroxide-based adsorbents for lithium removal, highlighting their advantages, limitations, and potential for practical implementation [1, 7].

By conducting an in-depth analysis of the literature and experimental findings, this study seeks to contribute to the understanding of lithium removal from aqueous solution and provide insights into the optimization of adsorption processes utilizing aluminum hydroxide. Developing efficient and eco-friendly methods to treat lithium-contaminated is crucial for reducing environmental risks and promoting sustainable management across various industries.

Materials Methods

Batch Study. In this investigation, commercial adsorbents including $Al(OH_3)$ sourced from Segam Malaysia were utilized, alongside another reagent procured from the same supplier. A Li synthesis solution of 10% concentration was prepared using high-grade $LiCO_3$ reagent mixed with distilled water prior to initiating the adsorption experiments.

In the batch adsorption experiments, various adsorbents and initial concentrations of lithium were introduced into 500 mL volumetric flasks, using a 10% stock solution with concentrations ranging from 300 mg/L to 2000 mg/L. The dosage of adsorbent and concentration were adjusted accordingly. pH adjustments of the lithium solutions were made using diluted HCl and NaOH to assess pH effects. The flasks were then agitated on an orbital shaker at 150 rpm for different durations. After the specified time periods, a filter membrane was used to separate the adsorbent from the liquid phase. The remaining lithium concentration was determined using an Automated Adsorption Spectrophotometer employing the carmine method.

Kinetic and Isotherm. Three different concentrations of 300, 1150, and 2000 mg/L were selected for adsorption kinetic and isotherm with three different contact time of 30, 255, 480 minutes.

Table 1: Both model for isotherm and kinetic

Model	linear	Ref.
Langmuir Isotherm	$\frac{C_e}{q_e} = \frac{1}{q_m b} + \frac{C_e}{q_m}$	[8]
Pseudo-first order Kinetic	$ln(q_e - q_t) = lnq_e - k_1 t$	[9]

Results and Discussions

This study presents the findings derived from a batch investigation employing AL(OH3) as the adsorbent material.

Impact of pH on Li Recovery. For The pH of the solution significantly impacts lithium recovery from produced water using aluminum hydroxide $Al(OH)_3$ as an adsorbent, with the pH range tested between 2 and 11 figure 1 illustrate the effect of pH to Li recovery.

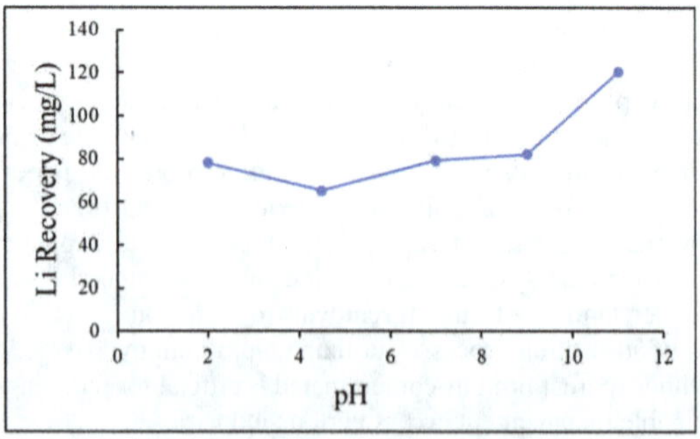

Fig.1: impact of pH on Li Recovery

The optimum pH found to be at 4.5 with lithium recovery of 65 mg/L at constant lithium concentration and adsorbent dosage of 1150 mg/L. This influence arises from its effect on the speciation of lithium ions and the surface chemistry of the adsorbent material. At lower pH values, lithium ions exist predominantly as monomeric species, while at higher pH values, lithium hydroxo complexes become prevalent, influencing their adsorption behavior onto the surface of aluminum hydroxide [10]. Additionally, the surface charge of aluminum hydroxide is pH-dependent, with a positively charged surface at low pH and a negatively charged surface at higher pH, affecting the electrostatic interaction between the adsorbent and lithium ions. Typically, the optimal pH range for lithium recovery using aluminum hydroxide lies within the neutral to slightly alkaline range, where the surface charge promotes the adsorption of lithium hydroxo complexes. pH control is essential in practical applications to ensure optimal conditions for lithium recovery, often involving adjustments using NaOH and HCl depending on the initial pH of the produced water. Understanding the interplay between pH, lithium speciation, and surface chemistry is crucial for optimizing adsorption conditions and maximizing lithium recovery efficiency [11].

Impact of Adsorption Dosage on Li Recovery. The effect of adsorption dosage on lithium (Li) recovery from produced water using aluminum hydroxide $Al(OH)_3$ as an adsorbent, within the range of 300 mg/L to 2000 mg/L, is crucial for optimizing the efficiency of the adsorption process figure 2 illustrate the effect of adsorbent dosage to Li recovery.

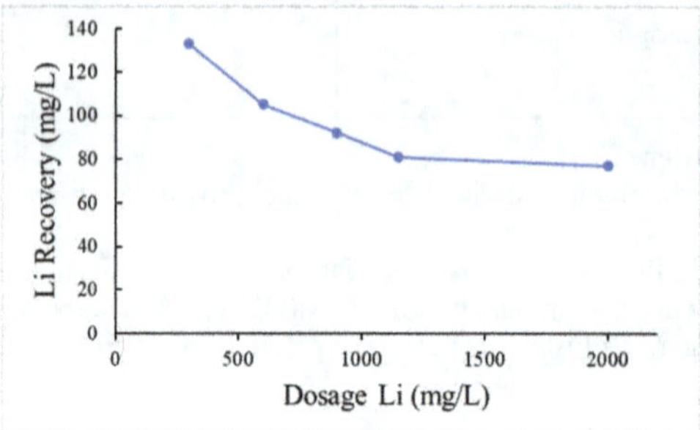

Fig. 2: Impact of Adsorption Dosage on Li Recovery

The optimum dosage for Li recovery was found to be 77 mg/L with adsorbent dosage of 2000 mg/L at pH of 4.5 and lithium concentration of 1150 mg/L. As the dosage increases within this range, the available adsorption sites on the aluminum hydroxide surface also increase, leading to enhanced Li recovery due to greater adsorbent utilization. This rise in adsorption sites allows for more lithium ions to be captured from the produced water, thereby increasing the overall efficiency of Li removal. However, beyond a certain dosage, typically at higher levels nearing 2000 mg/L, the adsorption sites may become saturated, resulting in diminishing returns in Li recovery. At this point, further increases in dosage may not significantly enhance Li removal and could lead to inefficient use of the adsorbent material, potentially increasing operational costs [12, 13]. Additionally, the pH of the solution may influence the impact of adsorption dosage on Li recovery, as higher dosages might require adjustments to maintain optimal pH conditions for adsorption. Therefore, careful consideration and optimization of adsorption dosage within the specified range are essential to achieve maximum Li recovery efficiency while minimizing operational costs and waste generation in the treatment of produced water [14]. Furthermore, the kinetics and equilibrium behavior of adsorption may be affected by the adsorption dosage, with higher dosages potentially accelerating the initial rate of adsorption but prolonging the time required to reach equilibrium. Thus, optimizing adsorption dosage involves striking a balance between maximizing Li recovery efficiency and minimizing operational costs, waste generation, and environmental impacts through experimental testing and modeling [15].

Impact of Li Concentration on Li Recovery. The impact of lithium (Li) concentration in produced water on Li recovery using aluminum hydroxide $Al(OH)_3$ as an adsorbent, within the range of 300 mg/L to 2000 mg/L, with constant dosage of 1150 and pH 4.5 is significant for optimizing the efficiency of the adsorption process figure 3 illustrate the effect of concentration to Li recovery.

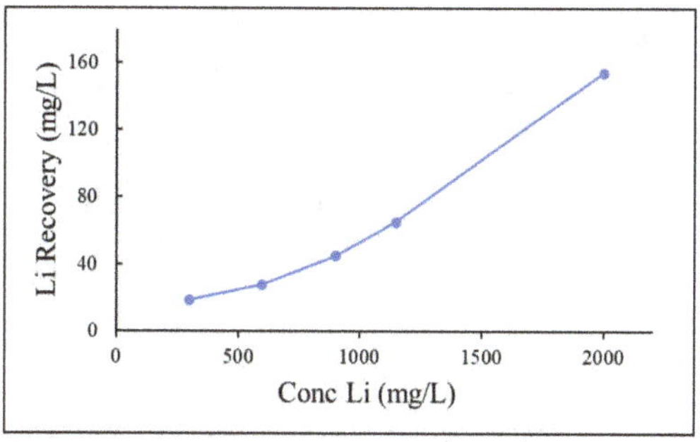

Fig.3: Impact of Li Concentration on Li Recovery

The optimum concentration for Li recovery was found in between 900 to 1150 mg/L with the Li recovery of 45, and 65 respectively. As the Li concentration increases within this range, there are more Li ions available in the solution for adsorption onto the surface of aluminum hydroxide. Consequently, higher Li concentrations typically result in greater Li recovery due to the increased availability of Li ions for adsorption. However, it's important to note that there may be a point of diminishing returns where further increases in Li concentration do not significantly enhance Li recovery and may even lead to inefficient utilization of the adsorbent material [16]. Therefore, careful consideration and optimization of the Li concentration within the specified range are crucial to achieve maximum Li recovery efficiency while minimizing operational costs and waste generation in the treatment of produced water [17].

Pseudo first Order kinetic Model, and Langmuir isotherm Model. The Pseudo First Order kinetic model and Langmuir isotherm model are commonly used to describe adsorption processes in batch studies.

In batch studies, the Pseudo First Order kinetic model and Langmuir isotherm model are frequently employed to characterize adsorption processes.

The Pseudo First Order kinetic model suggests that the adsorption rate depends on the discrepancy between the maximum adsorption capacity and the amount of solute adsorbed at a specific moment. On the other hand, the Langmuir isotherm model hypothesizes a single layer of adsorption onto a homogeneous surface with a restricted number of similar adsorption sites [18, 19].

In this investigation, we examined adsorption kinetics and isotherms using three distinct concentrations: 300 mg/L, 1150 mg/L, and 2000 mg/L, across three different contact durations: 30 minutes, 255 minutes, and 480 minutes. The kinetic model displayed an R^2 value of 0.9171, indicating a strong agreement between the experimental data and the Pseudo First Order kinetic model. Similarly, the isotherm model exhibited an R^2 value of 0.910, suggesting a favorable fit of the testing data to the Langmuir isotherm model [20]. These findings suggest that both models accurately depict the adsorption characteristics observed in the system being investigated [21, 22].

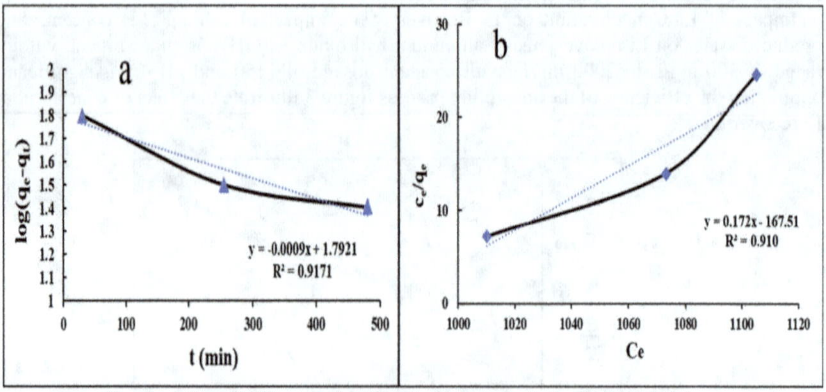

Fig.4: a) Pseudo First Order Kinetic b) Langmuir Isotherm Model

Conclusion

The adsorbent $Al(OH)_3$ demonstrated considerable potential for lithium recovery from produced water. Experimental findings revealed that $Al(OH)_3$ exhibited a notable adsorption capacity for lithium ions, resulting in a substantial reduction of lithium concentration in the solution. The dosage of adsorbent emerged as a crucial parameter influencing lithium recovery. Augmenting the dosage of $Al(OH)_3$ led to increased efficiency in lithium removal up to an optimal dosage. However, surpassing this threshold, further increments in adsorbent dosage did not yield significant improvements in lithium elimination, indicating saturation of available adsorption sites. Optimal lithium removal was attained at a pH level of 4.5, with a dosage and concentration of $Al(OH)_3$ both set at 1150 mg/L.

Acknowledgment

The authors sincerely appreciate the support extended by UTP and YUTP through grant cost center 015LC0-361.

References

[1] B. N. S. Al-dhawi et al., "Lithium adsorption from aqueous solution using aluminum hydroxide: Characterization, optimization by response surface methodology, kinetic modelling, and isotherm studies," Case Studies in Chemical and Environmental Engineering, vol. 7, p. 100350, 2023. https://doi.org/10.1016/j.cscee.2023.100350

[2] A. H. Birniwa et al., "Membrane technologies for heavy metals removal from water and wastewater: A mini review," Case Studies in Chemical and Environmental Engineering, p. 100538, 2023. https://doi.org/10.1016/j.cscee.2023.100538

[3] N. M. Almahbashi, S. R. Kutty, A. Jagaba, A. Al-nini, B. Al-Dhawi, and U. Rathnayake, "Phenol removal from aqueous solutions using rice stalk-derived activated carbon: Equilibrium, kinetics, and thermodynamics study," Case Studies in Chemical and Environmental Engineering, vol. 8, p. 100471, 2023. https://doi.org/10.1016/j.cscee.2023.100471

[4] A.-B. A. Al-Mekhlafi, A. S. N. Isha, A. N. Al-Tahitah, A. F. Kineber, B. N. S. Al-Dhawi, and M. Ajmal, "Modelling the impact of driver work environment on driving performance among oil and gas heavy vehicles: SEM-PLS," Safety, vol. 9, no. 3, p. 48, 2023. https://doi.org/10.3390/safety9030048

[5] Y. Boroumand and A. Razmjou, "Adsorption-type aluminium-based direct lithium extraction: The effect of heat, salinity and lithium content," Desalination, p. 117406, 2024. https://doi.org/10.1016/j.desal.2024.117406

[6] J. Farahbakhsh et al., "Direct lithium extraction: A new paradigm for lithium production and resource utilization," Desalination, p. 117249, 2023. https://doi.org/10.1016/j.desal.2023.117249

[7] A. Al-Yaari et al., "The effects of nanofluid thermophysical properties on enhanced oil recovery in a heterogenous porous media," Case Studies in Chemical and Environmental Engineering, vol. 9, p. 100556, 2024. https://doi.org/10.1016/j.cscee.2023.100556

[8] I. J. J. o. t. A. C. s. Langmuir, "The adsorption of gases on plane surfaces of glass, mica and platinum," Journal of the American Chemical society, vol. 40, no. 9, pp. 1361-1403, 1918. https://doi.org/10.1021/ja02242a004

[9] S. K. J. S. V. H. Lagergren, "About the theory of so-called adsorption of soluble substances," Sven. Vetenskapsakad. Handingarl, vol. 24, pp. 1-39, 1898.

[10] B. N. S. Al-dhawi et al., "Phosphorus Removal from Synthetic Wastewater by Using Palm Oil Clinker as Media in Continuous Activated Sludge," in Sustainability Challenges and Delivering Practical Engineering Solutions: Resources, Materials, Energy, and Buildings: Springer, 2023, pp. 41-44. https://doi.org/10.1007/978-3-031-26580-8_8

[11] P. Akbari Ziarani and A. Molaei Dehkordi, "Lithium Extraction from an Aqueous Medium through In Situ Synthesis of Aluminum Hydroxide: A Comprehensive Study on Adsorption and Desorption Processes, Kinetics, Isotherm Models, and Thermodynamic Parameters," Industrial & Engineering Chemistry Research, vol. 63, no. 1, pp. 445-458, 2023. https://doi.org/10.1021/acs.iecr.3c03069

[12] A. H. Jagaba et al., "Trend and current practices of coagulation-based hybrid systems for pulp and paper mill effluent treatment: mechanisms, optimization techniques and performance evaluation," Journal of Cleaner Production, p. 139543, 2023. https://doi.org/10.1016/j.jclepro.2023.139543

[13] B. N. S. Al-dhawi et al., "Boron adsorption from aqueous solutions through column study: desorption mechanisms, regeneration techniques, and kinetic insights," Desalination and Water Treatment, p. 100586, 2024. https://doi.org/10.1016/j.dwt.2024.100586

[14] B. N. S. Al-Dhawi et al., "Treatment of synthetic wastewater by using submerged attached growth media in continuous activated sludge reactor system," International Journal of Sustainable Building Technology and Urban Development, vol. 13, no. 1, pp. 2-10, 2022.

[15] L. Tian, W. Ma, and M. Han, "Adsorption behavior of Li+ onto nano-lithium ion sieve from hybrid magnesium/lithium manganese oxide," Chemical Engineering Journal, vol. 156, no. 1, pp. 134-140, 2010. https://doi.org/10.1016/j.cej.2009.10.008

[16] A. H. Jagaba et al., "Toxicity of carbon nanotubes in wastewater treatment," in Water Treatment Using Engineered Carbon Nanotubes: Elsevier, 2024, pp. 365-411. https://doi.org/10.1016/B978-0-443-18524-3.00004-0

[17] A. Kumar, H. Fukuda, T. A. Hatton, and J. H. Lienhard, "Lithium recovery from oil and gas produced water: a need for a growing energy industry," ACS Energy Letters, vol. 4, no. 6, pp. 1471-1474, 2019. https://doi.org/10.1021/acsenergylett.9b00779

[18] B. N. S. Al-dhawi et al., "Optimal parameters for boron recovery in a batch adsorption study: synthesis, characterization, regeneration, kinetics, and isotherm studies," Case Studies in Chemical and Environmental Engineering, vol. 8, p. 100508, 2023. https://doi.org/10.1016/j.cscee.2023.100508

[19] B. N. S. Al-dhawi et al., "Boron removal from produce water through adsorption," in BIO Web of Conferences, 2023, vol. 73: EDP Sciences, p. 02002. https://doi.org/10.1051/bioconf/20237302002

[20] K. VİCKY et al., "Kinetics study of nutrients removal from synthetic wastewater using media as submerged in continuous activated sludge system," Materials Research Foundations, vol. 29, 2022. https://doi.org/10.21741/9781644902516-12

[21] B. N. S. Al-dhawi et al., "Adsorptive removal of boron by DIAION™ CRB05: characterization, kinetics, isotherm, and optimization by response surface methodology," Processes, vol. 11, no. 2, p. 453, 2023. https://doi.org/10.3390/pr11020453

[22] N. Al-mahbashi, S. R. Kutty, A. Jagaba, A. Al-Nini, B. Al-dhawi, and A. Noor, "Preparation of Rice Paddy Stalks-Based Activated Carbon to Remove Phenol from Aqueous Solutions," in International Conference on Dam Safety Management and Engineering, 2023: Springer, pp. 675-685. https://doi.org/10.1007/978-981-99-3708-0_46

 https://doi.org/10.21741/9781644903575-7

Durian seed starch-nanostructured-based fillers modification to enhance film appearance

ASYIFA Andey[1,a], HIROSHI Uyama[2,b] and SYAZANA Abdullah Lim[1,c*]

[1]Food Science and Technology Programme, School of Applied Science and Mathematics, Universiti Teknologi Brunei, Tungku Highway, Gadong, BE1410, Brunei Darussalam

[2]Department of Applied Chemistry, Graduate School of Engineering, Osaka University, 2-1 Yamadaoka, Suita, Osaka 565-0871, Japan

[a]m20220019@student.utb.edu.bn, [b]uyama@chem.eng.osaka-u.ac.jp, [c]syazana.lim@utb.edu.bn

Keywords: Durian Seed Starch, Nanoparticles, High Pressure Homogenization, Green Application

Abstract. Durian seeds (*Durio zibethinus Murr*) are frequently discarded, adding to biomass waste. This research presents a green, economical, and high practicality approach for the fabrication of durian seed starch nanoparticles (DSS-NPs) derived from durian seed (*Durio Zibethinus Murr*) employing high pressure homogenizer to enhance its visual properties of film in sustainable applications. DSS-NPs suspension films were developed, and non-nanoparticle durian seed starch (DSS) suspension film was used as comparison. Morphological and structural aspects, including thickness and transparency of the suspension films were assessed. Thickness measurements for DSS and DSS-NPs films were recorded as 0.199 mm and 0.125 mm, respectively. The transparency of film revealed that it increased in DSS-NPs film, exhibiting higher absorbance values compared to DSS film. This indicated that DSS-NPs films exhibited enhanced compatibility in terms of morphology and appearance when compared to non-nanoparticle film which suggests significant potential for the application of DSS-NPs in diverse fields such as packaging materials and biomedical applications.

Introduction

Starch has emerged as the most viable substitute for traditional petroleum-based materials, owing to its physical advantages and additional benefits attributing to the simplicity of acquiring and processing starch [1]. Starch, as a natural, renewable, and biodegradable polymer, is typically extracted from traditional sources including corn, cassava, potato, wheat, and rice. It is widely utilized in a multitude of industries, including food, cosmetics, and pharmaceuticals, owing to its versatile properties. Nevertheless, native starch cannot be utilized directly owing to its inadequate thermo-mechanical properties and high-water absorption capacity [2]. Consequently, modifications are required prior to its application. Common techniques for starch modification encompass genetic, enzymatic, physical, and chemical methods, with chemical modification being extensively utilized in industrial settings. Recent studies have explored unconventional sources of starch, primarily derived from seeds and legumes, which exhibit comparable properties to traditional starch sources. These novel starch sources include durian (*Durio zibethinus L.*), lychee (*Litchi chinensis Sonn.*), loquat (*Eriobothrya japonica*), mango (*Mangifera indica*), jackfruit (*Artocarpus heterophyllus*), pine (*Araucaria angustifolia*) and tamarind (*Tamarindus indica L.*) [3]. Employing starch sourced from a variety of plants, including cassava, potatoes, yams, corn, rice, and durian seeds (*Durio zibethinus Murr*) presents a promising avenue for sustainable applications such as starch-based bioplastics [4]. The utilization of starch as a raw material offers notable advantages including cost-effectiveness and renewability. This makes them a viable option for utilizing food waste to create valuable products.

Durian seed (*Durio zibethinus Murr.*) has a relatively high starch content, amounting to 76.65% [5]. Besides their abundant starch content, durian seeds are readily accessible and still largely disposed of by the community. Moreover, given their significant potential for biomaterial development due to the abundance of biopolymers such as starch and cellulose, they offer a cost-effective and environmentally friendly alternative. Despite the high nutritional potential and relatively high starch, amylopectin content, and gum yield of durian seeds [6], they are known for their low mechanical strength and poor water resistance when used to develop biodegradable plastic films [7]. Hence, to enhance the properties of durian seed starch-based films, the starch matrix must be value-added or modified by incorporating various additives such as plasticizers, crosslinking agents and functional nanoparticles [8].

Amid rising apprehension regarding the environmental ramifications associated with traditional chemical approaches, there has been a surge in interest within the field of nanobiotechnology towards adopting environmentally sustainable methodologies, often referred to as green chemistry, for nanoparticle synthesis. Nanoparticles are solid or colloidal particles typically sized between 1 and 100 nm [9]. They possess distinctive physicochemical attributes, characterized by their small size effect, surface area, biological compatibility, water solubility, food pharmaceuticals, and water treatment [10]. Starch nanoparticles (SNPs) has the potential to enhance functional attributes of starch in which they are being explored for diverse prospects aimed at improving the quality of product development. Thus, widens its application especially in food systems such as bioplastic filler, encapsulating agent, and antimicrobial agent. Recent research has proven that SNPs can be derived from starch granules, exhibiting distinctive functional properties.

Transparency is a crucial technical parameter for evaluating and validating specific food packaging materials and usually accompanied with thickness value of the resultant film. The range of thicknesses for packaging material typically exhibit thickness value within the range of 50–100 μm depending on the polymer utilized [11].

There are many studies on the development of film from durian seed have been done, and mostly the resulting films displayed a brownish film attributed to the inherent color of the seeds and a textured surface. For instance, studies which have been conducted and resulted brownish film attributed to the inherent color of the seeds [12] and textured surface [13] coming from the incomplete homogenization or large particles of DSS.

The modification of DSS to nano particle-sized has the potential to improve the appearance of film in terms of its color which is related to transparency of the film as well as its texture. In this study, although there were many previous studies have been carried out to develop starch-based bioplastic film using DSS, as far as we are aware, there have been no investigations conducted on the utilization of durian seed starch nanoparticles, which possess unique functional characteristics as well as to improve its appearance and holds promise for diverse applications.

Experimental

Materials and instruments

The materials used to obtain durian seed starch (DSS) were durian seed waste which was collected from local supermarkets and deionized water.

The testing equipment used for this study were hot air oven, high-pressure homogenizer (Star Burst Mini, Sugino Machine Limited, Japan), scanning electron microscope (SU3500, Hitachi, Tokyo, Japan) and UV-Vis spectrophotometer (V-750, JASCO Corp., Japan).

Extraction of durian seed starch

2 kg of durian seeds were weighed and washed, followed by peeling and cutting them into small pieces. After washing and cleaning, the seeds were then oven-dried at 105°C for 24 h. The oven-dried seeds were then blended with water in a ratio of 1:2, squeezed and the resulting filtrate was collected. The filtrate was allowed to settle for 12 h to yield a precipitate, which was then oven-dried at 105°C for 24 h. Subsequently, the dried precipitate was processed in a blender and sifted through 100mm mesh sifters, and DSS was obtained. The starch was kept in room temperature in an air-tight container.

Preparation of durian seed starch-nano particles

DSS (5 wt %) (w/w DSS) was homogenized in deionized water at room temperature for 48 hours. DSS-NPs were subsequently prepared using a high-pressure homogenizer operating at 200 MPa for five passes. After each pass, the resulting emulsion was cooled for 15 minutes before advancing to the subsequent cycle. The final emulsion underwent stirring and ultrasonication for 1 hour to ensure complete dispersion of all DSS-NPs.

Preparation of durian seed starch suspension film

DSS-NPs suspension films were prepared by casting method as described by [14] with slight modification. The concentration of the DSS-NPs was set at 10% (w/v) of DSS. 1 g of glycerol was added, mixed on heater for 15 mins at 90 °C. A mixture of 1 g carboxymethyl cellulose (CMC) and 10 mL of deionized water was added, then the power was reactivated at the same temperature for an additional 10 min. After cooling, films were casted and dried overnight in an oven at 38 °C. The method for producing non-nanoparticle durian seed starch (DSS) suspension film was conducted in a similar manner, with the only difference being that DSS powder was used instead of DSS-NPs and mixed with deionized water.

Characterization of suspension films

Thickness of film

Thickness of the suspension film samples was assessed at three distinct points using a micrometer with a precision of 0.01 mm. The average measurement obtained represents the thickness of the film. This measurement procedure was conducted on all produced samples.

Transparency test

Film transparency was evaluated following the protocol established by [15] using UV-Vis at a wavelength to determine the absorbance value passed through the suspension films. Suspension films were cut into 1 cm x 3 cm to correspond the height and width of cuvette and then affixed to the side of the cuvette. Air was used as reference. The absorbance was measured across a spectrum ranging from 800 nm to 200 nm.

Scanning electron microscopy

Suspension film samples were subjected to morphological characterization using scanning electron microscopy (SEM). Prior to analysis, all films were trimmed into 1 cm x 1 cm and submerged in nitrogen liquid and double-sided adhesive was used to secure the samples to the specimen holder then attached on metal stub. The mounts were coated with gold after being dried under a strong vacuum. Scrutiny of these specimens was made at 500X magnifications.

Statistical analysis

The data underwent analysis through one-way analysis of variance (ANOVA) conducted with the Statistical Package for the Social Sciences (SPSS). Mean values were compared using Tukey's test at a significance level of 5%. The results were presented as the mean of triplicate values ± standard deviation.

Result and discussion

Thickness and transparency

Fig. 1 illustrates the thickness of DSS film and DSS-NPs film. These two suspension films yielded diverse thickness results, with measurements of 0.199 mm and 0.125 mm, respectively. Notably, the sample with DSS nanoparticles exhibited the lowest thickness. Discrepancies in thickness may be attributed to variations in mold area, solution volume, and total solids content. The increased thickness of DSS film is primarily attributed to its textured surface, resulting from the incomplete homogenization of starch granules.

Transparency testing of the suspension films shown in Fig. 2 revealed that the highest absorbance value was achieved in the film containing DSS-NPs at each wavelength (200–700 nm). Increased starch concentration in the film corresponded to higher absorbance values. Variations in absorbance values can be attributed to differences in the size of granule starch, amylopectin content, and amylose percentage of DSS and DSS-NPs. Film transparency can also be assessed visually by observing the clarity of an image behind the film, as illustrated in Fig. 3. The DSS-NPs suspension film depicted in Fig. 3b shows a much clearer image compared to the DSS suspension film shown in Fig. 3a. This visual clarity corresponds with higher transparency values, indicating a linear relationship between transmittance and image clarity. Additionally, there are significant differences in color and texture between the two films, the DSS film appears brownish with a textured surface, while the DSS-NPs film is clearer and has a smooth surface.

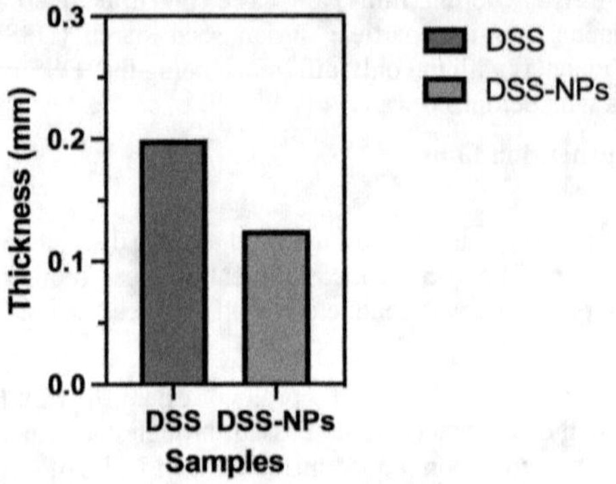

Figure 1. *Thickness of DSS and DSS-NPs films.*

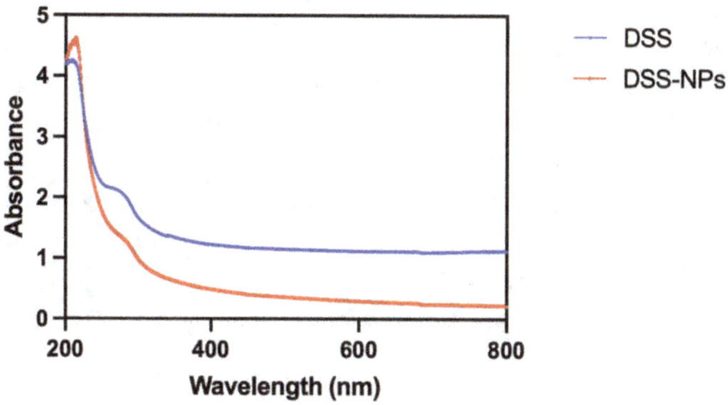

Figure 2. *Transparency of DSS and DSS-NPs suspension films.*

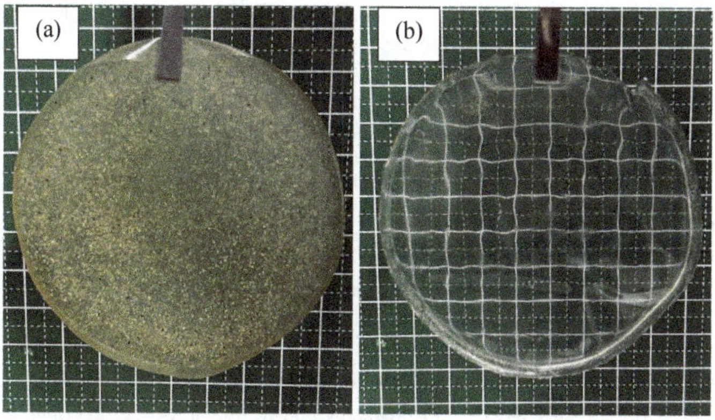

Figure 3. *Suspension films of (a) DSS and (b) DSS-NPs.*

Morphological studies of suspension films

SEM micrographs presented in Fig. 4a-b serve to visualize and analyze the surface morphology and microstructure of the DSS and DSS-NPs suspension film samples. Fig. 4a, representing the non-nanoparticle DSS film, illustrates surface features characterized by granules, indicating incomplete gelatinization of starch during the formation process. Conversely, Fig. 4b reveals well-homogenized DSS-NPs suspension film, hence resulting in smooth textured film as well as its transparency.

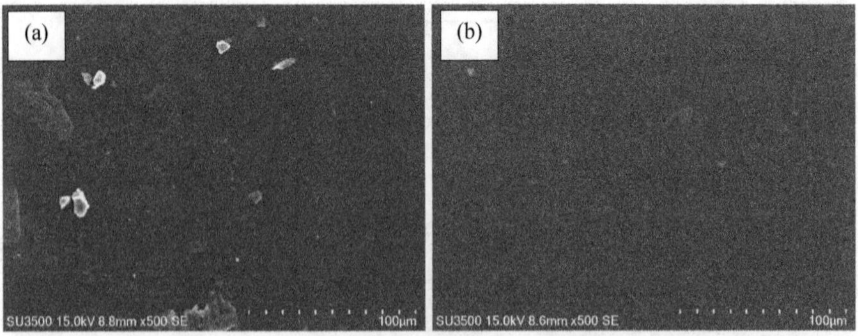

Figure 4. *SEM images of (a) DSS suspension film and (b) DSS-NPs suspension film at 500X magnifications.*

Conclusion

This study highlights the versatility of DSS-NPs in producing films with enhanced diverse properties such as the visual appearance, enhancing their applicability across various applications such as food packaging and encapsulating agent for drug delivery. Additionally, the modification of starch showcases the potential of high-pressure homogenization as a viable technique for large-scale industrial production of nanoparticles. The environmentally friendly nature of DSS-NPs films positions it as a viable alternative to petroleum-based plastics. While this research shows promising outcomes, it is crucial to acknowledge certain limitations, such as the need for further optimization of formulations and application methods to maximize effectiveness.

References

[1] M H S Ginting, R. H. Effect of Hydrochloric Acid Concentration as Chitosan Solvent on Mechanical Properties of Bioplastics from Durian Seed Starch (Durio Zibethinus) with Filler Chitosan and Plasticizer Sorbitol. IOP Conference Series: Materials Science and Engineering, 2017.

[2] Amaraweera, S.M.; Gunathilake, C.; Gunawardene, O.H.P.; Fernando, N.M.L.; Wanninayaka, D.B.; Dassanayake, R.S.; Rajapaksha, S.M.; Manamperi, A.; Fernando, C.A.N.; Kulatunga, A.K.; et al. Development of Starch-Based Materials Using Current Modification Techniques and Their Applications: A Review. Molecules 2021, 26, 6880. https://doi.org/10.3390/molecules26226880

[3] Caicedo Chacon, W. D., Gaviria, Y. A. R., Zapelini de Melo, A. P., Verruck, S., Monteiro, A. R., Valencia, G. A. Physicochemical Properties and Potential Food Applications of Starches Isolated from Unconventional Seeds: A Review. Biosynthesis Nutrition Biomedical 2023. https://doi.org/10.1002/star.202200228

[4] Marichelvam, M.K.; Jawaid, M.; Asim, M. Corn and Rice Starch-Based Bio-Plastics as Alternative Packaging Materials. Fibers 2019, 7, 32. https://doi.org/10.3390/fib7040032

[5] Handayani, P. A., Devi, A. L., and Ganisha, N. A., "Optimization of the Elongation of Bioplastic from Durian Seed Starch Using Response Surface Methodology", in IOP Conference Series: Earth and Environmental Science 2023, vol. 1203, no. 1. https://doi.org/10.1088/1755-1315/1203/1/012002

[6] N D Permatasari, J E Witoyo, E Ni'maturohmah, M Masruri, S S Yuwono, and S B Widjanarko. Potential of durian seed (*Durio zibenthinus Murr.*) flour as the source of eco-friendly plastics materials: a mini-review. 2021 https://doi.org/10.25181/icoaas.v2i2.2483

[7] Retnowati, Diah & Ratnawati, Ratnawati & Purbasari, Aprilina. (2015). A biodegradable film from jackfruit (Artocarpus heterophyllus) and durian (durio zibethinus) seed flours. Scientific Study and Research: Chemistry and Chemical Engineering, Biotechnology, Food Industry. 16. 395-404.

[8] M. Nikolic, Z. V. (2021). Metal oxide nanoparticles for safe active and intelligent food packaging. Trends in Food Science & Technology, 655-668. https://doi.org/10.1016/j.tifs.2021.08.019

[9] Qiang Yang, X. H. Preparation of starch nanoparticles by jet cavitation and enzymatic hydrolysis and their application in starch films. LWT. 2023 https://doi.org/10.1016/j.lwt.2023.115138

[10] Liu M, Guo X, Ma X, Xie Z, Wu Y, Ouyang J. Physicochemical properties of a novel chestnut porous starch nanoparticle. Int J Biol Macromol. 2024 Mar;261(Pt 2):129920. doi: 10.1016/j.ijbiomac.2024.129920. Epub 2024 Feb 2. PMID: 38311128.

[11] Coles, R., & Kirwan, M. J. (2011). Food and beverage packaging technology. John Wiley & Sons. https://doi.org/10.1002/9781444392180

[12] Jannah, N. R., Jamarun, N., & Putri, Y. E. (2021). Production of Starch-Based Bioplastic from Durio zibethinus Murr Seed Using Glycerol as Plasticizer. Jurnal Riset Kimia, 12(2). https://doi.org/10.25077/jrk.v12i2.398

[13] Rahmawati, S., Aulia, A., Hasfah, N., Nuryanti, S., Abram, P.H., Ningsih, P. (2021). The utilization of durian seeds (Durio Zibethinus Murr) as a base for making edible film. International Journal of Design & Nature and Ecodynamics, Vol. 16, No. 1, pp. 77-84. https://doi.org/10.18280/ijdne.160110

[14] Pimpa, W. & Pimpa, Chakriya & Junsangsree, P. Development of Biodegradable Films Based on Durian Seed Starch. Advanced Materials Research 2022, 506. 311-314. https://doi.org/10.4028/www.scientific.net/AMR.506.311

[15] Noryawati Mulyano, M. T. (2015). Development of bioplastic based on Cassava flour and its starch derivatives for food packaging. Journal of Harmonized Reserach in Applied Sciences, 125-132.

Computational study on amino acid base ionic liquids (AAILs) for gas separation using COSMO-RS

Mehtab Ali Darban[1,2,a], Serene Sow Mun Lock[1,2,b *], Zubair Ahmed Laghari[2,c]

[1]CO2 Research Centre (CO_2RES), Universiti Teknologi PETRONAS, Seri Iskandar 32610, Malaysia

[2]Department of Chemical Engineering, Universiti Teknologi PETRONAS, Seri Iskandar 32610, Malaysia

[a]mehtab_22011506@utp.edu.my, [b]sowmunlock@utp.edu.my, [c]zubair_22008648@utp.edu.my

Keywords: CO_2 Capture & Separation, COSMO-RS; Amino Acid Base Ionic Liquids; Solubility and Absorption. Computational Chemistry

Abstract. Greenhouse gases, especially carbon dioxide (CO_2), are a major contributor to global warming and poses a significant threat to the environment. Amino Acid based Ionic Liquids (AAILs) show strong potential as durable solvents with high CO_2 capture capacity, making them well-suited for long-term, large-scale applications to address this issue. This work is focused on the absorption study of various gases, such as methane (CH_4), nitrogen (N_2), oxygen (O_2), and CO_2 in AAILs using computational chemistry and molecular simulations. Capturing more gases involves trapping them within molecular structures, stabilizing them for future use or storage, and enabling efficient purification for various applications. Hereby, eighteen types of food grade AAILs, including 1-Butyl-3-methylimidazolium [BMIM], Tetrabutylphosphonium [$P(C_4)_4$] cations along with nine amino acids [AAs] (Alanate [ALA], Histidinate [HIS], Glycinate [GLY], Prolinate [PRO], Arginate [ARG], Lysinate [LYS], Leucinate [LEU], Valinate [VAL] and Methionate [MET]) anions are selected. Henry constant (K_H), activity coefficient (Υ), and sigma (σ) profiles and potentials are predicted through the Conductor-like Screening Model for Real Solvents (COSMO-RS) model for potential solvents using simulations studies. The value of K_H is inverse to the solubility of gases in AAILs. COSMO-RS results suggest that the ILs with cation [$P(C_4)_4$] exhibit improved CO_2 capture capacity (with $K_H = 62\ to\ 72$) as compared to [BMIM] cation ($K_H = 98\ to\ 130$) used in AAILs. The top three [$P(C_4)_4$] based ILs are selected to predict the solubility of a mixture of gases, comprising of CH_4, N_2, O_2, and CO_2. The phosphonium-based ILs possess a significant capacity for CO_2 capture and can be effectively utilized in various industrial sectors. The package of COSMO-RS enables a reliable pre-experimental forecast within 1–10% relative deviation (RD), which also satisfies the practical application in research and development.

Introduction

CO_2 emission to the environment is continuously increasing global warming and causes climate change. Consequently, countries around the world channelize the CO_2 emission applying Carbon Capture, Utilization, and Storage (CCUS) program to reduce its footprints and emission [1]. Conventional techniques employed in industries resulted in the difficulty of discovering a proficient and energy-efficient process for CO_2 capture [2].

AAILs having various cations and anions are used to tackle CO_2 capture, owing to their versatile characteristics, like non-volatility, controlled polarity, and exceptional stability [3]. A substantial number of research towards CO_2 capture and separation through ILs systems has been reported. It is estimated that using single cation to anion, the number of ILs can be anticipated by order of 10^{18} [4]. Enhanced performance and compatibility are observed in the CO_2 solubility data of ionic liquids based on pyrrolidinium, pyridinium, quaternary ammonium, tetra alkyl phosphonium, and imidazolium . N-(3-aminopropyl) aminoethyl tributylphosphonium amino acid salts ([apaeP444] [AA]) using different anions has been reported, where [LYS] discovered with highest capacity of CO_2 absorption with 1.54 mmol/g of each sorbent and [GLY] demonstrates superior long-term durability [5]. AAILs as designer solvents with vast properties are altered by introducing cations, anions which changes the chemical structures and use for various applications. Computational and numerical simulations and theoretical study proves concise screening and prediction of these molten salts in gas capture and its separation. It is better to perform numerical and computational simulations instead of going through the experimental work which save too much time and consumables. Similarly, the exploration of combination of different cations and anions in real world experiment are not feasible in novel AAILs production [6].

Therefore, it is crucial to consider theoretical or computational approaches as valuable counterparts in investigating a broader range of novel solutes and solvents. Molecular simulation has become an essential tool in modern materials science, enabling the use of theoretical models to study the properties and behavior of materials at the atomic scale [7]. COSMO-RS offers a novel, straightforward, and computationally efficient route from quantum chemistry (QC) to fluid phase thermodynamics, allowing the researchers to take advantage of the broad applicability and predictive power of contemporary quality control methods [8]. This computational tool has the capability to forecast thermodynamic characteristics, including solubilities , chemical potentials, and activity coefficients of fluids for a multitude of systems mostly for the AAILs [9].

In this work, various types of AAILs, including, [BMIM], $[P(C_4)_4]$ cations and nine different types of amino acid anions ([ALA], [HIS], [VAL], [GLY], [PRO], [ARG], [LYS], [LEU], and [MET]) are selected for this study. The COSMO-RS package is used in this work to predict the K_H, Υ, and σ profiles and potentials for the prediction of CO_2 solubility in 18 different types of AAILs. The results obtained for [BMIM] ILs are compared with the reported work. The top three best cation-anion combinations with highest CO_2 solubility are used to predict solubility for a mixture of gases, comprising of CH_4, N_2, O_2, and CO_2.

Methodology

Equimolar combinations of cations and anions are used to study AAILs. COSMO-RS files were produced for cations and anions employing Perdew–Burke–Ernzerhof (PBE) with triple zeta valence potential (TZVP) basis set via the TURBOMOLE (Tmole-X) add-on package [10]. The COSMO-Therm program is employed for all COSMO-RS calculations, providing a highly active and efficient performance.

Computational Details

Tmole-X was used to optimize the geometries and generate COSMO-RS files for the various ions. Simulations were performed on each ion using maximum iterations and high-quality settings. The structures and sigma surfaces of all ions are presented in Fig. 1.

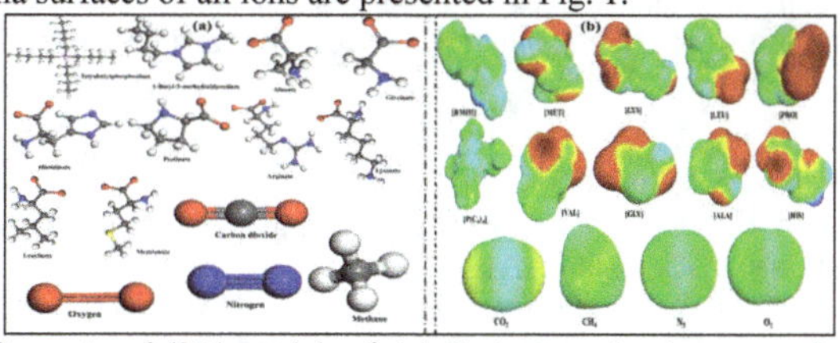

Figure 1: Three-dimensional (3D) models of, (a) Cations, and anions, gas phases and (b) Sigma iso-surfaces highlighting the electronegative forces.

The dark red shades on the iso-surfaces of the molecules imply higher positive screening charge density (SCD), dark blue shades signify higher negative SCD, and green color denotes neutrality of the molecules. To forecast CO_2, CH_4, N_2 and O_2 solubility, K_H, ϒ, and σ-profiles and potentials were considered from the property module of COSMO-RS. The estimation of interaction energies between two surfaces involves the energy associated with the Electrostatic-misfit energy (E_{misfit}), Hydrogen Bonding Energy (E_{HB}), and the van der waals Energy (E_{vdw}). The computational details associated with these three distinct forms of interaction energies are elucidated in Eq. 1, Eq. 2, and Eq. 3 [9]. Ultimately, the summation of these three energies results in the total interaction energy, denoted as ($E_{T_{INT}}$) Eq. 4.

$$E_{HB} = a_{eff} C_{HB} \min(0; \sigma_{donor} + \sigma_{HB}) * \max(0; \sigma_{acceptor} - \sigma_{HB}) \quad (1)$$

$$E_{misfit} = a_{eff} \frac{\alpha}{2} (\sigma + \sigma')^2 \quad (2)$$

$$E_{vdw} = a_{eff} (\tau_{vdw} + \tau'_{vdw}) \quad (3)$$

$$E_{T_{INT}} = E_{misfit} + E_{HB} + E_{vdw} \quad (4)$$

Where a_{eff} is effective contact area, σ_{HB} is threshold for hydrogen bond, C_{HB} strength of hydrogen bond, and τ_{vdw} specific van der Waals interaction. The chemical potential(μ) hereby can be applied to calculate the thermodynamic properties of system. For instance, the activity coefficient of "*i*" component at infinite dilution ($\Upsilon^{\infty}{}_{i}$) is calculated using the Eq. 5.

$$ln\Upsilon^{\infty}{}_{i} = \frac{\mu^{\infty}{}_{i} - \mu^{0}{}_{i}}{RT} \quad (5)$$

Where $\mu^{\infty}{}_{i}$, $\mu^{0}{}_{i}$ are chemical potentials of "*i*" component at infinite dilution. Henry's Law constant (K_H) utilized to predict the solubility of gas "*i*" in a solvent estimated using the Eq. 6 [11].

$$K_H = \frac{P}{x_i} \quad (6)$$

Where P is pressure and x_i is the solubility of the solute (gas) in the solvent expressed in mole fraction. The value of K_H demonstrates a direct relationship with the activity coefficient (ϒ). Where ϒ is inverse in relation to gas solubility. The percentage uncertainties in the estimated desired properties (X) are ascertained through the computation of the relative deviation (RD), as defined by Eq. (8) [12].

$$\% \, RD = \frac{1}{N} \sum \frac{(X_{obt} - X_{exp})}{X_{exp}} * 100 \quad (7)$$

Where, N is data number used, X_{obt} is the obtained desired property, X_{exp} is the corresponding previously reported experimental value.

Results and Discussions

Combination of cations and anions were first confirmed using COSMO-RS and following the forecast of solubility of gas mixture containing CH_4, N_2, O_2, and CO_2 is confirmed. The model validation and the determined values of K_H, ϒ, and σ-profiles and potentials are provided in the substituent sections.

Validation of COSMO-RS Method

Following the research methodology, this work finds comparable results regarding its predictive performance on CO_2 solubilities in AAILs. Hence, it becomes imperative to indorse the methodology by leveraging accessible experimental data [13], COSMO-RS calculations are carried out using the equimolar combinations of a cation and an anion in AAILs. The computed CO_2 solubility in [BMIM][AA], at a temperature of 298 K and a pressure of 1 bar is calculated using Eq. 6 and is compared with the corresponding experimental data presented [14].

Table 1. The anticipated density and CO_2 capture capacity of all [BMIM] based ILs exhibit a commendable alignment with the experimental data, exhibiting an error range of 1 - 10%. So, the COSMO-RS method can be used to find promising AAILs for capturing CO_2. This methodology is applied on different other ILs to predict their capacity to absorb a gas and separation of various gas phases through their amorphous phases. The results show minimum of 0.66% of error in [BMIM][GLY] and highest error 10.2% resulted in [BMIM][LEU]. [BMM][ARG] are observed with highest of CO_2 absorption and least CO_2 absorption is observed in [BMIM][PRO]. Overall, the trend is consistent with published data.

Table 1. Validation of the densities and CO_2 absorption capacity of AAILs.

S. No.	IL ([BMIM][AA])	Density [g/cm³]			M. Mass [g/mol]	K_H [per kpa]	CO_2 absorption capacity [mol CO_2/mol IL]		
		Exp.	Sim:	% Error			Exp.	Sim:	% Error
1	ARG	1	1.07	7.00	312	98.23	0.62	0.59	4.12
2	LYS	1.07	1.04	2.80	284	107.57	0.48	0.51	5.97
3	HIS	1.1	1.13	2.73	293	105.39	0.45	0.49	9.23
4	MET	1	1.10	9.90	287	115.89	0.42	0.45	7.31
5	ALA	1.05	1.07	1.90	227	117.62	0.39	0.36	7.81
6	VAL	1	1.03	3.00	255	128.02	0.39	0.38	1.40
7	GLY	1.05	1.09	3.81	213	102.82	0.38	0.38	0.66
8	LEU	0.96	1.02	6.25	269	125.34	0.38	0.42	10.21
9	PRO	1.13	1.11	1.77	253	130.70	0.32	0.35	9.13

Henry's constant (KH)

The solubility of gases, determined using K_H, was estimated via COSMO-RS and compared to experimental data. The K_H values for CO_2 in imidazolium and phosphonium-based ionic liquids are shown in Fig. 2(a) . The results demonstrate that phosphonium cations deprived of aromatic structure and anions including [MET] and [LYS] are efficacious combinations for the separation of CO_2. A lower K_H indicates improved solubility, aiding in the separation of these mixtures. The negatively charged group of the anion forms a strong hydrogen bond with the positively charged "*C*" atom in CO_2 due to the additional charge and high electronegativity. These combinations have superior solubility and are effective for separation because of their low K_H values. A study on the of CO_2 solubility predicts that [PH] cations, as opposed to [IM] and [PYRI], that form superior IL pairs with [BF_4], [PF_6], and [Tf_2N] anions are expected to operate as effective ILs [9].

Fig. 2 (b) presents the K_H for CO_2, CH_4, N_2, and O_2 in [P(C4)4]-[MET], [LYS], and [LEU]. Lowest K_H values are observed for O_2 supporting high O_2 solubility in ILs and highest K_H values, highlighting low solubility for CH_4. Although, O_2 is often not very soluble in ionic liquids however some ionic liquids have functional groups that interact with O_2 more effectively, resulting in increased O_2 solubility [15]. Based on the K_H results for N_2, [$P(C_4)_4$][LYS] are superior combination for N_2 capture. According to a study on N_2 solubility, the N_2 solubility is significantly higher for [PH] cation coupled with [eFAP] (0.28 mg/g) than for [IM] cation (0.2 mg/g). This is attributed to the stronger interactions that occur between the cation and N_2 [16].

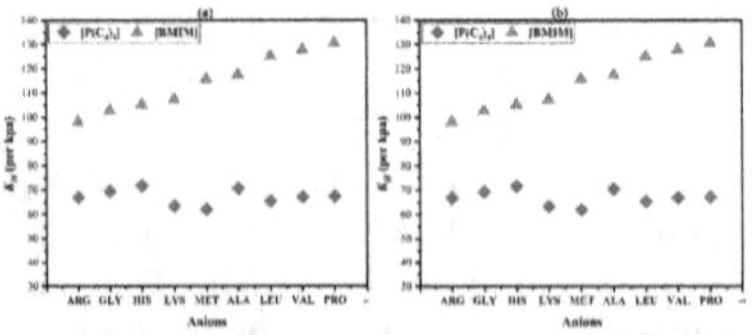

Figure 2: Henry's constants (K_H) at 298K and 1bar for (a) CO_2 in 18 ILs (b) CO_2, CH_4, N_2, and O_2 in [P(C₄)₄]- [MET], [Lys], and [LEU].

Activity coefficient (ϒ) at infinite dilution for gases

Activity coefficients are essential in selecting AAILs for gas absorption, as they indicate the solvent's solubilizing ability. Lower ϒ values signify stronger solute-solvent interactions, improving solubility. To identify suitable ILs for CO_2 capture, the COSMO-RS model was used to predict activity coefficients for CO_2 in all 18 ILs (Fig. 3(a)). Results show that all anions paired with [P(C4)4] had lower ϒ values than paired with [BMIM]. [P(C_4)$_4$][MET] and [P(C_4)$_4$][LYS] showed the lowest values, attributed to strong interactions between the [P(C_4)$_4$] cation and CO_2, while [MET] and [LYS] anions enhance CO_2 absorption. Higher γ values for [ALA] and [HIS] with [P(C_4)$_4$] are due to non-coordinating properties and steric hindrance. For CO_2 capture, [BMIM] cation shows weaker interaction due to its conjugated structure, though anions like [ARG], [LYS], and [HIS] perform well when paired with [BMIM]. Experimental findings suggest [ARG] is more effective with [BMIM] for CO_2 solubility [14].

In Fig. 3(b), CO_2 shows the lowest γ values in [P(C_4)$_4$][MET], indicating high solubility, while [P(C_4)$_4$][LYS] is moderate, and [P(C_4)$_4$][LEU] is lower. CH_4 displays high γ values, making [MET], [LYS], and [LEU] unsuitable for CH_4 due to their polarity [9]. For N_2, [MET] and [LEU] are ineffective, while [P(C_4)$_4$] has stronger interactions with O_2. This suggests that these AAILs are effective for CO_2/CH_4 separation.

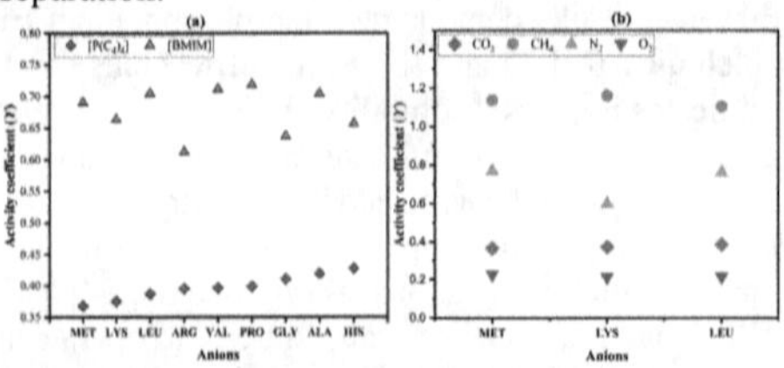

Figure 3: Activity coefficient at infinite dilution at 25 °C and 1 bar for (a) Activity coefficient of CO_2 in AAILs, and (b) Activity coefficient CO_2, CH_4, N_2 and O_2

Sigma(σ) Profiles and potentials

The influence of structural distinctions in the ILs can be identified by analysing their σ-profiles. profiles provide the relative amount of surface polarity for the solute and solvent under consideration. The σ-potentials of solvents can be utilised to assess the solvent's attraction to a solute. The shape, size, and positive and negative surface areas of ILs that interact with CO_2 determine the final solubility trend. The σ profiles for the chosen cations, anions, and gases are displayed in Fig. 4. The negative region represents the hydrogen bond donor (HBD <1), whereas the positive portion represents the hydrogen bond acceptor (HBA >1). The region between -1 and 1 is non-polar (N.P). In Fig. 4(a), the peaks of [P4444] and [BMIM] are located near H-Bond donor (HBD) regions and mainly in non-polar region. The peak (blue) for CO_2 on the other hand is near both regions HBD and HBA. The main peaks for anions lie in the non-polar (N.P) region as well as in H-Bond Acceptor (H.B.A) region that shows anions are electronegative and the highest peak of [ARG] shows it is more electronegative. This indicates that CO_2 can interact effectively with [ARG] anion. The results also indicate that [P4444] with [ARG] are very compatible with CO_2, shown in Table 1. As the peak for [P4444] is higher and near HBD region, this is why the CO_2 sorption capacity of ILs having [P4444] cation is higher with all anions [9]. The both HBD and HBA-natured ILs will improve CO_2 solubility, making them superior separation agents [17,18]. The peak CH_4 is negative side showing that the hydrogen bond donor region is suitable for CH_4 while the peak for O_2 and N_2 is in neutral region, so the more neutral cations are favorable for N_2 and O_2.

The σ-potential of the cations, anions, and gases are shown in Fig. 4 (b). The findings indicate that the positive values of σ-potential trigger CO_2 as HBA. This outcome is consistent with the authors' report that HBAs would be far better for CO_2 [19]. Most of the CO_2 surface molecules will have the highest favorable interaction with [P4444][MET], which possesses the highest negative sigma potential among the three options, followed by [P4444][LYS]. Therefore, the activity coefficients exhibit a decrease in [P4444][MET] [11].

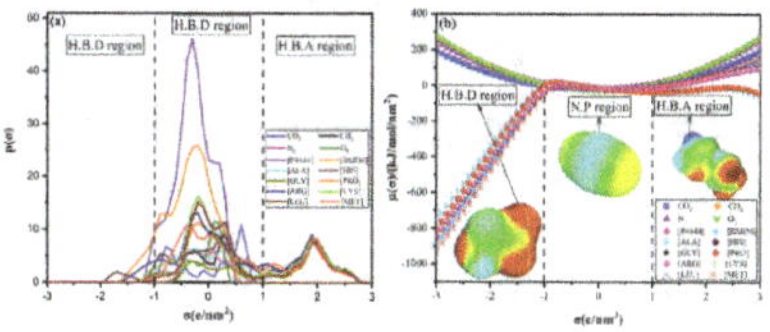

Figure 4. For selected cations, anions, and gases (a)Sigma(σ) Profiles (b)Sigma(σ) potentials

Conclusion

This study effectively developed a modern computational screening approach for the molecular design of 18 AAILs with 2 cations and 9 anions, based on their interactions with CO_2, CH_4, N_2, and O_2 gases. Henry's law constant, activity coefficients and sigma (σ) profiles and potentials are determined for the gas to liquid phase (AAILs) interaction using COSMO-RS. Major objective of this work was to analyse the performance of AAILs in separation of CO_2 from mixture of gases. The screening results show that cations have significant impact on the solubility performance of IL because of their evenly distributed charge density. AAILs with $[P(C_4)_4]$ cation and anions such as [MET], [LYS], and [LEU] are being observed as best for CO_2 solubility with highest capacity. The COSMO-RS predictions closely match experimental results, with a relative deviation (RD) ranging from 1% to 10%. Hence, it is concluded that COSMO-RS is robust predictive tool for CO_2 capture by screening and designing new potential AAILs. The methodology designed in this work can work as an impactful tool for calculating these parameters before setting up an experiment at laboratory level.

Acknowledgement:

We want to acknowledge the financial support provided by the Ministry of Higher Education, Malaysia, under Fundamental Research Grant Scheme (FRGS) (Ref. No.: FRGS/1/2022/TK09/UTP/03/3, Cost Centre: 015MA0-153) and YUTP-Fundamental Research Grant (Cost Centre: 015LC0-498). Moreover, we would also like to acknowledge the technical support from Department of Chemical Engineering, Universiti Teknologi PETRONAS for providing technical support and access to the simulation laboratories to perform this work.

References

[1] IPCC, Global Warming of 1.5°C, Cambridge University Press, 2022. https://doi.org/10.1017/9781009157940.

[2] Z. Bashir, S.S.M. Lock, N. e. Hira, S.U. Ilyas, L.G. Lim, I.S.M. Lock, C.L. Yiin, M.A. Darban, A review on recent advances of cellulose acetate membranes for gas separation, RSC Adv 14 (2024) 19560–19580. https://doi.org/10.1039/d4ra01315h.

[3] B.S. Lee, S.T. Lin, Screening of ionic liquids for CO_2 capture using the COSMO-SAC model, Chem Eng Sci 121 (2015) 157–168. https://doi.org/10.1016/j.ces.2014.08.017.

[4] J.D. Holbrey, K.R. Seddon, Ionic Liquids, Springer-Verlag, 1999. https://doi.org/10.1007/s100980050036.

[5] J. Ren, L. Wu, B.G. Li, Preparation and CO2 sorption/desorption of N -(3-aminopropyl)aminoethyl tributylphosphonium amino acid salt ionic liquids supported into porous silica particles, Ind Eng Chem Res 51 (2012) 7901–7909. https://doi.org/10.1021/ie2028415.

[6] A. Klamt, F. Eckert, W. Arlt, COSMO-RS: An alternative to simulation for calculating thermodynamic properties of liquid mixtures, Annu Rev Chem Biomol Eng 1 (2010) 101–122. https://doi.org/10.1146/annurev-chembioeng-073009-100903.

[7] M.A. Darban, S.S.M. Lock, S.U. Ilyas, D.-Y. Kang, M.H.D. Othman, C.L. Yiin, S. Waqas, Z. Bashir, Molecular simulation of [P8883][Tf_2N] ionic liquid decorated silica in 6FDA-ODA based mixed matrix membrane for enhanced CO2/CH4 separation, RSC Adv 14 (2024) 22894–22915. https://doi.org/10.1039/D4RA02851A.

[8] S. Ahmed Shahani, H. Warsi Khan, A. Mohd Shaiff, M. Goto, M. Moniruzzaman, Screening of ionic liquids for the solubility enhancement of quinine using COSMO-RS, J Mol Liq 409 (2024) 125388. https://doi.org/10.1016/j.molliq.2024.125388.

[9] N. Islam, H. Warsi Khan, A.A. Gari, M. Yusuf, K. Irshad, Screening of ionic liquids as sustainable greener solvents for the capture of greenhouse gases using COSMO-RS approach: Computational study, Fuel 330 (2022). https://doi.org/10.1016/j.fuel.2022.125540.

[10] M.S. Aminuddin, M.A. Bustam, K. Johari, Evaluation of amine-based ionic liquids as potential solvents for hydrogen sulfide absorption using COSMO-RS: Computational and experimental validation, Chemical Engineering Research and Design 203 (2024) 721–730. https://doi.org/10.1016/j.cherd.2024.01.054.

[11] K.Z. Sumon, A. Henni, Ionic liquids for CO_2 capture using COSMO-RS: Effect of structure, properties and molecular interactions on solubility and selectivity, Fluid Phase Equilib 310 (2011) 39–55. https://doi.org/10.1016/j.fluid.2011.06.038.

[12] N. Ab Manan, C. Hardacre, J. Jacquemin, D.W. Rooney, T.G.A. Youngs, Evaluation of gas solubility prediction in ionic liquids using COSMOthermX, J Chem Eng Data 54 (2009) 2005–2022. https://doi.org/10.1021/je800857x.

[13] M. Diedenhofen, F. Eckert, A. Klamt, Prediction of infinite dilution activity coefficients of organic compounds in ionic liquids using COSMO-RS, J Chem Eng Data 48 (2003) 475–479. https://doi.org/10.1021/je025626e.

[14] Y.S. Sistla, A. Khanna, CO_2 absorption studies in amino acid-anion based ionic liquids, Chemical Engineering Journal 273 (2015) 268–276. https://doi.org/10.1016/j.cej.2014.09.043.

[15] M.M. Hossain, A. Rawal, L. Aldous, Aprotic vs protic ionic liquids for lignocellulosic biomass pretreatment: Anion effects, enzymatic hydrolysis, solid-state NMR, distillation, and recycle, ACS Sustain Chem Eng 7 (2019) 11928–11936. https://doi.org/10.1021/acssuschemeng.8b05987.

[16] F. Zhou, L.M. Azofra, M. Ali, M. Kar, A.N. Simonov, C. McDonnell-Worth, C. Sun, X. Zhang, D.R. Macfarlane, Electro-synthesis of ammonia from nitrogen at ambient temperature and pressure in ionic liquids, Energy Environ Sci 10 (2017) 2516–2520. https://doi.org/10.1039/c7ee02716h.

[17] K.J. Fraser, D.R. MacFarlane, Phosphonium-based ionic liquids: An overview, Aust J Chem 62 (2009) 309–321. https://doi.org/10.1071/CH08558.

[18] X. He, J. Zhu, H. Wang, M. Zhou, S. Zhang, Surface functionalization of activated carbon with phosphonium ionic liquid for CO_2 adsorption, Coatings 9 (2019). https://doi.org/10.3390/COATINGS9090590.

[19] R. Sulaiman, M.K. Hadj-Kali, S.W. Hasan, S. Mulyono, I.M. AlNashef, Investigating the solubility of chlorophenols in hydrophobic ionic liquids, Journal of Chemical Thermodynamics 135 (2019) 97–106. https://doi.org/10.1016/j.jct.2019.03.026.

Materials Research Proceedings 53 (2025) 96-104 https://doi.org/10.21741/9781644903575-9

Correlation analysis of different solvent component in absorption-based acid gas removal unit

RAFI Jusar Wishnuwardana[1,a *], MADIAH Binti Omar[1,b], HASLINDA Binti Zabiri[1,c], MOCHAMMAD Faqih[1,d], MUHAMMAD Hasif Bin Mohd Zaid[1,e], and NURUL Izni Binti Ismail[1,f]

[1]Chemical Engineering, Universiti Teknologi PETRONAS, Perak, Malaysia

[a]rafi_24002236@utp.edu.my, [b]madiah.omar@utp.edu.my, [c]haslindazabiri@utp.edu.my, [d]mochammad.faqih@utp.edu.my, [e]muhammad_20001741@utp.edu.my, [f]nurul_19001247@utp.edu.my

Keywords: Acid Gas Removal Unit, Solvent Component, Correlation Analysis

Abstract. Natural gas is widely used in various applications including power generation and chemical industry. However, the natural gas from upstream contains several hazardous acid gases such as Hydrogen Sulfide (H_2S) and Carbon Dioxide (CO_2) which pose risks to both human health and the environment. Therefore, acid gas removal process is pivotal for purifying the natural gas before being delivered to the end-users. This paper aims to investigate the colleration between different solvent components used in absorption-based acid gas removal units. The analysis employs three correlation approaches: Pearson, Spearman, and Kendall. The data was simulated using a Hysys flowsheet by varying the type of solvent blends and compositions. The results indicate that the solvent blends MDEA+PZ and MDEA+DEA exhibit the highest correlation with CO2 removal, achieving a correlation value of -0.91. For H2S removal, the MDEA+SFL blend shows the highest correlation, with a value of -0.74. These findings suggest that specific solvent blending can significantly enhance the reduction of H2S and CO2 concentrations in sweet gas, underlining the importance of solvent selection in the acid gas removal process.

Introduction

Natural gas takes a quarter part of global electric production and emits less carbon compared than most other fossil fuels [1]. However natural gas contains large amount of dangerous acid gases such as H2S and non-flammable gases (CO2) [2]. Taking account to the human health, environmental impact and pipeline regulation, H2S and CO2 concentration should be kept at 4-20 ppm and less than 3% mol respectively [3]. Hence the removal of H2S and CO2 from natural gas is a critical concern in gas processing [4]. Absorption or tripping of Amine solvent has been dominant technique used in Acid Gas Removal Unit (AGRU) [5].

The main challenge of AGRU is to remove acid gases from natural gases to as low as possible [3]. In order to increase its removal efficiency methods such as amine blending have been used in previous researches. Esmaili et al. presented a 5 different amine solution are applied in AGRU resulting in the blend of Sulfolane and MDEA as the most appropriate blend with several advantages such as higher selectivity of H2S over CO2, and has the highest H2S loading [6]. Alkhatib et al discovery of hybrids solvent for AGRU showing that the blend of Physical Solvent including NMP or Sulfolane and amine solvent show an acceptable compromise between reduced absorption capability and reduced energy regeneration compared to typical MEA solvent [7]. Farooqi et al reveal that the blending of Triethanolamine (TEA) and diisopropanolamine (DIPA) decrease the Composition of CO2 and H2S in the sweet gas [8].

This paper aims to investigate the influence of the utilization of different solvent components with the H2S and CO2 composition in the sweet gas. There will be six amine blends with different certain concentration ranges selected in this study. This study will be using ASPEN HYSYS simulation to retrieve the data. Furthermore, the simulation data will be analyzed using three different Correlation Technique: Pearson, Spearman, and Kendall's Tau.

Methodology

The methods used to determine the correlation of AGRU parameter are considered in this section, as represented by Fig.1. Pearson correlation, Kendall rank correlation and Spearman correlation were used to determine each parameter that affected AGRU performance in absorbing H2S and CO2.

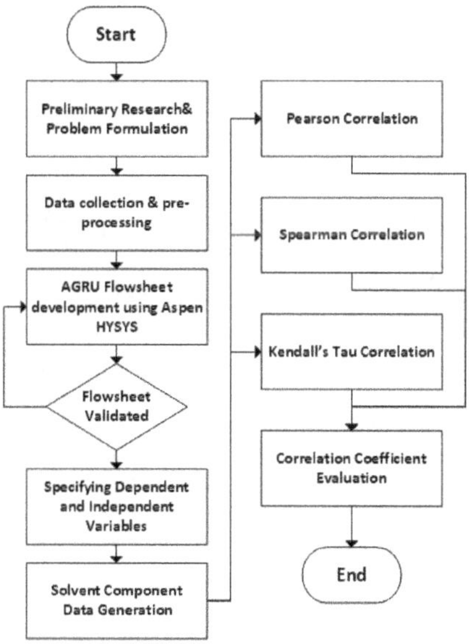

Fig.1. Research flowchart.

Acid Gas Removal Unit Flow Diagram

Absorption-based Acid gas removal using Amine as a solvent was used in this paper considering its dominance as technique for converting sour gas into sweet gas[5] . The absorption of CO2 and H2S reaction that occur in the Acid Gas Removal Unit is as follows[9]:

$$2(RNH2) + H2S \rightleftharpoons (RNH3)2S \quad (1)$$

$$2(RNH2) + CO2 \rightleftharpoons (RNH2)2H2CO3 \quad (2)$$

Where R is mono-, di- or tri-ethanol, while N, H, S, C, and O are nitrogen, hydrogen, Sulphur, carbon and oxygen, respectively.

Fig.1 illustrates the general process of Acid Gas Removal. The sour gas containing H2S, and CO2 enters the system through inlet separator (scrubber) to eliminate liquid and/or solid. The gas stream is then introduced at the bottom of the absorber while undergoing countercurrent contact with the lean solvent, where the lean solvent absorbs acid gas from the sour gas and converts it into sweet gas. The resulting sweet gas is discharged through the tip of the absorber and flows through an outlet separator before finally flowing to dehydration unit before being sold to the market.

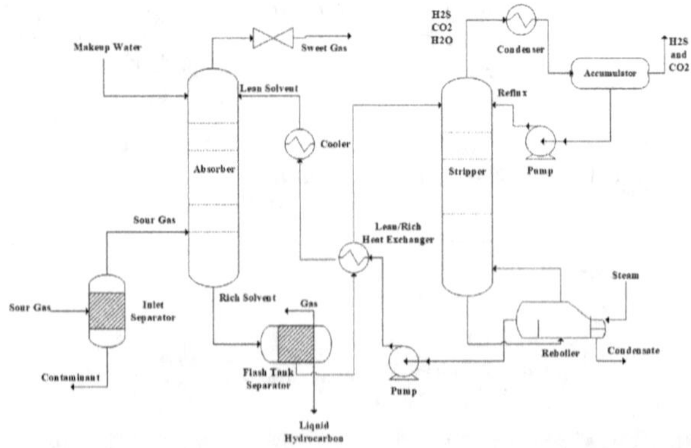

Fig.1. Typical process flow diagram of absorption-based acid gas removal unit.

Furthermore, the rich solvent flows from the base of the absorber to a flash tank in order to recover remaining hydrocarbons in the solvent. The purified rich solvent is then preheated before penetrating the top of the stripper operating at low pressure. The amine-amine heat exchanger acts as a heat conservation device to reduce total heat demand for the process. A portion of the absorbed acid gases will be discharged from the heated rich solvent at the top of the stripper. The rest of the solvent will stream downward through the stripper while undergoing countercurrent contact with the vapor generated by the reboiler for the purpose of stripping acid gases from rich solvent. The acid gases and the steam flashed from the top of the stripper and went to condenser, where most of the steam is cooled and condensed. The condensed vapor is returned to the top of the stripper as reflux. The acid gases are separated in the accumulator and directed to the flare or to further processing.

Lean solvent from the bottom of the stripper/regenerator, after precooling in the amine-amine heat exchanger, went to the amine surge tank as the solvent storage for solvent makeup. Lean solvent from the surge tank is then cooled in the cooler because the demerit of the high-temperature lean solvent would reduce the solvent volume because due to vaporization and affect the lowering of the gas carrying capacity. The chilled solvent is pumped by the circulating pump and returned to the absorber.

Flowsheet Input Parameter. Flowsheet was built based on industrial data at previous literature [10]. Initial variables for the several conditions are used such as sour gas flow, pressure, and temperature with the value of 7154.8 kmol/h, 54.03 bar and 52 C respectively. MDEA was used as a solvent with the concentration of 48% referring to common range usage of MDEA between 30-50%. Lean solvent inlet pressure and temperature were 54.32 bar and 55 C respectively. The sour feed gas composition is shown at Table 1.

Table 1. Specification of the sour gas.

Composition	Mole Fraction
H2O	0.00234
i-Pentane	0.0052
n-Pentane	0.0056
i-Butane	0.01243
n-Butane	0.0215

CO2	0.03001
H2S	0.00024
Methane	0.72583
n-Ethane	0.11717
Propane	0.07377

Flowsheet Validation

To validate the flowsheet industrial data from [10] was used as reference. The flowsheet was developed in Aspen HYSYS 12. Fig.2 shows the developed flowsheet for acid gas removal unit.

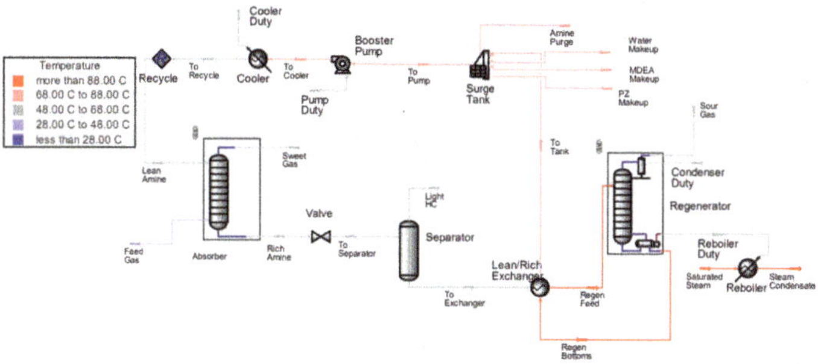

Fig.2. Flowsheet diagram of absorption-based acid gas removal unit.

Table 2 shows the data comparison between industrial and flowsheet simulation with the error value calculated using Equation (3).

$$\%error = \frac{|Industrial - Simulation|}{Industrial} x100 \tag{3}$$

Table 2. Comparison between industrial and simulation data.

Properties	Industrial [10]	Simulation	Error (%)
H2S (kmol/h)	0.00959	0.0101	5.31
CO2 (kmol/h)	170.87	171.09	0.13
Total Flow (kmol/h)	7031.80	6974.00	0.82

The comparison of the flowrate of H2S, CO2 and Total flow shows error value show a good result below. Therefore, the flowsheet can be used for data generation.

Data Generation. From the flowsheet, the data will be generated to define the correlation between the variables that occur in the AGRU system. Different solvent components applied to this paper are tabulated in Table 3, and the composition range is shown in Table 4. The data will be generated based on tables value and the result will be applied on three correlation technique: Pearson, Spearman, and Kendall's Tau.

Table 3. Variation of amine-based solvent blends.

Amine Blend	Mass Fraction (wt.%)	Reference
MDEA + PZ	40.5%+5%	[6]
MDEA + SFL	40.5%+5%	[6]
MDEA + DEA	27.5%+17.5%	[7]
MDEA + DGA	40%+3.25%	[11], [12]
DEA +PZ	20%+10%	[13]
TEA + DIPA	42.5%+7.5%	[8]

Table 4. Composition range of amine-based solvent blends.

Parameters		Lower	Base	Upper	Variables
MDEA (wt%)		35.5	40.5	45.5	Independent
PZ (wt%)		0	5	10	
MDEA (wt%)		35.5	40.5	45.5	
SFL (wt%)		0	5	10	
MDEA (wt%)		20	27.5	35	
DEA (wt%)		5	17.5	30	
MDEA (wt%)		35	40	45	
DGA (wt%)		1.5	3.25	5	
DEA (wt%)		15	20	25	
PZ (wt%)		5	10	15	
TEA (wt%)		35	42.5	50	
DIPA (wt%)		0	7.5	15	
Absorber Temperature (°C)		38.5	55.0	71.5	
Absorber Pressure (bar)	Feed	37.9	57.45	77.0	
	Lean	54.0	66.8	79.5	
	Top	37.3	52.85	68.41	
	Bottom	37.64	53.33	69.03	
Concentration of H_2S (ppm) in sweet gas					Dependent
Concentration of CO_2 (mol%) in sweet gas					

Correlation Technique. Correlation is defined as a measure of association between two variables. Considering variable a and b, there will be 2 scenarios: either (1) when variable an increases, variable b will follow (positive correlation); or (2) as the value of variable an increases, variable b will decrease (negative correlation). This paper were using 3 most commonly used correlation which are Pearson, Kendall, and Spearman[14][15]. The distinctive difference between these three techniques is that Pearson correlation measures linear association between two random variables, while Spearman and Kendall's Tau more suitable for measuring nonlinear monotonic association[16].

Pearson Correlation. The Pearson correlation is a method of measuring linear association between two normally distributed random variables[17]. The Pearson correlation coefficient is defined as follows [18] :

$$r_p = \frac{\sum_{i=1}^{n}(X_i-\bar{X})(Y_i-\bar{Y})}{\sqrt{\sum_{i=1}^{n}(X_i-\bar{X})^2 \sum_{i=1}^{n}(Y_i-\bar{Y})^2}} \tag{4}$$

Materials Research Proceedings 53 (2025) 96-104 https://doi.org/10.21741/9781644903575-9

With $\bar{X} = \sum_{i=1}^{n} X_i/n$ the sample average of X variable and $\bar{Y} = \sum_{i=1}^{n} Y_i/n$ is the sample average of Y variable. The value of the correlation coefficient is ranges from -1 to +1, with -1 showing a strong negative relationship and +1 showing strong positive relationship, and the closer value is to 0, the more it defines the deficiency of relationship between 2 variables.

Spearman Correlation. The Spearman correlation defines the monotonic relationship between two variables [17]. Its correlation coefficient is defined by equation below[19] :

$$r_s = \frac{\sum_{i=1}^{n}(rank(x_i)-\overline{rank(x)})(rank(y_i)-\overline{rank(y)})}{\sqrt{\sum_{i=1}^{n}(rank(x_i)-\overline{rank(x)})^2 \sum_{i=1}^{n}(rank(y_i)-\overline{rank(y)})^2}} \tag{5}$$

Where $rank(x_i)$ and $rank(y_i)$ are ranks of observations in the sample of the variable. The Spearman coefficient values varies from -1 to 1, describing the strength of the monotonic relationship. The closer value to the 0, the weaker the relationship between two variables [20], [21].

Kendall's Tau Correlation. Kendall's tau correlation is purpose to perceive the association between two ordinal or interval variables. The Kendall correlation coefficient equation is shown as follows [22] :

$$r_K = \frac{\sum_{i-1}^{n}\sum_{j=1}^{n} sgn(x_i-x_j)sgn(y_i-y_j)}{n(n-1)} \tag{6}$$

Where:

$$sgn(x_i - x_j) = \begin{cases} 1 \; if(x_i - x_j) > 0 \\ 0 \; if(x_i - x_j) = 0 \\ -1 \; if(x_i - x_j) < 0 \end{cases} ; sgn(y_i - y_j) = \begin{cases} 1 \; if(y_i - y_j) > 0 \\ 0 \; if(y_i - y_j) = 0 \\ -1 \; if(y_i - y_j) < 0 \end{cases} \tag{7}$$

This coefficient explains the difference between the number of concordant and discordant pairs. Similar to previous correlation coefficients, Kendall's tau correlation coefficient value ranges from -1 to +1, and if the value reaches 0, the relationship of concordant and discordant pairs will be weaker [20], [21].

Result and Discussion

Three different correlation techniques such as Pearson, Spearman and Kendall's Tau Correlation are utilized to correlate different solvent components, operating temperature, and pressure with H2S and CO2 composition on the sweet gas. Based on the data generated the amines that has been used in this correlation are MDEA + PZ, MDEA + SFL, MDEA + DEA, MDEA+DGA, DEA+PZ, TEA+DIPA.

(a) Pearson Correlation

(b) Spearman Correlation

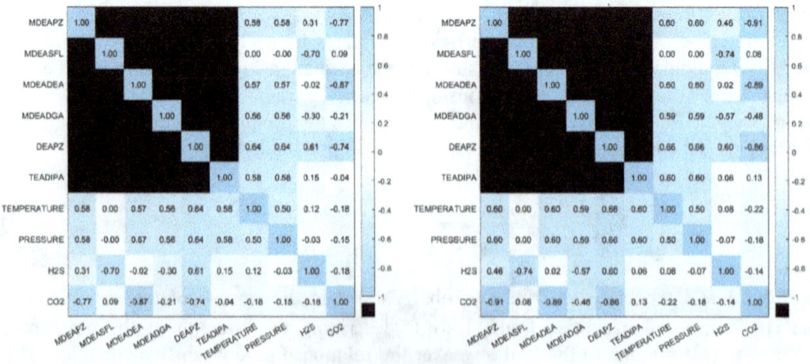

(c) Kendall Correlation

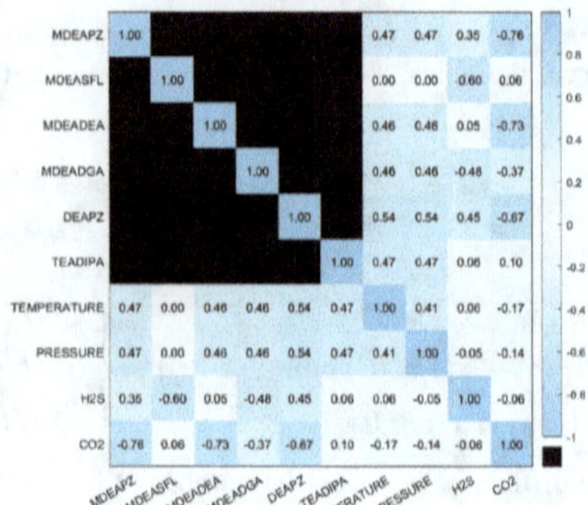

Fig.3 Correlation test results from Pearson (a), Spearman (b), and Kendall (c).

The correlation chart can be seen in Fig.3. It can be observed that different correlation techniques the correlation coefficient varies. Ignoring correlation between same variable the highest correlation value is show by Spearman correlation (-0.91) for MDEA PZ and CO2. The least correlation value is observed for MDEA SFL and Temperature Pressure in the value of 0 Three correlations confirm that MDEA PZ, MDEA DEA, MDEA DGA and DEA PZ negatively correlates with CO2 on the sweet gas. While MDEA SFL, and MDEA DGA shows negative correlation with H2S on the sweet gas.

Because the principal challenge of the AGRU is to remove acid gases and sulphur component from the natural gas as low as possible [3], hence negative relationship between different solvent component and H2S and CO2 composition in sweet gas is expected.

First Pearson correlation shows the strongest negative relationship between solvent component and sweet gas composition displayed at MDEA DEA and CO2 and MDEA SFL and H2S at the value of -0.74 and -0.70. While the weakest relationship displayed at MDEA DEA and H2S and MDEA SFL and CO2 at the value at -0.02 and 0.09 respectively.

Spearman correlation displayed different maximum value of coefficient. The highest value is at MDEA PZ and CO2 at -0.91 and MDEA SFL and H2S in -0.74 same agreement as Pearson correlation. The weakest value is shown at MDEA DEA and H2S at value of 0.02 and MDEA SFL and CO2 at the value of 0.08.

Lastly, Kendall Correlation shows the same conclusion with Spearman value for the highest and the lowest coefficient value. Kendall and Spearman basically have the same approach. Based on the correlation test amine selectivity can be identified by its correlation value. In this case the MDEA PZ, MDEA DEA, MDEA DGA and DEA PZ shows high selectivity to CO2 at the same time MDEA + SFL shows high selectivity to H2S. However, MDEA DGA shows the negative relationship in both H2S and CO2 with the weak relationship concluding that MDEA DGA cannot significantly decrease the amount of H2S and CO2.

Conclusion

This paper addressed the correlation analysis of different solvents against H2S and CO2 in acid gas removal unit. Based on the results, MDEA + PZ and MDEA DEA are highly correlated with the value of CO2 in the sweet gas while MDEA +SFL is the most affecting solvent component in H2S composition value. Kendall Tau Correlation and Spearman correlation show the same conclusion while Pearson Correlation outcome a different solvent component.

Acknowledgement

The authors are thankful to Yayasan Universiti Teknologi PETRONAS for the support in carrying this research through grant YUTP-FRG (015LC0-362).

References

[1] Carl Greenfield, "Natural Gas," International Energy Agency, 2023. https://www.iea.org/energy-system/fossil-fuels/natural-gas#

[2] L. Addington and C. Ness, "An evaluation of general 'rules of thumb' in amine sweetening unit design and operation," GPA Annu. Conv. Proc., vol. 1, pp. 119–135, 2010.

[3] S. Mokhatab and W. A. Poe, "Chapter 7 - Natural Gas Sweetening," S. Mokhatab and W. A. B. T.-H. of N. G. T. and P. (Second E. Poe, Eds. Boston: Gulf Professional Publishing, 2012, pp. 253–290. doi: https://doi.org/10.1016/B978-0-12-386914-2.00007-8

[4] A. A. Abd and S. Z. Naji, "Comparison study of activators performance for MDEA solution of acid gases capturing from natural gas: Simulation-based on a real plant," Environ. Technol. Innov., vol. 17, p. 100562, 2020. https://doi.org/10.1016/j.eti.2019.100562

[5] A. Pudi, M. Rezaei, V. Signorini, M. P. Andersson, M. G. Baschetti, and S. S. Mansouri, "Hydrogen sulfide capture and removal technologies: A comprehensive review of recent developments and emerging trends," Sep. Purif. Technol., vol. 298, no. June, p. 121448, 2022. https://doi.org/10.1016/j.seppur.2022.121448

[6] A. Esmaeili, T. Yoon, T. A. Atsbha, and C. J. Lee, "Rate-based modeling and energy optimization of acid gas removal from natural gas stream using various amine solutions," Process Saf. Environ. Prot., vol. 177, no. January, pp. 643–663, 2023. https://doi.org/10.1016/j.psep.2023.07.030

[7] I. I. I. Alkhatib, O. Khalifa, D. Bahamon, M. R. M. Abu-Zahra, and L. F. Vega, "Sustainability criteria as a game changer in the search for hybrid solvents for CO2 and H2S removal," Sep. Purif. Technol., vol. 277, no. August, 2021. https://doi.org/10.1016/j.seppur.2021.119516

[8] A. S. Farooqi, R. M. Ramli, S. S. M. Lock, N. Hussein, A. S. Farooqi, and S. M. Wajahat, "Simulation of acid gas removal unit using DIPA+TEA amine solvent," IOP Conf. Ser. Mater. Sci. Eng., vol. 1257, no. 1, p. 012033, 2022. https://doi.org/10.1088/1757-899x/1257/1/012033

[9] S. Y. Khan, M. Yusuf, and A. Malani, “Petroleum & Petrochemical Engineering Journal Selection of Amine in Natural Gas Sweetening Process for Acid Gases Removal : A Review of Recent Studies,” Pet. Petrochemical Eng. J., vol. 1, no. 3, pp. 1–7, 2017.

[10] A. Sharifi and E. Omidbakhsh Amiri, “Effect of the Tower Type on the Gas Sweetening Process,” Oil Gas Sci. Technol., vol. 72, no. 4, 2017. https://doi.org/10.2516/ogst/2017018

[11] U. Zahid, “Simulation of an acid gas removal unit using a DGA and MDEA blend instead of a single amine,” Chem. Prod. Process Model., vol. 15, no. 4, pp. 1–17, 2020. https://doi.org/10.1515/cppm-2019-0044

[12] U. Zahid, A. Sakheta, and C. J. Lee, “Techno-economic analysis of acid gas removal from associated and non-associated sour gas using amine blend,” Int. J. Greenh. Gas Control, vol. 98, no. November 2019, p. 103078, 2020. https://doi.org/10.1016/j.ijggc.2020.103078

[13] Y. C. Chang, R. B. Leron, and M. H. Li, “Equilibrium solubility of carbon dioxide in aqueous solutions of (diethylenetriamine + piperazine),” J. Chem. Thermodyn., vol. 64, pp. 106–113, 2013. https://doi.org/10.1016/j.jct.2013.05.005

[14] Albert W. Marshall, “Copulas, Marginals, and Joint Distributions,” Distrib. with Fixed Marginals Relat. Top., vol. 28, 1996, [Online]. Available: https://www.jstor.org/stable/4355894

[15] Nian Shong Chok, “Pearson’s versus Spearman’s and Kendall’s correlation coefficients for continuous data,” University of Pittsburgh, 2010. [Online]. Available: http://d-scholarship.pitt.edu/8056/1/Chokns_etd2010.pdf

[16] E. van den Heuvel and Z. Zhan, “Myths About Linear and Monotonic Associations: Pearson’s r, Spearman’s ρ, and Kendall’s τ,” Am. Stat., vol. 76, no. 1, pp. 44–52, 2022. https://doi.org/10.1080/00031305.2021.2004922

[17] P. Schober and L. A. Schwarte, “Correlation coefficients: Appropriate use and interpretation,” Anesth. Analg., vol. 126, no. 5, pp. 1763–1768, 2018. https://doi.org/10.1213/ANE.0000000000002864

[18] P. K. Pearson, F. G, and F. R. S. Received, “VII . Note on Regression and Inheritance in the Case of Two Parents,” Proceeding R. Soc. London, vol. 58, pp. 347–352, 1895.

[19] C. Spearman, “The Proof and Measurement of Association Between Two Things.,” Stud. Individ. Differ. search Intell., vol. 15, no. 1, pp. 45–58, 1904. https://doi.org/10.1037/11491-005

[20] J. D. Gibbons, Nonparametric measures of association. Sage, 1993.

[21] K. P. Kotz, Samuel, Tomasz Kozubowski, The Laplace distribution and generalizations: a revisit with applications to communications, economics, engineering, and finance. Springer Science & Business Media, 2012.

[22] M. G. Kendall, “A New Measure of Rank Correlation,” Biometrika, vol. 30, pp. 81–93, 1938. https://doi.org/10.1093/biomet/30.1-2.81

Investigation of physicochemical and magnetic properties of zinc ferrite ($ZnFe_2O_4$) with different ultrasonic power for enhancement of thermal conductivity of refrigeration oils

Poppy Puspitasari[1,2,a*], Avita Ayu Permanasari[1,2,b], Boby Wahyudi[1,c], Diki Dwi Pramono[1,d], Sri Hastuty[3,e]

[1]Department of Mechanical and Industrial Engineering, Universitas Negeri Malang, Indonesia

[2]Centre of Advanced Material and Renewable Energy, Universitas Negeri Malang, Indonesia

[3]Department of Mechanical Engineering, Universitas Pertamina, Indonesia

[a]poppy@um.ac.id, [b]avita.ayu.ft@um.ac.id, [c]boby.wahyudi.1905146@students.um.ac.id, [d]dikidwipramono@gmail.com, [e]sri.hastuty@universitaspertamina.ac.id

Keywords: Ultrasonic Power, Zinc Ferrite, Sonochemical, Nanolubricant, Polyolester Oil

Abstract. Zinc ferrite ($ZnFe_2O_4$) nanoparticles have gained significant attention globally, particularly due to their superior chemical properties, magnetic properties, and heat transfer referring to their particle size. The sonochemical technique is one of the most applied methods to prepare zinc ferrite nanoparticles as it offers faster reaction and the ability to break large aggregates into smaller aggregates up to nanoscale. This research uses sonochemical method with power variations of 25%, 50%, and 75% to obtain zinc ferrite nanoparticles, zinc ferrite nanoparticles obtained are then analyzed for morphology, phase, crystallite size, functional groups and magnetic properties. Then the obtained zinc ferrite nanoparticles were synthesized using 0.3% Polyester Oil (POE), this method was employed to analyze the nanolubricant as a refrigerant fluid additive. The nanolubricant was characterized using thermal conductivity. The smallest crystal, 11.51 nm, was found in the sample with 50% power. The functional groups confirmed the presence of Fe-O and Zn-O cation bonds in the octahedral and tetrahedral regions. Magnetic properties shows that zinc ferrite resulting from power variation 25%, 50%, and 75% is superparamagnetic. The characterization results show that the obtained zinc ferrite can be used as a nanolubricant in refrigerant fluid.

Introduction

Zinc ferrite ($ZnFe_2O_4$) is a ferrite material that is widely used in various applications, especially in electronics, magnetics, and energy processing [1]. As a ferrite material with ferromagnetic and semiconductor properties, $ZnFe_2O_4$ has a unique combination of physicochemical properties, such as magnetic ability, thermal stability, and good thermal conductivity [2]. These properties have led to research on potential applications of $ZnFe_2O_4$, including its utilization as battery lithium ion, catalysts, solar cells [3], sensors [4], and nanofluids as cooling fluids [5]. Therefore, $ZnFe_2O_4$ is an attractive candidate for development in thermal applications, including as an additive in cooling lubricating oils.

In refrigeration systems, thermal efficiency and heat transfer are key factors in maintaining optimal performance. The addition of nanoparticles to cooling lubricating oils has been shown to increase thermal conductivity, thereby improving the efficiency of the cooling system [6–8]. Additionally, Polyolester (POE) Oil is a lubricating fluid commonly used in refrigerant systems. Several studies have used POE as a base fluid or solvent for nanoparticles, such as in previous research that mixed POE with ZnO nanopowder to improve thermophysical properties [9]. However, the physicochemical and magnetic properties of $ZnFe_2O_4$ nanoparticles are strongly influenced by the synthesis method and process parameters, such as the ultrasonic power used during the nanoparticle production process.

In addition, the physicochemical and magnetic properties of $ZnFe_2O_4$ nanoparticles are strongly influenced by the synthesis method and process parameters, such as the ultrasonic power used during the nanoparticle production process known as sonochemistry [10]. This method offers a number of advantages, including faster reaction and the ability to break down larger crystals to smaller sizes, reaching the nanoscale [11]. This method involves an ultrasonic wave range of up to 1 MHz [12]. The sonochemical method utilizes ultrasonic waves radiated into a solution, leading to collisions as well as friction between solution particles under high pressure. Further, the colliding particles make up the solution, which will result in a local temperature increase of up to 5000K with a cooling rate of 1011 K/s [12]. Additionally, a study reported that the sonochemical method can influence particle size up to 6nm by using power variations [13]. The use of ultrasonic power in nanoparticle synthesis can result in a more uniform particle distribution, as well as improve the quality and properties of the resulting particles [14]. However, not many studies have deeply explored the effect of ultrasonic power variation on the physicochemical, magnetic, and thermal conductivity properties of $ZnFe_2O_4$, particularly in the context of its application as an additive to cooling lubricant oils.

Therefore, this study aims to investigate the effect of different ultrasonic power in the synthesis process of $ZnFe_2O_4$ on the physicochemical, magnetic, and thermal conductivity enhancement properties of refrigeration lubricating oils. With a better understanding of the correlation between synthesis parameters and material properties, it is expected that the results of this study can contribute to the development of more efficient and environmentally friendly refrigeration systems through the optimized use of $ZnFe_2O_4$ nanoparticles.

Method

Zinc Ferrite Synthesis

This study uses the sonochemical method to synthesize zinc ferrite by ultrasonic power variation shown in figure 1. Synthesis was started by mixing 2 gr of ZnO (Sigma-Aldrich, USA) and 3.95 gr of Fe_2O_3 (Sigma-Aldrich, USA) with a magnetic stirrer at 80°C temperature, with a pH control of 11. The process was continued using the Ultrasonic Cell Disruptor (UCD-2000, BIOBASE) with power variations of 25%, 50%, and 75%. This sonication process was performed at a time interval of 15s on and 5s off for 30 minutes for each sample. The synthesis was followed by the washing process after the sample solution was deposited using a centrifuge (LC-04R, Oregon) with 1500 ml of distilled water. The sample was then given a sintering treatment. After washing, the sample was dried at 100°C temperature, followed by dried crushing for 1 hour and the sintering process for 2.5 hours using Muffle Furnace (KSL-1750X-KA3, MTI Corporation). Once the sintering process was complete, the sample was crushed again for 1 hour to set nanopowder Zinc Ferrite

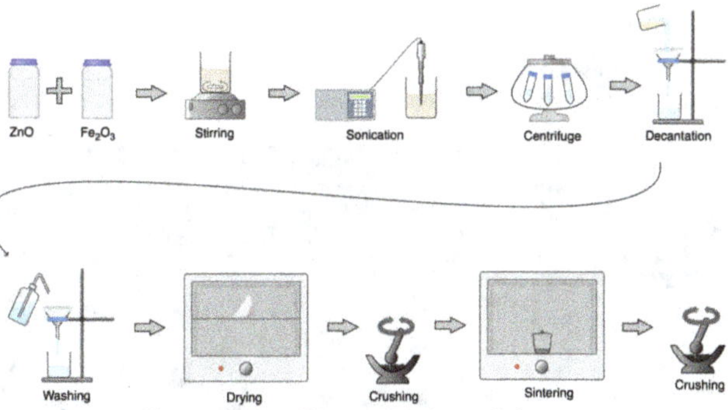

Figure 1. Zinc Ferrite Synthesis Scheme

Zinc Ferrite Characterization

Material characterization was carried out to identify the properties of the material. This research used three types of characterization, including SEM (Scanning Electron Microscopy) (Inspect-S50, FEI), to identify the surface morphology of zinc ferrite [15]. The XRD (X'Pert PRO, PANalytical) was used to determine the crystalline phase structure of nanoparticles by investigating the the material [16]. Lastly, the FTIR (Fourier Transform Infrared) (IRPrestige 21, Shimadzu) was performed to find the functional group using infrared [17]. The VSM (Vibration Sample Magnetometer) (Cryogen-free 3 Tesla, VersaLab™) was used to identification the magnetic properties of the zinc ferrite.

Preparation Nanolubricant

Figure 2 shows the synthesis process of nanolubricant with zinc ferrite additive. Zinc Ferrite nano-powder was synthesized with Polyolester (POE) Oil (SUNISO, Belgium) at a nanoparticle concentration of 0.3%. Zinc Ferrite nanoparticles were mixed using a magnetic stirrer for 1 hour and sonicated for 3 hours using an Ultrasonic Homogenizer (CGOLDENWALL, USA).

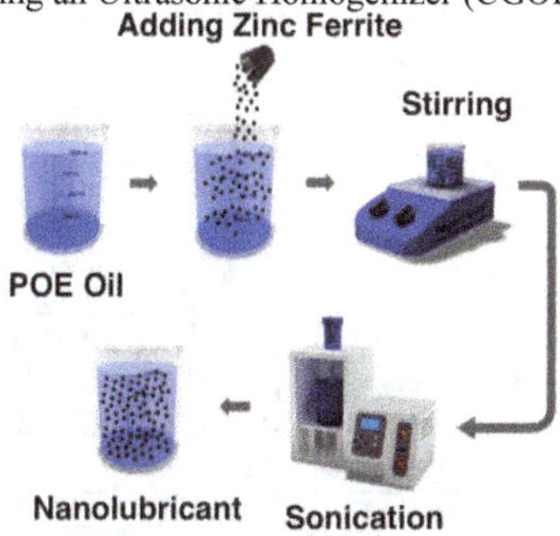

Figure 2. Scheme synthesized of POE Oil and Additive Zinc Ferrite

The Referigerant oil with additive zinc ferrite has thermal conductivity tests, the test was carried out to determine the conduction heat transfer in the cutting fluid using a KD2 pro thermal properties analyzer in Laboratorium Rekayasa Pangan dan Alat Mesin Pertanian, Universitas Jember

Result And Discussion

Morphology of Zinc Ferrite

The analysis of morphology was carried out using an SEM test. This test enabled us to determine the morphological differences of synthesized $ZnFe_2O_4$ with various UCD power.

a) b) c)

Figure 3. SEM Morphology of a) $ZnFe_2O_4$ 25% Ultrasonic Power, b) $ZnFe_2O_4$ 50% Ultrasonic Power, c) $ZnFe_2O_4$ 75% Ultrasonic Power

Figure 3. Samples of zinc ferrite ($ZnFe_2O_4$) nanopowder UCD power 25%, 50%, and 75% showing magnification of 50,000x SEM so that the particles of the sample can be seen. The micrograph shows that the particles tend to have cubic crystallites [18]. The particle sizes of the 25% and 75% UCD power samples look larger and more irregular when compared to the 50% UCD power sample.

Samples with 50% UCD power appear to have a better average particle size distribution compared to samples with 25% and 75% UCD power, the particle size distribution is due to the ultrasonic UCD power used in line with the sonochemical zinc ferrite ($ZnFe_2O_4$) synthesis method. High and not excessive ultrasonic wave intensity produces strong cavitation and acoustic streaming effect (liquid melting due to acoustic pressure gradient) [9]. Agglomeration in Figure 1 shows that in the 75% UCD power sample less agglomeration occurs, this is in line with the fact that high ultrasonic power can reduce the agglomeration of nanoparticles [19].

Phase and Crystallite size of Zinc Ferrite

The XRD test on the $ZnFe_2O_4$ was carried out to identify the phase and crystallite size [20]. The crystallite size was calculated using the Scherrer [21], Formula shown in 1.

$$d = \frac{k.\lambda}{\beta \cos \theta} \quad (1)$$

Table 1. Crystallite Size of $ZnFe_2O_4$.

Sampel	Pos. (°2Theta)	Height [cts]	FWHM [°2Th.]	d-spacing [Å]	Crystallite size [nm]
$ZnFe_2O_4$ 25%	35.51	324.227	0.4443	2.5281	38.98
$ZnFe_2O_4$ 50%	36.08	165.107	1.0000	2.4872	11.51
$ZnFe_2O_4$ 75%	35.43	396.242	0.6000	2.5317	31.32

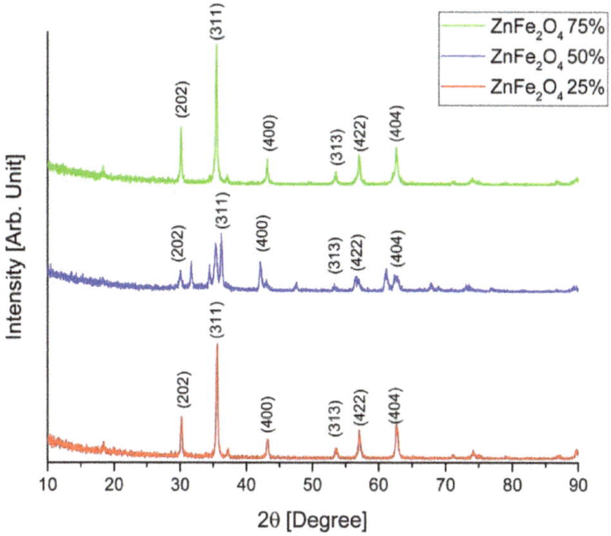

Figure 4. XRD Result of $ZnFe_2O_4$ synthesized using various ultrasonic power.

The results of phase identification for zinc ferrite (ZnFe2O4) nanopowder using the sonochemical method with variations in UCD power are presented in Figure 4 and table 1. Similar diffraction patterns were observed at the [202], [311], [400], [313], [422], and [404] peaks. These diffraction peaks indicate that the structure of the zinc ferrite ($ZnFe_2O_4$) nanopowder sample is face-centered cubic (FCC) [22]. The three research samples produce different sizes of 38.98 nm for UCD Power of 25%, 11.51 nm for 50% UCD Power, and 31.32 nm for 75% UCD Power. The ($ZnFe_2O_4$) nanopowder sample with UCD Power 50% has the smallest size among the others. Samples with 50% UCD power produce the smallest particle size, as suggested by the Scherrer equation that the greater the FWHM value results in a smaller obtained crystallite size [23]. If the ultrasonic power is too low, the sonication intensity may be insufficient for proper bubble nucleation [24]. Further, the ultrasonic power is directly proportional to the ultrasonic wave and amplitude value. Thus, if the ultrasonic power is too high, turbulence may occur due to excessive stirring, which can disrupt bubble growth and hinder the degradation process [25].

Functional Groups of Zinc Ferrite

The analysis and comparison of functional groups were completed using the FTIR test. Through this analysis, we identified the functional group bonds of $ZnFe_2O_4$ synthesized using various power UCD. The results of functional group identification are illustrated in Figure 3.

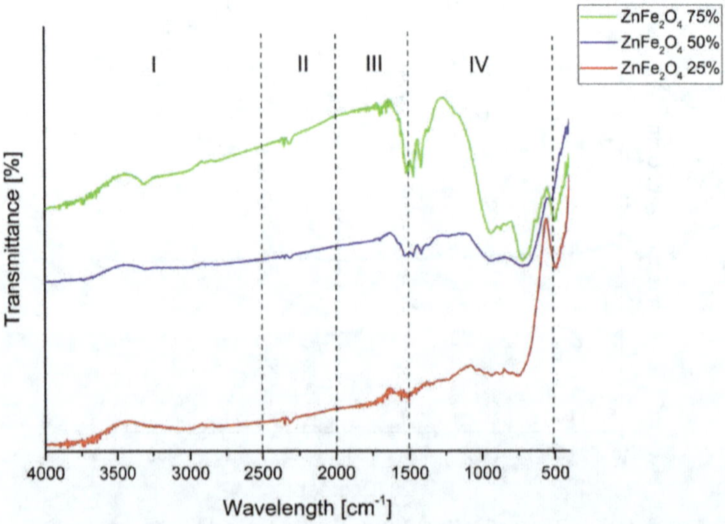

Figure 5. FTIR Result of $ZnFe_2O_4$ synthesized using various ultrasonic power.

Figure 5 shows a graph of the spectrum of a zinc ferrite ($ZnFe_2O_4$) sample with variations in UCD Power. The spectrum graph presents a similar shape except for the transmittance axis. The peak at 3400 cm^{-1} indicates O-H bonds that occur due to water vapor in the sintering process [18]. Vibration spectra at 2301 cm^{-1} indicate the presence of ferrite [26], while the vibration peak at 2346 cm^{-1} indicates the presence of a metal ferrite complex [27]. The spectrum observed at 2366 cm^{-1} is because of the indication of the CO_2 compound, specifically the C=O bond, resulting from the combustion reaction during sintering [28]. Additionally, the peaks at 1626 cm^{-1} and 1627 cm^{-1} are caused by bonding (O-H) and bending vibration, which respectively indicate the presence of water molecules absorbed and bound to $ZnFe_2O_4$ [29]. The UCD Power 75% Zinc Ferrite ($ZnFe_2O_4$) sample shows a marked decrease in frequency with a peak due to the formation of H bridges in the precursor, which disappears because of thermal treatment. This phenomenon influences the decreasing peak, as reported in a study [30]. The main band vibrations with strong absorption peaks at low frequencies (around 550 cm^{-1} and 415 cm^{-1}) can be attributed to the stretching vibrations of the Zn-O bond in the tetrahedral position and the Fe-O bond in the octahedral position [29] [18].

Magnetic Properties of Zinc Ferrite

The magnetic property of a material can be identified through VSM (*Vibrating Sample Magnetometer*) test. Some important things that can be observed from the results of VSM testing are magnetic saturation (M_s), magnetic remanence (M_r), coercivity (H_c), and remanence ratio (R) [31]. When the sample is exposed to an external magnetic field, the magnetic moment becomes aligned with the direction of the external magnetic field until it reaches the maximum saturation point. Magnetization in this process is known as magnetic saturation. When the external magnetic field is removed, the magnetic material sample retains the memories of the applied magnetic field, this is called magnetic remanence. The coercive force required to reduce the magnetization of the magnetic material to zero must be applied to close the hysteresis curve, this is called coercivity. The ability of the magnetic material to replace the curve in the hysteresis loop after the magnetic field is removed is called the remanance ratio [32].

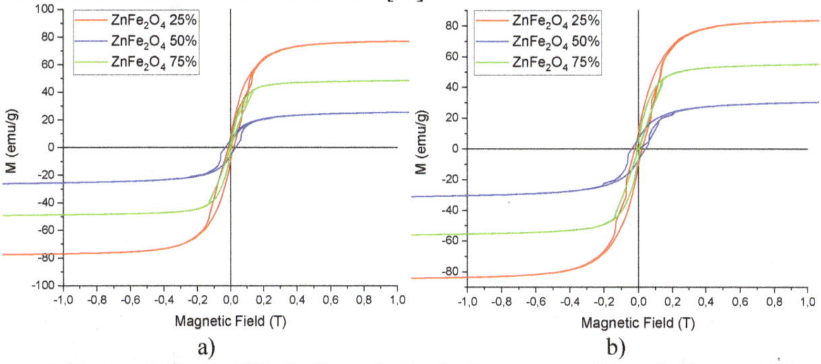

Figure 6. Hysteresis Curve of $ZnFe_2O_4$ synthesized using various ultrasonic Power at (a) room temperature, (b) at -100°C

Table 2. VSM Test Results of Zinc Ferrite ($ZnFe_2O_4$)

Sampel	Temp. (°C)	M_s (emu/g)	M_r (emu/g)	R	H_c (T)	H_c (Oe)
$ZnFe_2O_4$ 25%	30	77.02	8.71	0.113	0.0186	186
	-100	83.58	10.63	0.127	0.0205	205
$ZnFe_2O_4$ 50%	30	25.43	5.38	0.211	0.0300	300
	-100	30.31	6.61	0.218	0.0405	405
$ZnFe_2O_4$ 75%	30	48.45	5.05	0.104	0.0054	54
	-100	55.38	5.34	0.096	0.0077	77

Based on figure 6 and table 2, it can be seen that the samples with UCD Power 25% and 75% have a larger magnetization value (M_s) than the 50% sample. The decrease in M_s value in the sample with UCD power 50% is caused by the presence of Fe_2O_4 phase and ZnO phase which is antiferromagnetic, it causes a small magnetization value [33]. While the sample with UCD power 50% has the largest area ratio and is supported by a small Ms value, this indicates that in this sample there are more other impurity phases, the statement supports that the higher the degree of crystallinity, the more other impurity phases in the sample with UCD power 50% [34]. To analyze the results of the hysteresis curve against particle size can be seen in Figure 7. as follows.

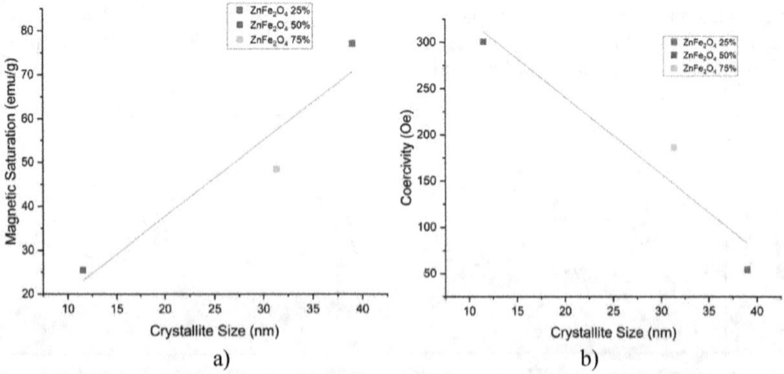

Figure 7. Relationship between particle size and (a) magnetic saturation and (b) coercivity

The M_s value is directly proportional to the particle size. Therefore, the larger crystallite size of the sample signifies a greater magnetization value (M_s) being obtained. The sample with $ZnFe_2O_4$ 25% UCD Power and crystallite size of 39.98 nm attains the largest M_s value among the other samples, while the sample $ZnFe_2O_4$ with 50% UCD Power and crystallite size of 11.51 nm has the smallest M_s value. The smaller the crystallite size, the smaller the M_s value obtained will be [35]. The coercivity (H_c) value of the sample $ZnFe_2O_4$with UCD power of 50% is high. Furthermore, the high H_c value observed in the sample $ZnFe_2O_4$ with UCD power of 50% does not align with the particle size. This is believed to be due to significant agglomeration in this particular sample [34]. Besides, indications of impurities in the sample also affect the level of coercivity and loop area in the sample.

The sample $ZnFe_2O_4$ with a UCD power of 50% presents different results from other samples due to the increase of particle size enhancing the coercive field. However, it should be noted that the H_c value does not increase linearly with particle size in the $ZnFe_2O_4$ 50% UCD power sample. As a result, the loop area of the 50% UCD power sample is larger due to the need for a field. The outer magnet is larger to make the net magnetization 0 [34]. Accordingly, the H_c value in each sample is not influenced by the crystallite size, while the level of coercivity (H_c) is affected by the uniformity of particle size and particle distribution or the presence of agglomeration [34]. A smaller H_c value indicates the sample's superparamagnetic nature. A small H_c value, which is close to zero in the sample, suggests that $ZnFe_2O_4$ particles with variations in UCD power are classified as superparamagnetic [18]. The shape of the loop area in the sample also confirms the superparamagnetic feature of the sample [36]. Additionally, the high M_s value also signifies the sample is a superparamagnetic material [37]. Superparamagnetic materials have high magnetic properties when exposed to an external magnetic field, but if the magnetic field disappears, then the average magnetization value is zero [37]. The classification of magnetic properties as either hard or soft is determined by the H_c value. A sample is considered hard magnetic if its H_c value is greater than 420 Oe. Conversely, if the H_c value is below 420 Oe, the sample is classified as soft magnetic [38]. In this study, all three sample variations are classified as soft magnetic materials.

Enhancement Thermal Conductivity of Refrigerant Oil

Figure 8 shows a histogram of the thermal conductivity test results using the KD2 Pro tool. Figure 8 suggests that the addition of zinc ferrite ($ZnFe_2O_4$) nanoparticles using the synthetic sonochemical method can increase the thermal conductivity of base fluid polyester oil (POE). The thermal conductivity test results for polyester oil SL68 showed a value of 0.127 W/m.K. The addition of nanolubricant to each sample resulted in an increase in thermal conductivity value, with POE/$ZnFe_2O_4$UCD Power 25% and 75% showing a value of 0.128 W/m.K. The highest thermal conductivity value was observed for POE/$ZnFe_2O_4$UCD Power 50%, reaching a value of 0.129 W/m.K. One of the factors that influences the thermal conductivity of a fluid is particle size [39]. Following the obtained phase identification results, the zinc ferrite ($ZnFe_2O_4$) sample with a UCD power of 50% presents the smallest particle size among the other samples. The greater the concentration of nanoparticles in a fluid increases the conductivity value of the fluid [40]. Other influencing factors for the thermal conductivity of nanolubricants include nanoparticle concentration, particle size, particle shape, thermal conductivity of the nanoparticles, type of base fluid, temperature, and preparation techniques of nanolubricant [41].

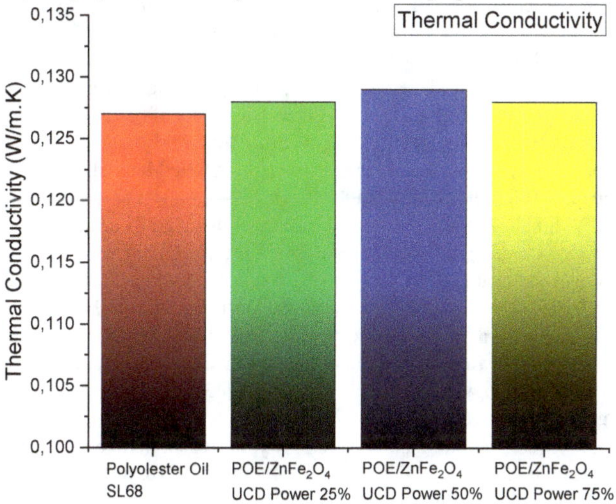

Figure 8. Histogram of Thermal Conductivity Test Nanolubricant Zinc Ferrite ($ZnFe_2O_4$)

Summary

This study successfully synthesized $ZnFe_2O_4$ nanoparticles using different ultrasonic power. Morphological analysis using SEM on zinc ferrite ($ZnFe_2O_4$) nanopowder samples synthesized using sonochemical methods with variations in UCD power shows that the samples have the shape of like-cubic crystallites. Phase analysis using XRD on the sample nanopowder zinc ferrite ($ZnFe_2O_4$) results show the sample has a shape Face Centered Cubic (FCC). The 50% UCD power variation has the smallest crystal size of 11.51 nm, the 25% variation has a crystal size of 38.98 nm, and the 75% variation has a crystal size of 31.32 nm. Functional group analysis using FTIR shows that the peak at m 550 cm^{-1} and 415 cm^{-1} can be assigned to the stretching vibrations of the Zn-O bond in the tetrahedral position and the Fe-O bond in the octahedral position. Following the analysis of the magnetic properties, the H_c value and loop area of the samples suggest its superparamagnetic nature. Synthesize Polyolester (POE) Oil/Zinc Ferrite ($ZnFe_2O_4$) Nanolubricant fluid with 0.3% concentration by weight has higher thermal conductivity than base fluid.

Acknowledgment

Authors would like to thank Universitas Negeri Malang for Thesis Publication Grant (4.4.279/UN32.14.1/LT/2024) in 2024.

References

[1] A.V. Humbe, J.S. Kounsalye, S.B. Somvanshi, A. Kumar, K.M. Jadhav, Cation distribution, magnetic and hyperfine interaction studies of Ni–Zn spinel ferrites: role of Jahn Teller ion (Cu^{2+}) substitution, Mater. Adv. 1 (2020) 880–890. https://doi.org/10.1039/D0MA00251H

[2] P.P. Goswami, H.A. Choudhury, S. Chakma, V.S. Moholkar, Sonochemical Synthesis of Cobalt Ferrite Nanoparticles, International Journal of Chemical Engineering 2013 (2013) 1–6. https://doi.org/10.1155/2013/934234

[3] S. Fabbiyola, L.J. Kennedy, T. Ratnaji, J.J. Vijaya, U. Aruldoss, M. Bououdina, Effect of Fe-doping on the structural, optical and magnetic properties of ZnO nanostructures synthesised by co-precipitation method, Ceramics International 8842 (2015) 1–32. https://doi.org/10.1016/j.ceramint.2015.09.110

[4] M.A.A. Kerroum, A. Essyed, C. Iacovita, W. Baaziz, D. Ihiawakrim, O. Mounkachi, M. Hamedoun, A. Benyoussef, M. Benaissa, O. Ersen, The effect of basic pH on the elaboration of ZnFe 2 O 4 nanoparticles by co-precipitation method : structural , magnetic and hyperthermia characterization Abstract, Journal of Magnetism and Magnetic Materials 8853 (2019) 1–18. https://doi.org/10.1016/j.jmmm.2019.01.081

[5] M. Gupta, V. Singh, Z. Said, Heat transfer analysis using zinc Ferrite/water (Hybrid) nanofluids in a circular tube: An experimental investigation and development of new correlations for thermophysical and heat transfer properties, Sustainable Energy Technologies and Assessments 39 (2020) 100720. https://doi.org/10.1016/j.seta.2020.100720

[6] Ç.V. Yıldırım, M. Sarıkaya, T. Kıvak, Ş. Şirin, The effect of addition of hBN nanoparticles to nanofluid-MQL on tool wear patterns, tool life, roughness and temperature in turning of Ni-based Inconel 625, Tribology International 134 (2019) 443–456. https://doi.org/10.1016/j.triboint.2019.02.027

[7] H. Hegab, U. Umer, M. Soliman, H.A. Kishawy, Effects of nano-cutting fluids on tool performance and chip morphology during machining Inconel 718, International Journal of Advanced Manufacturing Technology 96 (2018) 3449–3458. https://doi.org/10.1007/s00170-018-1825-0

[8] A.K. Sharma, A.K. Tiwari, A.R. Dixit, Effects of Minimum Quantity Lubrication (MQL) in machining processes using conventional and nanofluid based cutting fluids: A comprehensive review, Journal of Cleaner Production 127 (2016) 1–18. https://doi.org/10.1016/j.jclepro.2016.03.146

[9] Y. Yang, J. Lan, X. Li, Study on bulk aluminum matrix nano-composite fabricated by ultrasonic dispersion of nano-sized SiC particles in molten aluminum alloy, Materials Science and Engineering: A 380 (2004) 378–383. https://doi.org/10.1016/j.msea.2004.03.073

[10] M. Sivakumar, T. Takami, H. Ikuta, A. Towata, K. Yasui, T. Tuziuti, T. Kozuka, D. Bhattacharya, Y. Iida, Fabrication of Zinc Ferrite Nanocrystals by Sonochemical Emulsification and Evaporation: Observation of Magnetization and Its Relaxation at Low Temperature, J. Phys. Chem. B 110 (2006) 15234–15243. https://doi.org/10.1021/jp055024c

[11] P. Adamou, E. Harkou, S. Hafeez, G. Manos, A. Villa, S.M. Al-Salem, A. Constantinou, N. Dimitratos, Recent progress on sonochemical production for the synthesis of efficient photocatalysts and the impact of reactor design, Ultrasonics Sonochemistry 100 (2023) 106610. https://doi.org/10.1016/j.ultsonch.2023.106610

[12] D. Lestari, Preparasi Nanokomposit ZnO/TiO_2 Dengan Sonokimia Serta Uji Aktivitasnya Untuk Fotodegradasi Fenol, Indonesian Journal of Chemical Science 1 (2011).

[13] H.A. Choudhury, A. Choudhary, M. Sivakumar, V.S. Moholkar, Mechanistic investigation of the sonochemical synthesis of zinc ferrite, Ultrasonics Sonochemistry 20 (2013) 294–302. https://doi.org/10.1016/j.ultsonch.2012.06.006
[14] N. Pokhrel, P.K. Vabbina, N. Pala, Sonochemistry: Science and Engineering, Ultrasonics Sonochemistry 29 (2016) 104–128. https://doi.org/10.1016/j.ultsonch.2015.07.023
[15] S.K. Sharma, ed., Handbook of Materials Characterization, Springer International Publishing, Cham, 2018. https://doi.org/10.1007/978-3-319-92955-2
[16] G. Hübschen, I. Altpeter, R. Tschuncky, H.-G. Herrmann, eds., Materials characterization using nondestructive evaluation (NDE) methods, Elsevier/WP, Woodhead Publishing, Amsterdam Boston Cambridge Heidelberg, 2016
[17] K.E. Ramohlola, E.I. Iwuoha, M.J. Hato, K.D. Modibane, Instrumental Techniques for Characterization of Molybdenum Disulphide Nanostructures, Journal of Analytical Methods in Chemistry 2020 (2020) 1–29. https://doi.org/10.1155/2020/8896698
[18] P. Puspitasari, U.A. Rizkia, S. Sukarni, A.A. Permanasari, A. Taufiq, A.B.N.R. Putra, Effects of various sintering conditions on the structural and magnetic properties of zinc ferrite (ZnFe2O4), Materials Research 24 (2021). https://doi.org/10.1590/1980-5373-MR-2020-0300
[19] M. Esmaeili-Zare, M. Salavati-Niasari, A. Sobhani, Simple sonochemical synthesis and characterization of HgSe nanoparticles, Ultrasonics Sonochemistry 19 (2012) 1079–1086. https://doi.org/10.1016/j.ultsonch.2012.01.013
[20] A.A. Bunaciu, E. gabriela Udriştioiu, H.Y. Aboul-Enein, X-Ray Diffraction: Instrumentation and Applications, Critical Reviews in Analytical Chemistry 45 (2015) 289–299. https://doi.org/10.1080/10408347.2014.949616
[21] D.D. Pramono, P. Puspitasari, A.A. Permanasari, S. Sukarni, A. Wahyudiono, Effect of sintering time on porosity and compressibility of calcium carbonate from Amusium pleuronectes (scallop shell), in: Malang, Indonesia, 2023: p. 050003. https://doi.org/10.1063/5.0120990
[22] S.A. Morrison, C.L. Cahill, E.E. Carpenter, S. Calvin, R. Swaminathan, S.A. Morrison, C.L. Cahill, T. George, V.G. Harris, Magnetic and structural properties of nickel zinc ferrite nanoparticles synthesized at room temperature Magnetic and structural properties of nickel zinc ferrite nanoparticles synthesized at room temperature, 6392 (2004) 1–5. https://doi.org/10.1063/1.1715132
[23] P.R. Chaudhari, V.M. Gaikwad, S.A. Acharya, Role of mode of heating on the synthesis of nanocrystalline zinc ferrite, Applied Nanoscience (Switzerland) 5 (2015) 711–717. https://doi.org/10.1007/s13204-014-0367-5
[24] K. Thangavadivel, K. Okitsu, G. Owens, P.J. Lesniewski, R. Nishimura, Influence of sonochemical reactor diameter and liquid height on methyl orange degradation under 200 kHz indirect sonication, Journal of Environmental Chemical Engineering 1 (2013) 275–280. https://doi.org/10.1016/j.jece.2013.05.005
[25] N.L. Arifin, Pengaruh Sonikasi Bertahap dalam Proses Degradasi Kitosan terhadap Komposisi Dan Properti Produk, (2015) 110.
[26] K. Patil, S. Kadam, P. Lokhande, S. Balgude, P. More, The effects of cobalt and magnesium co-doping on the structural and magnetic properties of ZnFe2O4 synthesized using a sonochemical process, Solid State Communications 337 (2021). https://doi.org/10.1016/j.ssc.2021.114435
[27] S.E. Ali, Influence of preparation method on phase formation, structural and magnetic properties of BiFeO3, Journal of Electroceramics 48 (2022) 95–101. https://doi.org/10.1007/s10832-021-00276-1

[28] M. Sivakumar, T. Takami, H. Ikuta, A. Towata, K. Yasui, T. Tuziuti, T. Kozuka, D. Bhattacharya, Y. Iida, Fabrication of zinc ferrite nanocrystals by sonochemical emulsification and evaporation: Observation of magnetization and its relaxation at low temperature, Journal of Physical Chemistry B 110 (2006) 15234–15243. https://doi.org/10.1021/jp055024c

[29] V. Ramasamy Raja, A. Karthika, S. Lok Kirubahar, A. Suganthi, M. Rajarajan, Sonochemical synthesis of novel ZnFe 2 O 4 /CeO 2 heterojunction with highly enhanced visible light photocatalytic activity, Solid State Ionics 332 (2019) 55–62. https://doi.org/10.1016/j.ssi.2018.12.016

[30] C.A. Ladole, Preparation and characterization of spinel zinc ferrite ZnFe 2O 4, International Journal of Chemical Sciences 10 (2012) 1230–1234.

[31] M. Houshiar, F. Zebhi, Z.J. Razi, A. Alidoust, Z. Askari, Synthesis of cobalt ferrite (CoFe2O4) nanoparticles using combustion, coprecipitation, and precipitation methods: A comparison study of size, structural, and magnetic properties, Journal of Magnetism and Magnetic Materials 371 (2014) 43–48. https://doi.org/10.1016/j.jmmm.2014.06.059

[32] S. Kanagesan, M. Hashim, S. Tamilselvan, N.B. Alitheen, I. Ismail, M. Syazwan, M.M.M. Zuikimi, Sol-gel auto-combustion synthesis of cobalt ferrite and it's cytotoxicity properties, Digest Journal of Nanomaterials and Biostructures 8 (2013) 1601–1610.

[33] B.S. a dan Y. Mashadi, R. Andriyani Putri, Sintesis Bahan Magnetik $Zn_XFe_{(3-X)}O_4$ Dengan Metode Ko-Presipitasi Sebagai Penyerap Gelombang Mikro Synthesis Of The $Zn_XFe_{(3-X)}O_4$ Magnetic Materials As Microwave Absorber By Co-Precipitation Methods, Majalah Ilmiah Pengkanian Industri (MIPI) 13 (2019) 179–186.

[34] M. Pauzan, T. Kato, S. Iwata, E. Suharyadi, Pengaruh Ukuran Butir dan Struktur Kristal terhadap Sifat Kemagnetan pada Nanopartikel Magnetit (Fe 3 O 4), Prosidinga Pertemuan Ilmiah XXVII HFI (2013) 24–28.

[35] M. Afshari, A.R. Rouhani Isfahani, S. Hasani, F. Davar, K. Jahanbani Ardakani, Effect of apple cider vinegar agent on the microstructure, phase evolution, and magnetic properties of CoFe2O4 magnetic nanoparticles, International Journal of Applied Ceramic Technology 16 (2019) 1612–1621. https://doi.org/10.1111/ijac.13224

[36] S. Husain, M. Yusup, N.H. Haryanti, Suryajaya, M. Saukani, Rodiansono, S. Arjo, A. Riyanto, Characteristics of zinc ferrite nanoparticles (ZnFe 2O4) from natural iron ore, IOP Conference Series: Earth and Environmental Science 758 (2021). https://doi.org/10.1088/1755-1315/758/1/012001

[37] S. Tebriani, Analisis vibrating sample magnetometer (VSM) pada hasil elektrodeposisi lapisan Tipis magnetite menggunakan aruscontinue direct current, Natural Science Journal 5 (2019) 722–730.

[38] P.J. Wasilewski, 20 (1973) 67-72. north-holland publishing company, 20 (1973) 67–72.

[39] R.K. Asmoro, P. Puspitasari, A.A. Permanasari, M.I.H.C. Abdullah, Identification of Thermophysical and Rheological Properties of SAE 5w-30 with Addition of Hexagonal Boron Nitride, Transmisi 19 (2023) 41–48. https://doi.org/10.26905/jtmt.v19i1.9639

[40] W. Mulya, H. Dani Gustaman Syarif, T. Nuklir Nasional Bandung, Nanopartikel , Hasil Green Synthesis Dan Aplikasinya Di Mesin Pendingin Nanoparticle, a Green Synthesis Product and Its Application on Cooling Machines, Maret 5 (2018) 894

[41] M.A. Bin Saufi, H. Bin Mamat, A Review on Thermophysical Properties for Heat Transfer Enhancement of Carbon-Based Nanolubricant, Advanced Engineering Materials 23 (2021) 1–20. https://doi.org/10.1002/adem.202100403

Decarbonization Technology: ICDT2024 Materials Research Forum LLC
Materials Research Proceedings 53 (2025) 117-124 https://doi.org/10.21741/9781644903575-11

Biodiesel production from waste cooking oil in presence of modified waste eggshell catalyst via microwave heating

Siti Aminah Mohd Johari[1,2,a], Neshwign Chandrasehgaran[1,2,b], Muhammad Ayoub[1,2,c*], Chai Yee Ho[1,2,d] and Yoshito Ando[3,4,e]

[1]Centre for Biofuel and Biochemical Research, Universiti Teknologi PETRONAS, 32610, Seri Iskandar, Perak, Malaysia

[2]Department of Chemical Engineering, Universiti Teknologi PETRONAS, 32610, Seri Iskandar, Perak, Malaysia

[3]Department of Biological Function Engineering, Graduate School of Life Science and System Engineering, Kyushu Institute of Technology, 2-4 Hibikino, Wakamatsu, Kitakyushu, Fukuoka 808-0196, Japan

[4]Collaborative Research Centre for Green Materials on Environmental Technology, Kyushu Institute of Technology, 2-4 Hibikino, Wakamatsu, Kitakyushu, Fukuoka 808-0196, Japan

[a]aminah.mjohari@utp.edu.my, [b]neshwign_19001115@utp.edu.my, [c]muhammad.ayoub@utp.edu.my, [d]yeeho.chai@utp.edu.my, [e]yando@life.kyutech.ac.jp

Keywords: Biodiesel, Calcium Oxide, Eggshell, Hydrotalcite, Microwave

Abstract. This study compares the efficiency of producing biodiesel from waste cooking oil (WCO) utilizing modified eggshell based calcium oxide (CaO) catalysts improved with magnesium-aluminum hydrotalcite to pure eggshell based CaO by microwave-heated transesterification. The study optimized process parameters for a methanol to oil molar ratio of 9:1 at 65°C for 30 minutes, with the goal of enhancing catalyst stability and basicity for effective biodiesel synthesis. The study demonstrated that the eggshell based CaO/Mg-Al catalyst was better in biodiesel yield (89.44%) than the pure eggshell based CaO and exhibited improved stability and catalytic activity through gas chromatography with mass spectrometry (GC-MS) analysis for fatty acid methyl ester (FAME) composition and biodiesel yield.

Introduction

Among the most severe environmental problems the world is currently facing are the twin challenges of global warming and fossil fuel depletion. The global warming has been more alarming recently as the surrounding land environment felt hotter day by day and also affected the ocean ecosystem [1]. There was a report by the Intergovernmental Panel on Climate Change (IPCC) stated that when the average global temperature elevates the margin of minimal of 1.5°C, the chances of around one million species to extinct is higher which shows the critical of the global warming issue [2]. One of the primary factors contributing to the rise in temperature is the emission of greenhouse gases (GHGs) from deleterious activities, including the combustion of fossil fuels [3], which serves to meet the ever-increasing energy demand. This phenomenon prompts the community to persistently seek out eco-friendly and renewable alternatives for resources of energy such as biomass, biogas, biohydrogen and biodiesel.

The production of biodiesel has been implemented in numerous countries including Europe, with its annual output rising from 6.13 million tons in 2007 to 14.11 million tons in 2018 [3]–[5]. Therefore, this green fuel is not uncommon in terms of the manufacturing and application. In the realm of biodiesel feedstock, palm oil dominates with a 39% utilization rate, followed by soybean oil at 25%, rapeseed oil at 14% and waste cooking oil (WCO) at a mere 11% [6]. The WCO holds considerable potential as a feedstock due to its high content of free fatty acids (FFA), which is the

Materials Research Proceedings 53 (2025) 117-124 https://doi.org/10.21741/9781644903575-11

required source to transform into fatty acid methyl esters (FAME) and moreover, it is readily accessible and cost-effective, as it is commonly discarded [7]. Several earlier studies have reported biodiesel production from WCO including Erchamo et al. [8] where they obtained maximum 92% of biodiesel yield in presence of mixture of methanol and ethanol (8:4) within 2 hours of reaction time. In addition, the properties of the biodiesel from WCO also showed high quality within the range of ASTM limit. Nazir et al. [9] demonstrated biodiesel production from WCO with 18:1 methanol to oil molar ratio with 15 min of reaction time by using microwave heating method. They generated a maximum of 89.2% which is slightly higher than the above study most probably due to its heating method. Another earlier study by Tan et al. [10] reported similar result of biodiesel yield (89.5%) from the same feedstock but proceeded with longer reaction time (2 hours). Based on these previous works, WCO have the potential to generate a good quality biodiesel with cheaper cost.

Catalysts play a crucial role in the transesterification process for biodiesel production, significantly speeding up the reaction and improving the yield of the product [11]. Various type of catalysts is employed such as homogenous, heterogenous and enzyme catalysts with acid or base properties depending on the nature of the feedstock. Many researchers favor the use of heterogeneous catalysts because they are easily separated from the samples and can be reused several times, making them cost-effective [12], [13]. The base heterogenous sample can be prepared from waste biomass such as animal bones, mollusks shells and eggshell. A study by Pavlovic et al. [13] tested the performance of CaO from chicken eggshell modified with zeolite in biodiesel production and obtained 97.8% yield within 30 minutes of reaction with catalyst amount of 6 wt.%. Another earlier work applied CaO synthesized from combination of chicken and fish bones and achieved 89.5% with catalyst concentration of 2 wt.% [10]. Furthermore, other authors utilized snail shell as the source of CaO and managed to yield 98.9% of biodiesel with catalyst loading of less than 1 wt.% [14]. Drawing on the data laid by prior studies, the waste biomass has many potentials to be used as catalyst instead of catalyst from chemical which is not sustainable and green to the environment.

In recent years, microwave (MW) heating has been adopted to enhance reaction rates in processes such as organic and inorganic chemical synthesis, outperforming conventional methods [9], [15], [16]. The ability of microwaves to shorten reaction times lies in the direct momentum transfer of microwave energy that helps surpass the activation energy barrier, thus accelerating the reaction. Earlier research revealed that employing microwave heating in transesterifying macroalgae for biodiesel production significantly cuts down on reaction time compared to traditional techniques [17]. Applications of the MW heating also span to catalytic processes, encompassing environmental clean-up and biomass transformation. Catalysts prepared through MW techniques have shown impressive performance in both the breakdown of pollutants and the generation of biofuels [18]. A recent study by Yadav et al. [19] investigated esterification of oleic acid via microwave heating method with power of 50 W. The biodiesel yield achieved maximum of 99.6% within 1 hour at 80°C. Johari et al. [20] reported 90.5% biodiesel production from dairy waste scum oil within 25 minutes at 65°C. These studies show that high biodiesel generation is possible within a short reaction time and hence more eco-friendly and lower cost.

This study focuses on producing biodiesel from waste cooking oil (WCO) using a microwave-assisted transesterification process, aiming to improve efficiency and sustainability. By employing microwave heating, the process is expected to be more efficient, potentially reducing reaction time and energy consumption. The research compared the effectiveness of eggshell-based calcium oxide (CaO) and modified eggshell-based CaO catalysts in enhancing biodiesel yield.

Experimental

Materials and reagents

Magnesium (II) nitrate hexahydrate (HmbG chemicals with 99.9% purity), methanol (Merck with 99.9% purity), waste cooking oil was collected from local restaurants and waste eggshell was collected from household kitchen.

Preparation of CaO catalyst from eggshell

Initially, the eggshells underwent a thorough cleaning process to strip away any surface impurities, followed by air drying at room temperature to eliminate moisture. Subsequently, these eggshells were manually fragmented into finer pieces with a stone mortar. Afterward, a rinsing step in tap water was employed to further cleanse the surface, succeeded by a warm water bath to remove lingering contaminants. The preparation continued with the eggshells being dehydrated in a microwave reactor for 30 minutes at a temperature of 120°C, after which they were further broken down into smaller fragments. The final step involved calcining the eggshell in a furnace, where they were subjected to a temperature of 900°C for three hours, ensuring the conversion of $CaCO_3$ to CaO [21]. Lastly, the eggshell based CaO catalyst was kept in a desiccator.

Preparation of Mg-Al hydrotalcite ($Mg_6Al_2CO_3(OH)_{16}\cdot 4H_2O$) by using co-precipitation method

Aluminum nitrate nonahydrate ($Al(NO_3)_3\cdot 9H_2O$) and magnesium nitrate hexahydrate ($Mg(NO_3)_2\cdot 6H_2O$) were dissolved in degassed, deionized water to achieve a Mg to Al molar ratio of 2:1, all at room temperature. The solution was then vigorously mixed using a mechanical stirrer, with small quantities of NaOH and Na_2CO_3 added concurrently to adjust the solution's pH to above 10, under continuous, vigorous stirring. After thorough mixing, the solution was dried at 65°C for five hours, leading to the formation of a precipitate that was subsequently filtered and rinsed with distilled water at 50°C. This precipitate was then dried in a microwave reactor at 110°C for 30 minutes, sieved to ensure the particles were of a uniform size, and finally calcined in a furnace at 700°C for five hours [22]. The prepared eggshell based CaO/Mg-Al catalyst was stored in desiccator.

Preparation of modified eggshell based CaO with hydrotalcite ($Mg_6Al_2CO_3(OH)_{16}\cdot 4H_2O$)

CaO derived from waste eggshells and synthesized magnesium-aluminum hydrotalcite were precisely measured accordingly. These materials, assisted by microwave processing, were then dissolved in distilled water at ambient temperature. The solution underwent heating with magnetic stirring at 80°C until it dried. Half of the resulting precipitates were subsequently dried in a solar microwave reactor for 30 minutes at 110°C [23]. These precipitates were then pulverized and sieved to achieve particles of uniform size [23]. Following this, the particles were calcined in a furnace at 700°C for five hours. The process from the drying step to the calcination was repeated for the remaining precipitates, using a hot air oven for drying over 24 hours at 110°C.

Microwave-assisted transesterification of WCO for biodiesel production.

A total of 5.5 g of methanol and 25 g of WCO were measured and placed into separate beakers, maintaining an alcohol to oil molar ratio of 9:1. The methanol was then mixed with a catalyst at a 12 wt.% loading. This mixture was stirred at room temperature using a magnetic stirrer and covered with aluminum foil. Concurrently, the WCO was agitated with 400 rpm at 65°C [15]. Both of the heated WCO and the methanol-catalyst mixture were then combined in a round-bottom flask and subjected to microwave heating while stirring at 65°C for 30 minutes. Subsequently, the catalyst was filtered out from the mixture. The remaining liquid was transferred into a vial and centrifuged to facilitate the separation of biodiesel and glycerin. The FAME content in the biodiesel was analyzed using Agilent gas chromatography with mass spectrometry (GC-MS), After 15 minutes of centrifugation, the biodiesel was collected, measured, and its yield calculated manually based on the Eq. 1 [24]:

$$Biodiesel\ yield\ (\%) = \frac{mass\ of\ biodiesel\ produced\ (g)}{mass\ of\ WCO\ used\ (g)} \times 100\ \% \qquad (1)$$

Results and discussion

Influence of variance of catalyst in FAME composition via microwave heating transesterification

Fig. 1 and Table 1 present the gas chromatography-mass spectrometry (GC-MS) chromatogram and the identified components in the biodiesel, respectively, when produced using an eggshell based CaO catalyst. The FAME composition of biodiesel from WCO contained a majority of mixture of saturated and unsaturated esters which were methyl palmitate, methyl stearate and methyl linoleate and methyl elaidate respectively. This result is similar with previous study by Farooq and Ramli [25] where they reported their FAME composition from low free fatty acid WCO in presence of chicken bone based catalyst. Additionally, the findings are corroborated by previous research that examined the free fatty acid (FFA) content in WCO before transesterification, identifying the presence of linoleic, palmitic, and stearic acids [26]. As revealed by the GC-MS analysis, microwave heating transesterification can facilitate the conversion of oil into methyl ester, also known as biodiesel, in a manner comparable to conventional assisted process.

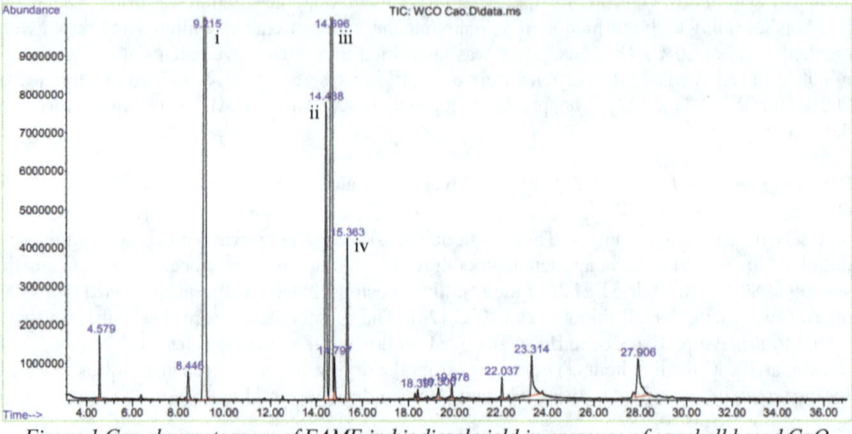

Figure 1 Gas chromatogram of FAME in biodiesel yield in presence of eggshell based CaO catalyst

Table 1 Identified FAME of WCO biodiesel in presence of eggshell based CaO catalyst

Peak Label	Area (%)	IUPAC Name	Common Name
i	35.13	Methyl hexadecanoate	Methyl palmitate
ii	12.10	Methyl 9,12-Octadecadienoate	Methyl linoleate
iii	34.36	Methyl (E)-octadec-9-enoate	Methyl elaidate
iv	4.93	Methyl octadecanoate	Methyl stearate

Fig. 2 and Table 2 display the chromatogram from GC-MS and type of components presence in biodiesel from GC-MS in presence of eggshell based CaO/Mg-Al catalyst respectively. The FAME composition was the same as the biodiesel production in presence of eggshell based CaO alone with slightly different in percentage area. The CaO/Mg-Al catalyst showed better catalytic activity as the FAME content in its biodiesel is higher than eggshell based CaO catalyst. This data can be supported by previous work [27] where CaO impregnated with Mg generated higher biodiesel yield compare to pure CaO.

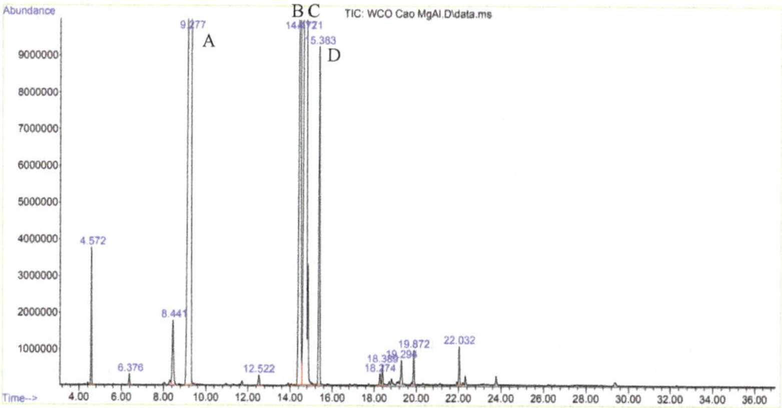

Figure 2 Gas chromatogram of FAME in biodiesel yield in presence of eggshell based CaO/Mg-Al catalyst

Table 2 Identified FAME of WCO biodiesel in presence of eggshell based CaO/Mg-Al catalyst

Peak Label	Area (%)	IUPAC Name	Common Name
A	34.55	Methyl hexadecanoate	Methyl palmitate
B	15.60	Methyl 9,12-Octadecadienoate	Methyl linoleate
C	36.20	Methyl (E)-octadec-9-enoate	Methyl elaidate
D	5.73	Methyl octadecanoate	Methyl stearate

Influence of variance of catalyst in biodiesel yield via microwave heating transesterification

Determining the yield of biodiesel generally involves accurately measuring the conversion of triglycerides found in the selected feedstock, such as waste cooking oil, into fatty acid methyl esters (FAME), which are the essential components of biodiesel. Based on Fig. 3, biodiesel yield from eggshell based CaO/Mg-Al catalyst was higher with value of 89.44% compared to eggshell based CaO alone (88.04%). Even though the difference is small most probably due to low percentage of hydrotalcite impregnated with CaO, it is still worthy to note that the modified catalyst is more stable and has higher basicity [27], [28]. Hsiao et al. [26] also reported high biodiesel yield (98.20%) from WCO when applying CaO modified with bromooctane via similar heating method. They achieved a higher biodiesel yield than the present study, most likely due to the longer reaction time of 75 minutes compared to only 30 minutes in this study. These outcomes make it clear that modified CaO catalyst, as opposed to pure eggshell based CaO catalyst, provide a more efficient and environmentally friendly option for producing biodiesel from WCO.

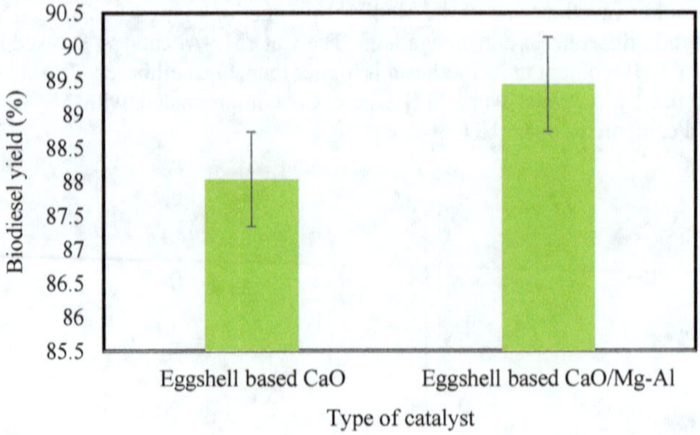

Figure 3 Biodiesel yield from WCO in presence of different type of catalyst

Conclusion

In short, the present work demonstrated that modification of eggshell based CaO with hydrotalcites hold better stability and higher catalytic activity than pure eggshell based CaO. The findings showed that biodiesel production from WCO in presence of the modified catalyst was 89.44% and better FAME composition under conditions of methanol to oil molar ratio of 9:1, at 65°C within 30 minutes of reaction time. Plus, microwave heating aided in shorter reaction time and lower energy consumption. Hence, the usage of modified CaO catalyst and microwave energy can offer enhanced stability, improved catalytic performance, and environmental benefits that not solely improve biodiesel yield and quality, but also aid in the development of more environmentally friendly biodiesel production processes.

Acknowledgement

This research was made possible by the kind support of the National Collaborative Research Fund (015MD0-190) which provided the financial resources needed to carry out this investigation.

References

[1] M. Allen, M. Babiker, and Y. Chen, "Summary of Policymakers. In: Global Warming of 1.5°C. An IPCC Special Report on the impacts of global warming of 1.5°C above pre-industrial levels and related global greenhouse gas emission pathways, in the context of strengthening the

global response to ," 2018. [Online]. Available: https://www.cambridge.org/core/product/identifier/9781009157940%23prf2/type/book_part

[2] M. Allen, M. Babiker, and Y. Chen, "Summary of Policymakers. In: Global Warming of 1.5°C. An IPCC Special Report on the impacts of global warming of 1.5°C above pre-industrial levels and related global greenhouse gas emission pathways, in the context of strengthening the global response to ," 2018.

[3] M. K. Pasha, L. Dai, D. Liu, M. Guo, and W. Du, "An overview to process design, simulation and sustainability evaluation of biodiesel production," *Biotechnol. Biofuels*, vol. 14, no. 1, pp. 1–23, 2021. https://doi.org/10.1186/s13068-021-01977-z

[4] UFOP, "Biodiesel 2014/2015: Report on the current situation and prospects," 2016.

[5] UFOP, "Report on Global Market Supply: 2019/2020," 2020. [Online]. Available: https://www.ufop.de/files/4815/4695/8891/WEB_UFOP_Report_on_Global_Market_Supply_18-19.pdf

[6] UFOP, "Report on Global Market Supply: 2019/2020," 2020.

[7] Z. Ullah, M. A. Bustam, and Z. Man, "Biodiesel production from waste cooking oil by acidic ionic liquid as a catalyst," *Renew. Energy*, vol. 77, pp. 521–526, 2015. https://doi.org/10.1016/j.renene.2014.12.040

[8] Y. S. Erchamo, T. T. Mamo, G. A. Workneh, and Y. S. Mekonnen, "Improved biodiesel production from waste cooking oil with mixed methanol–ethanol using enhanced eggshell-derived CaO nano-catalyst," *Sci. Rep.*, vol. 11, no. 1, pp. 1–12, 2021. https://doi.org/10.1038/s41598-021-86062-z

[9] M. H. Nazir *et al.*, "Development of lignin based heterogeneous solid acid catalyst derived from sugarcane bagasse for microwave assisted-transesterification of waste cooking oil," *Biomass and Bioenergy*, vol. 146, 2021. https://doi.org/10.1016/j.biombioe.2021.105978

[10] Y. H. Tan, M. O. Abdullah, J. Kansedo, N. M. Mubarak, Y. S. Chan, and C. Nolasco-Hipolito, "Biodiesel production from used cooking oil using green solid catalyst derived from calcined fusion waste chicken and fish bones," *Renew. Energy*, vol. 139, no. November 2014, pp. 696–706, 2019. https://doi.org/10.1016/j.renene.2019.02.110

[11] B. H. H. Goh *et al.*, "Progress in utilisation of waste cooking oil for sustainable biodiesel and biojet fuel production," *Energy Convers. Manag.*, vol. 223, no. May, 2020. https://doi.org/10.1016/j.enconman.2020.113296

[12] E. E. Çakırca, G. N Tekin, O. İlgen, and A. N Akın, "Catalytic activity of CaO-based catalyst in transesterification of microalgae oil with methanol," *Energy Environ.*, vol. 30, no. 1, pp. 176–187, 2019. https://doi.org/10.1177/0958305X18787317

[13] S. M. Pavlović *et al.*, "A CaO/zeolite-based catalyst obtained from waste chicken eggshell and coal fly ash for biodiesel production," *Fuel*, vol. 267, no. January, p. 117171, 2020. https://doi.org/10.1016/j.fuel.2020.117171

[14] K. N. Krishnamurthy, S. N. Sridhara, and C. S. Ananda Kumar, "Optimization and kinetic study of biodiesel production from Hydnocarpus wightiana oil and dairy waste scum using snail shell CaO nano catalyst," *Renew. Energy*, vol. 146, pp. 280–296, 2020. https://doi.org/10.1016/j.renene.2019.06.161

[15] S. A. Mohd Johari *et al.*, "Utilization of Dairy Scum Waste as a Feedstock for Biodiesel Production via Different Heating Sources for Catalytic Transesterification," *ChemBioEng Rev.*, no. 6, pp. 1–28, 2022. https://doi.org/10.1002/cben.202200003

[16] S. A. Mohd Johari, M. A. Ahmad Farid, M. Ayoub, N. A. Rashidi, and Y. Andou, "Optimization and kinetic studies for biodiesel production from dairy waste scum oil via microwave assisted transesterification," *Environ. Technol. Innov.*, vol. 34, p. 103580, 2024. https://doi.org/10.1016/j.eti.2024.103580

[17] A. Cancela, R. Maceiras, S. Urrejola, A. Sanchez, C. Lagoas-marcosende, and E. N. Militar, "Microwave-Assisted Transesterification of Macroalgae," *Energies*, pp. 862–871, 2012. https://doi.org/10.3390/en5040862

[18] W. Xu, J. Zhou, Z. Su, Y. Ou, and Z. You, "Microwave catalytic effect: A new exact reason for microwave-driven heterogeneous gas-phase catalytic reactions," *Catal. Sci. Technol.*, vol. 6, no. 3, pp. 698–702, 2016. https://doi.org/10.1039/c5cy01802a

[19] G. Yadav, N. Yadav, and M. Ahmaruzzaman, "Microwave-assisted sustainable synthesis of biodiesel on Oryza sativa catalyst derived from agricultural waste by esterification reaction," *Chem. Eng. Process. - Process Intensif.*, vol. 187, no. December 2022, p. 109327, 2023. https://doi.org/10.1016/j.cep.2023.109327

[20] S. A. Mohd Johari, M. A. Ahmad Farid, M. Ayoub, N. A. Rashidi, and Y. Andou, "Optimization and kinetic studies for biodiesel production from dairy waste scum oil via microwave assisted transesterification," *Environ. Technol. Innov.*, vol. 34, p. 103580, 2024. https://doi.org/10.1016/j.eti.2024.103580

[21] M. Khatibi, F. Khorasheh, and A. Larimi, "Biodiesel production via transesterification of canola oil in the presence of Na–K doped CaO derived from calcined eggshell," *Renew. Energy*, vol. 163, pp. 1626–1636, 2021. https://doi.org/10.1016/j.renene.2020.10.039

[22] N. Kurnia Julianti, T. Kusuma Wardani, I. Gunardi, and A. Roesyadi, "Effect of Calcination at Synthesis of Mg-Al Hydrotalcite Using co-Precipitation Method," *J. Pure Appl. Chem. Res.*, vol. 6, no. 1, pp. 7–13, 2017. https://doi.org/10.21776/ub.jpacr.2017.006.01.280

[23] M. Ayesha *et al.*, "Sorption enhanced steam reforming of methane over waste-derived CaO promoted MgNiAl hydrotalcite catalyst for sustainable H2production," *J. Environ. Chem. Eng.*, vol. 10, no. 3, p. 107651, 2022. https://doi.org/10.1016/j.jece.2022.107651

[24] V. Kavitha, V. Geetha, and P. J. Jacqueline, "Production of biodiesel from dairy waste scum using eggshell waste," *Process Saf. Environ. Prot.*, vol. 125, pp. 279–287, 2019. https://doi.org/10.1016/j.psep.2019.03.021

[25] M. Farooq and A. Ramli, "Biodiesel production from low FFA waste cooking oil using heterogeneous catalyst derived from chicken bones," *Renew. Energy*, vol. 76, pp. 362–368, 2015. https://doi.org/10.1016/j.renene.2014.11.042

[26] M. C. Hsiao, J. Y. Kuo, S. A. Hsieh, P. H. Hsieh, and S. S. Hou, "Optimized conversion of waste cooking oil to biodiesel using modified calcium oxide as catalyst via a microwave heating system," *Fuel*, vol. 266, no. October 2019, p. 117114, 2020. https://doi.org/10.1016/j.fuel.2020.117114

[27] M. Fan, Y. Liu, P. Zhang, and P. Jiang, "Blocky shapes Ca-Mg mixed oxides as a water-resistant catalyst for effective synthesis of biodiesel by transesterification," *Fuel Process. Technol.*, vol. 149, pp. 163–168, 2016. https://doi.org/https://doi.org/10.1016/j.fuproc.2016.03.029

[28] W. Xie, H. Peng, and L. Chen, "Calcined Mg-Al hydrotalcites as solid base catalysts for methanolysis of soybean oil," *J. Mol. Catal. A Chem.*, vol. 246, no. 1–2, pp. 24–32, 2006. https://doi.org/10.1016/j.molcata.2005.10.008.

Optimizing CO_2 storage efficiency using formic acid intervention for salt precipitation management: insights from static batch experiments and core flooding studies

VICTOR Darkwah-Owusu[1,2 a *], MD YUSOF Muhammad Aslam[1,2,b *], NGUKU Anita[1,2,c]

[1]Petroleum Engineering Department, Faculty of Geoscience and Petroleum Engineering, Universiti Teknologi PETRONAS, 32610 Bandar Seri Iskandar, Perak Darul Ridzuan, Malaysia

[2]Centre of Reservoir Dynamics (CORED), Universiti Teknologi PETRONAS, 32610 Bandar Seri Iskandar, Perak Darul Ridzuan, Malaysia

[a]victor_22009917@utp.edu.my, [b]aslam.myusof@utp.edu.my, [c]anita_22001594@utp.edu.my

Keywords: Geological CO_2 Storage, CO_2 Injectivity Improvement, Salt Precipitation, Formation Damage

Abstract. Geological CO_2 storage is a vital technique to reduce the global carbon footprint. However, this technique is hindered by frequent salt precipitation due to the CO_2 injection. This study experimentally evaluates the impact of formic acid as salt precipitation mitigation products to improve CO_2 injectivity, through core flooding experiments on selected core samples, and static batch experiments. The main hypothesis is to verify the claim if formic acid can improve CO_2 injectivity in salt-induced formation damage. The experiments were conducted in two stages using a 7wt% brine and 17wt% brine at conditions of 1600psi and 60 °C. The results were compared with freshwater, which is a conventional technique used to dissolve the salts. Also, SEM images were taken at each of the stages to understand the morphological and petrographical changes pre-and-post treatment. Results from this study showed salt precipitation led to a reduction in porosity and permeability. Specifically, porosity decreased by 25-30%, while permeability decreased by up to 72% across all tested scenarios post CO_2 injection. Also, the salting-out effect was more pronounced in high salinity brine compared to low salinity brine. Injection of remediation fluids during tests effectively opened pore spaces and throats in core samples, resulting in increased permeability. After using reactive fluids, permeability improved by up to 16% compared to the sample's initial permeability. Based on the SEM analysis of the core samples pre-and-post treatment, it was found that, salt precipitation not only blocked the pore spaces within the core matrix, but it also accumulated significantly around the rock grains, forming aggregates of salts. Also, after flooding with the treatment fluid, the SEM results reveal increased pore throats and salt dissolution. This observation shows the possibility of using formic acid as a treatment fluid for the reduction of salt precipitation occurring in both the pore spaces and pore throats. By understanding the intricate interplay between salt precipitation and rock behavior after using reactive fluids, the efficiency and sustainability of geological CO_2 storage can be enhanced.

Introduction

The combustion of fossil fuels for energy generation releases greenhouse gases, primarily carbon dioxide (CO_2), which is currently the major contributor to global warming and climate change [1]. To combat this pressing issue, Carbon capture and storage (CCS) is a transformative technology that has been identified for reducing atmospheric CO_2. To bolster efforts to decarbonize using CCS technology, selecting suitable geological formations for CCS is crucial for its effective implementation. However, during injection into these geological reservoirs, injectivity issues arise due to one or more of the following factors: (a) Brine vaporization near the wellbore leading to precipitation of salts, thus hindering further CO_2 flow [2] (b) convective multiphase flow, leading to acidified brine [3, 4] (c) Geochemical CO_2-brine-rock interactions which can result in fines clogging, causing further injectivity decline due to the coupled effect of salt precipitation and fines migration. [5-7].

In an ideal CO_2 injection case, CCS project operators aim to inject CO_2 at maximum injection rates in order to store more representative volumes of CO_2 in the subsurface geological reservoir [8, 9]. Most at times, the aim is to reduce the number of CO_2 injection wells since CO_2 sequestration in its current stages is not economically feasible. This practice, if not curtailed at the early stages of injection could lead to salt precipitation. This phenomenon arises from the complex interplay between gas-liquid interactions and mineral crystallization processes. As CO_2 interacts with formation brine, water molecules undergo evaporation and subsequent diffusion into the CO_2 phase. This phenomenon triggers the precipitation of previously dissolved salts within the formation water [10, 11]. The precipitated salts then migrate under the influence of the injected CO_2 flow, accumulating within pore throats near the wellbore [12]. This accumulation significantly reduces reservoir permeability, hindering CO_2 injectivity. While accurately predicting salt distribution and quantifying reservoir alterations remain ongoing challenges, a thorough understanding of this precipitation mechanism is critical for optimizing CO_2 injection strategies in saline formations.

Currently, most studies on salt precipitation during CO_2 injection in saline aquifers have predominantly focused on elucidating mechanisms and quantifying the resulting damage [2, 11, 13-15]. However, relatively few have delved into effective mitigation strategies for this challenge. Furthermore, existing mitigation methods have often demonstrated only temporary efficacy, with injectivity experiencing a decline upon the resumption of CO_2 injection. For instance, Kleinitz et al. [16] attempted to enhance injectivity through a freshwater treatment method incorporating 3% butyl glycol. However, this approach failed to yield the anticipated results, with some treatments proving entirely ineffective. Similarly, Smith et al. [17] pursued injectivity improvement by employing fresh water injection, reporting a remarkable 263% permeability enhancement post-treatment. Nevertheless, injectivity deteriorated rapidly upon the resumption of CO_2 injection, hinting at potential salt recrystallization issues. In a different approach, Grude et al. [18] proposed a weekly injection strategy utilizing a 90:10 mixture of methyl ethyl glycol and water to mitigate salt precipitation. Although their laboratory experiments initially showed promise, they observed a subsequent decrease in permeability. Additionally, early-stage injection pressure data suggested the possibility of near-well plugging, indicating potential challenges with this approach. These studies underscore the complexity of addressing salt precipitation during CO_2 injection and highlight the need for further research to develop robust and sustainable mitigation strategies that can effectively maintain injectivity over extended operational periods.

As previously elucidated, existing studies have investigated mechanism of salt precipitation due to CO_2 injection. However, a critical knowledge gap remains regarding core-scale investigation of injectivity improvement techniques. Understanding the core-scale dissolution characteristics will go a long way in helping to design improved injection scenarios. Thus, this study addresses this gap by evaluating the effects of these reactive fluids on permeability changes during the treatment process. Our research aims to assess the impact of the use of the reactive fluids on improving CO_2 injectivity flow using both static and dynamic core-flooding approach.

Experimental Approach

Materials. Because this is a preliminary study to verify the idea of using formic acid to mitigate salt precipitation in saline aquifers, two Berea sandstone cores were used to investigate the effect of formic acid on the salt precipitation and another two Berea core samples were used as the benchmark. In total 4 core samples were used for the experimental core flood. Also, four core plugs (with a 10mm by 25 mm dimension) were used for the static experiment. The main treatment fluid used in this study is formic acid (10wt%). Two different brine salinities of 17wt% and 7wt NaCl brine were used as the formation fluid and also tor measuring the rock permeability. A pure CO_2 with 99.5% purity was used as the working fluid. The CO_2 was injected in the supercritical phase at 1600psi and 60C.

Experimental Procedure. For the experimental procedure, a summary of this process is shown in Fig 1.

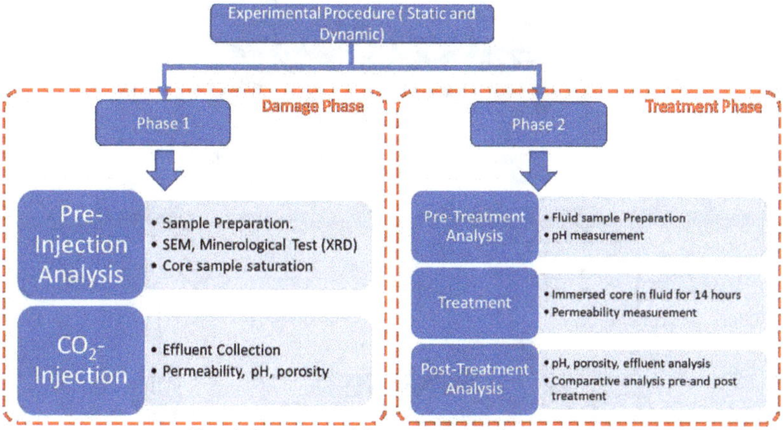

Fig. 1 Overview of the experimental procedure

Static Test. For the static tests, the average core plug dimensions were 10 mm in width and 25 mm in length. These core plugs underwent vacuum saturation in the designated brine concentration using a vacuum desiccator at a pressure of about 800 psi for 24 hours in order to ensure complete saturation with the desired fluid. Following saturation with the intended brine samples, the core plugs were placed in a Fann aging cell, where supercritical CO_2 ($scCO_2$) was injected to induce damage. A schematic of the aging cell test setup is depicted in Fig 3. After core damage was induced, Field Emission Scanning Electron Microscopy (FESEM) imaging was employed to assess the extent of pore blockage by salt precipitates. Subsequently, the core plugs underwent a 12-14 hour saturation process with various treatment fluids using the previously described methodology. A subsequent FESEM analysis was conducted to evaluate the impact of the acids on the intricate pore-scale dissolution behavior resulting from acid-salt interactions and potential in-situ geochemical reactions involving the acid, salt, and rock components. The Fann aging cells placed in the oven are illustrated in Fig 2.

Materials Research Proceedings 53 (2025) 125-135 https://doi.org/10.21741/9781644903575-12

Fig 2 Fann Aging Cells used for the experiment

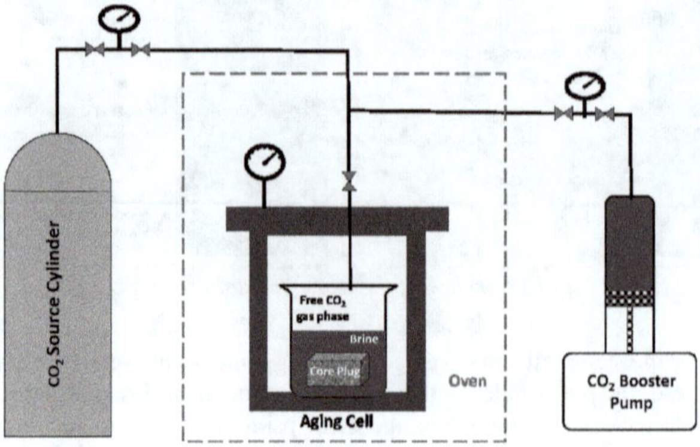

Fig. 3 Schematic of the aging cell for the static test

Core flooding Experiment. The Relative Permeability System coreflood equipment was used to carry out the injectivity experiments in this study. A schematic of the apparatus used in this study is shown in Fig 4. As mentioned earlier, two Berea sandstone cores were used to investigate the effect of formic acid on salt precipitation, and another two Berea core samples were used as benchmarks. The experimental phase was done in two phases- (a) a phase to create the damage and (b) the phase to treat the damage. The clean core sample was saturated inside the NaCl brine for about 24 hours in order to ensure full saturation, in a vacuum desiccator at 800 psi for about 4 hours. The sample was left overnight to ensure full saturation. The core sample was then wrapped in rubber sleeve and aluminum foil to prevent leakage, and mounted into the core holder.

In the damage phase, the NaCl brine saturated core is mounted into the core holder. The injection scheme consisted of the following procedures:

- The initial gas permeability of the clean, dry core sample was determined using the poro-perm unit.
- Brine concentrations of 7wt% and 17wt% were prepared, depending on the specific experimental run. The initial brine permeability was assessed by injecting 2PV brine

(approximately 2 hours of brine injection) and calculating the resulting liquid permeability using the Darcy equation.

- In order to replicate the convective mixing of CO_2 and brine, carbonated water with an 80% brine to 20% CO_2 ratio was injected at a controlled rate of 0.5 cc/min.
- Following this, the saturated core sample underwent flooding with supercritical CO2 at a regulated rate of 2 cm^3/min to displace the aqueous phase. The differential pressure was continuously monitored using pressure transducers throughout the core flooding experiment.
- The final permeability and injectivity of the affected core were evaluated using brine. This assessment was conducted once the introduced brine achieved a stabilized pressure drop, not exceeding a maximum of 2 PV.

For the treatment phase, the main treatment fluid used was formic acid. The samples were soaked for 14 hours within a vacuum desiccator to comprehend the dissolution dynamics. Pump pressure was regulated to 800psi to facilitate the migration of acids into the pore spaces. Upon completion of the soaking period, the core was reinstated into the coreholder and re-pressurized to its original conditions. Liquid permeability was then measured to assess the degree of injectivity improvement. Additionally, the core was dried subsequent to the measurement of liquid permeability post-treatment. To enable comparison with the initial gas and liquid permeability, the gas permeability of the dried core was gauged. A comparative analysis was conducted between the alterations in porosity and permeability of the treated core and the damaged core to ascertain the efficacy of the dissolver. Furthermore, the mass of the dried core was determined and juxtaposed with the initial undamaged core.

To quantify the extent of dissolution during CO_2-rock-brine and acidizing experiments, the mass change of each core sample was measured. This approach allows for a direct correlation between the pre- and post-experiment porosity changes and the amount of material dissolved. The pH of the brine samples was assessed before and after the experiment to monitor any potential changes.

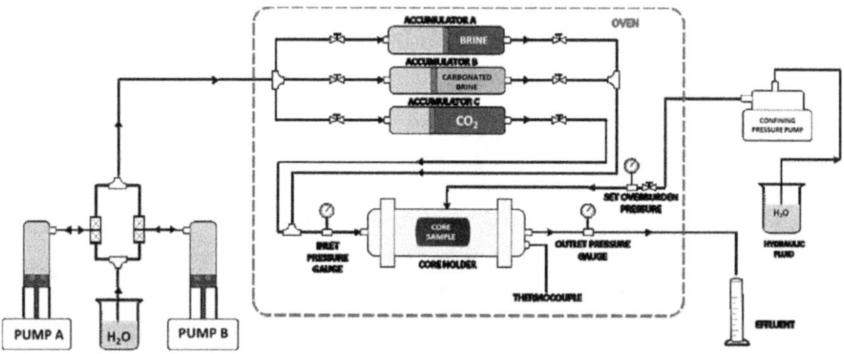

Fig. 4 Schematic of the coreflooding apparatus

Results and Discussion

Mineralogical and Petrographical Assessment. The comprehensive characterization of the core samples was achieved through a synergistic combination of analytical techniques, including X-ray Fluorescence (XRF), X-ray Diffraction (XRD), and Field Emission Scanning Electron Microscopy

(FESEM). XRF analysis facilitated the elucidation of the elemental composition of the core samples, offering precise quantification of constituent elements. This method enabled not only the identification of major elements but also the detection of trace elements that may play crucial roles in various geochemical processes. Also, XRD analysis provided invaluable insights into the mineral phases present within the core samples and their relative distributions. By utilizing oxide concentrations, as presented in Table 1 and Fig 5, mineral phases were identified, shedding light on the mineral components of the Berea sandstone cores used. The XRD results highlighted the prevalence of minerals such as Quartz, Kaolinite, Illite, Dolomite, Siderite, alongside various clay minerals and trace elements. Moreover, FESEM imaging offered detailed examinations of surface morphology and pore size distribution within the core samples. These high-resolution images provided insights into the microstructural features of the rocks, unveiling intricate pore networks, and the grain boundaries and other morphological features. The FESEM analysis, as illustrated in Fig 6, complemented the XRF and XRD data, enhancing our understanding of the petrophysical properties of the Berea sandstone samples used in this study.

Table 1. Mineralogical Composition of The Core Sample Based on XRF/XRD Data

Oxide Concentration								
Formula (by oxide concentration)	SiO_2	Al_2O_3	K_2O	Fe_2O_3	P_2O_5	TiO_2	CaO	Other mineral oxides
Concentration (%)	84.82	6.74	2.64	2.11	1.30	0.84	0.75	< 0.80

Fig. 3 Result from XRD intensity plots

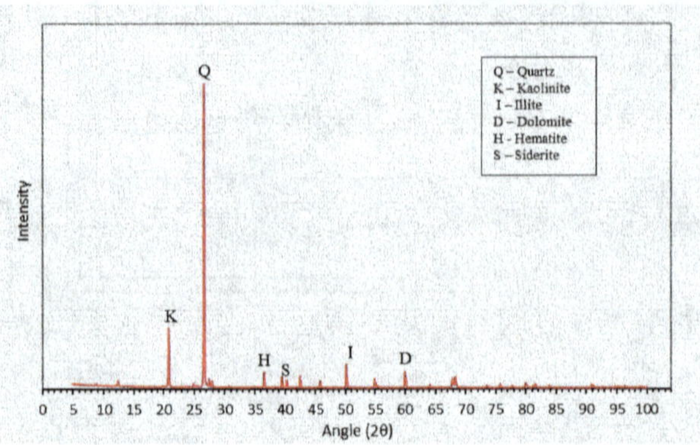

Fig. 6 SEM images prior to CO_2 injection

Results from Static Test – Insights derived from FESEM Results. Fig 7 shows the FESEM results post CO_2 injection which shows the evidence of salt precipitation post-CO_2 injection. The presence of salt precipitation in the porous media suggests that CO_2 injection has led to chemical reactions with the NaCl brine in the presence of the rock minerals. The complex structures observed may include salt crystals, mineral deposits, as highlighted in red. These crystals seem to form on the rock grain surface and sometime comingle with the rock minerals.

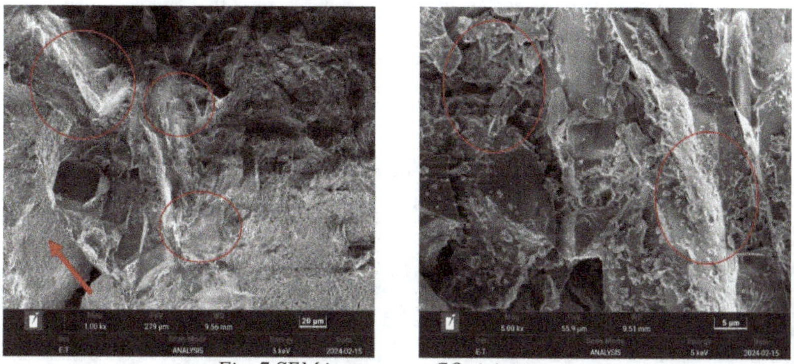

Fig. 7 SEM images post CO_2 injection

After this stage, the damaged cores were treated with the formic acid. FESEM results from this stage is shown in Fig 8. The arrows indicate areas of improved flow. This suggests that the formic acid may have helped to dissolve or remove some of the salt precipitates, potentially enhancing the rock's permeability. This observation proves the initial hypothesis that can potentially dissolve some of the precipitated salts, particularly those more susceptible to dissolution. This could lead to the reopening of previously blocked pore throats and an improvement in permeability.

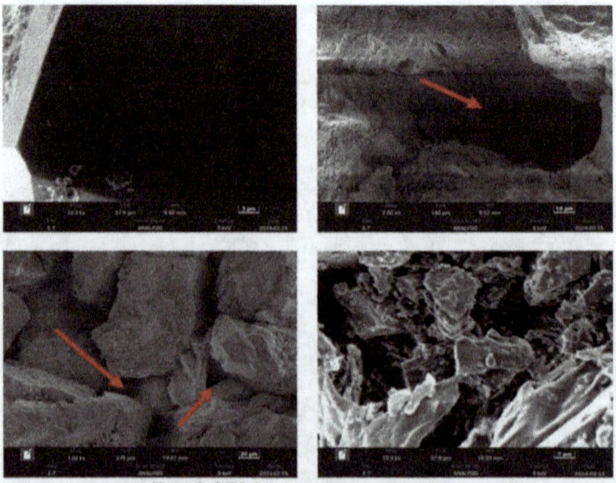

Fig. 8 SEM images post treatment

Results From Core Flooding Experiment. The summary of the dynamic test is presented in Table 2. The initial brine permeability ranged from 154 mD to 474 mD across the different core samples. After CO_2 injection, there was a notable reduction in permeability observed in all samples, with the percentage of permeability damage ranging from 45% to 75%. This reduction can be attributed to the damage caused by the drying effect due continuous CO_2 drying of the in-situ brine, causing salt precipitation when the saturation limit is reached. This interaction of the CO_2-saturated brine crystal along with the rock matrix, resulted in pore blockage or alteration of pore structure. Post-treatment with formic acid (FA) resulted in a significant improvement in permeability compared to the damaged. The permeability post-treatment ranged from 66 mD to 474 mD, indicating varying degrees of improvement across the samples. The percentage improvement compared to damage ranged from 54% to 317%, demonstrating the effectiveness of FA in enhancing permeability by mitigating the damage caused by CO_2 injection.

Table 2. Summary of the Various Coreflooding Results Pre CO_2 Injection, Pre-Treatment and Post-Treatment

Core sample number	Initial rock properties				Rock properties post damage		Properties post treatment				Analysis post-treatment	
	Mass	por (%)	Brine salinity (wt%)	k_i (mD)	k_d (mD)	% damage	Treatment Fluid	k_f (mD)	% improved compared to damage	% compared to initial k_i	dry mass (mD)	% change
S-1	188.175	20.52	17	154	39	75	FW	66	71	-57	185.5	-1.42
S-2	186.736	20.83	7	234	87	63	FW	133	54	-43	165.716	-11.3
S-3	179.22	20.40	17	334	94	72	FA (10wt%)	390	317	17	183.34	2.30
S-4	181.39	21.30	7	474	271	45	FA (10wt%)	474	75	0.00	182.25	0.48

K_i – initial brine permeability, K_d – permeability post CO_2 injection, K_f – Permeability post treatment, FW – Freshwater, FA – Formic Acod

A comparative analysis of freshwater (FW) and FA treatment reveals the superior performance of FA over FW in enhancing injectivity. Samples treated with formic acid exhibited remarkable improvements, with permeability values surpassing not only the damaged state but also the initial brine permeability. This is particularly evident in samples S-3 and S-4 treated with formic acid, where the permeability post-treatment surpassed the initial injectivity values by 17% and 0%, respectively. In contrast, samples treated with freshwater (S-1 and S-2) showed moderate improvements, although less pronounced compared to formic acid treatment. This phenomenon underscores the potential of formic acid treatment to not only restore but also improve the intrinsic permeability properties of the rock matrix after salt precipitation induced damage, especially in highly saline formations. Furthermore, the injectivity plots in Fig 9 confirm the flow behavior in each stage of the experiment. From Fig 9, there is a high pressure drop (dP) at the initial stages of injection due to the presence of salt precipitation. The dP decreases with time due to the fact that brine is used for the permeability measurement. The brine dissolves some of the solid cystals, thus reducing the pressure drop.

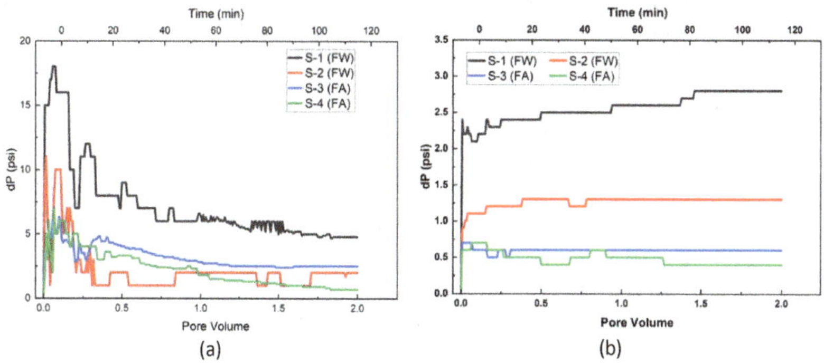

Fig 9. pressure drop profile of the various core samples (a) damage phase (b) treatment

Conclusion

The present study investigates the potential of using formic acid to improve injectivity during CO_2 injection. Key findings related to injectivity improvement are summarized below:

a) CO_2 injection led to salt precipitation, which caused a decline in injectivity due to geochemical reactions between the CO_2-brine and rock minerals.
b) The formic acid treatment applied to the core samples after CO_2 injection appears to have improved flow characteristics. We hypothesize that due to the reactive nature of formic acid, it is able to contact the plugged pore spaces, dissolves the minerals and thereby reducing capillary pressure. This in turn led to a greater contact area and volumetric sweep thus improving the overall flow.
c) The formic acid treatment was particularly effective in high salinity conditions, suggesting that its effectiveness as a halite dissolver is salinity-dependent (i.e., the initial brine salinity has a greater impact on the dissolution kinetics).

By restoring and even surpassing initial permeability levels in some cases, formic acid treatment demonstrates a promising approach to mitigate the negative effects of CO_2 injection on injectivity. Improved injectivity facilitates enhanced CO_2 storage capacity, providing a pathway to inject more CO_2 volumes to achieve net-zero emissions targets.

Acknowledgement
The first author would like to express gratitude to Universiti Teknologi PETRONAS for awarding the Graduate Assistantship. Special thanks are extended to Yayasan Universiti Teknologi PETRONAS (YUTP) for funding this research through the YUTP-FRG grant (015LC0-572).

Appreciation is also extended to the Centre of Reservoir Dynamics (CORED) under the Institute of Hydrocarbon Recovery (IHR) at UTP for providing the academic resources and computing facilities essential for the successful completion of this study. The World Engineering, Science & Technology Congress (ESTCON) committee is acknowledged for the opportunity to present this manuscript and for organizing the conference.

References

[1] IEA, "CCUS in Clean Energy Transitions, IEA, Paris," 2020. [Online]. Available: https://www.iea.org/reports/ccus-in-clean-energy-transitions

[2] Y. Peysson, L. André, and M. Azaroual, "Well injectivity during CO2 storage operations in deep saline aquifers-Part 1: Experimental investigation of drying effects, salt precipitation and capillary forces," (in English), Int J Greenh Gas Con, vol. 22, pp. 291- 300, Mar 2014, https://doi.org/10.1016/j.ijggc.2013.10.031

[3] H. Emami-Meybodi, H. Hassanzadeh, C. P. Green, and J. Ennis-King, "Convective dissolution of CO2 in saline aquifers: Progress in modeling and experiments," (in English), Int J Greenh Gas Con, vol. 40, pp. 238-266, Sep 2015, https://doi.org/10.1016/j.ijggc.2015.04.003.

[4] C. W. MacMinn, M. L. Szulczewski, and R. Juanes, "CO2 migration in saline aquifers. Part 2. Capillary and solubility trapping," (in English), J Fluid Mech, vol. 688, pp. 321-351, Dec 2011, https://doi.org/10.1017/jfm.2011.379

[5] M. A. Md Yusof et al., "Predictive Modelling of CO2 Injectivity Impairment due to Salt Precipitation and Fines Migration During Sequestration," in International Petroleum Technology Conference, 2021, vol. Day 1 Tue, March 23, 2021, D012S045R156, doi: 10.2523/iptc-21483-ms. [Online]. Available: https://doi.org/10.2523/IPTC-21483-MS

[6] M. Torsæter and P. Cerasi, "Geological and geomechanical factors impacting loss of near-well permeability during CO2 injection," Int J Greenh Gas Con, vol. 76, pp. 193-199, 2018. https://doi.org/10.1016/j.ijggc.2018.07.006

[7] F. Othman, M. Yu, F. Kamali, and F. Hussain, "Fines migration during supercritical CO2 injection in sandstone," (in English), J Nat Gas Sci Eng, vol. 56, pp. 344-357, Aug 2018, https://doi.org/ 10.1016/j.jngse.2018.06.001

[8] G. Cui, Z. Hu, F. Ning, S. Jiang, and R. Wang, "A review of salt precipitation during CO2 injection into saline aquifers and its potential impact on carbon sequestration projects in China, Fuel, vol. 334, pp. 126615-126615, 2023, https://doi.org/10.1016/j.fuel.2022.126615

[9] B. Wei, B. W. Wang, X. Li, M. Aishan, and Y. Ju, "CO2 storage in depleted oil and gas reservoirs: A review," (in English), Advances in Geo-Energy Research, vol. 9, no. 2, pp. 76-93, Aug 2023, https://doi.org/10.46690/ager.2023.08.02

[10] S. H. Hajiabadi, P. Bedrikovetsky, S. Borazjani, and H. Mahani, "Well Injectivity during CO2 Geosequestration: A Review of Hydro-Physical, Chemical, and Geomechanical Effects," (in English), Energ Fuel, vol. 35, no. 11, pp. 9240-9267, Jun 3 2021, https://doi.org/10.1021/acs.energyfuels.1c00931

[11] J. Ren and D. Feng, "Sensitivity Analysis Strategy to Assess the Salting-Out Problem during CO2 Geological Storage Process," (in English), Shock and Vibration, vol. 2021, pp. 1-14, Dec 21 2021, https://doi.org/10.1155/2021/2809467

[12] Y. A. Sokama-Neuyam et al., "Experimental and theoretical investigation of the mechanisms of drying during CO(2) injection into saline reservoirs," Sci Rep, vol. 13, no. 1, p. 9155, Jun 6 2023, https://doi.org/10.1038/s41598-023-36419-3

[13] S. M. Roels, H. Ott, and P. U. Zitha, "μ-CT analysis and numerical simulation of drying effects of CO2 injection into brine-saturated porous media," (in English), Int J Greenh Gas Con, vol. 27, pp. 146-154, Aug 2014, https://doi.org/10.1016/j.ijggc.2014.05.010

[14] R. Shaibu, Y. A. Sokama-Neuyam, and J. R. Ursin, "A Theoretical Study of the Effect of Salt Precipitation on CO2 Injectivity," presented at the Day 2 Thu, February 08, 2018, 2018. [Online]. Available: https://doi.org/10.2118/189470-MS

[15] M. Nooraiepour, H. Fazeli, R. Miri, and H. Hellevang, Salt Precipitation during Injection of CO2 into Saline Aquifers: Lab-on-chip Experiments on Glass and Geomaterial Microfluidic Specimens. 2018.

[16] W. Kleinitz, G. Dietzsch, and M. Köhler, "Halite Scale Formation in Gas-Producing Wells," Chemical Engineering Research and Design, vol. 81, no. 3, pp. 352-358, 2003, https://doi.org/0.1205/02638760360596900

[17] N. Smith et al., "Quest CCS facility: Halite damage and injectivity remediation in CO2 injection wells," (in English), Int J Greenh Gas Con, vol. 119, pp. 103718-103718, Sep 2022, https://doi.org/10.1016/j.ijggc.2022.103718

[18] S. Grude, M. Landro, and J. Dvorkin, "Pressure effects caused by CO2 injection in the Tubaen Fm., the Snohvit field," (in English), Int J Greenh Gas Con, vol. 27, pp. 178-187, Aug 2014, https://doi.org/10.1016/j.ijggc.2014.05.013

Materials Research Proceedings 53 (2025) 136-145 https://doi.org/10.21741/9781644903575-13

Removal of phosphate from aqueous solution through synergistic extractive extraction

Izzat Naim Shamsul Kahar[1,a], Shuhada A. Idrus-Saidi[1,2,b], Musyazrina Alya Musa[1,c], Norasikin Othman[1,2,d*] and Aishah Rosli[1,2,e]

[1]Faculty of Chemical & Energy Engineering, Universiti Teknologi Malaysia, 81310 Skudai, Johor, Malaysia

[2]Centre of Lipids Engineering & Applied Research (CLEAR), Ibnu Sina Institute for Scientific and Industrial Research, Universiti Teknologi Malaysia, 81310 Skudai, Johor, Malaysia

[a]izzatnaimsky@gmail.com, [b]shuhada.atika@utm.my, [c]musyazrinaalya@graduate.utm.my, [d]norasikin@cheme.utm.my, [e]aishahrosli@utm.my

Keywords: Phosphate, Removal, Synergistic Extractive Extraction, Green Diluent

Abstract. This study presents the utilization of a synergistic extractant for phosphate removal from an aqueous solution using extractive extraction. In this study, the synergistic extractant for phosphate removal was formulated and its concentration was investigated. Additionally, the stripping agent concentration was studied to strip phosphate from its loaded organic phase. The extraction and stripping mechanisms of phosphate were also examined. The results show that more than 97% of phosphate was effectively extracted using a synergistic extractant of 0.1 M TOMAC/0.002 M octanol dissolved in a green diluent mixture of palm oil and sunflower oil. The synergistic coefficient of the extraction is 36.76. Meanwhile an approximately 73% of phosphate was stripped using 0.5 M NaOH. Therefore, the developed formulation of synergistic extraction demonstrates a promise for removing phosphate from agricultural effluents.

Introduction

A significant amount of phosphorus stemming from diverse human activities (e.g., application of fertilizers, herbicides, manure, and pesticides) has been discharged into aquatic ecosystems in the form of phosphate [1]. This promotes the proliferation of algae and large aquatic plants which leads to a reduction in dissolved oxygen, a phenomenon known as eutrophication [2]. Eutrophication can have a negative impact on the aquatic ecosystems and biodiversity. Therefore, the removal of phosphate is crucial to prevent and mitigate the harmful effects of phosphate pollution.

Extractive extraction is an attractive method that is widely used in many industrial applications as it can effectively separate and remove or recover a wide range of compounds and elements from complex mixtures [3,4]. Hence, this method was utilized for removing phosphate from aqueous solution. In this method, a certain amount of extractant (dissolved in diluent) is required to extract the targeted solute through complexation in the organic phase until equilibrium is reached [5]. Previously, phosphate extraction from aqueous solution was investigated using a sole extractant. For instance, Ilyas and colleagues used benzyl dimethyl amine (BDMA) to extract 94% of phosphate with 0.5 M BDMA concentration [6]. Meanwhile, a study by Sulaiman and co-workers reported that almost 100% of phosphate can be extracted using a 0.3 M of tricaprylmethylammonium chloride [7]. However, employing a sole extractant in the extractive extraction comes with certain disadvantages, notably the necessity for a high concentration of extractant because of the low solute loading capacity in the organic phase [8,9]. Hence, the utilization of a synergistic extractant is introduced to overcome this limitation.

Materials Research Proceedings 53 (2025) 136-145 https://doi.org/10.21741/9781644903575-13

The synergistic extractant refers to a mixture of two or more extractants to improve extraction performance over their single capabilities [10]. This process involves combining various types of extractants in which their complementary mechanisms work together to enhance the efficiency and selectivity of the extraction process [11,12]. The effectiveness of synergistic extractive extraction is indicated when the synergistic coefficient (SC) exceeds one. Conversely, an SC less than one suggests antagonistic extraction, whereas an SC of one indicates no synergistic extraction [13]. Several studies have successfully utilized the synergistic extraction are for compounds such as turquoise blue dye [14], folic acid [15], pantothenic acid [16], amoxicillin [17], lactic acid [18], and phenol [19]. The utilization of a synergistic extractant from readily available sole extractants offers greater advantages compared to the laborious process of developing new extractants.

To the best of our knowledge, no study has been reported for the removal of phosphate from aqueous solution by employing a synergistic extractant. Therefore, in this study, synergistic extractant was formulated using different types of extractants. Besides, the effect of synergistic extractant and stripping agent concentration were investigated. The extraction and stripping mechanisms of phosphate were also studied.

Materials and Methods

Chemical Reagents. The feed solution was prepared by using potassium dihydrogen phosphate (KH_2PO_4, (≥99% purity, Merck). Several extractants were utilized, including tricaprylmethylammonium chloride (TOMAC, 95% purity, Sigma-Aldrich), di(2-ethylhexyl)phosphoric acid (D2EHPA, ≥95% purity, Merck), and octanol (≥99% purity, Merck). These extractants were diluted in a mixture (50:50) of cooking palm oil (100% purity, Buruh) and cooking sunflower oil (100% purity, Sunlico) that were purchased from a local market. Meanwhile, the stripping agent which is sodium hydroxide (NaOH, ≥99% purity) was procured from Merck. Then, the reagents used for phosphorus analysis were ammonium molybdate (≥99% purity, Sigma-Aldrich), antimony potassium tartrate (≥99% purity, Sigma-Aldrich), ascorbic acid (≥99% purity, Merck) and sulfuric acid (98% purity, Merck).

Extractive Extraction. The extractive extraction of phosphate was instigated by mixing the equal volume (10 ml) of simulated feed solution (50 ppm of phosphate) at pH of 6.82 and organic solution (extractant in green diluent) in a conical flask. The mixed solution was agitated in an incubator shaker with a rotation of 300 rpm for 18 hours to achieve equilibrium. After the extraction process, the mixture was poured into a separating funnel and left for 30 minutes to ensure complete phase separation. The concentration of phosphate in the aqueous solution was analyzed using an ultraviolet-visible spectrophotometer (λ = 880 nm) with the molybdenum blue method [20]. The synergistic effect between the different types of extractants was also investigated. The single extractant providing the highest extraction was termed the base extractant, while the other was referred to as synergist extractant. Subsequently, the organic solution loaded with phosphate was used for the stripping process, following the same procedure as the extraction process. Furthermore, the extraction and stripping mechanism was determined by using stoichiometry study as proposed by Sulaiman et al. [7]. The parameter and operating conditions for the extractive extraction process are tabulated in Table 1.

Table 1. Parameter and operating conditions

Parameter	Extraction Process	
[Phosphate]	50 ppm	
Diluent	Palm oil and sunflower oil (50:50)	
[Extractant]	Single Extractant	Mixed Extractant
	0.1 M	0.1 M base + 0.001 M synergist
		0.1-0.3 M base + 0.001 M synergist
		0.1 M base + 0.001-0.003 synergist
		0.003-0.1 M base + 0.002 synergist
	Stripping Process	
[Phosphate]	48.86 ppm	
[Stripping agent]	0.1-1.0 M	
	Operating Conditions	
Organic to aqueous ratio	1:1 (10 ml:10 ml)	
Agitation speed	300 rpm	
Agitation time	18 hr	
Temperature	26 °C	

Both processes were carried out using an incubator shaker with a temperature controller fixed at 26 °C. The extraction (E), distribution ratio (D), SC, and stripping (S), were determined using Eqs. 1-4, respectively:

$$E\ (\%) = \frac{C_i - C_{aq}}{C_i} \times 100 \tag{1}$$

$$D = \frac{C_{org}}{C_{aq}} \tag{2}$$

$$SC = \frac{D_{mixture}}{D_{base} + D_{extractant}} \tag{3}$$

$$S\ (\%) = \frac{C_s}{C_{org}} \times 100 \tag{4}$$

where C_i is the initial phosphate concentration (in ppm) in the feed phase for the extraction process. C_{aq} and C_s are the phosphate concentration (in ppm) of the aqueous phase and stripping phase after the extraction and stripping process, respectively. Meanwhile, C_{org} denotes the phosphate concentration (in ppm) in the organic phase after the extraction process. $D_{mixture}$ represents the distribution ratio of the synergistic extraction process, while D_{base} and $D_{synergist}$ refer to the distribution ratios obtained using a base and synergist extractant, respectively. The experimental data for extractive extraction experiments were reported as triplicate with standard deviation ±1%.

Materials Research Proceedings 53 (2025) 136-145 https://doi.org/10.21741/9781644903575-13

Results and Discussion

Effect of Single Extractant. The extraction performance of phosphate at different types of extractant at 0.1 M is shown in Table 2. The result shows that only TOMAC ($NR_4^+Cl^-$)as a basic extractant provide an extraction of phosphate around 53.78%. Meanwhile, there is no extraction of phosphate for D2EHPA (acidic) and octanol (solvating).

Table 2. Effect of extractant types of phosphate extraction (Experimental conditions: [Phosphate] = 50 ppm, [extractant] = 0.1 M, diluent = palm oil and sunflower oil (50:50), aqueous:organic ratio 1:1, agitation speed = 300 rpm, agitation time = 18 hr, temperature = 26 °C)

Extractant	E (%)
D2EHPA	0.00
TOMAC	53.78
Octanol	0.00

In this research, the phosphate existed primarily as $H_2PO_4^-$ due to the slightly acidic nature of the feed solution [6]. Concurrently, the extraction of anion complexes in the solution occurred because of the alkaline nature of the amine components present in TOMAC leading to an ion exchange process [7]. In this medium, phosphate necessitates an extractant possessing a positive charge to facilitate the formation of a complex between phosphate and the extractant within the organic phase. This complex formation is represented by Eq. 5. However, the use of a sole extractant during the extraction process resulted in a limited loading capacity of phosphate into the organic phase. To enhance the extraction of phosphate, TOMAC, which has been identified as the sole extractant for phosphate was functioned as the base extractant while D2EHPA and octanol served as synergist extractants.

$$H_2PO_4^-{}_{(aq)} + NR_4^+Cl^-{}_{(org)} \rightarrow (NR_4)H_2PO_{4(org)} + Cl^-{}_{(aq)} \quad (5)$$

Effect of Extractant Mixture. To assess the synergistic extraction of phosphate, a small proportion of synergist extractant (0.001 M) was mixed with TOMAC (0.1 M) and the result is shown in Table 3. The mixture of TOMAC with D2EHPA and octanol increased the extraction of phosphate to 74.74% and 85.52%, respectively. The SC value for both mixtures exceed one, signifying the successful synergism between the mixed extractants. During extraction, D2EHPA acts as a protective shield, inhibiting the interaction between TOMAC and triglycerides found in vegetable oils [21]. This suggests that TOMAC has the capability to form complexes with phosphate, potentially enhancing the efficiency of phosphate extraction processes. In the same vein, octanol serve as a protective shield and catalyst which helps TOMAC to extract phosphate more effectively [22]. In addition, both D2EHPA and octanol possess oxygen atoms within their molecular structure, exhibiting higher electronegativity compared to the nitrogen atoms present in TOMAC's structure [23]. Thus, they demonstrate a high affinity to form hydrogen bonds with the triglyceride compounds in palm oil, thereby augmenting the effectiveness of TOMAC in phosphate extraction. In this research, 0.1 M TOMAC/0.001 M octanol was chosen as the best mixture of extractants since it provides higher extraction with a higher SC compared to 0.1 M TOMAC/0.001 M D2EHPA.

Table 3. Effect of mixed extractants of phosphate extraction (Experimental conditions: [Phosphate] = 50 ppm, [base extractant] = 0.1 M, [synergist extractant] = 0.001 M, diluent= palm oil and sunflower oil (50:50), aqueous:organic ratio 1:1, agitation speed = 300 rpm, agitation time = 18 hr, temperature = 26 °C)

[TOMAC] (M)	[Synergist Extractant] (M)	E (%)	SC
0.1	**D2EHPA**		
	0.00	0	-
	0.001	74.74	2.55
	Octanol		
	0.00	0	-
	0.001	85.82	5.09

Effect of Extractant Mixture Concentration. The concentration of the mixture extractant was also studied by manipulating one extractant concentration while the other concentration was fixed as shown in Table 4. By increasing the TOMAC concentration from 0.1 to 0.3 M at fixed octanol concentration (0.001 M), phosphate extraction decreases as well as the SC value. This decrease can be attributed to the increased viscosity of TOMAC which hindered the mass transfer of phosphate during the extraction process [24]. Hence, 0.1 M of TOMAC concentration was preferred as the best concentration to synergize with 0.001 M of octanol.

Table 4. Effect of various concentration of TOMAC at fixed octanol concentration on phosphate extraction (Experimental conditions: [Phosphate] = 50 ppm, [octanol] = 0.001 M, diluent= palm oil and sunflower oil (50:50), aqueous:organic ratio 1:1, agitation speed = 300 rpm, agitation time = 18 hr, temperature = 26 °C)

[TOMAC] (M)	[Octanol] (0.001 M)	
	E (%)	SC
0.1	85.52	5.09
0.2	79.13	2.58
0.3	73.68	1.01

In the meantime, increasing the octanol concentration from 0.001 to 0.002 M at a fixed TOMAC concentration (0.1 M) increased phosphate extraction from 85.52 to 97.71% as depicted in Table 5. Aside from that, the SC value rose from 5.09 to 36.76 indicating the effective synergism of the mixed extractant. This is because the increase of octanol facilitates the solubilization of polar molecules [25], thus enhancing phosphate extraction. Meanwhile, the increase of octanol concentration to 0.003 M decreases the extraction of phosphate to 57.13%. This is attributed to the potential emulsification of octanol when the concentration of octanol is elevated beyond its concentration limit [26], leading to a decrease in phosphate extraction. Therefore, the mixed extractant of 0.1 M TOMAC/0.002 M octanol is preferable to obtain high extraction performance of phosphate.

Table 5. Effect of various concentration of octanol at fixed TOMAC concentration on phosphate extraction (Experimental conditions: [Phosphate] = 50 ppm, [TOMAC] = 0.1 M, diluent: palm oil and sunflower oil (50:50), aqueous:organic ratio 1:1, agitation speed = 300 rpm, agitation time = 18 hr, temperature = 26 °C)

[TOMAC] (0.1 M)	[Octanol] (M)					
	0.001		0.002		0.003	
	E (%)	SC	E (%)	SC	E (%)	SC
	85.52	5.09	97.71	36.76	57.13	1.15

Effect of Stripping Agent Concentration. In this study, NaOH was chosen as the stripping agent due to its high basicity, which enables the creation of chemical gradient between the organic and aqueous stripping phases [27]. This gradient promotes efficient decomplexation of phosphate-TOMAC at the interface of the organic-stripping phase. The effect of different concentration of NaOH on phosphate stripping is depicted in Table 6. The NaOH concentrations varied from 0.1 M to 1.0 M to strip phosphate from the phosphate loaded organic solution. Stripping performance exhibited an increasing trend from 5.96 to 73.27% with the increase in NaOH concentration from 0.1 to 0.5 M. However, increasing the NaOH concentration up to 1.0 M made it impossible to measure phosphate stripping due to saponification. At high concentrations of NaOH, a reaction occurs between NaOH and triglycerides in palm oil which produces soap, glycerol, and free fatty acids [28]. Therefore, 0.5 M NaOH was selected as the optimal stripping agent concentration to strip phosphate from its loaded organic solution.

Table 6. Effect of different concentration of NaOH towards phosphate stripping (Experimental conditions: (Experimental conditions: [Phosphate] = 48.86 ppm, [TOMAC] = 0.1 M, diluent= palm oil and sunflower oil (50:50), aqueous:organic ratio 1:1, agitation speed = 300 rpm, agitation time = 18 hr, temperature = 26 °C)

[NaOH] (M)	S (%)
0.1	5.96
0.3	34.73
0.5	73.27
1.0	-

Extraction and Stripping Mechanism. In order to determine the stoichiometry of phosphate extraction, the concentration of TOMAC was varied at a fixed concentration of octanol (0.002 M). The result is presented in Table 7. In this study, only TOMAC was found to react with phosphate. Meanwhile, octanol served a as a protective shield and a catalyst, thereby contributing to the enhancement of phosphate extraction.

Table 7. Effect of various concentration of TOMAC at fixed octanol concentration on phosphate extraction (Experimental conditions: [Phosphate] = 50 ppm, [octanol] = 0.002 M, diluent= palm oil and sunflower oil (50:50), aqueous:organic ratio 1:1, agitation speed = 300 rpm, agitation time = 18 hr, temperature = 26 °C)

[TOMAC] (M)	E (%)
0.03	34.65
0.04	36.05
0.05	51.72
0.1	97.71

The stochiometric reaction between phosphate and TOMAC was determined by plotting the Log D versus Log [TOMAC] as presented in Fig. 1. The slope of the graph is approximately 3.90, indicating that 4 moles of TOMAC are required to form a complex with one mole of phosphate. Therefore, Eq. 6 provides the chemical equilibrium responsible for phosphate extraction involving TOMAC. For the stoichiometry study of the stripping of phosphate, a graph of Log D versus Log [NaOH] was plotted as depicted in Fig. 2. The slope of the graph yields a value of 2.28, signifying that two moles of NaOH are required for one mole of phosphate-TOMAC complexes for the stripping process. As such, Eq. 7 represents the reaction of loaded phosphate-TOMAC complexes with NaOH.

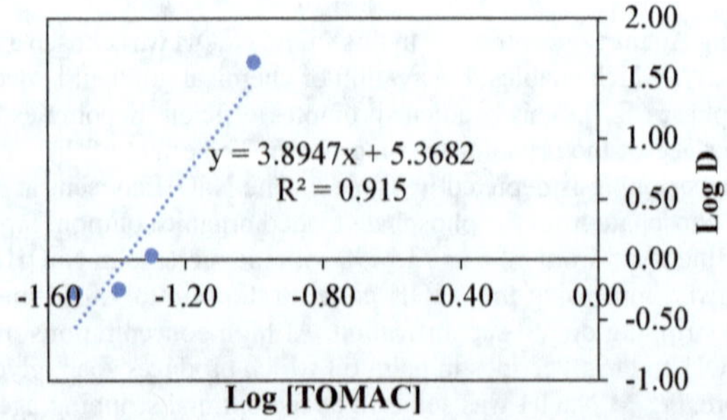

Fig. 1. Stoichiometry plot for the equilibrium extraction of phosphate

$$H_2PO_4^-{}_{(aq)} + 4NR_4^+Cl^-{}_{(org)} \rightarrow (NR_4)_4H_2PO_{4(org)} + 4Cl^-{}_{(aq)} \quad (6)$$

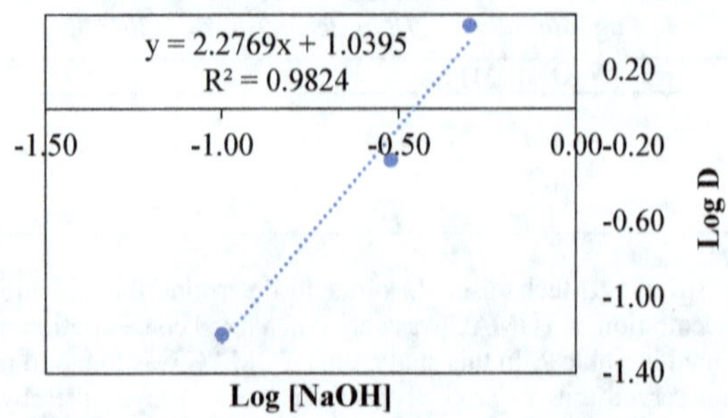

Fig. 2. Stoichiometry plot for the equilibrium stripping of phosphate

$$(NR_4)_4H_2PO_{4(org)} + +2NaOH_{(aq)} \rightarrow (Na)_2H_2PO_{4(aq)} + 2(NR_4)_2OH_{(org)} \quad (7)$$

Conclusion

In conclusion, the combination of 0.1 M TOMAC and 0.002 M octanol dissolved in a mixture of palm oil and sunflower oil (50:50) as a green diluent demonstrates synergistic effects in enhancing phosphate extraction from an aqueous solution. Over 97% of phosphate was effectively extracted, achieving a SC value of 36.76. Meanwhile, about 73% of phosphate was stripped using 0.5 M NaOH. Hence, the developed synergistic extractive extraction method shows great potential for phosphate removal from agricultural effluents.

Acknowledgement

The authors would like to sincerely thank the Ministry of Higher Education for financial support through the Fundamental Research Grant Scheme (FRGS/1/2023/TK05/UTM/02/3) and CLEAR for providing facilities to enable this research.

References

[1] S.P. Hong, H. Yoon, J. Lee, C. Kim, S. Kim, J. Lee, C. Lee, J. Yoon, Selective phosphate removal using layered double hydroxide/reduced graphene oxide (LDH/rGO) composite electrode in capacitive deionization, J. Colloid Interface Sci. 564 (2020) 1–7. https://doi.org/10.1016/j.jcis.2019.12.068

[2] N. Che, J. Qu, J. Wang, N. Liu, C. Li, Y. Liu, Adsorption of phosphate onto agricultural waste biochars with ferrite/manganese modified-ball-milled treatment and its reuse in saline soil, Sci. Total Environ. 915 (2024) 169841. https://doi.org/10.1016/j.scitotenv.2023.169841

[3] N.F.M. Noah, N. Othman, Y. V Arumugam, I.N.S. Kahar, S.S. Suliman, Extractive extraction for platinum recovery using synergistic organic formulation of Aliquat 336 / TBP mixtures, Mater. Today Proc. 96 (2024) 17–22. https://doi.org/10.1016/j.matpr.2023.08.156

[4] F.M. Antony, K.L. Wasewar, The Sustainable Approach of Process Intensification in Biorefinery Through Reactive Extraction Coupled with Regeneration for Recovery of Protocatechuic Acid, Appl. Biochem. Biotechnol. (2023) 1570–1591. https://doi.org/10.1007/s12010-023-04659-8

[5] M.E. Marti, H. Zeidan, Using eco-friendly alternatives for the recovery of pyruvic acid by reactive extraction, Sep. Purif. Technol. 312 (2023) 123309. https://doi.org/10.1016/j.seppur.2023.123309

[6] S. Ilyas, R.R. Srivastava, H. Kim, Liquid-liquid extraction of phosphorus from sulfuric acid solution using benzyl dimethyl amine, Int. J. Miner. Metall. Mater. 28 (2021) 367–372. https://doi.org/10.1007/s12613-020-2151-8

[7] R.N.R. Sulaiman, A. Rosli, N. Othman, M.A.A. Zaini, Novel Extraction of Phosphorus using Interaction of Tricaprylylmethylammonium Chloride (TOMAC) with Biodegradable Mixture of Vegetable Oils, Malaysian J. Fundam. Appl. Sci. 19 (2023) 769–780. https://doi.org/10.11113/mjfas.v19n5.3053

[8] G. Zhang, D. Chen, W. Zhao, H. Zhao, L. Wang, W. Wang, T. Qi, A novel D2EHPA-based synergistic extraction system for the recovery of chromium (III), Chem. Eng. J. 302 (2016) 233–238. https://doi.org/10.1016/j.cej.2016.05.063

[9] T.H. Nguyen, M.S. Lee, A review on the separation of lithium ion from leach liquors of primary and secondary resources by solvent extraction with commercial extractants, Processes. 6 (2018) 1–15. https://doi.org/10.3390/pr6050055

[10] H. Wang, S. Kuang, W. Liao, Synergistic extraction and separation of thorium from rare earths in chloride media using mixture of Cextrant 230 and Cyanex 923, J. Rare Earths. 42 (2024) 759–767. https://doi.org/10.1016/j.jre.2023.04.016

[11] A. Jackson, M. Nilsson, Synergism and Aggregation in Multi-Extractant Solvent Extraction Systems, Solvent Extr. Ion Exch. 37 (2019) 269–283. https://doi.org/10.1080/07366299.2019.1639367

[12] I.N.S. Kahar, N. Othman, N.F.M. Noah, S.S. Suliman, Recovery of copper and silver from industrial e-waste leached solutions using sustainable liquid membrane technology : a review, Environ. Sci. Pollut. Res. 30 (2023) 66445–66472. https://doi.org/10.1007/s11356-023-26951-0

[13] H.A. Rahman, N. Jusoh, N. Othman, M.B. Rosly, R.N.R. Sulaiman, N.F.M. Noah, Green formulation for synthetic dye extraction using synergistic mixture of acid-base extractant, Sep. Purif. Technol. 209 (2019) 293–300. https://doi.org/10.1016/j.seppur.2018.07.053

[14] N. Jusoh, N. Othman, N.M. Amin, R.N.R. Sulaiman, N.F.M. Noah, S. Nurulashikin, M.A.A. Zaini, Synergistic Formulation for Turquoise Blue Dye Extraction Using Mixed Trioctylamine/Cyanex 302 in Liquid-Liquid Extraction System, 200 (2021) 1–9. https://doi.org/10.2991/aer.k.201229.001

[15] D. Caşcaval, A.I. Galaction, L. Kloetzer, Synergic Extraction of Folic Acid with Di(2-ethylhexyl) Phosphoric Acid and Amberlite LA-2, Sep. Sci. Technol. 47 (2012) 834–841. https://doi.org/10.1080/01496395.2011.630056

[16] A.I. Galaction, M. Poştaru, A.S. Bompa, D. Caşcaval, Synergic extraction of pantothenic acid with two different extractants, Brazilian J. Chem. Eng. 33 (2016) 1031–1040. https://doi.org/10.1590/0104-6632.20160334s20150265

[17] C. Homsirikamol, N. Sunsandee, U. Pancharoen, K. Nootong, Synergistic extraction of amoxicillin from aqueous solution by using binary mixtures of Aliquat 336, D2EHPA and TBP, Sep. Purif. Technol. 162 (2016) 30–36. https://doi.org/10.1016/j.seppur.2016.02.003

[18] A. Kumar, A. Thakur, Reactive extraction of lactic acid using environmentally benign green solvents and a synergistic mixture of extractants, Sci. Iran. 26 (2019) 3456–3467. https://doi.org/10.24200/sci.2019.52233.2610

[19] Y. Wang, Q. Zhang, M. Liu, H. Liu, Y. Chen, X. Hu, Simulation and dynamic control of removal phenols from coal gasification wastewater by synergistic extraction, Sep. Purif. Technol. 285 (2022) 120347. https://doi.org/10.1016/j.seppur.2021.120347

[20] J.M. Racicot, T.L. Mako, A. Olivelli, M. Levine, A paper-based device for ultrasensitive, colorimetric phosphate detection in seawater, Sensors. 20 (2020) 2776.

[21] R.N.R. Sulaiman, H.A. Rahman, N. Othman, M.B. Rosly, N. Jusoh, N.F.M. Noah, Extraction of reactive dye via synergistic Aliquat 336/D2EHPA using emulsion liquid membrane system, Korean J. Chem. Eng. 37 (2020) 141–150. https://doi.org/10.1007/s11814-019-0418-4

[22] L. Kloetzer, R.A. Ilica, D. Cascaval, A.I. Galaction, Separation of fumaric acid by amine extraction without and with 1-octanol as phase modifier, Sep. Purif. Technol. 227 (2019) 115724. https://doi.org/10.1016/j.seppur.2019.115724

[23] N. Jusoh, M.B. Rosly, N. Othman, H.A. Rahman, N.F.M. Noah, R.N.R. Sulaiman, Selective extraction and recovery of polyphenols from palm oil mill sterilization condensate using emulsion liquid membrane process, Environ. Sci. Pollut. Res. 27 (2020) 23246–23257. https://doi.org/10.1007/s11356-020-07972-5

Materials Research Proceedings 53 (2025) 136-145 https://doi.org/10.21741/9781644903575-13

[24] A. Swarnkar, A. Keshav, A.B. Soni, Recovery of Methacrylic Acid from the Aqueous Phase Using Trioctylmethylammonium Chloride (TOMAC) in Different Diluents, J. Chem. Eng. Data. 61 (2016) 1412–1420. https://doi.org/10.1021/acs.jced.5b00553

[25] R.G. Lazar, A.C. Blaga, E.N. Dragoi, A.I. Galaction, D. Cascaval, Application of reactive extraction for the separation of pseudomonic acids: Influencing factors, interfacial mechanism, and process modelling, Can. J. Chem. Eng. 100 (2022) S246–S257. https://doi.org/10.1002/cjce.24124

[26] D. Luo, J. Huang, Y. Zhang, H. Liu, P. Hu, Efficient and environment-friendly vanadium (V) extraction from vanadium shale leachate using tri-n-octylmethylammonium chloride, Sep. Purif. Technol. 237 (2020) 116482. https://doi.org/10.1016/j.seppur.2019.116482

[27] M.B. Rosly, N. Jusoh, N. Othman, H.A. Rahman, N.F.M. Noah, R.N.R. Sulaiman, Synergism of Aliquat336-D2EHPA as carrier on the selectivity of organic compound dyes extraction via emulsion liquid membrane process, Sep. Purif. Technol. 239 (2020) 116527. https://doi.org/10.1016/j.seppur.2020.116527

[28] N.C. Firsta, R.D. Mentari, E.S. Salafiah, T. Estiasih, Preparation of unsaponifiable fraction from crude palm oil: A short review, IOP Conf. Ser. Earth Environ. Sci. 475 (2020). https://doi.org/10.1088/1755-1315/475/1/012032

Materials Research Proceedings 53 (2025) 146-154 https://doi.org/10.21741/9781644903575-14

Optimization of fly ash-*g*-polyacrylic acid hydrogel via response surface methodology for improved amine loading capacity

ALI SHAH Syed Farman[1,a], RABAT Nurul Ekmi[1,b,*], JOHARI Khairiraihanna[1,c], MT GHANI Siti Musliha[1,d] and ABDUL Sami[1,e]

[1]Chemical Engineering Department, Universiti Teknologi PETRONAS, 32610 Bandar Seri Iskandar, Perak, Malaysia

[a]syed_22009368@utp.edu.my, [b]nurulekmi.rabat@utp.edu.my, [c]khairiraihana.j@utp.edu.my [d]musliha.ghani@utp.edu.my, [e]abdul_22000064@utp.edu.my

Keywords: Hydrogel, Fly Ash, Amine Uptake, Response Surface Methodology

Abstract. Amine solvent has shown great sour gases separation capacity from natural gas. This paper studied the optimization of monoethanolamine loading capacity on the fly ash grafted polyacrylic acid hydrogel using response surface methodology. The central composite design was used to determine the effect of three independent variables particularly fly ash as a filler, N, N"-methylene bis acrylamide as a crosslinker, and ammonia peroxide sulphate as an initiator. The optimization study revealed the significance of factors in the following order: cross-linker > fly ash > initiator, with an R^2 value of 0.99. The fly ash grafted hydrogel achieved a maximum amine loading of up to 1550 wt.% with the addition of 1.1 g of fly ash, 0.07 g of cross-linker, and 0.07 g of initiator. It was investigated that varying the concentration of crosslinker was significantly enhanced the amine loading. Furthermore, SEM and FTIR were performed to analyze the morphology and chemical changes in the optimum fly ash grafted hydrogel. The results depicted that the fly ash grafted hydrogel has porous structure and ample space for amine uptake inside the pores. The fly ash-grafted hydrogel exhibits significant potential for efficient gas separation from natural gas. Its enhanced porosity, hydrophilicity, and capacity to incorporate functional groups, such as amines, enable the selective adsorption of impurities.

Introduction

Recently polymer hydrogel has been widely used as an adsorbent to separate sour gases from natural gas. Due to the inherent functional groups including amide (-NH-), carboxylic (COO-), and OH group having high affinity with sour gases. In addition, the hydrophilic nature of polymer hydrogel also plays an important role in improving the functionality of polymer hydrogel for toxic gas separation applications [1]. However, pure hydrogel has low porosity and limited adsorption capacity [2]. As a result, researchers have developed a composite hydrogel that by incorporating inorganic fillers particularly zeolite, silica, carbon nanotubes, oil palm empty fruit bunch (OPEFB), and fly ash, to enhance the polymer hydrogel porosity for hazardous gases from natural gas separation [3,4 ,5].

Fly ash, the solid waste from power stations with low utilization rates and primarily disposed of in landfills can be reused. By directly mixing fly ash with a polymer component to create a hydrogel porous support, an adsorbent can be produced to treat toxic gas from natural gas, thus contributing to low environmental concern. Ghani et al [6] investigated the adsorption performance of the toxic gases with and without the inclusion of fly ash as an additive. By adding fly ash filler as in a polymer hydrogel, 54% increment in adsorption capacity was attained. Thus, due to its readily available and considerably a low-cost resource, the utilization of may offer a substantial benefit preventing the environmental issues associated with the disposal of unused materials [7].

For practical application, the hydrogel adsorption capacity for toxic gas removal can be enhanced by impregnating with monoethanolamine (MEA) on the fly ash grafted polymer hydrogel. In the literature, Rabat et al have proven the effect of amine impregnation for removing pollutants. In this study, the MEA impregnations on the surface of the adsorbent drastically increased the adsorption capacity by three times as compared to a plain adsorbent. A comparable effect was also investigated by Kim et al [8]. In addition, MEA-functionalized hydrogel can accelerate the adsorption kinetics. The above-indicated technique was also utilized by investigating the use of water and ethylene glycol (EG) as solvents for the amine. Siti Musliha et al [9] revealed that a mixture of water and amine (MEA) 30 wt.% solution, integrated into the hydrogel matrix, acted as the activator for adsorbent with the ability to absorb a maximum of 1150% amine solution in their matrix.

Response Surface Methodology (RSM) was applied to optimize the synthesis parameter of fly ash grafted polyacrylic acid hydrogel in enhancing the amine uptake as an inverse suspension polymerization to check the effect of filler in water uptake extensively elaborated on the screening phase of the previous study utilizing full factorial experimental design [10]. It was found that fly ash predominantly affected the polymer hydrogel due to enhanced porosity. Therefore, optimizing the process parameters is essential. Although most studies on polymer hydrogels have used the conventional one-variable-at-a-time (OVAT) approach for parameter optimization, this method has notable limitations, including its inability to account for factor interactions. Therefore, an alternative approach utilizing statistical design, particularly RSM is suggested. Despite several advantages of RSM, there is a gap in the literature regarding the optimization of amine-infused grafting of fly ash onto polymer hydrogels.

This research work was performed to addresses gap in the literature by introducing a novel approach to enhance the efficiency and effectiveness of the adsorbent. By focusing on this underexplored area, the study aimed to provide new insights and methodologies that can improve the performance and applications of synthesis of FA-g-PAA hydrogel by solution polymerization via RSM for enhancing amine absorbance. The result RSM showed the amine absorbance capability by 27% increase using various parameters in Design Expert software.

Material and Methodology

Materials. Acrylic acid (AAc) (99%), the cross-linker; N, N'- methylene bis (acrylamide) (NNMBA) (99%), and the initiator ammonium persulfate (APS) (98%), and mono ethylene amine as functionalization were purchased from Sigma-Aldrich. The fly ash waste was collected from a coal power plant in Manjung, Perak. A synthesis grade of monomers and cross-linkers was used as-received without further purification. Deionized water was used as a solvent polymerization medium throughout this research.

Synthesis of fly ash-grafted-polyacrylic acid (FA-g-PAA) hydrogel. FA-g-PAA was synthesized via a solution polymerization by adding monomer AA, and water in a two-neck round bottom flask equipment with a reflux condenser for venting and an inlet for nitrogen gas. Initially, 3.3 g (14 w/v %) of AA was added into 20 ml of deionized water to produce a monomer solution and mixed for 10 minutes. The solution was put into a water bath at 60 °C and continuously stirred [6]. According to RSM using Central Composition Design (CCD), the suggested amounts of cross-linker, initiator and fly ash were added at 0.07 g (1 w/w %), 0.07 g (1 w/w %), and 1.1 g (15 w/w %), respectively. The reaction mixture was continuously stirred in order to allow a stable formation polymer hydrogel. The polymer hydrogel obtained was washed several times using deionized water to ensure the complete removal of the unconverted monomers prior to cutting into small pieces and dried at 60 °C for 18 h in the vacuum oven.

Optimization using response surface methodology. RSM is defined as a set of mathematical and statistical tools that is effectively applied to develop, improve, and optimize the process. It proposes a mathematical framework and illustrates outcomes through two-dimensional or three-dimensional visualizations, employing tailored experiments to identify optimal conditions for crucial factors. Second-order quadratic models are widely preferred and used in the context of

CCD effective in approximating the authentic response surface [11]. The response is examined through Analysis of Variance (ANOVA), considering the p-value at a confidence level of 95% [12].

In this study, a three-factor CCD was employed to optimize the experimental parameters of cross-linker (NNMBA), initiator (APS), and fly ash concentration to maximize the swelling percentage of the amine solution [13]. To analyze the response and produce response charts, each variable was evaluated at five levels (-α, -1, 0, +1, and +α) as shown in Table 1. The resulting model eq (1) is provided below.

$$Y = b_o + \sum_{i=1}^{k} b_i\, x_i + \sum_{i=1}^{j=1} \sum_{j=1}^{k} b_{ij}\, x_i\, x_j + \sum_{i=j}^{k} b_{ij}\, x_i^2 \tag{1}$$

Where Y is the swelling percentage of amine uptake, b_o,b_i, b_{ii}, and b_{ij} are the linear, squared, and interaction effect coefficients, respectively. *xi* and *xj* are also the coded values of the variables *i* and *j*, respectively.

Table 1. Independent factor and level runs in the Central Composition Design

Name	Units	low	high	-alpha	+alpha
Cross-linker	w/w %	1	2	0.659104	2.3409
Initiator	w/w %	1	3	0.318207	3.68179
Fly-ash	w/w %	5	15	1.59104	18.409

Amine Solution Absorption Analysis.Swelling characteristics of the hydrogel in 30 wt.% amine solution were studied over time until it reached equilibrium. For each experiment, 0.5 g of the polymer was placed in 20 ml of amine 30 wt. % solution at room temperature. Subsequently, the sample was put onto tissue paper to remove excess amine from the gel surface, and the weight of the gel-absorbed amine (expanded hydrogel) was measured using analytical balance with an accuracy of ± 0.001g. The amine uptake quality was evaluated according to the formula [14].

$$Amine\ uptake\ Q = \left(\frac{g(Amine)}{g\ \text{wet sampl}} x100\right) = \frac{m_f - m_i}{m_f} x\ 100 \tag{2}$$

Where m_i is the initial weight of dry hydrogel before being immersed in amine solution and m_f is the final amine uptake of the hydrogel.

Characterization

Fourier Transform Infrared Spectroscopy (FTIR). The FTIR spectra of fly ash grafted polyacrylic acid (FA-g-PAA) hydrogel were used to determine the chemical structure of the grafted fly ash on the maximum amine adsorbent hydrogel as well as its functional groups. This procedure was conducted in a KBr disc to obtain FTIR spectra results with a wavelength range spanning from 4000 cm^{-1} to 400 cm^{-1} [10].Scanning Electron Microscopic (SEM). Scanning electron microscopic analysis was carried out to determine the pore structure of fly ash grafted polyacrylic acid polymer hydrogel. The image was recorded by a microscope operated at an acceleration voltage of 5 to 10 kV [6]. Prior to the analysis, the sample was coated by a gold layer in order to avoid any discharging on sample surface. The pores on the surface were examined.

Results and Discussion

Characterization of FTIR Analysis. The result of the optimum sample amine MEA 30 % FA-g-PAA, pure polyacrylic acid hydrogel, and fly ash powder was characterized by FTIR spectrum. Fig.1 shows the structural property of gel of MEA amine functionalization under an optimization. The band stretching variation of -OH was between 3670 cm^{-1} to 3030 cm^{-1}. The spectral region between 3,700 and 3,280 cm^{-1} corresponded to the stretching vibration of the NH_2 and NH bond in monoethanolamine (MEA), and the stretching spectrum was observed at 1720 cm^{-1} corresponding to the bending mode of C=O carboxylic acid molecules, while the absorbency peak was detected in a range of 1250 to 770 cm^{-1} that represented the C-O stretching. The spectra of Si-O-Al were observed with different bandgaps at 1010, 770, and 680 cm^{-1} that indicated the presence of Si-O-Al. This finding explained these peaks showed the fly ash, was effectively grafted onto the polyacrylic acid. The peak of fly ash observed in this study aligns with the findings of previous research [5] [15].

Table 2. The runs in CCD and the result

Std	Run	Factor 1: Cross-linker w/w %	Factor 2: Initiator w/w %	Factor 3: Fly ash w/w %	Response: Amine uptake %	
					Actual	Predicted
18	1	1.5	2	10	1350	1349
11	2	1.5	0.318	10	1467	1472
17	3	1.5	2	10	1350	1349
1	4	1	1	5	1484	1477
5	5	1	1	15	1540	1536
6	6	2	1	15	1452	1447
20	7	1	2	10	1350	1349
19	8	1	2	10	1350	1349
10	9	2.34	2	10	1382	1374
13	10	1.5	2	1.59	1395	1390
12	11	1.5	3.68	10	1423	1418
3	12	1	3	5	1420	1424
8	13	2	3	15	1430	1436
7	14	1	3	15	1505	1498
14	15	1.5	2	18.40	1493	1498
4	16	2	3	5	1363	1366
15	17	1.5	2	10	1350	1349
2	18	2	1	5	1386	1392
16	19	1.5	2	10	1350	1349
9	20	0.659	2	10	1490	1497

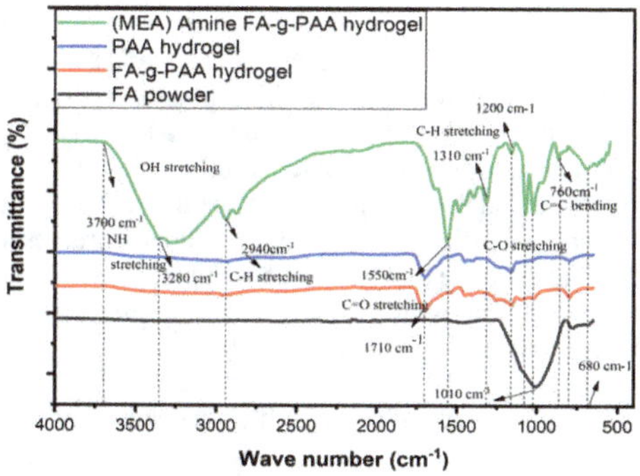

Fig. 1 FTIR results of MEA amine functionalization under optimization conditions compare with fly ash powder, plain PAA, and with the addition of fly ash

Scanning electron microscope (SEM). Fig. 2 shows the morphology of polymer hydrogel where the pore structure of hydrogel in the presence ((a)2 and (b)2) and absence ((a)1 and (b)1) of fly ash. The surface analysis of both PAA and FA-*g*-PAA was carried out at 100x and 500x magnification. The spherical small holes were identified on the FA hydrogel due to the addition of FA as shown in Fig. 2 (b1, b2). Large holes on the surface of hydrogel showed shrinkage due to the interlinkage of matrices concealed during the drying process which was indicated in encircled Figure 2 [16]. Additionally, the FA-g-PAA hydrogel showed a better surface pore than the plain PAA hydrogel. This surface condition could offer easier attachment of due to the addition of the fly ash. Thus, in conclusion, the addition of FA into the hydrogel improved the porosity and hydrophilicity of the polymer hydrogel, thereby potentially improving its amine loading capacity.

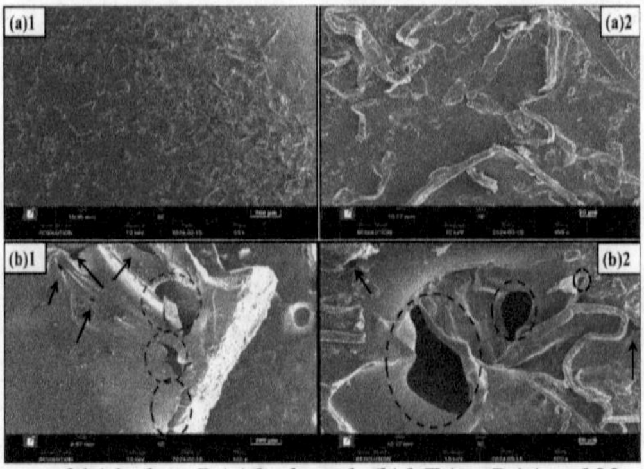

Fig. 2 SEM images of (a)1 plain PAA hydrogel, (b)1 FA-g-PAA at 100x magnification, (a)2 plain PAA hydrogel and (b)2 FA-g-PAA at 500x magnification

Analysis of the response surface. In this study the developed model R^2-value is 0.99 as shown in Table 3, which suggested the model was an acceptable and agreed with the predicted value. Furthermore, the margin between R^2 and adjusted R^2 was under 0.2, indicating a significant alignment between predicted R^2 and adjusted R^2, affirming the precision of the proposed model, as the R^2 value approaching 1, quadratic polynomial equation proven to be highly effective in capturing the impact of all independent variables on the response variable. This model can navigate the design space due to its high value of adeq: precision ratio of 38.8. which can be defined as the ratio (> 4: ideal). The model showed a significant F-value of 180.42 and a p-value of less than 0.0001, indicating a 0.01% chance of noise resulting in a large F-value. In this study, the non-significant terms of lack of fit including the interactions of AB, BC, and AC. The significant terms were only achieved by single factor effect of crosslinker (A^2), fly ash (B^2) and initiator (C^2). According to the ANOVA, a quadratic response surface model applied to amine uptake Table 4, the model was considerably significant and good agreement as represented by the p-value. The regression of the model evaluated the importance of each parameter and uses p-values to determine the intensity of its interactions. The ANOVA analysis evaluated the importance of each factor and used p-values to measure the intensity of their interactions.

$$Y = 2018.82 - 178.29\,A - 21.11\,B + 13.5\,C + 121.85\,A^2 + 33.64\,B^2 + 1.33\,C^2 \quad (3)$$

Where Y is the swelling percentage of amine, which depends on the initial concentration w/w% of cross-linker, initiator, and fly ash.

Table 3. R^2 value from ANOVA analysis

Std. Dev.	6.77	R^2	0.9939
Mean	14.1.50	Adjusted R^2	0.9884
C.V. %	0.4781	Predicted R^2	0.9514
		Adeq Precision	38.8602

Table 4. A complete ANOVA for a quadratic response surface model applied to amine uptake

Source	Sum Of Squares	df	Means square	F-value	p-value	
Model	74462.42	9	8273.60	180.42	< 0.0001	Significant
A-cross-linker	18279.01	1	18279.01	398.60	< 0.0001	
B-initiator	3479.83	1	3479.83	75.88	< 0.0001	
C-fly_ash	14099.82	1	14099.82	307.47	< 0.0001	
AB	364.50	1	364.50	7.95	0.0182	
AC	8.00	1	8.00	0.1745	0.6850	
BC	112.50	1	112.50	2.45	0.1484	
A^2	13373.74	1	13373.74	291.64	< 0.0001	
B^2	16313.52	1	16313.52	355.74	< 0.0001	
C^2	15972.47	1	15972.47	348.30	< 0.0001	
Residual	458.58	10	12.86			
Lack of Fit	88.58	5	12.72	1.79	0.156	Non-significant
Pure Error	0.0000	5	0.0000			
Cor Total	74921.00	19				

Data optimization using CCD. The importance of the design expert software as a substitute for many experiments in the full factorial design to reduce the cost, time, and resources just run the 20 experiments performed and screening significant variables for maximizing amine uptake and the pore structure of FA-g-PAA hydrogel. The high porous material commonly has a high amine uptake of 1550 % ratio was achieved at the (cross-linker 1 w/w %), (initiator 1 w/w %), and (fly-ash 15 w/w %). Fig.3 (a) demonstrates the correlation between the normal and residual values, revealing that the data points were closely distributed along a straight line. This alignment suggested a suitable dispersion of data and a substantial lack of fit model. The normal plot of residuals in Fig. *3* confirmed that the errors were normally distributed, affirming the ability of model to accurately forecast response within the experimental range. Fig. 4 shows a three-dimensional quadratic graph between the cross-liker concentration inversed relationship and direct relationship with the fly ash content. Increasing cross-linker decreased the amine uptake and increasing the fly ash increased the amine uptake [10].

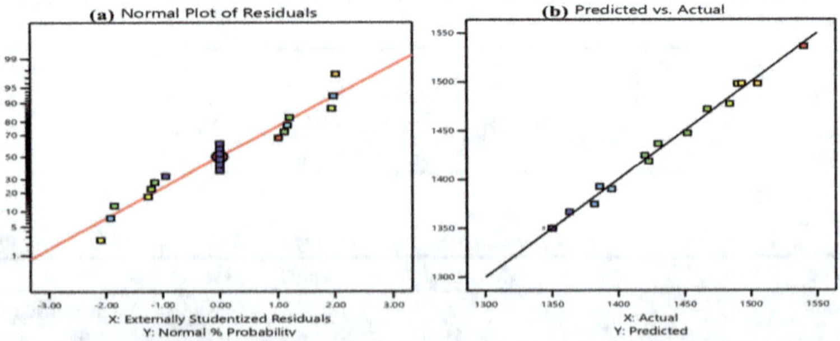

Fig. 3 (a)Normal probability plot of residual and (b) Residual versus predicted

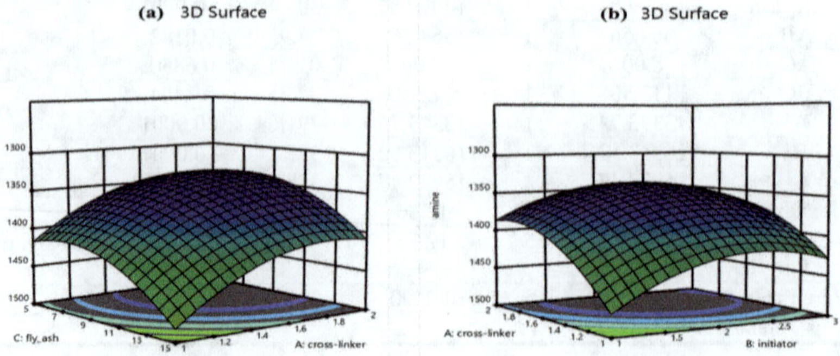

Fig. 4 The 3D graph between fly ash and cross-linker 3D (b) The graph between cross-linker and initiator

Validation with Experiment. Validation of the experimental trials was conducted using the optimal conditions. The outcomes of these validation runs were presented in Table 5. The mean amine uptake achieved was 1450 %. The result optimization model yielded the maximum amine uptake of 1435 %, the error, which was approximately 1.7 %. Comparatively, the experimental amine uptake value suggested the optimization model was validated with a confidence level of 95% in the design expert software. This condition markedly indicated satisfaction through the manipulation of the processing conditions of NNMBA concentration, fly ash content, and initiator. This study claimed that optimized material offered 27 % increase in amine uptake as compared to the most reported studies.

Table 5. Comparison of existing and experimental data

Run	Cross-linker w/w%	Initiator w/w%	Fly ash w/w%	Predicted Amine uptake	Obtained amine uptake	Error %	Desirability
1	0.09	0.137	0.97	1402	1385	1.3	1
2	0.079	0.017	1	1457	1435	1.6	1
3	0.07	0.09	0.43	1437	1423	1	1

Conclusion

This study focused on optimizing the amine uptake on fly ash grafted polyacrylic acid using design expert software at different levels of parameter. The maximum swelling of 30 wt.% MEA was achieved on FA-*g*-PAA up to 1550% with an addition of 1.1 g fly ash, 0.07 g of the crosslinker, and 0.07 g of the initiator with a R^2 value of 0.99. In this study, the highest significance factor was crosslinker followed by fly ash and initiator. Furthermore, chemical and morphological analyses were conducted to compare the properties of FA-*g*-PAA with the plain PAA hydrogel. The FTIR data revealed the presence of hydroxyl, alkanes, carboxyl groups, and silica oxide groups. The morphological analysis revealed a rough surface and pores in the FA-*g*-PAA hydrogel structure, suggesting a larger contact area for amine molecules as compared to the plain PAA hydrogel. In conclusion, this study effectively achieved synthesized FA-*g*-PAA by solution polymerization process with both significantly enhanced properties of amine uptake and pore structure. In the future, this research could be performed to explore the practical applications of this optimized material for utilization as an efficient adsorbent for hazardous gas separation.

Acknowledgements

The author would like to thank the financial support from Yayasan Universiti Teknologi PETRONAS (YUTP) under cost center 015LC0-359. The authors would like to thank the Chemical Engineering Department, Universiti Teknologi PETRONAS, Malaysia for providing technical services and facilities.

References

[1] E. M. Ahmed, "Hydrogel: Preparation, characterization, and applications: A review," *J Adv Res*, vol. 6, no. 2, pp. 105–121, 2015. https://doi.org/10.1016/j.jare.2013.07.006

[2] N. M. F. M. Yasin, N. H. Meri, N. Talib, W. A. W. A. K. Ghani, Z. A. Rashid, and A. B. Alias, "Breakthrough Analysis of Empty Fruit Bunch-Based Hydrogel Biochar Composite (EFB-HBC) for Hydrogen Sulphide (H 2 S) Adsorption Study Removal," vol. 200, no. ICoST, pp. 216–225, 2021. https://doi.org/10.2991/aer.k.201229.030

[3] R. K. N. E. R. M. A. M. A. S. F. N. A. X. E. C. Siti Musliha Mat Ghani, "Fly ash grafted poly(acrylic acid-co-acrylamide) composite hydrogel as the carbon dioxide adsorbent," in *INTERNATIONAL SYMPOSIUM ON ADVANCED MATERIALS & PROCESSING*, AIP Publishing, 2024.

[4] S. Musliha, M. Ghani, X. Enna, R. Kumeressan, and N. Ekmi, "Improved Hydrogel as Potential Carbon Dioxide Adsorbent," *The Italian Association of Chemical Engineering Online*, vol. 97, no. August, pp. 457–462, 2022. https://doi.org/10.3303/CET2297077

[5] E. E. B. Campbell. A. V. Gromov, A. Kulur, J.A.A. Gibson, E. Mangano, S. Brandani, "Carbon Nanotube/PVA Aerogels Impregnated with PEI: Solid Adsorbents for CO_2 Capture A.," *Sustainable Energy & Fuel*, 2018. https://doi.org/10.1039/C8SE00089A

[6] A. A. S. and R. K. Siti Musliha Mat Ghani, Nurul Ekmi Rabat, Abdul Rahman Abdul Rahim, Khairiraihanna Johari, "Amine Infused Fly Ash Grafted Acrylic Acid / Acrylamide," *Gels*, 2023.

[7] I. Cretescu, G. Soreanu, and M. Harja, "A low-cost sorbent for removal of copper ions from wastewaters based on sawdust/fly ash mixture," *International Journal of Environmental Science and Technology*, vol. 12, no. 6, pp. 1799–1810, Jun. 2015. https://doi.org/10.1007/s13762-014-0596-x

[8] H. Kim, J. Y. Jung, K. H. Park, P. Linga, Y. Seo, and C. D. Wood, "Enhanced Kinetic Performance of Amine-Infused Hydrogels for Separating CO2from CH4/CO2Gas Mixture,"

Energy and Fuels, vol. 35, no. 17, pp. 13889–13899, 2021. https://doi.org/10.1021/acs.energyfuels.1c01501

[9] C. White, E. Adam, Y. Sabri, M. B. Myers, B. Pejcic, and C. D. Wood, "Amine-Infused Hydrogels with Nonaqueous Solvents: Facile Platforms to Control CO2Capture Performance," *Ind Eng Chem Res*, vol. 60, no. 41, pp. 14758–14767, 2021. https://doi.org/10.1021/acs.iecr.1c02525

[10] S. N. A. Mazlan, S. Abd Rahim, S. Ghazali, and S. S. Jamari, "Optimization of N, N'-methylenebis(acrylamide), and ammonium persulfate content in carbonaceous/acrylic acid-co-acrylamide superabsorbent polymer," *Mater Today Proc*, vol. 57, pp. 1088–1094, 2022. https://doi.org/10.1016/j.matpr.2021.09.393

[11] B. S. Kaith, Saruchi, R. Jindal, and M. S. Bhatti, "Screening and RSM optimization for the synthesis of a Gum tragacanth-acrylic acid-based device for in situ controlled cetirizine dihydrochloride release," *Soft Matter*, vol. 8, no. 7, pp. 2286–2293, 2012. https://doi.org/10.1039/c2sm07033b

[12] A. S. M. Ghumman, R. Shamsuddin, M. M. Nasef, W. Z. Nisa Yahya, and A. Abbasi, "Optimization of synthesis of inverse vulcanized copolymers from rubber seed oil using response surface methodology," *Polymer (Guildf)*, vol. 219, no. December 2020, p. 123553, 2021. https://doi.org/10.1016/j.polymer.2021.123553

[13] S. N. Ainda Mazlan and S. S. Jamari, "Factorial experimental design for superabsorbent carbonaceous polymer through inverse suspension polymerization method," *IOP Conf Ser Mater Sci Eng*, vol. 523, no. 1, 2019. https://doi.org/10.1088/1757-899X/523/1/012021

[14] J. Ren, N. Zhan, J. Yang, and X. Xiang, "Regulated the Swelling Properties of Poly(acrylamide-co-acrylic acid) Hydrogel by Changing the Charge Density," *Macromol Chem Phys*, vol. 224, no. 22, pp. 1–5, 2023. https://doi.org/10.1002/macp.202300266

[15] A. A. S. and R. K. Siti Musliha Mat Ghani, Nurul Ekmi Rabat, Abdul Rahman Abdul Rahim, Khairiraihanna Johari, "Amine Infused Fly Ash Grafted Acrylic Acid / Acrylamide," *Gels*, 2023.

[16] A. Sami *et al.*, "Development of an amine-impregnated polymer aerogel for CO_2 capture," *Colloids Surf A Physicochem Eng Asp*, vol. 690, no. January, p. 133778, 2024. https://doi.org/10.1016/j.colsurfa.2024.133778

[17] A. Sami, K. Johari, F. F. Hilmi, and M. Rashid, "A new initiators-free technique for synthesizing stable amine-impregnated polymeric aerogel using electron beam radiation for CO2 capture," *Polymer (Guildf)*, p. 126987, 2024. https://doi.org/10.1016/j.polymer.2024.126987

[18] D. Nikjoo and F. Akhtar, "Structured emulsion-templated porous copolymer based on photopolymerization for carbon capture," *Journal of CO2 Utilization*, vol. 21, no. May, pp. 473–479, 2017. https://doi.org/10.1016/j.jcou.2017.08.007

Materials Research Proceedings 53 (2025) 155-163 https://doi.org/10.21741/9781644903575-15

Rubber waste tyre: Effect of temperature on pyrolysis char yield and properties

HO Tze Jing[1,a], RAMLI Raihan Mahirah[1,b*], MOHD HASHIM Muhammad Firdaus[1,c], SAIDIN Khairul Afizan[1,d]

[1]Universiti Teknologi PETRONAS, 32610, Bandar Seri Iskandar, Perak Darul Ridzuan, Malaysia

[a]tzejingfirst123@gmail.com, [b]raihan.ramli@utp.edu.my, [c]mohdfirdaus@utp.edu.my, [d]khairulafizan@utp.edu.my

Keywords: Waste Tyre Pyrolysis; Pyrolysis, Temperature, Char Yield, Char Properties

Abstract. Waste tyre pyrolysis is a viable technology to convert tyres into useful products such as pyrolysis oil, fuel and gas. Most of the research focus on investigating the effect of pyrolysis operating parameters on the oil product. With the increasing applications of the solid char produced, it is found that no detailed investigation on the effect of pyrolysis temperature on the char yield and properties. Thus, this project is aimed to determine the effect of pyrolysis temperature on tyre- derived char yield and properties. A total of three char samples were produced from rubber crumb, namely Char400, Char550 and Char700, corresponding to their pyrolysis temperature at 400°C, 550°C and 700°C respectively. Rubber crumb and these 3 char samples were sent to Fourier Transform Infrared Spectroscopy (FTIR), Elemental analysis of carbon, hydrogen, nitrogen and sulphur (CHNS), Scanning Electron Microscopy (SEM) and Brunauer-Emmett-Teller (BET) analysis for characterization purposes. It was found that increasing temperature from 400°C to 700°C led to significant char yield reduction from 76% to 36.25%. In terms of composition, Char400 has additional alkane and phenol functional groups indicating the main constituents in rubber crumb did not decompose completely at low temperatures. Furthermore, char produced at higher temperatures has lower carbon content, higher sulphur content, higher specific surface area, lower average pore width and more structured surface morphology. Overall, temperature affects char yield and properties significantly.

Introduction

More than three billion tyres are produced worldwide annually to support the growing demand for tyres in line with the accelerated development in the automobile industry [1]. When the tyre has reached its maximum usable period or when its tread has worn out, the tyre can no longer be used and will be disposed of. Statistics show that around one billion rubber waste tyres are produced worldwide every year, which accounted for 2% of the total amount of solid waste [2]. Landfilling and burning of waste tyres are not environmentally friendly waste tyre management methods as these methods release greenhouse gases and cause water and land pollution when the toxic chemicals leach from the waste tyres. In addition, the waste tyre is non-biodegradable, which will continue to pose health hazards to the community such as acting as breeding grounds for mosquitoes and life-threatening diseases if it is not managed properly [3].

These concerns had led to the development of more environmentally sustainable waste tyres management methods such as pyrolysis process. Applying pyrolysis technology, the waste tyres were heated in an oxygen-free environment to be converted into pyrolysis oil, solid char and non-condensable gas. Pyrolysis oil can be recovered as fuel, pyrolysis gas which has high heating value can serve as energy source and the solid char can be modified into commercial carbon black, activated carbon, catalyst support and filler in the modifying asphalt [1]. The yield and properties of the pyrolysis products are highly impacted by the process operating conditions thus it is crucial to comprehend how the operating parameters like temperature, heating rate and residence time influencing the yield and characteristics of products. Extensive research reported the effect of operating parameters on the yield of pyrolysis oil over the years due to its high potential to be fuel, but less attention on investigating the effects of operating parameters towards the char yield and properties. With the advancements in the application of char in various fields, it is significant to investigate the most important operating parameter of pyrolysis process which is temperature towards the char yield and properties.

Methodology

Char Production. Waste tyre was obtained in the rubber crumb form ranging from 0.6 mm to 1.18 mm sizes. In a typical pyrolysis process, 2g of rubber crumb was placed in a quartz boat and then the boat was transferred into the tube furnace (model number: MTI OTF-1200X-S-50-LVT). To study the impact of temperature on the char yield and properties, pyrolysis temperature of the process was varied and set to 400°C, 550°C, and 700°C respectively. Heating rate was set at 35 °C/min and residence time was set at 120 minutes following suitable range reported by previous studies[1,4]. Nitrogen was provided throughout the pyrolysis process to ensure heating was conducted in inert atmosphere. The solid char produced at 400°C, 550°C, 700°C were named as Char400, Char550 and Char700, respectively. The char yield and char yield percentage are calculated using Eq. 1 and Eq. 2, respectively.

$$W_{char} = W_{boat+char} - W_{boat} \tag{1}$$

$$Char\ Yield\ (wt\%) = \frac{W_{char}}{2\,g} \times 100\ \% \tag{2}$$

Where W_{char}, $W_{boat+char}$, W_{boat} and char yield (wt%) are weight of char produced after pyrolysis (g), total weight of sample boat and the char after pyrolysis (g), weight of sample boat without waste tyre before pyrolysis (g), and char yield in weight percentage, respectively.

Char Characterization. Four characterization tests including Fourier Infrared Spectroscopy (FTIR), Brunauer-Emmett- Teller (BET) surface area analysis, CHNS and scanning electron microscopy (SEM) and were all performed on rubber crumb (RC), Char400, Char550 and Char700. Characterization was done on rubber crumb too to provide better insights on how properties have changed when the rubber crumb was pyrolyzed to produce char. FTIR tests were performed using PerkinElmer Frontier FT-IR spectrometer with ATR method to determine the compositions and the functional groups present in the samples. The element analysis of the samples in terms of carbon, hydrogen, nitrogen and sulphur were conducted performed using Elementar VarioMicro CHNS analyzer. Micrometric TriStar II 3020 was used to perform nitrogen absorption tests on the samples to find out the surface area and porosity of the samples. The surface morphology of the samples were analysed using Phenom Pro (X) G6 Desktop SEM.

Results and Discussion

Effect of Temperature on Char Composition and Functional Groups. The similarities and differences between FTIR spectra of RC, Char400, Char550 and Char700 can be observed in Figure 1. From Figure 1, it was found that RC had many unique peaks that are not detected in the other 3 spectra. This indicated the rubber crumb had decomposed and been converted to another substance with different functional groups during the pyrolysis process. These peaks were at 3269.03 cm^{-1}, 2915.25 cm^{-1}, 2847.20 cm^{-1}, 2113.69 cm^{-1}, 1994.67 cm^{-1},1891.4 cm^{-1},1536.03 cm^{-1}, 1220.82 cm^{-1}, 823.05 cm^{-1}, 783.36 cm^{-1} and 686.07 cm^{-1}. The peaks found in this FTIR spectrum of rubber crumb were aligned to the reported natural rubber and styrene- butadiene rubber related functional groups for waste tyre rubber crumb [6]. For natural rubber, wavenumber between 1372 cm^{-1} and 1450 cm^{-1} were allocated for C−H bending of $-CH_2$ and $-CH_3$, peaks found at 1387.05 cm^{-1} and 1361.7 cm^{-1} in the spectrum were within the range and can be matched to functional groups related to natural rubber. Moreover, peak at 1600 cm^{-1} is attributed to C=C in 1,4-cis-isoprene(natural rubber) and butadiene (styrene-butadiene rubber). Peak detected at 738.36 cm^{-1} can be matched to monosubstituted benzene in SBR which has wavenumber between 698 cm^{-1} and 804 cm^{-1} [6]. Essentially, FTIR spectrum of rubber crumb showed that the raw material used for the experiment is waste tyre.

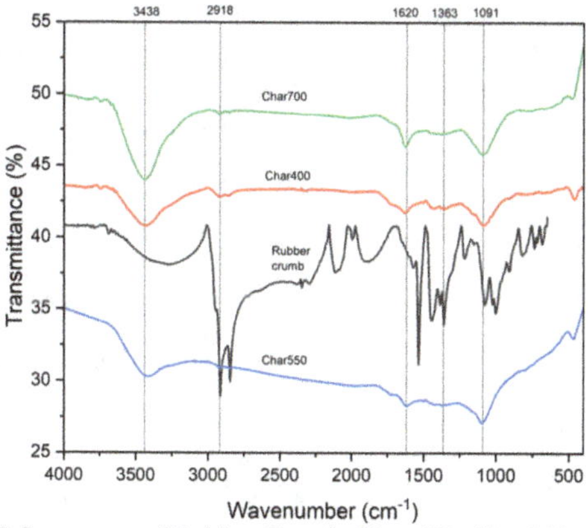

Fig. 1: FTIR Spectrums of Rubber Crumb, Char400, Char550 and Char700

From Figure 1, it can be clearly observed that Char400, Char550 and Char700 had very similar patterns in the FTIR analysis except the unique detection of peaks at 2918.3 cm^{-1} and 1363.14 cm^{-1} for Char400C. From this observation, it was suggested that pyrolysis temperature affects the composition and functional groups present in the char. Peak found at 2918.3 cm^{-1} and 1363.14 cm^{-1} indicated the presence of alkane (C−H stretching between 2840 cm^{-1} and 3000 cm^{-1}) and phenol (O−H bending between 1310 cm^{-1} and 1390 cm^{-1}) respectively[5]. This happened may be due to at lower pyrolysis temperatures, the pyrolysis was incomplete, C−H stretching and O−H bending were still not broken, leaving higher content of phenols, alcohols and other substance on the surface of char[1]. In terms of similarities between the chars, peaks were found at 3438.17 cm^{-1}, 3414.3 cm^{-1} and 3437.46 cm^{-1} for Char400, Char550 and Char700 respectively indicates presence of alcohol (wavenumber 3200 cm^{-1} and 3550 cm^{-1}). Strong peaks found at 1631.74 cm^{-1}, 1615.86 cm^{-1} and 1628.22 cm^{-1} for Char400, Char550 and Char700 respectively indicate the presence of C=C for conjugated alkene (1600 cm^{-1} and 1650 cm^{-1}). Conjugated alkene functional group was found in natural rubber (NR) and Styrene-Butadiene Rubber (SBR), showing that certain components in the rubber crumb will not decompose even after heating at high temperatures till 700°C. Furthermore, peaks found at 1088.1 cm^{-1}, 1097.6 cm^{-1} and 1091.28 cm^{-1} for Char400, Char550 and Char700, respectively which indicate the presence of C−O stretching for aliphatic ether (wavenumber 1085 cm^{-1} and 1150 cm^{-1}), is also found in the rubber crumb. Compared to the rubber crumb, monosubstituted benzene in SBR with wavenumber between 698 cm^{-1} and 804 cm^{-1} was not detected in the chars, indicating that the benzene had decomposed during the pyrolysis process. To summarize, based on the FTIR spectra obtained, temperature influenced the functional groups present because at lower temperature like 400°C, NR, SBR and volatiles did not decompose completely so there will still be C−H and O−H bonds not broken in Char400. However, at

Materials Research Proceedings 53 (2025) 155-163 https://doi.org/10.21741/9781644903575-15

temperature above 550°C, the effect of temperature was very minimal, no obvious differences were noticed between Char550 and Char700, this might be due to most components already degrade fully at high temperatures, leaving some important functional groups remains even at the highest pyrolysis temperature.

Effect of Temperature on Char CHNS Percentage. From Table 1, it was found that the carbon, hydrogen, nitrogen and sulphur content percentage in the char varies with the temperature change. For carbon content, rubber crumb had a high carbon content of 79.16%, Rising temperature from 400°C to 550°C resulted a decrease in the carbon content which was from 76.12% to 69.57%, this could be due to the decomposition of NR and SBR. The possible reason of Char400 having slightly higher carbon content than Char550 might be the pyrolysis is still incomplete at 400°C, which leads to more substances not decomposed yet, as supported by its FTIR spectrum. For Char700, its carbon content was the highest, this might be due to complete pyrolysis that had resulted the other components in very small amount and causing the percentage of carbon to be large. For hydrogen content, Char400 has lower hydrogen content compared to rubber crumb and Char700 also showed minimal amount of hydrogen left in the char. However, at 550°C, the hydrogen content is even higher than the hydrogen content in rubber crumb, further investigation should be carried out to find out the reasons. For nitrogen content, it can be observed that different pyrolysis temperatures did not lead to significant differences in all three char samples but it was proven that nitro compound present in rubber crumb had degraded in pyrolysis process. For sulphur content, all 3 char samples had higher sulphur content compared to rubber crumb. When temperature was increased, the sulphur content will be generally higher because the char yield is less and sulphur might react with zinc oxide (constituent added for vulcanization of tyres) to produce ZnS [7,8,9,10]. To conclude, as temperature rises, a decrease in carbon content and increase in sulphur content on the char will be expected.

Table 1: Composition of carbon, hydrogen nitrogen and sulphur in different samples

Sample	C (%)	H (%)	N (%)	S (%)
Rubber Crumb	79.16	7.427	2.26	0.062
Char400	76.12	3.627	0.02	1.828
Char550	69.57	7.472	0.01	0.292
Char 700	78.73	0.85	0.03	3.041

Effect of Temperature on Char Surface Area and Average Pore Width. From Table 2, rubber crumb has the smallest specific surface area and smallest average pore size. A positive increasing trend was observed in the specific surface area of char with the increase of temperature from 400°C to 700°C. Specific surface area of Char550 was around 30 m^2/g higher than Char400. Char700 had the highest specific surface area. Increase in specific surface area might be due to the pores were more expanded at temperature higher than 500°C, leading to much higher specific surface area in Char550 and Char700 compared to Char400. While at temperature lower than 500°C, the pores were not fully expanded and volatiles still attached on the surface, causing Char400 had much smaller specific surface area[11]. In addition, the average pore size reduced with the rising of temperature from 400°C to 700°C. Char400 has the highest pore width which is 2.306 nm whereas Char700 has the lowest pore width which is 9.166 nm. Similar to the specific surface parameter, the effect of temperature is more obvious when the temperature was increased from 400°C to 550°C, increasing above 550°C did not bring significant impact to the pore size. This trend can be explained by the blockage of char pores due to sintering at high temperatures. When the char pores were block, there will be reduction in micropore and mesopore volumes[12]. Nevertheless, all chars were found to be mesopore as they had pore size between 2nm and 50nm. To conclude, temperature affects the char's specific surface area and average pore width significantly.

Table 2: BET Specific Surface Area and Average Pore Width of different samples

Sample	BET Specific Surface Area (m^2/g)	Average pore width (nm)
Rubber Crumb	0.3857	2.30604
Char400	1.6180	26.18209
Char550	31.7418	10.15486
Char700	58.6786	9.16624

Effect of Temperature on Char Morphology. Figure 2 shows the SEM images of rubber crumb, Char400, Char550 and Char700 at 1000x and 10000x magnification. Effect of temperature on the changes on surface morphology can be observed through SEM Images. For rubber crumb, the sample had a compact and dense structure with very minimal pores at 1000x. When the magnification was increased to 10000x, the sample displayed an uneven structure with noticeable structures of irregular shapes on it. The irregular structures were suspected to be the other components present in the waste tyre such as tyres additives. For Char400, it showed compact structure as rubber crumb did at 1000x, very small pores were found at the surface indicating the pore starts to develop during the pyrolysis process but not fully expanded. When the sample was further magnified to 10000x, the surface looked smoother, exhibited less number of irregular structures on it and had more pores. For Char550, it can be observed that there were more small-sized fragments appearing on the surface at 1000x, the sample did not look as compact as in rubber crumb and in Char400.At 10000x, flat fragments with irregular shapes were found and these fragments appeared to be stacking with each other on the surface of the char. For Char700C, the sample appeared to have a rough but flat surface at 1000x. Noticeable number of small pores were found on the surface which further supported the increase in surface area of char at high temperature. Increasing the magnification till 10000x, flake-shaped structures of various sizes were arranged in an irregular pattern over the surface aligned to the findings about char will be arranged in a more structured geometry at temperature about 800°C [1]. This morphology had not been observed in any other char samples, indicating the morphology of the char changed significantly at high pyrolysis temperature. In overall, the sample started to develop pores when temperature was increased. The higher the temperature, the pore appeared to be more expanded. In addition, higher temperatures facilitated the breakdown of the original structures on the surface and promoted growth of flat and irregular shaped sheets on the char surface. Based on the results, it was suggested to set the pyrolysis temperature to be at least higher than 700°C in order to change the morphology of the char to be significantly different from the rubber crumb.

Table 3: SEM Images of different samples at different magnifications. SEM image of (a) Rubber Crumb at 1000x magnification; b) Rubber Crumb at 10000x magnification; c) Char400 at 1000x magnification; d) Char400 at 10000x magnification; e) Char550 at 1000x magnification; f) Char550 at 10000x magnification; g) Char700 at 1000x magnification; h)Char700 at 10000x magnification

Magnification	1000x	10000x
Rubber Crumb		
Char400		
Char550		

Effect of Temperature on Char Yield. From Figure 3, it can be observed that char yield decreased when the pyrolysis temperature was increased. When the temperature is 400 °C, the char yield is 76%. When the temperature was increased to 550 °C, the char yield reduced to 48.2%. Further increasing the temperature to 700 °C, the char yield continued to show a decreasing trend and recorded 36.25% yield only. The overall decreasing tread shown in the char yield against temperature can be explained by the its thermo- gravimetric analysis(TGA) which demonstrated the mass change of the sample over time when temperature was increased[1].

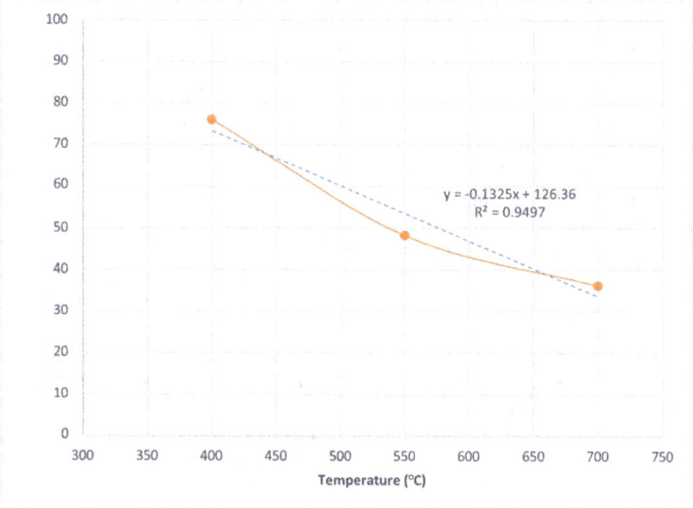

Fig. 3: Graph of char yield vs temperature

From TGA graph, waste tyres started to decompose around 150 °C. At 150°C to 200 °C, the plasticizer and additives in the tyres started to decompose. When the temperature was further increased to higher than 250 °C, some of the waste tyres decomposed to produce pyrolysis oil and gas. Natural rubber (NR) and Styrene-Butadiene rubber (SBR) which are the most abundant component in the waste tyres degraded at temperatures between 250 °C and 500 °C. Furthermore, the volatiles decomposed fully at temperatures around 500 °C to 600 °C. When the temperature is above 600 °C, the products started to undergo cracking to produce small molecular products [1]. From table 4, the char yield decreases significantly when the temperature was increased from 400 °C to 550 °C, this phenomena might be due to NR and SBR did not decompose fully at 400 °C, but at 550 °C, all the NR and SBR have decomposed and converted to pyrolysis oil and gas, thus leaving less amount of solid product. Moreover, the small char yield reduction when the temperature was increased from 550 °C to 700 °C may be due to the decomposition of volatiles and cracking of larger molecules to smaller molecules. In overall, char yield decreased following the increase in pyrolysis temperature. Equation describes the relationship between these two variables is as followed:

$$Char\ Yield = -0.1325 \times Temperature + 126.36 \quad (3)$$

This equation is only valid from 400 °C to 700 °C. It has high R^2 value too which is 0.9497.

Table 4: Char yield produced at different temperatures

Sample	W_{char} (g)	Char Yield (%)
Char400	1.520	76.00
Char550	0.964	48.20
Char 700	0.725	36.25

Conclusion

Effect of temperature on the char produced from waste tyre pyrolysis was studied in terms of char yield and properties such as the composition, functional groups, elemental composition, specific surface are, average pore width and surface morphology. To summarize, increase in pyrolysis temperature cause the char yield to decrease. For the composition and functional groups, characterization results shown that all chars produced have similar pattern except Char400 still have some alkane and phenols not being decomposed at low temperatures. In CHNS analysis, it was found that increasing temperature will reduce the carbon content and increase the sulphur content of the char. Moreover, when temperature was increased, the specific surface area increased too but a decline in average pore width was observed as well. To determine the surface morphology, SEM images of the char were investigated and it was found out that the char had less pores at low temperatures because the pores are not fully expanded. High temperature caused the morphology to change significantly, leading to more structures with ordered arrangement covering the surface of the char. To conclude, temperature is a crucial operating parameter that will affect the yield and properties of char produced from waste tyre pyrolysis. Researchers that wish to modify char obtained from waste tyre pyrolysis for applications like activated carbon are recommended to run the pyrolysis at optimum temperature that favours the production of char that suits the needs of the application. This research has also enhanced knowledge on recovering high yield of char with high quality. The insights can be utilized to improve char utilization that further reduced environmental impacts of solid waste tyre and also facilitating circular economy of tyres.

Acknowledgement

The authors acknowledge financial support from Universiti Teknologi PETRONAS, Malaysia under University Internal Research Funding (URIF) (cost centre 015LB0-123).

References

[1] N. Gao, F. Wang, C. Quan, L. Santamaria, G. Lopez, and P. T. Williams, 'Tire pyrolysis char: Processes, properties, upgrading and applications', *Prog. Energy Combust. Sci.*, vol. 93, p. 101022, Nov. 2022. https://doi.org/10.1016/j.pecs.2022.101022

[2] S. Dabic-Miletic, V. Simic, and S. Karagoz, 'End-of-life tire management: a critical review', *Environ. Sci. Pollut. Res.*, vol. 28, no. 48, pp. 68053–68070, Dec. 2021. https://doi.org/10.1007/s11356-021-16263-6

[3] H. Chittella, L. W. Yoon, S. Ramarad, and Z.-W. Lai, 'Rubber waste management: A review on methods, mechanism, and prospects', *Polym. Degrad. Stab.*, vol. 194, p. 109761, Dec. 2021. https://doi.org/10.1016/j.polymdegradstab.2021.109761

[4] G. I. Danmaliki and T. A. Saleh, 'Influence of conversion parameters of waste tires to activated carbon on adsorption of dibenzothiophene from model fuels', *J. Clean. Prod.*, vol. 117, pp. 50–55, Mar. 2016. https://doi.org/10.1016/j.jclepro.2016.01.026

[5] 'Infrared Spectroscopy Absorption Table', Chemistry LibreTexts. Accessed: Feb. 25, 2024. [Online]. Available: https://chem.libretexts.org/Ancillary_Materials/Reference/Reference_Tables/Spectroscopic_Reference_Tables/Infrared_Spectroscopy_Absorption_Table

[6] P. C. Tsipa, M. M. Phiri, S. A. Iwarere, N. M. Mkhize, M. J. Phiri, and S. P. Hlangothi, 'The effect of pre-pyrolysis chemical treatment of waste tyre rubber crumbs: comparison between pre-treated and conventional waste tyre-derived oil', *J. Chem. Technol. Biotechnol.*, vol. 98, no. 5, pp. 1279–1289, 2023. https://doi.org/10.1002/jctb.7346

[7] F. Cataldo, 'Preparation of Pyrolytic Carbon Black from Scrap Tire Rubber Crumb and Evaluation in New Rubber Compounds', *Macromol. Mater. Eng.*, vol. 290, no. 5, pp. 463– 467, May 2005. https://doi.org/10.1002/mame.200400388

[8] A. Chaala, H. Darmstadt, and C. Roy, 'Acid-base method for the demineralization of pyrolytic carbon black', *Fuel Process. Technol.*, vol. 46, no. 1, pp. 1–15, Jan. 1996. https://doi.org/10.1016/0378-3820(95)00044-5

[9] S.-Q. Li, Q. Yao, Y. Chi, J.-H. Yan, and K.-F. Cen, 'Pilot-Scale Pyrolysis of Scrap Tires in a Continuous Rotary Kiln Reactor', *Ind. Eng. Chem. Res.*, vol. 43, no. 17, pp. 5133–5145, Aug. 2004. https://doi.org/10.1021/ie030115m

[10] E. Manchón-Vizuete, A. Macías-García, A. Nadal Gisbert, C. Fernández-González, and V. Gómez-Serrano, 'Preparation of mesoporous and macroporous materials from rubber of tyre wastes', *Microporous Mesoporous Mater.*, vol. 67, no. 1, pp. 35–41, Jan. 2004. https://doi.org/10.1016/j.micromeso.2003.10.002

[11] M. F. Laresgoiti, B. M. Caballero, I. de Marco, A. Torres, M. A. Cabrero, and M. J. Chomón, 'Characterization of the liquid products obtained in tyre pyrolysis', *J. Anal. Appl. Pyrolysis*, vol. 71, no. 2, pp. 917–934, Jun. 2004. https://doi.org/10.1016/j.jaap.2003.12.003

[12] Y. Alhassan, N. Kumar, and I. M. Bugaje, 'Catalytic upgrading of waste tire pyrolysis oil via supercritical esterification with deep eutectic solvents (green solvents and catalysts)', *J. Energy Inst.*, vol. 89, no. 4, pp. 683–693, Nov. 2016. https://doi.org/10.1016/j.joei.2015.05.003

Enhanced CO_2 adsorption kinetics with hierarchical zeolite 5A

SYED ABDUL Moiz Hashmi[1,a], NG Xiuxin[1,b], CHUAH Chong Yang[2,c,*], JOHARI Khairiraihana[1,3,d], POON Wai Ching[4,e]

[1]Department of Chemical Engineering, Universiti Teknologi PETRONAS, Bandar Seri Iskandar, 32610 Perak, Malaysia

[2]Department of Chemical Engineering, Faculty of Engineering, University of Malaya, Kuala Lumpur, 50603, Malaysia

[3]Centre of Carbon Capture, Utilisation, and Storage, Institute of Sustainable Energy & Resources, Universiti Teknologi PETRONAS, Bandar Seri Iskandar, 32610 Perak, Malaysia

[4]School of Management & Marketing, Faculty of Business & Law, Taylor's University, Selangor, 47500, Malaysia

[a]syed_22009552@utp.edu.my, [b]xiu_18002461@utp.edu.my, [c]chongyang.chuah@utp.edu.my, [d]khairiraihana.j@utp.edu.my, [e]waiching.poon@utp.edu.my

Keywords: Zeolites, Mesoporous Materials, Carbon Capture, Hierarchical Zeolite

Abstract. In this work, hierarchical zeolite LTA (H-4A and H-5A) was synthesized using dimethyloctadecyl[3-(trimethoxysilyl)propyl]ammonium chloride surfactant and characterized through various analyses. N_2 physisorption analysis shows that H-4A and H-5A exhibited a hierarchical structure with coexisting microporous and mesoporous domains. The existence of mesoporosity within the crystals is accountable for the decreased diffraction peaks as observed in the XRD analysis of H-4A and H-5A. This observation is also consistent with the FESEM images, where H-4A and H-5A with hierarchical structures display as spherical particles while bulk zeolite LTA (B-4A and B-5A) showcase as well-defined and uniform cubic shape crystals. Investigation of CO_2 and N_2 adsorption has shown that the synthesized zeolites displayed a significant CO_2 adsorption capacity compared to N_2. Furthermore, by measuring the CO_2 fractional uptake, it is concluded that H-5A improves CO_2 adsorption kinetics significantly as compared to B-5A. This attractive characteristic is expected to be able to be applied effectively in the carbon capture and storage process as an improved overall processing rate of adsorbents under repetitive adsorption-desorption cycling can be achieved.

Introduction

Zeolites are microporous crystalline aluminosilicates formed by silicon-oxygen tetrahedron (SiO_4) and aluminium-oxygen tetrahedron (AlO_4), which show widespread applications in many industrial processes such as adsorption, separation, and catalysis[1]. Their high cation-exchange capacity, high adsorption, and hydration-dehydration properties are because of the aluminosilicate's framework. Adsorbent applications of zeolites are widely used in the chemical industry's purification and separation processes, including as CO_2 removal from natural gas (CH_4), sweetening of petroleum gas, and sulfur compound removal[2]. Zeolites can be found in nature as a constituent in volcanic and basaltic rock cavities, but their crystallinity and purity are insufficient for most industrial purposes[3]. Hence, they are usually synthesized through synthetic means to improve their properties such as crystallinity, purity, surface area, selectivity, porosity, and thermal stability.

Numerous zeolites have been applied for gas adsorption with various molecular sizes such as Type X, Type Y, ZSM-5, and LTA (Zeolite Type A or Zeolite A). Zeolites X and Y belong to Faujasite (FAU-type zeolite) group which is widely synthesized industrially. Zeolite X exhibits a lower silica-alumina ratio, as well as less thermal and hydrothermal stability compared to zeolite Y. On the other hand, Zeolites Socony Mobil-5 (ZSM-5) is a synthetic zeolite that has a pore diameter of 5 Å (angstroms) with Si/Al ratios higher than five[4]. Zeolite LTA is usually present in the sodium- exchanged form, which can be expressed under the chemical formula of $Na_{12}[(AlO_2)_{12}(SiO_2)_{12}] \cdot 27H_2O$, where sodium ions are the exchangeable extra-framework cations. It is also frequently referred to as zeolite 3A, 4A, or 5A if the material is potassium, sodium, or calcium ion-exchanged respectively, depending on the exchangeable cation in the zeolite structure. Among the various zeolite types, zeolite LTA has been utilized extensively for gas adsorption due to its superior properties and well-established production method. It is reported that zeolite Ca-A has remarkable performance features for post-combustion CO_2 capture, including highly selective CO_2 uptake and rapid adsorption kinetics[5].

It has been well noted that other than the synthesis method, tunable pore sizes of zeolite adsorbents' structure also play a vital role in effective gas adsorption. For instance, zeolite LTA with various micropore sizes (3A, 4A, and 5A) can be created via ion exchange. Through a detailed literature review, the majority of studies only focus on varying the synthesis method of zeolite, with limited investigation towards tuning the mesoporosity of zeolite adsorbents[3]. The enormous pore volumes created by the mesopores can be useful for improving the accessibility of gas molecules. Currently, only a few mesoporous materials (MCM-41, SBA-15, and COK-12) have been successfully performed through a surfactant templating strategy[3]. Hence, the goal of this project is the development of zeolite LTA adsorbent by creating hierarchical structures (i.e., micropores with mesoporous domains) through a soft templating technique.

Materials and Methods

Materials: Dimethyl-octadecyl [3-(trimethoxysilyl)propyl] ammonium chloride solution (TPOAC, 42 wt% in methanol), calcium nitrate tetrahydrate ($Ca(NO_3)_2 \cdot 4H_2O$), sodium metasilicate pentahydrate ($Na_2O_3Si \cdot 5H_2O$), sodium hydroxide (NaOH) and sodium aluminate ($AlH_4O_4 \cdot Na$) were purchased from Sigma-Aldrich, Merck and it was used without additional purifications.

Synthesis of hierarchical zeolite:[6, 7] Initially, the precursor gel was prepared at the molar composition of $Al_2O_3/Na_2O/SiO_2/H_2O$/TPOAC of 20/80/30/4800/2.5. Solution 1 ($Na_2O_3Si \cdot 5H_2O$, NaOH & TPOAC) and solution 2 ($AlH_4O_4 \cdot Na$) were prepared with distilled water using two separate polypropylene bottles. After vigorously agitating the solutions for 2 hours, two solutions were mixed and shaken for 5 minutes. Next, the resultant solution was heated at 100°C for 4 hours. The products were collected using vacuum filtration and washed with a large amount of distilled water to eliminate unreacted raw materials and contaminants. The final product was dried overnight at 100°C in a convection oven to purify the product by removing volatile substances and moisture. The product was then calcinated at 550°C in a furnace with a heating rate of 1°C/min. This step aided in the removal of TPOAC, which is used to create mesoporosity, thus allowing the formation of hierarchical zeolite 4A (indicated as H-4A herein). Subsequently, ion exchange from H-4A to hierarchical zeolite 5A (indicated as H-5A herein) was performed by preparing 50 mL of 0.5 M of calcium nitrate solution. The solution was vigorously stirred at 60°C for 12 hours. The resultant product was collected by vacuum filtration and rinsed with a large amount of deionized water. The samples were dried at 80°C for 24 hours to remove excess moisture.

Synthesis of bulk zeolite: The synthesis of bulk zeolite 4A (B-4A) and 5A (B-5A) with only the microporous domain was performed using the same technique as hierarchical zeolite above but without the inclusion of TPOAC.

Characterization: X-ray diffraction (XRD), which was performed to verify the crystal structures was performed at ambient conditions using a Bruker D2 phaser diffractometer that is allocated with Cu-Kα (1.5418 Å) source. The 2θ range of 5 - 40° is selected for this investigation.

Next, the structural morphology of the crystals was performed using field-emission scanning electron microscopy (FESEM). The samples were first coated with platinum before the measurements were taken, based on the acceleration voltage of 5 kV. The porosity properties of the synthesized zeolite were measured using N_2 physisorption isotherm at 77 K under P/P_o range from 0 to 1 by using an automated volumetric gas sorption analyser (Quantachrome Autosorb). The samples were outgassed at 250°C for 24 hours to ensure effective removal of the residual solvents. Barrett-Joyner-Halenda (BJH) algorithm was used to analyse the samples' mesopore size distribution, which was obtained from the desorption branch of the isotherms. The isotherms' data were used to compute the surface areas and pore volumes of the zeolite samples. Lastly, the CO_2 and N_2 adsorption properties of the zeolite samples were measured using a gas sorption analyser (Quantachrome IsorbHP1) by conducting the measurements at 25°C and 40°C at the pressure range from 0 to 1 bar. The CO_2 adsorption kinetics were measured at 25°C and 40°C under the fixed dosing pressure of 1 bar to monitor the pressure drop in the adsorption cell.

Result and Discussion

Synthesis of bulk and hierarchical zeolite 4A and 5A: The porosity properties of B-4A, B-5A, H-4A, and H-5A were described using N_2 physisorption isotherms at 77 K. **Figure 1 (a)** clearly shows that B-5A has high N_2 sorption at low P/Po, followed by H-5A. Furthermore, based on the IUPAC classification of adsorption isotherms, the bulk and hierarchical zeolite 5A adsorption isotherms are categorized as type I and type IV, respectively. As observed, type I isotherms are provided by microporous materials with a small external surface area, which is B-5A while the N_2 adsorption isotherm indicated partially no adsorption in B-4A; N_2 is almost completely excluded at temperatures below 173 K in microporous zeolite 4A[8]. On the contrary, type IV(a) isotherms are acquired by both H-4A and H-5A. This is because mesoporous structures were successfully created in both hierarchical zeolites using soft templating. According to **Figure 1 (b)**, the adsorption amount when $P/P_o < 0.6$ increases moderately with increasing relative pressure; this continuous increase in adsorption amount indicates multilayer N_2 adsorption in the mesopores. Next, the adsorption of N_2 of H-4A and H-5A increases steeply in the $P/P_o > 0.6$ range, and hysteresis develops, which could be due to multilayer formation near the saturation vapor pressure, followed by capillary condensation in mesopores[9].

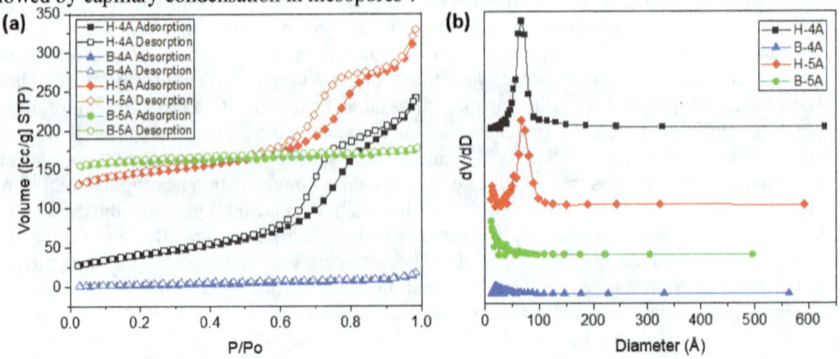

*Figure 1 **(a)** N_2 physisorption isotherms of B-4A, B-5A, H-4A, and H-5A (shaded and unshaded symbols describe the adsorption and desorption isotherms); **(b)** BJH pore distribution of B-4A, B-5A, H-4A, and H-5A.*

Table 1 summarizes the surface areas and pore volumes of the synthesised samples by utilizing N_2 physisorption isotherms. The Brunauer-Emmett-Teller (BET) and Langmuir surface areas of

Materials Research Proceedings 53 (2025) 165-173 https://doi.org/10.21741/9781644903575-16

zeolite 5A were observed to be greater than zeolite 4A. This is because the pore size of zeolite 5A (approximately 5Å) is larger than that of zeolite 4A (approximately 4Å), a greater amount of N_2 can be adsorbed by the zeolite 5A with larger pore size. This resulted in zeolite 5A having a larger total pore volume than zeolite 4A. Furthermore, when compared to B-4A (BET surface area: 6.69 m^2/g and Langmuir surface area: 11.8 m^2/g), the BET and Langmuir surface areas of H-4A were significantly improved to 143 and 199 m^2/g, respectively. On the other hand, **Table 1** shows that the BET and Langmuir surface area of H-5A is lower than that of B-5A, which may be due to the formation of mesoporous structures that reduce the amount of zeolite LTA. Since zeolite 4A contains very little mesoporosity, therefore the available micropore apertures are too narrow for effective diffusion of N_2 gas[10]. As a result, the micropore area and volume for both B-4A and H-4A were found to be zero. Aside from that, the micropore surface area and volume (510 m^2/g and 0.237 cc/g) observed in B-5A decreased significantly in H-5A (336 m^2/g and 0.152 cc/g), indicating a loss in microporosity while increasing mesoporosity. It should be noted that the external surface area of zeolite 4A is similar to BET surface area. On the contrary, BET surface area of zeolite 5A (B-5A: 534 m^2/g, H-5A: 492 m^2/g) is greater than its external surface area (B-5A: 24.5 m^2/g, H-5A: 156 m^2/g) as their microporosity and mesoporosity structures have occupied the zeolite domains, resulting in small external surface area[11].

***Table 1** Surface areas and pore volumes of bulk and hierarchical zeolite 4A and 5A, based on N_2 sorption isotherm at 77 K*

Sample	S_{BET}[a] (m^2/g)	$S_{Langmuir}$[a] (m^2/g)	A_{micro}[b] (m^2/g)	$S_{external}$[b] (m^2/g)	V_{micro}[b] (cc/g)	V_{total}[c] (cc/g)
B-4A	6.69	11.8	0	6.69	0	0.0258
H-4A	143	199	0	143	0	0.373
B-5A	534	705	510	24.5	0.237	0.274
H-5A	492	652	336	156	0.152	0.509

Note: [a] BET and Langmuir surface area (SBET and SLangmuir) were determined at P/Po = 0.05-0.2; [b] Micropore surface area, external surface area, and micropore volume (Amicro, Sexternal, and Vmicro) were determined at P/Po = 0.4-0.6 using t-plot method; [c] Total pore volume (Vtotal) was determined at P/Po = 0.99.

Moreover, the XRD patterns were examined to verify the crystallinity of the B-4A, B-5A, H-4A, and H-5A. **Figure 2 (a)** shows that despite the zeolites' material and the use of surfactant, the overall structural crystallinity of all the zeolite LTA samples remains. Further verification with the simulated XRD pattern (**Figure 2 (b)**)[12] indicated that the reported samples were indeed comparable. The diffraction peaks of the samples are consistent with the diffraction peaks reported. Nevertheless, the diffraction peaks of H-4A and H-5A were comparatively lower than that of B-4A and B-5A. This is because the presence of intracrystalline mesoporosity in the zeolites 4A and 5A with hierarchical structures decreases the number of zeolite domains in the sample[13].

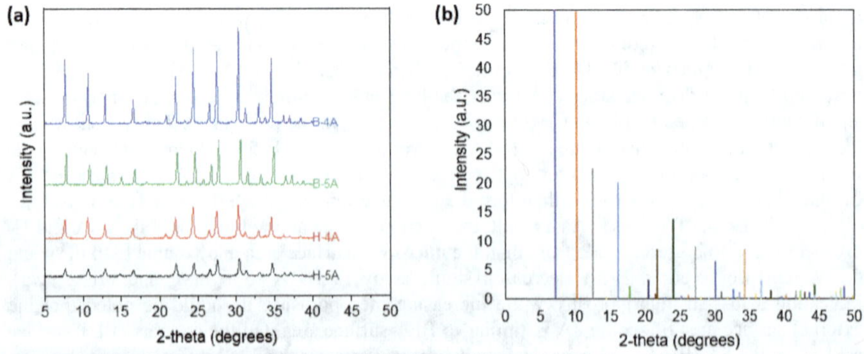

Figure 2 (a) XRD patterns of B-4A, B-5A, H-4A and H-5A, (b) Simulated XRD pattern for Zeolite Linde Type A (LTA)[12]

Figure 3 shows the microstructure or morphology of the B-4A, H-4A, B-5A and H-5A respectively. Field emission scanning electron microscopy (FESEM) was used to examine the morphology of the zeolite LTA. According to **Figure 3 (a)** and **(c)**, both zeolites B-4A and B-5A have well-defined and uniform cubic shape crystals which confirms the samples' pure phase and high crystallinity[14]. According to the FESEM images, the B-4A and B-5A's particle sizes are approximately 3 μm. When a hierarchical structure is formed, small crystals assemble into larger spherical particles and thus form secondary pores with less crystallinity (**Figure 3 (b)** and **(d)**), presumably in the mesopore range. The presence of intracrystalline mesoporosity in zeolites 4A and 5A with surfactant addition causes the size of the crystal domains to moderately decrease. As a result, the repeating crystalline structure is disrupted, which corresponds to the low intensity of diffraction peaks observed in the XRD pattern (**Figure 2 (a)**) for both hierarchical zeolites.

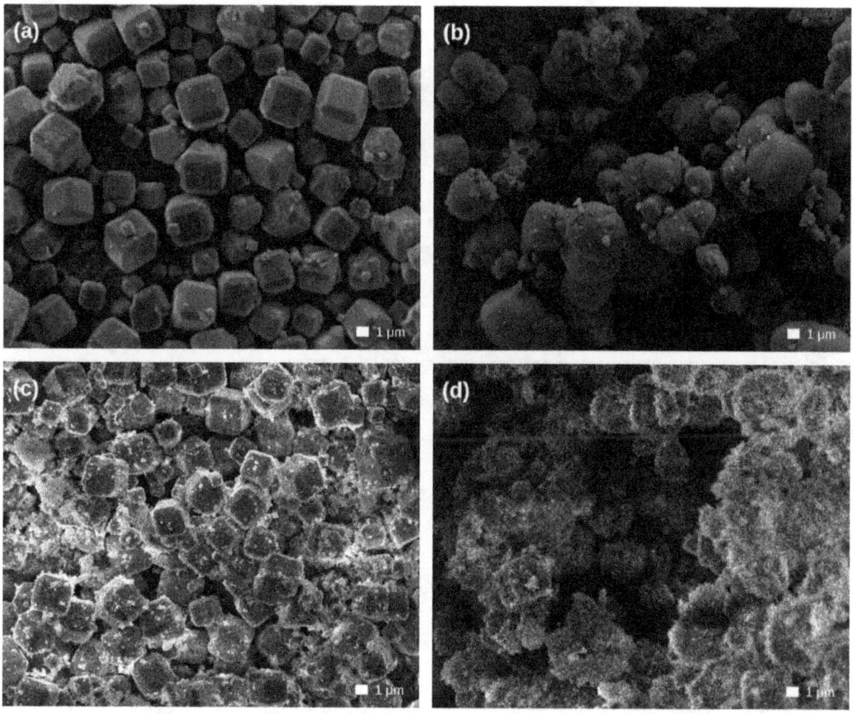

***Figure 3** FESEM images of (a) B-4A; (b) H-4A; (c) B-5A; (d) H-5A*

CO_2 and N_2 adsorption of bulk and hierarchical zeolite 5A: In the subsequent investigation, B-5A and H-5A with greater surface areas and pore volumes compared to zeolite 4A (**Table 1**) have been chosen as the representative samples for the pure component gas adsorption process to evaluate their affinities towards CO_2 and N_2. The difference in CO_2 and N_2 adsorption capacities is desirable for post-combustion CO_2 capture, namely CO_2 removal from flue gas[15]. **Figure 4 (a)** and **(b)** shows the CO_2 adsorption isotherm while **Figure 4 (c)** and **(d)** presents the N_2 adsorption isotherm for both B-5A and H-5A, measured at 25 and 40°C, respectively. Based on **Figure 4**, the CO_2 and N_2 adsorption were greater on B-5A than on H-5A, which is due to the introduction of mesoporous domains in the zeolitic framework of H-5A that causes a slight reduction in the number of available active sites. Hence, comparatively, less quantity of CO_2 or N_2 molecules can be adsorbed on H-5A as compared to the B-5A. Apart from that, it was observed that as the temperature rose from 25 to 40°C, the quantity of CO_2 and N_2 molecules that can be absorbed decreases. Accordingly, as the temperature rises, the internal energy and mobility of the adsorbed molecules in the pores of the zeolite increase, causing the adsorbate's interaction with the surface to decrease. Hence, this reduces the adsorption strength of the gas molecules on the zeolite adsorbent surface.

Materials Research Proceedings 53 (2025) 165-173 https://doi.org/10.21741/9781644903575-16

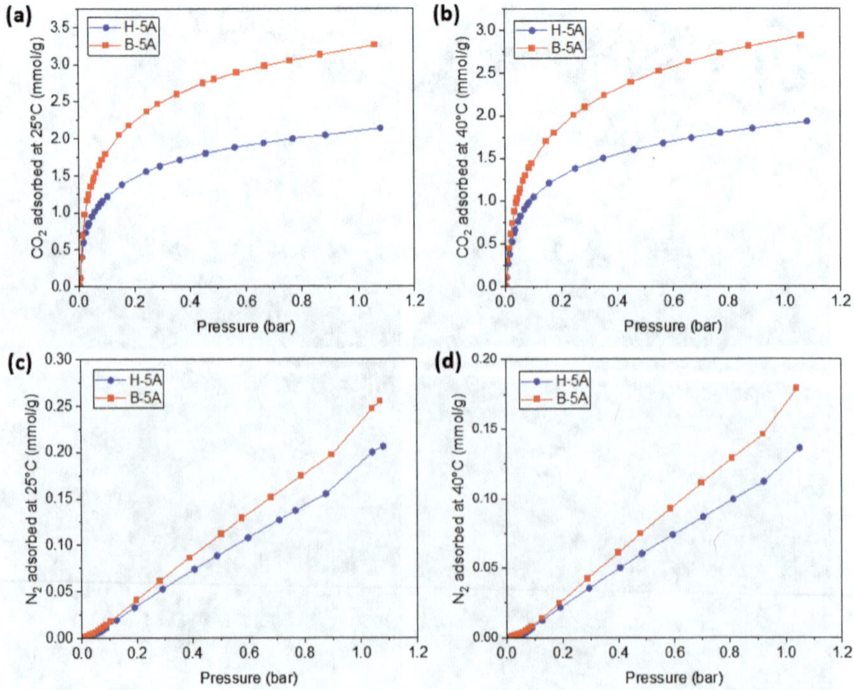

***Figure 4 (a, b)** CO_2 and **(c, d)** N_2 adsorption isotherm of B-5A and H-5A measured at 25 and 40°C, respectively*

Apart from that, CO_2 adsorption kinetics in an adsorbent is critical as it dictates the overall processing rate of a particular adsorbent under repetitive adsorption/desorption cycling process. Hence, the CO_2 uptake rate was measured using the dosing pressure of 1 bar at two different temperatures, with the findings illustrated in **Figure 5**. According to **Figure 5**, despite temperature variations, H-5A allows impressively rapid CO_2 uptake in a shorter time compared to B-5A. At 25 and 40°C, the H-5A required roughly 0.66 and 0.53 min, respectively, to reach a fractional uptake of 0.9. In general, higher CO_2 fractional uptake was observed at higher temperatures, which could be attributed to the expansion of the zeolite structure due to thermal energy. In contrast, B-5A required 0.90 and 0.94 min to reach the same CO_2 fractional uptake at 25 and 40°C respectively. As a result, creating micropores and mesopores in zeolite 5A is very beneficial in improving CO_2 adsorption kinetics.

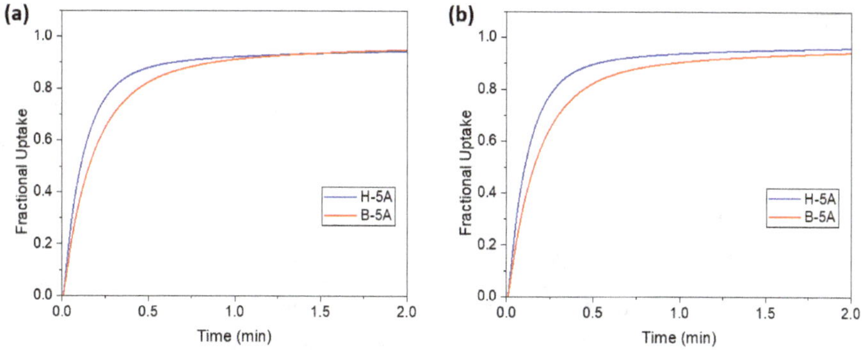

Figure 5 *CO_2 adsorption kinetics of B-5A and H-5A at 25 and 40°C, respectively*

Conclusion

In conclusion, hierarchical zeolite LTA has been successfully synthesized hydrothermally using TPOAC surfactant. This was verified through characterization such as N_2 physisorption isotherm, XRD analysis, FESEM, and gas adsorption analysis. Both H-4A and H-5A with mesoporous structures obtained type IV(a) isotherms while achieving significantly high N_2 adsorption. N_2 physisorption studies show that B-4A and B-5A are completely microporous, whereas H-4A and H-5A exhibit a hierarchical structure with both microporous and mesoporous domains coexisting. Furthermore, all synthesized zeolites displayed significant CO_2 adsorption capacity compared to N_2, indicating its usefulness in various applications such as carbon capture and storage, where the goal is to selectively capture and store CO_2 from the industrial processes. Additionally, CO_2 fractional uptake rate which is desirable for evaluating the performance features of zeolite 5A for post-combustion CO_2 capture was measured. Despite temperature variations, H-5A allows impressively rapid CO_2 uptake at a much shorter time compared to B-5A. This study concluded that the development of both microporous and mesoporous domains in zeolite 5A improves CO_2 adsorption kinetics significantly.

Acknowledgment

This work is supported by the Yayasan Universiti Teknologi PETRONAS-Fundamental Research Grant (YUTP-FRG, Cost Centre: 015LC0-486 and 015LC0-508). The authors would also like to thank University of Malaya for the funding support.

References

[1] Faust, S. D.; Aly, O. M., Adsorption processes for water treatment. Elsevier: 2013.

[2] Indarto, A.; Putra, I.; Noersalim, S.; Hartanto, Y.; Handojo, L. In Zeolites as adsorbent materials for decolorization of crude terpineol, IOP Conference Series: Materials Science and Engineering, IOP Publishing: 2019; p 012021. https://doi.org/10.1088/1757-899X/599/1/012021

[3] Chuah, C. Y.; Goh, K.; Yang, Y.; Gong, H.; Li, W.; Karahan, H. E.; Guiver, M. D.; Wang, R.; Bae, T.-H., Harnessing filler materials for enhancing biogas separation membranes. Chem. Rev. 2018, 118 (18), 8655-8769. https://doi.org/10.1021/acs.chemrev.8b00091

[4] Sadeghbeigi, R., Fluid catalytic cracking handbook: An expert guide to the practical operation, design, and optimization of FCC units. Butterworth-Heinemann: 2020.

[5] Bae, T.-H.; Hudson, M. R.; Mason, J. A.; Queen, W. L.; Dutton, J. J.; Sumida, K.; Micklash, K. J.; Kaye, S. S.; Brown, C. M.; Long, J. R., Evaluation of cation-exchanged zeolite adsorbents

for post-combustion carbon dioxide capture. Energy Environ. Sci. 2013, 6 (1), 128-138. https://doi.org/10.1039/C2EE23337A

[6] Nguyen, T. H.; Kim, S.; Yoon, M.; Bae, T. H., Hierarchical zeolites with amine-functionalized mesoporous domains for carbon dioxide capture. ChemSusChem 2016, 9 (5), 455-461. https://doi.org/10.1002/cssc.201600004

[7] Li, W.; Goh, K.; Chuah, C. Y.; Bae, T.-H., Mixed-matrix carbon molecular sieve membranes using hierarchical zeolite: A simple approach towards high CO2 permeability enhancements. Journal of membrane science 2019, 588, 117220. https://doi.org/10.1016/j.memsci.2019.117220

[8] Cho, K.; Cho, H. S.; de Menorval, L.-C.; Ryoo, R., Generation of mesoporosity in LTA zeolites by organosilane surfactant for rapid molecular transport in catalytic application. Chem. Mater. 2009, 21 (23), 5664-5673. https://doi.org/10.1021/cm902861y

[9] Mousavi, H.; Darian, J. T.; Mokhtarani, B., Enhanced nitrogen adsorption capacity on Ca2+ ion-exchanged hierarchical X zeolite. Sep. Purif. Technol. 2021, 264, 118442. https://doi.org/10.1016/j.seppur.2021.118442

[10] Panda, D.; Kumar, E. A.; Singh, S. K., Introducing mesoporosity in zeolite 4A bodies for Rapid CO2 capture. J. CO2 Util. 2020, 40, 101223. https://doi.org/10.1016/j.jcou.2020.101223

[11] Nguyen, T. H.; Gong, H.; Lee, S. S.; Bae, T. H., Amine-Appended Hierarchical Ca-A Zeolite for Enhancing CO2/CH4 Selectivity of Mixed-Matrix Membranes. ChemPhysChem 2016, 17 (20), 3165-3169. https://doi.org/10.1002/cphc.201600561

[12] Tounsi, H.; Mseddi, S.; Djemel, S., Preparation and characterization of Na-LTA zeolite from Tunisian sand and aluminum scrap. Physics Procedia 2009, 2 (3), 1065-1074. https://doi.org/10.1016/j.phpro.2009.11.064

[13] Sachse, A.; Garcia-Martinez, J., Surfactant-templating of zeolites: from design to application. Chem. Mater. 2017, 29 (9), 3827-3853. https://doi.org/10.1021/acs.chemmater.7b00599

[14] Bae, T.-H.; Liu, J.; Thompson, J. A.; Koros, W. J.; Jones, C. W.; Nair, S., Solvothermal deposition and characterization of magnesium hydroxide nanostructures on zeolite crystals. Micropor. Mesopor. Mater. 2011, 139 (1-3), 120-129. https://doi.org/10.1016/j.micromeso.2010.10.028

[15] Khoramzadeh, E.; Mofarahi, M.; Lee, C.-H., Equilibrium adsorption study of CO2 and N2 on synthesized zeolites 13X, 4A, 5A, and beta. J. Chem. Eng. Data 2019, 64 (12), 5648-5664. https://doi.org/10.1021/acs.jced.9b00690

Decarbonization Technology: ICDT2024 Materials Research Forum LLC
Materials Research Proceedings 53 (2025) 173-185 https://doi.org/10.21741/9781644903575-17

Kinetics modeling for the extraction of phenolic compound from palm kernel cake using microwave-assisted method

ROSLI Ellisa Natasya[1,a,*], MOHD ARIFF Mohd Azahar[1,b], MOHAMAD NOR Norhusna[1,2,c], OSMAN Mohamed Syazwan[1,3,d], ISMAIL Nor Fariza[1,2,e], and MAT YAMAN Siti Hawa[1,f]

[1]Chemical Engineering Studies, College of Engineering, Universiti Teknologi MARA, Cawangan Pulau Pinang, 13500 Permatang Pauh, Pulau Pinang, Malaysia

[2]Waste Management and Resource Recovery (WeResCue) Group, Chemical Engineering Studies, College of Engineering, Universiti Teknologi MARA, Cawangan Pulau Pinang, 13500 Permatang Pauh, Pulau Pinang, Malaysia

[3]Lab EMZI-UiTM Nanoparticles Colloids, Universiti Teknologi MARA, Cawangan Pulau Pinang, 13500 Permatang Pauh, Pulau Pinang, Malaysia

[a]ellisanatasya@gmail.com, [b]azahar.ariff@uitm.edu.my, [c]norhusna811@uitm.edu.my, [d]syazwan.osman@uitm.edu.my, [e]norfariza5031@uitm.edu.my, [f]sitihawamatyaman@gmail.com

Keywords: Antioxidant Activity (AOA), Microwave-Assisted Extraction (MAE), Palm Kernel Cake (PKC), Total Phenolic Content (TPC)

Abstract. This study investigates the influence of microwave heating power and extraction time on the extraction of total phenolic content (TPC) from palm kernel cake (PKC). Additionally, it explores the antioxidant activity (AOA) of the extracted TPC under various microwave heating powers. Microwave-assisted extraction (MAE) is utilized as a non-conventional method for phenolic compound extraction, offering benefits such as reduced extraction time, minimal solvent usage, cost-effectiveness, and simplicity, while achieving high TPC and AOA yields. The parameters for microwave extraction include extraction times of 10, 15, 20, and 25 minutes, and microwave power of 300, 450, and 600 W. The TPC and AOA were analysed using UV-vis spectrophotometer at wavelength of 765 nm and 517 nm respectively. The highest TPC and AOA obtained were 1.715 mg GAE/g dry oil and 186.03 g AAE/mL respectively at 25 minutes and 300 W while the lowest TPC and AOA were 0.931 mg GAE/g dry oil and 99.68 g AAE/mL respectively at 10 minutes and 600 W. Low level of DPPH correlates with lower absorbance at 517 nm that indicates high scavenging activity, and consequently, greater antioxidant activity. FTIR study reveals that the phenol component of both the extracted and pure solid PKC is visible in the wavenumber region of 3500–3200 cm^{-1}. Finally, from the kinetic study, at 600 W of microwave power, the second-order extraction rate constant, k, was 0.6292 $Lg^{-1}min^{-1}$, the extraction capacity (Cs) was 1.0761 gL^{-1}, and the greatest coefficient of determination, R^2 was 0.9782.

Introduction

The oil palm, known scientifically as *Elaeis guineensis* is the most significant species in the genus Elaeis which belongs to the Palmae family from west Africa [1]. Production of oil palms is a significant part of Malaysia's agricultural sector, and it has emerged as the nation's most important agriculture-based business in recent decades [2]. In 2022, Malaysia possessed a cumulative count of 450 palm oil mills, together capable of processing 119.36 million tons of fresh fruit bunches (FFB) every year [3]. As a result, Malaysia has overtaken Indonesia to become the world's second-largest producer of palm oil [4]. 1000 kg of oil palm yields 250 kg of oil, consisting of 200 kg of crude palm oil and 50 kg of crude palm kernel oil. As a result, the byproducts and solid waste such

Materials Research Proceedings 53 (2025) 173-185 https://doi.org/10.21741/9781644903575-17

as palm kernel cake (PKC), palm kernel shell (PKS), empty fruit bunch (EFB), and mesocarp fibre (MF) that result from the production of palm oil weigh roughly up to 750 kg [5]. As the byproduct generated from palm oil industry increases, solid waste disposal also increases that might offer threats to the environment.

Phenolic compounds are widely used in various industries like food preservation, cosmetics, packaging, and textiles due to their natural antioxidant, antibacterial, and coloration properties [6]. In the food industry, phenolic extracts from sources like *Litchi chinensis Sonn* are utilized for preservation purposes, showing effectiveness comparable to synthetic antioxidants like butylated hydroxyl toluene (BHT). Cosmetics also employ phenolic chemicals for their UV absorption properties, aiding in sun protection and skin health. Moreover, phenolics derived from fruits are used in therapeutic applications for cardiovascular health and cancer prevention [6]. In packaging, phenolic compounds such as anthocyanins help prevent lipid oxidation in oils. Textile industry utilizes biodegradable phenolic dyes from sources like oak bark and tea extracts for coloration and UV protection of fabrics.

Phenolic compounds, terpenes, and sterol are the most common types of phytochemicals found in palm-based products [6]. The phenolic compounds, most of which are phenolic acids, flavonoids, tannins, and lignans, make up the largest and most interesting category of phytochemicals [6]. Phenolic molecules, which are found all over the place as byproducts of secondary metabolism, may exist in either a free or bound state. They have extraordinary antioxidant activity because of their capacity to serve as chelators of divalent cations, which enables them to scavenge free radicals and interrupt radical chain reactions. The antioxidant activity of phenolic compounds is mostly due to their redox characteristics, which enable them to operate as reducing agents, hydrogen donors, and singlet oxygen quenchers [7].

Palm kernel Cake (PKC) is a solid byproduct of oil extraction from palm kernels in order to produce palm kernel oil [8]. Due to high levels of protein and fibre, PKC is an interesting protein source since it is often utilized as a feed supplement for ruminants [5]. Additionally, 6.98% of PKC is made up of important amino acids such as arginine, leucine, valine, phenylalanine, and threonine [5]. The crude protein (CP) content of PKC varies between 10.0 to 19.8% while the crude fiber (CF) content ranging between 13 to 20% [9]. PKC was found to include hemicellulose, cellulose, lignin, proteins, amino acids, oligosaccharides, phenolics, and phytosterols according to chemical studies. Furthermore, it has been shown that PKC contains organic acids, water-soluble vitamins, and phenolic antioxidants which the phytochemical compound such flavonoids, polyphenols, and phenolic acids [10]. Tungchaisin et al. [5] stated that phenolic compound extracted from PKC is 32.46% yield of extraction.

There are two types of extraction techniques for phenolic compounds which are conventional and non-conventional techniques. Maceration, Soxhlet extraction, and hot reflux extraction are three conventional methods that are often used for recovering phenolics from solid samples. The examples of the non-conventional method are ultrasound-assisted extraction (UAE), microwave-assisted extraction (MAE), ultrasound-microwave-assisted extraction (UMAE), supercritical fluid extraction (SFE), sub-critical water extraction (SCWE) and high hydrostatic pressure processing (HHPP). These techniques reduce harmful pollutant emission by lowering organic solvent usage and are generally straightforward to implement [11]. Among them, MAE is a relatively novel technology for phenolic compound extraction. Figure 1 shows the mechanism of heat and mass transfer of MAE extraction. The process acceleration and high extraction yield seen in microwave-assisted hydrodistillation can be attributed to the synergistic interaction between two transport processes which are heat and mass gradients, operating in parallel [12]. In contrast, typical extractions involve mass transfer from the interior to the exterior, while heat transfer takes place from the exterior to the interior of the substrate. Moreover, the yellow arrow shows heat transfer from the outer layers to the center, occurring when initial microwave energy causes the outer layers to heat up and transfer heat inward [13]. Whereas, the red arrow involves heat moving from the center outward. This is due to microwaves generating internal heat by interacting with polar molecules, causing uniform heat distribution from the inside out. This central heating ensures rapid and even heating throughout the material, enhancing the efficiency of the microwave-assisted hydrodistillation process [13]. According to Martino (2006), the optimum extraction of phytocomponents from Melilotus officinalis flowering tops was achieved using 50% aqueous ethanol with MAE [7].

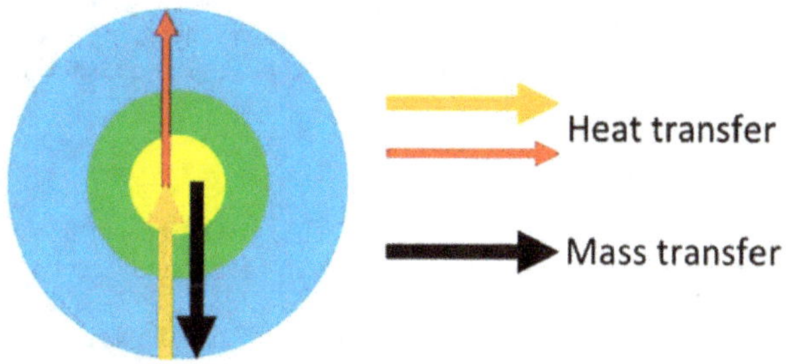

Figure 1: Mechanism of heat and mass transfer in microwave-assisted extraction

The MAE method has the benefits of taking less time, using fewer solvents, producing greater extraction rates, and producing better products at a lesser cost. [7]. The main factors affecting the optimum yield of the phenolic compound are solvent choice, temperature, liquid-to-solid ratio, microwave power and extraction time [14]. According to Ali Khoddami (2013), the best extraction was accomplished by using 25 mL of 90% ethanol for 25 min at 110 °C with 1,000 W of microwave power [11]. As reported by Djenar et al., from their research, they found that the highest yield of pandan leave extract was 23.53% by using MAE method[15].

In this study, non-conventional technique which is the MAE will be conducted to extract phenolic compound from the palm kernel cake (PKC). Utilizing MAE for phenolic compound extraction from palm kernel cake can contribute to more efficient, cost-effective, and better selectivity to enhance the overall effectiveness of this study. Although previous studies have explored the extraction of phenolic compounds from different sources employing MAE, such as fruits or medicinal plants, there appears to be a lack of literature on the precise conditions necessary for achieving effective extraction from PKC.

However, this study highlights several potential limitations and sources of error. A significant limitation is the inhomogeneous microwave power distribution, which causes uneven heating and affects both extraction efficiency and reproducibility. Consistent temperature maintenance during MAE is challenging, with fluctuations potentially impacting extraction kinetics and the stability of phenolic compounds [16]. Additionally, variations in the composition and moisture content of PKC samples can influence extraction efficiency and kinetics, making standardized sample preparation essential. The choice of solvent is also critical, as different solvents uniquely interact with phenolic compounds and other PKC constituents, potentially limiting extraction efficiency. Measurement inaccuracies in temperature, microwave power, solvent volume, and sample weight can introduce errors in kinetic modeling [17]. Not only that, analytical errors in UV-vis spectrophotometry also can affect the result reliability. Standardizing sample preparation,

Materials Research Proceedings 53 (2025) 173-185 https://doi.org/10.21741/9781644903575-17

optimizing solvent selection, improving measurement accuracy, and ensuring multiple replicates can mitigate these issues. By adopting these measures, the study can enhance the reliability and applicability providing valuable insights into the efficient and sustainable extraction of phenolic compounds from PKC using MAE.

This study aims to examine the impact of several factors, including microwave power and extraction time on the efficacy and characteristics of phenolic compounds obtained from PKC. This study has the potential to offer significant insights into improving the exploitation of PKC as a sustainable source of phenolic compounds. By doing so, it can contribute to the advancement of scientific knowledge and find practical applications in various industries. Thus, this study was conducted with the aim to investigate the effect of microwave heating power and extraction time on total phenolic content (TPC), antioxidant activity (AOA) and kinetic study extracted from PKC.

Methodology

Chemicals
Chemicals that have been used were palm kernel cake (PKC) bought from a local seller from Terengganu, ethanol (QRec, 96%), silica gel, Folin-Ciocalteu reagent (QRec, ≥95%), 2,2-diphenyl-1-picryhydrazyl (DPPH) free radicals (Sigma-Aldrich, ≥97), sodium carbonate (R&M, Anhydrous), deionized water (100%), gallic acid (QRec, 97.5-102.5%) and ascorbic acid (HmbG, Food Additive).

Preparation of Palm Kernel Cake (PKC)
PKC was dried for 48 hours in an oven at 40°C, and the drying was terminated when a consistent mass was attained. Then, the PKC ground using a grinder and sieved to a particle size of between 0.5 and 1 mm. Next, the PKC was weighed at 10 g and inserted in an empty tea bag.

Microwave-Assisted Extraction (MAE) of Phenolics from PKC
The extraction procedure was carried out by using microwave oven (Samsung, MS23K3513A/SM) using microwave-assisted extraction (MAE) method as shown in Figure 2. A 10 g of PKC sample and 300 mL of ethanol as a co-solvent were added into the microwave digestion vessel. The mixture then was allowed to stand for 5 minutes. The process will be varied at different parameters including extraction time for 10, 15, 20 and 25 minutes, and microwave power of 300, 450 and 600 W. Extracts containing phenolic compounds will be collected.

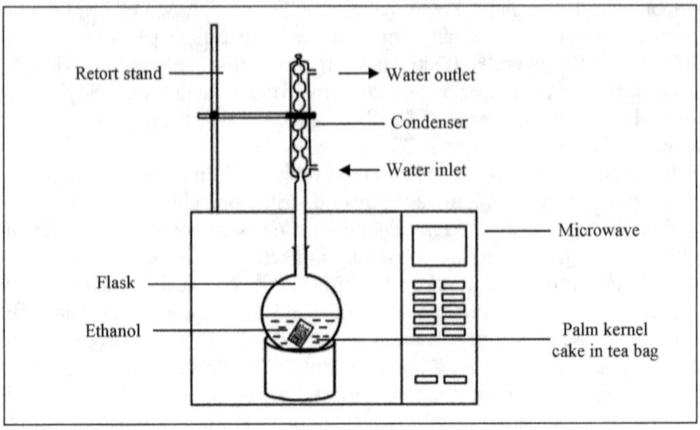

Figure 2: Schematic diagram of MAE

Total Phenolic Content (TPC)

Folin-Ciocalteu colorimetric method was used to determine total phenolic content (TPC). Dissolving 0.5 g gallic acid in 10 mL ethanol produced a typical gallic acid solution. The solution was transferred to a 100 mL volumetric flask and filled with distilled water to 100 mL. Standard gallic acid solutions in ethanol were made at 50, 100, 250, and 500 mg/L. For experiments preparing gallic acid at different concentrations, use different micropipette tips. Next, 0.2 mL of 0.2N Folin-Ciocalteu reagent, 3.16 mL of distilled water, 0.6 mL of sodium carbonate (7.5%, w/v), and 40μL of gallic acid at varied concentrations were added to 5 test tubes. Then, the mixture was rapidly mixed and incubated for 2 hours at dark room temperature. Gallic acid was replaced with ethanol for blank sample. Then, the gallic acid was replaced with the extracted PKC for sample analysis. For accuracy, testing was done in duplicate. A double-beam UV-Vis spectrophotometer (PEAK INSTRUMENTS INC, T-9200) measured absorbance at 765 nm and compared it to a gallic acid calibration curve. Absorbance was evaluated against an ethanol blank. Results were reported as mg of gallic acid equivalents (GAE) per L [11].

Determination of Antioxidant Activity (AOA)

The 2,2-diphenyl-1-picryl-hydrazyl (DPPH) free radicals were used to assess the sample's capacity to scavenge radicals. Stock DPPH solution was prepared by diluting 0.1183 g of DPPH in 500 mL of ethanol to produce 0.6 mM solution of DPPH. Then, the DPPH solution was shaken vigorously for well mixing and covered with aluminium foil. The DPPH solution was stored in a dark place for 30 minutes. In addition, a conventional ascorbic solution was prepared through the dissolution of 10 mg of ascorbic acid in 10 mL of ethanol. The solutions of ascorbic acid in ethanol were prepared using the standard solution, with concentrations of 10, 20, 40, 60, 80 and 100 ppm. For blank sample, 2.5 mL of distilled water was diluted into ethanol in a 25 mL volumetric flask. After 30 minutes, 1 mL of ascorbic acid is diluted into 9 mL of stock DPPH solution in a conical flask. The solution is mixed well and covered with aluminium foil before stored in a dark room for 30 minutes. For sample analysis, the ascorbic acid was replaced with an extracted sample. Each sample was triplicate to get an accurate result and the calibration curve was determined using the average absorbance values obtained at different concentrations of ascorbic acid. For control solution, 1 mL of blank sample was diluted into 9 mL of DPPH solution and covered with aluminium foil. The control solution was stored in a dark place for 30 minutes. After 30 minutes, the samples were ready to be analyzed by using double-Beam UV-Vis Spectrophotometer (PEAK INSTRUMENTS INC, T-9200) at wavelength of 517 nm.

Determination of Fourier Transform Infrared Spectroscopy (FTIR) Analysis

The process of analysing phenolic chemicals from palm kernel cake using FTIR (Thermo Nicolet 6700) analysis comprises many essential stages. Firstly, the palm kernel cake samples are coarsely crushed and then mixed with 300 mL of ethanol solvent to extract phenolic compound. Next, the FTIR analysis is performed by depositing the solid palm kernel cake, extracted sample with highest and lowest TPC. Spectra are then captured within the mid-infrared range, generally spanning from 4000 to 500 cm^{-1}. The FTIR spectra of the palm kernel cake are examined to identify and analyze the distinct peaks that correspond to the functional groups of phenolic compounds, including hydroxyl groups, aromatic rings, and other important vibrations. This analysis identifies the presence of phenolic compounds in the palm kernel cake. The FTIR approach offers unique insights into the chemical composition of the material, facilitating the qualitative and quantitative evaluation of phenolic compounds in palm kernel cake.

Determination of Kinetic Study

A second-order mechanism model is a kinetics model that precisely conveys the rate of the extraction process. This model proposes that the extraction rate is directly proportional to the square of the phenolic compound's concentration or the product of the concentrations of two reactants participating in the extraction process [12]. A second-order mechanism model refers to the occurrence of extraction via two simultaneous processes. The rate of oil extraction first experiences substantial increases, followed by a steady decrease until the end of the extraction process [12]. The dissolving rate of the phenolic compound in the solid-to-solution system may be described by below equation:

$$C_t = k(C_s - C_t)^2 \quad \text{(Eq 1)}$$

where k is the second-order extraction rate constant (Lg-^{1}min^{-1}), C_s is the concentration of phenolic compound at saturation (gL^{-1}), C_t is the concentration of phenolic compound at any time (min).

The integral rate law for a second-order extraction was derived by considering the beginning and boundary conditions, where t = 0 to t, and C_t = 0 to C_t:

$$C_t = \frac{C_s^2 kt}{1 + C_s kt} \quad \text{(Eq 2)}$$

Equation 2 may be converted into a linear form, represented by equation (3), which allows us to express the extraction rate as Equation 4:

$$\frac{t}{C_t} = \frac{1}{kC_s^2} + \frac{t}{C_s} \quad \text{(Eq 3)}$$

$$\frac{C_t}{t} = \frac{1}{\left(\frac{1}{kC_s^2}\right) + \left(\frac{t}{C_s}\right)} \quad \text{(Eq 4)}$$

The initial extraction rate, h, as C_t/t when t approaches 0. Can be described as:

$$h = kC_s^2 \quad \text{(Eq 5)}$$

The concentration of phenolic compound at any time can be expressed as:

$$C_t = \frac{t}{\left(\frac{1}{h}\right) + \left(\frac{t}{C_s}\right)} \quad \text{(Eq 6)}$$

Where the values of initial extraction rate, h, the extraction capacity, Cs, and the second-order extraction rate constant, k, can be determined from the slope and intercept by plotting the graph of t/C_t against t.

Results and Discussion

Total Phenolic Content of Oil Extracted from PKC

Figure 3 shows the TPC of PKC versus time at different microwave power and extraction time. The TPC increases as the extraction time increases from 10 to 25 minutes. It has been observed that the duration of the extraction process typically has a positive impact on the yield of the extracted product. For microwave power, the TPC decreases as the microwave power increases due to the denaturation of protein. This is because excessive microwave power has the potential to elevate the temperature of the product to an undesirable level, leading to reduced extraction yield due to product damage or compound breakdown. From the graph, the highest TPC obtained is 1.715 mg GAE/g dry oil at 25 minutes and 300 W while the lowest TPC obtained is 0.931 mg

Materials Research Proceedings 53 (2025) 173-185 https://doi.org/10.21741/9781644903575-17

GAE/g dry oil at 10 minutes and 600 W. This result is consistent with previous study which stated that as extraction time increases, the TPC is also increases [18]. This is due to the fact that the phenolic compound became more soluble in the solvent as the microwave extraction period increased. As a result, it has been discovered that the extraction duration typically has a favorable impact on the yield of the extracted material.

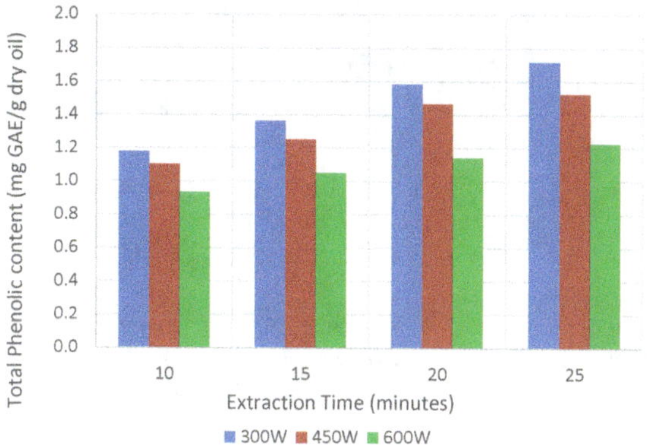

Figure 3: Total Phenolic Content (TPC) of PKC at different MW power (300 W, 450 W and 600 W) and extraction time

Various research have examined the influence of extraction parameters on total phenolic content (TPC) in different plant materials, revealing comparable patterns reported in the current study. The positive correlation between extraction time and TPC corresponds with the findings of Osorio-Tobon et al. (2020), who found that increasing the extraction time enhances the solubility of phenolic compounds in the solvent, thereby yielding higher extraction amounts [14]. This is because phenolic compounds are exposed to the extraction solvent for a longer period, leading to increased solubilization. Furthermore, previous research conducted by Aida, has yielded comparable results about the adverse effects of excessive microwave power on total phenolic content (TPC) [19]. These findings attribute these effects to the denaturation and breakdown of phenolic compounds that occur under elevated temperatures.

Antioxidant Activity of Oil Extracted From PKC

Figure 4 shows the AOA versus time at different microwave power. The AOA is increases as the extraction time increases from 10 to 25 minutes. For microwave power at 10 minutes, the AOA decreases as the microwave power increases due to the denaturation of phenolic. This is because phenol compounds are very sensitive to heat so an increase of microwave power from 300, 450 and 600W reduces the antioxidant activity of the sample. Based on previous research, it has been observed that when the microwave power exceeds 200 W, there is a decrease in antioxidant activity [18]. This decline can be attributed to excessive microwave power, which has the capacity to raise the temperature of the product to an undesirable extent. Consequently, this elevated temperature can result in product damage or breakdown of compounds [18]. From the graph, the highest AOA obtained is 186.03 g AAE/mL at 25 minutes and 300 W while the lowest AOA obtained is 99.68 g AAE/mL at 10 minutes and 600 W. Low level of DPPH correlates with lower absorbance at 517 nm that indicates high scavenging activity and consequently, greater antioxidant activity.

Materials Research Proceedings 53 (2025) 173-185 https://doi.org/10.21741/9781644903575-17

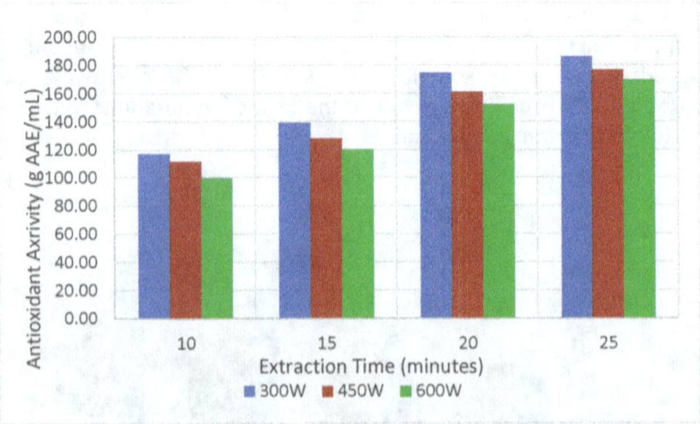

Figure 4: Antioxidant activity (AOA)

In a literature study conducted by Moraes et al. [20], it was observed that an excessive application of microwave power had a detrimental impact on the antioxidant activity of phenolic compounds that were extracted. The decrease in antioxidant activity was ascribed to the deterioration or denaturation of phenolic compounds that are susceptible to heat, which occurs when heat is applied at high temperatures due to increased levels of microwave power. Moreover, previous research conducted by Perussello et al. [21] and Prabhakar et al. [22] has yielded similar results, indicating a decline in antioxidant activity beyond a specific microwave power threshold as a result of thermal degradation. The present investigation provides support for the apparent negative correlation between microwave power and AOA, as evidenced by the collective findings of these studies. Furthermore, the AOA values obtained are consistent with the ranges reported in the existing literature, providing additional support for the findings.

Surface Functional Groups

Fourier Transform Infrared Spectroscopy (FTIR) is a powerful analytical technique that offers a distinct and identifiable representation of the chemical components within a given sample [23]. This is achieved by examining the molecular vibrations exhibited by these substances. FTIR also can be used for qualitative analysis in order to ascertain the presence of particular compounds or to validate the existence of phenolic functional groups within the sample. Figure 5 shows the Fourier Transform Infrared (FTIR) spectra of both the pure solid PKC and the extracted PKC. Multiple peaks can be observed, indicating the presence of functional groups and modes of vibration associated with the individual components. The presence of alcohols, carboxylic acids, esters, and ethers can be inferred from the distinct peaks observed in the range of 1200-1000 cm^{-1} [24]. The observed prominent peak within the wavenumber range of 3500-3200 cm^{-1} can be ascribed to the existence of phenols and alcohols [24]. However, the solid form of PKC shows a relatively lower peak compared to the extracted PKC, which contain both highest and lowest total phenolic content (TPC).

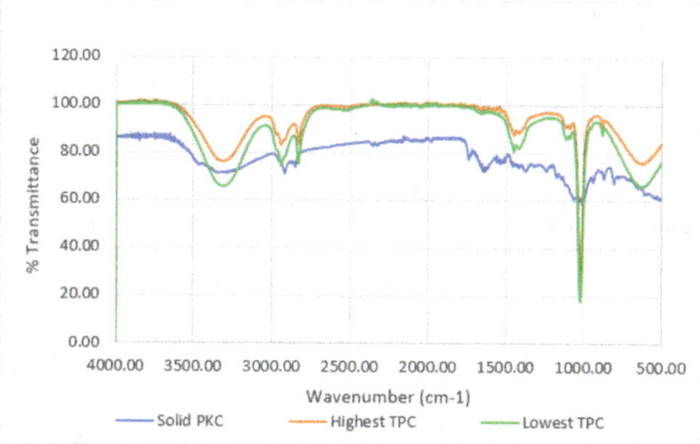

Figure 5: FTIR spectra of pure solid PKC and Extracted PKC

The FTIR analysis in this study reveals functional groups such as alcohols, carboxylic acids, esters, and ethers in PKC, with peaks in the 1200-1000 cm^{-1} range, which is consistent with the findings of Osorio-Tobon et al. (2020) [14]. The prominent peak within the 3500-3200 cm^{-1} range, attributed to phenols and alcohols, aligns with Zhang et al. (2021), who observed similar O-H stretching vibrations in phenolic compounds [25]. These comparisons validate our findings and underscore the reliability of FTIR in identifying phenolic compounds in PKC.

Kinetic Study

Figure 6 shows the concentration of phenolic compound in the solution at any time versus time for different microwave power. Based on figure, it shows that as the time of extraction increases, the rate of extraction also increases. The experimental result was analyzed using second-order model by plotting t/Ct versus time for microwave power of 300 W, 450 W and 600 W as shown in Fig 7 (a), (b) and (c) respectively.

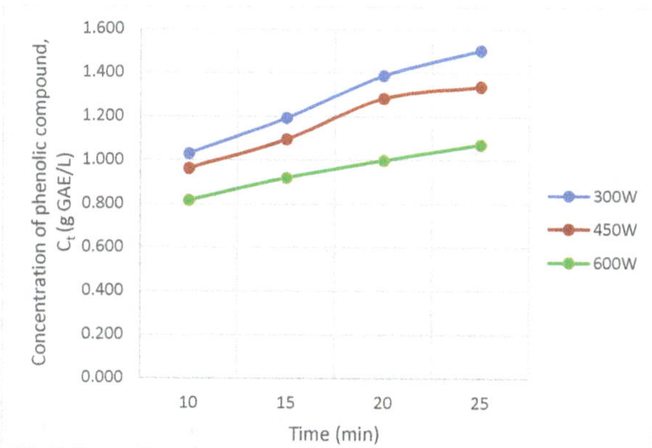

Figure 6: The concentration of phenolic compound in the solution at any time, C_t (gL^{-1}) versus time (min) at various microwave power

According to Figure 6, the rate of extraction was fast from time 0 to 25 minutes for all of the microwave power. Not only that, it is shows that microwave power at 300 W has the highest concentration of phenolic compound at any time compared to other microwave power. Our study's application of a second-order model to examine the extraction kinetics at 300 W, 450 W, and 600 W aligns with the findings of Wu et al. (2019) who reported similar rapid initial extraction phases transitioning into slower diffusion-controlled stages [26]. The extraction process occurs in three distinct stages: an equilibrium phase including solubilization and partition phenomena, during which the substrate is extracted from the outer surface of the particle at a relatively constant speed. Subsequently, an intermediate transitional phase to diffusion ensues. The beginning of resistance to mass transfer occurs at the interface between the solid and liquid phases. During this stage, mass transport is primarily driven by convection and diffusion. During the last stage, the solute has to surpass the interactions that hold it together with the matrix and spread out into the extraction solvent. The extraction rate during this stage is rather modest since it is primarily driven by the diffusion mechanism for extract removal. This particular point represents a phase in the extraction process that cannot be undone; it is often considered to be the step that sets the limit for the whole operation. The diffusion rate declined as the extraction time increased, owing to the elevated solute content in the liquid during the third stage. Although the extraction time increased after the extraction of the maximum phenolic component, there were no observable changes or substantial differences in the quantity of oil removed.

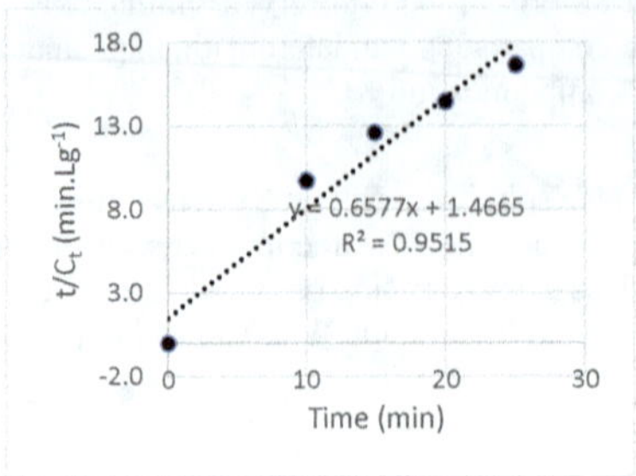

(a)

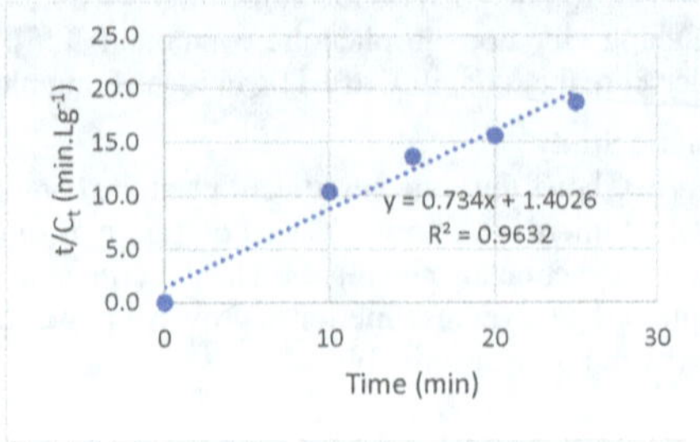

(b)

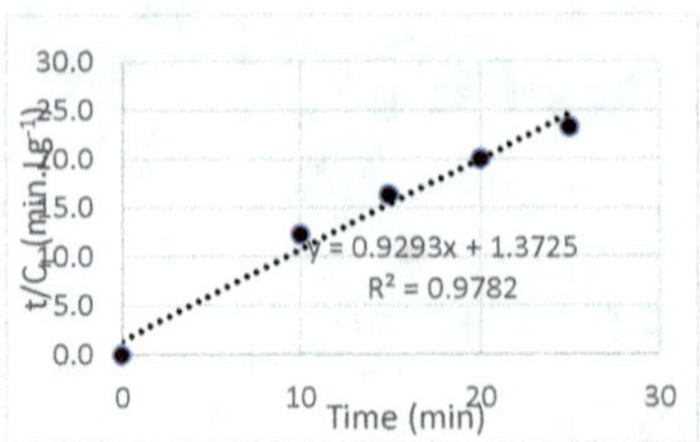

(c)

Figure 7: Second-order extraction kinetics of phenolic compound at microwave power of (a) 300 W, (b) 450 W and (c) 600 W

The values for the initial extraction rate (h), extraction rate capacity (C_s), second-order extraction rate constant (k), and coefficient of determination (R^2) were determined from experimental calculations using the linear curve shown in Figure 7 (a), (b) and (c). The slope of the t/C_t against time graph is equal to $1/C_s$, and the intercept is equal to 1/h. The data is shown in Table 1.

Table 1: Linearization of second-order kinetic model of microwave-assisted extraction of phenolic compound

Microwave power (W)	Cs (gL^{-1})	K ($Lg^{-1}min^{-1}$)	h ($gL^{-1}min^{-1}$)	R^2
300	1.5205	0.2950	0.6819	0.9515
450	1.3624	0.3841	0.7130	0.9632
600	1.0761	0.6292	0.7286	0.9782

From Table 1, the highest value coefficient of determination, R^2 for this research is 0.9782. This shows that there is a strong linear relationship between the independent and dependent variables. From previous research, Kusuma et al. [12] applied MAE for oil extraction from sandalwood and the final result obtained for R^2 was 0.9597 which is less 0.0185 from extraction of palm kernel cake using MAE method [12]. This result shows that MAE is an efficient method for oil extraction.

Conclusions

In conclusion, this study has been successfully performed to investigate the effect of microwave heating power and extraction time on total phenolic content extracted from palm kernel cake and to study the AOA towards TPC extracted from palm kernel cake at different microwave heating power effect on MAE parameters on antioxidant. Among the two parameters studied, which are microwave power and extraction time, the most important factors that contributed to the TPC is extraction time followed by microwave power. Microwave power is not really affecting the result production, in fact microwave power can degrade the phenolic compound as the compound is very sensitive to the heat. It was found that the highest TPC and AOA of the extracted sample were 2.107 mg GAE/g dry oil and 186.03 g AAE/mL respectively at 25 minutes and 300 W while the lowest TPC and AOA were 0.505 mg GAE/g dry oil and 99.68 g AAE/mL respectively at 10 minutes and 600 W. In addition, the phenol compound of the pure solid PKC and extracted PKC can be observed within the wavenumber range between 3500-3200 cm^{-1} from the FTIR analysis. Lastly, form the kinetic study, the highest coefficient of determination, R^2 was 0.9782 with initial extraction rate, h was 0.7286 $gL^{-1}min^{-1}$, the extraction capacity, Cs, was 1.0761 gL^{-1} and the second-order extraction rate constant, k, was 0.6292 $Lg^{-1}min^{-1}$ at microwave power of 600 W.

Acknowledgement

The authors express their gratitude to Universiti Teknologi MARA (UiTM) Cawangan Pulau Pinang for the facilities provided and financial support through the Fundamental Research Grant Scheme (Grant NO. FRGS/1/2022/TK0/UITM/02/100) that enabled them to complete this research project.

References

[1] H. Teoh Cheng, "The Palm Oil Industry In Malaysia From Seed to Frying Pan," 2002.

[2] N. Saifuddin, A. Saltanat, and H. Refal, "Enhancing the Removal of Phenolic Compounds from Palm Oil Mill Effluent by Enzymatic Pre-treatment and Microwave-Assisted Extraction," *Chem Sci Trans*, Jun. 2014. https://doi.org/10.7598/cst2014.797.

[3] G. K. A. Parveez *et al.*, "OIL PALM ECONOMIC PERFORMANCE IN MALAYSIA AND R&D PROGRESS IN 2022," *J Oil Palm Res*, vol. 35, no. 2, pp. 193–216, Jun. 2023. https://doi.org/10.21894/jopr.2023.0028.

[4] N. M. Saifuddin, S. Uskenbayeva, and M. Y. Ong, "On increase of the efficiency of extracting phenolic compounds from palm oil mill effluent Development of Bio-Hydrogen Production From Algae : A Review View project adsorption and waste water treatment View project References 3," 2018. [Online]. Available: https://www.researchgate.net/publication/322674594

[5] T. Tungchaisin, S. Sae-Tan, and K. Rattanaporn, "Optimization of Microwave-Assisted Extraction of Palm Kernel Cake Protein," *E3S Web of Conferences*, vol. 355, p. 02018, 2022. https://doi.org/10.1051/e3sconf/202235502018.

[6] B. R. Albuquerque, S. A. 8. Heleno, M. B. P. P. Oliveira, L. Barros, and I. C. F. R. Ferreira, "Phenolic compounds: Current industrial applications, limitations and future challenges," Jan. 07, 2021, *Royal Society of Chemistry*. doi: 10.1039/d0fo02324h.

[7] M. Gallo, R. Ferracane, G. Graziani, A. Ritieni, and V. Fogliano, "Microwave assisted extraction of phenolic compounds from four different spices," *Molecules*, vol. 15, no. 9, pp. 6365–6374, Sep. 2010. https://doi.org/10.3390/molecules15096365.

[8] H. Y. Saw, J. Janaun, S. Kumaresan, and C. M. Chu, "Characterization of the physical properties of palm kernel cake," *Int J Food Prop*, vol. 15, no. 3, pp. 536–548, May 2012. https://doi.org/10.1080/10942912.2010.492543.

[9] Sharmila, Alimon, Azhar, and Samsudin, "Malaysian Society of Animal Production 1 Improving Nutritional Values of Palm Kernel Cake (PKC) as Poultry Feeds: A Review," 2014. [Online]. Available: http://www.etawau.com/OilPalm/Elaeis_guineensis.htm

[10] E. Oskoueian *et al.*, "Cytoprotective effect of palm kernel cake phenolics against aflatoxin B1-induced cell damage and its underlying mechanism of action," *BMC complementary and alternative medicine, 15(1), 1-12*, 2015.

[11] A. Khoddami, M. A. Wilkes, and T. H. Roberts, "Techniques for analysis of plant phenolic compounds," Feb. 2013. doi: 10.3390/molecules18022328.

[12] H. S. Kusuma and M. Mahfud, "Preliminary study: Kinetics of oil extraction from sandalwood by microwave-assisted hydrodistillation," in *IOP Conference Series: Materials Science and Engineering*, Institute of Physics Publishing, May 2016. doi: 10.1088/1757-899X/128/1/012009.

[13] H. Kusuma and M. Mahfud, "Microwave-assisted Hydrodistillation for Extraction of Essential Oil from Patchouli (Pogostemon cablin) Leaves," *Periodica Polytechnica Chemical Engineering*, vol. 61, Jul. 2016. https://doi.org/10.3311/PPch.8676.

[14] J. F. Osorio-Tobón, "Recent advances and comparisons of conventional and alternative extraction techniques of phenolic compounds," *Journal of Food Science and Technology, 57, 4299-4315*, 2020.

[15] N. S. Djenar, E. W. S. Mulyono, and T. R. Saputra, "The effect of microwave power variations on phytochemical characteristic of pandan leaves (Pandanus Amaryllifolius) using the Microwave-Assisted Extraction (MAE)," in *Journal of Physics: Conference Series*, Institute of Physics Publishing, Mar. 2020. doi: 10.1088/1742-6596/1450/1/012007.

[16] M. A. Rostagno and J. M. Prado, "Natural Product Extraction Principles and Application," 2017, pp. P001–P004. doi: 10.1039/9781849737579-fp001.

[17] W. Routray and V. Orsat, "Microwave-Assisted Extraction of Flavonoids: A Review," Feb. 2012. doi: 10.1007/s11947-011-0573-z.

[18] Mashuni *et al.*, "The determination of total phenolic content of cocoa pod husk based on microwave-assisted extraction method," in *AIP Conference Proceedings*, American Institute of Physics Inc., Jun. 2020. doi: 10.1063/5.0001364.

[19] W. Aida, "Effect of ethanol concentration, extraction time and extraction temperature on the recovery of phenolic compounds and antioxidant capacity of Orthosiphon stamineus extracts phenolic content (TPC), total flavonoid content (TFC), condensed tannin content (CTC), 2,2'-Azino-bis(3-ethylbenzothiazoline-6-sulphonic acid) (ABTS) radical-scavenging capacity, 2,2-Diphenyl-1-picrylhydrazyl (DPPH) radical-scavenging capacity," *Int Food Res J*, vol. 18, no. 4, pp. 1427–1435, 2011.

[20] D. P. Moraes *et al.*, "Effect of Microwave Hydrodiffusion and Gravity on the Extraction of Phenolic Compounds and Antioxidant Properties of Blackberries (Rubus spp.): Scale-Up Extraction," *Food Bioproc Tech*, vol. 13, no. 12, pp. 2200–2216, Dec. 2020. https://doi.org/10.1007/S11947-020-02557-Z.

[21] C. A. Perussello, Z. Zhang, A. Marzocchella, and B. K. Tiwari, "Valorization of Apple Pomace by Extraction of Valuable Compounds," *Compr Rev Food Sci Food Saf*, vol. 16, no. 5, pp. 776–796, Sep. 2017. https://doi.org/10.1111/1541-4337.12290.

[22] P. Prabhakar *et al.*, "Optimization of microwave-assisted extraction (MAE) of key phenolic compounds from pigeon pea (Cajanus cajan L.), their characterization, and measurement of their anti-diabetic and cytotoxic potential," *Journal of Food Measurement and Characterization*, vol. 17, no. 6, pp. 5697–5720, Dec. 2023. https://doi.org/10.1007/S11694-023-02082-5.

[23] M. Lucarini *et al.*, "Grape seeds: Chromatographic profile of fatty acids and phenolic compounds and qualitative analysis by FTIR-ATR spectroscopy," *Foods*, vol. 9, no. 1, 2020. https://doi.org/10.3390/foods9010010.

[24] K. Nisa Mahmud *et al.*, "Evaluation on Efficiency of Pyroligneous Acid from Palm Kernel Shell as Antifungal and Solid Pineapple Biomass as Antibacterial and Plant Growth Promoter (Penilaian terhadap Keberkesanan Asid Piroligneus daripada Tempurung Isirung Sawit sebagai Antikulat dan Sisa Pepejal Nanas sebagai Antibakteria dan Promoter Pertumbuhan Tumbuhan)," 2016.

[25] R. Zhao *et al.*, "Efficient removal of phenol and p-nitrophenol using nitrogen-doped reduced graphene oxide," *Colloids Surf A Physicochem Eng Asp*, vol. 611, p. 125866, 2021. https://doi.org/https://doi.org/10.1016/j.colsurfa.2020.125866.

[26] Z. Wu, Y. Li, B. Zhang, W. Yang, and B. Yang, "Co-pyrolysis behavior of microalgae biomass and low-rank coal: Kinetic analysis of the main volatile products," *Bioresour Technol*, vol. 271, pp. 202–209, Jan. 2019. https://doi.org/10.1016/J.BIORTECH.2018.09.076.

Optimization of viscosity of diamond and boron nitride based nanofluids for enhanced thermal management and pumping efficiency

Aymn Abdulrahman[1,*]

[1]Chemical Engineering Department, University of Jeddah, Jeddah 23890, Kingdom of Saudi Arabia

aabdulrahman@uj.edu.sa

Keywords: Nanofluids, Viscosity Optimization, Response Surface Methodology, Thermal Management, Optimization

Abstract. The present study investigates optimizing the viscosity of hybrid nanofluids composed of diamond and boron nitride in thermal oil for high-temperature heat transfer applications. Recognizing that higher viscosity equates to increased pumping costs, the study seeks to minimize this critical parameter to enhance the overall efficiency of thermal systems. Employing Response Surface Methodology (RSM) and a desirability function approach, explored the effect of temperature (25°C to 65°C), concentration (0.2 wt.% to 0.6 wt.%), and shear rate (1 to 1000 1/s) on the viscosity of nanofluids. The optimization process pinpointed conditions that yield the minimum viscosity with the highest desirability score, signifying the most advantageous operating scenario. The results showed an optimal balance at a higher temperature and mid-range concentration, culminating in a nanofluid demonstrating Newtonian behavior at low shear rates and shear-thinning attributes at elevated temperatures and shear rates. These findings explain the potential of tailored nanofluids to substantially cut pumping costs while maintaining excellent heat transfer properties, making them ideal candidates for advanced cooling systems in demanding thermal management settings.

Introduction

The advancement of heat transfer technology through nanofluids has emerged as a focal point in engineering research, spurred by their potential to enhance thermal properties significantly. Nanofluids, created by dispersing nanoparticles such as diamond and boron nitride within base fluids like oil, water, or ethylene glycol, have shown promise in revolutionizing thermal management systems due to their superior thermal conductivities and stability, particularly in high-temperature applications [1]. The literature affirms the impact of nanoparticles not only on thermal conductivity but also on specific heat, density, and viscosity, which collectively influence the thermal performance of a system[2].

Exploration of nanofluids for heat transfer applications has been an area of intense research due to their remarkable potential to enhance thermal properties. Nanofluids, by definition, consist of a dispersion of nanoparticles, with sizes typically between 1 and 100 nm, within a base fluid[3]. These nanoparticles have been found to significantly improve the system's thermal transport properties, which is beneficial for various engineering applications where heat transfer is a critical process[4, 5].

Nanofluids' capability to enhance the thermal properties of a system is not solely attributed to their high thermal conductivity but also to the combined effects of other critical properties such as specific heat, density, and viscosity[6]. While the literature is replete with studies examining the thermal performance of nanofluids in water and ethylene glycol, the exploration of thermal oil-based nanofluids remains relatively uncharted. This gap presents a significant research opportunity, especially considering the primary use of thermal oil in heat exchange applications, where its thermal stability and non-Newtonian behavior can be advantageously leveraged [7].

However, introducing nanoparticles into a base fluid often increases viscosity, a parameter of paramount importance due to its direct impact on pumping power and heat transfer effectiveness. Viscosity, the measure of a fluid's resistance to flow, can significantly affect a system's operational costs and efficiency. Several studies have observed that as the concentration of nanoparticles increases or the temperature decreases, the viscosity of nanofluids tends to increase, corroborating the intricate relationship between these parameters[8, 9].

The rheological behavior of nanofluids, too, cannot be overlooked. This behavior describes the response of fluid to applied shear or stress, a characteristic that significantly influences the fluid's pumping and heat transfer capability[10]. Previous studies have reported that nanofluids can exhibit shear-thinning behavior, where the fluid viscosity decreases with increasing shear rate, indicating non-Newtonian characteristics[11]. These behaviors necessitate carefully examining the conditions under which nanofluids are utilized to optimize their performance in heat transfer applications.

This research focuses on thermal oil-based hybrid nanofluids containing diamond and boron nitride nanoparticles. Thermal oil is an ideal carrier fluid due to its high thermal stability and ability to maintain stable rheological properties even at elevated temperatures. The diamond and boron nitride nanoparticles are selected for their exceptional thermal properties and compatibility with the thermal oil.

Ali et al. investigated the thermal and rheological properties of diamond and boron nitride hybrid nanofluids in thermal oil, revealing improved thermal conductivity and specific heat with increasing temperature and nanoparticle concentration. Their findings suggest these nanofluids could reduce pumping costs while enhancing thermal management efficiency, especially at higher temperatures and moderate concentrations, positioning them as suitable for advanced cooling systems [12]. Subsequently, in 2024, Ali et al. applied machine learning to predict the rheological behavior of BN-diamond/thermal oil hybrid nanofluids. They found Gradient Boosting Regression to be the most accurate among the tested algorithms, underscoring machine learning's potential in nanofluid research, albeit with a reliance on substantial experimental data for precise predictions [13]. While comprehensive in exploring nanofluid synthesis and predictive modeling, the research did not extend to optimization studies.

Recent studies have highlighted the role of viscosity in influencing pumping power and heat transfer effectiveness. Increased fluid viscosity can significantly affect the heat transfer rate's effectiveness, impacting the system's overall performance [14-16]. Consequently, the need for an optimal balance between improved thermal properties and manageable fluid viscosity has been a driving force behind recent research efforts.

Optimizing the viscosity for enhanced thermal management cannot be overstated. The primary goal of this study is to minimize the viscosity within the constraints of maintaining or improving the thermal efficiency of the nanofluid. RSM was employed alongside optimization techniques to achieve this. This systematic approach allows for determining the optimal combination of temperature, nanoparticle concentration, and shear rate that will result in the minimum possible viscosity while ensuring the nanofluids' thermal conductivity and specific heat capacity remain at levels suitable for high-performance applications.

The findings of this study are expected to fill the research gap in the design and optimization of the thermophysical properties of nanofluids. This research contributes to the broader goal of developing efficient and cost-effective thermal management solutions by presenting a systematic approach to optimizing these properties, particularly focusing on viscosity. The optimized nanofluids have the potential to offer significant benefits in various industrial applications, including but not limited to electronics cooling, energy storage, and automotive thermal systems. Therefore, a case study has been subjected to investigation where the viscous property of a

Materials Research Proceedings 53 (2025) 186-197 https://doi.org/10.21741/9781644903575-18

nanofluid, comprised of BN and diamond nanoparticles and heat transfer oil, is optimized based on the three input parameters.

Materials and Methodology

The methodology for this research was grounded in utilizing the detailed nanofluid preparation and characterization process as outlined in previous work [12]. This study adopted a meticulous approach to developing nanofluids, involving the acquisition of high-purity spherical diamond nanoparticles and hexagonal boron nitride. Diamond nanoparticles (spherical) and BN (hexagonal) with an average diameter of 3–10 nm and 0–80 nm, respectively, were utilized in preparing a highly stable dispersion in heat transfer oil. These nanoparticles underwent a comprehensive characterization process, employing techniques such as Transmission Electron Microscopy (TEM), Field Emission Scanning Electron Microscopy (FESEM), X-ray Diffraction (XRD), along with elemental and spectroscopic analysis to determine their precise properties. The nanofluids were prepared using a two-step method, which included mechanical mixing of the nanoparticles in a 1:1 ratio and their subsequent dispersion in high thermal oil (HTO) to achieve stable suspensions at varying concentrations. This preparation addressed potential nanoparticle agglomeration, incorporating an ultrasonic probe-type homogenizer for de-agglomeration and the surfactant Span-85 to enhance stability. This study conducted optimization analyses, not on preparing nanofluids anew but leveraging the existing rheological data from Ali et al. [12]. The focus was squarely on utilizing the established dataset to optimize these nanofluids' thermal and rheological behavior, thereby contributing insights into their enhanced application potential within thermal management systems.

This study employed the RSM to model and optimize the thermal and rheological properties of diamond and boron nitride-based hybrid nanofluids. RSM significantly reduces the need for extensive experimental runs by modeling the interactions between key variables such as temperature, nanoparticle concentration, and shear rate, which influence the nanofluids' viscosity and, consequently, the thermal performance. This method enables the identification of optimal conditions for minimized viscosity without detracting from the nanofluid's thermal efficiency. RSM's capability to predict the nanofluid behavior through a generated mathematical model and its graphical tools, like contour and surface plots, aids in visually identifying the conditions for achieving the desired responses. Thus, RSM facilitates a streamlined approach to enhancing thermal management by optimizing nanofluid properties with precision and efficiency.

Temperature, nanofluid concentration, and shear rate were the three key parameters examined to understand their collective impact on the viscosity of nanofluids. A face-centered Central Composite Design (CCD) incorporating six center points, six face-centered, and eight corner points was utilized to explore these input parameters comprehensively. The experimentation followed the Design of Experiments (DoE) framework, with RSM employed for statistical analysis to establish an empirical relationship between the inputs and the viscosity response. RSM's robustness lies in its ability to discern the effects of multiple inputs on a single output, making it highly effective for both modeling and optimization purposes. This approach led to the formulation of a second-order polynomial equation, capturing the interactions among temperature, concentration, and shear rate and their collective influence on nanofluid viscosity, thereby facilitating the optimization process.

Results and Discussion

Table 1 presents the results of an experimental design based on a face-centered Central Composite Design (CCD) to investigate the effects of temperature, nanofluid concentration, and shear rate on the viscosity of nanofluids containing diamond and boron nitride in thermal oil for cooling applications. The table is organized into five columns, listing the serial number (Sr. No) of each

Materials Research Proceedings 53 (2025) 186-197 https://doi.org/10.21741/9781644903575-18

run, the temperature (in °C), the nanofluid concentration (in weight percent, wt.%), the shear rate (in 1/s), and the measured viscosity (in Pascal-seconds, Pa.s, multiplied by 1000 for readability).

The experimental conditions vary systematically to cover a range of temperatures (25°C, 45°C, 65°C), nanofluid concentrations (0.2 wt.%, 0.4 wt.%, 0.6 wt.%), and shear rates (1 s^{-1}, 500.5 s^{-1}, 1000 s^{-1}). The table includes 20 runs, showcasing a wide span of viscosity measurements from as high as 0.058812 (Pa.s) under low temperature and concentration with minimal shear to as low as 0.011701 (Pa.s) at higher temperatures and the same low concentration but also minimal shear.

Table 1: Experimental Results of Viscosity Measurements for Nanofluids Containing Diamond and Boron Nitride in Thermal Oil Across Varied Conditions

Sr. No	Temperature (°C)	Nanofluid Concentration (wt. %)	Shear Rate (1/s)	Viscosity (Pa.s)
1	25	0.2	1	0.058812
2	45	0.6	500.5	0.02103
3	65	0.2	1	0.011701
4	65	0.2	1000	0.01256
5	45	0.4	500.5	0.025581
6	65	0.4	500.5	0.012934
7	45	0.4	500.5	0.02558
8	45	0.4	500.5	0.025581
9	45	0.4	500.5	0.02557
10	25	0.6	1000	0.014266
11	45	0.4	500.5	0.025581
12	25	0.2	1000	0.057436
13	45	0.2	500.5	0.034415
14	65	0.6	1000	0.014266
15	45	0.4	1000	0.025483
16	65	0.6	1	0.018663
17	25	0.4	500.5	0.030591
18	45	0.4	1	0.024681
19	25	0.6	1	0.018663
20	45	0.4	500.5	0.025583

RSM was leveraged to forge an empirical linkage between the input variables—temperature, nanofluid concentration, and shear rate—and the viscosity response. After the equation's construction, the Analysis of Variance (ANOVA) technique was employed for response optimization, scrutinizing the data through a quadratic polynomial model. ANOVA was helpful in assessing the model's effectiveness and the significance of its parameters. This analytical method enabled a deep dive into the dynamics between the parameters, their interactions, and the influence of their squared values.

Table 2: ANOVA results for the viscosity

Source	Sum of Squares	Degree of freedom	Mean Squares	F Values	P values
Model	0.0031	9	0.0003	115.99	< 0.0001
A	0.0012	1	0.0012	406.45	< 0.0001
B	0.0008	1	0.0008	262.03	< 0.0001
C	7.24E-06	1	7.24E-06	2.45	0.1488
A^2	0.0011	1	0.0011	357.6	< 0.0001
B^2	6.25E-07	1	6.25E-07	0.2112	0.6557
C^2	8.56E-06	1	8.56E-06	2.89	0.1197
AB	0	1	0	8.08	0.0175
AC	0	1	0	8.43	0.0157
BC	3.78E-07	1	3.78E-07	0.1279	0.7281
Residual	0	10	2.96E-06		
Lack of Fit	0	5	5.92E-06	2.71E+05	< 0.0001
Pure Error	1.09E-10	5	2.19E-11		
Cor Total	0.0031	19			

Table 2 presents the ANOVA results for the viscosity model, detailing the influence of temperature (T), nanofluid concentration (C), shear rate (S), their squared terms, and interaction terms (TC, TS, CS) on viscosity. The overall model, with 9 degrees of freedom, shows a substantial sum of squares (0.0031), highlighting a robust model fit as evidenced by an F-value of 115.99 and a highly significant P-value (< 0.0001). Temperature (T) and its squared term (T^2) significantly affect viscosity, with F-values of 406.45 and 357.6, respectively, and P-values below 0.0001, underscoring their crucial impact. Nanofluid concentration (C) also significantly influences viscosity, demonstrated by an F-value of 262.03. In contrast, shear rate (S) and its squared term (S^2), while included in the model, did not achieve statistical significance, as their P-values (0.1488 and 0.1197, respectively) suggest.

The interaction between temperature and nanofluid concentration (TC) and temperature and shear rate (TS) are statistically significant, with P-values of 0.0175 and 0.0157, indicating that the combined effects of these parameters significantly affect viscosity. However, the interaction between nanofluid concentration and shear rate (CS) does not significantly impact viscosity, as indicated by a P-value of 0.7281. The residual and lack of fit tests support the model's adequacy, showing no significant lack of fit and affirming the model's capability and reliability in capturing the nuanced relationship between the input variables and the viscosity response.

Equation (1) models the viscosity of nanofluids by incorporating the effects of temperature, concentration, and shear rate, highlighting the nuanced interplay between these factors.

$$\begin{aligned} Viscosity = {} & 0.1162 - 0.001049 \times T - 0.2284 \times C - 3.077 \times 10^{-7} \times S \\ & + 0.002875 \times T \times C + 2.797 \times 10^{-8} \times T \times S - 1.036 \times 10^{-5} \times C \times S \\ & - 7.371 \times 10^{-6} \times T^2 + 0.0753 \times C^2 \\ & + 1.487 \times 10^{-9} \times S^2 \qquad (1) \end{aligned}$$

Fig. 1 illustrates a Normal Plot of Residuals from a statistical analysis, typically used to assess the normality of residuals in regression models. Residuals are the differences between observed and predicted values of the dependent variable, and their distribution is expected to be approximately normal (bell-shaped) if the model is appropriate for the data. In the plot, externally studentized residuals are plotted against the normal percentage probability. The majority of the data points fall along the reference line, which indicates a good fit of the model to the data. The

Materials Research Proceedings 53 (2025) 186-197 https://doi.org/10.21741/9781644903575-18

color differentiation of the points could signify various experimental groups or conditions. The presence of a single data point (the red square) far from the reference line might suggest an outlier or an influential observation that could warrant further investigation. The distribution of the residuals here appears to largely support the assumption of normality, an essential consideration in the validation of the regression model.

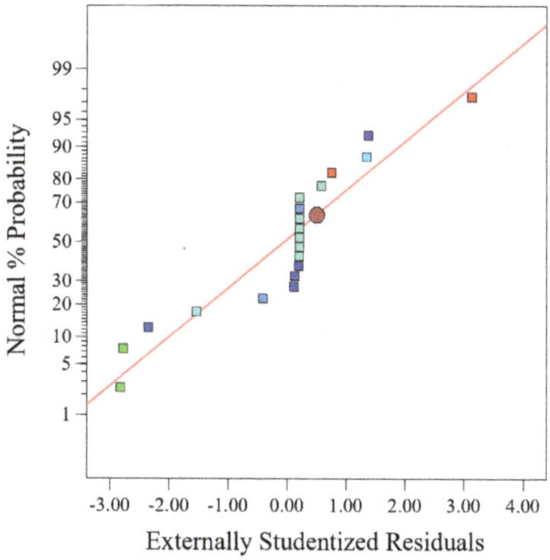

Fig. 1: Normal Probability Plot of Residuals for the viscosity model.

Fig. *2* shows a plot of externally studentized residuals against the predicted viscosity values, a key diagnostic tool in regression analysis to assess the homoscedasticity of the residuals. The plot's color gradient represents the range of viscosity values, providing a visual correlation between the magnitude of residuals and the predicted values. Data points are mostly centered around the horizontal axis, which is desirable, suggesting that the residuals are fairly evenly distributed. However, the presence of two points marked with red squares, significantly deviating from the rest, indicates potential outliers or influential points that could affect the regression model's accuracy. These points warrant further investigation to ensure the robustness of the model and to understand the underlying causes of these deviations. The red horizontal lines represent the thresholds beyond which the residuals are considered statistically significant outliers, given the commonly used cutoffs in studentized residuals analysis.

Fig. 3 shows a Cook's Distance plot, a diagnostic tool for identifying influential points in a statistical model. On the x-axis, each point represents a different run number from the set of experiments. At the same time, the y-axis measures the Cook's Distance, indicating the influence of each run on the model's parameters. Most data points are well below the threshold line (marked in red), suggesting they have little influence on the model. However, one data point markedly exceeds the threshold, denoted by a red square, signifying it as a potential influential point. This could indicate that this particular run disproportionately impacts the model and may need further scrutiny to determine if it is an outlier or an indication of a model anomaly or special case in the experimental conditions.

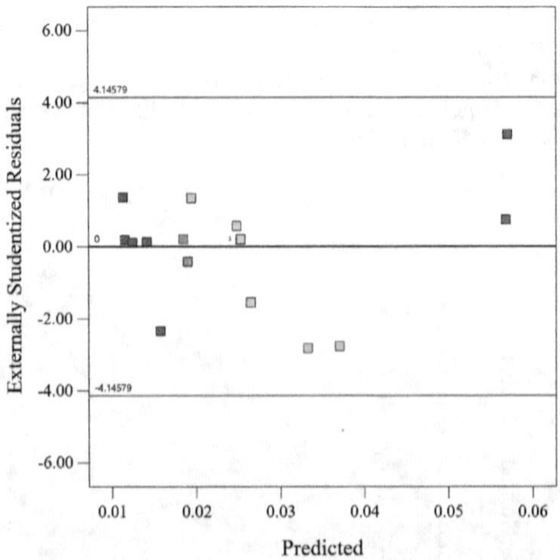

Fig. 2: Residuals vs. Predicted Viscosity Plot showing the distribution of residuals and highlighting potential outliers in the viscosity prediction model.

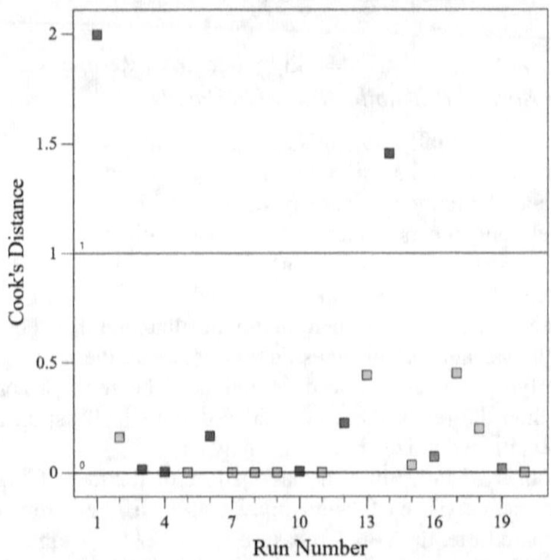

Fig. 3: Cook's Distance Plot identifying the influence of individual runs on the regression model.

Fig. 4 is an actual vs. predicted plot, which is a fundamental tool for assessing the performance of a regression model. The data points represent individual observations where the actual viscosity values are plotted against the predicted viscosity values derived from the model. Ideally, if the model predictions are perfect, all points would lie on the diagonal line, indicating a perfect correlation between actual and predicted values. The color-coding of the points may signify different experimental groups or ranges of viscosity values. The points generally follow the line closely, demonstrating the model's accuracy in predicting viscosity. However, the red square, which lies off the line, indicates a discrepancy between the actual and predicted values, suggesting a potential area for model improvement or a specific case where the model does not perform as well.

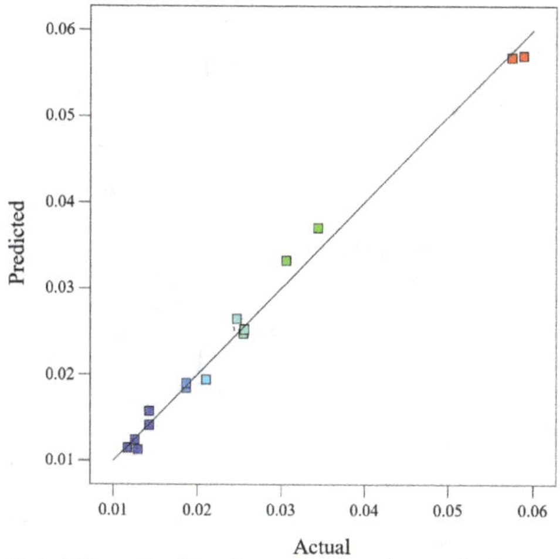

Fig. 4: Actual vs. Predicted Viscosity Plot demonstrating the predictive accuracy of the model.

Fig. 5 illustrates a three-dimensional surface plot, which is a graphical representation of the relationship between viscosity (Pa.s), temperature (°C), and concentration. The axes represent temperature and concentration, while the surface color gradient and height indicate the viscosity levels. The plot shows how viscosity changes with variations in both temperature and concentration, with the color gradient providing an intuitive visualization of the viscosity range. The red points on the plot likely represent actual data points from experiments, which appear to lie close to the surface created by the model, suggesting the model's predictions are reasonably consistent with the empirical data. The shape of the surface suggests that viscosity decreases as temperature increases, which is typical for most fluids, and that there might be a nonlinear relationship between viscosity and concentration.

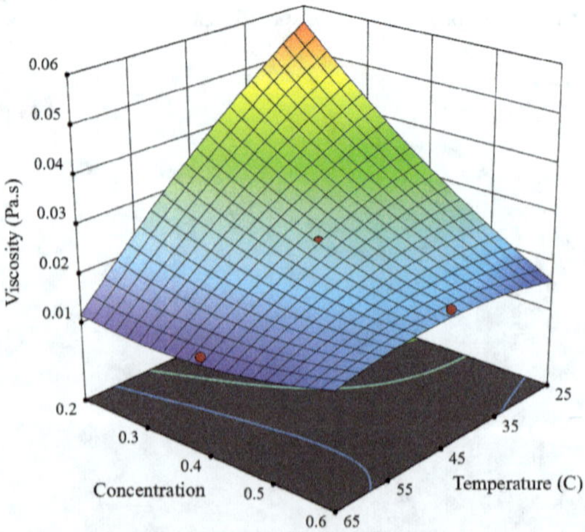

Fig. 5: 3D Surface Plot demonstrating the combined effect of temperature and concentration on the viscosity of the nanofluid.

The results presented focus on optimizing the parameters within specified ranges to achieve the minimum viscosity for the nanofluid. The optimization process considered three key variables: temperature, with a range of 25°C to 65°C; concentration, with a range of 0.2 wt.% to 0.6 wt.%; and shear rate, with a range of 1 1/s to 1000 1/s. Each parameter was weighted equally in terms of importance, and all were set to have a high impact on the optimization outcome, as indicated by an importance level of 3 (Table 3).

Table 3: Optimization constraints

Name	Goal	Lower Limit	Upper Limit	Lower Weight	Upper Weight	Importance
A:Temperature	is in range	25	65	1	1	3
B:Concentration	is in range	0.2	0.6	1	1	3
C:Shear rate	is in range	1	1000	1	1	3
Viscosity	minimize	0.0117	0.0588	1	1	3

The goal was to minimize the viscosity, with the target range for viscosity being from 0.0117006 Pa.s to 0.0588121 Pa.s. The optimization algorithm has identified a solution within these constraints, achieving a desirability score of 1.000, which suggests an optimal solution was found within the defined parameter space. The red points on the optimizer's ramp plots for each variable indicate the optimal levels for temperature, concentration, and shear rate that lead to the minimum viscosity, as defined by the boundaries and goals of the study.

Fig. 6 represents the optimization results for a process with the objective of minimizing viscosity. The ramp graphs for temperature (A), concentration (B), and shear rate (C) show the optimal levels within their respective ranges that contribute to the lowest viscosity outcome. The optimal temperature is shown at the upper limit of 65°C, the concentration at a mid-range value of 0.4, and the shear rate at a value of 500.5. These conditions align with the objective to minimize the viscosity, which is represented in the bottom right plot. The final viscosity achieved under these optimal conditions is marked with a blue dot, indicating a value of 0.0113191, within the lower bound of the target viscosity range. The desirability function value of 1.000 implies that these conditions are the most favorable within the experimental bounds for achieving the goal of minimal viscosity.

Fig. 7 shows two contour plots representing a nanofluid's viscosity optimization. The plot on the left details the desirability function, with warmer colors indicating higher desirability; the highest desirability aligns with higher temperatures and mid-range concentrations. The plot on the right maps the actual viscosity prediction, with cooler colors representing lower viscosity levels, which is the optimization goal. The optimal point, marked on both plots, corresponds to a temperature of around 65°C and a concentration of 0.4, yielding the lowest viscosity and desirability. These plots effectively visualize the combined effect of temperature and concentration on the optimization process, guiding the selection of the best conditions to achieve the desired outcome.

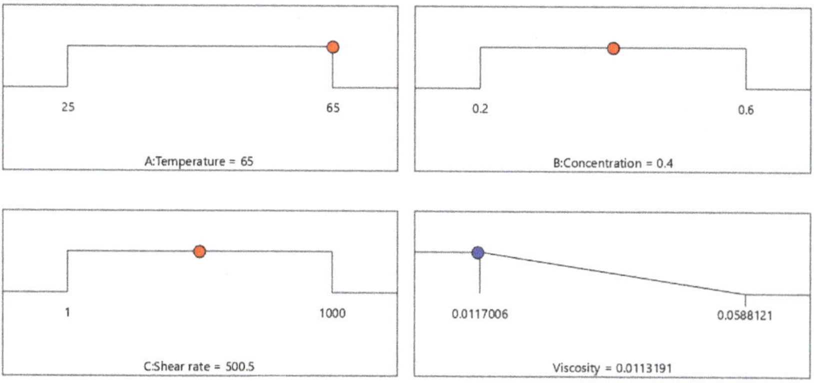

Fig. 6; Optimization Output: Ramp graphs illustrating the optimal temperature, concentration, and shear rate for minimizing viscosity in a nanofluid.

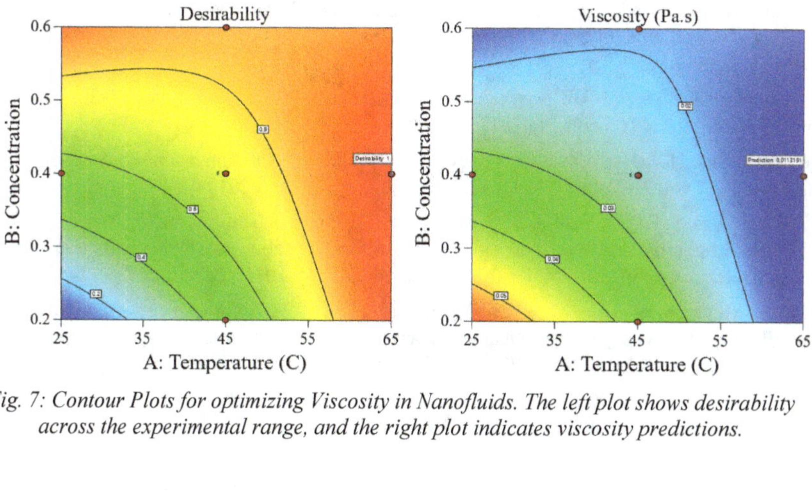

Fig. 7: Contour Plots for optimizing Viscosity in Nanofluids. The left plot shows desirability across the experimental range, and the right plot indicates viscosity predictions.

Conclusions

The present study conducted a comprehensive optimization analysis on hybrid nanofluids containing diamond and boron nitride particles suspended in thermal oil. The investigation was focused on achieving an optimal reduction in viscosity, which is a critical factor in enhancing the efficiency of thermal management systems, particularly in high-temperature applications. The optimization process was meticulously carried out through RSM), where the influences of temperature, nanoparticle concentration, and shear rate on the viscosity of the nanofluids were examined. The optimized conditions were determined to be a temperature of 65°C, nanoparticle concentration of 0.4 wt.%, and a shear rate of 500.5 1/s. Under these parameters, the viscosity of the nanofluids was significantly reduced to 0.0113191 Pa.s, demonstrating the potential of these hybrid nanofluids in reducing operational costs associated with pumping power in heat transfer applications. The model's accuracy was substantiated by the analysis of variance (ANOVA), which presented significant model terms and a high Model F-value, indicating a strong model with an almost negligible probability of the results occurring due to random chance. With an R^2 value of 0.9905, the model's predictive accuracy is well-established, assuring that the optimization approach is both reliable and repeatable.

Acknowledgment

This work was funded by the University of Jeddah, Jeddah, Saudi Arabia, under grant No. (UJ-23-DR-39). The authors, therefore, acknowledge with thanks the University of Jeddah technical and financial support.

References

[1] S. U. Ilyas, R. Pendyala, N. J. C. E. Marneni, and Technology, "Preparation, sedimentation, and agglomeration of nanofluids," vol. 37, no. 12, pp. 2011-2021, 2014. https://doi.org/10.1002/ceat.201400268

[2] S. K. Das, S. U. Choi, and H. E. J. H. t. e. Patel, "Heat transfer in nanofluids-a review," vol. 27, no. 10, pp. 3-19, 2006. https://doi.org/10.1080/01457630600904593

[3] Y.-j. Hwang et al., "Stability and thermal conductivity characteristics of nanofluids," vol. 455, no. 1-2, pp. 70-74, 2007. https://doi.org/10.1016/j.tca.2006.11.036

[4] S. U. Ilyas, R. Pendyala, M. Narahari, L. J. E. c. Susin, and management, "Stability, rheology and thermal analysis of functionalized alumina-thermal oil-based nanofluids for advanced cooling systems," vol. 142, pp. 215-229, 2017. https://doi.org/10.1016/j.enconman.2017.01.079

[5] S. U. Ilyas, R. Pendyala, M. J. C. Narahari, S. A. Physicochemical, and E. Aspects, "Stability and thermal analysis of MWCNT-thermal oil-based nanofluids," vol. 527, pp. 11-22, 2017. https://doi.org/10.1016/j.colsurfa.2017.05.004

[6] W. Yu and H. J. J. o. n. Xie, "A review on nanofluids: preparation, stability mechanisms, and applications," vol. 2012, pp. 1-17, 2012. https://doi.org/10.1155/2012/435873

[7] S. U. Ilyas, S. Ridha, F. A. A. J. C. Kareem, S. A. Physicochemical, and E. Aspects, "Dispersion stability and surface tension of SDS-Stabilized saline nanofluids with graphene nanoplatelets," vol. 592, p. 124584, 2020. https://doi.org/10.1016/j.colsurfa.2020.124584

[8] J. P. Meyer, S. A. Adio, M. Sharifpur, and P. N. J. H. T. E. Nwosu, "The viscosity of nanofluids: a review of the theoretical, empirical, and numerical models," vol. 37, no. 5, pp. 387-421, 2016. https://doi.org/10.1080/01457632.2015.1057447

[9] S. S. Murshed, P. J. R. Estellé, and S. E. Reviews, "A state of the art review on viscosity of nanofluids," vol. 76, pp. 1134-1152, 2017. https://doi.org/10.1016/j.rser.2017.03.113

[10] N. A. C. Sidik, H. Mohammed, O. A. Alawi, S. J. I. C. i. H. Samion, and M. Transfer, "A review on preparation methods and challenges of nanofluids," vol. 54, pp. 115-125, 2014. https://doi.org/10.1016/j.icheatmasstransfer.2014.03.002

[11] M. Hojjat, S. G. Etemad, R. Bagheri, J. J. I. C. i. H. Thibault, and M. Transfer, "Rheological characteristics of non-Newtonian nanofluids: experimental investigation," vol. 38, no. 2, pp. 144-148, 2011. https://doi.org/10.1016/j.icheatmasstransfer.2010.11.019

[12] A. Ali et al., "Thermal and Rheological Behavior of Hybrid Nanofluids Containing Diamond and Boron Nitride in Thermal Oil for Cooling Applications," pp. 1-18, 2023.

[13] A. Ali et al., "Application of Machine Learning Algorithms in Predicting Rheological Behavior of BN-diamond/Thermal Oil Hybrid Nanofluids," vol. 9, no. 1, p. 20, 2024. https://doi.org/10.3390/fluids9010020

[14] A. B. Mahfouz, A. Ali, M. Mubashir, A. S. Hanbazazah, M. Alsaady, and P. L. J. F. Show, "Optimization of viscosity of titania nanotubes ethylene glycol/water-based nanofluids using response surface methodology," vol. 347, p. 128334, 2023. https://doi.org/10.1016/j.fuel.2023.128334

[15] R. Prasher, D. Song, J. Wang, and P. J. A. p. l. Phelan, "Measurements of nanofluid viscosity and its implications for thermal applications," vol. 89, no. 13, 2006. https://doi.org/10.1063/1.2356113

[16] I. M. Mahbubul, R. Saidur, M. A. J. I. J. o. H. Amalina, and M. Transfer, "Latest developments on the viscosity of nanofluids," vol. 55, no. 4, pp. 874-885, 2012. https://doi.org/10.1016/j.ijheatmasstransfer.2011.10.021

Materials Research Proceedings 53 (2025) 198-207 https://doi.org/10.21741/9781644903575-19

Preparation of NiCu-PANI/N-rGO nanocomposite for hybrid supercapacitor

RAHMAN Mohammad Obaidur[1,a *], M NOR Nursyarizal B[1,b], ROMLIE Mohd Fakhizan[1,c], SAKIDIN Hamzah[2,d], SAMSUDIN Dafrizal[3,e], MOHD KASIM Abdul Rahman[4,f], ADYTIA Didit[5,g] and KANNAN Ramani[1,h]

[1]Department of Electrical and Electronic Engineering, Universiti Teknologi PETRONAS, 32610 Seri Iskandar, Perak, Malaysia

[2]Department of Fundamental & Applied Science, Universiti Teknologi PETRONAS, 32610 Seri Iskandar, Perak, Malaysia

[3]Universitas Islam Riau Communication, Riau 28284, Indonesia

[4]Centre for Mathematical Sciences, Universiti Malaysia Pahang, Kampung Melayu Gambang, 26300 Gambang, Pahang, Malaysia

[5]Universitas Telkom, Computing, Terusan Buah Batu 40257 Kabupaten Bandung, Indonesia

[a]mohammad_18003517@utp.edu.my, [b]nursyarizal_mnor@utp.edu.my, [c]fakhizan.romlie@utp.edu.my, [d]hamzah.sakidin@utp.edu.my, [e]dafrizal@comm.uir.ac.id, [f]rahmanmohd@ump.edu.my, [g]adytia@telkomuniversity.ac.id, [h]ramani.kannan@utp.edu.my

Keywords: Hybrid Supercapacitor, Nanocomposite, Solvothermal Synthesis, Polyaniline, Heteroatom Doping, Graphene

Abstract. A highly efficient nickel/copper oxide-PANI grafted nitrogen-doped reduced graphene oxide (NiCu-PANI/N-rGO) nanocomposite electrode material was developed for hybrid supercapacitors (HSCs). The prepared HSC achieved an excellent specific capacitance of 290 F g^{-1} at the current density of 1.5 A g^{-1}. The rate capability of the device was recorded as 37% at 10 A g^{-1} current density and the calculated energy density was also recorded as 32.6 W h kg^{-1} at the power density of 670 kW kg^{-1}. After 8000 charge-discharge cycles in the KOH electrolyte solution, an excellent capacitance retention (82%) was realized. These findings effectively illustrated the possible uses of the manufactured hybrid supercapacitor to mitigate upcoming energy demands.

Introduction

The trade names or well-known colloquial terms for capacitors include electrochemical capacitors (ECs), supercapacitors (SCs), ultracapacitors (UCs), power capacitors, and gold capacitors. SCs are such kinds of electrochemical energy storage technologies that are going through comprehensive evaluation [1]. They also have been given much attention as like as battery in designing and manufacturing processes from both the commercials and academics owing to their identical properties such as low maintenance cost, large power density, fast charge-discharge rate, excellent rate capability, long cycle life, batter safety, wider temperature range. SCs are known as power supply systems with 4–5 W h kg^{-1} energy density and 10–20 kW kg^{-1} higher power density and these figures can be improved soon. Additionally, they can fill the power and energy gap that exists between battery fuel cells and electrostatic capacitors [2]. Even though SCs have lower specific energy compared to batteries they have several merits than those of batteries, high energy storage density, high power density, rapid charge-discharge rate, long cycle life, no need for maintenance, and wide thermal stability range [3].

There are currently different types of supercapacitors available for energy storage systems and the classification is growing along with the use of electroactive materials and electrolytes in supercapacitor device configuration, which is necessary to produce commercially viable supercapacitors. Based on how they store charge, SCs are categorized into main three parts: hybrid supercapacitors, pseudocapacitors, and electrical double-layer capacitors (EDLC) [4]. While pseudocapacitors possess a larger specific capacitance along with excellent specific energy, EDLCs have superior cycle stability and high power density [5]. By combining electrical double-layer and pseudocapacitor processes, a hybrid supercapacitor may achieve higher energy densities and higher working potential windows without sacrificing power density. The concept of a hybrid supercapacitor gained traction as a means of improving energy density. Because of its asymmetric nature, it produces a working potential and specific capacitance that are two to three times higher than those of ordinary capacitors, EDLCs, and pseudocapacitors [6, 7].

Choosing the right electrode materials and their proper design are essential to improve the capacitive performance of SCs. Thermal stability, high SSA, corrosion resistance, high electrical conductivity, proper chemical stability, and acceptable surface wettability are all required of SC electrodes [8]. Graphene, a two-dimensional allotrope of carbon with a single atom thick, is bonded in a hexagonal honeycomb lattice of sp^2 carbons. It is a promising electrode material for supercapacitor electrodes due to its combination of several unique features, including a large surface area (2630 $m^2\ g^{-1}$), high mechanical strength (42 N/m breaking strength), excellent electrochemical stability, and good electrical conductivity (104 S/cm) [9], great mechanical strength (42 N/m breaking strength), exceptional electrochemical stability, superior electrical conductivity (104 S/cm), high thermal conductivity (5,000 W $m^{-1}\ K^{-1}$), and excellent electron mobility (250 000 $cm^2\ V^{-1}s^{-1}$) at ambient temperature[10]. However, graphene-based supercapacitors are unable to meet the current energy demand due to the low energy density of existing SC electrodes caused by the high agglomeration of graphene sheets during the synthesis process. The rational construction of a graphene-based composite was tried to resolve this awkward contradiction, and it worked well in practice [11]. In many different forms, such as reduced graphene oxide, functionalized graphene, graphene doped with heteroatoms like nitrogen or iodine, and graphene composites with transition metal oxides or polymers, graphene has been widely designed and investigated as supercapacitor electrodes. A hollow-structured N-doped graphene composite made of polyaniline, polystyrene, and graphene oxide (GO) was also reported on by the Fan group [12]. The material with a hollow spherical structure doped with N showed remarkable cycle stability and a noteworthy capacitance of 381 F g^{-1} at 1 A g^{-1} current density. Lei et al. have developed an N-doped graphene-based hybrid material for electrochemical SCs [13]. Furthermore, current improvements in compositing graphene with conductive polymers (CPs) exhibit the auspicious properties of those materials for ultracapacitor appliances. Graphene and CP-based composites can produce huge electrochemical and mechanical facilities that finally enhance both cyclability and specific capacitance [14]. The process of creating a reduced graphene oxide (rGO/CP) composite involves distributing well-soluble graphene oxide (GO) in an aqueous polymerizable monomer solution instead of less soluble graphene and then reducing the in-site polymerized GO/CP. Hydrazine hydrate is frequently used as a reductant in the production of rGO/CP composites because of its remarkable efficiency in eliminating oxygen-containing groups from GO. Nevertheless, hydrazine is extremely hazardous and unstable, and it is particularly intensely reductive, which causes the polymer's π–π conjugated system to be destroyed [11]. The cycle stability and rate capability of the electrodes can be impacted by the weak electronic conductivity and poorer ion-transport kinetics that are frequently observed in supercapacitors consisting of carbon materials, conducting polymers, and metal oxides. Moreover, a small working voltage window is a downside of metal/oxide-based electrodes, despite their many advantages over carbon-based materials [15]. Consequently, there has been a lot of interest in the improvement of

TMOs/CPs and carbon/TMOs-based electrodes, and has the potential to lead to revolutionary applications in the electrochemical area. Because they integrate the benefits of both materials in a mutually beneficial way [16] bimetallic oxides and composites are becoming more and more significant. They perform better in electrochemical tests than single metal oxides [17]. Moreover, bimetallic nanoparticles of Ni–Cu are excellent hydrophobic and exhibit outstanding dispersion capability [18].

In our previous work, we produced N-rGO by a one-pot solvothermal synthesis process with the assistance of an organic solvent ethylene glycol (EG) [19]. In this study, NiCu-PANI/N-rGO nanocomposite electrode material which is intended for use in supercapacitor technology is developed following the same procedure.

Materials and Methods

Synthesis of GO/PANI Nanocomposite

Initially, 30 ml of aqueous GO mixture (2mg/ml) was diluted in 50 ml of DI water by sonication for 0.5 h. After that, 0.2 ml of aniline was added to the above solution and subsequently, a specific amount of APS solution (APS to aniline molar ratio:1:1) was mixed drop by drop into the solution while vigorously stirring. At 0 °C, the polymerization process lasted for 4 h. After filtering and washing the reacted mixture with DI water and ethanol, the composite was then vacuum-dried for 24 h at 60 °C.

Synthesis of NiCu-PANI/N-rGO Nanocomposite by EG

After 30 minutes of ultrasonication and magnetically stirring, 50 ml of GO/PANI aqueous solution was mixed with 3 g of urea, 5 ml hydrazine hydrate, 1 mmol $Ni(NO_3)_2 \cdot 6H_2O$, 2 mmol $Cu(NO_3)_2 \cdot 6H_2O$, and 50 ml of EG. A round-bottom flask was filled with the suspension and refluxed for 6 hours at 120 °C. The final composite was dried in a vacuum oven at 60 °C for 24 hours after being cleaned twice with ethanol and water.

Fabrication of Hybrid Supercapacitors

For the electrode preparation, the mass ratio of active materials (NiCu-PANI/N-rGO), PVDF, and conductive carbon were 80%, 15%, and 5% respectively. All substances were taken to a beaker and mixed with N-Methyl-pyrrolidone (NMP) solvent as required followed by stirring for 4 h to make a homogeneous slurry. After that, the mixture was loaded onto the pre-cleaned nickel foam (1x1 cm^2 area). For the cleaning purpose of nickel (Ni) foam, HCl (1 M), DI water, and ethanol were used to wash for 10 min each following ultrasonication to remove surface oxidation of the contaminated layer. Subsequently, the coated Ni foam was dried at 80 °C using a hotplate. The mass loading of active materials was determined by the difference in mass of post-loaded Ni foam and pre-loaded Ni foam. Following the same procedure, an activated carbon (AC)-based electrode was also prepared and finally, a hybrid supercapacitor was fabricated with the assistance of a stainless steel two-electrode split test cell system.

Physical and Electrochemical Characterization

SEM (Evo LS15 VPSEM, Zeiss) was used to describe the morphologies of both N-GO and N-rGO/PANI. X-ray diffraction (XRD, Xpert3 Powder, Panalytical) and XPS (K-Alpha, Thermo) were used to examine the sample structures. Using an electrochemical workstation (Autolab PGSTAT204), electrochemical impedance spectroscopy (EIS), galvanostatic charging/discharging (GCD), and cyclic voltammetry (CV) were performed on the as-prepared samples.

A traditional strategy to obtain the capacitance of an electrode using CV measurement is to assume that the charge q is stored in the electrode. If the applied voltage V, then the total capacitance (c_s) can be defined as [20]:

$$C_s = \frac{q}{V} \quad (1)$$

The specific capacitance of any material would be assessed by the below equation and the area under CV curves.

$$C_s = \frac{\int_{V_1}^{V_2} I(V)\, dV}{mv(V_2 - V_1)} \quad (2)$$

Where active materials mass is denoted by m, $V_1$and V_2 are the lower and upper potentials, respectively, and the scan rate is v. This value ($\int_{V_1}^{V_2} I\, dV$), reflects the area under the CV curve, and the following equation [291] can be used to compute the specific capacitance based on GCD curves.

$$C_s = \frac{I \times t}{V \times m} \quad (3)$$

Where I is referred to as the current and t is the discharge time. The quantity of energy that a material can store per unit volume is known as energy density. The usual units of measurement for volumetric capacitance and gravimetric capacitance are watt-hours per liter (W h L^{-1}) and watt-hours per kilogram (W h kg^{-1}), respectively [21]. The specific energy and specific power can be determined as below equations.

$$E = \frac{1}{2} C_s V^2 x \frac{1}{3600} \quad (4)$$

$$P = \frac{E}{t} x3600 \quad (5)$$

Results and Discussion

The morphological nanostructure of previously prepared N-rGO and NiCu-pANI/N-rGO materials from SEM analysis are depicted in Fig. 1. For N-rGO, GO was reduced and doped with N molecules via the solvothermal method. The corrugations and wrinkles realized in Fig. 1 (a) indicate the dopant elements (N) and oxygen-related functional groups. As depicted in Fig. 1 (b), the SEM image in the scale of 2 μm from the NiCu-PANI/N-rGO-based sample represents the morphological properties with hierarchic microstructure. Metal nanoparticles (Ni & Cu) are equally distributed across graphene nanosheets, as seen by the SEM illustration, and well-agglomerated NiO and CuO nanoparticles which are globular-like crystalline structures are also seen in the visual indicating mesoporous structure of the sample [22]. Because of the repulsive force and van der Waals force, a small percentage of PANI nuclei may have first formed at the GO surface in the polymerization of aniline on non-functionalized GO; however, many more nuclei were involved in the solution. As a result, nucleation on the PANI nuclei that had formed in the solution or close to the GO happened concurrently with the polymerization, resulting in the creation of two different morphologies (Fig. 1 (b) – PANI/N-rGO sheets and PANI aggregates) [23].

Figure 2 displays the XRD spectrum graph obtained from NiCu-PANI/N-rGO and N-rGO nanocomposites. Following the post-treatment work by solvothermal process, broad and small peaks concentrated at 2θ=24.9° (reduction of GO) and 2θ=43.10° for N-rGO materials were depicted, consisting of the 002 and 001 planes, respectively, rather than the sharp peak located at 2θ=11.10° for GO, indicating that the GO had been restacked and exfoliated (intercalated) at the doping process due to electrostatic and van der Waals [24].

Materials Research Forum LLC
https://doi.org/10.21741/9781644903575-19

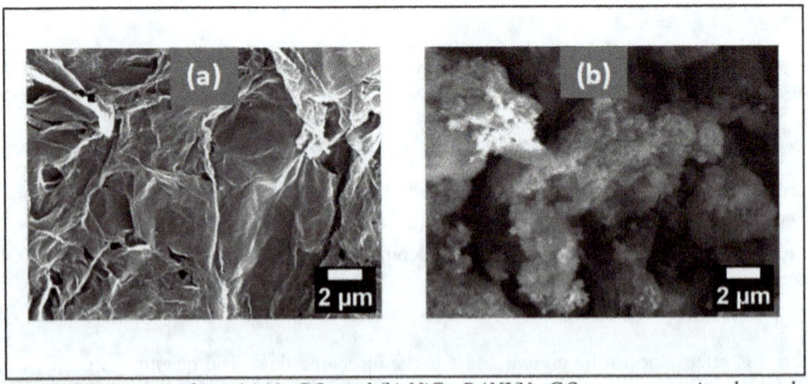

Fig. 1. SEM images from (a) N-rGO, and (b) NiCu-PANI/N-rGO nanocomposite electrode material for hybrid supercapacitors.

The outcome shows that the graphene-decorated Ni-Cu bimetallic composite is obtained. The crystal planes (111, 200, 020, 202 220, and 311) of the structure are represented by the diffraction peaks at 44.03°, 44.5°, 51.1°, 51.5°, 75.09°, and 75.5°, respectively in the XRD pattern of NiCu-PANI/N-rGO [25]. The PANI is in a crystalline structure, as indicated by the distinctive π/π (C=C) intense peak at 15.3° and strong intense peak at 25° (d-spacing 7.2 and 3.5 A) [26].

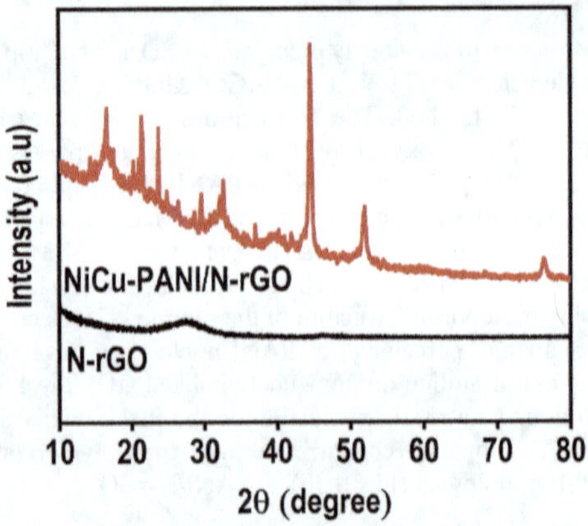

Fig. 2. XRD plot of (a) N-rGO, (b) NiCu-PANI/N-rGO nanocomposite materials.

Figure 3 displays the FTIR spectra of the NiCu-PANI/N-rGO and N-rGO composites. The C=C stretching vibrations of the quinoid and benzenoid rings, the C–N stretching of the secondary aromatic amine, and the C–H stretching are responsible for the peaks of PANI at 1581, 1494, 1305, 1140, and 802 cm^{-1}. On the other hand, the benzene ring's in-plane and out-of-plane C–H bending vibrations are the source of the bands at 1140 and 802 cm^{-1}. The peak exhibits a small shift and produces a red shift when compared to PANI. This is due to the rGO/Ni-Cu doping, which can improve the electronic cloud distribution and electron delocalization between the π electrons in graphene and the electric charge in the PANI. The outcome unequivocally demonstrates the covalent link between PANI [27].

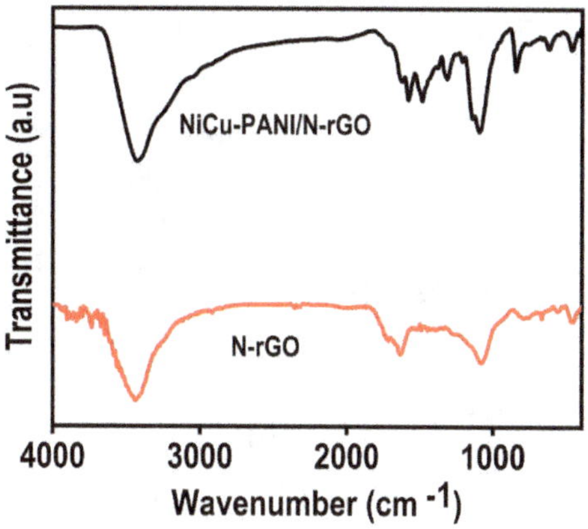

Fig. 3. FTIR graph from NiCu-PANI/N-rGO nanomaterials.

A hybrid supercapacitor from NiCu-PANI/N-rGO and activated carbon electrodes was fabricated by a split test cell where 3 M KOH was applied as electrolyte solution to evaluate the electrochemical performance of NiCu-PANI/N-rGO material. CV curves presented in Fig. 4 (a) with the chosen potential window of 1.1 V (-0.6 + 0.5) and the scan rates ranging from 10 mV/s to 100 mV/s have the physical structure of battery type storage system. The quasi-rectangular shape of the CV graphs indicates the mixed features of pseudo-capacitance and EDLC implying an improved electrochemical characteristic [28]. GCD curves of NiCu/N-rGO//AC hybrid supercapacitor with 2 – 8 A g^{-1} current density range are shown in Fig. 4 (b) having faradaic reaction characteristics [29]. The obtained specific capacitance was 290 A g^{-1} at the current density of 1.5 A g^{-1} which was calculated using equation (3). The highest energy and power densities were also recorded as 32.6 W h kg^{-1} and 670 W kg^{-1}, respectively. The Ragone plot shows the capacitive performance of the device at various current densities having a rate capability of 37% at the current density of 10 A g^{-1} (Fig. 4 (c)) which suggests that this device has better capacitance performance.

EIS is a crucial characterization technique for researching the supercapacitor's fundamental behavior such as electrical resistance and ion transport characteristics. With an applied open circuit potential, electrode/electrolyte interfacial resistance (series resistance) was measured in the frequency range of 100 kHz – 0.1 Hz. The EIS graph from the fabricated hybrid device creates an indistinct semicircle in the high-frequency zone and a line in the low-frequency area as displayed in Fig. 4 (d). The real axis (Z′) cut-off shows the combined resistance of electrode material and electrolyte solution in the high-frequency region. The slope of the curves in the low-frequency zone indicates the Warburg impedance which is related to proton diffusion in host materials and electrolyte diffusion in porous electrodes. The higher slope of the EIS line with the supercapacitor assigns the higher diffusion rate of electrolyte and electrode efficiency. The line angle with the real axis of the EIS data indicates lower diffusion and contact resistance [30, 31].

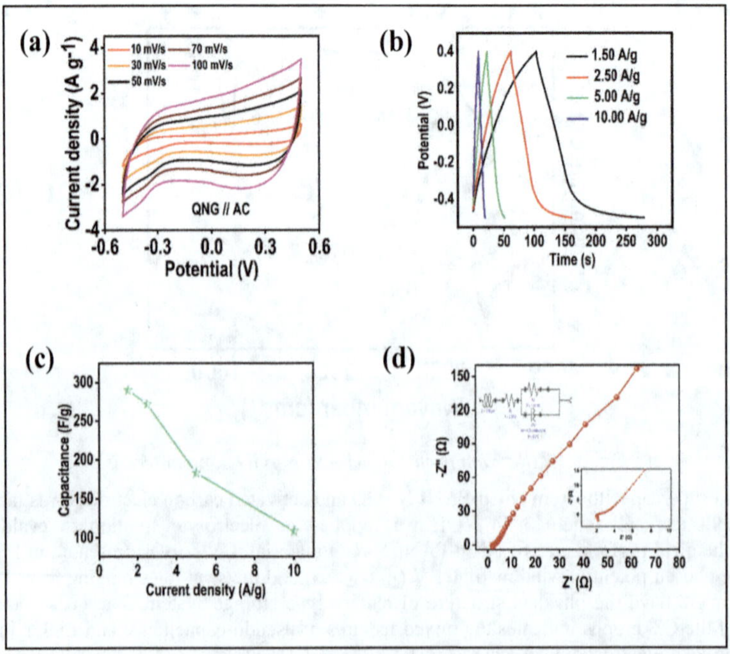

Fig. 4. (a) CV curve, (b) GCD curve, (c) rate performance, and (d) Nyquist plot of NiCu-PANI/N-rGO//AC hybrid supercapacitor.

The hybrid device showed superb cyclic stability since the specific capacitance of the electrode remained at 82% after 8000 cycles. It could occur because of the optimized material's large surface area where N-rGO is responsible for achieving the surface area and can directly touch the electrolyte to lower the diffusion resistance. The capacitive performance gradually reduced until 1500 cycles after that it remained constant up to 7000 cycles.

Conclusion

By solvothermal synthesis process, a high-performance nanocomposite electrode material was developed for supercapacitor applications, and a hybrid storage device was fabricated. The performance of the device was thoroughly assessed by an electrochemical workstation. Notably, the NiCu-PANI/N-rGO//AC hybrid supercapacitor device demonstrated the highest capacitance value of 290 F g^{-1} at a current density of 1.5 A g^{-1}. An impressive energy density of 32.6 W h kg^{-1} at the power density of 670 W kg^{-1} was also observed. The device holds an excellent capacitance retention of 82% after 8000 charge-discharge cycles. Additionally, the EIS outcomes revealed low resistivity of the electrodes indicating the material's excellent charge transfer capability. This proposed approach, with its outstanding results, indicates an efficient route for developing carbon-metal alloys for the application of supercapacitor technology.

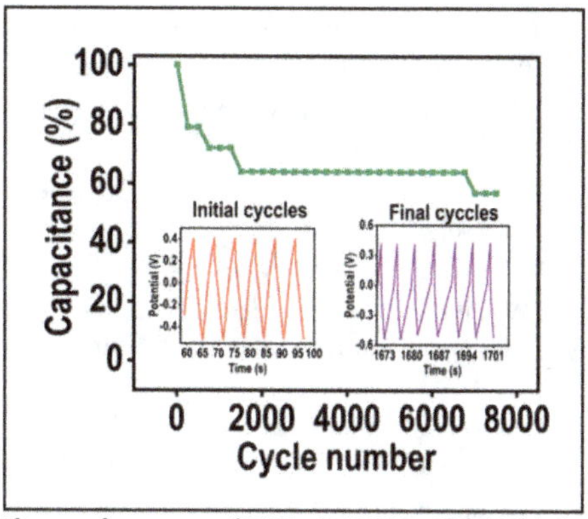

Fig. 5. Cyclic stability performance of NiCu-PANI/N-rGO//AC hybrid supercapacitor.

Acknowledgment
This work is fully supported by the Collaboration Research Funding (CRF) under the grant scheme 015ME0-336.

References

[1] R. Kötz and M. Carlen, "Principles and applications of electrochemical capacitors," Electrochim. Acta, vol. 45, no. 15, pp. 2483-2498, 2000/05/03/ 2000, doi: https://doi.org/10.1016/S0013-4686(00)00354-6.

[2] V. K. Mariappan, K. Krishnamoorthy, P. Pazhamalai, S. Sahoo, S. S. Nardekar, and S.-J. Kim, "Nanostructured ternary metal chalcogenide-based binder-free electrodes for high energy density asymmetric supercapacitors," Nano Energy, vol. 57, pp. 307-316, 2019/03/01/ 2019, doi: https://doi.org/10.1016/j.nanoen.2018.12.031.

[3] A. González, E. Goikolea, J. A. Barrena, and R. Mysyk, "Review on supercapacitors: Technologies and materials," Ren. & Sus. Eng. Rev., vol. 58, pp. 1189-1206, 2016/05/01/ 2016, doi: https://doi.org/10.1016/j.rser.2015.12.249.

[4] M. Tomy, A. Ambika Rajappan, V. Vm, and X. Thankappan Suryabai, "Emergence of Novel 2D Materials for High-Performance Supercapacitor Electrode Applications: A Brief Review," Energy & Fuels, vol. 35, no. 24, pp. 19881-19900, 2021/12/16 2021, doi: 10.1021/acs.energyfuels.1c02743.

[5] Z. S. Iro, C. Subramani, and S. Dash, "A brief review on electrode materials for supercapacitor," Int. J. Electrochem. Sci, vol. 11, no. 12, pp. 10628-10643, 2016.

[6] H. Liu, X. Liu, S. Wang, H.-K. Liu, and L. Li, "Transition metal based battery-type electrodes in hybrid supercapacitors: A review," Energy Storage Materials, vol. 28, pp. 122-145, 2020.

[7] C.-y. Hui, C.-w. Kan, C.-l. Mak, and K.-h. Chau, "Flexible energy storage system—an introductory review of textile-based flexible supercapacitors," Processes, vol. 7, no. 12, p. 922, 2019.

[8] P. Forouzandeh, V. Kumaravel, and S. C. Pillai, "Electrode Materials for Supercapacitors: A Review of Recent Advances," Catalysts, vol. 10, no. 9, p. 969, 2020. [Online]. Available: https://www.mdpi.com/2073-4344/10/9/969.

[9] Y. Zhu et al., "Graphene and graphene oxide: synthesis, properties, and applications," Adv. Mater., vol. 22, no. 35, pp. 3906-3924, 2010.

[10] K. S. Novoselov et al., "Two-dimensional gas of massless Dirac fermions in graphene," nature, vol. 438, no. 7065, pp. 197-200, 2005.

[11] Y. Liu, Y. Zhang, G. Ma, Z. Wang, K. Liu, and H. Liu, "Ethylene glycol reduced graphene oxide/polypyrrole composite for supercapacitor," Electrochim. Acta, vol. 88, pp. 519-525, 2013/01/15/ 2013, doi: https://doi.org/10.1016/j.electacta.2012.10.082.

[12] W. Fan, Y.-Y. Xia, W. W. Tjiu, P. K. Pallathadka, C. He, and T. Liu, "Nitrogen-doped graphene hollow nanospheres as novel electrode materials for supercapacitor applications," J. Power Sources, vol. 243, pp. 973-981, 2013.

[13] H. Lei, J. Tu, D. Tian, and S. Jiao, "A nitrogen-doped graphene cathode for high-capacitance aluminum-ion hybrid supercapacitors," New J. Chem., vol. 42, no. 19, pp. 15684-15691, 2018.

[14] W. Raza et al., "Recent advancements in supercapacitor technology," Nano Energy, vol. 52, pp. 441-473, 2018.

[15] H. S. Yaddanapudi, K. Tian, S. Teng, and A. Tiwari, "Facile preparation of nickel/carbonized wood nanocomposite for environmentally friendly supercapacitor electrodes," Scientific Reports, vol. 6, no. 1, p. 33659, 2016.

[16] T. E. Balaji, H. Tanaya Das, and T. Maiyalagan, "Recent Trends in Bimetallic Oxides and Their Composites as Electrode Materials for Supercapacitor Applications," ChemElectroChem, vol. 8, no. 10, pp. 1723-1746, 2021/05/11 2021, doi: https://doi.org/10.1002/celc.202100098.

[17] M. Yewale et al., "Sphere-Shaped CuCo2O4 Nanostructures battery type electrode for supercapacitor via Hydrothermal Synthesis Approach," Colloids and Surfaces A: Physicochemical and Engineering Aspects, p. 132541, 2023.

[18] L. Chen, H. Xu, H. Cui, H. Zhou, H. Wan, and J. Chen, "Preparation of Cu–Ni bimetallic nanoparticles surface-capped with dodecanethiol and their tribological properties as lubricant additive," Particuology, vol. 34, pp. 89-96, 2017/10/01/ 2017, doi: https://doi.org/10.1016/j.partic.2016.12.006.

[19] M. O. Rahman et al., "One-Step Solvothermal Synthesis by Ethylene Glycol to Produce N-rGO for Supercapacitor Applications," Nanomaterials, vol. 13, no. 4, doi: 10.3390/nano13040666.

[20] M. O. Rahman et al., "One-Step Solvothermal Synthesis by Ethylene Glycol to Produce N-rGO for Supercapacitor Applications," Nanomaterials, vol. 13, no. 4, p. 666, 2023. [Online]. Available: https://www.mdpi.com/2079-4991/13/4/666.

[21] S. V. Mulik, S. A. Jadhav, P. S. Patil, and S. D. Delekar, "4 - Transition metal oxide–conducting polymer nanocomposites and metal-organic framework-based composites for supercapacitor application," in Advances in Metal Oxides and Their Composites for Emerging Applications, S. D. Delekar Ed.: Elsevier, 2022, pp. 135-185.

[22] M. Shoeb, M. Mobin, M. A. Rauf, S. M. Adnan, and M. Y. Ansari, "Graphene nickelcopper nanocomposite (Gr@ NiCu NCs) as a binder free electrode for high energy density supercapacitor and antimicrobial application," Journal of Materiomics, vol. 7, no. 4, pp. 815-827, 2021.

[23] Z.-F. Li, H. Zhang, Q. Liu, Y. Liu, L. Stanciu, and J. Xie, "Covalently-grafted polyaniline on graphene oxide sheets for high performance electrochemical supercapacitors," Carbon, vol. 71, pp. 257-267, 2014.

[24] Y. Zhu, W. Peng, Y. Li, G. Zhang, F. Zhang, and X. Fan, "Multiple roles of a heterointerface in two-dimensional van der Waals heterostructures: insights into energy-related applications," J of Mater. Chem. A, vol. 7, no. 41, pp. 23577-23603, 2019.

[25] H. Siddiqui, M. R. Parra, M. S. Qureshi, M. M. Malik, and F. Z. Haque, "Studies of structural, optical, and electrical properties associated with defects in sodium-doped copper oxide (CuO/Na) nanostructures," Journal of Materials Science, vol. 53, no. 12, pp. 8826-8843, 2018/06/01 2018, doi: 10.1007/s10853-018-2179-6.

[26] S. Sravya, D. RamaDevi, N. Belachew, K. E. Rao, and K. Basavaiah, "Vitamin C assisted synthesis of rGO–Ag/PANI nanocomposites for improved photocatalytic degradation of pharmaceutical wastes," RSC advances, vol. 11, no. 20, pp. 12030-12035, 2021.

[27] D. Luan et al., "Fabrication and electrochemical properties of graphene/copper–nickel solid solution reinforced polyaniline composite," Journal of Materials Science: Materials in Electronics, vol. 28, no. 19, pp. 14738-14746, 2017/10/01 2017, doi: 10.1007/s10854-017-7342-3.

[28] R. Wang et al., "Flexible ultrathin all-solid-state supercapacitors," Rare Metals, vol. 37, pp. 536-542, 2018.

[29] S. C. Sekhar, G. Nagaraju, and J. S. Yu, "High-performance pouch-type hybrid supercapacitor based on hierarchical NiO-Co3O4-NiO composite nanoarchitectures as an advanced electrode material," Nano Energy, vol. 48, pp. 81-92, 2018.

[30] G. Liu, S. Lu, W. Xu, G. He, Y. Zheng, and Y. Cheng, "Fabrication of graphene/copper–nickel foam composite for high performance supercapacitors," New J. Chem., 10.1039/C8NJ01017J vol. 42, no. 12, pp. 9455-9462, 2018, doi: 10.1039/C8NJ01017J.

[31] D. Shi, L. Zhang, X. Yin, T. Huang, and H. Gong, "A one step processed advanced interwoven architecture of Ni (OH) 2 and Cu nanosheets with ultrahigh supercapacitor performance," J of Mater. Chem. A, vol. 4, no. 31, pp. 12144-12151, 2016.

TBP-D2EHPA system for synergizing phenol extraction in emulsion liquid membrane process

OTHMAN Norasikin[1,2,a*], YAN Kok Jia[1,b], NOAH Norul Fatiha Mohamed[1,c], KAHAR Izzat Naim Shamsul [1,d]

[1]Faculty of Chemical and Energy Engineering, Universiti Teknologi Malaysia, 81310 Skudai, Johor Bahru, Malaysia

[2]Centre of Lipids Engineering and Applied Research (CLEAR), Ibnu Sina Institute for Scientific and Industrial Research, Universiti Teknologi Malaysia, 81310 Skudai, Johor Bahru, Malaysia

[a]norasikin@cheme.utm.my, [b]kokyan@graduate.utm.my, [c]norulfatiha.mn@gmail.com, [d]izzatnaimsky@gmail.com

Keywords: Synergist Formulation, Phenol Recovery, Emulsion Liquid Membrane, Emulsion Stability, TBP-D2EHPA

Abstract. Phenol in wastewater causes adverse pollution problems and could affect aquatic and human life. Emulsion liquid membrane (ELM) had a high potential possibility to extract phenol from wastewater. The new ELM formulation that consists of Tributyl phosphate (TBP) and Di-(2-ethylhexyl) phosphoric acid (D2EPHA) in kerosene as a synergistic carrier is investigated. Several operating parameters affecting the phenol extraction and stability of W/O/W emulsion through breakage or swelling were studied, such as extraction time, agitation speed, and treat ratio. At the optimum condition of a treat ratio of 1:5, an agitation speed of 300 rpm, and an extraction time of 3 minutes, a stable ELM process yielded an 89 % phenol extraction. This research highlights the potential synergistic effect of the ELM formulation of the TBP-D2EHPA system in enhancing emulsion stability and improving phenol extraction efficiency.

Introduction

Phenol, though valuable for its role in various products, presents a significant environmental threat due to its toxicity and carcinogenicity. Wastewater from refineries and processing plants often contains traces of this harmful pollutant. The noxious odor of phenol, even at low concentrations, induces respiratory irritation. Inhalation may result in adverse effects such as anorexia, weight loss, and headaches [1]. Phenol-laden wastewater poses a significant threat to water ecosystems if left untreated, as it can contaminate water bodies and jeopardize aquatic life [2]. Legislation such as the Malaysia Environmental Quality Act 1974 mandates stringent limits on phenol concentration in discharged effluents, set at 1 ppm. However, industrial processes can yield effluents with phenol concentrations ranging from 2.8 to 6900 ppm [3]. Thus, imperative measures are requisite to recover phenol from industrial effluents prior to discharge into water bodies.

Fortunately, emulsion liquid membrane (ELM) technology has become a promising solution. This efficient process offers several advantages, including low chemical consumption, high selectivity, non-dependence on equilibrium consideration and high interfacial area for extraction, [4]. ELM have found utility in the retrieval of metals and organics, yet their broader adoption is impeded by challenges related to emulsion instability and swelling. The development of an ELM for phenol extraction involves incorporating a carrier for facilitating phenol extraction. Commonly utilized single carriers, including TBP, TOA and D2EHPA, are employed for phenol extraction through an ion exchange mechanism [5–7].

However, employing a single carrier tends to yield suboptimal extraction efficiencies [4]. Achieving higher extraction efficiencies may necessitate elevated concentrations of a single carrier. Consequently, researchers have increasingly focused on the utilization of synergistic carriers, comprising combinations of two or more carriers. This approach has garnered substantial interest due to its potential to enhance solute extraction, ion transport efficiency, and process selectivity compared to single carrier [8–10]. Furthermore, synergistic carriers have been observed to augment the solubility of extracted solutes within the liquid membrane phase while preventing the formation of a tertiary phase [11]. By embracing this innovative technology, the environment can be safeguarded and a cleaner future can be paved.

To the best of our knowledge, to date, there is a notable absence of research reports detailing the extraction of phenol using a synergistic solvating-acidic carrier within the formulation of ELM. The present study aims to assess the efficacy of the synergistic carrier combination of tributyl TBP and D2EHPA in the ELM process. The investigation involves varying affecting parameters of phenol extraction such as treat ratio, extraction time and agitation speed.

Materials and Methods

Chemicals and Reagents. Kerosene and phenol crystal (99 % assay) were purchased from Sigma-Aldrich. Span 80 (60 % oleic acid), sodium hydroxide pellets (NaOH, 98 % purity), 1-butyl-3-methyl-imidazolium bis(trifluoro-methhylsulfonyl)imide ([BMIM]+ [PF6]-) and di-(2-ethylhexyl) phosphoric acid (D2EHPA, 95 % purity) were obtained from Merck. Tributyl phosphate (TBP, 99 % purity) was obtained from Across Organic. All reagents utilized were of analytical grade and were employed without additional purification.

Preparation of Simulated Phenol Solution. The aqueous solution of phenol feedstock (100 ppm) was prepared by dissolving the appropriate amount of crystallized phenol in distilled water. To establish calibration, the 100 ppm phenol solution was diluted to concentrations of 10, 20, 30, 40, and 50 ppm. Phenol content in wastewater was measured using a UV-spectrophotometer (JASCO V-730 Spectrophotometer) at a wavelength (λ) of 270 nm **[12].**

Single and Synergistic Carrier for Phenol Extraction in Liquid-Liquid Extraction. The investigation included conducting a Liquid-Liquid Extraction (LLE) process to examine the impact of single and synergistic carriers on phenol extraction. In a 100 mL conical flask, a 1:1 volume ratio of simulated phenol solution and organic liquid membrane (carrier in kerosene) was added, and the mixture was shaken for 18 hours at 300 rpm using an incubator shaker (Sastec, Canada). Following the extraction procedure, the mixture was transferred into a separating funnel and allowed to phase separation for 30 minutes at room temperature (25 ± 1 °C). The UV spectrophotometer was used to ascertain the amount of phenol in the treated aqueous solution. The synergistic effect was quantified through the extraction and synergistic coefficient (S.C) which can be defined using Equations 1 and 2.

$$D = \frac{[C]_{org}}{[C]_{aq}} \quad (1)$$

$$S.C = \frac{D_{mix}}{D_{synergist} + D_{carrier}} \quad (2)$$

Where the phenol concentrations in the organic and aqueous phases represents by $[C]_{org}$ and $[C]_{aq}$, respectively. Meanwhile, the distribution ratios obtained using each carrier separately represents by $D_{carrier}$ and $D_{synergist}$ are and the distribution ratio with the mixture represents by D_{mix}.

Phenol recovery using Emulsion Liquid Membrane (ELM) Process. The extraction study was carried out using the optimum operating parameters which have been studied in previous work [13]. The organic liquid membrane was prepared by mixing Span 80, $[BMIM]^+$ $[PF6]^-$stabilizer and 0.025M synergistic carrier in kerosene, while the aqueous stripping phase was prepared by diluting NaOH in distilled water at 0.1 M. Then, water in oil (W/O) emulsion was prepared by homogenizing the liquid membrane phase and aqueous stripping phase at a ratio of 3:1. The solution was homogenized at 8000 rpm for 3 minutes using a high-speed homogenizer (Silent Crusher, Heidolph).

After preparing the primary emulsion, it was dispersed into 100 ppm phenol solution as a feed phase. The mixture of the feed and emulsion phase was stirred using a digital mixer (ColeParmer EW-50006-00). After the extraction process, the mixture of emulsion and aqueous was immediately poured into a separating funnel and was left for 30 minutes for the phase separation of the emulsion and feed phase. The treated phenol solution (feed phase) was taken for phenol content analysis using the UV-Spectrophotometer.

In order to find the optimum parameters for the ELM process with the presence of a synergist carrier, the extraction study was carried out by varying the agitation speed (200-500 rpm), extraction time (1-7 minutes) and treat ratio of emulsion to feed phase (1:3 - 1:10). The study was carried out based on the one-factor-at-a-time (OFAT) method. All the experiments were conducted at room temperature (25 ± 1 °C). Equations 3 and 4 were employed to assess both the efficiency of phenol extraction and the extent of emulsion breakage, respectively.

$$\text{Phenol extraction percentage } (\%) = \frac{P_f - P_i}{P_i} \times 100 \quad (3)$$

$$\text{Breakage/Swelling } (\%) = \frac{V_f - V_i}{V_i} \times 100 \quad (4)$$

Where P denotes the concentration of phenol in the feed phase (ppm) while f and i denote the final and initial, respectively. V_i represents the initial volume of the emulsion and V_f represents the final volume of the emulsion after extraction.

Result and Discussion

Effect of Single and Synergistic Carrier. Table 1 tabulates the extraction and distribution ratio of phenol using single and synergistic carrier. The result shows that the phenol extraction percentages using single TBP and D2EHPA carriers are reported as 67.91 % and 58.53 %, respectively, with corresponding distribution ratios of 2.12 and 1.41. Single TBP demonstrates higher extraction efficiency compared to single D2EHPA, attributable to its capacity to interact with polar phenol molecules and dissolve into the diluent. The TBP structure comprising four oxygen atoms facilitates the formation of hydrogen bonds with phenol molecules [13]. On the other hand, the extraction of phenol using D2EHPA was lower owing to weaker molecular interactions. Due to its polarity, phenol can be effectively extracted using various basic and solvating carriers.

However, a more promising performance is observed with the combination of TBP and D2EHPA, where a significantly higher extraction percentage of 93.51 % is achieved. Additionally, the system yielded high D = 14.41 and S.C = 25.70 values. These findings indicate a synergistic effect between TBP and D2EHPA, resulting in more efficient transport of phenol molecules compared to when either TBP or D2EHPA is used individually. This formulation was used in the next ELM process.

Table 1. Distribution ratio and extraction of phenol using single and synergist carrier (Experimental conditions: agitation speed : 300 rpm; [Phenol] : 100mg/L; pH of feed phase : 8.50 ± 0.01; agitation time : 18 h)

Carrier	Extraction (%)	Distribution Ratio
0.025M TBP	67.91	2.12
0.025M D2EHPA	58.53	1.41
0.0125M TBP + 0.0125M D2EHPA	93.51	14.41

Effect of Treat Ratio. The effect of treat ratio on phenol extraction and distribution ratio is present in Table 2. The result indicates that within treat ratios of 1:3 to 1:5, there is a significant increase in phenol extraction percentage from 19.88 % to 83.40 %. This indicates that an increase in the treat ratio has a positive impact on the membrane and internal phase capacity, consequently enhancing the extraction of phenoxide ions. This improvement is attributed to the expansion of the feed aqueous phase within the emulsion's overall volume as the treat ratio increases, leading to the formation of numerous emulsion globules that increase the surface area for efficient solute mass transfer into the internal phase. The result is parallel to Sulaiman et al. [14].

However, as the treat ratio is further increased to 1:7 and 1:10, the phenol extraction efficiency begins to decline, reaching values of 66.12 % and 44.94 %, respectively. This decrease is linked to the diminished ability of the emulsion phase to interface with the feed phase, leading to diminished extraction of phenoxide ions into the internal phase. This phenomenon arises due to the abundance of water molecules in the external feed phase, thereby creating a significant osmotic pressure disparity relative to the emulsion phase [15]. Consequently, the external feed phase undergoes permeation into the internal droplet, resulting in swelling and a tendency to rupture the emulsion.

The swelling percentage, on the other hand, did not significantly change when the treat ratio increased, initially rising from 8 to 12 % and subsequently decreasing to 10 %. The study concludes that a 1:5 treat ratio emerges as the optimal condition, delivering the highest extraction efficiency while maintaining the stability of the ELM process, making it a suitable choice for future research.

Table 2. Effect of treat ratio on the extraction performance of phenol (Experimental conditions: O/I ratio : 3:1; emulsification time : 3 min; [Phenol] : 100mg/L; pH of feed phase: 8.50 ± 0.01; [Carrier] : 0.025 M; homogenizer speed : 8000 rpm; surfactant composition in the liquid membrane : 3 % (w/v); [NaOH] : 0.1M; [ionic liquid] : 0.107 % (w/v); extraction time : 3 min; agitation speed : 200 rpm).

Treat Ratio (emulsion: feed phase)	Extraction (%)	Swelling (%)
1:3	19.88	8
1:5	83.40	12
1:7	66.12	12
1:10	44.94	10

Effect of Agitation Speed. The effect of phenol extraction and swelling percentage with agitation speed of 200 to 500 rpm range are shown in Table 3. The study observed a progressive increase in extraction percentage, rising from 83.40 % to 89.01 %, as the agitation speed increased from 200 to 300 rpm. This enhancement is attributed to the exposure of the emulsion to the agitator's shear, generating numerous smaller emulsion globules and thereby providing a larger mass transfer area for efficient phenoxide ion extraction between phases [7]. In addition, the swelling percentage was unchanged at 10 %. This suggests that the stability of the ELM system can be maintained at up to 300 rpm agitator speed.

Upon increasing agitation speeds to 400 rpm and 500 rpm, a declining trend in extraction was observed, dropping to 51.62 % and 50.93 %, respectively. As the agitation speed rises, there is a noticeable reduction in extraction efficiency, primarily due to the formation of larger emulsion globules. These larger globules diminish the available interface for mass transfer, thereby impeding the extraction process. The inefficiency arises from certain internal droplets failing to participate in the reaction due to longer diffusion paths. These results corroborate those reported by Zaulkiflee et al. [7]. In addition, the swelling percentage is raised from 12 % to 22 %. This phenomenon arises from surpassing the critical threshold (up to 400 rpm), triggering the entrainment swelling phenomenon. Repetitive coalescence events among the globules induce the generation of secondary inner droplets within the globules. Consequently, the volume of the internal phase increases, culminating in the formation of larger-sized globules. This finding is in line with Suliman et al. [15]. Agitation speed of 300 rpm was selected since it produced the highest phenol extraction percentage.

Table 3. Effect of agitation speed on the extraction performance of phenol (Experimental conditions : O/I ratio : 3:1; [Phenol] : 100mg/L; treat ratio : 1:5; pH of feed phase : 8.50 ± 0.01; [Carrier] : 0.025 M; homogenizer speed : 8000 rpm; emulsification time : 3 min; surfactant composition in the liquid membrane : 3 % (w/v); [NaOH]: 0.1M; [Ionic liquid] : 0.107 % (w/v); extraction time : 3 min).

Agitated Speed (rpm)	Extraction (%)	Swelling (%)
200	83.4	12
300	89.01	12
400	51.62	22
500	50.93	8

Effect of Extraction Time. Table 4 depicts the impact of extraction time on both phenol extraction and emulsion swelling. The findings reveal diminished extraction efficiency at a 1 minute extraction time, suggesting inadequate time for the transport of phenoxide ions from the feed phase to the stripping phase. Additionally, a shorter extraction time proves insufficient for emulsion dispersion to facilitate the extraction process effectively, leading to decreased extraction efficiency.

As the extraction time increased to 3 minutes, an increase in phenol extraction to 89.01 % indicating the high reaction kinetics of phenol by ELM. This improvement is attributed to the thorough dispersion observed facilitates the formation of a stable water-in-oil-in-water (W/O/W) emulsion, aiding in the reduction of emulsion globule size. [8]. The generation of smaller globules enhances solute mass transfer, ultimately leading to a higher extraction efficiency in the phenol extraction process.

Table 4. Effect of extraction time on the extraction performance of phenol (Experimental conditions: O/I ratio : 3:1; [Phenol] : 100mg/L; treat ratio : agitation speed : 300 rpm; homogenizer speed : 8000 rpm; 1:5; pH of feed phase : 8.50 ± 0.01; [Carrier] : 0.025 M; emulsification time : 3 min; surfactant composition in the liquid membrane : 3 % (w/v); [NaOH] : 0.1M; [ionic liquid] : 0.107 % (w/v)).

Extraction time (min)	Extraction (%)	Swelling (%)
1	67.32	10
3	89.01	12
5	53.96	14
7	40.401	20

On the other hand, the phenol extraction demonstrates a decrease to 53.96 % and 40.40 % as the extraction time extends to 5 and 7 minutes, respectively. This is due to beyond optimum conditions, an extended contact time between the impeller vessel wall and emulsion potentially lead to emulsion instability due to reduces membrane thickness [7]. In addition, an increase in the extraction time generates large globules and reduces the interfacial mass transfer area. These findings align with those of Khadem et al. [16], indicating that extended extraction time encourage greater coalescence of emulsion globules. This phenomenon facilitates increased water transfer into the internal phase, thereby inducing membrane swelling and potentially triggering emulsion phase rupture. This assertion is reinforced by the data presented in Table 4, where the swelling percentage steadily escalates to 20% with prolonged extraction time.

Conclusion

In this research, the LLE of phenol presents itself as one of the most straightforward methods for conducting preliminary investigations of the ELM process in order to find a good liquid membrane formulation. The use of synergistic carrier in the LLE process manages to extract about 93.51 % of phenol. To further optimize ELM performance, the ELM process then focused on three key parameters: extraction time, treat ratio and agitation speed. The results shows that a 3 minutes extraction time, 300 rpm agitation speed, 1:5 treat ratio and yielded the highest ELM extraction efficiency of 89.01 %. These findings provide valuable insights for further development and optimization of ELM as a promising technique for effective phenol removal.

Acknowledgements

This study received financial support from the Ministry of Higher Education Malaysia through the Fundamental Research Grant Scheme (FRGS), registration no: (FRGS/1/2022/STG05/UTM/01/2).

References

[1] M.B. Rosly,, N. Jusoh,, N. Othman,, H.A. Rahman,, R.N.R. Sulaiman,, N.F.M. Noah, Chemical Engineering and Processing - Process Intensification 148(December 2019) (2020) 107790. https://doi.org/1016/j.cep.2019.107790

[2] P. Xiang,, P. Ma,, Q. He,, Z. Song,, Z. Miao, Bioresour Technol 394 (2024) 130207. https://doi.org/1016/J.BIORTECH.2023.130207

[3] G. Busca,, S. Berardinelli,, C. Resini,, L. Arrighi, J Hazard Mater (2008) 265–88. https://doi.org/1016/j.jhazmat.2008.03.045

[4] I.N.S. Kahar,, N. Othman,, N.F.M. Noah,, M.H.M. Sofi,, S.S. Suliman, Journal of Industrial and Engineering Chemistry (2023). https://doi.org/1016/j.jiec.2023.10.035

[5] Y. Qi,, S. Zhang, Can J Chem Eng 101(2) (2023) 1027–42. https://doi.org/1002/CJCE.24421

[6] A.R. Raval,, H.P. Kohli,, O.K. Mahadwad, Chemical Engineering Journal Advances 12 (2022) 100398. https://doi.org/1016/J.CEJA.2022.100398

[7] N.D. Zaulkiflee,, A.L. Ahmad,, J. Sugumaran,, N.F.C. Lah, ACS Omega 5(37) (2020) 23892–7. https://doi.org/1021/ACSOMEGA.0C03142

[8] M.B. Rosly,, N. Jusoh,, N. Othman,, H.A. Rahman,, N.F.M. Noah,, R.N.R. Sulaiman, Sep Purif Technol 239 (2020). https://doi.org/1016/j.seppur.2020.116527

[9] R.N.R. Sulaiman,, H.A. Rahman,, N. Othman,, M.B. Rosly,, N. Jusoh,, N.F.M. Noah, Korean Journal of Chemical Engineering 37(1) (2020) 141–50. https://doi.org/1007/s11814-019-0418-4

[10] A. Kumar,, A. Thakur,, P.S. Panesar, Sep Purif Technol (2019). https://doi.org/1016/j.seppur.2018.09.048

[11] V. Eyupoglu,, R.A. Kumbasar, Sep Sci Technol 49(16) (2014) 2485–94. https://doi.org/1080/01496395.2014.938167

[12] M.B. Rosly,, N. Othman,, H.A. Rahman, Malaysian Journal of Analytical Sciences 22(4) (2018) 702–14. https://doi.org/17576/mjas-2018-2204-17

[13] C. Guo,, Q. Cao,, B. Chen,, S. Yang,, Y. Qian, J Clean Prod 211 (2019) 380–6. https://doi.org/1016/J.JCLEPRO.2018.11.227

[14] R.N.R. Sulaiman,, N. Othman,, N.H. Harith,, H.A. Rahman,, N. Jusoh,, N.F.M. Noah,, M.B. Rosly, IOP Conf Ser Mater Sci Eng 736(2) (2020) 022031. https://doi.org/1088/1757-899X/736/2/022031

[15] S.S. Suliman,, N. Othman,, N.F.M. Noah,, I.N.S. Kahar, J Mol Liq 374 (2023) 121261. https://doi.org/1016/J.MOLLIQ.2023.121261

Decarbonization Technology: ICDT2024 Materials Research Forum LLC
Materials Research Proceedings 53 (2025) 215-222 https://doi.org/10.21741/9781644903575-21

Catalytic performance of calcium oxide-modified copper- Zinc-alumina catalysts in methanol synthesis from CO2 hydrogenation

AGNES Ng Xin Hui[1,a], Syed Muhammad WAJAHAT UL HASNAIN[1,2,b], Ahmad Salam FAROOQI[3,c], and Bawadi ABDULLAH[1,2,d*]

[1]Department of Chemical Engineering, Universiti Teknologi PETRONAS, Seri Iskandar, 32610, Perak, Malaysia

[2]CO_2 Research Centre (CO2RES), Institute of Contaminant Management, Universiti Teknologi PETRONAS, Bandar Seri Iskandar, 32610 Perak, Malaysia

[3]Interdisciplinary Research Center for Hydrogen Technologies and Carbon Management (IRC-HTCM), King Fahd University of Petroleum and Minerals, Dhahran, 31261, Saudi Arabia

[a]agnes_19001155@utp.edu.my, [b]syed_22000103@utp.edu.my, [c]ahmad.farooqi@kfupm.edu.sa, [d]bawadi_abdullah@utp.edu.my

Keywords: Catalyst Development, Alumina-Based Catalysts, Methanol Production, CO_2 Hydrogenation

Abstract. Proposed as a carbon utilization method, the catalytic hydrogenation with carbon dioxide (CO_2) as feedstock to produce methanol is a viable approach for utilizing excess CO_2 and countering climate change. The current work aims to enhance methanol production from CO_2 hydrogenation over a CaO-promoted Cu/ZnO-based catalyst supported on Al_2O_3. The catalysts, prepared via impregnation method with 10 wt.% active metal loading and 1.0 wt% CaO promoter, were analyzed for physicochemical properties using N_2 physical adsorption-desorption and XRD. It was found that the addition of a CaO promoter improved the textural properties of the catalyst, where a larger BET surface area (214.9 m^2/g) and smaller crystallite size (7.76 nm) were obtained. The 1.0 wt% CaO-promoted catalyst showed the highest CO_2 conversion (19.14%) and methanol yield (4.66%) due to its excellent textural properties and enhanced basicity.

Introduction

In the 1980s, greenhouse gases (GHGs) were identified as the primary cause of global warming, with carbon dioxide being the most prevalent. By 2022, global carbon dioxide emissions had risen by 0.9%, reaching a record high of 36.8 Gigatons [1]. Addressing this issue has become urgent to meet climate goals, with carbon capture and utilisation (CCU) emerging as a key strategy to reutilize 7 to 32 percent of emitted carbon dioxide, particularly in industries like methanol and urea production [2–4].

Methanol serves as a versatile feedstock for various chemicals like tert-butyl methyl ether and dimethyl ether, with its production expected to double by 2030 due to increased usage in fuel and diverse chemical applications. Utilizing carbon dioxide hydrogenation for methanol production offers numerous advantages, including high energy density, easy storage at low pressure and room temperature, low toxicity, and environmental friendliness [5,6]. The process, initiated in the 1940s, employs copper-zinc oxide catalysts and operates at temperatures of 150°C to 300°C and pressures of 5 to 10 MPa with syngas containing less than 5% CO_2 to enhance methanol selectivity.

Balancing temperature and pressure is crucial for optimal reactant conversion and methanol selectivity while preventing side reactions like reverse water gas shift [7].

Heterogeneous catalysts are favored for methanol synthesis from CO_2 hydrogenation due to their easy removal, handling, and recovery from various reactors. However, the conventional Cu-ZnO/Al_2O_3 catalyst faces challenges such as low CO_2 activation, catalyst poisoning by water adsorption, and production of undesired by-products. To address these issues, various methods, including promoter introduction, synthesis optimization, and reactor type variation, have been explored. Promoters, like metal oxides, enhance catalyst activity by modifying active sites and physicochemical characteristics. However, challenges persist, including unsatisfactory catalytic performance, catalyst deactivation, and gaps in understanding catalyst-promoter interactions. This study aims to investigate the impact of calcium oxide as a promoter on catalyst performance, particularly its physicochemical properties like basicity, which have not been extensively studied.

Materials and methods

Catalyst Synthesis. A catalyst's properties are dependent on the method of preparation and raw materials used. The most common methods of catalyst preparation were found to be precipitation and impregnation methods. In this work, the incipient wetness impregnation method (IWI) was used to synthesize the catalysts. Alumina (Al_2O_3) was used as the support for the catalyst of the CO_2 hydrogenation reaction to methanol for its outstanding properties and positive effects on the Cu-ZnO-based catalysts in active site dispersion, catalyst stability, and Cu-ZnO interactions. The catalyst support was prepared using γ-Al_2O_3 from PURALOX ScFa 230 provided from SASOL. Before impregnation with precursor solution, the γ-Al_2O_3 was pre-treated using argon flow at temperature of 400°C for 5 h to remove its impurities and moisture.

A Cu-ZnO methanol synthesis catalyst with 1.0wt% CaO promoter (CCZA) and an unpromoted Cu-ZnO catalyst (CZA) supported by Al_2O_3 were synthesized. The Cu metal loading was kept constant and the ratio of Cu:Zn remained at approximately 60:30 for decent catalytic activity. To produce the precursor solution, the required amount of $Cu(NO_3)_2{\cdot}3H_2O$ and $Zn(NO_3)_2{\cdot}6H_2O$ were dissolved in 63.7 mL of deionized water to form a precursor solution with a concentration of 0.5M. The solution was stirred using a magnetic stirrer on a hotplate stirrer at 90°C for 1 hour to produce a homogeneous solution. The pH value of the mixture was controlled to remain at pH 7 using 10% ammonia solution or 10% hydrochloric acid solution while the precursor was added. The mixture was stirred for 24 h, followed by filtration and then washing using deionized water. Afterward, the paste produced was dried in an oven at a temperature of 120 °C for 12 h, then calcinated using a furnace at 600 °C for 4 h. CaO-promoted Cu-ZnO/Al_2O_3 catalysts were synthesized using the impregnation method. The synthesis procedure for CaO- promoted catalysts is the same as a base catalyst, except for preparing the precursor solution. The required amounts of the precursor of $Cu(NO_3)_2{\cdot}3H_2O$, $Zn(NO_3)_2{\cdot}6H_2O$, and additionally, $Ca(NO_3)_2{\cdot}4H_2O$) were dissolved in 71.5 mL of deionized water to form precursor solution with 0.5M concentration. Following that, the steps of catalyst synthesis are identical to those detailed in Section 3.3.2 previously.

Catalyst Characterization. At 77 K, a gas adsorption analyzer (Tristar 3020, Micromeritics) was used to measure the textural properties of catalysts using an N_2 physisorption isotherm. The BET method by relative pressure range (p/po) of 0.05 – 0.20 was used to calculate particular surface areas. While the amount of N_2 absorbed in a relative pressure range (p/po) was used to calculate pore volumes. XRD was used to determine the composition and structure of the catalyst. Powder X-ray diffraction patterns with a Bruker D8B Advance X-ray diffractometer are reported on the calcined catalysts in which Cu-Kα at a wavelength of 1.54Å was used. With an exposure time of 3 seconds and a step size of 1°, the XRD had a scanning range of 10-90°.

Catalyst Evaluation. The catalytic performance of the synthesized catalysts were tested using the Microactivity Effi fixed bed reactor. 0.1g of the catalyst was inserted into the reactor between glass wool. First, the reactor was purged with He gas at flowrate of 20 mL/min for 30 min to eliminate any impurities. After purging, the catalyst was reduced at 250 °C with H_2 at flowrate of 20 mL/min for 2 h. After that, the reactor tube was pressurized to reaction conditions of temperature 250 °C and pressure of 30 bar with He gas at flow rate of 25mL/min for 1 h. Following that, the reactants gases H_2 and CO_2 were introduced in to the reactor at a ratio of 3:1. The total gas hourly space velocity (GHSV) was determined to be 3000 mL/g_{cat} h.

The effluent gas from the reactor was analyzed using the online gas chromatographer (GC) provided by Agilent Technologies (model Agilent 7890A) which is equipped with a thermal conductivity detector (TCD) and flame ionization detector (FID). The X_{CO2} and the Y_{CH3OH} were calculated using formulas shown in Equation (1) and Equation (2) respectively, where F_{CO2in} is the inlet molar flowrate of CO_2, F_{CO2out} is the outlet molar flowrate of CO_2, F_{CH3OH} is the molar flowrate of methanol produced in the reactor, and $F_{CH3OH,T}$ is the theoretical molar flowrate for methanol production assuming no side reactions occurred.

$$X_{CO_2}(\%) = \left(\frac{F_{CO_{2,in}} - F_{CO_{2,out}}}{F_{CO_{2,in}}}\right) x100 \quad (1)$$

$$CO_2(\%) = \left(\frac{F_{CH_3OH_{actual}}}{F_{CH_3OH_T}}\right) x100 \quad (2)$$

Results and Discussion

N_2 Physisorption Analysis. The textural properties of the base catalyst (CZA) and the CaO-promoted catalyst (CCZA) have been analyzed by using N_2-physiosorption analysis and results are reported in Table 4.1. The BET surface area increased upon adding a CaO promoter to the catalyst (from 170.1 m^2/g to 214.9 m^2/g). This is attributed to the larger size of crystallites formed in the unpromoted base catalyst compared to the CaO-promoted catalyst as shown in Table 1. The decrease in crystallite size with the addition of an alkaline CaO promoter was in agreement with the findings from the literature report [8]. The increased BET surface area and basicity of the CaO-promoted catalyst increased conversion and selectivity through enhanced adsorption and activation of reactant molecules [6,9].

N_2 adsorption-desorption isotherm and pore size distribution are presented in Fig. 1. The graph showed Type IV isotherm with H_2 type hysteresis loop which represents characteristics of mesoporous material. The steep desorption branch at higher relative pressures (p/p^o) of over 0.6 is due to the pore-blocking or percolation in small ranges of pore necks or could be caused by cavitation-induced evaporation as well [10,11].

Table 1 Textural properties of 5Ni/KCC-1 catalyst

Catalyst	Surface Area (m^2/g)	Pore Volume (cm^3/g)	Pore Size (nm)	Crystallite Size (nm)
CZA	170.1	0.46	9.15	7.93
CCZA	214.9	0.55	8.60	7.76

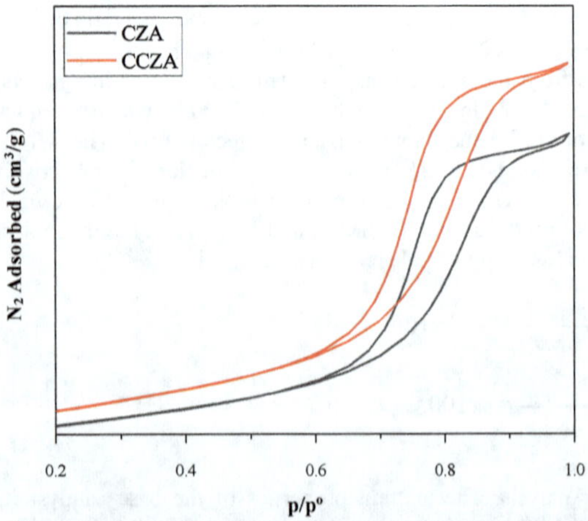

Fig. 1 N_2 adsorption-desorption isotherms of synthesized catalysts

XRD Analysis. The XRD profiles of the base catalyst and the CaO-promoted catalyst are presented in Fig. 4.3. The existence of γ-phase Al_2O_3 in both catalysts is confirmed by the presence of two significant diffraction peaks at 2θ values of 46.0° and 66.6° [12]. On the other hand, CuO diffraction peaks were detected at 2θ of 32.1°, 35.7°, 38.9°, 48.7°, 58.1°, and 61.5° [13,14]. Furthermore, the addition of a CaO promoter increases the intensity of the CuO diffraction peaks at 35.7° and 38.9° which indicates a higher level of crystallinity in CuO [15,16]. Meanwhile, ZnO diffraction peaks were identified at 32.1° and 48.7°. However, no diffraction peaks assigned to CaO species were detected for the promoted catalyst. This could be due to the low concentration of the promoter introduced, good dispersion of CaO species or that the promoter was dispersed in an amorphous state [12,17].

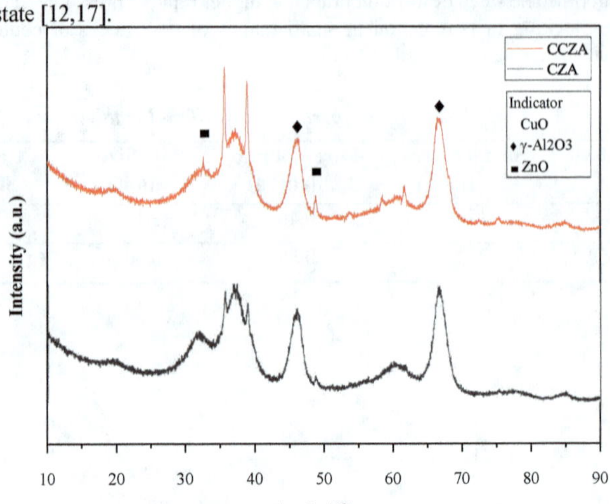

Fig. 2 XRD results of synthesized catalysts.

Catalytic Performance. The catalytic performance tests of CZA and CCZA catalysts were done through the reaction of CO_2 hydrogenation to CH_3OH at a temperature of 250°C and pressure of 30 bar in a Microactivity Effi fixed bed reactor with fixed inlet ratio of reactant gases H_2 and CO_2 for 5 h. The produced gases were analyzed by gas chromatography every 20 minutes. The data obtained was organized to evaluate the catalyst performance based on the CO_2 conversion and CH_3OH yield.

First, as observed from Fig. 3 below, the CO_2 conversion of CZA and CCZA showed a similar trend of sharply decreasing conversion rate in the first 2 h of the reaction followed by its stabilization after the 2-hour mark. This may be due to the unstable flow rate of reacting materials at the start of the test due to pressure build-up in the reactor. Moreover, CCZA achieved a higher CO_2 conversion than CZA. This is due to the higher pore volume and BET surface area of the CCZA catalyst compared to CZA. This can also be explained by the increase in basicity and total basic sites on the catalyst with the addition of alkaline earth metal promoter of CaO which led to an enhancement of the CO_2 adsorption ability, improving the CO_2 conversion [8,18,19].

Furthermore, Fig. 4 below compares the methanol yield of CZA and CCZA catalysts. The methanol yield of CCZA was found to be higher than CZA due to enhanced basicity from the CaO promoter, possibly leading to the formation of strongly basic sites, which increased the likeliness of intermediates to be converted to methanol and prevented the formation of by-products such as methane [20,21]. The methanol yield is also likely to be improved from the increase in copper dispersion caused by the weakening of Cu and Al_2O_3 interaction with the addition of a CaO promoter, as well as from the decreased crystallite size that was found to increase selectivity [11]. Additionally, the CO_2 conversion and CH_3OH yield of the CZA catalyst observed in Fig. 3 and 4 were found to be in good agreement with those described in previous literature [22].

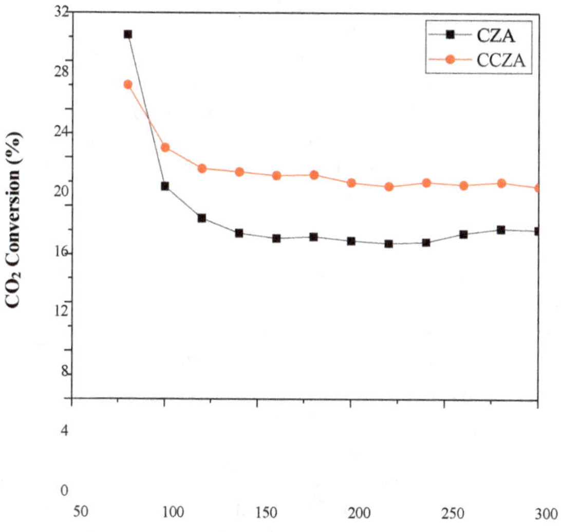

Fig. 3 %CO_2 Conversion for synthesized catalysts.

Materials Research Proceedings 53 (2025) 215-222 https://doi.org/10.21741/9781644903575-21

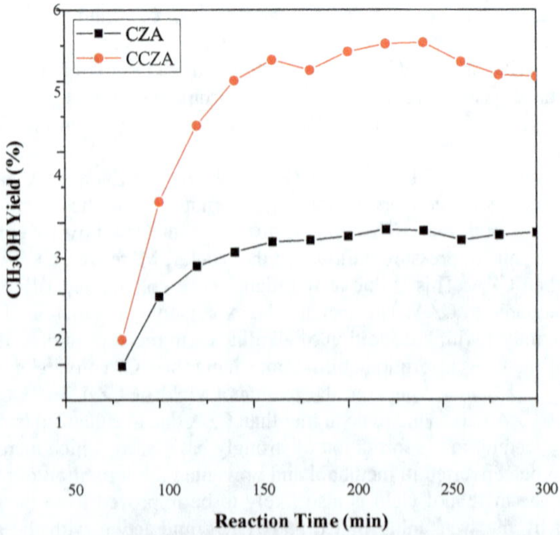

Fig. 4 CH_3OH yield for synthesized catalysts.

Table 2 and Fig. 5 below present the performance of the CZA and CCZA catalysts from the current study in comparison with other Cu-ZnO-based catalysts for CH_3OH production from CO_2 hydrogenation. It was observed that amongst both singly promoted catalysts and tri-promoted catalysts, the CCZA catalyst showed the best catalytic performance with a CO_2 conversion of 19.14% and a CH_3OH yield of 4.66%. CCZA has a higher BET surface area than the MgCZA, TCZA, and MnNZrCZA catalysts, in which a higher CO_2 conversion and CH_3OH yield were observed. Meanwhile, MnNZrCZS showed agglomeration of particles on the outer surface of the SBA-15 instead of inside, unlike the dispersion of particles in the CCZA catalyst which was likely to be improved by the CaO promoter addition, leading to superior catalytic performance [23].

Table 2 Performance comparison of the catalysts studied with literature.

Catalyst	Promoter	Support	T (°C)	P (bar)	X_{CO2}	Y_{CH3OH}	Ref.
CZA	-	Al_2O_3	250	30	15.49	2.52	Current work
CCZA	CaO				19.14	4.66	
MgCZA	MgO		190	30	4.40	3.50	[24]
TCZA	TiO_2		260	26	16.10	4.07	[25]
MnNZrCZA	Mn, Nb,		250	22.5	2.61	2.03	[26]
MnNZrCZS	Zr	SBA-15			0.48	0.45	

Therefore, from the catalytic performance tests carried out, it can be concluded that the addition of a CaO promoter to the conventional CZA catalyst to form the CCZA catalyst leads to an enhancement of the CO_2 conversion and CH_3OH yield for the reaction of CO_2 hydrogenation to methanol. However, it may be necessary to vary the promoter loading of the catalyst to further study its effect and achieve the optimum catalyst performance.

Acknowledgment
The authors acknowledge financial support from Universiti Teknologi PETRONAS, Malaysia, under the YUTP grant number 015LC0-392.

Conclusions

This study investigated the characterization and performance of CaO-promoted Cu-ZnO/Al_2O_3 catalysts for CO_2 hydrogenation to methanol. Two catalysts were prepared via impregnation method. Characterization using N_2 physical adsorption-desorption and X-ray diffraction revealed that the addition of CaO increased the BET surface area and reduced the crystallite size of the catalyst. The CaO-promoted catalyst exhibited superior performance in CO_2 conversion (19.14%) and CH_3OH yield (4.66%) compared to the conventional unpromoted catalyst, attributed to its enhanced textural properties and increased basicity.

References

[1] Wajahat ul Hasnain SM, Farooqi AS, Singh O, Hidayah Ayuni N, Ayodele BV, Abdullah B. Response surface optimization of hydrogen-rich syngas production by the catalytic valorization of greenhouse gases (CH4 and CO2) over Sr-promoted Ni/SBA-15 catalyst. Energy Convers Manag X 2023;20:100451. https://doi.org/10.1016/j.ecmx.2023.100451

[2] Hasnain SMW ul, Farooqi AS, Ayodele BV, Farooqi AS, Sanaullah K, Abdullah B. Advancements in Ni and Co-based catalysts for sustainable syngas production via Bi- reforming of methane: A review of recent advances. J Clean Prod 2024;434:139904. https://doi.org/10.1016/j.jclepro.2023.139904

[3] Salam Farooqi A, Mahirah Binti Ramli R, Sow Mun Lock S, Salam Farooqi A, Wajahat ul Hasnain SM, Anis Farooqi S. Absorption of acid gases (CO2, H2S) from natural gas using a ternary blend of N-methyldiethanolamine (MDEA), 2-amino-2-methyl-1-propanol (AMP), and Sulfolane. Mater. Res. Proc., 2022.

[4] Sharifianjazi F, Esmaeilkhanian A, Bazli L, Eskandarinezhad S, Khaksar S, Shafiee P, et al. A review on recent advances in dry reforming of methane over Ni- and Co-based nanocatalysts. Int J Hydrogen Energy 2022;47:42213-33. https://doi.org/10.1016/j.ijhydene.2021.11.172

[5] GUO X ji, LI L min, LIU S min, BAO G ling, HOU W hua. Preparation of CuO/ZnO/Al2O3 catalysts for methanol synthesis using parallel-slurry-mixing method. Ranliao Huaxue Xuebao/Journal Fuel Chem Technol 2007;35:329-33. https://doi.org/10.1016/S1872-5813(07)60023-1

[6] Budi CS, Wu HC, Chen CS, Saikia D, Kao HM. Ni Nanoparticles Supported on Cage-Type Mesoporous Silica for CO2 Hydrogenation with High CH4 Selectivity. ChemSusChem 2016;9:2326-31. https://doi.org/10.1002/cssc.201600710

[7] Al-Swai BM, Osman N, Alnarabiji MS, Adesina AA, Abdullah B. Syngas Production via Methane Dry Reforming over Ceria-Magnesia Mixed Oxide-Supported Nickel Catalysts. Ind Eng Chem Res 2019;58:539-52. https://doi.org/10.1021/acs.iecr.8b03671

[8] Chen F, Gao W, Wang K, Wang C, Wu X, Liu N, et al. Enhanced performance and stability of Cu/ZnO catalyst by introducing MgO for low-temperature methanol synthesis using methanol itself as catalytic promoter. Fuel 2022;315:123272. https://doi.org/10.1016/j.fuel.2022.123272

[9] Choi SM, Kang YJ, Kim SW. Effect of Γ-alumina nanorods on CO hydrogenation to higher alcohols over lithium-promoted CuZn-based catalysts. Appl Catal A Gen 2018;549:188- 96. https://doi.org/10.1016/j.apcata.2017.10.001

[10] Thommes M, Kaneko K, Neimark A V., Olivier JP, Rodriguez-Reinoso F, Rouquerol J, et al. Physisorption of gases, with special reference to the evaluation of surface area and pore size distribution (IUPAC Technical Report). Pure and Applied Chemistry, 87(9-10) | 10.1515/pac-2014-1117. Pure Appl Chem 2015;87:1051-69. https://doi.org/10.1515/pac-2014-1117

[11] Din IU, Alotaibi MA, Alharthi AI, Al-Shalwi MN, Alshehri F. Green synthesis approach for preparing zeolite based Co-Cu bimetallic catalysts for low temperature CO2 hydrogenation to methanol. Fuel 2022;330:125643. https://doi.org/10.1016/j.fuel.2022.125643

[12] Ren H, Xu CH, Zhao HY, Wang YX, Liu J, Liu JY. Methanol synthesis from CO2 hydrogenation over Cu/γ-Al2O3 catalysts modified by ZnO, ZrO2 and MgO. J Ind Eng Chem 2015;28:261-7. https://doi.org/10.1016/j.jiec.2015.03.001
[13] Berahim NH, Asmawati N, Zabidi M, Ramli RM. The Activity and Stability of Promoted Cu / ZnO / Al 2 O 3 Catalyst. Processes 2023. https://doi.org/10.3390/pr11030719
[14] Zhang Y, Chen J. Controllable preparation of CeF3:Tb3+ nanostructures with different morphologies from an ionic liquid-based extraction system. Colloids Surfaces A Physicochem Eng Asp 2015;470:130-6. https://doi.org/10.1016/j.colsurfa.2015.01.076
[15] Al-Swai BM, Osman NB, Ramli A, Abdullah B, Farooqi AS, Ayodele BV, et al. Low-temperature catalytic conversion of greenhouse gases (CO2 and CH4) to syngas over ceria-magnesia mixed oxide supported nickel catalysts. Int J Hydrogen Energy 2021;46:24768- 80. https://doi.org/10.1016/j.ijhydene.2020.04.233
[16] Abd Ghani NA, Azapour A, Syed Muhammad SAF, Mohamed Ramli N, Vo D-VN, Abdullah B. Dry reforming of methane for syngas production over Ni-Co-supported Al2O3-MgO catalysts. Appl Petrochemical Res 2018;8:263-70. https://doi.org/10.1007/s13203-018-0218-5
[17] Sharma SK, Paul B, Srivastava A, Pal RS, Poddar MK, Khan TS, et al. The role of MgO during CO2 hydrogenation to methanol over Pd/ZnO catalyst. Sustain Chem Clim Action 2023;2. https://doi.org/10.1016/j.scca.2023.100019
[18] Zhou L, Guo Y, Chen J, Sakurai M, Kameyama H. Trace precious metal Pt doped plate-type anodic alumina Ni catalysts for methane reforming reaction. Fuel 2012;92:373-6. https://doi.org/10.1016/j.fuel.2011.06.039
[19] Dasireddy VDBC, Štefančič NS, Huš M, Likozar B. Effect of alkaline earth metal oxide (MO) Cu/MO/Al2O3 catalysts on methanol synthesis activity and selectivity via CO2 reduction. Fuel 2018;233:103-12. https://doi.org/10.1016/j.fuel.2018.06.046
[20] Yang L, Yu H, Wang S, Wang H, Zhou Q. Carbon Dioxide Captured from Flue Gas by Modified Ca-based Sorbents in Fixed-bed Reactor at High Temperature. Chinese J Chem Eng 2013;21:199-204. https://doi.org/10.1016/S1004-9541(13)60459-0
[21] Dang C, Yu H, Wang H, Peng F, Yang Y. A bi-functional Co-CaO-Ca12Al14O33 catalyst for sorption-enhanced steam reforming of glycerol to high-purity hydrogen. Chem Eng J 2016;286:329-38. https://doi.org/10.1016/j.cej.2015.10.073
[22] Ren S, Fan X, Shang Z, Shoemaker WR, Ma L, Wu T, et al. Enhanced catalytic performance of Zr modified CuO/ZnO/Al2O3 catalyst for methanol and DME synthesis via CO2 hydrogenation. J CO2 Util 2020;36:82-95. https://doi.org/10.1016/j.jcou.2019.11.013
[23] Ganesh I. Conversion of carbon dioxide into methanol - A potential liquid fuel: Fundamental challenges and opportunities (a review). Renew Sustain Energy Rev 2014;31:221-57. https://doi.org/10.1016/j.rser.2013.11.045
[24] Kowalik P, Próchniak W, Borowiecki T. The effect of alkali metals doping on properties of Cu/ZnO/Al 2O3 catalyst for water gas shift. Catal Today 2011;176:144-8. https://doi.org/10.1016/j.cattod.2011.01.028
[25] Zhang L, Zhang Y, Chen S. Effect of promoter SiO2, TiO2 or SiO 2-TiO2 on the performance of CuO-ZnO-Al2O 3 catalyst for methanol synthesis from CO2 hydrogenation. Appl Catal A Gen 2012;415-416:118-23. https://doi.org/10.1016/j.apcata.2011.12.013
[26] Berahim NH, Mohd Zabidi NA, Suhaimi NA. Co-Production of Methanol and Methyl Formate via Catalytic Hydrogenation of CO2 over Promoted Cu/ZnO Catalyst Supported on Al2O3 and SBA-15. Catalysts 2022. https://doi.org/10.3390/catal12091018

Comparison of antioxidative performance of synthetic and natural-derived antioxidants for oxidative stabilization of biodiesel

AMRAN Nurul Aini[1,2,a*], BELLO Usman[2,b], and
RUSLAN Muhammad Syafiq Hazwan[3,c]

[1]HICOE—Centre for Biofuels and Biochemical Research (CBBR), Institute of Sustainable Energy (ISE), Universiti Teknologi PETRONAS, Seri Iskandar 32610, Malaysia

[2]Chemical Engineering Department, Universiti Teknologi PETRONAS, Seri Iskandar 32610, Malaysia

[3]School of Chemical Engineering, College of Engineering, Universiti Teknologi MARA, Shah Alam 40450, Malaysia

[a]nurul.amran@utp.edu.my, [b]usmanbello088@gmail.com, [c]syafiqhazwan@uitm.edu.my

Keywords: Biodiesel Stabilization, Solventless Extraction, Induction Period, Rancimat

Abstract. In this work, liquid extracts of banana, mango, and fruit palm peels recovered via supercritical fluid extraction (SFE) were used as natural antioxidants whose antioxidative effectiveness was compared with those of their synthetic antioxidant counterparts namely quercetin, gallic acid, and beta-carotene on improving biodiesel's resistance to oxidation reaction. Blends of the two antioxidant types were tested in mono (single antioxidant) and binary (mixture of two antioxidants) dosages. The results show that a superior antioxidative effect was manifested by the natural antioxidants extracts as evidenced by having higher induction period (IP) values of 6.3 and 8.5 h for the mono and binary-blended biodiesel respectively, compared to 4.5 and 6.7 h for the corresponding synthetic antioxidants-blended peers. However, the presence of water-soluble polyphenols in the natural extracts was identified as the driving factor for the enhanced solubility of the natural antioxidant extracts giving rise to faster reaction by the hydrocarbon chains of the antioxidants and that in the biodiesel, and subsequent deceleration of the radical chain growth.

Introduction

The silent inversion of greenhouse gases (GHG) into the atmosphere and the entire ecosystems poses an increasing threat that discomforts the plight of human existence and the natural system. These global warmers especially carbon dioxide and methane are majorly derived from excessive fossil combustion which subsequently, accumulates and traps heat in the earth's atmosphere, leading to the planet's warm-up and climatic patterns' disruption. While the growing consequences of greenhouse effects are capturing headlines in recent years, mitigation strategies are continuously been explored among regions and countries to curb their long-standing effects [1]. Therefore, in pursuit of identifying optimal approaches for clamping down the rising GHG emissions and winding up their menace, embracing sustainable energy like biodiesel is considered a plausible route for strategically reducing fossil fuel consumption through decarbonization of the transport sector and simultaneously promoting the quest for ensuring cheap and sustainable clean energy as advocated by Sustainable Development Goals 7 (SDG-7) [2]. Consequently, biodiesel is becoming an increasingly demanded commodity, whose consumption and production are growing with many countries developing interest in it as an attractive substitute to conventional diesel [3]. Inevitably, the cost of purchasing biodiesel outweighs that of conventional diesel, and this is largely attributed to the inflated price of raw materials, mainly feedstocks. However, the feedstock employed has successively transitioned from edible oils to non-edible oils, microalgae, and waste from sludge as sustainable feedstocks [4]. Waste cooking oil (WCO) represents a cost-effective and environmentally friendly feedstock for biodiesel production, owing to its zero competition for use as food and being socially sustainable.

Besides, the inability of biodiesel to withstand long-term exposure to certain pro-oxidizing conditions like air, light, high temperature, humidity or moisture, and metal residues during production or storage, is unarguably viewed as a constraining factor that limits its mass production and broader commercialization [3]. In view of this problem, the use of antioxidants was considered a cost-effective approach for increasing the biodiesel's longevity. Notably, both natural and synthetic antioxidants were reported to be effective antioxidative candidates with proficiency in promoting the longevity of the stored biodiesel [4]. Most natural antioxidants from plants containing polyphenols are called primary antioxidants because they alter the radical chain growth by giving out hydrogen atoms to restrict the chain propagation [5]. Therefore, this work explored the potentials of fruit palm, banana, and mango peel extracts as natural antioxidants owing to their abundant polyphenolic contents and compared their antioxidative efficacies with synthetic-based mainly gallic acid, quercetin, and beta-carotene in promoting biodiesel's shelf life. Moreover, the work evaluates the efficacy of the mono, and binary dosages to assess the role of synergistic influence in promoting superior antioxidative effects among both antioxidant types.

Material and methods

The three (3) fruit peel samples were separately extracted using SFE as shown by the schematic 2D diagram represented by Figure 1, with the details of the procedure reported in our previous work [6]. Subsequently, the recovered extracts used as the natural antioxidants together with the solution of the synthetic antioxidants were blended with the biodiesel individually and in combined form (Table 1) to evaluate their proficiency in promoting the biodiesel's ability to withstand oxidation reaction using the Rancimat based on EN 14112 standard [7]. The IP was identified by observing a sudden rise in conductivity upon the generation of volatile acids. A 10 g each of unblended and antioxidants-blended samples were analyzed using Rancimat with conductivity measurement apparatus containing 50 ± 0.1 mL of deionized water, maintained at a block temperature of 110 °C and a constant airflow of 10 L/h. The solution's conductivity was continuously monitored, resulting in an oxidation curve that represents the IP expressed as conductivity (µs/cm) against time (h).

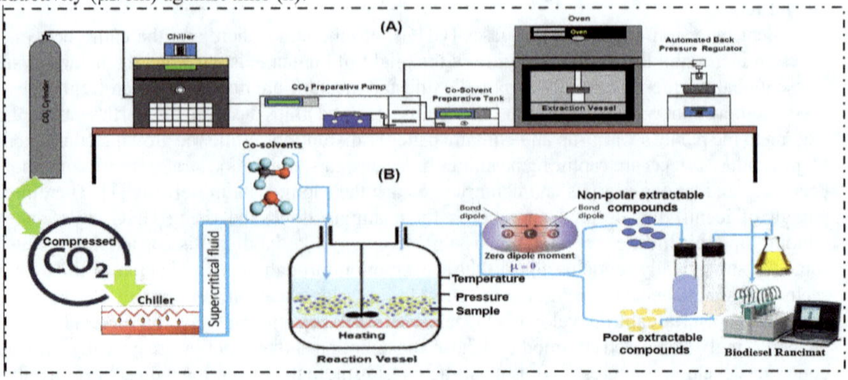

Fig. 1 (A) is the SFE assemble extraction set-up in 2D scheme, and (B) is the description of the processing mechanism.

Table. 1 Blending mixture formulation of synthetic and natural antioxidants

Sample's ID	Synthetic antioxidants			Natural antioxidants (Peel extracts)			
	Gallic acid	Quercetin	Beta-carotene	Mango peel	Banana peel	Fruit palm peel	Blending type
S.1	1	0	0	1	0	0	Mono blend
S.2	0	1	0	0	1	0	
S.3	0	0	1	0	0	1	
S.4	0	½	½	0	½	½	Binary blend
S.5	½	0	½	½	0	½	
S.6	½	½	0	½	½	0	

Results and Discussion

Fatty acid composition profile

The biodiesel was produced using WCO feedstock by esterification reaction followed by base-catalyzed transesterification with the aid of a microwave synthesis reactor (Dixson 103L) as described in our previously published work [8]. The fatty acid composition profile for the raw feedstock (WCO) and the resulting produced biodiesel was analyzed using GC-FID. Figure 2 shows the FAME profile in which eight fatty acids were quantified in the WCO-biodiesel compared to four in the raw WCO, indicating the conversion process. The presence of C16 and C18 fatty acids in the raw WCO confirmed its suitability as a high-quality biodiesel feedstock. However, for the raw WCO, the FAME composition profile indicates 27.2% saturated fatty acids (SFAs) and 72.7% unsaturated fatty acids (UFAs), whereas the synthesized WCO-biodiesel showed 43.82% SFAs and 56.2% UFAs. Given this, it can be claimed that some of the UFAs present in raw WCO were transformed into SFAs in synthesized biodiesel. These compositions closely align with those reported by Malaphong et al. [9] for similar feedstocks produced under conditions. However, the high degree of unsaturation in WCO makes the produced biodiesel susceptible to quick oxidation due to the presence of reactive bis-allylic moieties in the unsaturated carbon chains.

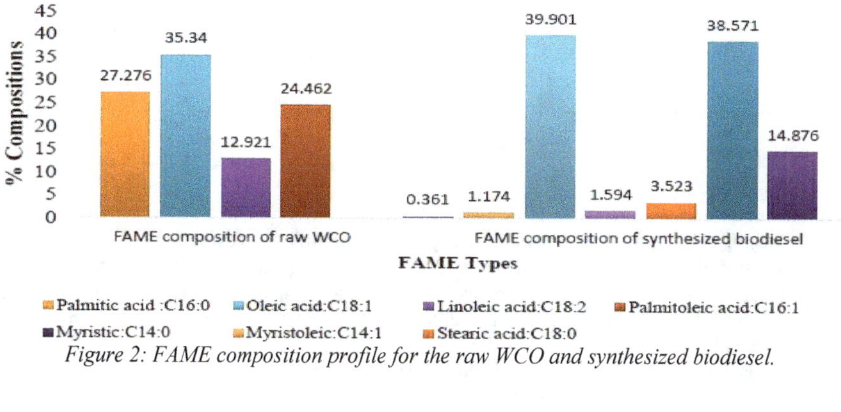

Figure 2: FAME composition profile for the raw WCO and synthesized biodiesel.

Oxidative stability testing

The ability of the biodiesel to withstand oxidation reaction upon exposure to the stream of purified air through the Rancimat analyzer also called the IP was evaluated. Based on the result, the unblended biodiesel (without antioxidants) presented an IP value of 2.43 h (Table 2). This value is far below the earmarked 6.0 h EN 14112 standard prescribed for biodiesel [10]. However, this is presumably expected considering the proportion of UFAs in raw WCO as indicated by the GC-FID results, thereby crediting the feedstock to susceptibly undergo oxidation [11]. For the antioxidant-blended biodiesel, the mono-doped samples (S.1 – S.3) for both synthetic and natural antioxidants, show an appreciable increase in IP from the initial 2.43 h to a maximum of 6.3 h. However, a comparably alike BDE for gallic (71.8 kcal/mol) and quercetin (72.2 kcal/mol), might be responsible for similar IP values presented by samples S.1 and S.2 of the synthetic-blended samples. Consequently, relatively higher IP values were presented by the natural antioxidants blended samples compared to their synthetic-based and this was largely attributed to the enhanced solubility of the former owing to their water-soluble containing phenolic OH groups [12-14]. Among the binary blends, sample S.6 outperforms other blends by presenting IP values of 8.5 h and 6.7 h for the corresponding natural and synthetic antioxidants blended samples respectively. For the natural antioxidants-blended samples, the blend containing a mixture of banana and mango peel extracts (S.6) exhibited the highest protective effect (Table 2). Hence, shared analogous compounds (polyphenolics) present in the sample may be considered responsible for this strong antioxidative effect [15]. The oxidative properties linked to beta-carotene found in palm fruit extracts might contribute to the reduction in the IP values observed in samples containing blends.

Table 2. Result of Rancimat-analyzed IP values for the antioxidants-blended biodiesel.

Sample's ID	Synthetic antioxidants				Natural antioxidants (Peel extracts)			
	Gallic acid	Quercetin	Beta-carotene	IP (h)	Mango peel	Banana peel	Fruit palm peel	IP (h)
S.1	1	0	0	4.5	1	0	0	4.1
S.2	0	1	0	4.3	0	1	0	6.3
S.3	0	0	1	3.8	0	0	1	3.9
S.4	0	½	½	5.8	0	½	½	5.2
S.5	½	0	½	6.0	½	0	½	6.4
S.6	½	½	0	6.7	½	½	0	8.5
Raw biodiesel								2.43

Conclusion

In conclusion, this study evaluates the efficacy of natural antioxidant extracts, obtained through supercritical fluid extraction of banana, mango, and fruit palm fruit peel, in enhancing the resistance of biodiesel to oxidation reactions. Through comparative analysis with synthetic antioxidant counterparts mainly quercetin, gallic acid, and beta-carotene, it was demonstrated that biodiesel blended with natural antioxidants exhibits superior antioxidative effects, as evidenced by significantly higher IP values for mono and binary blends, respectively. The findings, therefore, underscore the promising prospects of utilizing peel extracts as a sustainable alternative source of natural antioxidants with the potency of replacing the synthetic-based in promoting the shelf life of biodiesel.

Acknowledgments
The financial assistance provided by the HI-COE Centre for Biofuel and Biochemical Research (CBBR) and Yayasan Universiti Teknologi Petronas (Cost Centre of 015MA0-052 and 015LC0-438) was sincerely appreciated. Also, the support extended by the Chemical Engineering Department, UTP, and the Centre Graduate Studies, CGS, through the HI-CoE award to CBBR was duly acknowledged.

References

[1] U. Bello, N. A. Amran, S. Samsuri, M. S. H. Ruslan " Fruit palm peel extract derived from supercritical CO2 extraction as a potent source of bioactive compounds" Sustainable Processes and Clean Energy Transition - ICSuPCET2022, Materials Research Proceedings 29 (2023) 35-42. https://doi.org/10.21741/9781644902516-5

[2] J. S. Rodrigues et al., "Comparative study of synthetic and natural antioxidants on the oxidative stability of biodiesel from Tilapia oil," Renew. Energy, vol. 156, no. 13, pp. 1100-1106, 2020. https://doi.org/10.1016/j.renene.2020.04.153

[3] D. Borsato et al, Oxidation kinetics of biodiesel from soybean mixed with synthetic antioxidants BHA, BHT, and TBHQ: Determination of activation energy, Fuel Process.Technol. 127 (2014) 111-116. https://doi.org/10.1016/j.fuproc.2014.05.033

[4] M.E.A.O. Souza, N. Mezzomo, L.C. Correa, M.S. Lima, L.C. Azevedo, S.R.S. Ferreira, Recovery of antioxidant compounds from mango peel by green extraction processes, Int. Food Res. J. 26 (2) (2019) 1845-1859.

[5] N. Kumar, "Oxidative stability of biodiesel: Causes, effects, and prevention," Fuel, vol. 190, pp. 328-350, 2017. https://doi.org/10.1016/j.fuel.2016.11.001

[6] U. Bello, N. Aini, and S. Samsuri, "Kinetics, thermodynamic studies, and parametric effects of supercritical CO2 extraction of banana peel wastes," Sustain. Chem. Pharm., vol. 31, no. December 2022, p. 100912, 2023, doi: 10.1016/j.scp.2022.10091
https://doi.org/10.1016/j.scp.2022.100912

[7] M. S. H. Ruslan, Z. Idham, L. Nian Yian, M. A. Ahmad Zaini, and M. A. Che Yunus, "Effect of operating conditions on catechin extraction from betel nuts using supercritical CO2-methanol extraction," Sep. Sci. Technol., vol. 53, no. 4, pp. 662-670, 2018.
https://doi.org/10.1080/01496395.2017.1406947

[8] U. Bello, N.A. Amran, M.S.H. Ruslan, H. Adamu, Characterization, bioactivity evaluation, thermo-kinetic studies of mango (Mangifera indica L.) peel extract, and its applicability in oxidative stabilization of biodiesel, Biomass Convers. Biorefinery 13 (4) (2023) pp.
https://doi.org/10.1007/s13399-023-03973-9

[9] C. Malaphong, A. Tangwanitchakul, S. Boriboon, and N. Tangtreamjitmun, "A simple and rapid HPLC method for determination of S-allyl-L-cystein and its use in quality control of black garlic samples," Lwt, vol. 160, no. February, p. 113290, 2022.
https://doi.org/10.1016/j.lwt.2022.113290

[10] J. Nagarajan and B. Narayanasamy, "Effects of natural antioxidants on the oxidative stability of waste cooking oil biodiesel," Biofuels, vol. 0, no. 0, pp. 1-10, 2020.

[11] A. C. Roveda, K. V. Tenório, A. R. L. Caires, V. S. Ferreira, C. T. de Carvalho, and M. A. G. Trindade, "Alternative binary blends and their synergistic effect on the stability of soybean biodiesel," Clean Technol. Environ. Policy, vol. 19, no. 10, pp. 2479-2489, 2017.
https://doi.org/10.1007/s10098-017-1445-7

[12] T. C. P. M. Ramos, E. F. de Souza, C. C. Pina, A. A. Cavalheiro, A. R. Fiorucci, and M. S. da Silva, "Evaluation of natural antioxidants action in oxidative stability of commercial biodiesel," Orbital, vol. 10, no. 1 Special Issue, pp. 26-30, 2018. https://doi.org/10.17807/orbital.v10i1.1027

[13] S. R. Sommano, P. Ounamornmas, M. Nisoa, S. Sriwattana, P. Page, and G. Colelli, "Characterisation and physiochemical properties of mango peel pectin extracted by conventional and phase control microwave-assisted extractions," Int. Food Res. J., vol. 25, no. 6, pp. 2657-2665, 2018.

[14] P. V. Gadkari and M. Balaramanl, "Mass transfer and kinetic modelling of supercritical CO2 extraction of fresh tea leaves (Camellia sinensis L.)," Brazilian J. Chem. Eng., vol. 34, no. 3, pp. 799-810, 2017. https://doi.org/10.1590/0104-6632.20170343s20150545

[15] M. P. Garcia-Mendoza, J. T. Paula, L. C. Paviani, F. A. Cabral, and H. A. Martinez-Correa, "Extracts from mango peel by-product obtained by supercritical CO2 and pressurized solvent processes," Lwt, vol. 62, no. 1, pp. 131-137, 2015. https://doi.org/10.1016/j.lwt.2015.01.026

Materials Research Proceedings 53 (2025) 229-240 https://doi.org/10.21741/9781644903575-23

Extraction of polyphenols using TBP/1-octanol/mixed diluent in reactive system via flat sheet supported liquid membrane process

SULIMAN Sazmin Sufi[1,a], OTHMAN Norasikin[1,2,b*], ROSLI Aishah[1,2,c], MOHAMED NOAH Norul Fatiha[1,2,d], SHAMSUL KAHAR Izzat Naim[1,e] and JAMIL Nur Shahira[1,f]

[1]Faculty of Chemical and Energy Engineering, Universiti Teknologi Malaysia, 81310 Skudai, Johor Bahru, Malaysia

[2]Centre of Lipids Engineering and Applied Research (CLEAR), Ibnu Sina Institute for Scientific and Industrial Research, Universiti Teknologi Malaysia, 81310 Skudai, Johor Bahru, Malaysia

[a]sazmin5821@gmail.com, [b]norasikin@cheme.utm.my, [c]aishahrosli@utm.my, [d]norulfatiha.mn@gmail.com, [e]izzatnaimsky@gmail.com, [f]00shahira@gmail.com

Keywords: Polyphenols, Carriers, Supported Liquid Membrane, Mixed Diluents

Abstract. Agricultural wastewaters are attractive sources in order to recover precious bioactive compounds particularly polyphenols. These compounds offer multiple advantages toward human health and are popular in several industries branches such as pharmaceutical, food, and cosmetic owing to their strong antioxidant properties. In this current study, flat sheet supported liquid membrane (FSSLM) was used to recover polyphenols from the simulated wastewater. Even though this process offers high selectivity and simple process, the application of FSSLM process is still limited in the industries because of the rapid membrane loss and short lifetime process. Therefore, this study exposed the feasibility of FSSLM process containing a mixture of tributyl phosphate (TBP) and 1-Octanol as a synergistic carrier, diluents constituting of kerosene, corn oil, sunflower and palm oil and sodium hydroxide (NaOH) as a stripping agent. Several parameters that commonly influence the extraction process were manipulated including feed and stripping phase flow rate, carrier concentration, and feed initial concentration. The results show that 60.13% of extraction with 32.25% of polyphenols recovery were achieved at the favorable condition of 1:1 ratio of palm oil: sunflower as a diluent, 100 mL/min feed phase flow rate, 100 mL/min stripping phase flow rate, 1M of TBP, 1M of 1-Octanol and initial concentration of polyphenols at 300 ppm.

Introduction

Polyphenols are one of the chemical substances that are mostly found in plants species with enormous structural diversities. Ranging from the simpler phenolic acids to the more complex compounds like tannins, they offer a wide spectrum of molecular structures. It is noted that the global polyphenols market size was valued at USD 1.68 billion in 2022 and is expected to grow at a compound annual growth rate (CAGR) of 7.4% from 2023 to 2030 [1]. This growth is driven by their extensive utility across cosmetic, food, and pharmaceutical sectors, primarily due to their potent antioxidant, antimicrobial, and antifungal properties. Moreover, polyphenols are linked to the potential reduction of certain cancers and cardiovascular diseases, further augmenting their demand. [2].

At present, polyphenols are mainly derived from primary sources like vegetables, fruits, and cereals. Nevertheless, there is an increasing emphasis on exploring agricultural wastewater as an alternative source for extracting various precious bioactive compounds, including polyphenols. This shift in focus is essential for reducing reliance on primary sources and enhancing the efficiency of product generation in agriculture. It is important to highlight the considerable potential of various wastewater streams that could be harnessed for polyphenols recovery including wastewater from citrus, palm oil, winery, and olive oil processing [3].

Numerous method have been exploited for the removal and recovery of polyphenols from wastewater including solvent extraction, adsorption, distillation and membrane filtration [3]. Nevertheless, each of the method consist of multiple drawbacks such as generation of secondary waste, high chemical consumption, and not eco-friendly [3]. Therefore, supported liquid membrane (SLM) process was introduced in conjunction with several unique advantages such as rapid extraction, high selectivity and recyclability of membrane components [4]. Despite its advantages, SLM encounters significant drawbacks such as process instability and a short lifespan, limiting its widespread implementation [5]. Nevertheless, these challenges can be mitigated through proper formulation of the membrane phase (diluent, carrier and stripping agent) and operating conditions (feed and stripping phase flowrate and stirring speed) [3]. Several experiments had been done to prove the feasibility of SLM process in order to extract and recover catechol and vanillin [6], lignosulfonate [7], and arsenites [8] with more than 90% of extraction performance.

Previously, the usage of the petroleum-based diluents including hexane, kerosene, chloroform are generally used as a diluent in SLM formulation. However, they are considered as a toxic and harmful to environment. Therefore, many researchers have turn their attention to the implementation of a greener and sustainable diluent such as ionic liquid, vegetable oil and deep eutectic solvent (DES) [9]. Among of them, vegetable oil has been highlighted due to its inexpensiveness, abundance in the market and no additional treatment required. There are many vegetable oils which has highly potential that could be used as a diluent in SLM process, for example, mustard oil [10], sunflower oil [11], palm oil [5] and coconut oil [12]. The aforementioned vegetable oils are rich with triglycerides fatty acid which make them non-polar and suitable to be used as a diluent.

To the best of our knowledge, there have been no prior studies examining the extraction and recovery of gallic acid from aqueous solutions using a synergistic carrier dissolved in mixed vegetable oil as the organic liquid membrane in FSSLM process. In addition, several crucial factors influencing the extraction of phenolic compounds were investigated, including the concentrations of the carrier, synergist carrier, and stripping agents, along with the flow rates of the feed and stripping phases.

Materials and Methods

Chemicals and Reagents

Gallic acid (98% purity) as a feed phase was procured from Sigma-Aldrich. Cooking palm oil, sunflower oil, corn oil as a diluents were supplied from a local manufacturer in Malaysia. Meanwhile, kerosene was supplied by Sigma-Aldrich. Tributyl phosphate (TBP, 99% purity) and 1-Octanol as carriers were obtained from Acros Organic and Sigma-Aldrich, respectively. Sodium hydroxide (NaOH, 99% purity) as a stripping phase was provided by Sigma-Aldrich. In this study, a commercial hydrophobic polyvinylidene fluoride (PVDF) has been selected as a solid membrane support in the SLM process. The PVDF membrane was provided by Millipore and possesses several advantages for the separation process including resistance to high temperature, high porosity and wide chemical compatibility. The physical characteristics of the PVDF solid membrane support are average pore size (0.22 μm), thickness (125 μm) and porosity (75 %). All chemicals and reagents utilized in this study were employed without any additional modification.

Reactive Extraction

A simulated polyphenols solution was used as an external feed phase. The solution was prepared by dissolving the required amount of gallic acid into 1 litre of distilled water. The organic liquid membrane was prepared based on the optimal conditions identified in a prior investigation conducted by Jusoh et al [13]. The carrier concentration was fixed at 1 M of TBP/1 M of 1-Octanol (8:2 by volume) and dissolved in kerosene. Also, the concentration of NaOH as the stripping agent was fixed at 0.3M. The mixture was shaken using a mechanical shaker with the agitation speed and agitation time was fixed at 320 rpm and 18 hours, respectively. After that, the solution was separated using a separatory funnel and left for about 30 minutes. In order to analyse the polyphenols concentration, Folin-Ciocalteu calorimetric method was used [13]. About 0.5 mL of Folin-Ciocalteu reagent was mixed with 0.1 mL of supernatant, followed by the addition of 7 mL of deionized water, and left for 5 minutes. Then, 1.5 mL of Na2CO3 (7.5% w/v) solution was added, and the mixture was allowed to stand for 2 hours in a dark place. The absorbance of the samples was then measured using a UV-Visible spectrophotometer (Jenway, model 7305) at 760 nm against the blank. This method was repeated for the next type of diluent including palm, sunflower, corn and mixed vegetable oil of sunflower and palm oil at fixed ratio of 50:50. The best diluent giving the highest extraction performance was selected for further investigation.

FSSLM Process

Initially, the FSSLM process was conducted by immersing the flat sheet membrane support in organic liquid membrane containing TBP, 1-Octanol and mixed vegetables oil for almost 24 hours [5]. It was assumed that all the empty voids were filled with the organic liquid membrane. The presence of the synergistic carrier in the membrane support's pore would enhance the transportation of gallic acid the from feed phase into the stripping phase. Filter paper was used to remove the excess organic liquid membrane on the PVDF membrane support's surface. Next, the PVDF membrane support impregnated with organic liquid membrane was then placed in between two compartments of the membrane cell (stagnant sandwich) as shown in Fig. 1 [5]. About 300 mL of feed and strip solutions were added into cell 1 and cell 2, respectively. The range of each parameters involved in this study were summarized in Table 1. All experiments were performed at ambient pressure and temperature.

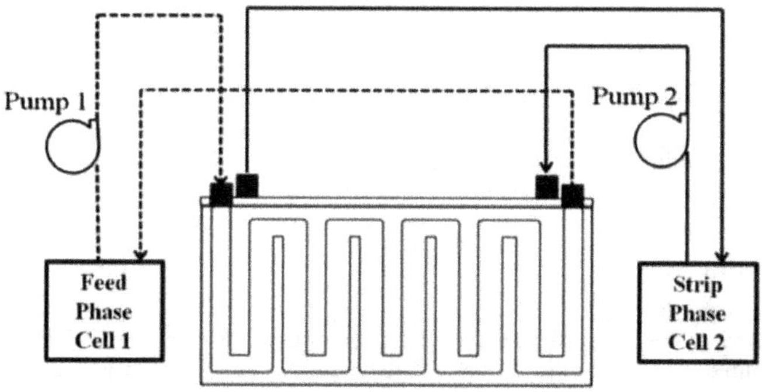

Fig 1. Schematic diagram of FSSLM process

Table 1. The range of each parameter involved in FSSLM process

Parameters	Range of study
Feed phase flowrate (mL/min)	50, 100, 150
Stripping phase flowrate (mL/min)	50, 100, 150
Carrier concentration (M)	0.5, 2.0, 2.5, 3.0
Initial feed concentration (ppm)	150, 300, 450

Analysis of Data

For the reactive extraction, the performance was calculated using Eq. (1). Meanwhile, for the FSSLM process, the extraction and recovery performance of polyphenols were obtained using the Eqs. (2) and (3) [5], respectively:

$$\text{Reactive extraction (\%)} = \frac{C_0 - C_f}{C_0} X\,100 \quad (1)$$

$$\text{Extraction (\%)} = \frac{C_0 - C_t}{C_0} X\,100 \quad (2)$$

$$\text{Recovery (\%)} = \frac{C_S}{C_0} X\,100 \quad (3)$$

Where, C_0 represents initial polyphenols content in the feed phase (ppm), C_f refers to the polyphenols concentration after extraction (ppm), C_t is the polyphenols concentration at the elapsed time (t) (ppm) and C_S refers to the polyphenols content in NaOH solution (stripping phase) (ppm).

Organic Liquid Membrane Loss

The loss of the organic liquid membrane was determined by measuring the weight of dry, wet and used membrane support represented as m_o, m_1 and m_2, respectively. Dry PVDF membrane refers to the membrane support without the impregnation of the organic liquid membrane. On the other hand, wet membrane support indicates the PVDF support that has been impregnated with the organic liquid membrane. The PVDF membrane support used in the extraction process was acquired upon completion. Next, the membrane support was dried in a gel vessel until its mass stabilized. Subsequently, three identical pieces with a fixed area were cut, and the average weight of these pieces, denoted as m_2 was used to determine the organic liquid membrane loss per unit area [14]. The percentage of organic liquid membrane loss, $\Delta m(\%)$ is therefore expressed as Eq. (4).

$$\text{Organic Liquid Membrane Loss}, \Delta m(\%) = \frac{m_1 - m_2}{m_1 - m_o} X\,100 \quad (4)$$

Where m_o and m_1 and m_2 is the weights of the dry, wet and used PVDF membrane support, respectively. Regarding the stability investigation, the loss of organic liquid membrane was assessed by weighing the impregnated membrane support after each cycle. The support was not dried in a desiccator in order to prevent the loss of organic liquid membrane.

Results and Discussion

Selection of Diluents

Fig. 2 shows the extraction performance of polyphenols using different type of diluents which are kerosene, palm oil, sunflower oil, corn oil, mixed vegetable oil of sunflower and palm oil at ratio of 50:50.

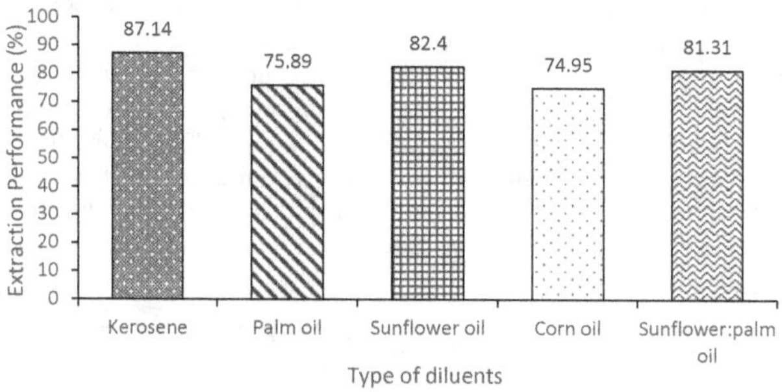

Fig. 2 Extraction performance of polyphenols using different type of diluents (Experimental conditions: Feed phase concentration = 300 mg GAE/L; Carrier concentration = 1.0 M TBP and 1.0 M 1-Octanol (8:2 by volume); Agitation time: 18 hours; Agitation speed: 320 rpm)

It can be seen that kerosene exhibited the highest extraction performance up to 87.14% compared to the green diluents. However, kerosene is considered as a toxic substance. In this research, palm oil, sunflower and corn oil were used to replace kerosene as a diluent. Sunflower oil shows the highest extraction performance up to 82.4% followed by mixed diluent (81.31%), palm oil (75.89%) and corn oil (74.95%). This is because sunflower oil offers lower viscosity (44mPa.s) providing low resistance for the solute to diffuse. On the contrary, palm oil exhibits slightly higher extraction performance compared to corn oil even though the viscosity of palm oil (77mPa.s) is higher than corn oil (47mPa.s). This is because palm oil consist of lower dielectric constant (3.225) compared to corn oil (3.86). This is in agreement with Admawi et al. [15] who stated that lower dielectric constant increased the extraction ability and solute-carrier complex stability. Despite the higher extraction performance offered by sunflower oil, however, it consists of a lot of unsaturated fatty acid which can easily react with water via hydration and caused co-extraction of water into organic liquid membrane [16]. The mixed vegetable diluents consisting of palm and sunflower oil could enhanced the extraction since the major constituent of palm oil are saturated fatty acid (palmitic and stearic acid) [17]. This could enhance the solute-carrier complex stability and reduced the co-extraction of water in organic liquid membrane through hydrophobic interaction [18]. Therefore, mixed sunflower and palm oil at ratio 50:50 was selected for subsequent study.

Effect of Feed Phase Flowrate

The feed phase flow rate is one of the parameters that influences the stability of FSSLM. Fig. 3(a) and (b) illustrate the polyphenols extraction and recovery at different flow rate of feed phase from 50 to 150 mL/min after 180 min of operation, respectively.

The results shows that the highest extraction and recovery performances of polyphenols were obtained up to 70.32% and 67.48% at 150 mL/min, respectively, within 60 mins. Nevertheless, a significant drop of recovery performance (39.48%) was observed by extending the period of time up to 180 min. This is caused by the organic liquid membrane loss at higher feed flowrate. From the experiment, the organic liquid membrane loss obtained after 1 cycle for 150 mL/min was high up to 10.75% compared to 100 mL/min (0.76%) and 50 mL/min (1.36%). High organic liquid membrane loss is unfavourable since it needs to be reimpregnated in order to achieved high extraction and recovery performance in subsequent cycles. Mei et al. [19] also observed similar trend where the extraction performance declined as the flowrate was increased beyond the critical limit. The lowest extraction and recovery was achieved at 50 mL/min with 60.04 and 27.73%, respectively, within 180 min. This is due to the higher thickness of the boundary layer near the feed-membrane interface, offering high resistance for the solute to diffuse. When the flow rate of the feed phase was increased up to 100 mL/min, the extraction and recovery improved up to 60.13 and 32.25%, respectively. In this case, the reduction of the boundary layer thickness and minimal diffusion resistance was achieved, subsequently, enhancing the extraction and recovery performance. This is in line with the findings of Sulaiman et al. [20]. Even though, 100 mL/min of feed phase flowrate resulted in slightly low extraction and recovery performance, however, low organic liquid membrane loss is preferred since it could be reused many times without being reimpregnated, hence, reducing the operating cost. Therefore, 100 mL/min was selected as the best condition for this study.

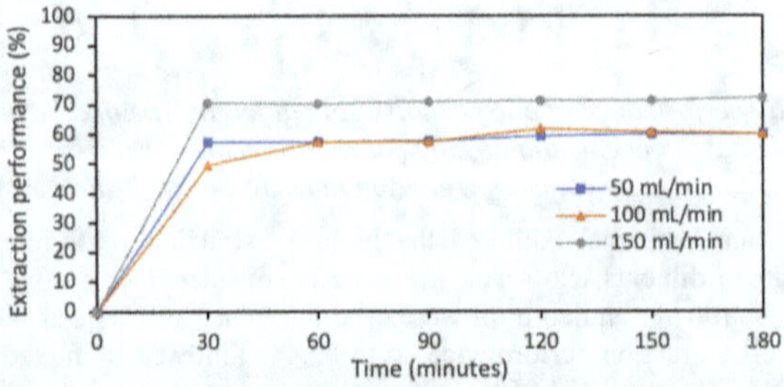

a) Extraction performance of polyphenols

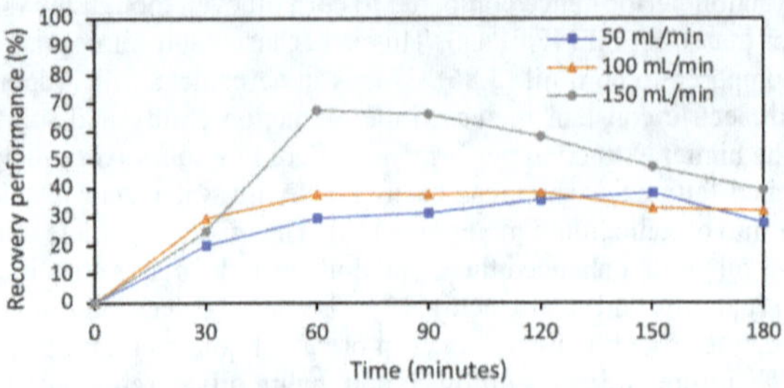

b) Recovery performance of polyphenols

Fig. 3 The effect of feed phase flowrate on the a) extraction and b) recovery of polyphenols via FSSLM process (Experimental conditions: Feed phase concentration = 300 mg GAE/L; Carrier concentration = 1.0 M TBP and 1.0 M 1-Octanol (8:2 by volume); Stripping agent concentration = 0.3 M NaOH)

Effect of Stripping Phase Flowrate

The flowrate of the stripping phase is important since it could affects the polyphenols extraction and recovery performance. Fig. 4(a) and (b) show the experimental results at different stripping flowrates (50– 150 mL/min).

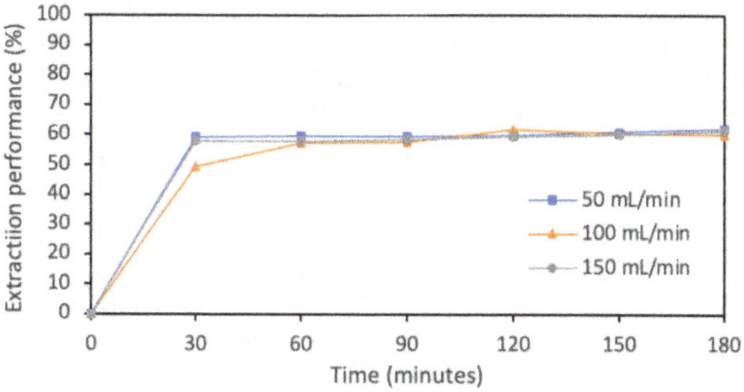

a) Extraction performance of polyphenols

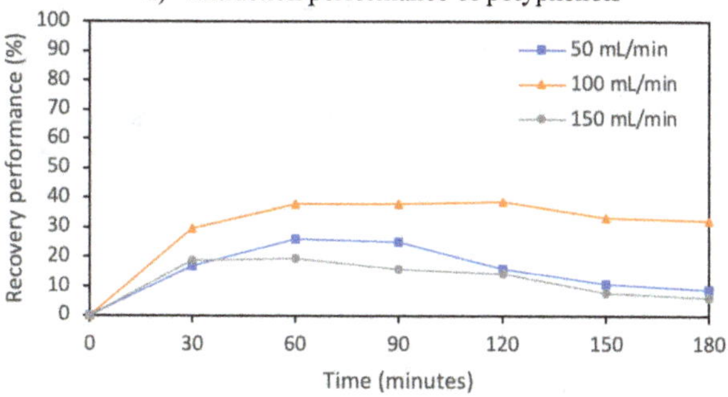

b) Recovery performance of polyphenols

Fig. 4 The effect of stripping flowrate on the a) extraction and b) recovery of polyphenols via FSSLM process (Experimental conditions: Feed phase concentration = 300 mg GAE/L; Carrier concentration = 1.0 M TBP and 1.0 M 1-Octanol (8:2 by volume); Stripping agent concentration = 0.3 M NaOH; Feed phase flowrate: 100 mL/min)

Fig. 4(a) clearly shows that varying the stripping flowrate from 50 to 150 mL/min had no significant effects on the polyphenols extraction performance. Nevertheless, the recovery of polyphenols demonstrated contrary results as shown in Fig. 4(b). The recovery performance was increased from 8.76 to 32.25% as the stripping flowrate increased from 50 mL/min to 100 mL/min after 180 min. With the increases of stripping flowrate, the mass transfer driving force increases, subsequently, increasing the transport efficiency [21]. However, further increase of the stripping flowrate up to 150 mL/min resulted in the lowest polyphenols recovery up to 6.05%. This can be attributed to the improper reaction of the stripping agent with solute-carrier complex at the membrane-stripping interface. As a result, saturation occurred at the stripping interface and consequently blocking the membrane pore [20]. Also, higher flowrate exhibited the highest organic liquid membrane loss up to 19.5% which significantly act as a major contributor to the reduction of the recovery performance. These results are in agreement with those of Xu et al. [22] for rare earths removal using SLM process. Thus, stripping flow rate at 100 mL/min was selected as optimum stripping flowrate and was used for the subsequent study.

Effect of Carrier Concentration

Another crucial factor in designing a stable FSSLM process is choosing the adequate carrier concentration. It can be seen that the recovery of polyphenols in the stripping phase using FSSLM process in the previous studies (Fig. 3(b) and Fig. 4 (b)) was low, therefore, the carrier concentration was varied from 0.25M TBP/0.25M 1-Octanol to 1.5M TBP/ 1.5M 1-Octanol at fixed 8:2 by volume ratio. Fig. 5 depicts the effect of synergist carrier concentration in the membrane phase on the extraction and recovery performance.

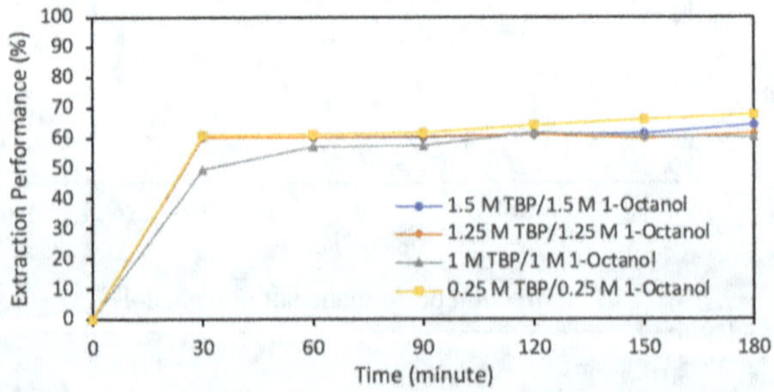

a) Extraction performance of polyphenols

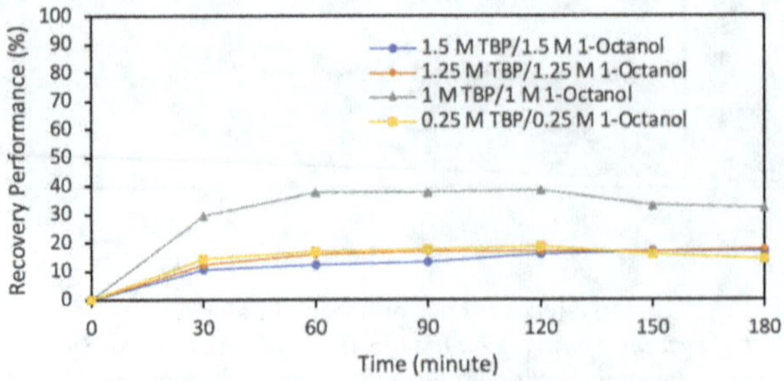

b) Recovery performance of polyphenols

Fig. 5 The effect of carrier concentration on the a) extraction and b) recovery of polyphenols via FSSLM process (Experimental conditions: Feed phase concentration = 300 mg GAE/L; Stripping agent concentration = 0.3 M NaOH; Feed phase flowrate: 100 mL/min; Stripping phase flowrate = 100 mL/min)

Based on the graph in Fig. 5(a), it can be seen that the extraction performance was plateaued for every concentration after 30 min of extraction time, but performed at different recovery rates. The recovery performance increased from 14.18 to 32.25% as the carrier concentration was increased from 0.25 M TBP/0.25 M 1-Octanol to 1 M TBP/1 M 1-Octanol after 180 min. The recovery reached a maximum at about 1 M TBP/1 M 1-Octanol and then the recovery decreased to 16.89% with increase in carrier concentration up to 1.5 M TBP/1.5 M 1-Octanol. This is probably due to the increment of the membrane phase viscosity at higher carrier concentration [23]. This subsequently led to the extra resistance to the diffusion of solute-carrier complex. The experimental data is parallel with the equation proposed by Stokes-Einstein, where the diffusion coefficient of the solute is inversely proportional to the viscosity [23, 24]. As a results, low recovery performance in the stripping phase was observed. Also, the increment of the carrier concentration in the membrane phase accelerated the organic liquid membrane loss. More than 9.83% of organic liquid membrane loss was obtained at higher carrier concentration (>1 M TBP/1 M 1-Octanol) after 180 min of operation time. According to Kazemi et al. [25], the solubility of

TBP in aqueous solution is higher than organic liquid membrane at higher concentration, hence, the organic liquid membrane loss escalated. Hence, 1 M TBP/1 M 1-Octanol was selected as the best condition in this study.

Effect of the Initial Feed Concentration

Initial concentration in the feed phase also affected the transportation of polyphenols across the membrane phase. The effect of initial feed concentration on the extraction and recovery performance are presented in Fig. 6(a) and (b), respectively.

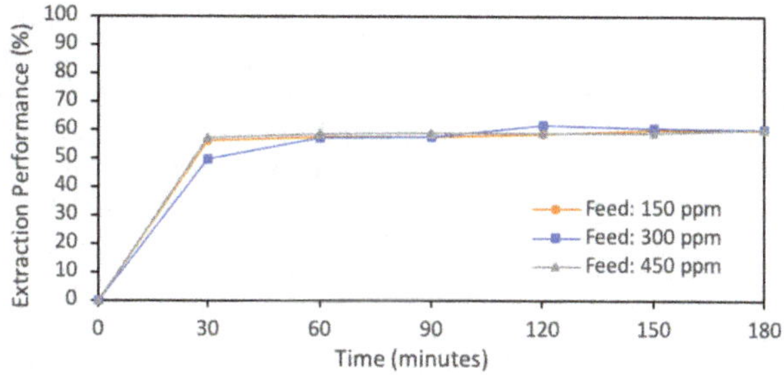

a) Extraction performance of polyphenols

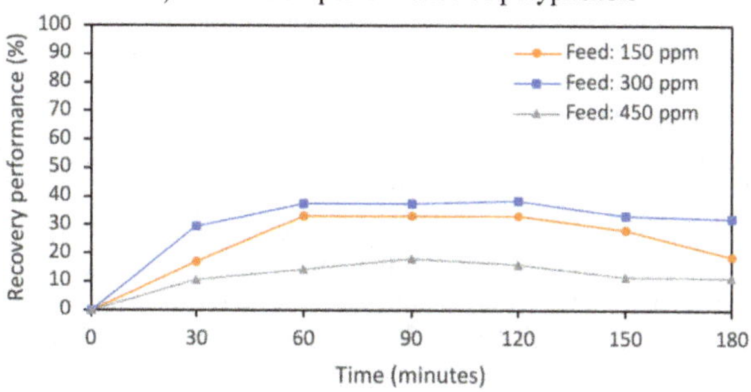

b) Recovery performance of polyphenols

Fig. 6 The effect of initial feed concentration on the a) extraction and b) recovery of polyphenols via FSSLM process (Experimental conditions: Feed phase concentration = 300 mg GAE/L; Carrier concentration = 1.0 M TBP and 1.0 M 1-Octanol (8:2 by volume); Stripping agent concentration = 0.3 M NaOH; Feed phase flowrate: 100 mL/min; Stripping phase flowrate = 100 mL/min)

From Fig. 6(a), it can be seen that the curve corresponding to 150, 300 and 450 ppm are close together and have the same trend where the highest extraction performance was up to 60.00% at 30 min and then plateaued until 180 min of operation time. However, the recovery performance in Fig. 6(b) show that there is a significant difference between the polyphenols concentration in the feed phase and the stripping phase. The recovery performance for 150, 300 and 450 ppm was up to 18.66, 32.25 and 11.15%, respectively. Increase in the initial feed concentration resulted in an increment of recovery performance. This is because the feed flux is proportional to polyphenols concentration in the feed phase [26]. Therefore, with an increase in the polyphenols concentration in the feed phase, feed flux also increased. This is strongly supported by Ali et al. [27]. Also, with increasing of polyphenols concentration in the feed phase beyond 300 ppm (at 450 ppm) the recovery performance decreased. This trend may be attributed to the reduced effective membrane area resulting from membrane saturation [28]. This indicates that the membrane pores become saturated with polyphenols, leading to the formation of a layer on the membrane interface, thereby reducing the recovery performance [28]. Bhatluri et al. [29] also observed the similar trend for the recovery of lead (II) and cadmium (II) via SLM process. Hence, the best optimum initial feed concentration is at 300 ppm.

Conclusion

In conclusion, the use of palm oil and sunflower oil at composition of 50:50 and 1 M of TBP/1 M of 1-Octanol (8:2 by volume) as carrier has been proven to achieve the extraction performance 81.31% of polyphenols in liquid-liquid extraction process. The extraction and recovery of polyphenols were also studied in this research using FSSLM process. From the result obtained, 60.13 and 32.25% of polyphenols were successfully extracted and recovered after 3 hrs of operation time under optimum conditions. The optimum conditions for treating 300 ppm (pH 3.74) of initial feed concentration are feed phase flow rate of 100 mL/min, stripping phase flow rate of 100 mL/min, and 1 M TBP/ 1 M 1-Octanol of carrier concentration.

Acknowledgement

The authors would like to acknowledge the financial support from the Ministry of Higher Education (MoHE) Malaysia under Fundamental Research Grant Scheme (FRGS) (FRGS/1/2022/STG05/UTM/01/2), Geran Kursi Premier Sarawak (PY/2024/00707), and Universiti Teknologi Malaysia (UTM) under UTMNexus Scholarship.

References

[1] Grand View Research, Polyphenols Market Size, Share & Trends Analysis Report By Product (Grape Seed, Green Tea, Apple, Cocoa), By Application (Beverages, Food, Feed) By Region And By Segment Forecasts, 2022-2030, 2023. https://www.grandviewresearch.com/industry-analysis/polyphenols-market-analysis.

[2] Rathod, N. B., Elabed, N., Punia, S., Ozogul, F., Kim, S. K., & Rocha, J. M. (2023). Recent developments in polyphenol applications on human health: a review with current knowledge. Plants, 12(6), 1217. https://doi.org/10.3390/plants12061217

[3] Suliman, S. S., Othman, N., Noah, N. F. M., & Kahar, I. N. S. (2023). Separation of phenolic compounds from fruit processing wastewater using liquid membrane technology: A short review. Biochemical Engineering Journal, 109096. https://doi.org/10.1016/j.bej.2023.109096

[4] Amini, M., Rahbar-Kelishami, A., Alipour, M., & Vahidi, O. (2018). Supported liquid membrane in metal ion separation: An overview. Journal of Membrane Science and Research, 4(3), 121-135.

[5] Sulaiman, R. N. R., Jusoh, N., Othman, N., Noah, N. F. M., Rosly, M. B., & Rahman, H. A. (2019). Supported liquid membrane extraction of nickel using stable composite SPEEK/PVDF support impregnated with a sustainable liquid membrane. Journal of hazardous materials, 380, 120895. https://doi.org/10.1016/j.jhazmat.2019.120895

[6] Pavón, S., Blaesing, L., Jahn, A., Aubel, I., & Bertau, M. (2020). Liquid membranes for efficient recovery of phenolic compounds such as vanillin and catechol. Membranes, 11(1), 20. https://doi.org/10.3390/membranes11010020

[7] Chakrabarty, K., Saha, P., & Ghoshal, A. K. (2009). Separation of lignosulfonate from its aqueous solution using supported liquid membrane. Journal of Membrane Science, 340(1-2), 84-91. https://doi.org/10.1016/j.memsci.2009.05.016

[8] Ali, N., Azeem, S., Khan, A., Khan, H., Kamal, T., & Asiri, A. M. (2020). Experimental studies on removal of arsenites from industrial effluents using tridodecylamine supported liquid membrane. Environmental Science and Pollution Research, 27, 11932-11943. https://doi.org/10.1007/s11356-020-07619-5

[9] Harun, M. H. Z. M., & Ahmad, A. L. (2022). Sustainable and effective contaminants transport beyond stability barrier of supported liquid membrane: How crucial and vital is the green membrane system. Journal of Water Process Engineering, 49, 103172. https://doi.org/10.1016/j.jwpe.2022.103172

[10] Mondal, S. K., & Saha, P. (2018). Separation of hexavalent chromium from industrial effluent through liquid membrane using environmentally benign solvent: A study of experimental optimization through response surface methodology. Chemical Engineering Research and Design, 132, 564-583. https://doi.org/10.1016/j.cherd.2018.02.001

[11] Ashraf, M. W., Abulibdeh, N., & Salam, A. (2019). Selective removal of malachite green dye from aqueous solutions by supported liquid membrane technology. International Journal of Environmental Research and Public Health, 16(18), 3484. https://doi.org/10.3390/ijerph16183484

[12] Chakrabarty, K., Saha, P., & Ghoshal, A. K. (2010). Separation of mercury from its aqueous solution through supported liquid membrane using environmentally benign diluent. Journal of Membrane Science, 350(1-2), 395-401. https://doi.org/10.1016/j.memsci.2010.01.016

[13] Jusoh, N., Othman, N., Geeta, G., & Bukhari, M. (2020). Emulsion liquid membrane extraction of polyphenols compound from palm oil mill effluent. Malaysian Journal of Fundamental and Applied Sciences, 16(1), 96-101. https://doi.org/10.11113/mjfas.v16n1.1443

[14] Zhao, W., He, G., Nie, F., Zhang, L., Feng, H., & Liu, H. (2012). Membrane liquid loss mechanism of supported ionic liquid membrane for gas separation. Journal of Membrane Science, 411, 73-80. https://doi.org/10.1016/j.memsci.2012.04.016

[15] Admawi, H. K., & Mohammed, A. A. (2023). Extraction of cadmium from aqueous solutions by emulsion liquid membrane (ELM) using three different carriers dissolved in a 70: 30 ratio of sunflower oil to kerosene. Journal of the Indian Chemical Society, 100(9), 101081. https://doi.org/10.1016/j.jics.2023.101081

[16] Mabrouk, A. F., & Dugan Jr, L. R. (1961). Solubility of linoleic acid in aqueous solutions and its reaction with water. Journal of the American Oil Chemists' Society, 38(1), 9-13. https://doi.org/10.1007/BF02633110

[17] Chang, S. H. (2014). Vegetable oil as organic solvent for wastewater treatment in liquid membrane processes. Desalination and Water Treatment, 52(1-3), 88-101. https://doi.org/10.1080/19443994.2013.782829

[18] Bryant, C. M., & McClements, D. J. (1998). Molecular basis of protein functionality with special consideration of cold-set gels derived from heat-denatured whey. Trends in Food Science & Technology, 9(4), 143-151. https://doi.org/10.1016/S0924-2244(98)00031-4

[19] Mei, X., Li, J., Jing, C., Fang, C., Liu, Y., Wang, Y., ... & Ding, Y. (2020). Separation and recovery of phenols from an aqueous solution by a green membrane system. Journal of Cleaner Production, 251, 119675. https://doi.org/10.1016/j.jclepro.2019.119675

[20] Sulaiman, R. N. R., Othman, N., Noah, N. F. M., & Jusoh, N. (2018). Removal of nickel from industrial effluent using a synergistic mixtures of acidic and solvating carriers in palm oil-

based diluent via supported liquid membrane process. Chemical Engineering Research and Design, 137, 360-375. https://doi.org/10.1016/j.cherd.2018.07.034

[21] Ren, Z., Zhang, W., Liu, Y., Dai, Y., & Cui, C. (2007). New liquid membrane technology for simultaneous extraction and stripping of copper (II) from wastewater. Chemical Engineering Science, 62(22), 6090-6101. https://doi.org/10.1016/j.ces.2007.06.005

[22] Xu, D., Shah, Z., Sun, G., Peng, X., & Cui, Y. (2019). Recovery of rare earths from phosphate ores through supported liquid membrane using N, N, N′, N′-tetraoctyl diglycol amide. Minerals Engineering, 139, 105861. https://doi.org/10.1016/j.mineng.2019.105861

[23] Kamran Haghighi, H., Irannajad, M., Fortuny, A., & Sastre, A. M. (2019). Non-dispersive selective extraction of germanium from fly ash leachates using membrane-based processes. Separation Science and Technology, 54(17), 2879-2894. https://doi.org/10.1080/01496395.2018.1555170

[24] Panja, S., Ruhela, R., Misra, S. K., Sharma, J. N., Tripathi, S. C., & Dakshinamoorthy, A. (2008). Facilitated transport of Am (III) through a flat-sheet supported liquid membrane (FSSLM) containing tetra (2-ethyl hexyl) diglycolamide (TEHDGA) as carrier. Journal of Membrane Science, 325(1), 158-165. https://doi.org/10.1016/j.memsci.2008.07.019

[25] Kazemi, P., Peydayesh, M., Bandegi, A., Mohammadi, T., & Bakhtiari, O. (2014). Stability and extraction study of phenolic wastewater treatment by supported liquid membrane using tributyl phosphate and sesame oil as liquid membrane. Chemical engineering research and design, 92(2), 375-383. https://doi.org/10.1016/j.cherd.2013.07.023

[26] Jalilvand, P., Rahbar-Kelishami, A., Mohammadi, T., & Shayesteh, H. (2020). Optimizing of malachite green extraction from aqueous solutions using hydrophilic and hydrophobic nanoparticles. Journal of Molecular Liquids, 308, 113014. https://doi.org/10.1016/j.molliq.2020.113014

[27] Ali, N., Azeem, S., Khan, A., Khan, H., Kamal, T., & Asiri, A. M. (2020). Experimental studies on removal of arsenites from industrial effluents using tridodecylamine supported liquid membrane. Environmental Science and Pollution Research, 27, 11932-11943. https://doi.org/10.1007/s11356-020-07619-5

[28] Kamran Haghighi, H., Irannajad, M., Fortuny, A., & Sastre, A. M. (2019). Non-dispersive selective extraction of germanium from fly ash leachates using membrane-based processes. Separation Science and Technology, 54(17), 2879-2894. https://doi.org/10.1080/01496395.2018.1555170

[29] Bhatluri, K. K., Manna, M. S., Ghoshal, A. K., & Saha, P. (2015). Supported liquid membrane based removal of lead (II) and cadmium (II) from mixed feed: Conversion to solid waste by precipitation. Journal of Hazardous Materials, 299, 504-512. https://doi.org/10.1016/j.jhazmat.2015.07.030

Decarbonization Technology: ICDT2024 Materials Research Forum LLC
Materials Research Proceedings 53 (2025) 241-254 https://doi.org/10.21741/9781644903575-24

Optimization of supercritical fluid extraction of phenolic-compounds from palm kernel cake and its antioxidant activity

Siti Hawa Mat Yaman[1,a], Norhusna Mohamad Nor[1,2,b*], Ellisa Natasya Rosli[1,c], Mohd Azahar Mohd Ariff[1,2,d], Mohamed Syazwan Osman[1,3,e] and Nor Fariza Ismail[1,2,f]

[1]Chemical Engineering Studies, College of Engineering, Universiti Teknologi MARA, Cawangan Pulau Pinang, 13500 Permatang Pauh, Pulau Pinang, Malaysia

[2]Waste Management and Resource Recovery (WeResCue) Group, Chemical Engineering Studies, College of Engineering, Universiti Teknologi MARA, Cawangan Pulau Pinang, 13500 Permatang Pauh, Pulau Pinang, Malaysia

[3]Lab EMZI-UiTM Nanoparticles Colloids, Universiti Teknologi MARA, Cawangan Pulau Pinang, 13500 Permatang Pauh, Pulau Pinang, Malaysia

[a]sitihawamatyaman@gmail.com, [b]norhusna811@uitm.edu.my, [c]ellisyanatasya@gmail.com, [d]azahar.ariff@uitm.edu.my, [e]syazwan.osman@uitm.edu.my, [f]norfariza5031@uitm.edu.my

Keywords: Antioxidant Activity, Design Expert, Gallic Acid Equivalent, Palm Kernel Cake, Supercritical Fluid Extraction, Total Phenolic Content

Abstract. Palm Kernel Cake (PKC), a byproduct of the palm oil industry, is rich in phytochemical compounds, including phenolic acids, and has various potential applications. This study aimed to analyse the Total Phenolic Content (TPC) and Antioxidant Activity (AOA) of PKC extracts using Supercritical Fluid Extraction (SFE). The phenolic compounds extracted from the PKC were optimised using the Response Surface Methodology (RSM) approach. Three parameters, pressure (150-300bar), temperature (40-80°C), and extraction time (20-60 minutes), were varied to optimise the TPC extraction. The results, displayed through response surface plots, showed that the TPC ranged from 1.99 to 3.05 mg of Gallic Acid Equivalent (GAE)/g bio-oil, with an R^2 value of 0.9890, indicating high accuracy. The optimised conditions of a temperature of 80°C, pressure of 300bar, and reaction time of 42 minutes led to a TPC of 3.035mg of GAE per g of bio-oil. Additionally, the highest AOA value, 68.11g of Ascorbic Acid Equivalent (AAE)/mL solvent, was obtained at a condition pressure of 300bar, temperature of 60°C and extraction time of 40 minutes. The study highlights the potential of SFE for extracting valuable compounds from PKC, which could find applications in industries such as cosmetics and medicine due to their antioxidant, anti-aging, anti-inflammatory, and antiproliferative properties.

Introduction

The palm oil industry is one of the most significant contributors to the global supply of edible oil. Indonesia and Malaysia are responsible for generating 46.8 million metric tonnes (MT), which accounts for 58% of the total supply, and 20.9 million MT, which accounts for 26% of the world's palm oil production, respectively [1]. Due to the vast production of the palm oil industry, biomass wastes are significantly generated. As a result, this may be a factor in the accumulation of solid waste disposal, where the untreated biomass may pose potential environmental threats. Some examples of this include palm kernel cake (PKC), palm kernel shell (PKS), empty fruit bunch (EFB), and mesocarp fibre (MF), which are produced annually at 17.0 [2], 1.10, 4.42, and 2.71 tonnes/hectare [3], respectively. The palm kernel oil is obtained by pressing or solvent extracting of the kernels, which comes after the separation and cracking of the palm fruits to extract the kernels [4].

The phytochemicals obtained from the PKC are mainly composed of terpenes, phenolic compounds, and sterols [5]. It has been acknowledged that these phenolic compounds can potentially be utilized in preventing neurological disorders, malignancies, cardiovascular diseases, and diabetes [6]. The phenolic compounds, which include primarily phenolic acids, flavonoids, tannins, and lignans, are the most numerous and alluring of these phytochemicals. Secondary metabolic products known as phenolic compounds can exist in free and bound forms [5]. As a result of their ability to scavenge free radicals, break radical chains, and act as chelators of divalent cations, they exhibit remarkable antioxidant activity. Due to wide range of biological activity and potential uses of phenolic compounds, numerous research has explored this abundant reservoir found in different natural sources. Among these, Nuchdang et al. [7] conducted a study that illuminated the complex composition of palm kernel cake (PKC) by concentrating on phenolic chemicals. Interestingly, their research revealed the existence of five distinct types of phenolic compounds: 3-pyrodinol, aminophenol, vanillin lactoside, phenol, and maltol. Furthermore, a study by Tsuoko et al. [5] extracted almost 10 different kinds of phenolic compounds from PKC. The study additionally claimed that PKC was the most promising because it had the highest amount of total phenolics and could be used as a preservative in foods, medicines, and cosmetics because it had the highest amount of 4-hydroxybenzoic acid [5]. The finding of several phenolic compounds inside PKC by both Nuchdang et al. [7] and Tsuoko et al. [5] emphasizes the extensive antioxidant potential of this agricultural byproduct, highlighting its usefulness for various applications in food, medicine, and cosmetics.

Conventional and non-conventional extraction technologies exist to obtain phytochemicals from plants, and both can be used to get high-quality results. Maceration, percolation, reflux, and Soxhlet extraction are typical traditional techniques [8]. Supercritical fluid extraction (SFE), pressurised liquid extraction (PLE), microwave-assisted extraction (MAE), deep eutectic solvents (DES), and combinations thereof are also examples of non-conventional extraction technologies [9], [10]. The goals of both methods are the same: to extract the valuable chemical compounds in plants so they can be further utilised in medicine, food, or even just a relaxing cup of tea. SFE is a green technology with the advantages of high efficiency, selectivity, and quick extraction [11]. This contrasts with traditional solvent extraction processes, which take a long time, use toxic solvents, have low efficiency, and pollute the environment.

According to previous studies [11], [12], [13], [14], [15], SFE was the most preferable extraction approach due to its ability to increase the extraction yield of chemical compounds from biomasses. These previous studies were conducted to obtain an optimum yield of the compound based on different variations of operating parameters such as CO_2 flow rates, pressure, temperature, and extraction time. Gu et al. [15] obtained an optimum yield of bio-oil (56.47%) of extraction using SFE at 30 MPa of pressure, 50 °C of temperature, 1.5 hours of extraction time, and 0.16 L/min of CO_2 flow rate by using *Xanthoceras sorbifolia* Bunge (yellow horn) seed. Studies also stated that co-solvents, such as methanol, ethanol, and water, and the combination of ethanol-water and methanol-water would help the solvent, CO2, increase the ability to dissolve polar compounds. Moreover, Costa et al. [12] reported in their investigation that the Soxhlet extraction of PKC oil produced a yield of 7.92%, which was comparable to the maximum oil yield that SFE (7.82%) at 80°C and 350bar attained. The comparison of the oil yields from the SFE and Soxhlet procedures was conducted by Zaidul et al. [16], who confirmed the high degree of similarity between both. On the other hand, the Soxhlet extraction method has several drawbacks, including the consumption of flammable and hazardous liquid organic solvents (e.g., hexane), the emission of hazardous gases during the extraction process, the absence of selective extraction, and the operating need that is highly time-consuming [17]. On the other hand, SFE produce minimum waste.

This work proposes a novel method for extracting TPC from PKC by optimizing the operating conditions of SFE with ethanol as a co-solvent. This is done by using the Response Surface Method (RSM). This approach provides improved selectivity and efficiency in extracting phenolic compounds from PKC compared to traditional extraction methods. The research thoroughly assesses the quality and bioactivity of the isolated phenolic compounds by combining total phenolic content and antioxidant activity analysis. The work uses Design Expert simulation to find the best SFE parameters for maximizing both TPC and oil output from PKC, aiming to improve sustainable and economically feasible extraction techniques. This study contributes to the current knowledge on extracting and using natural products, providing information on the potential uses of phenolic compounds produced from PKC in several sectors.

Methods

Materials and Chemicals. Throughout the study, various chemicals and materials were identified and used to extract palm kernel cake (PKC), analyse total phenolic content (TPC), and determine antioxidant activity. The PKC was bought from a local seller from Terengganu. The chemicals that were employed in these studies are silica gel, ethanol (QRec, AR Grade, 96%), Carbon Dioxide (Alpha Gas, 99.9%), Folin-Ciocalteu reagent (QRec, ≥95%), Sodium Carbonate (R&M Chemicals, Anhydrous), 2,2-diphenyl-1-picryhydrazyl free radicals (DPPH, Sigma-Aldrich), Gallic acid (Sigma-Aldrich, 97.5-102.5% (titration)), and Ascorbic acid (HmbG Chemicals, Food Additive).

Optimisation Study using RSM. The RSM method was used to design the optimum SFE parameters using Design Expert V13 software. In this study, Central Composite Design (CCD) was chosen to perform the optimisation for extracting the highest value of TPC from the PKC. Three manipulated variables used for this study were (A) pressure, (B) temperature and (C) extraction time.

Table 1 The variables was manipulated for the experiment

Factors	Parameters	Unit	Low	High
A	Pressure	bar	150	300
B	Temperature	°C	40	80
C	Extraction Time	t	20	60

Table 2 Design experiment provided by Design Expert V13.

Run	Factor 1: Pressure (bar)	Factor 2: Temperature (°C)	Factor 3: Extraction time (minutes)
1	150	80	80
2	350	80	20
3	350	60	50
4	250	60	50
5	350	40	80
6	250	60	50
7	150	60	50
8	350	40	20
9	250	60	20

10	150	40	20
11	250	60	80
12	350	80	80
13	250	80	50
14	150	40	80
15	250	40	50
16	250	60	50
17	250	60	50
18	250	60	50
19	150	80	20
20	250	60	50

Preparation of PKC. Firstly, the PKC was washed and rinsed with tap water to remove the fine debris affecting the experiment. After that, the PKC was rinsed for another two rounds with distilled water. The PKC was dispersed equally into trays and dried at 45°C for 2 days or until it reached constant mass. The PKC was weighed before and after the drying process, which is 2 days. The moisture content in the PKC sample was calculated, and it should be lower than plus or minus 10 to ensure no difficulty occurs during the extraction process [12]. Eq. 1 shows the formula to calculate the moisture content in the PKC sample.

$$\text{Moisture content (\%)} = \frac{M_1 - M_2}{M_1 - M_0} \times 100\% \qquad \text{(Eq. 1)}$$

where M_1 is the mass of the PKC sample before drying, M_2 is the mass of the sample and tray after drying, and M_0 is the mass of the tray. After that, the dried PKC was ground and sieved into particle size. The sieved PKC was stored in a plastic container with silica gel for further use. For this experiment, the particle size of PKC used is 1.0mm to 2.0mm.

Extraction of PKC using Supercritical Fluid Extraction (SFE). PKC extractions were performed utilising Supercritical Fluid Extraction (SFE) equipment from Thar Technologies, Inc. The essential components of this equipment are a liquid CO_2 cylinder, CO_2 pump, co-solvent reservoir, co-solvent pump, oven, pressure control valve and regulatory valves. The SFE parameters, sourced from Design Expert version 13, included pressure (150-300 bar), temperature (40-80°C), extraction time (20-60 minutes), and using ethanol as a co-solvent. The extraction process involved employing 15g of dried PKC with a flow rate of 25mL/min of carbon dioxide and ethanol as a co-solvent. The extracts were depressurised and then collected in an amber bottle before being stored in the fridge at ≤10°C.

Materials Research Forum LLC
https://doi.org/10.21741/9781644903575-24

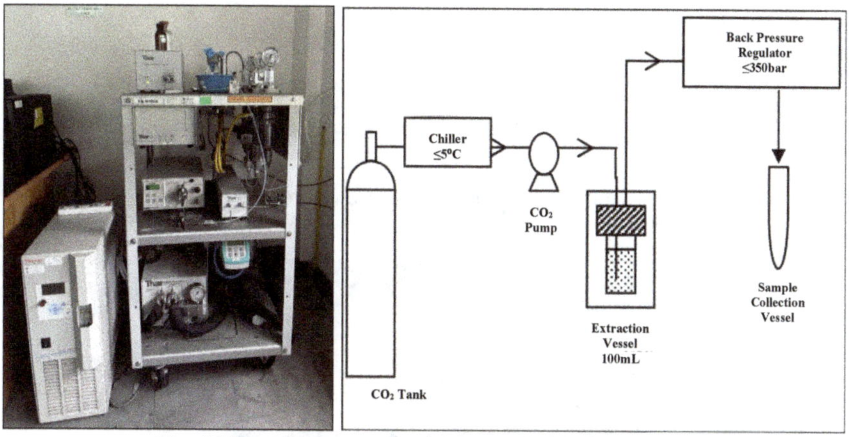

Fig. 1 The equipment was used to extract PKC which is SFE.

Total Phenolic Contents (TPC) Analysis. The Folin–Ciocalteu colorimetric method was used to quantify total phenolic contents (TPC) in extracts from PKC, with modifications to the protocol described by Burapan et al. [18]. A gallic acid solution of standard concentration was made and diluted to concentrations of 50, 100, 250, and 500mg/L. Samples were prepared in triplicate by mixing 0.04mL of PKC's oil sample, 0.2mL of Folin-Ciocalteu reagent, 3.16mL distilled water, and 0.6mL sodium carbonate solution (7.5 %, w/v) in amber bottles. The sample mixture was incubated at room temperature for two hours, then measure absorbance at 765 nm using a UV-visible spectrophotometer (Portable Peak USA T9200, 190-1100nm). Ethanol and gallic acid were used for the blank sample and calibration curve, respectively. The Eq. 2 to represent the results of TPC as mg of gallic acid equivalent (GAE) per grammes of bio-oil extract were as follows:

$$\text{C (mg GAE/g bio-oil)} = \frac{c \times V}{m} \qquad \text{(Eq. 2)}$$

where C is the total phenolic content mg GAE/g dry extract, c is the concentration of gallic acid obtained from the calibration curve in mg/L, V is the volume of extract in ml, and m is the mass of extract in grams.

Antioxidant Activity (AOA) Analysis. The in vitro antioxidant activities of the extracts were assessed using the DPPH free radicals method, as Phuyal et al. [19] outlined with some adjustments. A 0.6mM DPPH solution was created by diluting 0.1183g of DPPH in ethanol in a 500 mL volumetric flask. The solution was then incubated and shaken in darkness at room temperature for 30 minutes. A standard ascorbic acid solution was made by dissolving 10 mg of ascorbic acid in 10 mL of ethanol to achieve a 1 mg/mL concentration. Concentrations of ascorbic acid in ethanol were generated at 25, 50, 75, and 100 g/mL using the provided standard solution. Each concentration sample was prepared by combining 9 mL of DPPH solution with 1 mL of the corresponding ascorbic acid solution to reach 10 mL. The blank sample was created by diluting 2.5 mL of distilled water with ethanol in a 25 mL volumetric flask. The control solution was made by diluting 9 mL of DPPH with 1 mL of the blank solution. The ascorbic acid was substituted with the oil sample to examine oil-extracted samples. The absorbance of the ascorbic acid sample, as illustrated in Fig. 2, was measured at a wavelength of 517nm using a UV-Vis Spectrophotometer after shaking and incubation for 30 minutes at room temperature. The oil extract's free radical scavenging activity was determined using Eq. 3.

$$\text{Antioxidant activity (\%)}=\frac{A_c - A_s}{A_c}\times 100\% \qquad \text{(Eq. 3)}$$

where A_s is the absorbance of the sample and A_c is the absorbance of the control solution.

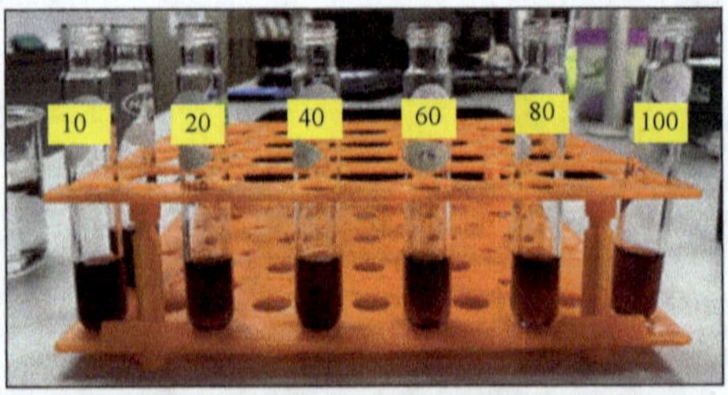

Fig. 2 Ascorbic Acid used in the construction of calibration curve.

Results and Discussion

Model Fitting and Statistical Significance Analysis. The findings of the Analysis of Variance (ANOVA) for the Total Phenolic Content of PKC extract are shown Table 3. The p-values were used to assess the importance of each coefficient, which reflected the interaction patterns between independent variables [20]. p-values below 0.05 indicate that the model terms are significant, while p-values more than 0.1000 show that the model terms are not significant [19], [20].

The ANOVA findings shown in Table 3 give a very significant overall model for the TPC model, with a F-value of 100.07 and a p-value of 1.32×10^{-08}. This indicates that the model adequately accounts for the variability in the response variable, TPC. The analysis reveals that A-pressure, B-Temperature, C-extraction time, and the quadratic terms A^2 and B^2 all have p-values below 0.05, indicating their significant influence on TPC. This suggests that changes in these characteristics substantially affect the observed variances in TPC.

In this analysis, the Lack of Fit F-value of 4.29, coupled with a corresponding p-value of 0.068, suggests a lack of significance, indicating a 6.80% probability that the model's lack of fit is attributed to random noise rather than systematic inadequacy. This finding is advantageous as it suggests that the model adequately captures the relationship between the predictors and the response variable. Consequently, the quadratic regression model for TPC in PKC extracted derived from CCD design and expressed in terms of coded factor is shown in Eq. 4.

$$\begin{aligned} TPC = {} & 2.13 + 0.291A + 0.0820B + 0.055C + 0.015AB \\ & -0.035AC - 0.0175BC + 0.4382A^2 + 0.0832B^2 \\ & -0.0118C^2 \end{aligned} \qquad \text{(Eq. 4)}$$

where A was pressure, B was temperature, and C was extraction time.

Table 3 ANOVA for regression model

Source	Sum of Squares	df	Mean Square	F-value	p-value	
Model	2.14	9	0.2373	100.07	< 0.0001	significant
A-Pressure	0.8468	1	0.8468	357.14	< 0.0001	
B-Temperature	0.0672	1	0.0672	28.36	0.0003	
C-Extraction Time	0.0303	1	0.0303	12.76	0.0051	
AB	0.0018	1	0.0018	0.7591	0.4040	
AC	0.0098	1	0.0098	4.13	0.0695	
BC	0.0025	1	0.0025	1.03	0.3334	
A^2	0.5280	1	0.5280	222.69	< 0.0001	
B^2	0.0190	1	0.0190	8.02	0.0178	
C^2	0.0004	1	0.0004	0.1620	0.6958	
Residual	0.0237	10	0.0024			
Lack of Fit	0.0192	5	0.0038	4.29	0.0680	not significant
Pure Error	0.0045	5	0.0009			
Cor Total	2.16	19				

The regression scores generated for the TPC model demonstrate a highly robust and dependable forecasting capacity. A higher R^2 value indicates a more robust fit between the empirical model and the actual data [19]. Table 4 shows a coefficient of determination (R^2) of 0.989, indicating that the model accounts for over 98.9% of the variability in TPC, demonstrating a strong association between the predictors and the response variable. The modified R^2 value of 0.979 confirms the model's dependability, accounting for the number of predictors included. The estimated R^2 value of 0.9297 indicates the model's success in generalising to new data, showing its practical relevance beyond the training dataset. Accuracy above 28 indicates a strong signal-to-noise ratio, confirming the model's ability to differentiate genuine impacts from random fluctuations effectively.

Table 4 Regression score for TPC analysis

Std. Dev.	0.0487	R^2	0.9890
Mean	2.38	Adjusted R^2	0.9791
C.V. %	2.05	Predicted R^2	0.9297
		Adeq Precision	28.0806

Response Surface of Total Phenolic Content (TPC) Extracted from PKC. The Palm Kernel Cake extract was analysed for phytochemicals, revealing the presence of phenolics. The Folin–Ciocalteu method was employed to quantify total phenolic content, reported in mg of Gallic Acid equivalent (GAE) per g of bio-oil. A graphical image generated by the design expert programme gives information regarding the relationship between the responses and experimental levels of each variable and the sort of interactions that occur between the variables being evaluated to ascertain the optimal condition. The 3D response surfaces were plotted using Eq. 4. Figs. 3(a), 3(b), and 3(c) display the graph.

Materials Research Forum LLC
https://doi.org/10.21741/9781644903575-24

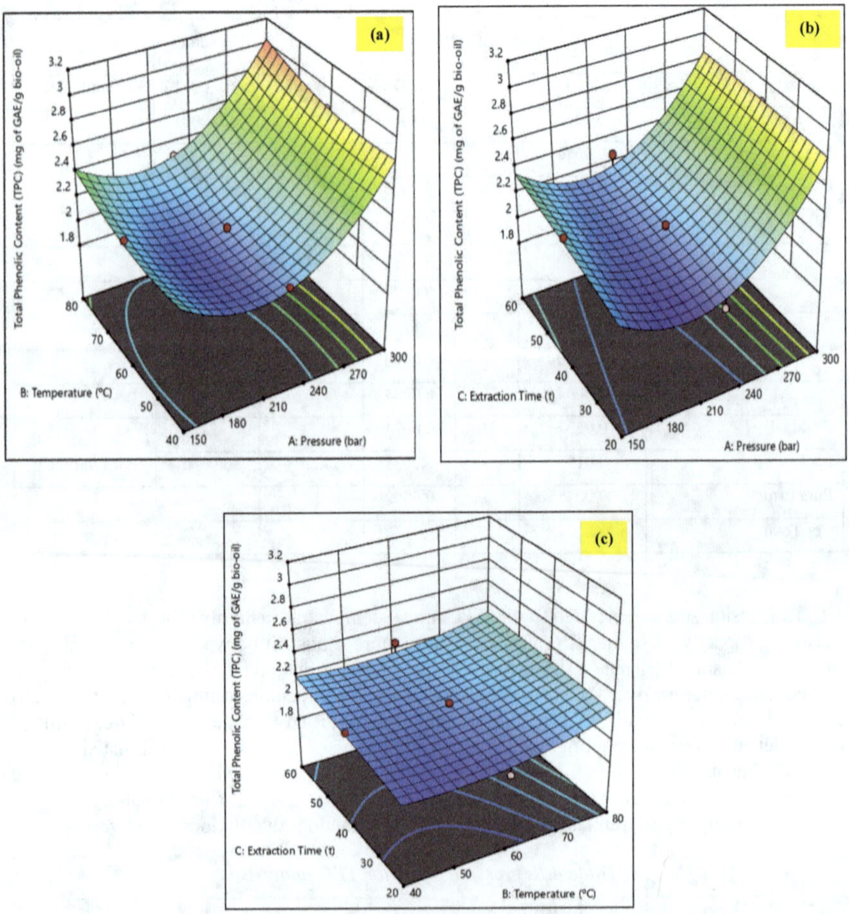

Fig. 3 Response surface plot on the effect of parameters towards the TPC: (a) effect of pressure and temperature, (b) effect of pressure and extraction time, and (c) effect of extraction time and temperature towards TPC.

Fig. 3 (a) illustrates the effect of pressure and temperature towards Total Phenolic Content (TPC) extracted from PKC. As pressure increases from 150 to 300bar, the extracted TPC increases. This result is consistent with the previous study, which stated that increasing the pressure will increase the TPC [19], [21], [22]. Furthermore, in this finding, the highest peak of TPC occurs at a pressure of 300 bar and a temperature of 80°C, with a value of 3.01 mg GAE/g bio-oil. This causes the diffusion of solutes to be slowed down. The density of supercritical CO_2 ($ScCO_2$) is increased by extraction at pressures of more than 300 bar, which also results in a rise in the solubility of the solutes it contains [21], [22]. In addition, raising the pressure setting will cause the supercritical CO_2 ($scCO_2$) to have a high density, which will cause its solvating power to rise.

Because of this, the solution to the solute in the sample will be more effective. The correlation between the solvating power and the density will impact the amount of material that can be extracted. Therefore, an increase in solvating power because of an increase in density will result in an increase in extraction yield because more extract may be flushed out [19].

Furthermore, in Fig. 3(a) and Fig. 3(c), it can be seen that as the temperature increases, the TPC also increase. This finding is in line with what was found in the earlier research [19], [23], [24]. It was found that greater temperatures had a beneficial impact on both solubility and diffusion. It indicates that elevated temperatures enhance both solubility and diffusion. This phenomenon can be elucidated by the hypothesis that elevated temperatures elevate vapour pressure and improve phenolic solubility in the solvent [24]. Nevertheless, excessively high temperatures may cause the elimination of undesired impurities and the decomposition of heat-sensitive phenolic compounds, which are susceptible to oxidation [23]. As a result, it can reduce the amount of oil that can be extracted. This assertion is corroborated by the findings of a study conducted by Azahar et al. [50], in which the researchers investigated the influence of temperature on the accumulation yield of Red Butterfly Wing Leaves. According to the data from Azahar et al. [50], the extraction yield gradually drops after 50°C, going from 2.94% at 50°C to 2.58% at 60°C. However, in this study, it does not show this adverse effect.

Determining the extraction time is of utmost importance because it will impact the extraction process and may result in an incomplete extraction if the extraction time is too short [25]. As shown in Fig. 3(b) where TPC reading is at its lowest, 2.35 mg GAE/g bio-oil when the extraction time is 20 minutes, and the pressure is 150 bar. This is compared to a TPC reading of 2.46 mg GAE/g bio-oil at the same pressure of 150 bar, but with an extended extraction period of 60 minutes. However, the extraction process required excessive time, which would result in the waste of both time and the solvent that was employed, in addition to the degradation of the bioactive component [25].

Optimisation Study of TPC Extracted from PKC. The study showed that the three independent variables analysed during the extraction of PKC utilising SFE had a substantial impact on the TPC results [20]. One crucial consequence is identifying the best parameters to enhance the TPC outputs. Table **5** displays the optimum conditions for Total Protein Content (TPC) values, using the conditions shown in CCD Design Expert version 13 and confirmed by a validated experiment.

Table 5 Summary of optimisation study.

Parameters			TPC (mg of GAE/ g bio-oil)		
Pressure (bar)	Temperature (°C)	Extraction time (minutes)	Predicted value	Experimental value	Percentage error (%)
300	80	42	3.035	2.985	1.65

Numerical optimisation was used to predict the maximum TPC achievable from PKC by studying the relationships among three variables: temperature, pressure, and extraction time. The optimised value indicated by the RSM was the temperature of 80°C, pressure of 300 bar, and extraction time of 42 minutes, which led to a TPC of 3.035mg GAE/g bio-oil. The analysis revealed a 1.65% discrepancy between the experimental data and expected values, which is considered insignificant. The inaccuracy may stem from interference during sample measurement or unforeseeable faults in the experiment.

Antioxidant Activity (AOA) of Phenolic Compounds Extracted from PKC. Firstly, increasing the pressure in SFE generally enhances Antioxidant Activity (AOA), as shown in Fig. 4(a). The highest AOA is observed at 300bar, while the lowest is at 150bar, under constant conditions of 60°C for 50 minutes, which are 68.11 and 38.07 g of AAE/mL bio-oil, respectively. This aligns with findings in a recent publication by Santos et al. [26], where higher pressure (200bar) resulted in 5mmol Trolox Equivalent (TE)/kg oil compared to 4mmol TE/kg oil at 30MPa, using Favela seeds. Elevating extraction pressure can boost AOA [27]. However, it may also increase the solubility of non-antioxidant components, like colours, waxes, or contaminants, which can adversely affect antioxidants. Extracting these non-antioxidant components alongside antioxidants might reduce the overall antioxidant activity of the extract.

Furthermore, as extraction time increases, Fig. 4(b) shows that Antioxidant Activity (AOA) also increases. The highest AOA of 56.59 g AAE/L bio-oil is recorded at 20 minutes, while the lowest, 44.74 g AAE/L bio-oil, is at 60 minutes. This supports prior studies [19], [25], [28]. The lower AOA at 20 minutes is due to the shorter extraction time, which reduces the recovery of antioxidants. This means not all available antioxidants are fully extracted, resulting in a lower AOA reading. Inadequate extraction time also lowers the overall concentration of antioxidants, reducing their effectiveness. Moreover, with longer extraction times, more antioxidant compounds are extracted. This is because antioxidants have a better chance to diffuse from the source material into the solvent over an extended period. Therefore, a longer extraction time leads to a higher concentration of antioxidants in the extract, contributing to a higher AOA reading.

Fig. 4(c) indicates that the Antioxidant Activity (AOA) value is highest at 80°C and lowest at 20°C. This trend mirrors the Total Phenolic Content (TPC) trend, reinforcing the consistency of the findings with earlier research [19], [25], [28]. Higher temperatures provide molecules with incredible kinetic energy, facilitating their movement and diffusion through a substance. Consequently, the extraction process becomes more effective in retrieving antioxidants from the source material. This is because, at elevated temperatures, the molecular motion is heightened, allowing for a more efficient transfer of antioxidants into the extraction medium. In simpler study proves, higher temperatures enhance the extraction process by increasing the mobility of the molecules, leading to a higher AOA value.

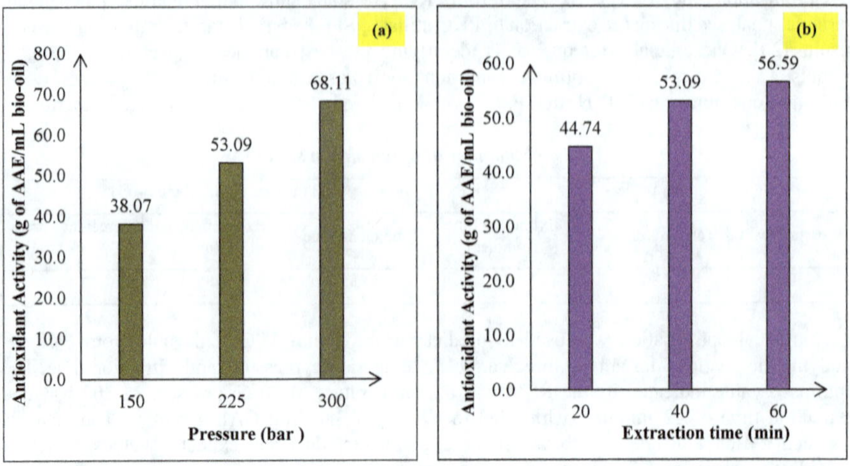

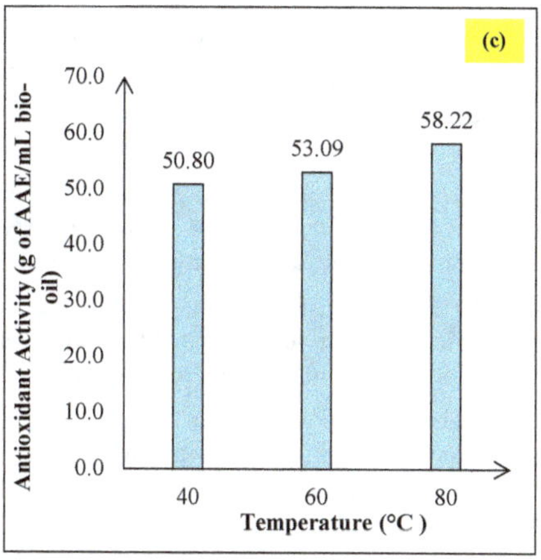

Fig. 4 Effect of parameters towards AOA: (a) effect pressure at conditions 60°C and 40 minutes, (b) effect extraction time at conditions 60°C and 225 bar, and (c) effect temperature at 225 bar and 40 minutes.

Conclusions

In summary, the purpose of this research publication is to give an optimisation study on the Total Phenolic Content (TPC) of PKC through the use of Supercritical Fluid Extraction (SFE), the Response Surface Methodology (RSM), and the Design Expert software. The Central Composite Design (CCD) approach proposed 20 experimental designs. The ideal parameters for achieving the highest Total Polar Compound (TPC) value were identified as 80°C temperature, 300 bar pressure, and a reaction time of 42 minutes. The highest AOA value obtained was 68.11g of Ascorbic Acid Equivalent (AAE)/mL bio-oil under 300 bar pressure, 60°C temperature, and 40 minutes extraction time. The lowest AOA value obtained was 38.07g of Ascorbic Acid Equivalent (AAE)/mL bio-oil, attained under 150 bar pressure, 60°C temperature, and a 40-minute extraction. The CCD model forecasted that the highest TPC value would be 3.035mg of GAE per g of bio-oil. Experimental results demonstrated a removal of 2.985mg of GAE per g of bio-oil, revealing a 1.65% discrepancy between the predicted and experimental values. The error is considerably below the acceptable threshold of less than 5%, demonstrating the model's accuracy.

This work shows that Supercritical Fluid Extraction (SFE) successfully extracts TPC from PKC. The optimised circumstances serve as a beneficial reference for professionals in the same industry. The CCD model's accuracy highlights its dependability and appropriateness for additional optimisation research in the SFE extraction process.

Acknowledgements

This research was supported by the Ministry of Higher Education (MOHE) through the Fundamental Research Grant Scheme (Grant No. FRGS/1/2022/TK0/UITM/02/100). The authors are grateful for the facilities Universiti Teknologi MARA (UiTM) Cawangan Pulau Pinang provided in completing this work.

References

[1] M. A. Hassan, M. A. A. Farid, M. R. Zakaria, H. Ariffin, Y. Andou, and Y. Shirai, "Palm oil expansion in Malaysia and its countermeasures through policy window and biorefinery approach," *Environ Sci Policy*, vol. 153, Mar. 2024. https://doi.org/10.1016/j.envsci.2024.103671

[2] I. L. Gold, C. E. Ikuenobe, O. Asemota, and D. A. Okiy, "Palm and Palm Kernel Oil Production and Processing in Nigeria," *Palm Oil: Production, Processing, Characterization, and Uses*, pp. 275–298, Jan. 2012. https://doi.org/10.1016/B978-0-9818936-9-3.50013-7

[3] G. Su, N. W. Mohd Zulkifli, H. C. Ong, S. Ibrahim, Q. Bu, and R. Zhu, "Pyrolysis of oil palm wastes for bioenergy in Malaysia: A review," *Renewable and Sustainable Energy Reviews*, vol. 164, no. May, 2022. https://doi.org/10.1016/j.rser.2022.112554

[4] C. C. Nnaji, A. E. Agim, C. N. Mama, P. G. C. Emenike, and N. M. Ogarekpe, "Equilibrium and thermodynamic investigation of biosorption of nickel from water by activated carbon made from palm kernel chaff," *Sci Rep*, vol. 11, no. 1, pp. 1–20, 2021. https://doi.org/10.1038/s41598-021-86932-6

[5] E. Tsouko, M. Alexandri, K. Vieira Fernandes, D. Maria Guimarães Freire, A. Mallouchos, and A. A. Koutinas, "Extraction of Phenolic Compounds from Palm Oil Processing Residues and Their Application as Antioxidants," *Food Technol Biotechnol*, vol. 57, no. 1, 2019. https://doi.org/10.17113/ftb.57.01.19.5784

[6] A. Sadeghi, A. Rajabiyan, N. Nabizade, N. Meygolinezhad, and A. Z. Ahmady, "Seaweed-derived phenolic compounds as diverse bioactive molecules: A review on identification, application, extraction and purification strategies," *Int J Biol Macromol*, p. 131147, Mar. 2024. https://doi.org/10.1016/J.IJBIOMAC.2024.131147

[7] S. Nuchdang *et al.*, "Soxhlet, microwave-assisted, and room temperature liquid extraction of oil and bioactive compounds from palm kernel cake using isopropanol as solvent," *Ind Crops Prod*, vol. 176, p. 114379, Feb. 2022. https://doi.org/10.1016/J.INDCROP.2021.114379

[8] Q. W. Zhang, L. G. Lin, and W. C. Ye, "Techniques for extraction and isolation of natural products: A comprehensive review," *Chinese Medicine (United Kingdom)*, vol. 13, no. 1, pp. 1–26, 2018. https://doi.org/10.1186/s13020-018-0177-x

[9] L. Panzella, F. Moccia, R. Nasti, S. Marzorati, L. Verotta, and A. Napolitano, "Bioactive Phenolic Compounds From Agri-Food Wastes: An Update on Green and Sustainable Extraction Methodologies," *Front Nutr*, vol. 7, no. May, pp. 1–27, 2020. https://doi.org/10.3389/fnut.2020.00060

[10] S. Tahir *et al.*, "Deep eutectic solvents as alternative green solvents for the efficient desulfurization of liquid fuel: A comprehensive review," *Fuel*, vol. 305, no. July, 2021. https://doi.org/10.1016/j.fuel.2021.121502

[11] V. H. Pise and B. N. Thorat, "Supercritical fluid extraction of dried Surangi flowers (Mammea suriga)," *Ind Crops Prod*, vol. 186, no. April, 2022. https://doi.org/10.1016/j.indcrop.2022.115268

[12] W. A. da Costa *et al.*, "Supercritical CO 2 extraction and transesterification of the residual oil from industrial palm kernel cake with supercritical methanol," *Journal of Supercritical Fluids*, vol. 147, no. October 2018, pp. 179–187, 2019. https://doi.org/10.1016/j.supflu.2018.10.012

[13] H. Ahangari, J. W. King, A. Ehsani, and M. Yousefi, "Supercritical fluid extraction of seed oils – A short review of current trends," *Trends Food Sci Technol*, vol. 111, no. February, pp. 249–260, 2021. https://doi.org/10.1016/j.tifs.2021.02.066

[14] K. Patel, N. Panchal, and P. Ingle, "Review of Extraction Techniques Extraction Methods: Microwave, Ultrasonic, Pressurized Fluid, Soxhlet Extraction, Etc," *International Journal of Advanced Research in Chemical Science*, vol. 6, no. 3, pp. 6–21, 2019. https://doi.org/10.20431/2349-0403.0603002

[15] L. B. Gu *et al.*, "Comparative study on the extraction of Xanthoceras sorbifolia Bunge (yellow horn)seed oil using subcritical n-butane, supercritical CO2, and the Soxhlet method," *Lwt - Food Science and Technology*, vol. 111, pp. 548–554, 2019. https://doi.org/10.1016/j.lwt.2019.05.078

[16] I. S. M. Zaidul, N. A. Nik Norulaini, A. K. Mohd Omar, and R. L. Smith, "Supercritical carbon dioxide (SC-CO2) extraction of palm kernel oil from palm kernel," *J Food Eng*, vol. 79, no. 3, pp. 1007–1014, Apr. 2007. https://doi.org/10.1016/J.JFOODENG.2006.03.021

[17] M. D. Luque de Castro and F. Priego-Capote, "Soxhlet extraction: Past and present panacea," *J Chromatogr A*, vol. 1217, no. 16, pp. 2383–2389, Apr. 2010. https://doi.org/10.1016/J.CHROMA.2009.11.027

[18] S. Burapan, M. Kim, Y. Paisooksantivatana, B. E. Eser, and J. Han, "Thai Curcuma Species: Antioxidant and Bioactive Compounds," *Foods 2020, Vol. 9, Page 1219*, vol. 9, no. 9, p. 1219, Sep. 2020. https://doi.org/10.3390/FOODS9091219

[19] M. Azahar *et al.*, "Supercritical CO2 extraction of red butterfly wing leaves: process parametric study towards extraction yield," *MATEC Web of Conferences* , vol. 154, 2018. https://doi.org/10.1051/matecconf/201815401015

[20] Mohd Azahar Mohd Ariff, Mohamed Syazwan Osman, Suzana Tukiman, Noor 'Aina Abdul Razak, and Jefri Jaapar, "Optimization of Supercritical Fluid Extraction for Mariposa Christia Vespertilionis Leaves Towards Yield by Using Response Surface Methodology," *J Phys Conf Ser*, vol. 1349, no. 12054, 2019. https://doi.org/10.2174/2405520412666190807141906

[21] A. Sharma, A. Ray, and R. S. Singhal, "Co-extraction of turmeric (Curcuma longa L.) and dried coconut shreds by supercritical fluid extraction (SFE): Chemical and bioactivity profile," *J Clean Prod*, vol. 382, Jan. 2023. https://doi.org/10.1016/j.jclepro.2022.135313

[22] D. Alwazeer, M. A. Elnasanelkasim, S. Çiçek, T. Engin, A. Çiğdem, and E. Karaoğul, "Comparative study of phytochemical extraction using hydrogen-rich water and supercritical fluid extraction methods," *Process Biochemistry*, vol. 128, pp. 218–226, May 2023. https://doi.org/10.1016/j.procbio.2023.01.022

[23] I. S. Che Sulaiman, M. Basri, H. R. Fard Masoumi, W. J. Chee, S. E. Ashari, and M. Ismail, "Effects of temperature, time, and solvent ratio on the extraction of phenolic compounds and the anti-radical activity of Clinacanthus nutans Lindau leaves by response surface methodology," *Chem Cent J*, vol. 11, no. 1, pp. 1–11, Jun. 2017. https://doi.org/10.1186/S13065-017-0285-1/TABLES/6

[24] S. A. Radzali, M. Markom, and N. Md Saleh, "Parameter Effects and Optimisation in Supercritical Fluid Extraction of Phenolic Compounds from Labisia pumila," *Separations 2022, Vol. 9, Page 385*, vol. 9, no. 12, p. 385, Nov. 2022. https://doi.org/10.3390/SEPARATIONS9120385

[25] N. A. Aris *et al.*, "The Effect of Fluid Flow Rate and Extraction Time in Supercritical Carbon Dioxide," *Journal of Advanced Research in Applied Sciences and Engineering Technology Journal homepage*, vol. 16, pp. 26–34, 2019, Accessed: Jan. 28, 2023. [Online]. Available: www.akademiabaru.com/araset.html

[26] K. A. Santos, E. A. da Silva, and C. da Silva, "Supercritical CO2 extraction of favela (Cnidoscolus quercifolius) seed oil: Yield, composition, antioxidant activity, and mathematical modeling," *Journal of Supercritical Fluids*, vol. 165, 2020. https://doi.org/10.1016/j.supflu.2020.104981

[27] M. Karamać, A. Kosińska, I. Estrella, T. Hernández, and M. Dueñas, "Antioxidant activity of phenolic compounds identified in sunflower seeds," *European Food Research and Technology*, vol. 235, no. 2, pp. 221–230, 2012. https://doi.org/10.1007/s00217-012-1751-6

[28] Mohd Azahar Mohd Ariff, Nurul Nadiah Mohd Nazri, Noor Aina Abdul Razak, Mohamed Syazwan Osman, and Jefri Jaapar, "Optimization of Supercritical Fluid Extraction for Mariposa Christia Vespertilionis Leaves Towards Yield by Using Response Surface Methodology," *Recent Innovations in Chemical Engineering*, vol. 12, Jul. 2019. https://doi.org/10.2174/2405520412666190807141906

Development of carbon dots from lignocellulose oil palm wastes: A potential CO_2 adsorbent

Aimi Solihah Zaul Kapri[1,a], Norhusna Mohamad Nor[1,2,b*], Nur Ainaa Mohd Hasyim Chan[1,c], Mohamed Syazwan Osman[1,3,d], Abdul Rahman Mohamed[4,e]

[1]Chemical Engineering Studies, College of Engineering, Universiti Teknologi MARA, Cawangan Pulau Pinang, 13500 Permatang Pauh, Pulau Pinang, Malaysia

[2]Waste Management and Resource Recovery (WeResCue) Group, Chemical Engineering Studies, College of Engineering, Universiti Teknologi MARA, Cawangan Pulau Pinang, 13500 Permatang Pauh, Pulau Pinang, Malaysia

[3]Lab EMZI-UiTM Nanoparticles Colloids, Universiti Teknologi MARA, Cawangan Pulau Pinang, 13500 Permatang Pauh, Pulau Pinang, Malaysia

[4]School of Chemical Engineering, Universiti Sains Malaysia, 14300 Nibong Tebal, Pulau Pinang, Malaysia

[a]aimi998822@gmail.com, [b]norhusna811@uitm.edu.my, [c]ainaamimi20@gmail.com, [d]syazwan.osman@uitm.edu.my, [e]chrahman@usm.my

Keywords: Carbon Dots, Empty Fruit Bunches, Mesocarp Fibre, Oil Palm Waste, Palm Kernel Shell

Abstract. This paper explores the utilisation of oil palm wastes, specifically palm kernel shells (PKS), mesocarp fibres (MF), and empty fruit bunches (EFB), into valuable carbon dots (CDs) through hydrothermal synthesis. Through comprehensive chemical techniques, including alkaline bleaching treatment, cellulose was successfully isolated from oil palm waste, yielding 59.28%, 57.64%, and 55.98% from EFB, MF, and PKS, respectively. Morphological analysis revealed separated fibrils with pits and holes, indicating successful impurity removal during delignification. Fourier-transform infrared spectroscopy (FTIR) confirmed the removal of wax, lignin, and hemicelluloses, while thermogravimetric analysis (TGA) indicated cellulose's thermal stability at 360°C. Elemental analysis showed carbon as the predominant element. The novel aspect of this research lies in the subsequent hydrothermal synthesis of CDs from the extracted cellulose at 180°C for 24 hours, using a 1:1:3 weight ratio of cellulose, thiourea, and potassium hydroxide (KOH) as doping agents. Results indicated that MF-derived CDs exhibited the highest quantum yield (64.14%), followed by EFB (62.36%) and PKS (58.99%). Surface characterisation revealed functional groups enhancing CDs' hydrophobicity, with nitrogen and sulfur doping significantly improving their CO_2 adsorption properties. These findings highlight the potential of CDs derived from oil palm waste as a novel and efficient CO_2 adsorbent, contributing to sustainable waste management and environmental efforts for CO_2 capture applications. This research stands out by showcasing an innovative approach to waste valorisation, turning abundant oil palm wastes into high-value functional materials with environmental benefits, and demonstrating the practical applications and sustainability of CDs technology in CO_2 capture.

Introduction

The escalating trend of waste generation and its consequential disposal challenges in developing nations pose a significant concern. Malaysia, renowned for its proficient palm oil-based industries, annually produces substantial waste comprising empty fruit bunches (EFB), mesocarp fibres (MF), palm kernel shells (PKS), and palm oil mill effluent (POME). This substantial waste generation is

 Materials Research Forum LLC
https://doi.org/10.21741/9781644903575-25

attributed to the inefficiency of oil palm trees, utilising only 10% of the produced oil and consequently leading to a staggering 90% wastage [1]. As a rough estimation, 1 kg of palm oil produces about 4 kg of dry biomass [2]. According to Jafri et al., the production of oil palm waste increased from 25,041.19 metric tons in 2018 to 28,106.48 metric tons in 2019 and rose to 32,579.48 metric tons in 2020 [3]Hence, in response to this challenge and to encourage sustainable resource utilisation, significant research efforts have been dedicated to exploring the development of value-added products from the abundant biomass generated by the oil palm industry. Being rich in sources of lignocellulosic biomass (LCBs) obtained from palm oil processing, Malaysia has been a focal point for such investigations. LCBs, characterised by their elevated carbon content and inherent heteroatom doping features, have attracted attention as promising alternative carbon sources for diverse applications, including their conversion into carbon dots (CDs).

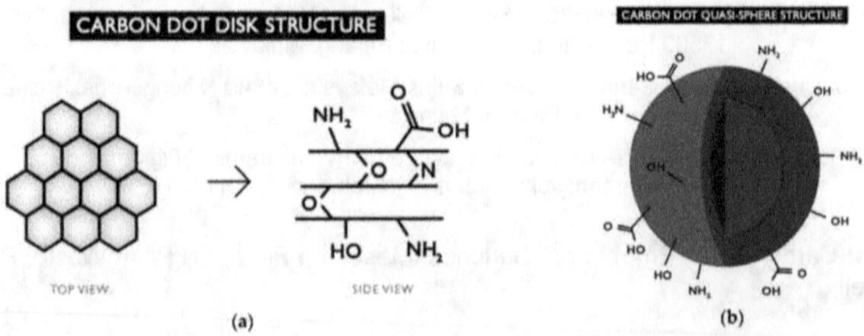

Fig 1 General structure of CDs where a) Top and side views of the disk-shaped morphology, Disk-shaped CDs with graphene-like sheets in a honeycomb structure, bound by dopants or any intermolecular forces. b)Typical quasi-spherical CDs with a crystalline carbon core (brown) and an amorphous shell (blue), with dopants forming surface functional groups on the CDs [4].

Carbon dots (CDs) are typically described as diminutive carbon nanoparticles dispersed in an aqueous solution [5]. CDs are characterised by their fluorescence and possess a diameter of less than 10 nm. These nanoparticles consist primarily of carbon atom aggregates, accompanied by trace elements such as nitrogen (N_2), oxygen (O_2), sulfur (S), and phosphorus (P), which were accidentally discovered by Xu et al. while purifying single-walled carbon nanotubes in 2004 [6,7]. The conventional CDs described to date typically comprise sp2 hybrid carbon cores and oxygen (O_2) and nitrogen (N) containing functional groups dispersed across their entire surface. These CDs possess an inherent hydrophobic nature within their carbon cores, encapsulated by a layer of hydrophilic groups. This configuration results in a stable aqueous dispersion of functionalised CDs [8,9]. In contrast to conventional semiconductor quantum dots (SQDs), commonly associated with significant production expenses, potential health risks, and environmental harm, CDs offer a more cost-effective, safer, and environmentally friendly alternative [10].

Two primary methodologies for synthesising CDs are top-down and bottom-up approaches. The top-down method predominantly involves a physical approach to producing CDs by breaking or peeling larger carbon material structures. Techniques under this category include arc discharge, electrochemical oxidation, chemical oxidation, and laser ablation [11]. However, this approach has several drawbacks, including expensive materials, harsh reaction conditions, prolonged reaction time and particles with unpredictable shapes and sizes [12,13]. The bottom-up method uses small molecules as precursors to generate CDs through chemical reactions. This approach includes various techniques such as combustion, microwave, and ultrasonic methods, as well as chemical solution synthesis and hydrothermal reactions [14]. Due to their straightforward operation and precise control over carbon precursors, bottom-up methodologies have become extensively utilised for generating CDs with consistent morphologies, narrow size distributions, and enduring characteristics [15]. Hydrothermal synthesis stands out as the preferred method for producing eco-friendly CDs owing to its versatility, cost-effectiveness, and resource-efficient nature. This process entails relatively brief reaction times within pressurised autoclave vessels, typically operating between 110°C and 220°C [16,17].

Various lignocellulosic biomass (LCBs) sources have been explored for synthesising carbon dots (CDs) through hydrothermal methods, each exhibiting unique quantum yield percentages. Dried lemon peel subjected to hydrothermal treatment at 200°C for 6 hours yielded CDs with a quantum yield of 11%. Palm kernel shells treated at temperatures ranging from 120 to 200°C for 3 hours demonstrated a higher quantum yield of 13.7%, while banana peels hydrothermally processed at 200°C for 24 hours exhibited a quantum yield of 16% [18–20]. Due to biomass composition, CDs produced from different biomass and synthesis methods exhibit non-identical quantum yields (QYs). It can be seen that the traditional hydrothermal method produced low CDs production with a yield of less than 20% even at optimal condition parameters.

These limitations hinder their large-scale production and future industrial applications. Thus, heteroatom doping, such as the addition of nitrogen (N_2), sulfur (S), phosphorus (P), or boron (B), can effectively regulate the surface properties of CDs and address these challenges. Doping can be achieved by incorporating chemical co-dopants during the one-step synthesis of CDs using the hydrothermal method. Heteroatom doping is a promising strategy for surface modification of CDs, allowing the manipulation of their intrinsic properties, including fluorescence enhancement. Nitrogen is particularly favoured among dopants because of its comparable atomic size to carbon and its capacity to alter CDs' electronic and optical characteristics by introducing additional electrons into the carbon structure. Nitrogen doping and chemical activation have demonstrated efficacy in enhancing the number of adsorption sites on CDs [21]. Sulfur, with an electronegativity similar to carbon, can also improve the surface functionality of CDs, expanding their potential applications as CO_2 adsorbents.

The research on solid CO_2 adsorbents in Malaysia lacks tailored development of materials suitable for the country's industrial and environmental needs. There is a lack of studies focusing on using abundant local waste materials, particularly those from the palm oil industry, for developing CO_2 adsorbents. Bridging these gaps is crucial for advancing sustainable CO_2 adsorbent development, aligning with global sustainability goals. Given the above, this research aims to convert the lignocellulose oil palm wastes into potential CO_2 adsorbents using hydrothermal methods to synthesise CDs. This study will employ a one-pot synthesis method to produce CDs using thiourea as the nitrogen (N_2), sulfur (S) source, and lignocellulose oil palm wastes as the carbon and oxygen sources. Concurrently, introducing KOH played a crucial role by providing surface functional groups to the CDs. KOH acted as a stabilising agent, contributing to synthesised CDs' structural integrity and stability. This methodological approach is hypothesised not only to harness the unique properties of thiourea and lignocellulose oil palm wastes but also to leverage the stabilising effects of KOH to facilitate the controlled synthesis of CDs under low-temperature conditions as a potential adsorbent for CO_2 removal. The exploration of alternative materials for adsorption, such as CDs adsorbent from EFB, MF and PKS were studied in this research to address the limitations associated with existing technologies, fostering more sustainable and efficient CO_2 capture methods.

Methodology

Materials and Chemicals. The samples of oil palm wastes (i.e. PKS, MF, EFB) were collected from Taclico Company at Padang Serai, Kedah. All chemicals were used as received which are sodium hydroxide (Emsure, 100%, NaOH), glacial acetic acid (Emsure, 100%, CH_3COOH), sodium chlorite (R&M Chemical, 80%, $NaClO_2$), thiourea (Sigma Aldrich, 99%, CH_4N_2S), potassium hydroxide (Merck, 30-32%, KOH), and ethanol (R&M Chemical, 96%, C_2H_6O).

Cellulose Extraction of Oil Palm Wastes. The alkaline treatment involved immersing 25.0 g of dried EFB, MF, and PKS in a 4% (w/v) NaOH solution at 80°C for 9 hours. After filtration, the samples were washed with distilled water until reaching a neutral pH (6-7). Subsequently, the treated samples underwent a soaking process with an equal part of acetate buffer (27g NaOH; 75ml glacial acetic acid diluted with distilled water to 1L), 1.7% (w/v) aqueous sodium chlorite, and purified water at 90°C for 12 hours. [22,23]. The obtained cellulose was filtered and washed with distilled water until pH 6-7. Then, it was dried at 80◦C for 24 hours.

Synthesis of CDs Using Hydrothermal Approach. The synthesis of ternary N/S/O-doped CDs involved a modified hydrothermal treatment of the extracted cellulose from lignocellulosic oil palm wastes. The process began by crushing the dried cellulose and mixing it with thiourea and KOH in a ratio of 1:1:3 (cellulose:thiourea: KOH). Subsequently, 5.0 g of the resulting mixture powder was dissolved in 125 mL of deionised (DI) water and stirred for 30 minutes. This solution was then transferred into a 200 mL Teflon autoclave reactor. The mixture underwent hydrothermal treatment at 180°C for 24 hours using an electric oven (Universal Oven, Memmert GmbH), then cooling to room temperature using a fan [24–26]. Then, the samples were centrifuged and freeze-dried in a freeze dryer (Labconco) for further use as potential CO_2 adsorbent.

Quantum Yield Measurements using Fluorescence Spectroscopy (FS). Absorbance Instrumentation/Measurements. A UV-VIS Scanning Spectrophotometer with automatic wavelength calibration and baseline correction was used to measure the solutions' absorbances. The solvent was put in two matched quartz spectrophotometer cells in the sample and blank cell positions. The baseline was "zeroed," and measurements were taken at 1-nm intervals. The observed baseline deviated from 0.000 by less than 0.001 absorbance units at no wavelength between 220 and 700 nm [27].

Absorbance/Fluorescence Solutions. The stock Quinine sulphate standard and sample solutions were prepared in volumetric flasks, and then the appropriate solvent (diluted sulphuric acid) was added to the mark. Dilutions were prepared volumetrically to produce solutions with absorbances of around 0.1 at the peaks employed for excitation in the fluorescence investigations using a fluorescence spectrophotometer (FS, Varian Cary Eclipse). A series of quinine standards from the 100 mg/L stock solution precisely were prepared. As a result, five diluted samples with concentrations of 20 mg/L, 10 mg/L, 8 mg/L, 5 mg/L, and 2.5 mg/L were obtained.

Quantum Yield Formula. The quantum yield (QY) of the CDs was calculated by the following equation:

$$Qx = Qs \frac{I_X A_S \eta_X^2}{I_S A_X \eta_S^2} \quad (1)$$

where Q is the fluorescence quantum yield and x and s are the samples and reference compounds. The reference was quinine sulphate dissolved in 0.1 M H_2SO_4 (Qs = 0.54), which is the refractive index of the aqueous solution (1.33). A is the absorbance at 365 nm, which is the excitation wavelength. I is the fluorescence emission spectrum's integrated fluorescence intensity [28].

CO_2 Removal using Continuous Fixed Bed Adsorption Setup

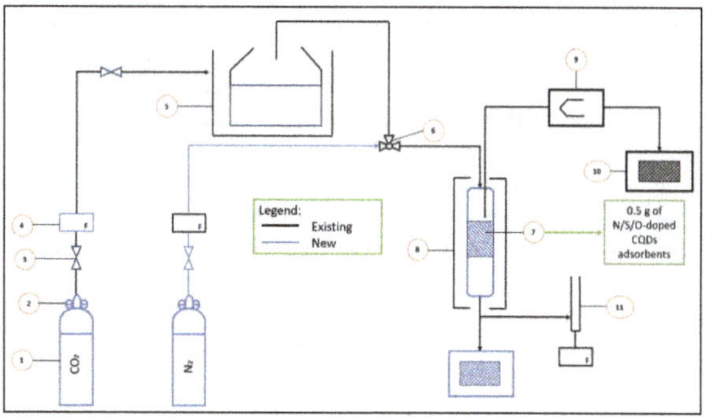

Fig 2 Schematic diagram of existing adsorption pilot plant, (1) Gas Cylinder, (2) Pressure Regulator, (3) Valve, (4) Mass Flow Controller, (5) Humidity System, (6) 3-Ways Valve, (7) Stainless-steel Column, (8) Heating Jacket, (9) K-Type Thermocouple, (10) Temperature Display, (11) Bubble Soap Flow Meter, (12) Gas Analyzer

The performance of N/S/O-doped CQDs adsorbents will be tested in a custom-designed CO_2 adsorption rig, as shown in Fig 2. A 0.5 g of N/S/O-doped CQDs adsorbents will be weighed and placed in the middle of a stainless-steel column with 250 mm length and 9.525 mm diameter and supported with 0.3 g borosilicate glass wool. At ambient temperature, a simulated gas containing CO_2 flow will be passed through the column at various flow rates. The composition of the simulated gas will be adjusted by controlling the mass flow meter (Aalborg). The inlet and outlet concentration of CO_2 will be monitored using a portable gas analyzer via an electrochemical sensor calibrated for syngas composition. The concentration of CO_2 will be recorded continuously for every minute, and the adsorption test will be stopped until it reaches the saturated concentration (Ci/Co = 1.0). The adsorption capacity of N/S/O-doped CQDs adsorbents will be calculated using the breakthrough time and flow rate as in equation 2.

$$q = \frac{Q_f t_f y_f}{m_c} \tag{2}$$

where q, is the adsorption capacity of fresh adsorbent (mg/g), Q_f is the volumetric feed flow rate (mL/min), y_f is the mole fraction of the adsorbate (mg/mL), t_f is the breakthrough time (minute) and m_c is the mass of adsorbent (g).

Physical and Chemical Properties of Cellulose Extracted and CDs. Field Emission Scanning Electron Microscope Energy Dispersive X-ray (FESEM-EDX). The extracted cellulose was morphologically examined using a Field Emission Scanning Electron Microscope (Nova NanoSEM 450). A small sample was deposited onto a carbon tape, and morphological features were observed by randomly selecting three fibres from the captured images. Additionally, elemental analysis was conducted using energy-dispersive X-ray (EDX) coupled with FESEM.

Fourier Transmission Infra-Red Spectroscopy (FTIR). Fourier Transform Infrared Spectroscopy (Thermo Scientific Nicolet iS10) was employed for the chemical analysis of functional groups in both the extracted cellulose and CDs. To prepare the sample disc, the sample was mixed with KBr at a ratio of 1:10 and then compressed. FTIR spectra were generated by scanning the sample covering a transmission wavelength range from 3500 to 1000 cm^{-1}.

Cellulose Quantification Analysis using Thermogravimetric Analysis (TGA). The extracted cellulose was thermally stabilised via Thermogravimetric Analysis (SDT Q600). Ten milligrams of the sample were subjected to TGA analysis within a temperature range of 30 to 500°C, utilizing an alumina crucible under a nitrogen flow of 100 ml/min and a heating rate of 10°C/min [29]. The typical decomposition temperatures for hemicellulose, cellulose, and lignin were 225°C, 325°C, and 425°C, respectively[29].

CHNS Elemental Analysis. Elemental analysis was conducted to determine the overall elemental composition of both isolated cellulose from lignocellulose oil palm wastes and CDs, utilising a CHNS elemental analyser (Perkin Elmer). The measured elements in the extracted cellulose and CDs, including carbon, hydrogen, nitrogen, and sulfur, were determined individually.

BET Analysis. To analyze the porosity and surface area of the extracted cellulose, nitrogen adsorption and desorption isotherm measurements were conducted. These measurements provided insights into the morphological and porous properties, including specific surface area, average pore diameter, and total pore volume of the extracted cellulose.

Result and Discussions

Yield of Extracted Cellulose and Quantum Yield of CDs Synthesis. The yield percentage of the extracted cellulose from lignocellulose oil palm wastes was then calculated. This parameter provides insight into the cellulose extraction efficiency of the alkaline-bleaching pretreatment. It quantifies the percentage of cellulose extracted relative to the initial biomass used. The pretreatment method can be evaluated by comparing the yields from EFB, MF and PKS, and the cellulose content variations can be determined. The percentage yield of cellulose extracted from LCBs was determined using Equation 2:

$$Percentage\ Yield = \frac{Weight_2}{Weight_1} \times 100\% \tag{2}$$

where $Weight_1$ represents the weight of the LCBs before treatment, and Weight2 represents the weight of the LCBs fibre post-treatment.

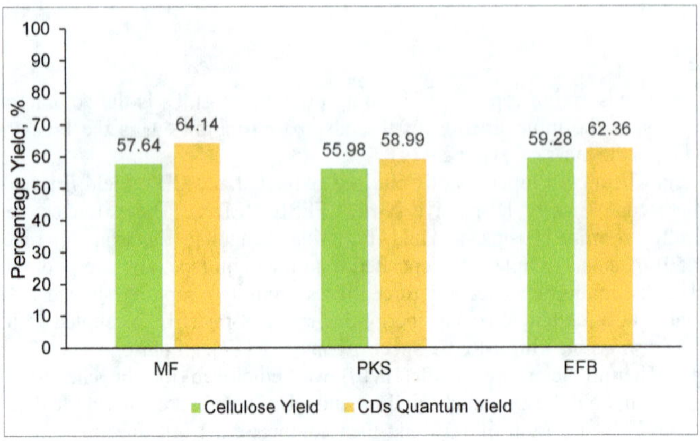

Fig 3 Percentage yield of extracted cellulose from lignocellulose oil palm wastes and quantum yield of CDs

Based on Fig 3, the percentage yield of extracted cellulose is computed, and EFB has the maximum percentage yield, followed by MF and PKS with respective contents of 59.28%, 57.64%, and 55.98%. This could be explained by each fibre's function. For instance, PKS is the seed's exterior layer in palm fruits and protects the kernel oil-containing seed [30]. At the same time, MF contains the hydrophobic fragment (lignin) that could aid in preserving the fruit's oil from dampness conditions [31]. Since PKS initially contained the most significant quantity of lignin, followed by MF, more hemicellulose fraction is covalently bound to lignin in PKS and MF than in EFB. Further delignification has effectively removed the lignin and hemicellulose fraction and produced cellulose. It is noted that the percentage yield for this study exhibited quite similar results as reported in previous research by Izzaha et al. that the content of cellulose is 65.78 %, 64.13 % and 61.65 % for oil palm frond (OPF), oil palm empty fruit bunches (EFB) and oil palm mesocarp fibre (MF) respectively [22,32]. When synthesising CDs, cellulose acts as a carbon source, providing the necessary carbon atoms for forming the carbonaceous core of the dots. A larger pool of cellulose precursor allows for a higher potential yield of CDs during the synthesis process. This is because more cellulose translates to more carbon for conversion into CDs. Thus, a higher yield of extracted cellulose provides a larger quantity of precursor material for CDs synthesis and potentially contributes to the production of CDs with improved optical characteristics and overall performance.

On the other hand, the percentage quantum yield (QY%) obtained for the CDs-derived EFB, CDs-derived MF, and CDs-derived PKS was 62.36%, 64.14% and 58.99%, respectively. A higher quantum yield denotes a more efficient emission. These numbers reflect the proportion of absorbed photons the CDs successfully transform into released photons. [33]. It is emphasised that a critical element in enhancing the properties of CDs is the addition of nitrogen, oxygen, and sulphur doping during their fabrication. The CDs' electronic structure and surface characteristics can improve optical properties, including higher quantum yield [34]. Furthermore, amination/amidation creates radiative recombination centres, including amine (–C–N) and amide (R2–NCOR) groups, which serve as electron-donating entities, potentially enhancing the fluorescence intensity of CDs. [35]. On top of that, incorporating S atoms into CDs will enhance their electron-donating capacity due to the electron-rich nature of S atoms, thereby modifying the energy levels and electronic structure of the CDs, resulting in enhanced light emission [36]. However, the simple synthesis (using water as the reaction medium) of CDs-derived PKS has a low quantum yield (58.99%), which may be due to incomplete carbonisation as a result of the low boiling point of water [37]. Based on a previous study, using a-lipoic acid as the carbon source and ethylenediamine as the nitrogen source, Ding et al. synthesised N, S-CDs via a hydrothermal method in 2014. N, S-CDs exhibited a high QY of 54.4% and an excellent fluorescence property, allowing for their successful application to $Fe3+$ detection and multicolour cell imaging [38]. Yang et al. reported that employing an ultrasonic-hydrothermal process without a co-catalyst or modification yields pure CDs with an average diameter of 3.4 nm and a relative QY of 2.42% [16]. Therefore, this suggests that the N/S-doped CDs derived from cellulose derived from lignocellulose oil palm wastes in this study exhibited higher quantum yield values than non-doped CDs.

The correlation between the percentage yield of extracted cellulose and the percentage quantum yield (QY%) of the derived CDs from different oil palm waste sources highlights an exciting relationship. While the percentage yield of extracted cellulose reflects the efficiency of cellulose extraction from the raw biomass, the percentage quantum yield of CDs indicates the efficiency of light emission from the synthesised CDs. Interestingly, despite EFB exhibiting the highest percentage yield of extracted cellulose, CDs derived from MF demonstrated the highest percentage quantum yield. This suggests that while a higher yield of extracted cellulose may contribute to a larger precursor pool for CDs synthesis, other factors such as the inherent composition and structure of the biomass source may influence the optical properties and emission efficiency of the resulting CDs.

Physical Characteristics. Surface Morphology of Cellulose. The FESEM analysis revealed the surface morphology of lignocellulosic oil palm wastes (i.e., MF) following the alkaline-bleaching treatment. This treatment effectively removed non-cellulosic and macromolecular components such as hemicelluloses, lignin, pectin, and wax, exposing the underlying cellulose structure. Isolated fibrils with disrupted lignocellulosic structures were observed, indicating the removal of impurities (i.e., silica bodies, residual oils, and proteins) and the dissolution of lignin and hemicellulose, contributing to the enhanced purity and porosity of the cellulose as agreed by Mohammad et al [39]. The extracted cellulose exhibited a rougher and corrugated surface with numerous holes and pits, suggesting enhanced porosity and surface area, as shown in Fig 4. The alkaline treatment by sodium hydroxide (NaOH) contributed to removing wax and cuticle from the fibre surface, further exposing the cellulose [40]. According to Al-Rajabi et al., after the delignification process, pinholes appeared on the surface of de-waxed oil palm waste fibres with a 'rotten' appearance [41]. The EDX spectra further validated the elemental composition of the extracted cellulose, which predominantly consisted of carbon (54.12 wt%), oxygen (42.94 wt%), and nitrogen (2.95 wt%). Besides, raw MF has a surface area of 1.7480 m^2/g while the extracted cellulose of MF shows a surface area of 10.7445 m^2/g, which is a significant increase indicating effective removal of non-cellulosic components through alkaline-bleaching treatment in enhancing the porosity and surface area of MF. Overall, the alkaline and bleaching treatments effectively disrupted the lignocellulosic complex, resulting in the solubilization of lignin and hemicellulose. This process also increased cellulose fibres' porosity and surface area, making them more accessible for further extraction procedures. The FESEM micrograph depicted a delignified oil palm waste cellulose network with increased cylindrical holes, highlighting the structural modifications induced by the treatment.

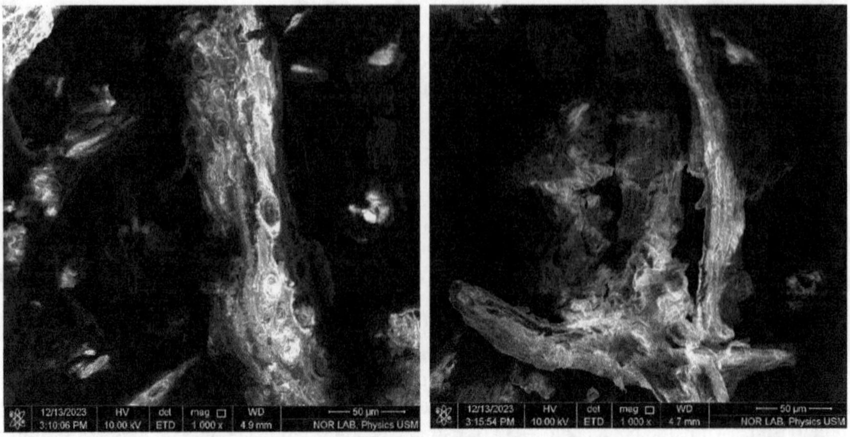

Fig 4 FESEM micrograph of extracted cellulose from MF

Table 1 EDX composition and surface area of extracted cellulose from MF

Sample	Carbon, Wt%	Oxygen, Wt%	Nitrogen, Wt%	BET Surface Area, m^2/g
Oil palm waste (MF)	54.12	42.94	1.98	10.7445

Thermal Stability of Cellulose Extracted from Oil Palm Waste. Using a TGA, the prediction of cellulose, lignin, and hemicellulose composition can be analysed in correlation with other analyses. The thermal stability and decomposition peak of the cellulose fibre are presented in Figure 5.

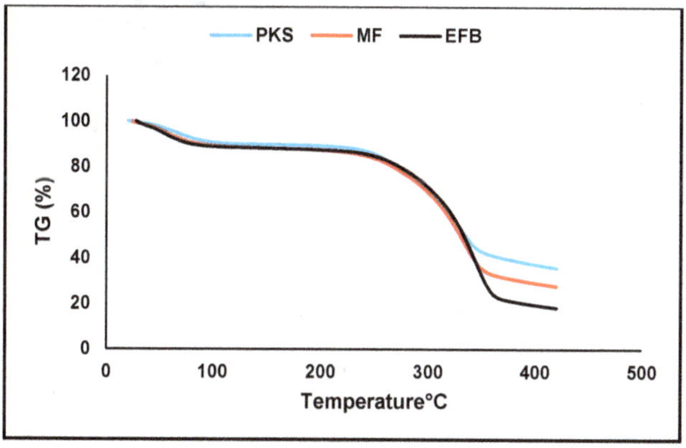

Fig 5 Thermal stability of extracted cellulose derived from lignocellulose oil palm wastes

The degradation process of lignocellulose oil palm wastes can be delineated into three distinct phases. In phase one, there is the dehydration and evaporation of lighter components. This phase is followed by phase two, which involves devolatilization cellulose and hemicellulose. Finally, in phase three, lignin decomposition occurs [42]. Based on the findings in Fig 5, the weight loss observed at a temperature near 100°C can be attributed to water in the samples. Subsequently, a significant weight loss was observed between 230°C and 360°C, indicating the degradation of cellulose as the primary component. The degradation of lignin and hemicellulose compounds occurred at higher temperatures, resulting in a slower and more consistent mass loss in the third phase at 360°C. Comparing the thermal stability of the samples, it was observed that MF exhibited the lowest stability, while EFB and PKS samples showed similar thermal stability. The amorphous structure of hemicellulose and lignin in MF reduces the crystallinity of isolated cellulose, impacting its mechanical and thermal properties. On the other hand, an increase in cellulose crystallinity enhances its thermal stability. The higher crystallinity observed in the EFB-cellulose sample, resulting from removing amorphous regions through hemicellulose extraction, could explain its higher thermal stability than MF. According to Eleanor et al., the analysis of OPEFB-cellulose, OPF-cellulose, and OPT-cellulose's thermal stability indicated a single-step degradation between 210°C and 380°C. Overall, understanding the thermal behaviour of these biomass sources is crucial for optimising their utilisation in synthesising CDs.

Chemical Characteristics of Extracted Cellulose and CDs. Surface Functional Group of Extracted Cellulose and CDs. FTIR spectroscopy was employed to analyse the functional groups and chemical bonds within the extracted cellulose and CDs, providing crucial details about their chemical composition and structural characteristics, as illustrated in Fig 6 and 7.

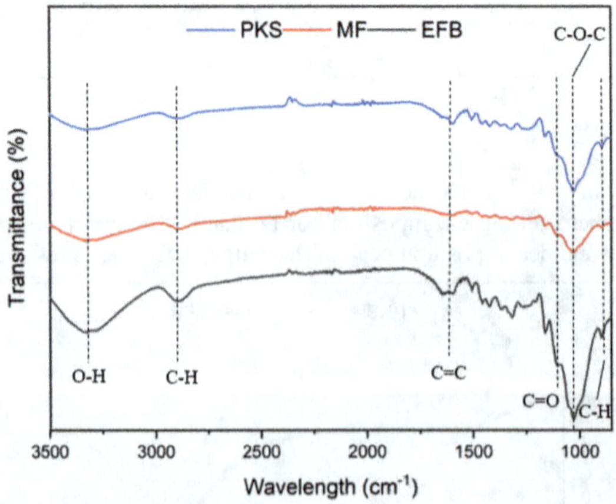

Fig 6 FTIR spectroscopy of cellulose derived from lignocellulose oil palm wastes

Results from FTIR spectroscopy show that EFB, MF, and PKS have similar chemical compositions. Broadband between 3200 and 3400 cm-1 reflects the tendency of all LCB samples linked with the O-H group, which is present in the glucose unit [43]. The peak intensity increased upon treatment with alkaline, attributed to the heightened concentration of O-H groups, as alkaline conditions reduced hydrogen bonding in cellulosic hydroxyl groups. The peak displays pronounced and distinct intensity due to the exposure of O-H groups and the prevalence of intermolecular O-H bonding within the cellulose structure on the fibre surface of EFB compared to MF and PKS. The next noticeable peak was observed around 2835-2945 cm^{-1}, indicating the stretching vibration of C-H methylene groups and confirming the presence of lignocellulose in all samples [22]. The corresponding adsorption peaks at 1592 cm^{-1} indicate the diminished aromatic skeletal C=C vibrations corresponded to the lignin were mostly absent in the extracted celluloses [44], and it is noted that the MF sample has the least lignin content compared to other samples. Then, the low intensity of peaks 1028 cm–1 (C-O-C, aryl-alkyl-ether) on all LCBs samples revealed the ether bond between the hydroxyl groups of lignin and the carbohydrates in the samples were removed after alkali-bleaching treatment, which also indicates the complete removal of the hemicelluloses [44]. Meanwhile, the peaks observed at 895 cm-1, 1160 cm–1, and 1313 cm–1 corresponded to the β-glycosidic linkages between the glucose monomers, the stretching and bending vibrations of C-H bonds, and the stretching vibrations of C=O bonds of hemicellulose, respectively. The presence of glycosidic linkages is evidence of a cellulose structure, as these linkages bond anomeric carbon atoms of saccharides to form polysaccharides [45].

Based on previous research by Apaydin Varol et al., the broad stretching vibration band observed between 3400 and 3000 cm-1 is similarly attributed to O-H groups. In contrast, the medium and weak band observed between 2800 and 3000 cm^{-1} indicates asymmetric and symmetric stretching C-H vibrations in xylose, cellulose, and lignin structures. The bands between 1740 and 1510 cm^{-1} in aromatic structures are owing to olefinic C=C and C=O vibrations. These findings align with previous research, which proved that the cellulose extraction from lignocellulose oil palm wastes has been successful.

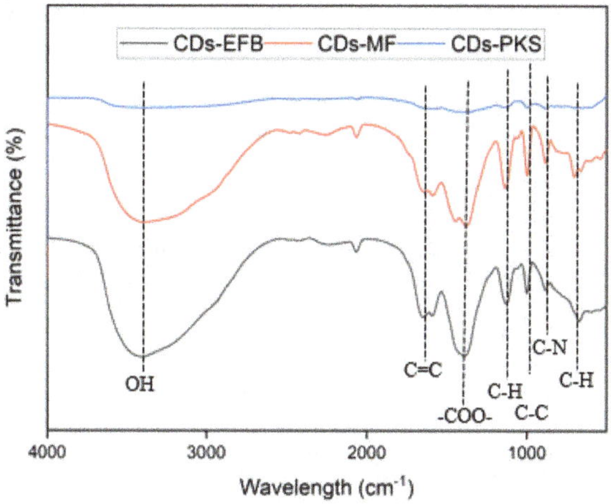

Fig 7 FTIR spectroscopy of freeze-dried CDs

Based on Fig 7, the results confirmed the presence of carboxyl, hydroxyl, and amino groups on the surface of the CDs samples. The broad absorption band observed at 3200-3400 cm^{-1} corresponded to the stretching vibrations of (O-H) bonds, and CDs-PKS had lower intensity peaks than CDs-EFB and CDs-MF. The broadening of the O-H peak is attributed to the formation of hydrogen bonds with the solvent during the synthesis process using the hydrothermal method [46]. A prior study by Brichkin et al. utilised FTIR analysis to investigate CDs-derived biomass and identified a similar absorption band at 3309 cm^{-1}, corresponding to the stretching vibration of O-H groups. The existence of C=C stretching vibrational modes was verified by the moderately broad absorption band centred at 1640 cm^{-1}. After the hydrothermal reaction, all CDs samples showed a noticeable peak corresponding to the amino group (C-N), indicating the presence of residual components from the surface modification process using a doping agent before hydrothermal synthesis of the CDs. The sharp intensity peaks in CDs-EFB and CDs-MF suggest adequate nitrogen doping into the CDs with thiourea. Besides, a clear peak at 1595 cm^{-1} indicated the presence of carboxylic groups (-COO-) on the surface of the CDs. The FTIR spectrum revealed the presence of oxygen-rich groups like carboxyl and hydroxyl on the surface, elucidating the CDs' excellent water solubility [47]. The absorption peaks at 1392 and 679 cm^{-1}, attributed to the C–H stretching vibrations of the methyl group C–H band, were identified. The peaks may arise from the bending vibration of the C–H groups in the untreated cellulose components within the lignocellulose biomass samples. Additional absorption bands at 1005 cm^{-1} were linked to transitions involving C-O stretching, indicating the presence of ether linkages or alcoholic functional groups. These findings highlight the successful fabrication of CDs with tailored surface functionalities from the extracted cellulose, demonstrating their potential for CO_2 adsorbent.

Elemental Analysis of Extracted Cellulose and CDs. The elemental composition of the lignocellulose oil palm waste samples is depicted in Table 1. Based on the result, the comparison between the weight percentages of carbon, hydrogen, nitrogen, and sulfur in lignocellulose oil palm wastes and freeze-dried CDs revealed significant differences. In lignocellulose oil palm wastes, carbon comprises the highest proportion at 40.99%, followed by hydrogen at 5.12%, nitrogen at 1.98%, and sulfur at 0.45%. Conversely, carbon constitutes a lower percentage in freeze-dried CDs at 9.62%, with hydrogen slightly higher at 5.98%. Notably, nitrogen content significantly increases to 4.44% in CDs compared to lignocellulose oil palm wastes, indicating intentional nitrogen doping during synthesis. Moreover, the sulfur content in CDs rises to 1.24%, suggesting incorporating sulfur-containing compounds during synthesis. Overall, the freeze-drying process and synthesis methods contribute to the altered composition of CDs compared to lignocellulose oil palm wastes, with deliberate modifications to enhance their properties for specific applications as CO_2 adsorbent. With its higher carbon content, oil palm waste might inherently possess some CO_2 adsorption capabilities. However, the deliberate modifications during synthesising CDs, including nitrogen and sulfur doping, likely enhance their CO_2 adsorption properties. Nitrogen and sulfur functionalities introduce active sites on the CDs' surface, facilitating stronger interactions with CO_2 molecules through chemisorption or physisorption mechanisms. Additionally, the increased presence of these heteroatoms may contribute to enhanced surface polarity and charge distribution, further improving the CDs' affinity towards CO_2 molecules. Therefore, the engineered properties of CDs, stemming from their elemental composition and surface modifications, could render them more effective adsorbents for CO_2 capture applications. Overall, the correlation between the EDX analysis of extracted cellulose and the CHNS elemental analysis of CDs highlights the deliberate modifications made during the synthesis process to tailor the CDs' elemental composition for improved CO_2 adsorption performance.

Table 2 Elemental analysis of extracted cellulose from oil palm wastes and CDs

Sample	Carbon, Wt%	Hydrogen, Wt%	Nitrogen, Wt%	Sulfur, Wt%
Oil palm waste (MF)	40.99	5.12	1.98	0.45
Freeze Dried CDs	9.62	5.98	4.44	1.24

CDs on CO_2 Adsorption Performance

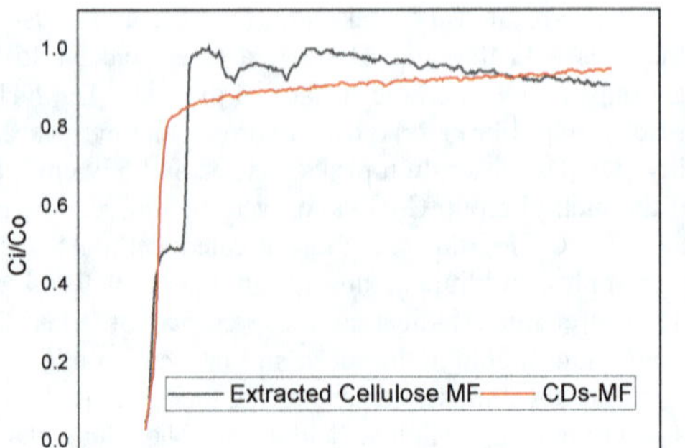

Fig 8 CO_2 Adsorption capacity over time for two different samples (i.e., extracted cellulose MF and CDs-MF)

Both samples exhibit quick initial adsorption rates, as seen in Fig. 8, with CDs-MF reaching a greater maximum adsorption capacity than isolated cellulose MF. According to equation 2's calculations, the isolated cellulose MF had an 11.09 mg/g CO_2 uptake. In the first few minutes, the raw MF adsorption curve increases swiftly, indicating rapid adsorption; nonetheless, it achieves its maximal adsorption capacity rather quickly. This reduced adsorption capability suggests that the isolated cellulose MF sample has a weaker affinity for CO_2 and fewer active sites. On the other hand, compared to isolated cellulose MF, CDs-MF showed a much higher CO_2 absorption of 49.02 mg/g. A faster initial climb to a considerably higher value is also shown in the adsorption curve for CDs-MF, suggesting a bigger surface area and more accessible active sites. The significant adsorption capacity difference between the two samples highlights CDs-MF's better CO_2 capture capabilities. According to Dziejarski et al, the typical values of CO_2 uptake using carbon nanoparticles (i.e., carbon nanotubes) adsorbent which are often lower than those of activated carbons and MOFs, usually in the range of 20-50 mg CO_2/g [48]. In contrast, it is reported that activated carbon capacities typically range from 50 to 150 mg CO_2/g under optimal conditions and MOFs ranging from 100 to 300 mg CO_2/g [49–51]. While CDs-MF may not be the highest capacity compared to all available adsorbents, it demonstrates significant potential, particularly for applications prioritizing sustainability and cost-effectiveness. Nevertheless, research on CDs for CO_2 capture is still in its early stages. This investigation is crucial in addressing, at least partially, the significant challenge of global warming.

CDs as A Potential CO_2 Adsorbent. Given the acidic nature of CO_2, optimising the surface properties of solid carbonaceous materials becomes paramount for enhancing their selectivity in capturing this pollutant. Despite typically possessing smaller surface areas and tuneable surface chemistry, CDs still hold promise in CO_2 adsorption applications. As per May et al., the adsorption characteristics of CDs are primarily governed by the surface area and the surface functional groups, which actively remove anionic or cationic adsorbates [52]. Doping CDs with heteroatoms is expected to enhance the charge distribution, thereby facilitating the interaction of the adsorbent with CO_2 molecules [53]. Primary nitrogen-containing groups are expected to promote the occupation of CO_2 in the adsorption sites of CDs by facilitating the selective adsorption of CO_2 through acid-base interactions. Conversely, a single sulfur electron can establish robust interactions between oxidised sulfur functions and CO_2. Oxygen functionalities can also modify the hydrophilicity of a CD's surface. It has been reported that carbon-based materials containing sulfur or nitrogen can capture CO_2 up to eight times more effectively than aqueous amines and are more efficient than any other known methods [54]. Moreover, surface polarity arising from the variation in electronegativity between the heteroatom and carbon could augment interactions between gas molecules and the carbon surface, thereby enhancing adsorption properties [55].

The findings demonstrate the potential of CDs derived from oil palm waste as efficient CO_2 adsorbents. The successful fabrication of CDs in this study with tailored surface functionalities using KOH and thiourea to further augments their properties for CO_2 removal application. Furthermore, the elemental analysis revealed deliberate modifications during synthesis, such as nitrogen and sulfur doping, which contribute to the enhanced CO_2 adsorption capacity of the CDs. The existence of surface functional groups like carboxyl, hydroxyl, and amino groups on the CDs amplifies their attraction towards CO_2 molecules. The CDs' tailored surface functionalities, combined with their high surface area and porous structure derived from cellulose, make them promising candidates for CO_2 capture applications. Hence, through physical and chemical modifications, LCBs can be transformed into promising CO_2 adsorbents with the potential for high adsorption capacities [56]. LCBs can be utilised to synthesise CDs as potential adsorbents due to their abundance, minimal cost, and high cellulose and carbon content. Through tailored surface modifications, CDs can be engineered to exhibit enhanced affinity and efficiency in capturing CO_2

molecules, thereby contributing to the mitigation of CO_2 emissions and environmental sustainability efforts.

Conclusion

This research introduced an innovative approach to extracting cellulose from lignocellulosic oil palm waste to synthesise CDs. The comprehensive chemical methods employed in this study, namely alkaline bleaching treatment, have successfully isolated cellulose from oil palm waste (i.e., EFB, MF, and PKS), which yielded 59.28%, 57.64%, and 55.98%, respectively. This pattern aligns with each fibre's intrinsic composition and roles; PKS, serving as the protective outer layer of the seed, harbours higher levels of lignin and hemicellulose residues, yielding less cellulose compared to MF and EFB. FESEM examination revealed that the structure of the extracted cellulose comprised separated fibres with pits and holes, indicative of impurities removed during the delignification process. The increase in BET surface area from 0.1748 m^2/g to 1.0744 m^2/g demonstrates the effectiveness of the alkaline-bleaching treatment in enhancing the porosity and surface area of MF. The alkaline-bleaching treatment effectively removed non-cellulosic impurities and macromolecular components from oil palm biomass fibres, exposing the cellulose structure and increasing surface porosity. Furthermore, FTIR analysis indicated the elimination of hemicellulose functionalities, lignin and wax. The FTIR spectroscopy results highlight the chemical composition similarities among various lignocellulosic biomass samples, with alkaline treatment affecting the intensity of specific peaks indicative of structural changes (i.e., 3200 and 3400 cm^{-1} linked with O-H group, 2835-2945 cm^{-1} indicated C-H methylene groups and low peaks at 1592 cm^{-1} means the diminished aromatic skeletal C=C vibrations). The TGA assessment highlighted the thermal stability of cellulose, demonstrating its purity within the temperature range of 230°C to 360°C. At the same time, CHNS elemental analysis proved that the extracted cellulose comprises the highest proportion at 40.99% of carbon, followed by hydrogen at 5.12%, nitrogen at 1.98%, and sulfur at 0.45%. These findings underscore the importance of understanding lignocellulosic oil palm biomass composition and structure in optimising cellulose extraction efficiency for various applications, including producing valuable materials such as CDs for CO_2 adsorption. Moreover, a practical and swift approach was developed for preparing modified CDs via hydrothermal treatment. CDs have been successfully synthesised by incorporating KOH and thiourea as doping agents and using EFB, MF, and PKS as carbon precursors to evaluate the effect of carbon sources on CDs' efficiency. Based on the experimental findings, quantum yield measurements reveal that CDs-MF exhibited the highest emission efficiency (64.14%) compared to those derived from EFB (62.36%) and PKS (58.99%). The surface characterisation of CDs derived from lignocellulose oil palm wastes revealed distinct functional groups, including carboxyl, hydroxyl, and amino groups, which enhance the hydrophobicity of the CDs. Last, the CHNS analysis of freeze-dried CDs proved that nitrogen and sulfur doping had been successfully incorporated due to an increase in weight percentage compared to the extracted cellulose. The deliberate modifications, including nitrogen and sulfur doping, during the synthesis of CDs enhance their CO_2 adsorption properties, potentially making them more effective adsorbents than raw oil palm waste. Comparative analysis of the adsorption capacities showed that raw MF achieved a CO_2 uptake of 11.09 mg/g, while the modified CDs-MF exhibited a substantially higher uptake of 49.02 mg/g. This significant improvement highlights the performance of CDs-MF in capturing CO_2, attributed to the increased number of accessible active sites and higher surface area.The engineered properties of CDs, stemming from their altered elemental composition and surface modifications, underscore their potential for CO_2 capture applications.

Acknowledgement

This research was supported by the Ministry of Higher Education (MOHE) through the Fundamental Research Grant Scheme (Grant No. FRGS/1/2022/TK0/UITM/02/55). The authors are grateful for the facilities provided by Universiti Teknologi MARA (UiTM) Cawangan Pulau Pinang in completing this work.

References

[1] K.E. Anyaoha, R. Sakrabani, K. Patchigolla, A.M. Mouazen, Critical evaluation of oil palm fresh fruit bunch solid wastes as soil amendments: Prospects and challenges, Resour Conserv Recycl 136 (2018) 399–409. https://doi.org/10.1016/j.resconrec.2018.04.022

[2] N.H.S. Jafri, D.N. Jimat, F.M. Azmin, S. Sulaiman, Y.A. Nor, The potential of biomass waste in Malaysian palm oil industry: A case study of Boustead Plantation Berhad You may also like The potential of biomass waste in Malaysian palm oil industry: A case study of Boustead Plantation Berhad, (2021). https://doi.org/10.1088/1757-899X/1192/1/012028

[3] N.H.S. Jafri, D.N. Jimat, N.F.M. Azmin, S. Sulaiman, Y.A. Nor, The potential of biomass waste in Malaysian palm oil industry: A case study of Boustead Plantation Berhad, IOP Conf Ser Mater Sci Eng 1192 (2021) 012028. https://doi.org/10.1088/1757-899x/1192/1/012028

[4] M. Jorns, D. Pappas, A review of fluorescent carbon dots, their synthesis, physical and chemical characteristics, and applications, Nanomaterials 11 (2021). https://doi.org/10.3390/nano11061448

[5] L. Cui, X. Ren, M. Sun, H. Liu, L. Xia, Carbon dots: Synthesis, properties and applications, Nanomaterials 11 (2021). https://doi.org/10.3390/nano11123419

[6] K.W. Chu, S.L. Lee, C.J. Chang, L. Liu, Recent progress of carbon dot precursors and photocatalysis applications, Polymers (Basel) 11 (2019). https://doi.org/10.3390/polym11040689

[7] S. Punyasamudram, R. Prasad Puthalapattu, S. Gumma, N.Y. Sreedhar, P. V Nagendra Kumar, Synthesis and characterization of high fluorescent engineered carbon dots, (n.d.). https://doi.org/10.1016/j.matpr.2023.02.412

[8] M. Wu, J. Zhan, B. Geng, P. He, K. Wu, L. Wang, G. Xu, Z. Li, L. Yin, D. Pan, Scalable synthesis of organic-soluble carbon quantum dots: Superior optical properties in solvents, solids, and LEDs, Nanoscale 9 (2017) 13195–13202. https://doi.org/10.1039/c7nr04718e

[9] H.A.S. Tohamy, M. El-Sakhawy, S. Kamel, Eco-friendly Synthesis of Carbon Quantum Dots as an Effective Adsorbent, J Fluoresc 33 (2023) 423–435. https://doi.org/10.1007/s10895-022-03085-z

[10] J. Gan, L. Chen, Z. Chen, J. Zhang, W. Yu, C. Huang, Y. Wu, K. Zhang, Lignocellulosic Biomass-Based Carbon Dots: Synthesis Processes, Properties, and Applications, Small (2023). https://doi.org/10.1002/smll.202304066

[11] J. Zhang, H. Wang, Y. Xiao, J. Tang, C. Liang, F. Li, H. Dong, W. Xu, A Simple Approach for Synthesizing of Fluorescent Carbon Quantum Dots from Tofu Wastewater, Nanoscale Res Lett 12 (2017). https://doi.org/10.1186/S11671-017-2369-1

[12] S. Whitaker, C. De Silva, A.A.M. Noor, M.M.A. Karim, B. Gunasekaran, S.A. Gani, M.A. Cabrera, S.A. Ahmad, The green synthesis and characterisation of silver nanoparticles from Serratia spp Síntesis verde y caracterización de nanopartículas de plata de Serratia spp, Rev Mex Ing Quim 19 (2020) 1327–1339. https://doi.org/10.24275/rmiq/Bio1059

[13] O.G. Rojas-Valencia, M. Regules-Carrasco, J. Hernández-Fuentes, C.M.R.S. Germán, M. Estrada-Flores, E. Villagarcía-Chávez, Synthesis of blue emissive carbon quantum dots from Hibiscus Sabdariffa flower: Surface functionalization analysis by FT-IR spectroscopy, Materialia (Oxf) 19 (2021) 101182. https://doi.org/10.1016/J.MTLA.2021.101182

[14] A. Sharma, J. Das, Small molecules derived carbon dots: synthesis and applications in sensing, catalysis, imaging, and biomedicine, Journal of Nanobiotechnology 2019 17:1 17 (2019) 1–24. https://doi.org/10.1186/S12951-019-0525-8

[15] S. Chahal, J.R. Macairan, N. Yousefi, N. Tufenkji, R. Naccache, Green synthesis of carbon dots and their applications, RSC Adv 11 (2021) 25354–25363. https://doi.org/10.1039/d1ra04718c

[16] K.W. Chu, S.L. Lee, C.J. Chang, L. Liu, Recent progress of carbon dot precursors and photocatalysis applications, Polymers (Basel) 11 (2019). https://doi.org/10.3390/polym11040689

[17] A.M. El-Shafey, Carbon dots: Discovery, structure, fluorescent properties, and applications, Green Processing and Synthesis 10 (2021) 134–156. https://doi.org/10.1515/gps-2021-0006

[18] X. Lin, M. Xiong, J. Zhang, C. He, X. Ma, H. Zhang, Y. Kuang, M. Yang, Q. Huang, Carbon dots based on natural resources: Synthesis and applications in sensors, Microchemical Journal 160 (2021) 105604. https://doi.org/10.1016/J.MICROC.2020.105604

[19] Y.N. Monday, J. Abdullah, N.A. Yusof, S.A. Rashid, R.H. Shueb, Facile Hydrothermal and Solvothermal Synthesis and Characterization of Nitrogen-Doped Carbon Dots from Palm Kernel Shell Precursor, Applied Sciences 2021, Vol. 11, Page 1630 11 (2021) 1630. https://doi.org/10.3390/APP11041630

[20] A. Su, D. Wang, X. Shu, Q. Zhong, Y. Chen, J. Liu, Y. Wang, Synthesis of Fluorescent Carbon Quantum Dots from Dried Lemon Peel for Determination of Carmine in Drinks, Chem Res Chin Univ 34 (2018) 164–168. https://doi.org/10.1007/S40242-018-7286-Z/METRICS

[21] H. Zhou, Y. Ren, Z. Li, W. He, Z. Li, Selective Detection of Fe3+ by Nitrogen–Sulfur-Doped Carbon Dots Using Thiourea and Citric Acid, Coatings 12 (2022). https://doi.org/10.3390/coatings12081042

[22] N.F. Sayakulu, S. Soloi, The Effect of Sodium Hydroxide (NaOH) Concentration on Oil Palm Empty Fruit Bunch (OPEFB) Cellulose Yield, J Phys Conf Ser 2314 (2022). https://doi.org/10.1088/1742-6596/2314/1/012017

[23] B.W. Chieng, S.H. Lee, N.A. Ibrahim, Y.Y. Then, Y.Y. Loo, Isolation and characterization of cellulose nanocrystals from oil palm mesocarp fiber, Polymers (Basel) 9 (2017) 1–11. https://doi.org/10.3390/polym9080355

[24] H. Zhou, Y. Ren, Z. Li, W. He, Z. Li, Selective Detection of Fe3+ by Nitrogen–Sulfur-Doped Carbon Dots Using Thiourea and Citric Acid, Coatings 12 (2022). https://doi.org/10.3390/coatings12081042

[25] Z. Fu, M. Yao, X. Niu, F. Cui, Facile synthesis of highly luminescent co-doped carbon nanodots for rapid, sensitive, and label-free detection of Hg2+, Sens Actuators B Chem 226 (2016) 486–494. https://doi.org/10.1016/j.snb.2015.12.030

[26] J. Xiao, Y. Wang, T.C. Zhang, S. Yuan, N,S-containing polycondensate-derived porous carbon materials for superior CO2 adsorption and supercapacitor, Appl Surf Sci 562 (2021) 150128. https://doi.org/10.1016/j.apsusc.2021.150128

[27] R.A. Velapoldi, H.H. Tønnesen, Corrected Emission Spectra and Quantum Yields for a Series of Fluorescent Compounds in the Visible Spectral Region, J Fluoresc 14 (2004)

[28] Z. Li, Q. Wang, Z. Zhou, S. Zhao, S. Zhong, L. Xu, Y. Gao, X. Cui, Green synthesis of carbon quantum dots from corn stalk shell by hydrothermal approach in near-critical water and applications in detecting and bioimaging, Microchemical Journal 166 (2021). https://doi.org/10.1016/j.microc.2021.106250

[29] M.A.A. Sisak, R. Daik, S. Ramli, Characterization of cellulose extracted from oil palm empty fruit bunch, in: AIP Conf Proc, American Institute of Physics Inc., 2015. https://doi.org/10.1063/1.4931295

[30] Characteristics of Palm Kernel Shell and Palm Kernel Shell-Polymer Composites: A Review, (n.d.). https://doi.org/10.1088/1757-899X/1107/1/012090

[31] Y. Yee, N.A. Ibrahim, N. Zainuddin, B.W. Chieng, H. Ariffin, W.M. Zin, W. Yunus, Alkaline peroxide & composites, Bioresources 10 (2015) 1730–1746

[32] N.S. Izzaha, A.N.A. Yahaya, N.Z.M. Zuhudi, N.A. Khalil, M. Zulkifli, Extraction of cellulose from oil palm empty fruit bunch using eco-friendly solvents for preparation of transparent cellulose thin film, IOP Conf Ser Mater Sci Eng 1195 (2021) 012062. https://doi.org/10.1088/1757-899x/1195/1/012062

[33] J.F.Y. Fong, Y.H. Ng, S.M. Ng, Carbon dots as a new class of light emitters for biomedical diagnostics and therapeutic applications, Fullerenes, Graphenes and Nanotubes: A Pharmaceutical Approach (2018) 227–295. https://doi.org/10.1016/B978-0-12-813691-1.00007-5

[34] Y. Lou, X. Hao, L. Liao, K. Zhang, S. Chen, Z. Li, J. Ou, A. Qin, Z. Li, Recent advances of biomass carbon dots on syntheses, characterization, luminescence mechanism, and sensing applications, Nano Select 2 (2021) 1117–1145. https://doi.org/10.1002/nano.202000232

[35] P. Wu, W. Li, Q. Wu, Y. Liu, S. Liu, Hydrothermal synthesis of nitrogen-doped carbon quantum dots from microcrystalline cellulose for the detection of Fe3+ ions in an acidic environment, RSC Adv 7 (2017) 44144–44153. https://doi.org/10.1039/c7ra08400e

[36] Y. Wang, L. Yan, G. Ji, C. Wang, H. Gu, Q. Luo, Q. Chen, L. Chen, Y. Yang, C.Q. Ma, X. Liu, Synthesis of N,S-Doped Carbon Quantum Dots for Use in Organic Solar Cells as the ZnO Modifier to Eliminate the Light-Soaking Effect, ACS Appl Mater Interfaces 11 (2019) 2243–2253. https://doi.org/10.1021/acsami.8b17128

[37] W.L. Ang, C.A.L. Boon Mee, N.S. Sambudi, A.W. Mohammad, C.P. Leo, E. Mahmoudi, M. Ba-Abbad, A. Benamor, Microwave-assisted conversion of palm kernel shell biomass waste to photoluminescent carbon dots, Sci Rep 10 (2020). https://doi.org/10.1038/s41598-020-78322-1

[38] S. Miao, K. Liang, J. Zhu, B. Yang, D. Zhao, B. Kong, Hetero-atom-doped carbon dots: Doping strategies, properties and applications, Nano Today 33 (2020). https://doi.org/10.1016/j.nantod.2020.100879

[39] I.N. Mohammad, C.M. Ongkudon, M. Misson, Physicochemical properties and lignin degradation of thermal-pretreated oil palm empty fruit bunch, Energies (Basel) 13 (2020). https://doi.org/10.3390/en13225966

[40] B.W. Chieng, S.H. Lee, N.A. Ibrahim, Y.Y. Then, Y.Y. Loo, Isolation and Characterization of Cellulose Nanocrystals from Oil Palm Mesocarp Fiber, Polymers 2017, Vol. 9, Page 355 9 (2017) 355. https://doi.org/10.3390/POLYM9080355

[41] M.M. Al-Rajabi, & Teow, Y. Haan, Green Extraction Method of Cellulose Fibers from Oil Palm Empty Fruit Bunches, (2022). https://doi.org/10.17576/jkukm-2022-34(5)-12

[42] G.T. Chala, Y.P. Lim, S.A. Sulaiman, C.L. Liew, Thermogravimetric analysis of empty fruit bunch, in: MATEC Web of Conferences, EDP Sciences, 2018. https://doi.org/10.1051/matecconf/201822502002

[43] A.W. Noorshamsiana, J. Nor Faizah, H. Kamarudin, A.O. Nur Eliyanti, I. Fatiha, A.A. Astimar, Extraction and characterisation of cellulose from the residue of oil palm empty fruit

bunch-xylan extraction, J Oil Palm Res 32 (2020) 610–620. https://doi.org/10.21894/jopr.2020.0052

[44] S.R.A.M. Rasli, I. Ahmad, A.M. Lazim, A. Hamzah, Extraction and characterizations of cellulose from agriculture residue, Malaysian Journal of Analytical Sciences 21 (2017) 1065–1073. https://doi.org/10.17576/mjas-2017-2105-08

[45] B. Yimlamai, W. Choorit, Y. Chisti, P. Prasertsan, Cellulose from oil palm empty fruit bunch fiber and its conversion to carboxymethylcellulose, Journal of Chemical Technology and Biotechnology 96 (2021) 1656–1666. https://doi.org/10.1002/jctb.6689

[46] A.S. Kamarol Zaman, T.L. Tan, Y. A/P Chowmasundaram, N. Jamaludin, A.R. Sadrolhosseini, U. Rashid, S.A. Rashid, Properties and molecular structure of carbon quantum dots derived from empty fruit bunch biochar using a facile microwave-assisted method for the detection of Cu2+ ions, Opt Mater (Amst) 112 (2021) 110801. https://doi.org/10.1016/J.OPTMAT.2021.110801

[47] N. Jamaludin, T.L. Tan, A.S.K. Zaman, A.R. Sadrolhosseini, S.A. Rashid, Acid-Free Hydrothermal-Extraction and Molecular Structure of Carbon Quantum Dots Derived from Empty Fruit Bunch Biochar, Materials 2020, Vol. 13, Page 3356 13 (2020) 3356. https://doi.org/10.3390/MA13153356

[48] B. Dziejarski, J. Serafin, K. Andersson, R. Krzyżyńska, CO2 capture materials: a review of current trends and future challenges, Materials Today Sustainability 24 (2023) 100483. https://doi.org/10.1016/J.MTSUST.2023.100483

[49] S. Mahajan, M. Lahtinen, Recent progress in metal-organic frameworks (MOFs) for CO2 capture at different pressures, J Environ Chem Eng 10 (2022) 108930. https://doi.org/10.1016/J.JECE.2022.108930

[50] J. Saleem, U. Bin Shahid, M. Hijab, H. Mackey, G. McKay, Production and applications of activated carbons as adsorbents from olive stones, Biomass Conversion and Biorefinery 2019 9:4 9 (2019) 775–802. https://doi.org/10.1007/S13399-019-00473-7

[51] S. Acevedo, L. Giraldo, J.C. Moreno-Piraján, Adsorption of CO2 on Activated Carbons Prepared by Chemical Activation with Cupric Nitrate, ACS Omega 5 (2020) 10423–10432. https://doi.org/10.1021/ACSOMEGA.0C00342/ASSET/IMAGES/LARGE/AO0C00342_0005.JPEG

[52] J. May, L. Thoe, N. Surugau, H. Lye, H. Chong, Application of Oil Palm Empty Fruit Bunch as Adsorbent: A Review, 2019. http://tost.unise.org/

[53] S. Giraldo, A.P. Ramirez, E. Flórez, N.Y. Acelas, Adsorbent materials obtained from palm waste and its potential use for contaminants removal from aqueous solutions, in: J Phys Conf Ser, Institute of Physics Publishing, 2019. https://doi.org/10.1088/1742-6596/1386/1/012036

[54] Current Research Paper Carbon dioxide capture by adsorption using sulfur doped carbon material, (n.d.)

[55] R. Busquets, S. Reljic, M. Martinez-Escandell, J. Silvestre-Albero, Effect of Porosity and Surface Chemistry on CO2 and CH4 Adsorption in S-Doped and S-/O-co-Doped Porous Carbons, C 2022, Vol. 8, Page 41 8 (2022) 41. https://doi.org/10.3390/C8030041

[56] H. Khurshid, M.R.U. Mustafa, M.H. Isa, Modified Activated Carbon Synthesized from Oil Palm Leaves Waste as a Novel Green Adsorbent for Chemical Oxygen Demand in Produced Water, Sustainability (Switzerland) 14 (2022). https://doi.org/10.3390/su14041986

Dynamic simulation investigation of e-methanol production from hydrogen and Co_2 capture using Zif-67 and ZIF-8 catalysts

Zulkifli Merican Aljunid Merican[1,2*], Judy Ng Min Hui[1,2], Muhammad Khairi Zakuwan Kamaruddin[1,2], Fatin Nadhrah Shamsul Azhar[1,2], Nur Arifah Mohd Zul Sharinan[1,2], Nik Nur Syamimi Nik Danya[1,2], Ali Khatib Juma[1,2] and Khairulazhar Jumbri [1,2]

[1]Department of Fundamental and Applied Sciences, Universiti Teknologi PETRONAS, 32610 Seri Iskandar, Perak Darul Ridzuan, Malaysia

[2]Institute of Contaminant Management, Universiti Teknologi PETRONAS, 32610 Seri Iskandar, Perak Darul Ridzuan, Malaysia

* zulkifli.aljunid@utp.edu.my

Keywords: Metal-Organic Frameworks, Photocatalysis, Molecular Simulation, CO_2/H_2 Capture, E-Methanol Production

Abstract. Malaysia and global stakeholders are committed to addressing climate change and the prevailing fuel crisis driven by non-renewable fossils. One promising solution is the photocatalytic reduction of methanol production (e-methanol). However, there are challenges with the utilization of ZIF-67 catalyst, and existing literature lacks sufficient data on its interaction with CO_2 and H2 during adsorption and e-methanol production. To address this gap, we conducted DFT simulations using BIOVIA Materials Studio software to study ZIF-67 and compare it with the extensively studied ZIF-8. The Sorption module revealed Type I adsorption isotherms for both ZIF-67 and ZIF-8 in CO_2 and e-methanol capture. Additionally, hydrogen molecules exhibited slow adsorption rates on both ZIFs. Notably, CO_2 showed stronger chemisorption interaction with ZIF-67 at 298 K, indicating its potential for CO_2 capture. Furthermore, the two-step reactions for e-methanol production on ZIF-67 demonstrated lower energy (-363.65 kJ/mol) compared to ZIF-8 (-312.17 kJ/mol), attributed to their different band gap energies (0.009 eV and 4.313 eV, respectively). These findings suggest ZIF-67's potential as a thermodynamically favored medium for e-methanol production.

Introduction

The world's largest energy source, nonetheless, still the heavily dependent fossil fuels. Fossil fuels being the non-renewable energy, predicted to be in scarcity by 2060 with the current rate of usage. Besides, United Nations and research reported that the fossil fuels, such as coal, oil, and gas, are significant contributors for more than 75% of greenhouse gas emissions and nearly 90% of CO_2 emissions. This phenomenon has been drawing the attention of many countries that signed the Net Zero GHG emission pledges, to tackle the alarming issues of GHGs and climate change that resulting in unparalleled warming and disturbances to ecosystems, in action with the SDG 13 (Climate Action) and SDG 7 (Affordable and Clean Energy) [1]. Furthermore, e-methanol has emerged as a crucial factor in the worldwide movement towards a low-carbon economy to alleviate the effects of climate change and reduce dependency on fossil fuels. This transformation is enabled by its production via electrochemical processes, which provide a low-carbon replacement for a wide range of industrial application and energy needs. Malaysia's energy policies, such as the Twelfth Malaysia Plan and the National Energy Policy 2022–2040, which emphasize the importance of sustainable and innovative energy solutions, are closely linked to the production of e-methanol [2].

Current alternatives include biomass and carbon capture, which aim to minimize reliance on natural gas yet encounter challenges. The study of the Zeolitic Imidazolate Framework (ZIF), a unique class of Metallic Organic Framework (MOF) that bears resemblance to zeolites in terms of porosity, mechanical strength, chemical and thermal stability, and flexibility in manipulating the structure during synthesis, is focused on CO_2 capture and photocatalysis for the production of e-methanol. The identified ZIF catalyst in this study - ZIF-67 provides selective CO_2 adsorption, effective solar-powered photocatalysis, and increased catalytic activity as compared to other MOF and ZIF, making ZIF-67 emerges as a possible solution to address the environmental impacts of catalytic methanol production efficiency and reducing reliance on fossil fuels by lower carbon footprint [3].

In this study, ZIF-67 and ZIF-8 will be compared to demonstrate that ZIF-67 is more efficient in CO_2 capture and e-methanol production. ZIF-8, chosen for its structural and application similarities to ZIF-67, has a framework of Zn(II) metal centres linked to nitrogen atoms within five-membered imidazolate rings, forming a sodalite topology [4]. The metal-substituted variant of ZIF-8, known as ZIF-67, replaces zinc with cobalt, resulting in stiffer Co-N bonds compared to Zn-N bonds [5]. While there is extensive research on various applications of ZIF-67, only a few studies have explored its potential in CO_2 capture. Due to these reasons, researchers are turning to MD, making it an emerging technology in the sector. MD is a computational technique for modelling and studying the behavior of molecular systems over time at the atomic or molecular level [6]. In this context, MD will be used to study the dynamic interactions between molecules, catalysts, and reaction intermediates throughout the complex process. Moreover, current research does not provide a thorough knowledge of the dynamic interactions inside the ZIF-67 catalyst during CO_2 capture and e-methanol production. Filling these gaps is crucial to increasing the efficiency and sustainability of e-methanol production. Precise understanding of molecular interactions can lead to improved catalyst designs and process conditions, making e-methanol a more viable option [7]. Therefore, the goal of this research is to use dynamic molecular simulation to bridge existing knowledge gaps in ZIF-67 catalyzed CO_2 capture and e-methanol production. It aims to understand molecular-level interactions and provide optimization solutions to improve catalyst performance.

Materials and Methods

The project focused on the computational studies, on quantum basis (density functional theory), to select suitable geometry and environment for the surface adsorption of substrates – CO_2 and H_2 onto the chosen heterogenous catalyst – ZIF-67 and compared with ZIF-8 catalyst. The adsorption isotherms and fixed loading at different temperatures using Sorption module in Materials Studio 17.1.0.48 version. Meanwhile, the reaction kinetics of photocatalytic reduction of CO_2 to produce e-methanol were calculated by DMol^3 module of Materials Studio 17.1.0.48 version. The software was operated in High Performance Computer (Intel(R) Xeon(R) Gold 5218R CPU @ 2.10GHz 2.10 GHz with 32GB RAM) at Universiti Teknologi PETRONAS.

MOF Unit Cell. The unit cell (1 x 1 x 1) of the selected MOF, including ZIF-67 and ZIF-8, were obtained from Cambridge Cystallographic Data Centre (CCDC), downloaded as. mol^2 file. The structures were modified to primitive space group using Build Crystal function in Materials Studio, with the crystal parameters of a=14.736 Å, b= 14.780 Å, c= 14.764 Å; α=109.61°, β=109.47°, γ=109.36°. The crystals were triclinic structure with unequal length of sides and angles.

Sorption of CO_2, H_2, and CH_3OH. Density functional theory (DFT) calculations integrated within Sorption package module of BIOVIA Materials Studio Accelrys Inc. (17.1.0.48 version) was used to identify the adsorption isotherms of sorbates involving in the photocatalytic reaction

Materials Research Proceedings 53 (2025) 273-278 https://doi.org/10.21741/9781644903575-25

for e-methanol production. Throughout the adsorption simulations, the unit cells ZIF-67 and ZIF-8 was assumed to be remained at constant size of crystal lattice (a=14.736 Å, b= 14.780 Å, c= 14.764 Å), and all metals and respective coordinating ligands were static at its original position without taking part in mass transport of sorbates. To study the adsorption isotherm and distribution densities of each sorbate – CO2, H2, CO2 and H2 pair, and CH3OH, major reactants and products of e- methanol production, the metropolis method with medium quality was used, simulated in the pressure range of 10 kPa to 1000 kPa at 273 K, 298 K and 323 K. The Universal forcefield was utilized, applying current charges of the atoms, and using the electrostatic Ewald and group summation method [8].

Band Gap Energy of ZIF-67 and ZIF-8. The band gap energy between valence band and conduction band of the ZIF-67's primitive cell, was analysed with the DMol^3 module in BIOVIA Materials Studio. The exchange-correlation of GGA-PW91 functional were chosen, for energy study of the band structure, and density of states with PDOS. The Brillouin zone path was identified and cleaved at g1, g2 and g3 planes [9].

Results and Discussion

Adsorption Isotherms on ZIF-67 Catalyst.

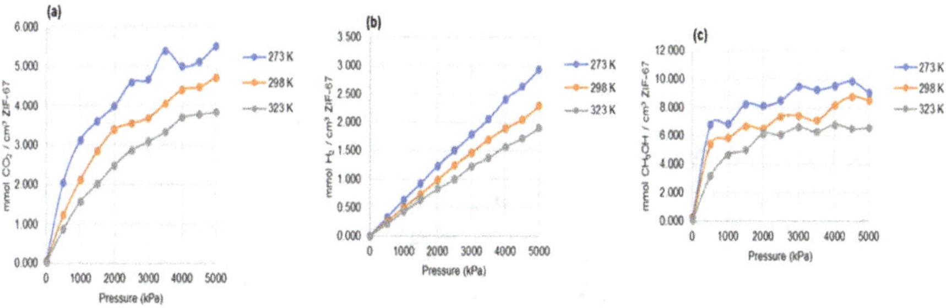

Fig. 1. Adsorption isotherm results of sorbates at different temperatures on ZIF-67, (a) CO2, (b) H2, (c) CH3OH.

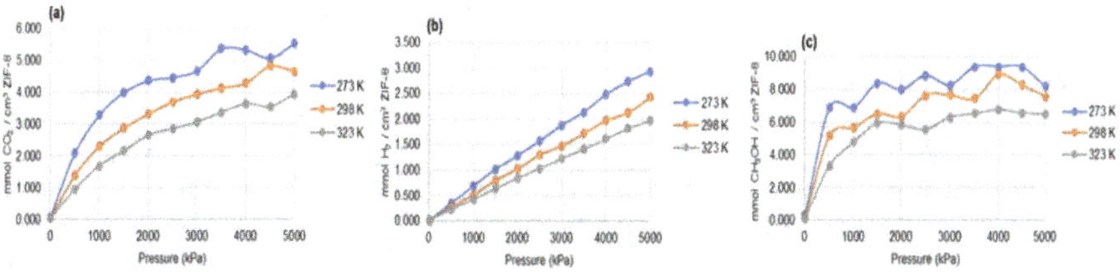

Fig. 2. Adsorption isotherm results of sorbates at different temperatures on ZIF-8, (a) CO2, (b) H2, (c) CH3OH.

Adsorption Energies of CO2/H2 Binary Mixture on ZIF-67 Catalyst. Based on the stoichiometric equation of reaction between CO2 and H2, every one mole of CO2 reacted with three moles of H2, therefore, this molar ratio that applied in fixed loading but at varied temperature was shown in Table 1.

Materials Research Proceedings 53 (2025) 273-278 https://doi.org/10.21741/9781644903575-25

Table 1. Adsorption energies on ZIF-67 catalyst

Temperature (K)	Isosteric Heat of Sorbates: Reactant (kcal/mol)		Overall Energy of System: Reactant (kcal/mol)	Isosteric Heat of Sorbates: Product (kcal/mol)	Overall Energy of System: Product (kcal/mol)
	CO2	H2		CH3OH	
273	3.167	1.489	163.822	6.449	136.830
298	3.215	1.503	163.923	5.701	103.237
323	3.196	1.551	163.984	5.443	86.768

Comparison: Adsorption energies of CO2/H2 Binary Mixture on ZIF-8 Catalyst. Based on the stoichiometric equation of reaction between CO2 and H2, every one mole of CO2 reacted with three moles of H2, therefore, this molar ratio that applied in fixed loading but at varied temperature was shown in Table 2.

Table 2. Adsorption energies on ZIF-8 catalyst

Temperature (K)	Isosteric Heat of Sorbates: Reactant (kcal/mol)		Overall Energy of System: Reactant (kcal/mol)	Isosteric Heat of Sorbates: Product (kcal/mol)	Overall Energy of System: Product (kcal/mol)
	CO2	H2		CH3OH	
273	3.220	1.450	163.886	6.190	124.373
298	3.173	1.476	164.048	5.504	97.319
323	3.189	1.501	164.160	5.232	80.694

Band Gap Energy. From the band structure graphs plots of ZIF-67 and ZIF-8 above, the identified band gap energies are 0.009 eV and 4.313 eV, respectively. In addition, as the hydrogenation of CO_2 in ZIF is a photocatalysis reaction, the band gap energy in relation to the charge transfer between metal and ligands, has decisive effect to the reaction enthalpies, whereby ZIF-67 with smaller band gap energy (0.009 eV) than ZIF-8 (4.313 eV) for electron discharge to conduction band, provide a lower reaction energy pathway selective for CO_2 hydrogenation conversion to e-methanol [10].

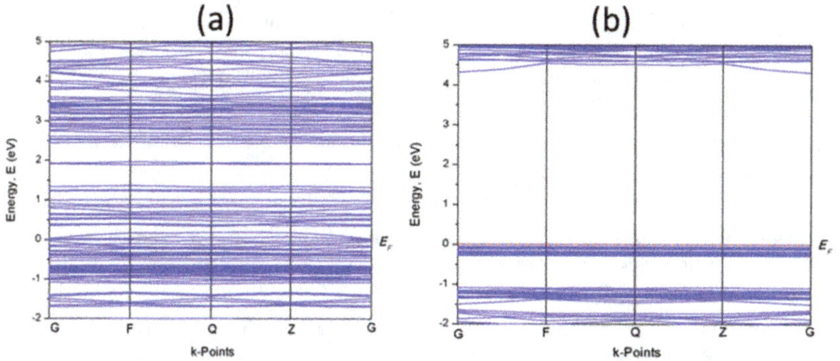

Fig. 8. Band structure graphs of catalysts, (a) ZIF-67, (b) ZIF-8

Conclusion

This study outlines a dynamic simulation framework for e-methanol production using ZIF-67 catalyst, showing better performance than ZIF-8. ZIF-67 exhibits superior methanol production at lower temperatures, with slightly lower CO_2 adsorption. Its smaller band gap leads to stronger methanol adsorption, making it a more suitable catalyst.

Acknowledgement

The authors would like to acknowledge the funding support from Universiti Teknologi PETRONAS and the Yayasan UTP (Np. FRG/1/2021/015LCC0–376). We would also like to express our gratitude to the UTP CO_2 Research Centre (CO_2RES) and the Institute of Contaminant Management, for their invaluable support throughout this study.

References

[1] A. I. Osman *et al.*, "Cost, environmental impact, and resilience of renewable energy under a changing climate: a review," *Environ. Chem. Lett.*, vol. 21, no. 2, pp. 741–764, 2023. https://doi.org/10.1007/s10311- 022-01532-8

[2] Z. Li, L. W. Leong, M. M. N Aldoseri, I. Muda, A. Abu-Rumman, and A. Al Shraah, "Examining the role of sustainability and natural resources management in improving environmental quality: Evidence from Asian countries," *Resour. Policy*, vol. 80, 2023. https://doi.org/10.1016/j.resourpol.2022.103136

[3] M. Wang *et al.*, "Metal-organic frameworks (ZIF-67) as efficient cocatalysts for photocatalytic reduction of CO2: The role of the morphology effect," *J. Mater. Chem. A*, vol. 6, no. 11, pp. 4768– 4775, 2018. https://doi.org/10.1039/c8ta00154e

[4] M. Barooah and B. Mandal, "Synthesis, characterization and CO2 separation performance of novel PVA/PG/ZIF-8 mixed matrix membrane," *J. Memb. Sci.*, vol. 572, pp. 198–209, 2019. https://doi.org/10.1016/j.memsci.2018.11.001

[5] S. Meshkat, S. Kaliaguine, and D. Rodrigue, "Comparison between ZIF-67 and ZIF-8 in Pebax® MH-1657 mixed matrix membranes for CO2 separation," *Sep. Purif. Technol.*, vol. 235, 2020. https://doi.org/10.1016/j.seppur.2019.116150

[6] Q. Mao, M. Feng, X. Z. Jiang, Y. Ren, K. H. Luo, and A. C. T. van Duin, "Classical and reactive molecular dynamics: Principles and applications in combustion and energy systems," *Prog. Energy Combust. Sci.*, vol. 97, 2023. https://doi.org/10.1016/j.pecs.2023.101084

[7] H. Park *et al.*, "Zeolitic imidazolate framework-67 (ZIF-67) rhombic dodecahedrons as full-spectrum light harvesting photocatalyst for environmental remediation," *Solid State Sci.*, vol. 62, pp. 82–89, 2016. https://doi.org/10.1016/j.solidstatesciences.2016.10.018

[8] X. Cheng, Y. Liao, Z. Lei, J. Li, X. Fan, and X. Xiao, "Multi-scale design of MOF-based membrane separation for CO2/CH4 mixture via integration of molecular simulation, machine learning and process modeling and simulation," *J. Memb. Sci.*, vol. 672, 2023. https://doi.org/10.1016/j.memsci.2023.121430

[9] A. Baghban, S. Habibzadeh, and F. Zokaee Ashtiani, "Bandgaps of noble and transition metal/ZIF-8 electro/catalysts: A computational study," *RSC Adv.*, vol. 10, no. 39, pp. 22929–22938, 2020. https://doi.org/10.1039/d0ra02943b

[10] L. E. Mphuthi, E. Erasmus, and E. H. G. Langner, "Metal Exchange of ZIF-8 and ZIF-67 Nanoparticles with Fe(II) for Enhanced Photocatalytic Performance," *ACS Omega*, vol. 6, no. 47, pp. 31632–31645, 2021. https://doi.org/10.1021/acsomega.1c04142

Optimization of ethylene oxide production via hydrogen peroxide process route

ABID Salam Farooqi[1,2,a], CELESTINE Wong Yi Ni[1,2,b], RAIHAN Mahirah Binti Ramli[1,2,c*], AMANI Hussein Sharaf Addin[1,2,d], AHMAD Salam Farooqi[4,e], BAWADI Abdullah[1,3,f]

[1]Department of Chemical Engineering, Universiti Teknologi PETRONAS, Bandar Seri Iskandar, 32610 Seri Iskandar, Perak, Malaysia

[2]Centre of Innovative Nanostructures & Nanodevices, Universiti Teknologi PETRONAS, Bandar Seri Iskandar, 32610 Seri Iskandar, Perak, Malaysia

[3]CO2 Research Center (CO2RES), Universiti Teknologi PETRONAS, 32610 Seri Iskandar, Malaysia

[4]Interdisciplinary Research Center for Hydrogen Technologies and Carbon Management (IRC HTCM), King Fahd University of Petroleum and Minerals, Dhahran, 31261, Saudi Arabia

[a]abid_20001276@utp.edu.my, [b]celestine0214@gmail.com, [c]raihan.ramli@utp.edu.my, [d]amani_22011516@utp.edu.my, [e]ahmad.farooqi@kfupm.edu.sa, [f]bawadi_abdullah@utp.edu.my

Keywords: Catalysis, Carbon Dioxide, Aspen HYSYS, Optimization

Abstract. This study investigates ethylene oxide (EO) production using a Titanium-Silicate (TS) catalyst in methanol, addressing the issue of excess hydrogen peroxide (H_2O_2). The objective of this study includes developing and validating a process model, evaluating key parameters, and proposing optimized conditions. Aspen HYSYS is used for modeling, and response surface (RSM) methods are applied for parameter analysis. The optimal conditions identified are a reactor temperature of 40°C, pressure of 8.5 bar, and H_2O_2 concentration of 2.326 wt.%. Results show high alignment between predicted and modeled EO selectivity (0.06%). Future work could explore recycling streams and improve separation systems to enhance product purity and minimize waste.

Introduction

The production of ethylene glycols (EG), ethoxylates, glycol ethers (GE), polyethylene terephthalate (PET), and ethanolamine are just a few of the chemicals that ethylene oxide (EO) is used to make. As a result, EO has garnered a lot of interest [1]. These compounds are used in many industries, such as detergents, anti-freeze, plastics, textiles, medicines, disinfection, and pest management. They are also used in the sterilization of surgical equipment [2]. Due to several factors, including the growing requirement for heat-sensitive medical equipment and the increased demand for anti-freeze solutions, the market for EO is experiencing substantial revenue growth [3]. The growing use of PET bottles and containers is also expected to drive further market growth throughout the projected timeframe. Also, it is anticipated that the spike will aid the market revenue growth trajectory in EO research activities [4].

Since 1931, the production of EO has relied on the reaction between ethylene and oxygen (O_2), facilitated by a heterogeneous silver-based catalyst, a method still widely employed in the industry today [5]. However, recent concerns have emerged regarding its substantial environmental impact. For instance, Zakkour [6] stated that producing 10.2 tonnes of EO results in a staggering 3.4 million tonnes of carbon dioxide (CO_2) gas emission, presenting a significant challenge to global sustainability goals to achieve net-zero carbon emissions. In response to these challenges, a team of researchers at the Center for Environmentally Beneficial Catalysis (CEBC) has developed an alternative, CO_2-free method for EO production [7]. This process utilizes hydrogen peroxide (H_2O_2) as the oxidant and methyl trioxorhenium (CH_3ReO_3, abbreviated as MTO) as a homogeneous catalyst [8]. However, the MTO catalyst is prone to deactivation due to water accumulation, and the high cost of rhenium (Re) metal used in its preparation, priced at $1995/kg [9], coupled with its unstable supply, poses additional challenges. Introducing a titanium-silicate catalyst instead of the MTO catalyst has been proposed as an alternate solution to these problems. However, to improve EO selectivity, the experiment uses extra H_2O_2, which results in unreacted H_2O_2 at the reactor outlet [10]. Despite this progress, a thorough literature on ethylene epoxidation employing H_2O_2 on the titanium silicate (TS-1) catalyst is still lacking [11], [12].

Consequently, this research aims to delve into the experiment by Matias Alvear et al. [11] by utilizing Aspen HYSYS simulation software. The primary objective is replicating their experimental setup within the software and generating simulated results. These results will then be rigorously validated by comparing them with the empirical data obtained from the original experiment. This validation process ensures that the simulation accurately reflects real-world conditions and outcomes. Following this validation, the next step will involve optimizing the operating conditions. This optimization will focus on refining the parameters and settings used in the experiment to enhance efficiency, performance, or other desired metrics within the experimental range. By doing so, the study aims to provide valuable insights and improvements based on the validated simulation data.

Methodology

Development of process model in Aspen HYSYS. To design the process model for EO production, Aspen HYSYS was used as the simulation software. This choice was made possible by Aspen HYSYS wide range of capabilities in process design, performance modeling, and steady-state and dynamic system simulation. The experimental configuration suggested was then used to build the process model for the manufacture of ethylene oxide. After developing the model which was built from published data available in the literature shown in Table 1, validation processes were carried out by contrasting the selectivity of EO found in published works [11].

Table 1: Experimental Optimal Operating Conditions for reaction of Ethylene with Hydrogen Peroxide on a Titanium-Silicate Catalyst in Methanol

Parameter	Optimal Operating Condition
Temperature (°C)	45
Pressure (bar)	4.5
Concentration of H_2O_2(wt.%)	2
Ethylene Oxide Selectivity (%)	90

After the process model was developed and validated, the process model was further modified by referring to the process model developed by Madhav Ghanta [8] using Aspen HYSYS. The process model was developed and confirmed by comparing the simulated EO selectivity with the 90% experimental selectivity reported in the literature [11]. Equation 1 below displays the formula used to calculate the percentage deviation. The difference between the literature value and the simulated selectivity had to be less than 5% to meet the validation criterion. If the variation was greater than 5%, the process model must be adjusted until the ethylene oxide selectivity deviation was within an acceptable range.

Materials Research Proceedings 53 (2025) 279-286 https://doi.org/10.21741/9781644903575-27

$$\text{Percentage deviation}=\left(\frac{\text{EO selectivity from process model-EO selectivity from literature review}}{\text{EO selectivity from literature review}}\right)\text{x100}$$

Optimization of process parameters using response surface method (RSM). Response Surface Methodology (RSM) in the design expert (DE) software was used to calculate the ideal temperature, pressure, and concentration values. Parameter ranges were first evaluated and then entered into the program to create a design of the experiment (DOE) table. The developed process model was then used to simulate EO selectivity for each combination of parameters by entering different combinations. The DE program examined the DOE table once filled out with EO selectivity values to provide a fit summary, ANOVA tables, model equations, and optimize EO selectivity. The EO selectivity computed from the model equations was compared with simulated values to verify that the model equations were accurate. After that, the process model was used to validate the optimization outcomes by comparing them to benchmarks. 90% was established as the standard for EO selectivity. Afterward, the process model was inserted with the optimized process parameters for validation. Equation 2 was used to determine percentage deviations of EO selectivity from the RSM optimization and the process model simulation. The findings were considered valid when the error between RSM optimization and process model simulation was less than 5%.

$$\text{Percentage deviation of EO selectivity}=\left(\frac{\text{EO selectivity from process model}-\text{EO selectivity from RSM}}{\text{EO selectivity from process model}}\right)\text{x100}$$

Result & Discussion

EO selectivity from process model

Validation of the process model. Based on the experimental setup outlined by Matias Alvear [11] for EO production using a TS-1 catalyst from the reaction between ethylene and H_2O_2, the setup was replicated in Aspen HYSYS, as illustrated in Figure 1. The system comprised three inlet streams: nitrogen (N_2), ethylene, and H_2O_2 + methanol. The operational pressure of these streams was raised from 1 bar to 3 bar by directing them into a pump (P-100) or compressors (K-100 & K-101). After being combined in a mixer (MIX-100), the streams were heated in a heater (E-100) to the operating temperature of 30°C. After heating, the mixture was added to a plug flow reactor (PFR-100), where, under the assumption that no side reactions took place, ethylene and H_2O_2 interacted to create EO. A reactor pressure of 3 bar, a temperature of 30°C, and a concentration of 1 wt.% of H_2O_2 were used to validate the process model. The reactor's ethylene flow rate was 1.32×10^{-5} kmol/h, and the rate at which EO was produced was 1.16×10^{-6} kmol/h under these circumstances. The selectivity of EOs attained by the process model was 90%, as shown below.

$$\text{EO selectivity for process model}=\frac{8.28\,x\,10^{-4}(0.0014)}{1.32\,x\,10^{-5}-8.28\,x\,10^{-4}(0.014)}x100=90.79\%$$

$$\text{Percentage deviation of EO selectivity}=\frac{90.79-90}{90}x100=0.87\%$$

Materials Research Proceedings 53 (2025) 279-286 https://doi.org/10.21741/9781644903575-27

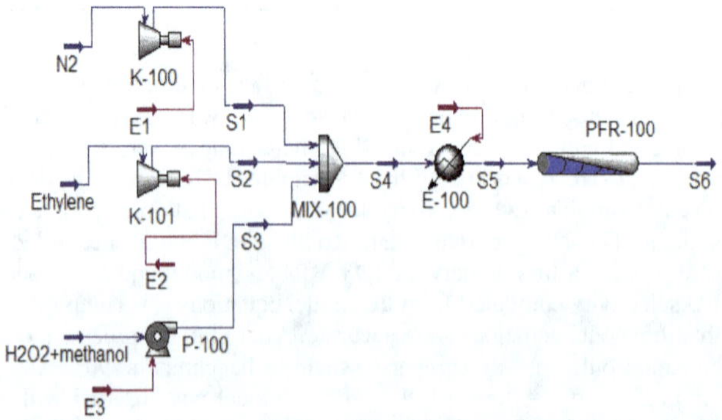

Figure 1: Process Model for EO Production based on Literature.

Modification. Using Aspen HYSYS, the process model which was first based on the literature was improved by adding components from the model created by Madhav Ghanta [8]. Figure 2 shows the system three inlet streams: N2, ethylene, and H2O2 + methanol. Pumps (P-100) or compressors (K-100 & K-101) were used to pressurize these streams from 1 bar to 6.5 bar. They were then combined in a mixer (MIX-100) and heated in a heater (E-100) to 46˚C. The resulting mixture was then put into a plug-flow reactor (PFR-100), where ethylene and H2O2 were combined to create EO.

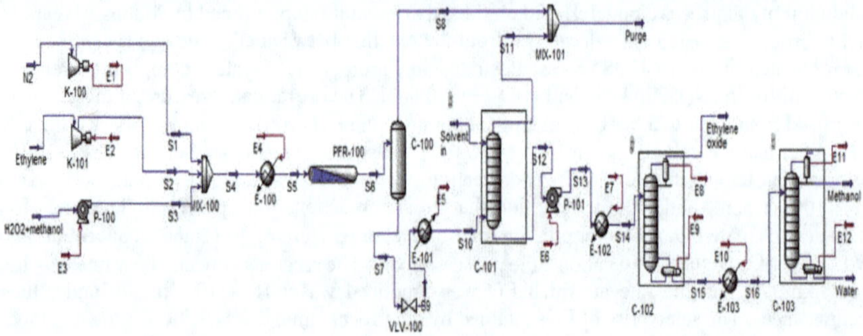

Figure 2: Process model for EO Production developed and modified based on the Process Model.

The material was first fed into a C-100 two-phase separator to extract the liquid mixture (S7) and the gas phase of ethylene and N_2. Venting was used to release the separated gases (S8). To prevent EO loss from vaporization, the liquid mixture (S7) was chilled and depressurized using a valve (VLV-100) and a cooler (E-101) to 40˚C and 4.5 bar. Subsequently, it was introduced into an absorber column (C-101), wherein a solvent absorbed an EO. The absorbed EO (S12) was transferred for purging after being separated using a mixer (MIX-101) from N_2, ethylene, and H_2O vapor (S11). To separate the EO mixture (S12) from H_2O and methanol, it was fed into an EO stripper (C-102) after being further compressed and heated to 50˚C and 3 bar using a pump (P- 101)

Materials Research Proceedings 53 (2025) 279-286 https://doi.org/10.21741/9781644903575-27

and a heater (E-102). The stripper (C-102) had 90% EO purity. In a methanol recovery column (C-103) operating at 87˚C and 10.13 bar, methanol and H_2O were separated. A reactor pressure of 6.5 bar, a temperature of 46˚C, and a concentration of 2 wt.% of H_2O_2 were used to validate the process model. In such conditions, the reactor's ethylene flow rate was 1.32×10^{-5}kmol/h, while the rate at which EO was produced was 1.13×10^{-5} kmol/h. The obtained EO selectivity was 85.64%, differing 4.84% from the 90% value reported in the literature.

$$\text{EO selectivity for process model} = \frac{1.256\,x\,10^{-5}(0.9)}{1.32\,x\,10^{-5}}x100 = 85.64\%$$

$$\text{Percentage deviation of EO selectivity} = \left|\frac{85.64 - 90}{90}\right|x100 = 0.87\%$$

Optimization of Process Parameters Using Response Surface Method (RSM). After entering the reactor temperature, pressure and H2O2 concentration into the RSM in the DE Software, Table 2 below displays the produced fit summary. The quadratic model, which produced an adjusted R^2 value of 0.9443, was the most accurate in explaining the effects of reactor temperature, reactor pressure, and H2O2 concentration on EO selectivity. This value is greater than the linear model corrected R^2 value of 0.8211. Thus, to maximize EO selectivity, the quadratic model was chosen.

Table 2: Fit Summary for EO Selectivity.

Source	Sequential p-value	Lack of Fit p-value	Adjusted R^2	Predicted R^2	
Linear	**< 0.0001**		**0.8211**	**0.7222**	**Suggested**
2FI	0.0914		0.8668	0.5840	
Quadratic	**0.0121**		**0.9443**	**0.6914**	**Suggested**
Cubic	< 0.0001		0.9993	0.8563	**Aliased**

The study produced an ANOVA table, which is shown in Table 3. The quadratic model has a p-value of less than 0.0001, indicating its importance ($p < 0.05$) in explaining the impact of process factors on EO selectivity. The ANOVA table also includes the specific p-values for every process parameter. The p-values demonstrate that the selectivity of EO is significantly influenced by the reactor temperature (A), pressure (B), and H2O2 concentration (C). For this reason, data reduction is not necessary. To summarise, using the quadratic model, the ANOVA results significantly support investigating EO selectivity with all three factors.

Table 3: ANOVA Table for EO Selectivity.

Source	Sum of Squares	df	Mean Square	F-value	p-value	
Block	0.0022	1	0.0022			
Model	16.61	9	1.85	34.91	< 0.0001	significant
A-Reactor Temperature	1.87	1	1.87	35.35	0.0002	
B-Reactor Pressure	12.02	1	12.02	227.33	< 0.0001	
C-Concentration of H2O2	0.6520	1	0.6520	12.34	0.0066	
AB	0.0190	1	0.0190	0.3597	0.5635	
AC	0.6786	1	0.6786	12.84	0.0059	
BC	0.3321	1	0.3321	6.28	0.0335	
A^2	0.1432	1	0.1432	2.71	0.1341	
B^2	0.5687	1	0.5687	10.76	0.0095	
C^2	0.5096	1	0.5096	9.64	0.0126	
Residual	0.4757	9	0.0529			
Lack of Fit	0.4757	5	0.0951			
Pure Error	0.0000	4	0.0000			
Cor Total	17.09	19				

Optimization of EO Selectivity. The coded equation (Equation 3) and the real equation (Equation 4) are the two types of model equations created. The real equation is used to predict EO selectivity, whereas the coded equation is used to compare the relative effects of variables. The coded equation showed that component AB had the least influence on EO selectivity, whereas factor B had a major impact. Though factor B had a beneficial effect, factors A, C, AB, AC, BC, A^2, B^2, and C^2 had a negative impact on EO selectivity. The process model was utilized to verify the equation and assess its accuracy. The two equations were modified by inserting three values for the reactor's temperature, pressure, and H_2O_2 content. The percentage differences between the process model and the EO selectivity derived from the real equation were computed. The findings in Table 4 demonstrated that the three trials percentage variations were all less than 5%, indicating that the equation prediction of EO selectivity was entirely accurate.

Coded Equation (Eq.3)

$$\text{EO selectivity} = 86.12 - 0.30.3699\text{A} + 0.938\text{B} - 0.2185\text{C} - 0.0487\text{AB} - 0.2912\text{AC} - 0.2037\text{BC} - 0.0997\text{A}^2 - 0.1987\text{B}^2 - 0.1881\text{C}^2$$

Actual Equation (Eq. 4)

$$\text{EO selectivity} = 44.0719 - 0.6328\text{A} + 3.7606\text{B} - 10.9314\text{C} - 0.0078\text{AB} - 0.1165\text{AC} - 0.3260\text{BC} - 0.0039\text{A}^2 - 0.1271\text{B}^2 - 0.7524\text{C}^2$$

Table 4: Calculation of percentage deviation between EO selectivity obtained from process model and actual equation.

Trial		1	2	3
Reactor Temperature (°C)		40	50	45
Reactor Pressure (bar)		6	6	7.25
Concentration of H_2O_2 (wt%)		2	3	2.5
EO Selectivity (%)	Model Equation	84.74	84.07	86.12
	Process Model	84.39	83.87	86.11
	Percentage deviation	0.41	0.23	0.01

After validating the EO selectivity equations, RSM in DE software optimized the reactor temperature, pressure, and H2O2 concentration. Table 5 displays the top 5 options recommended by RSM.

Table 5: Optimization solutions generated by RSM.

Sr. No	Reactor Temperature (°C)	Reactor Pressure (bar)	Concentration of H_2O_2 (wt.%)	EO Selectivity (%)	Desirability
1	**40**	**8.5**	**2.326**	**87.198**	**0.982**
2	40	8.5	2.326	87.198	0.982
3	40	8.5	2.337	87.198	0.982
4	40.008	8.5	2.324	87.198	0.982
5	40	8.5	2.342	87.198	0.982

Table 6 illustrates the 87.25% simulated EO selectivity obtained by substituting these variables into the process model created in Aspen HYSYS. The process model simulated EO selectivity and the EO selectivity predicted by the Response Surface Methodology (RSM) differed by a mere 0.06%. This minimal discrepancy, being significantly less than the acceptable 5% threshold, strongly suggests the reliability and accuracy of the model. Consequently, it can be inferred with confidence that the suggested optimized process parameters are valid. This close alignment between the simulated and predicted values underscores the robustness of the optimization approach used, ensuring that the process parameters are theoretically sound and practically applicable in real-world scenarios. The minor variance further reinforces the model credibility, highlighting its potential for consistent and reliable performance in the desired applications.

Table 6: Proposed optimized process parameters with EO selectivity and percentage deviation.

Parameters		Value
Reactor Temperature (°C)		40
Reactor Pressure (bar)		8.5
Concentration of H_2O_2 (wt.%)		2.326
EO Selectivity (%)	Predicted Value	87.198
	Simulated Value	87.25
	Percentage deviation	0.06

Conclusion

In summary, this project sought to increase the output of EO by changing the catalyst and refining the process variables. Initially, the process model derived from earlier research was verified with few modifications. Subsequently, the model underwent adjustments and was verified with excellent outcomes. Using RSM, the ideal circumstances were ascertained and found to be in line with the simulated values. About EO synthesis, these adjusted parameters offer insightful information for both industrial and future research uses.

Acknowledgment
The authors acknowledge financial support from Universiti Teknologi PETRONAS, Malaysia, under the Yayasan Universiti Teknologi PETRONAS (YUTP) grant number 015LC0-534.

References

[1] V. Russo, R. Tesser, E. Santacesaria, and M. Di Serio, "Kinetics of propene oxide production via hydrogen peroxide with TS-1," *Ind. Eng. Chem. Res.*, vol. 53, no. 15, pp. 6274–6287, Apr. 2014. https://doi.org/10.1021/IE404271K/ASSET/IMAGES/IE-2013-04271K_M048.GIF

[2] "Ethylene oxide - American Chemical Society." Accessed: Mar. 30, 2024. [Online]. Available: https://www.acs.org/molecule-of-the-week/archive/e/ethylene-oxide.html

[3] L. G. Pinaeva and A. S. Noskov, "Prospects for the Development of Ethylene Oxide Production Catalysts and Processes (Review)," *Pet. Chem.*, vol. 60, no. 11, pp. 1191– 1206, Nov. 2020. https://doi.org/10.1134/S096554412011016X/METRICS

[4] A. V. Sulimov, S. M. Danov, A. V. Ovcharova, A. A. Ovcharov, and V. R. Flid, "Kinetics of propylene epoxidation with hydrogen peroxide catalyzed by extruded titanium silicalite in methanol," *Kinet. Catal.*, vol. 57, no. 4, pp. 466–473, Jul. 2016. https://doi.org/10.1134/S0023158416040121/METRICS

[5] M. Mubashir, M. Ahsan, I. Ahmad, and M. Nouman Aslam Khan, "Process modeling and simulation of ethylene oxide production by implementing pinch and cost analysis," *Ain Shams Eng. J.*, vol. 13, no. 3, p. 101585, May 2022. https://doi.org/10.1016/J.ASEJ.2021.09.012.

[6] P. Zakkour, "Carbon Counts Report", Accessed: Mar. 30, 2024. [Online]. Available: www.unido.org

[7] H. J. Lee, M. Ghanta, D. H. Busch, and B. Subramaniam, "Toward a CO2-free ethylene oxide process: Homogeneous ethylene oxide in gas-expanded liquids," *Chem. Eng. Sci.*, vol. 65, no. 1, pp. 128–134, Jan. 2010. https://doi.org/10.1016/J.CES.2009.02.008

[8] M. Ghanta, T. Ruddy, D. Fahey, D. Busch, and B. Subramaniam, "Is the liquid-phase H2O2-based ethylene oxide process more economical and greener than the gas-phase O2- based silver-catalyzed process?," *Ind. Eng. Chem. Res.*, vol. 52, no. 1, pp. 18–29, Jan. 2013. https://doi.org/10.1021/IE301601Y/SUPPL_FILE/IE301601Y_SI_001.PDF

[9] "Rhenium Price - Historical Prices - 2024 Forecast - How to Buy." Accessed: Mar. 30, 2024. [Online]. Available: https://strategicmetalsinvest.com/rhenium-prices/

[10] H. Yao and D. E. Richardson, "Epoxidation of alkenes with bicarbonate-activated hydrogen peroxide [3]," *J. Am. Chem. Soc.*, vol. 122, no. 13, pp. 3220–3221, Apr. 2000. https://doi.org/10.1021/JA993935S/ASSET/IMAGES/LARGE/JA993935SF1.JPEG

[11] M. Alvear *et al.*, "Continuous Liquid-Phase Epoxidation of Ethylene with Hydrogen Peroxide on a Titanium-Silicate Catalyst," *Ind. Eng. Chem. Res.*, vol. 60, no. 26, pp. 9429–9436, Jul. 2021. https://doi.org/10.1021/ACS.IECR.1C01722/ASSET/IMAGES/LARGE/IE1C01722_0010.JPEG

[12] P. P. McClellan, "Manufacture and Uses of Ethylene Oxide and Ethylene Glycol," *Ind. Eng. Chem.*, vol. 42, no. 12, pp. 2402–2407, Dec. 2002. https://doi.org/10.1021/IE50492A013

Materials Research Proceedings 53 (2025) 287-297 https://doi.org/10.21741/9781644903575-28

Role of different pressure equalization strategies on cycle performance of dual-reflux pressure swing adsorption for biogas upgradation

MOHD HARDYIANTO Vai Bahrun[1,2,a], AWANG Bono[3,b], NORASIKIN Othman[1,2,c] and MUHAMMAD ABBAS Ahmad Zaini[1,2,d *]

[1]Department of Chemical Engineering, Faculty of Chemical and Energy Engineering, Universiti Teknologi Malaysia, 81310 UTM, Johor Bahru, Johor, Malaysia

[2]Centre of Lipids Engineering and Applied Research (CLEAR), Ibnu-Sina Institute for Scientific and Industrial Research (ISI-SIR), Universiti Teknologi Malaysia, 81310 UTM, Johor Bahru, Johor, Malaysia

[3]GRISM Innovative Solutions, Kota Kinabalu, Sabah, Malaysia

[a]mohdhardyiantovai@graduate.utm.my, [b]awangbono@gmail.com, [c]norasikin@cheme.utm.my, [d]abbas@cheme.utm.my

Keywords: Aspen Adsorption, Biogas Upgrading, DR-PSA, Numerical Simulation, Pressure Equalization

Abstract. Cycle sequence plays a crucial role in determining the performance of dual-reflux pressure swing adsorption (DR-PSA) process. Pressure equalization (PE) is one of the cycle modifications that is often implemented in DR-PSA, which can influence the performance of the process. In this study, numerical simulations of a 2-bed DR-PSA, with different PE strategies, were conducted in Aspen Adsorption in order to elucidate the role played by the PE step on the cycle performance. The process was designed for biogas upgradation from a binary gas mixture of a 45 mol% CO_2 + 55 mol% CH_4, with silica gel as solid adsorbent. The introduction of the PE step into the cycle sequence enhances the cycle performance in terms of product purity, recovery, and cycle work. The extent to which the cycle performance improves depends on the PE strategies being implemented.

Introduction

Dual-reflux pressure swing adsorption (DR-PSA) is an advanced PSA process, capable of simultaneously producing two high-purity gas components with high recovery from a binary feed mixture. Theoretical analysis under highly idealized conditions has established the possibility of perfect binary separation with 100% recovery for each gas component [1]. Furthermore, it has been demonstrated that the DR-PSA is capable of producing a nearly perfect binary separation of several gas separation applications through rigorous numerical simulation and/or experimental, *i.e.* air separation [2], [3], nitrogen rejection from natural gas [4], helium recovery [5], biogas upgrading [6], etc.

Biogas is anaerobically-produced gas that is mainly consisted of methane (CH_4) and carbon dioxide (CO_2), is regarded as a potential sustainable renewable energy source alongside hydrogen. Since its production consists of a considerable amount of CO_2, thus, hindered its application as energy source due to its low energy content [7]. In this regard, biogas upgrading is a step towards achieving CH_4 specifications matching that of pipeline quality of natural gas for sustainable fuel sources, by removing or separating CO_2 [8]. It is of highly important, during which the upgrading of biogas, if two products CH_4 and CO_2 can be separated efficiently and producing each with high purity and recovery. In this regard, the DR-PSA shows a promising prospect, whereby CH_4 can be obtained as fuel alternative to natural gas, and CO_2 can be captured for utilization.

Like in any conventional PSA processes, the performance of DR-PSA is influenced by its cycle sequence design. The pressure equalization (PE) step between two adsorption beds is a crucial aspect of many PSA processes [9] including DR-PSA. This step involves connecting two beds and allowing them to equalize in pressure. Open literature has investigated the effect of different PE strategies on the DR-PSA performance. For instance, Li et al. [10] investigated three PE methods on CO_2 and N_2 purities and recoveries, revealing the top-to-top bed PE (T-T PE) as giving the highest performance. Nevertheless, to the best of authors' knowledge, no study has investigated the impact of PE step on bed productivity and power requirements in the DR-PSA process. While the effects of PE strategies on bed productivity and power requirements have been the subject in several conventional PSA processes [11], their impact on DR-PSA's bed productivity and power requirements remains uncertain.

Therefore, the objective of this study is to analyze the influence of introducing a PE step with different PE strategies on the performance of DR-PSA in the context of biogas upgrading. The biogas feed mixture comprises a binary mixture of 45 mol% CO_2 + 55 mol% CH_4, utilizing CO_2-selective silica gel as the adsorbent packed in the DR-PSA. This is accomplished through rigorous phenomenological numerical model in Aspen Adsorption, considering three scenarios: the absence of a PE step (No PE), top-to-top bed PE step (T-T PE), and bottom-to-bottom bed PE step (B-B PE) on the DR-PSA performance. The performance is tracked by four key performance indicators (KPIs): products purity, products recovery, bed productivity, and cycle work. It is worth noting that this study represents an initial exploration of DR-PSA for biogas upgrading under various PE strategies. The optimal conditions for the DR-PSA have yet to be determined, thus the results reflect unoptimized performance – however, enough to reflect the influence played by the PE step on the cycle performance.

Methods

Process Description of a Basic 4-Step DR-PSA Cycle. In the first attempt, this study applied a basic 2-bed, 4-step P_H-A DR-PSA configuration with the absence of PE step, specifically consisting of half-cycle sequence of (1) feed (FE)/light reflux (LR) and (2) blowdown (BD)/pressurization (PR). A commercial simulation tool, which is the Aspen Adsorption (V12.1, AspenTech, USA) is utilized for numerical investigation of the DR-PSA cycle. Fig. 1 depicts the half-cycle of the process, with each step description given below:

Step 1: FE/LR – The feed gas with a constant flowrate of Q_{feed} is admitted at an intermediate position, Z_F of a high-pressure column (Bed 1), and a portion is used as flushing gas with a constant flowrate of Q_{LR} to regenerate adsorbent in a low-pressure column (Bed 2), while the remaining portion is withdrawn at the top as a light product. Meanwhile, the heavy product is desorbed from the bottom of Bed 2, a part of which is discharged with a flow rate of Q_{HP} as heavy product, and the remaining of which is recycled and sent to Bed 1 as a heavy reflux.

Step 2: BD/PR – The blowdown step and pressurization step concurrently proceed in Bed 1 and Bed 2, respectively, to initiate the pressure down and up to the low-and high-pressure conditions, which is achieved with the help of a compressor. The function is to desorb the adsorbed components by lowering the pressure in Bed 1, so that the adsorbent can be reused in the next cycle. While, for Bed 2, the function is to ready the bed for undergoing the next step, which is the feed adsorption. When one half of DR-PSA cycle is completed in Bed 1, a similar cycle sequence commences in Bed 2, and the cycles continue until a pre-determined cyclic steady-state conditions are achieved.

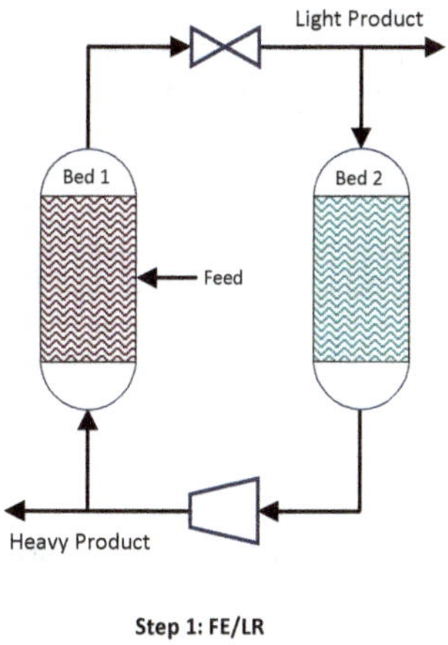

Fig. 1: Simplified diagram of a half-cycle of the P_H-A DR-PSA configuration. Step 1: FE/LR, Step 2: BD/PR.

Introduction of PE Step into the Basic 4-Step DR-PSA Cycle. The effect of different PE strategies on the DR-PSA cycle performance is investigated by introducing two strategies: (1) top-to-top (T-T PE) step, which connects the two beds through the topmost part of the columns via a T-T PE valve, and (2) bottom-to-bottom (B-B PE) step, which connects the two beds through the bottommost part of the columns via a B-B PE valve. The illustration of these two strategies is provided in Fig. 2. The PE step is added after Step 1: FE/LR in the half-cycle sequence.

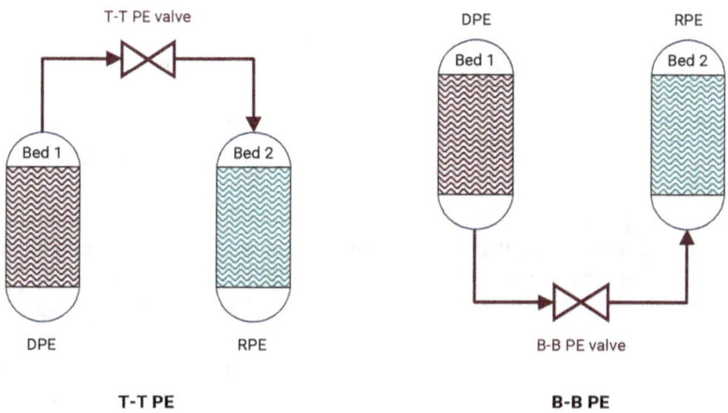

Fig. 2: The depiction of the PE step for T-T PE and B-B PE strategies.

Theoretical Assumptions and Mathematics of DR-PSA model. DR-PSA is a complex and advanced dynamic process, in which all variables change over time and space until the system reaches a condition whereby all the variables show no change between the current and the next cycle, known as the cyclic steady state (CSS). To elucidate the behaviour of the DR-PSA, numerical simulation is carried out within the simulation tool, incorporating rigorous non-isothermal model consisting of the mass, momentum, and energy balances. The following theoretical assumptions are adopted in constructing the model equation of the adsorption bed flowsheet [12], [13], [14].

1. The gas obeys the ideal gas law, $PV = nRT$.
2. There is no radial variation in both concentration and temperature.
3. The pressure drop across the bed is determined through the Ergun equation.
4. The competitive multi-component adsorption effects are represented by an Extended Langmuir 2 isotherm equation.
5. A rigorous heat equation consisting of gas-phase, solid-phase, and bed wall energy balances.
6. A linear driving force (LDF) approximation as a function of solid loading difference is employed to describe the mass transfer between the gas and solid phases.

Based on the theoretical assumptions, the complete governing equations (Eqs. 1 – 8) are presented in Table 1, mainly including mass, energy, and momentum transfer models. The detailed description of the equations used here is widely discussed and can be found in many refs. [12], [14], [15].

In the Eqs. 1 - 3 of the above Table 1, ε_b is the interparticle voidage, c_i is the gas-phase concentration of component i, v_g is the gas-phase superficial velocity, ε_p is the intraparticle voidage, ρ_s is the bulk solid density, q_i is the solid-phase loading of component i, z is the axial coordinate, and t is time. In Eq. 3, P is the pressure, μ is the gas-phase viscosity, M is the gas-phase molecular weight, ρ_g is the gas-phase mass density, and Ψ is the adsorbent particles' shape factor.

For Eq. 4, k_{ga} is the axial gas-phase thermal conductivity, T_g is the gas-phase temperature, C_{vg} is the specific gas-phase heat capacity at constant volume, HTC is the gas-solid heat transfer coefficient, a_p is the specific particle surface per unit volume bed, T_s is the solid-phase temperature, H_w is the gas-wall heat transfer coefficient, D_B is the bed diameter, and T_0 is the reference temperature. In Eq. 5, k_{sa} is the axial solid thermal conductivity, T_s is the solid-phase temperature, C_{ps} is the specific heat capacity of adsorbent, C_{pai} is the specific heat capacity of adsorbed phase, and ΔH_i is the heat of adsorption of component i, While for Eq. 6, k_w is the column wall thermal conductivity, T_w is the wall temperature, ρ_w is the wall density, C_{pw} is the specific heat capacity of column wall, W_T is the width of column wall, H_{amb} is the wall-ambient heat transfer coefficient, and T_{amb} is the ambient temperature. In Eqs. 7 and 8, k_{LDF} is the LDF mass transfer coefficient, q_i^* is the equilibrium solid-phase loading of component i, and $IP_1 – IP_4$ are the isotherm parameters of the Extended Langmuir 2 model.

The input for the simulation used in this study are tabulated in Table 2. This includes the details about the bed properties, adsorbent properties, process conditions, and the associated mass and heat transport parameters. Meanwhile, the Extended Langmuir 2 isotherm model is used to relate the multi-component adsorption behaviour of CO_2 and CH_4. Their respective isotherm parameters are given in Table 3, after being modified to suit the default equation used by the Aspen Adsorption simulator. The original isotherm parameters can be referred to the work by Shen et al. [7].

Table 1: Mathematical model equations for conservation of mass, momentum, and energy and mass transfer rate, as well as the adsorption isotherm model used for the simulation.

Items	Equations
Conservation of mass	Component i mass balance (Eq. 1) $$\upsilon_g\frac{\partial c_i}{\partial z}+[\varepsilon_b+(1-\varepsilon_b)\varepsilon_p]\frac{\partial c_i}{\partial t}+(1-\varepsilon_b)\rho_s\frac{\partial q_i}{\partial t}=0$$ Overall mass balance (Eq. 2) $$\frac{\partial c}{\partial z}+[\varepsilon_b+(1-\varepsilon_b)\varepsilon_p]\frac{\partial c}{\partial t}+(1-\varepsilon_b)\rho_s\sum\frac{\partial q_i}{\partial t}=0$$
Conservation of momentum	*Ergun's equation (Eq. 3)* $$\frac{\partial P}{\partial z}=-\left(\frac{1.50\times10^{-3}\mu(1-\varepsilon_b)^2}{\varepsilon_b{}^3(2r_p\Psi)^2}\upsilon_g+\frac{1.75\times10^{-5}M\rho_g(1-\varepsilon_b)}{\varepsilon_b{}^3 2r_p\Psi}\upsilon_g{}^2\right)$$
Conservation of energy	*Gas-phase (Eq. 4)* $$-k_{ga}\varepsilon_i\frac{\partial^2 T_g}{\partial z^2}+C_{vg}\upsilon_g\rho_g\frac{\partial T_g}{\partial z}+\varepsilon_i C_{vg}\rho_g\frac{\partial T_g}{\partial t}+P\frac{\partial \upsilon_g}{\partial z}+HTCa_p(T_g-T_s)+\frac{4H_w}{D_B}(T_g-T_0)=0$$ *Solid-phase (Eq. 5)* $$-k_{sa}\frac{\partial^2 T_s}{\partial z^2}+\rho_s C_{ps}\frac{\partial T_s}{\partial t}+\rho_s\sum_{i=1}^{n}(C_{pai}q_i)\frac{\partial T_s}{\partial t}+\rho_s\sum_{i=1}^{n}\left(\Delta H_i\frac{\partial q_i}{\partial t}\right)-HTCa_p(T_g-T_s)=0$$ *Wall phase (Eq. 6)* $$-k_w\frac{\partial^2 T_w}{\partial z^2}+\rho_w C_{pw}\frac{\partial T_w}{\partial t}-H_w\frac{4D_B}{(D_B+W_T)^2-D_B{}^2}(T_g-T_w)+H_{amb}\frac{4(D_B+W_T)^2}{(D_B+W_T)^2-D_B{}^2}(T_w-T_{amb})=0$$
Mass transfer rate	*LDF approximation (Eq. 7)* $$\frac{\partial q_i}{\partial t}=k_{LDF,i}(q_i^*-q_i)$$
Adsorption isotherm	*Extended Langmuir 2 (Eq. 8)* $$q_i=\frac{IP_{1i}e^{IP_{2i}/T}P_i}{1+\sum_k(IP_{3k}e^{IP_{4k}/T}P_k)}$$

Table 2: Parameters of the column, adsorbent, process conditions, and heat transport used for the simulation. Adapted from Shen et al. [7].

Parameter	Value	Parameter	Value
Bed properties		Adsorbent properties	
Column length, L (m)	1.00	Material	Silica gel
Column inner diameter, D (m)	0.06	Solid density, ρ_b (kg/m^3)	750
Wall thickness, W_t (m)	0.002	Particle diameter, d_p (m)	0.004
Wall density, ρ_w (kg/m^3)	7800	Interparticle voidage, ε_b (-)	0.35
Process conditions		Intraparticle voidage, ε_p (-)	0.37
Feed flowrate, Q_{feed} (SLPM)	15	Shape factor, ψ (-)	0.83
Feed CO_2/CH_4 mole fraction, y_{feed} (-)	0.45/0.55	Mass transfer parameters	
Feed temperature, T (K)	303	LDF coefficient of CO_2, $k_{LDF,CO2}$ (1/s)	0.0643
High pressure, P_H (bar)	5	LDF coefficient of CH_4, $k_{LDF,CH4}$ (1/s)	0.356
Low pressure, P_L (bar)	1.1		
Axial feed location, Z_{feed} (-)	0.5		
Heat transfer parameters			
Column wall thermal conductivity, k_w (W/m/K)	17	Specific heat capacity of gas, C_{pg} (kJ/kg/K)	0.758
Axial solid-phase thermal conductivity, k_s (W/m/K)	0.3	Gas-solid heat transfer coefficient, HTC ($W/m^2/K$)	219
Axial gas-phase thermal conductivity, k_g (W/m/K)	0.242	Gas-wall heat transfer coefficient, H_w ($W/m^2/K$)	94
Specific heat capacity of column wall, C_{pw} (kJ/kg/K)	0.502	Wall-ambient heat transfer coefficient, H_{amb} ($W/m^2/K$)	60
Specific heat capacity of adsorbent, C_{ps} (kJ/kg/K)	0.902		

Table 3: Isotherm parameters of Extended Langmuir 2 for CO_2 and CH_4 adsorption over silica gel. Modified from Shen et al. [7].

Component	CO_2	CH_4
IP_1 (kmol/kg/bar)	5.1706×10^{-8}	2.4710×10^{-7}
IP_2 (K)	3003.01	1788.91
IP_3 (1/bar)	8.6090×10^{-6}	7.5380×10^{-5}
IP_4 (K)	3003.01	1788.91

To model sequential cycles, the dynamic simulation is continued until steady-state conditions are achieved while the results approach the same amount with no changes from one cycle, *N*, to the next cycle, *N*+1, as mathematically expressed in Eq. 9. The performance of the DR-PSA would be reported at the CSS conditions.

$$e_{CSS} = \left| y_{\text{N cycle}} - y_{\text{N+1 cycle}} \right| \leq \varepsilon_{CSS} \quad \text{Eq. 9}$$

Here, y denotes the variables that are used to monitor the CSS, in this case are the material and temperature, and e_{CSS} is a small positive value that plays a critical role in evaluating the number of cycles to reach the CSS. In this study, the value of ε_{CSS} is set as 1×10^{-5}, small enough for ensuring the CSS [4], [16].

DR-PSA Performance Indicators. To assess the performance of the DR-PSA in separating CH_4 and CO_2 from biogas, key performance indicators (KPIs) are established, including the product purity, product recovery, bed productivity, and cycle work. The respective KPIs are defined as in Eqs. 10 – 13, respectively.

$$\text{Product purity (\%)} = \frac{\text{Moles of } CO_2 \text{ or } CH_4 \text{ in the heavy or light product stream}}{\text{Total moles of gas in that product stream}} \times 100 \quad \text{Eq. 10}$$

$$\text{Product recovery (\%)} = \frac{\text{Moles of } CO_2 \text{ or } CH_4 \text{ in the heavy or light product stream}}{\text{Moles of that component in the feed stream}} \times 100 \quad \text{Eq. 11}$$

$$\text{Bed productivity (mol } CH_4\text{/h/kg)} = \frac{\text{Total mole of } CH_4 \text{ in the light product stream}}{\text{Cycle time} \times \text{Total mass of adsorbent in two beds}} \quad \text{Eq. 12}$$

$$\text{Cycle work (kJ/mol } CH_4 \text{ captured)} = \frac{\text{Enery consumption of compressor}}{\text{Total mole of } CH_4 \text{ in the light product stream}} \quad \text{Eq. 13}$$

The numerical model and flowsheet used in this study were comprehensively validated against experimental data from literature [17], demonstrating their reliability in predicting the behaviour of the DR-PSA. The validation shows a combined error margin of less than 10% for light and heavy product purities across 24 experimental runs [18].

Results and Discussion

Effect of PE Step on Product Purity and Recovery. Table 4 presents the results of the effect of PE step on the purity of CH_4 and its recovery in LP, as well as the purity of CO_2 and its recovery in HP. Overall, the inclusion of the PE step within the cycle sequence enhances both product purity and recovery, with a slight improvement observed in the T-T PE compared to the B-B PE. This study suggests that the presence of the PE step improves both the purity and recovery of a DR-PSA process, especially at the heavy product-to-feed flowrate ratio equals to feed heavy component, aligning with a literature [10] – unlike in conventional PSAs system, the presence of the PE step only improves the recovery, not purity [19], [20].

Table 4: The products purity and products recovery of DR-PSA at different PE strategies.

PE strategies	Product purity (%)		Product recovery (%)	
	CH_4 in LP	CO_2 in HP	CH_4 in LP	CO_2 in HP
No PE	76.70	71.55	76.73	71.55
T-T PE	77.33	72.30	77.32	72.30
B-B PE	76.84	71.70	76.79	71.70

The underlying reason behind the observed phenomenon at different PE strategies can be explained through the historical profile of the heavy component CO_2 progression along the axial location of a bed column at the start until the end of the PE step, as depicted in Fig. 3. In the FE step (a step before DPE step), a significant amount of CO_2 was enriched at the lower part of Bed 1 (~70%), while a large quantity of CH_4 existed in the topmost part of Bed 1 (~65%) at the end of the FE step (and thus the start of the DPE step). During the PE step, the high-pressure gas from Bed 1 would flow into Bed 2. In both PE strategies, the progression of the CO_2 front induces the displacement of gas within the column. However, the direction of the CO_2 front movement differs between these two PE strategies.

In the T-T PE strategy, a greater quantity of CH_4 is liberated from the top of the column initially at the start of the DPE step. This phenomenon arises because CH_4, being the lighter product, has accumulated near the upper region of the column. As the gas is released, the pressure within the column decreases, which causes the desorption of CO_2 from the adsorbent into the gas phase, contributing to the net increase in CO_2 composition by the end of the step.

Conversely, in the B-B PE strategy, a larger proportion of CO_2 initially accumulates near the lower region of the column due to the previous step involving the injection of heavy reflux from the bottom of the column. As the gas is released from the bottom during the DPE step, CO_2 is rapidly expelled, and the pressure drop inside the column similarly promotes the desorption of CO_2 into the gas phase, leading to an overall increase in CO_2 composition by the end of the step. However, the net CO_2 composition in B-B PE is lower compared to T-T PE by the end of the DPE step. This is particularly due to CO_2 is being released more rapidly and in greater amounts from the bottom, where it was initially more concentrated, resulting in a faster depletion of CO_2 composition in B-B PE compared to T-T PE.

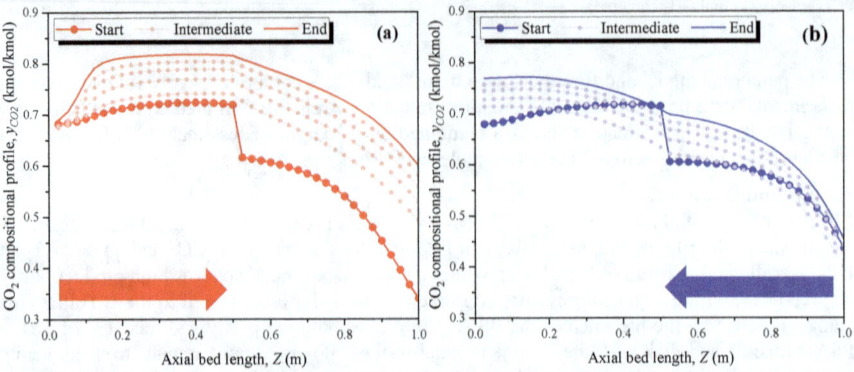

Fig. 3: Gas-phase CO_2 compositional profiles at the start to the end of DPE step for (a) T-T PE and (b) B-B PE. The arrows indicate the direction of flow during the step.

Effect of PE Step on Bed Productivity and Cycle Work. The effect of the PE step on bed productivity and cycle work is summarized in Table 5. It was observed that the presence of a PE step reduces bed productivity but enhances cycle work requirement. The No PE configuration achieved the highest bed productivity of 2.56 mol CH_4/h/kg adsorbent, while the B-B PE recorded the lowest cycle work, requiring only 64.95 kJ/mol CH_4 captured.

Bed productivity, as defined by the quantity of CH_4 produced per unit amount of adsorbent per unit time (Eq. 12), is directly influenced by the cycle time. Faster cycles correspond to higher bed productivity, whereas the inclusion of additional steps increases the total cycle time, thereby reducing bed productivity. In this study, the No PE configuration recorded the highest bed productivity at 2.56 mol CH_4/h/kg adsorbent, compared to configurations with PE. While there was a noticeable difference in bed productivity between No PE and with PE configurations (T-T PE or B-B PE), the difference between T-T PE and B-B PE was not clearly distinguishable.

It might be expected that bed productivity would be equal across different PE strategies (T-T PE and B-B PE). However, since the BD/PR step in this study was event-driven – meaning that $t_{BD/PR}$ was determined by the time required for the bed undergoing blowdown to reach a predetermined low-pressure value, P_L of 1.1 bar – the time for BD/PR was not explicitly set in the simulation. This led to the observed differences in bed productivity between the two PE strategies. Nevertheless, this difference can be considered negligible, as the magnitude of the difference in bed productivity was insignificant, with a magnitude difference of only 0.02 mol CH_4/h/kg, and that practical pilot- or industrial-scale operations would likely to considers such a small difference in $t_{BD/PR}$ as minor relative to the longer overall cycle time.

Furthermore, the cycle work reduces in the presence of the PE step, consistent with the underlying theoretical principle [11], [19]. This reduction occurs because the compressed gas from the high-pressure bed, instead of being vented, is utilized to partially pressurize the low-pressure column, leading to shorter period for the subsequent BD/PR step to complete its step. This explanation parallels the effect of PE step on bed productivity. A shorter period for completing the BD/PR step requires less compression duty for gas transfer, thus lowering the cycle work per mole CH_4 captured, with B-B PE showing the most efficient performance, followed by T-T PE, and finally, No PE. This observation suggests that B-B PE is a more efficient PE step for conserving cycle work, most likely due to its ability to distort the CO_2 front towards the direction of the subsequent BD/PR step, which pulling the gas fronts from the bottom of the column. Overall, it is evidence that the presence of PE step saves the energy consumption by requiring smaller amount of compression cycle work by as much as 8% in relative to the absence of PE step.

Table 5: The bed productivity and cycle work of DR-PSA at different PE strategies.

PE strategies	$t_{PR/BD}$ (s)	Bed productivity (mol CH_4/h/kg adsorbent)	Cycle work (kJ/mol CH_4 captured)
No PE	22	2.56	70.57
T-T PE	17	2.41	68.78
B-B PE	16	2.43	64.95

Conclusion

The presence of PE step generally improves DR-PSA performance in terms of products purity, recovery, bed productivity, and cycle work. However, the effectiveness of these improvements is contingent on the specific PE strategy employed. The presence of a PE step led to a maximum increase of 1% in product purities and recoveries, while bed productivity decreased by up to 6%, and cycle work improved by as much as 8%. Among the PE strategies evaluated, the T-T PE emerged as the most effective, offering a balanced improvement across product purity, recovery, bed productivity, and cycle work. Overall, this study demonstrates the relative benefits of incorporating a PE step in the DR-PSA process.

Acknowledgement

The authors are grateful for the financial support from Universiti Teknologi Malaysia under UTM Zamalah Research Fund No. 00N08.

References

[1] A. D. Ebner and J. A. Ritter, "Equilibrium theory analysis of dual reflux PSA for separation of a binary mixture," *AIChE J.*, vol. 50, no. 10, pp. 2418–2429, 2004. https://doi.org/10.1002/aic.10191

[2] C. Tian, Q. Fu, Z. Ding, Z. Han, and D. Zhang, "Experiment and simulation study of a dual-reflux pressure swing adsorption process for separating N2/O2," *Sep. Purif. Technol.*, vol. 189, pp. 54–65, 2017. https://doi.org/10.1016/j.seppur.2017.06.041

[3] Y. Wang, Y. An, Z. Ding, Y. Shen, Z. Tang, and D. Zhang, "Integrated VPSA Processes for Air Separation Based on Dual Reflux Configuration," *Ind. Eng. Chem. Res.*, vol. 58, no. 16, pp. 6562–6575, 2019. https://doi.org/10.1021/acs.iecr.8b06272

[4] W. Lu, G. Xiao, R. Weh, and E. F. May, "Rejecting N2 from natural gas by dual reflux pressure swing adsorption with activated carbon," *J. Nat. Gas Sci. Eng.*, vol. 81, no. July, p. 103457, 2020. https://doi.org/10.1016/j.jngse.2020.103457

[5] R. Weh, G. Xiao, E. S. Pouya, and E. F. May, "Helium recovery and purification by dual reflux pressure swing adsorption," *Sep. Purif. Technol.*, vol. 288, no. December 2021, p. 120603, 2022. https://doi.org/10.1016/j.seppur.2022.120603

[6] G. Spoorthi, R. S. Thakur, N. Kaistha, and D. P. Rao, "Process intensification in PSA processes for upgrading synthetic landfill and lean natural gases," *Adsorption*, vol. 17, no. 1, pp. 121–133, 2011. https://doi.org/10.1007/s10450-010-9302-6

[7] Y. Shen, W. Shi, D. Zhang, P. Na, and B. Fu, "The removal and capture of CO2 from biogas by vacuum pressure swing process using silica gel," *J. CO2 Util.*, vol. 27, no. May, pp. 259–271, 2018. https://doi.org/10.1016/j.jcou.2018.08.001.

[8] M. H. V. Bahrun, A. Bono, N. Othman, and M. A. A. Zaini, "Carbon dioxide removal from biogas through pressure swing adsorption - A review," *Chem. Eng. Res. Des.*, vol. 183, pp. 285–306, 2022. https://doi.org/10.1016/j.cherd.2022.05.012

[9] B. Fakhari-Kisomi, L. Erden, A. D. Ebner, and J. A. Ritter, "Equilibrium Theory Analysis of Pressure Equalization Steps in Pressure Swing Adsorption," *Ind. Eng. Chem. Res.*, vol. 60, no. 27, pp. 9928–9939, 2021. https://doi.org/10.1021/acs.iecr.1c01144

[10] D. Li *et al.*, "Experiment and simulation for separating CO2/N2 by dual-reflux pressure swing adsorption process," *Chem. Eng. J.*, vol. 297, pp. 315–324, 2016. https://doi.org/10.1016/j.cej.2016.03.075

[11] K. Warmuzinski, "Effect of pressure equalization on power requirements in PSA systems," *Chem. Eng. Sci.*, vol. 57, no. 8, pp. 1475–1478, 2002. https://doi.org/10.1016/S0009-2509(02)00060-X

[12] Z. Guan *et al.*, "Simulation and analysis of dual-reflux pressure swing adsorption using silica gel for blue coal gas initial separation," *Int. J. Hydrogen Energy*, vol. 46, no. 1, pp. 683–696, 2021. https://doi.org/10.1016/j.ijhydene.2020.09.209

[13] E. F. May, Y. Zhang, T. L. H. Saleman, G. Xiao, G. K. Li, and B. R. Young, "Demonstration and optimisation of the four Dual-Reflux Pressure Swing Adsorption configurations," *Sep. Purif. Technol.*, vol. 177, pp. 161–175, 2017. https://doi.org/10.1016/j.seppur.2016.12.007

[14] Y. Zhang, T. L. H. Saleman, G. K. Li, G. Xiao, B. R. Young, and E. F. May, "Non-isothermal numerical simulations of dual reflux pressure swing adsorption cycles for separating N2 + CH4," *Chem. Eng. J.*, vol. 292, pp. 366–381, 2016. https://doi.org/10.1016/j.cej.2016.02.018

[15] Z. Kamin, M. H. V. Bahrun, and A. Bono, "A Short Review on Pressure Swing Adsorption (PSA) Technology for Nitrogen Generation from Air," *AIP Conf. Proc.*, 2022.

[16] F. Brandani, P. Pullumbi, and C. Monereau, "Nitrogen rejection from landfill gas using Pressure Swing Adsorption," *Adsorption*, vol. 27, pp. 603–617, 2021. https://doi.org/10.1007/s10450-021-00304-0

[17] T. L. Saleman *et al.*, "Capture of low grade methane from nitrogen gas using dual-reflux pressure swing adsorption," *Chem. Eng. J.*, vol. 281, pp. 739–748, 2015. https://doi.org/10.1016/j.cej.2015.07.001

[18] M. H. V. Bahrun, A. Bono, N. Othman, and M. A. Ahmad Zaini, "Parametric optimization of dual-reflux pressure swing adsorption in biogas upgrading," *(submitted)*.

[19] S. Jain, A. S. Moharir, P. Li, and G. Wozny, "Heuristic design of pressure swing adsorption: A preliminary study," *Sep. Purif. Technol.*, vol. 33, no. 1, pp. 25–43, 2003. https://doi.org/10.1016/S1383-5866(02)00208-3

[20] M. Yavary, H. A. Ebrahim, and C. Falamaki, "The effect of number of pressure equalization steps on the performance of pressure swing adsorption process," *Chem. Eng. Process. Process Intensif.*, vol. 87, pp. 35–44, 2015. https://doi.org/10.1016/j.cep.2014.11.003

Materials Research Proceedings 53 (2025) 298-304 https://doi.org/10.21741/9781644903575-29

Morphological and crystallinity properties of bacterial nanocellulose foam reinforced by copper oxide nanoparticles

JIBRIL Maulana[1,a], HERU Suryanto[2,b*], AMINNUDIN Aminnudin[2,c], JOSEPH Selvi Binoj[3,d], MAVINKERE Rangappa Sanjay[4,e], SUCHART Siengchin[4,f], UUN Yanuhar[5,g]

[1]Faculty of Vocational, Universitas Negeri Malang, Jl. Semarang 5, Malang 65145, Indonesia

[2]Center of Excellence for Cellulose Composite (CECCom), Department of Mechanical and Industrial Engineering, Universitas Negeri Malang, Jl. Semarang 5, Malang 65145, Indonesia

[3]Institute of Mechanical Engineering, Saveetha School of Engineering, Saveetha Institute of Medical and Technical Sciences (SIMATS), Saveetha University, Tamilnadu 602105, India

[4]Natural Composites Research Group Lab, Department of Materials and Production Engineering, The Sirindhorn International Thai-German Graduate School of Engineering (TGGS), King Mongkut's University of Technology North Bangkok (KMUTNB), Bangkok 10800, Thailand

[5]Faculty of Fisheries and Marine Science, Brawijaya University, Jl Veteran Malang, East Java, Indonesia

[a]jibril.maulana.fv@um.ac.id, [b]heru.suryanto.ft@um.ac.id, [c]aminnudin.ft@um.ac.id, [d]binojlaxman@gmail.com, [e]mavinkere.r.s@op.kmutnb.ac.th, [f]suchart.s@tggs.kmutnb.ac.th, [g]doktoruun@ub.ac.id

Keywords: Bacterial Cellulose, Crystallinity, Copper Oxide, Foam, Nanoparticles

Abstract. Bacterial cellulose (BC) is a material with advantages such as high purity, mechanical strength, crystallinity, and unique physical properties while being biodegradable. However, in order to improve specific characteristics, cellulose should be combined with other materials. One type of metal that can be applied as an additive in bacterial nanocellulose (BNC) networks is copper oxide nanoparticles (CuO-NPs). Copper has great potential for application as a cellulose tissue filler. Cetyl Trimethyl Ammonium Bromide (CTAB) is required to assist in the distribution and bonding of BC. This research provided the use of BC, which is produced from pineapple peel. Afterward, the BC was homogenized with High-pressure homogenization to produce BNC, mixed with varied CuO-NPs (0, 0.25, 0.5, 1, 1.5, 2%). As a result, a BNC/CuO-NPs nanocomposite is synthesized. CTAB can successfully homogenize the distribution of CuO-NPs, according to morphological investigations. The addition of CTAB and CuO-NPs toward the crystallinity increased the intensity at diffraction angles of 21.3°, 35.5°, and 38.6°, thus according to XRD observations.

Introduction

Bacterial cellulose (BC) is composed principally of a glucose monomer with the general formula $(C_6H_{10}O_5)_n$, which is produced by *Acetobacter xylinum* or *Gluconobacter xylinus* [1, 2]. The synthesis of BC necessarily requires sucrose, glucose, and other nutrients. These substances have been used as a source of carbon and energy towards the single-cell protein fermentation medium [3]. Pineapple peels are one of the substances with a high glucose content. Regardless of the fact that it contains almost 50% pineapples, this material is one of the agro-waste underexploited [4]. However, most of the pineapple peels are discarded and no longer used. As a result, these kinds of unused agro-waste were reprocessed for further usage in different sectors.

BC, which has a three-dimensional structural matrix and is developed by hydrogen bonds having a diameter ranging from 20 to 50 nm [5]. The cellulose produced by bacteria has the same structure as cellulose produced by plants, except that it has higher purity, crystallinity, mechanical strength, unique physical properties, and biodegradability [1, 5–8]. BC has a variety of applications, such as high-quality paper, photocatalysts, sensor systems, tissue engineering, diet food, artificial blood vessels, biodegradable packaging, water filtration, and air purifiers [5, 8–12]. To improve certain properties, cellulose should form a composite in its application.

One type of metal that can be used as an additive in the BC network is economically available copper oxide nanoparticles (CuO-NPs). Copper can indeed form hydrogen bonds with tissues through the use of cellulose O-H bonds [13]. Copper has high suitability for use as a reinforcement material in cellulose tissue based on its properties. Since nanomaterials have high surface energy, making them difficult to bond with the material, a surfactant having the capability of facilitating the bonding of CuO-NPs and cellulose should be added. Surfactants have a chemical composition and an amphiphilic condition that aids in surface orientation and development, allowing the material to be more stable and enhancing the distribution of material and metal agglomeration [14]. Cetyl Trimethyl Ammonium Bromide (CTAB) was chosen as a surfactant because it has a better material distribution than Sodium Dodecylbenzene Sulfonate (SDS) and Polyethylene Glycol (PEG) surfactants [14]. CTAB could also create a strong bond between BC and copper oxide. The objective of this research is to explore the effects of CuO-NPs on the morphology and crystallinity of bacterial nanocellulose (BNC) foam composites. This study's contribution is to prepare an effective pretreatment for developing foam materials in order to be applied as foam in adsorbent for waste, in the future.

Material and Methods

Materials

BC is sourced from pineapple peel starter culture. *A. xylinum* is the bacteria used in the production of cellulose. Sugar ($C_{12}H_{22}O_{11}$), acetic acid, and urea have been used as bacterial fermentation reagents (CH_4N_2O). The chemical agent was procured the NaOH, CuO-NPs, and CTAB from Sigma Aldrich, Singapore.

Bacterial Cellulose Synthesis

A nitrogen-rich culture produced by step: 150 g of pineapple peel is blended using 1 liter of water, 75 g of sugar, and 10 g of urea. Then it was boiled, and acetic acid was added to adjust the pH to reach 4.5. During the cold extract, the culture was inoculated using *A. xylinum*, and then fermented for 10 days. Harvest time will indeed be based on forming a transparent cellulose pellicle that floats on top of the solution. The pellicle is washed after harvesting to remove bacteria and other impurities. In addition, to remove impurities, the pellicle was washed in a 1% NaOH solution for 2 h at 90°C. The pellicle was rinsed multiple cycles until pH 7.0.

Homogenization Process

50 g of cellulose pellicles were mashed in a blender (ICH-DS7, Fomac, China) with 1 liter of water for 5 min. The cellulose solution was also homogenized for about 5 cycles at 150 bar using High-Pressure Homogenization equipment (Berkley Scientific, AH-100D type, China) as well as to obtain BNC, BNC solution was filtered.

Composite Foam Synthesis

5 g of cellulose and 40 ml of dH_2O were added with 2% CTAB and various concentrations of CuO-NPs (0.25, 0.5, 1, 1.5, and 2%). After 45 min of stirring with a magnetic stirrer and 1 h of homogenizing with an ultrasonic homogenizer, the cellulose solution was vacuum freeze-dried (Berkley Scientific, China) at 1.0 Pascal and a freezing temperature of -62 °C for 5 days. The composite foam was kept dry by packing it in a plastic bag and placing it in a dry box.

Materials Research Proceedings 53 (2025) 298-304 https://doi.org/10.21741/9781644903575-29

Morphology Analysis

SEM (Scanning Electron Microscope) testing was done in a Hitachi (TM4000) instrument. Also, to increase the conductivity surface of BC was added with a carbon paper prior to testing. In addition, the morphological testing was done at magnification 50X with an accelerating voltage of 15kV and displayed a signal of scattered electrons.

Crystallinity Characterization

The crystallinity of BNC/CuO-NPs was studied using X-ray diffraction (XRD). XRD (PANanalytical, E'xpert Pro type) is used to analysis of crystalline properties such as the degree of crystallinity and crystallinity index of BNC. XRD examination was performed in the 2θ angle of 10° to 90° range, with CuKα (λ) radiation with wavelength 1.5406 Å, at 30.0 mA and 40.0 kV [15]. The Segal equation was used to determine a crystallinity index (CI):

$$CI = \frac{I_{(002)} - I_{(am)}}{I_{(002)}} x100\% \tag{1}$$

Where $I_{(002)}$ is the highest X-ray intensity at 22.5° and I(am) is the X-ray intensity at 18°.

Result and Discussions

Morphology Analysis

Figure 1 shows the surface morphology of BNC with the treatment of (a) CTAB surfactants and CuO-NPs of 0.25% (b), 0.5% (c), 1% (d), 1.5% (e), and 2% (f).

Surfactants can affect the bonds between cellulose, as shown in Figure 1. The SEM of BC control results in large porosity and the porosity that occurs is not homogeneous. Meanwhile, the morphology of the CTAB and CuO-NPs addition resulted in more even porosity. CTAB, when used as a surfactant, could enhance the homogeneity of the resulting porosity. CTAB will decompose and disappear from the material, activating mesoporous in the material [16]. CTAB also could diffuse into the cellulose bond, resulting in better material distribution [17]. CTAB not only greatly enhances the homogeneity of the resulting porosity but also can significantly influence the amount of porosity up to 92.63 % [18].

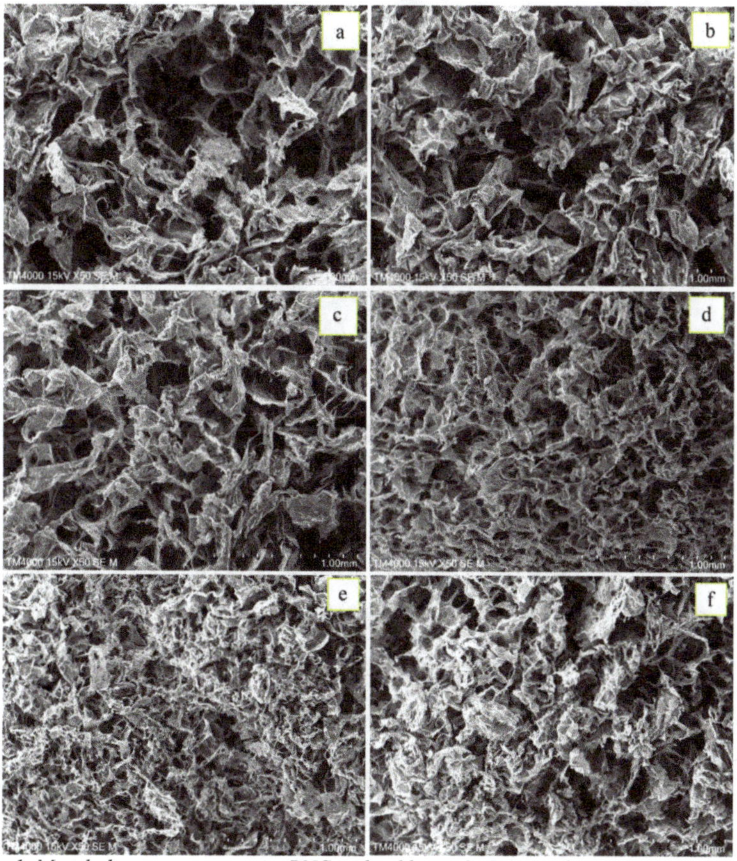

Figure 1. Morphology nanocomposite BNC with addition CuO-NPs of 0% (a), 0.25% (b), 0.5% (c), 1% (d), 1.5% (e), and 2% (f)

Crystallinity Characteristic

Figure 2 shows X-ray diffraction laboratory tests for BNC nanocomposites treated with surfactants and variations of CuO-NPs (0.25, 0.5, 1, 1.5, and 2%). Using equations 1, the crystallinity index (CI) was obtained from the XRD data analysis, as shown in Table 1.

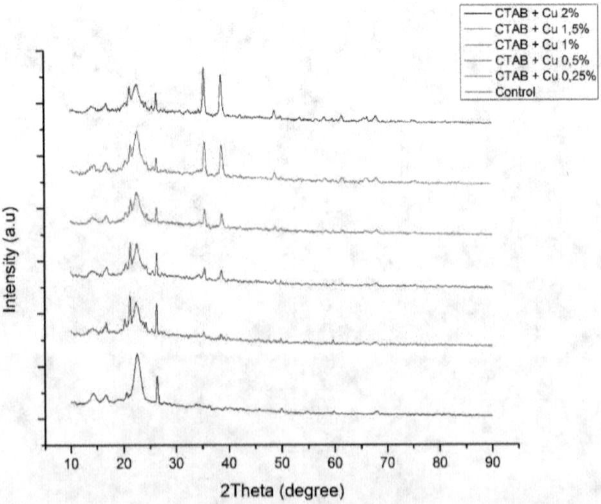

Figure 2. X-Ray Diffraction (XRD) Graphic of nanocomposite BNC with addition CuO-NPs of 0% (a), 0.25% (b), 0.5% (c), 1% (d), 1.5% (e), and 2% (f)

Table 1 Diffraction pattern values of BNC composite foams

Sample	The peak in diffraction pattern (Degree)						Crystalline index (%)
Control	14.32	16.64	-	22.57	-	-	77.43
BNC + CuO 0.25%	14.67	16.81	21.24	22.49	-	38.54	75.50
BNC + CuO 0.5%	14.39	16.84	21.38	22.58	35.47	38.64	71.92
BNC + CuO 1%	14.43	16,90	21.45	22.64	35.59	38.74	68.22
BNC + CuO 1.5%	14.53	16.86	21.46	22.67	35.53	38.77	70.83
BNC + CuO 2%	14.43	16.86	21.35	22.67	35.44	38.64	61.22

The characterization with X-ray diffraction results in a diffraction pattern, as shown in Figure 2. In Figure 2, four main peaks appear in the control sample: 14.5°, 16.8°, 22.6°, and 26°, where these peaks are indicated as cellulose I. In cellulose, I would most represent three main peaks at 14.5°, 16.8°, and 22.6°, while cellulose II will form at an angle of 12.1° and 20° [19]. Cellulose I have multiple polymorphs: triclinic structure (Iα) and monoclinic structure (Iβ), and BC is a type of cellulose I with crystal planes [110], [$1\bar{1}0$], and [200].

The presence of CTAB and CuO-NPs in BNC network resulted in new peaks at an angle of 21.3°. Also, the appearance of peaks at this angle is affected by bond changes in CuO bonds caused by interactions with CTAB surfactants, where they bind to oxygen to form Cu_2O. All the peaks at a lower angle caused by the addition of CTAB can be characterized as simple cubic unit cells, indicating that the lattice of decanethiolate-capped copper nanoparticles was based on the periodic arrangement of nanoparticles. Meanwhile, all of the reflections in the low-angle region were assigned to simple cubic unit cells [20]. The increase in CuO-NPs percentage increased the intensity of CuO-NPs peaks formed at 35.5° and 38.6° [13, 21]. These two peaks represent the type of bonding of monoclinic crystal planes with [$\bar{1}11$] and [111] planes on CuO-NPs [22].

Conclusions

Researchers succeeded in adding CuO-NPs into BNC bonds to obtain foam materials. Also, additional CTAB can affect in morphology and crystallinity of BNC. CTAB can make smaller and homogenize porosity BC/CuO; besides, CTAB can better distribute materials. XRD analysis showed some distinct peaks at 2θ of 14.5°, 16.8°, and 22.6° that belong to crystalline peaks [110], [$1\bar{1}0$], and [200], respectively. In addition, CTAB and CuO increased the intensity by 21.3°, 35.5°, and 38.6°, showing the existence of CuO-NPs. In the future, this treatment can be adopted to prepare bacterial cellulose-based foam material to be applied for wastewater treatment application.

References

[1] Z. Mohammedi, Structure, properties and medical advances for biocellulose applications: A review, Am. J. Polym. Sci. Technol., 3 (2017) 89-96. https://doi.org/10.11648/j.ajpst.20170305.12

[2] M. Zeng, A. Laromaine, and A. Roig, Bacterial cellulose films: influence of bacterial strain and drying route on film properties, Cellulose, 21 (2014) 4455–4469. https://doi.org/10.1007/s10570-014-0408-y

[3] H. Pawignya, Pembuatan protein sel tunggal dari limbah nanas dengan proses fermentasi, Pros. Semin. Nas. Tek. Kim., (2011) A05-1-A05-5.

[4] L.C. De L. Novaes, V. De C.S. Ebinuma, P.G. Mazzola, and A.P. Júnior, Polymer-based alternative method to extract bromelain from pineapple peel waste, Biotechnol. Appl. Biochem., 60 (2013) 527–535. https://doi.org/10.1002/bab.1121

[5] H. J. Song, H. Li, J. H. Seo, M. J. Kim, and S. J. Kim, Pilot-scale production of bacterial cellulose by a spherical type bubble column bioreactor using saccharified food wastes, Korean J. Chem. Eng., 26 (2009) 141–146. https://doi.org/10.1007/s11814-009-0022-0

[6] T. Mahsuli, A. Larasati, A. Aminnudin, and J. Maulana, Effect of the homogenization process on titanium oxide-reinforced nanocellulose composite membranes, J. Mech. Eng. Sci. Technol., 7 (2023) 137-146. https://doi.org/10.17977/um016v7i22023p137

[7] T.A. Agustin and A. Putra, The effect of addition of polyethylene glycol (PEG) on biodegradable plastic based on bacterial cellulose from coconut water (Coconus nucifera), Int. J. Progress. Sci. Technol., 17 (2019) 50–57. [Online]. Available: http://ijpsat.ijsht-journals.org/index.php/ijpsat/article/view/1398/742.

[8] M. Muhajir, H. Suryanto, Y.R.A. Pradana, and U. Yanuhar, Effect of homogenization pressure on bacterial cellulose membrane characteristic made from pineapple peel waste, J. Mech. Eng. Sci. Technol., 6 (2022) 34–39. https://doi.org/10.17977/um016v6i12022p034

[9] A.Y. Booshehri, R. Wang, and R. Xu, Simple method of deposition of CuO nanoparticles on a cellulose paper and its antibacterial activity, Chem. Eng. J., 262 (2015) 999–1008. https://doi.org/10.1016/j.cej.2014.09.096

[10] N. Kawee, N. T. Lam, and P. Sukyai, Homogenous isolation of individualized bacterial nanofibrillated cellulose by high-pressure homogenization, Carbohydr. Polym., 179 (2018) 394–401. https://doi.org/10.1016/j.carbpol.2017.09.101

[11] N. Daels, S. De Vrieze, I. Sampers, B. Decostere, P. Westbroek, A. Dumoulin, P. Dejans, K. De Clerck, S.W.H. Van Hulle, Potential of a functionalised nanofibre microfiltration membrane as an antibacterial water filter, Desalination, 275 (2011) 285–290. https://doi.org/10.1016/j.desal.2011.03.012

[12] J. Nemoto, T. Saito, and A. Isogai, Simple freeze-drying procedure for producing nanocellulose aerogel-containing, high-performance air filters, Appl. Mater. interfaces, (2015). https://doi.org/10.1021/acsami.5b05841

[13] K. Li, L.N. Skolrood, T. Aytug, H. Tekinalp, and S. Ozcan, Strong and tough cellulose nanofibrils composite films: Mechanism of synergetic effect of hydrogen bonds and ionic interactions, ACS Sustain. Chem. Eng., 7 (2019) 14341–14346. https://doi.org/10.1021/acssuschemeng.9b03442

[14] W. Wu, L. Zhang, X. Zhai, C. Liang, and K. Yu, Preparation and photocatalytic activity analysis of nanometer TiO_2 modified by surfactant, Nanomater. Nanotechnol., 8 (2018). https://doi.org/10.1177/1847980418781973

[15] H. Suryanto, A. Aminnudin, R.D. Bintara, A.F. Putra, F.M. Nashrullah, J.S. Binoj, and N. Panicker, Analysis of structure and functional group of filament product-based PLA/nanographite nanocomposite, J. Mech. Eng. Sci. Technol., 7 (2023) 129-136. https://doi.org/10.17977/um016v7i22023p129

[16] H.I. Lee, H.J. Park, Y.K. Park, J.Y. Hur, J.K. Jeon, and J.M. Kim, Synthesis of highly stable mesoporous aluminosilicates from commercially available zeolites and their application to the pyrolysis of woody biomass, Catal. Today, 132 (2008) 68–74. https://doi.org/10.1016/j.cattod.2007.12.029

[17] S.H. Yoon, M. Kang, W. Il Park, and H.J. Jin, Electrically conductive polymeric membranes by incorporation of carbon nanotubes, Mol. Cryst. Liq. Cryst., 464 (2007) 103/[685]-108/[690]. https://doi.org/10.1080/15421400601030449

[18] H. Suryanto, U. Yanuhar, J. Maulana, and Y.R.A. Pradana, Effect of addition surfactant on properties of bacterial cellulose based composite foam reinforced by copper oxide nanoparticle, J. Eng. Sci. Technol. Rev., 14 (2021) 102–106. https://doi.org/10.25103/jestr.146.11

[19] A. Vazquez, M.L. Foresti, P. Cerrutti, and M. Galvagno, Bacterial cellulose from simple and low cost production media by Gluconacetobacter xylinus, J. Polym. Environ., 21 (2013) 545–554. https://doi.org/10.1007/s10924-012-0541-3

[20] L. Chen, D. Zhang, J. Chen, H. Zhou, and H. Wan, The use of CTAB to control the size of copper nanoparticles and the concentration of alkyl thiols on their surfaces, Mater. Sci. Eng. A, 415 (2006) 156–161. https://doi.org/10.1016/j.msea.2005.09.060

[21] Z. Singh and I. Singh, CTAB surfactant assisted and high pH nano-formulations of CuO nanoparticles pose greater cytotoxic and genotoxic effects, Sci. Rep., 9 (2019) 1–13. https://doi.org/10.1038/s41598-019-42419-z

[22] H. Almasi, L. Mehryar, and A. Ghadertaj, Characterization of CuO-bacterial cellulose nanohybrids fabricated by in-situ and ex-situ impregnation methods, Carbohydr. Polym., 222 (2019). https://doi.org/10.1016/j.carbpol.2019.114995

Characteristics of carbon dots derived chitosan-based shrimp shells deacetylated at different temperatures as a potential CO_2 adsorbent

NUR AINAA Mohd Hasyim Chan[1,a], NORHUSNA Mohamad Nor[1,2,b*], AIMI SOLIHAH Zaul Kapri[1,c], MOHAMED SYAZWAN Osman[1,3,d] and ABDUL RAHMAN Mohamed[4,e]

[1]Chemical Engineering Studies, College of Engineering, Universiti Teknologi MARA, Cawangan Pulau Pinang, 13500 Permatang Pauh, Pulau Pinang, Malaysia

[2]Waste Management and Resource Recovery (WeResCue) Group, Chemical Engineering Studies, College of Engineering, Universiti Teknologi MARA, Cawangan Pulau Pinang, 13500 Permatang Pauh, Pulau Pinang, Malaysia

[3]Lab EMZI-UiTM Nanoparticles Colloids, Universiti Teknologi MARA, Cawangan Pulau Pinang, 13500 Permatang Pauh, Pulau Pinang, Malaysia

[4]School of Chemical Engineering, Universiti Sains Malaysia, 14300 Nibong Tebal, Pulau Pinang, Malaysia

[a]ainaamimi20@gmail.com, [b]norhusna811@uitm.edu.my, [c]aimi998822@gmail.com, [d]syazwan.osman@uitm.edu.my, [e]chrahman@usm.my

Keywords: Carbon Dots (CDs), Chitosan, CO_2 Adsorbents, Deacetylation Degree (DA), Hydrothermal Treatment (HT)

Abstract. This research addresses the need to efficiently utilize chitosan-based shrimp shells through synthesizing and characterizing CDs for potential CO_2 applications. It aims to study the effect of deacetylation temperature on chitosan extracted using the chemical extraction method, the effect of deacetylation temperature of chitosan towards the yield of CDs, and the physical and chemical properties of CDs as a potential CO_2 gas adsorbent. The deacetylation temperature was varied (55, 85, and 115 °C) to study its effect on the yield of CDs derived from chitosan-based shrimp shells modified with thiourea/KOH. The CDs were synthesized using the hydrothermal method. The synthesized CDs were characterized using various techniques, including FTIR, TGA, UV-vis spectrophotometry, fluorescence spectroscopy, CHNS/O elemental analysis, and FESEM-EDX. The best quantum yields of the synthesized CDs were determined to be 58.15% for 115 °C of deacetylation temperature. The results showed distinct peaks in the FTIR spectra, indicating the presence of C-H bonds, amide I and amide II bands, C-O-C, O-H, C-H, and C-H out-of-plane functional groups related to the chitosan structure. The thermal stability of chitosan was found to be at 300°C, where the surface morphology explained that the chitosan structure indicated potential modifications or surface alterations due to the synthesis process. The CHNS/O analysis results signify a notable alteration in chitosan to chitosan-based CDs, marked by a reduction in carbon content and the appearance of additional elements, suggesting effective carbonization and modification procedures. In summary, this research contributes to understanding CDs synthesized from chitosan-based shrimp shells, highlighting the importance of deacetylation temperature towards CDs' quantum yield, and underscoring the potential of these synthesized CDs as CO2 adsorbents.

Introduction

Shrimp shells are a byproduct of the seafood industry and contain chitin, protein, and mineral salts. They are generated in large quantities, with millions of tons produced annually, making their efficient utilization crucial for waste management and resource sustainability. The amount of chitin-containing waste from shrimp processing can be as high as 60-80% of the biomass [1]. After being discarded, shrimp shells offer valuable potential in numerous applications, including chitosan production, bioplastics manufacturing, animal feed, and fertilizer creation. Waste can be minimized by repurposing shrimp shells in these ways, and sustainable solutions can be developed [2]. Shrimp shells, often discarded as waste, hold significant value as a source of chitosan, the deacetylated form of chitin. This versatile compound finds applications across various industries, including pharmaceuticals, hair care, skincare, household products, the chemical industry, and the food industry. By extracting chitosan from shrimp shells, we can unlock their potential and contribute to developing sustainable solutions while reducing waste [3].

The reported extraction methods of chitosan are varied, consisting of chemical and biological extraction [4]. Chemical extraction is favored over biological extraction methods due to faster extraction time. The chitosan can be produced using demineralization, deproteination, and deacetylation steps in the chemical extraction method. Among the extraction steps, the deacetylation process is vital in determining the deacetylation degree (DA) of chitosan, which refers to the extent to which the acetyl groups attached to the N-acetylglucosamine units of chitin are removed, resulting in the conversion of chitin into chitosan [5]. As reported by Wozniak et al., chitosan is an excellent carbon precursor for CDs since it is a naturally occurring chitin derivative that contains a polymeric structure of carbon (C) components [6].

According to a study conducted by Alyani et al., comprehensive bibliographic research was employed to collect and study a total of 82 samples from various sources, including shellfish (crustaceans and molluscs), insects, and others, for their chemical extraction of chitin and its derivatives. Alyani et al. aimed to review the chitosan extraction method, specifically focusing on the optimal demineralization and deproteinization processes. The methods used involved comparing the chitin and chitosan contents of insects with those of commercial sources, such as crustaceans, to determine their viability as alternative sources [7]. Previous data from past studies were analyzed to demonstrate the comparable yield values of chitin and chitosan from insects, further supporting the interest in insects as alternative sources. Bombyx eri exhibited a chitosan content of 45%, while Periostracum cicadae (cicada sloughs) demonstrated a slightly lower yield at 36.6%. Chyrysomya megacephala was found to contain chitosan at a rate of 26.2%. These findings underscore the variability in chitosan content among different sources, providing insights into the potential viability of various organisms as alternative chitin sources for industrial applications.

Carbon dots (CDs) have piqued the interest of researchers due to several advantageous physicochemical properties, including exceptional and tuneable photoluminescence (PL), high quantum yield (QY), low toxicity, nano-scale size, numerous low-cost sources, unique optical properties, abundant functional groups, high stability, and providing significant applications [8]. CDs can be categorized into the nanomaterial group and fall into less than 10 nm in size range [9]. CDs can be derived from any carbon precursors, including palm kernel shells, eggshells, coconut shells, orange peels, sugarcane residue, and shrimp shells [10] by hydrothermal carbonization method. Chitosan extracted from shrimp shells can be synthesized into CDs and is abundantly available, with cheaper precursors, a simple synthesis approach, and naturally occurring multi-heteroatom dopants. The presence of several heteroatoms in CDs is a crucial component for future utilization. The existing nitrogen (N) and oxygen (O) contents in the chitosan can be added with additional dopants, such as sulfur-based (S) components, prior to hydrothermal synthesis [11]. Heteroatom doping contributes to forming small micropores and very polar adsorption sites on CDs [12].

There are two general ways to prepare CDs: the top-down approach and the bottom-up approach [13]. The top-down approach requires high pressure, temperature, and lengthy and difficult processes, such as pyrolysis and gasification, making them less preferable. Contrarily, the bottom-up approach is considered more environmentally friendly because of the utilization of existing carbon atoms abundantly available in the biomass precursor. In the bottom-up approach, a conventional hydrothermal synthesis method was reported to produce a high CDs yield. The hydrothermal synthesis method operates at a lower temperature (180°C) and pressure compared to the top-down approach [14]. Due to its straightforward experimental design and lack of requirement for highly toxic chemicals or high temperatures, the hydrothermal synthesis method may also be simple and effective [15].

Due to their unique attributes, bio-based CDs present a compelling avenue for developing novel CO2 removal adsorbents. These CDs offer a high surface area, enhancing their capacity to adsorb CO_2 molecules efficiently from gas streams. Additionally, their surface chemistry can be tailored by introducing heteroatoms like nitrogen, sulfur, and oxygen, optimizing interactions with CO_2 and improving adsorption efficiency. Derived from biomass sources such as shrimp shells, bio-based CDs offer a cost-effective alternative to conventional adsorbents while also promoting sustainability by repurposing waste materials. Furthermore, their production from renewable resources reduces reliance on fossil-based precursors and minimizes environmental impact.

Chitosan-based CDs represent a class of nanoparticles derived from chitosan, a natural polymer obtained from chitin, a significant component of crustacean shells. These CDs exhibit unique properties that make them promising candidates for various applications. Firstly, their synthesis from chitosan, a renewable and biocompatible material, offers an environmentally friendly approach. CDs typically possess high surface areas and tuneable surface chemistries, providing abundant active sites for molecule interactions. This property makes them highly efficient for adsorption applications, including CO_2 capture. Additionally, CDs can be modified with heteroatoms like nitrogen, sulfur, and oxygen, further enhancing their adsorption capabilities and selectivity for specific gases. Their small size, typically ranging from a few nanometers to tens of nanometers, allows for easy dispersion and penetration into various matrices, facilitating diverse applications in sensing, imaging, drug delivery, and environmental remediation.

Furthermore, CDs are generally biocompatible and exhibit low cytotoxicity, making them suitable for biomedical applications. For example, the study by Li et al. focuses on the preparation and characterization of biocompatible nanoparticles by encapsulating CdS quantum dots (QDs) with quaternized chitosan (HTCC) to improve their biocompatibility and stability. The properties of the resulting nanoparticles, referred to as HTCC/CdS QDs, can be controlled by varying the mass ratios of QD to HTCC. The quantum yield of CdS quantum dots (QDs) can be improved from 9% to 23% when encapsulated with quaternized chitosan (HTCC) at a mass ratio of 1:1. The encapsulation of CdS QDs with HTCC at a mass ratio of 1:0.5 resulted in a quantum yield of 18%. CdS QDs encapsulated with HTCC at a mass ratio of 1:2 exhibited a quantum yield of 15%. The encapsulation of CdS QDs with HTCC enhances their quantum yield and stability at room temperature, with the highest quantum yield observed at a mass ratio of 1:1 [16].

Figure 1 illustrates the development of antibacterial functional CDs utilizing chitosan (CS) and its derivatives as primary raw materials.

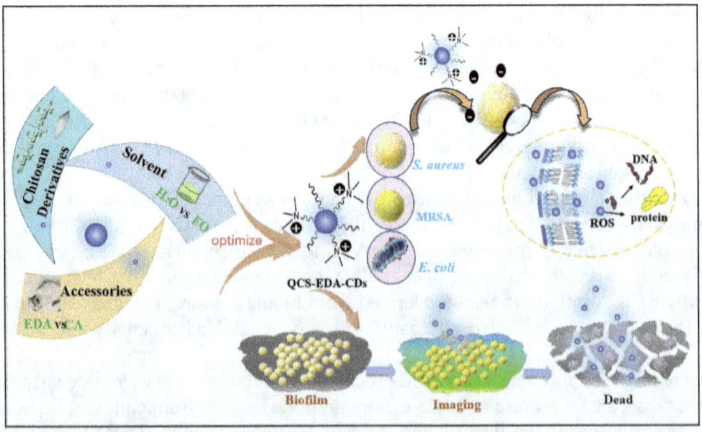

Fig. 1 The Development of Antibacterial Functional CDs Using Chitosan (CS) and Its Derivatives as Raw Materials [17]

The study by Zhao et al. [17] explained the development of antibacterial functional CDs using chitosan (CS) and its derivatives as starting materials. The objective is to produce highly effective antibacterial and antibiofilm nanomaterials to combat pathogenic bacteria, especially those forming biofilms. Through investigations into various experimental parameters, including the structures of raw materials, excipients, and solvents, the research evaluates their impact on the CDs' optical properties and antibacterial efficacy. The study concludes that these functional CDs exhibit promising potential as visual probes and therapeutic agents for treating infections caused by bacteria and biofilms. While Zhao et al. concentrate on developing antibacterial CDs for combating pathogenic bacteria and biofilms, the latter study investigates the potential of CDs derived from chitosan-based shrimp shells as CO_2 adsorbents. While both studies involve the utilization of chitosan-based CDs, they address distinct challenges and applications: antibacterial properties in one case and CO_2 capture potential in the other. These diverse applications showcase chitosan-based CDs' versatility and wide-ranging applicability in various fields, from healthcare to environmental remediation.

The available adsorbent materials typically used in CO_2 removal were activated carbon, biochar, zeolites, graphene, and carbon nanotubes [18]. The potential of CDs synthesized from chitosan-derived shrimp shells can be further explored, where the characteristics of the CDs' adsorbent can be modified. Hence, including N/S/O heteroatoms in the chitosan-derived shrimp shell is anticipated to assist the distribution charge of the CDs' surface [19], thus improving the adsorbent's interaction with the CO_2 molecules.

This study focuses on synthesizing CDs from chitosan-derived shrimp shells through hydrothermal synthesis at varying deacetylation degrees. Characterization techniques were employed to analyze the produced CDs, including FTIR spectroscopy, TGA, UV-vis spectrophotometry, fluorescence spectroscopy, CHNS/O analysis, and FE-SEM-EDX. Investigating the influence of deacetylation temperature on chitosan extraction from shrimp shells using a chemical extraction method, the research aims to explore the intricate relationship between deacetylation temperature, CD yield, and the resulting physical and chemical properties of CDs as potential CO_2 gas adsorbents. Specifically, the study examines the impact of different deacetylation temperatures (55, 85, and 115 °C) on the yield of CDs derived from chitosan-based shrimp shells modified with thiourea/KOH and synthesized using the hydrothermal method. The novelty of this work lies in its comprehensive analysis of how chitosan deacetylation conditions affect CDs properties, particularly in relation to CO_2 adsorption. Addressing a research gap in systematically exploring the influence of deacetylation temperature on CDs yield and CO_2 adsorption suitability, this research provides valuable insights for optimizing CDs synthesis processes, thereby advancing sustainable carbon capture technologies and aiding in climate change mitigation efforts.

Methodology

Chemicals. Shrimp shells were obtained from 105 Frozen Seafood & Nicole Cottage, Selangor. All chemicals are used as received, such as hydrochloric acid, HCl (Emsure, 37%), sodium hydroxide NaOH (Emsure, 98%), ethanol (Sigma Aldrich, 95%), acetic acid (R&M Chemicals, 99.8%), thiourea (Sigma Aldrich, 99%), and potassium hydroxide (Merck, 85%).

Extraction of Chitosan from Shrimp Shell Waste. The shrimp shells obtained from kitchen waste were extracted using a chemical extraction method via demineralization, deproteination, and deacetylation procedures to get a high chitosan yield. Deacetylation temperatures (55°C, 85°C, and 115°C) were varied to study the effect of deacetylation degree towards the yield of CDs. The deacetylation degree was determined using the method proposed by Sarbon et al. [20]. Using equation (1), the DA was calculated:

$$DA = \frac{(161.16\,x10^{-3} * (V_2 - V_1)N)}{W_1} \quad (1)$$

where 161.16 x10-3 is the constant mass of chitosan monomer, V1 and V2 are the volumes of NaOH solution used, N is the strength of NaOH solution (0.1 M), and W1 is the mass of the sample.

Synthesis of CDs derived from Chitosan-based Shrimp Shells. The extracted chitosan-derived shrimp shell was synthesized using a hydrothermal method. Before employing the hydrothermal synthesis, the extracted chitosan-derived shrimp shell would undergo a chemical modification using thiourea and KOH. Using a hydrothermal reactor, the hydrothermal synthesis was carried out at a doping ratio of chitosan: thiourea: KOH (1:1:3) [21], [22], synthesis time (24 hours), and synthesis temperature (180 °C) [14].

Characterizations. The physical and chemical properties of the synthesized CDs were further analyzed using Cary Eclipse Fluorescence Spectrophotometer (FS), Thermal Gravimetry Analysis was performed SDT Q600 Thermal Gravimetry Analysis (TGA), Fourier transform infrared spectra were determined by Thermo Scientific Nicolet iS10 Fourier-transform Infrared Spectroscopy (FTIR), CHNS/O Elemental Analyzer were obtained by using Perkin Elmer CHNS/O Elemental Analyzer, and high-resolution images were performed using Nova Nanosem 450 Field emission scanning electron microscopy (FESEM-EDX). The study comprehensively analyzed the characteristics of CDs, which relate to the chemical modification with thiourea/KOH and hydrothermal synthesis parameters, as well as the potential of CDs as CO_2 gas adsorbent. Fluorescence Quantum Yield, φ_x:

$$= \frac{Slope\ Sample}{Slope\ Standard} \times Quantum\ Yield\ Standard, \varphi_{std} \quad (2)$$

where the standard is the reference compound (Quinine sulphate dissolved in 0.1 M H_2SO_4), φstd is 0.54 (reference) [23], [24].

Results and Discussion

Effect of Deacetylation Temperature on Chitosan Yield. The deacetylation process involved removing acetyl groups from chitosan molecules was conducted at various temperatures. These temperatures included 55 °C, 85 °C, and 115 °C, each representing different conditions under which chitosan extraction from shrimp shells was performed. Figure 2 illustrates the yields of deacetylated products, and the deacetylation degree of chitosan obtained at these distinct deacetylation temperatures. The obtained yield values of the deacetylated product derived from shrimp shells indicate that as the deacetylation temperature increases, the yield of the deacetylated product decreases.

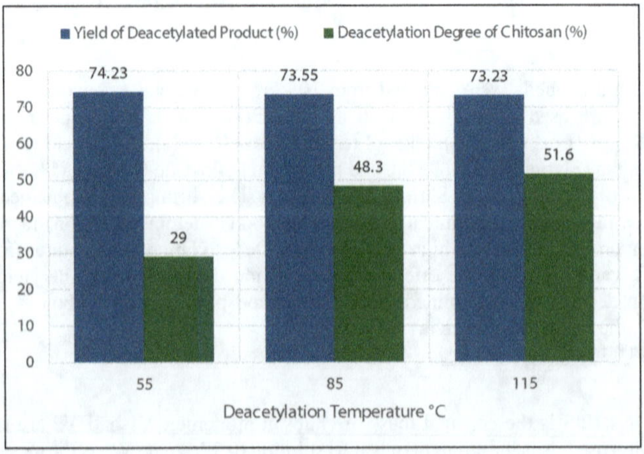

Fig. 2 The Yield of Deacetylated Product and the Deacetylation Degree of Chitosan at Different Deacetylation Temperature

Specifically, the yield values obtained were 74.23% for the sample synthesized at 55 °C, 73.55% for the sample synthesized at 85 °C, and 73.23% for the sample synthesized at 115 °C, as shown in Figure 2. This inverse relationship between deacetylation temperature and yield suggests that higher temperatures during the deacetylation process have a negative impact on the overall yield of the deacetylated product. One possibility is that higher temperatures may lead to increased degradation or loss of the chitosan structure, resulting in lower yields of the deacetylated product. Additionally, higher temperatures might accelerate the reaction kinetics, leading to incomplete deacetylation and the formation of byproducts that lower the overall yield. In summary, the inverse relationship between deacetylation temperature and yield of the deacetylated product derived from shrimp shells indicates that higher temperatures result in lower yields [25].

From Figure 2, the deacetylation degree was 29% for the sample deacetylated at 55 °C, 48.3% for the sample deacetylated at 85 °C, and 51.6% for the sample deacetylated at 115 °C. A higher deacetylation degree indicates a more excellent conversion of chitin into chitosan, with a higher proportion of acetyl groups being replaced by amino groups. As the deacetylation temperature increases from 55 °C to 115 °C, there is a noticeable increase in the deacetylation degree. The increase in deacetylation degree with higher deacetylation temperatures suggests that elevated temperatures promote a more thorough deacetylation process, resulting in a higher degree of conversion from chitin to chitosan. This can be attributed to the enhanced reactivity of the chitin structure at higher temperatures, facilitating the cleavage of acetyl groups and promoting the deacetylation reaction. The results obtained in this study highlight the impact of deacetylation temperature on the degree of deacetylation and provide insights for optimizing the deacetylation process [26], [27]. By carefully controlling the deacetylation temperature, it is possible to achieve the desired level of deacetylation and tailor the properties of the resulting chitosan-based material to meet specific requirements for various applications. In summary, determining the deacetylation degree for the different samples synthesized at varying deacetylation temperatures reveals the influence of temperature on the extent of deacetylation. The higher deacetylation degrees obtained at elevated temperatures demonstrate the potential for increasing chitin conversion to chitosan.

Effect of Deacetylation Temperature towards CDs' Yield. Figure 3 shows the fluorescence quantum yield and deacetylation degree for each sample. From Figure 3, the quantum yield represents the proportion of absorbed photons re-emitted as fluorescence by the CDs derived from chitosan-based shrimp shells.

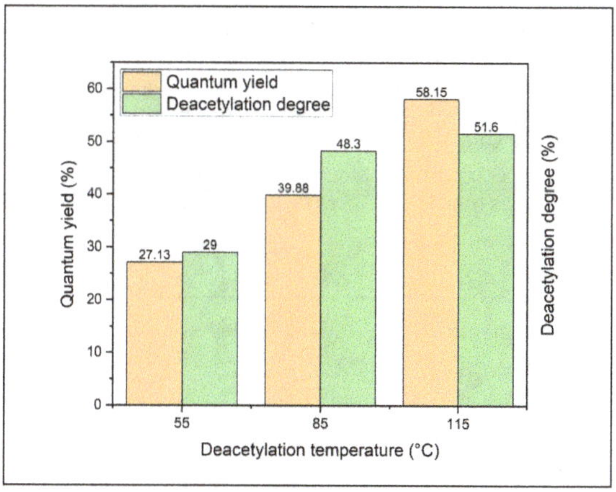

Fig. 3 The Fluorescence Quantum Yield and Deacetylation Degree for Each Sample

In this study, the fluorescence quantum yield values obtained were 27.13% for the sample synthesized at 55 °C, 39.42% for the sample synthesized at 85 °C, and 58.15% for the sample synthesized at 115 °C. These quantum yield values indicate the relative fluorescence efficiency of the samples and provide valuable information regarding their optical properties. Higher quantum yield values indicate a higher proportion of absorbed photons emitted as fluorescence, suggesting greater efficiency in the fluorescence emission process. The observed trend in the quantum yield values across the different deacetylation temperatures reveals the influence of temperature on the fluorescence efficiency of the CDs. As the deacetylation temperature increases from 55 °C to 115 °C, there is a noticeable increase in the fluorescence quantum yield. This suggests that higher deacetylation temperatures improve fluorescence performance and efficiency [28]. The increase in quantum yield with higher deacetylation temperatures could be attributed to several factors. One possibility is that higher temperatures during the deacetylation process result in a higher degree of deacetylation, leading to enhanced carbon dot formation and improved fluorescence properties. Additionally, increased deacetylation temperatures might facilitate better control of the synthesis process, resulting in the formation of CDs with more favorable structural and optical characteristics for efficient fluorescence emission [29]. The variation in quantum yield values among the samples synthesized at various deacetylation temperatures highlights the importance of optimizing the synthesis conditions to achieve desired fluorescence properties. The results indicate that a higher deacetylation temperature, such as 115 °C, might lead to synthesizing CDs with superior fluorescence efficiency, making them more promising for potential applications in fluorescence-based technologies. Overall, the obtained quantum yield values provide quantitative measures of the fluorescence efficiency of the CDs derived from chitosan-based shrimp shells synthesized at different deacetylation temperatures.

Thermal Stability (TGA) of Chitosan. Figure 4 shows the thermal stability of deacylated chitosan samples. The thermal stability analysis, depicted by the thermal decomposition graph of the deacetylated chitosan, as shown in Figure 4, revealed that the materials exhibited a similar onset of thermal decomposition at around 300°C. This observation aligns with recent studies investigating the thermal behavior of chitosan-based materials. In a study by Li et al., it was reported that the thermal decomposition of chitosan occurred in two stages, with the initial stage starting at approximately 300°C. This stage involves the degradation of the polysaccharide backbone, resulting in the release of volatile products. This research's onset temperature of thermal decomposition aligns with this reported temperature range, suggesting a similar degradation mechanism for the deacetylated chitosan. The comparable thermal decomposition behaviors observed in the deacetylated chitosan samples and the corroboration from recent studies indicate that these materials possess similar thermal stability characteristics. Understanding the thermal behavior of these materials is crucial for their potential applications, such as in thermal energy storage, catalysis, or pyrolysis processes, where thermal stability plays a vital role [30].

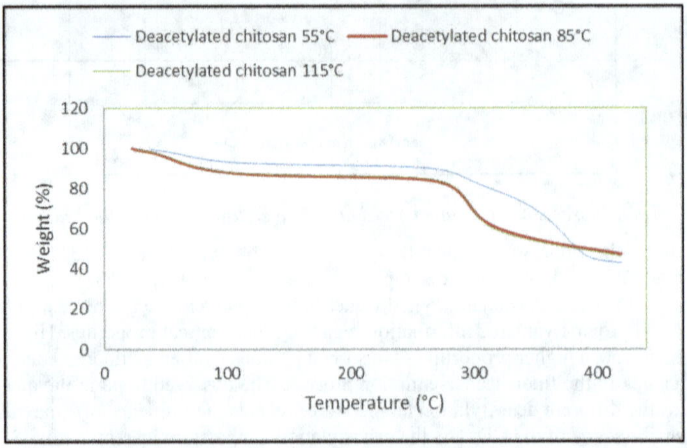

Fig. 4 Thermal Stability of Deacylated Chitosan

Surface Morphology of Chitosan. The FE-SEM analysis of the chitosan sample at varying magnifications offers detailed insights into its morphological characteristics and structural features. Figure 5 shows the analysis images of chitosan sample at x100, x1000, and x3000 magnifications.

https://doi.org/10.21741/9781644903575-30

a)

b)

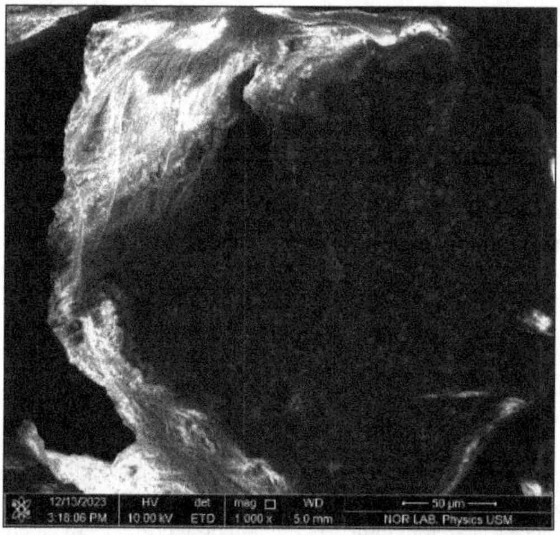

Materials Research Proceedings 53 (2025) 305-321 https://doi.org/10.21741/9781644903575-30

c)

Fig. 5 FESEM Analysis Images at varying magnification of x100 (a), x1000 (b), and x3000 (c)

The FESEM-EDX analysis of the chitosan sample provides a comprehensive understanding of its structural morphology at varying magnifications, offering a detailed understanding of its characteristics. At x100 magnification, the image reveals the overall structure of the shrimp shells, providing insights into their size, shape, and surface features. Upon increasing the magnification to x1000, distinct dots emerge on the chitosan structure, indicating potential modifications or surface alterations resulting from the synthesis process. However, it is at x3000 magnification that these dots are distinctly observed to be attached to the chitosan structure, suggesting a close association or integration between the dots and the chitosan matrix. This observation underscores the successful formation and attachment of the dots onto the chitosan framework, which holds promise for various applications.

The Energy Dispersive X-ray (EDX) analysis results as shown in table 1 indicate the elemental composition of the sample in weight percentage (wt%). In this case, the sample contains carbon (C) at 49.49%, nitrogen (N) at 6.44%, oxygen (O) at 39.24%, sulfur (S) at 0.24%, and calcium (Ca) at 4.6%. These elements are detected and quantified based on the characteristic X-rays emitted when the sample is bombarded with high-energy electrons. The presence of carbon, nitrogen, and oxygen is consistent with the composition of organic materials like chitosan, while the detection of sulfur suggests the presence of sulfur-containing compounds, possibly from impurities or additives. The presence of calcium indicates the potential presence of calcium-containing minerals or compounds, which may originate from the source material or be introduced during processing. Overall, the EDX results provide valuable information about the elemental composition of the sample, aiding in its characterization and analysis.

Table 1 The Energy Dispersive X-ray (EDX) analysis results

Element	Wt%
C	49.49
N	6.44
O	39.24
S	0.24
Ca	4.6

Surface Functional Groups of Chitosan and CDs. Figure 6 represents the FTIR Spectra of the Deacetylated Chitosan Sample. The FTIR spectra analysis reveals similarities among the deacetylated chitosan samples synthesized at various temperatures (55 °C, 85 °C, and 115 °C), indicating the presence of shared functional groups. Specifically, the observed peak at approximately 2880 cm-1 signifies the presence of C-H bonds in the chitosan structure, originating from the acetylated units of chitin that have undergone deacetylation to form chitosan. Additionally, peaks at around 1650 cm-1 and 1400 cm^{-1} correspond to the amide I and amide II bands, respectively, indicating the presence of amide bonds formed during the deacetylation process. Furthermore, the peak at approximately 1050 cm^{-1} corresponds to the glycosidic C-O-C stretching vibration characteristic of chitosan, confirming the presence of glycosidic linkages within the structure [31]. Despite subtle variations in peak intensity and shape, the overall similarity in the FTIR spectra suggests a consistent chemical composition among the deacetylated chitosan samples, providing a basis for further exploration of their structure-property relationships and potential applications. Figure 6 shows the FTIR Spectra of CDs Derived from Chitosan Sample.

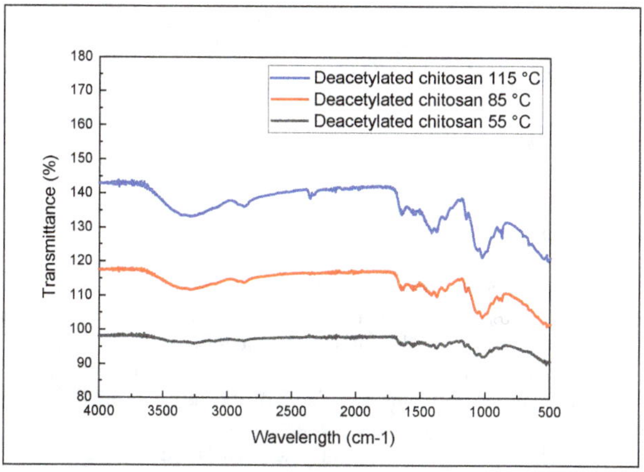

Fig. 6 FTIR Spectra of Deacetylated Chitosan Sample

The Fourier Transform Infrared (FTIR) spectra of CDs samples shown in Figure 7 were synthesized at 55, 85, and 115 °C exhibit similar trends, suggesting the presence of common functional groups [17]. Notably, peaks at around 3300 cm-1 indicate O-H stretching vibrations, typically associated with hydroxyl groups found in organic compounds like alcohols and phenols. These groups may originate from the chitosan precursor or other components involved in the synthesis. The presence of peaks at 1350 cm-1 signifies C-H bending vibrations, indicative of the carbon backbone of the CDs, possibly arising from the chitosan structure in the case of CDs synthesized from shrimp shells. Peaks at 700 cm-1 suggest C-H out-of-plane bending vibrations, reflecting the bending of carbon-hydrogen bonds in aromatic or aliphatic compounds within the CDs [32].

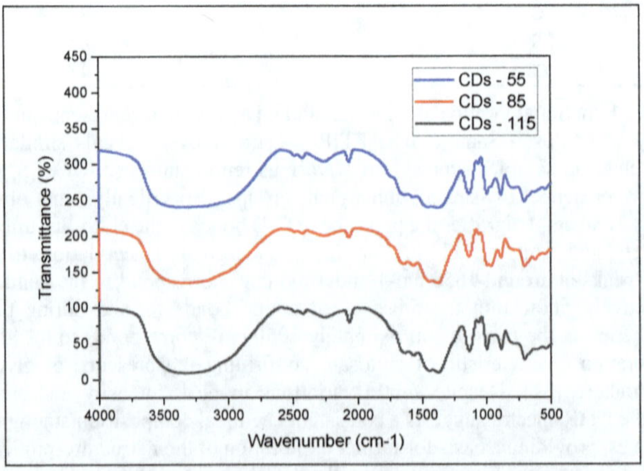

Figure 7 FTIR Spectra of CDs Derived from Chitosan Sample

The consistent presence of these peaks across varying deacetylation temperatures suggests stability in fundamental functional groups. However, subtle variations in peak characteristics could provide insights into factors such as the degree of deacetylation or other structural modifications. Further detailed analysis of peak assignments and their variations with temperature could offer a more comprehensive understanding of the synthesized CDs' chemical composition and structural changes.

CHNS/O Elemental Analysis of Chitosan and CDs. The CHNS analysis offers vital insights into materials' elemental composition and chemical structure, which is crucial for understanding their properties and potential applications. The study reveals notable differences in their elemental compositions in the context of chitosan and chitosan-based CDs. Chitosan, derived from shrimp shells, primarily consists of carbon, hydrogen, nitrogen, sulfur, and miscellaneous components. However, after synthesizing CDs, there is a significant alteration in the elemental composition. The CHNS/O analysis results reveal substantial differences between the composition of chitosan and chitosan-based CDs. In the chitosan sample, carbon comprises the most significant portion at 34%, followed by hydrogen (9.58%), nitrogen (7.99%), and sulfur (0.62%) [33]. However, in the chitosan-based CDs, the carbon content decreases substantially to 6.38%, while hydrogen (3.36%), nitrogen (4.62%), and sulfur (1.09%) also show reductions. These findings suggest a considerable transformation of the chitosan structure during the synthesis of CDs, likely due to carbonization and modification processes. The decrease in carbon content, the emergence of other elements, and a substantial increase in miscellaneous components suggest significant structural changes during carbonization and modification processes. Reducing carbon content indicates the successful conversion of chitosan into CDs, which typically have lower carbon content due to the formation of functional groups and heteroatoms on their surfaces. The presence of nitrogen and sulfur in the CDs may also enhance their adsorption properties, potentially improving their performance as CO_2 adsorbents. Thus, the CHNS analysis is a critical tool for elucidating the chemical transformations occurring during synthesizing CDs from chitosan and provides key understanding of their potential as CO_2 adsorbents.

Table 2 CHNS/O Elemental Analysis of Chitosan and Chitosan-based CDs

Sample Name	Weight (mg)	Carbon (%)	Hydrogen (%)	Nitrogen (%)	Sulfur (%)
Chitosan	2.236	34.00	9.58	7.99	0.62
Carbon Dots	1.760	6.38	3.36	4.62	1.09

CO_2 Adsorption Performance of Chitosan and CDs. The performance of chitosan and N/S/O-doped CQDs adsorbents was evaluated using a custom-designed CO_2 adsorption rig. In this setup, 0.5 grams of the adsorbent was accurately weighed and placed at the center of a stainless-steel column measuring 250 mm in length and 9.525 mm in diameter, supported by 0.3 grams of borosilicate glass wool. At ambient temperature, a simulated gas stream containing CO_2 was passed through the column at flow rates of 50 ml/min, with the gas composition controlled via a flow meter. The CO_2 concentrations at the inlet and outlet were continuously monitored using a portable gas analyzer equipped with an electrochemical sensor calibrated for syngas composition. CO_2 concentrations were recorded every 10 seconds, and the adsorption test continued until the system reached saturation (C_i/C_o = 1.0). Below is the graph depicting the CO_2 adsorption performance of the chitosan and N/S/O-doped CQDs adsorbents.

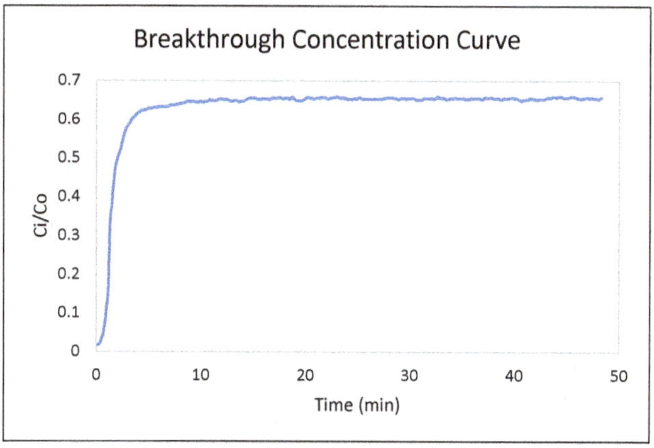

Fig. 8 CO_2 Adsorption Breakthrough Concentration Curve of Chitosan

The CO_2 adsorption performance of the chitosan sample, as indicated by the graph, shows that equilibrium is reached at approximately 23 minutes. The breakthrough point, where $C_i/C_o = 0.1$, is observed around one minute into the adsorption evaluation, suggesting that the adsorbent quickly becomes saturated and does not exhibit significant CO_2 adsorbent criteria for this sample. The total adsorption capacity is measured at 31.88 mg CO_2 per gram of adsorbent. These findings indicate a relatively rapid initial adsorption phase followed by a quick saturation, which implies limited adsorption efficiency for this sample. The quick breakthrough and lower total capacity suggest that while the adsorbent can initially capture CO_2, it quickly reaches its limit, which may be a limitation for extended or large-scale CO_2 capture applications.

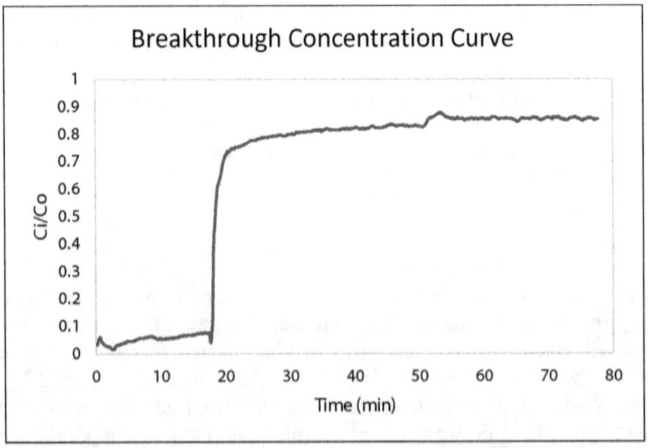

Fig. 9 CO_2 Adsorption Breakthrough Concentration Curve of N/S/O-doped CQDs

The CO2 adsorption performance of the N/S/O-doped CQDs adsorbents, as illustrated in the graph, shows that equilibrium is reached at approximately 53 minutes. This indicates the point at which the adsorbent can no longer uptake additional CO_2, signifying its maximum adsorption capacity. The breakthrough point, where $C_i/C_o = 0.1$, occurs around 18 minutes into the adsorption evaluation, highlighting the initial rapid adsorption phase where the adsorbent is most effective. The total adsorption capacity of this adsorbent is determined to be 60.38 mg CO_2 per gram of adsorbent, demonstrating a significant ability to capture CO_2. These findings underscore the efficiency of the N/S/O-doped CQDs as a potential CO_2 adsorbent, with a relatively quick initial adsorption phase followed by a sustained capacity until equilibrium is achieved.

In conclusion, the comparison of CO_2 adsorption performance between chitosan and N/S/O-doped CQDs adsorbents highlights distinct differences in their capabilities. The chitosan sample reaches equilibrium rapidly at 23 minutes with a breakthrough point at one minute, and a total adsorption capacity of 31.88 mg CO_2 per gram, indicating a quick saturation and limited efficiency for prolonged CO_2 capture. On the other hand, the N/S/O-doped CQDs adsorbent reaches equilibrium at 53 minutes, with a breakthrough point at 18 minutes, and a significantly higher total adsorption capacity of 60.38 mg CO_2 per gram. This demonstrates a more effective initial adsorption phase and a greater overall CO_2 capture ability. These findings suggest that while chitosan may offer faster initial adsorption, the N/S/O-doped CQDs are superior in terms of sustained adsorption capacity and efficiency, making them more suitable for extended or large-scale CO_2 capture applications.

The Potential of CDs as CO_2 Adsorbent. CDs exhibit promising potential as CO_2 adsorbents owing to their distinctive characteristics. These nanoparticles typically boast high surface areas and customizable surface chemistries, facilitating efficient CO_2 capture by providing ample active sites for adsorption. Moreover, renewable biomass sources, such as shrimp shells, for CDs synthesis aligns with sustainable practices, offering an environmentally friendly approach to CO_2 mitigation. Incorporating heteroatoms into CDs further augments their adsorption capabilities, enhancing selectivity and efficiency in capturing CO_2 from gas streams. Additionally, CDs' capacity for regeneration and multiple reuses enhances their cost-effectiveness and practicality in continuous CO_2 capture processes, reducing overall operational expenses. Overall, CDs' versatility and eco-friendly attributes position them as up-and-coming candidates for addressing the pressing challenges associated with CO_2 emissions and climate change mitigation, highlighting their potential significance in advancing sustainable carbon capture technologies.

Conclusion

In conclusion, the results demonstrated that the effect of deacetylation temperature on chitosan extraction from shrimp shells was feasible, with yields of 74.23% at 55 °C, 73.55% at 85 °C, and 73.23% at 115 °C. The deacetylation process yielded chitosan samples with deacetylation degrees ranging from 29% to 51.6%, depending on the deacetylation temperature. The synthesis of CDs resulted in varying fluorescence quantum yields of 27.13%, 39.88%, and 58.15% for deacetylation temperatures of 55°C, 85°C, and 115°C, respectively. The FTIR analysis revealed the presence of C-H bonds, amide I and amide II bands, C-O-C, O-H, C-H, and C-H out-of-plane functional groups related to the chitosan structure and the synthesized CDs. Chitosan exhibited thermal stability up to 300°C, with the surface morphology indicating possible alterations or modifications due to the synthesis process. The CHNS/O analysis suggests a significant transformation of chitosan into chitosan-based CDs, characterized by a decrease in carbon content and the emergence of other elements, suggesting successful carbonization and modification processes. The obtained CDs demonstrated potential as CO_2 adsorbents due to their fluorescence properties and the presence of functional groups suitable for CO_2 interactions. Overall, this research contributes to understanding CDs synthesis from chitosan-based shrimp shells and provides insights into the relationship between deacetylation temperature, fluorescence quantum yield, and CO_2 adsorption potential. The findings pave the way for further optimizing and exploring CDs derived from chitosan for various environmental and technological applications, particularly in CO_2 capture and mitigation.

Acknowledgement

This research was supported by the Ministry of Higher Education (MOHE) through the Fundamental Research Grant Scheme (Grant No. FRGS/1/2022/TK0/UITM/02/55). The authors are grateful for the facilities provided by Universiti Teknologi MARA (UiTM) Cawangan Pulau Pinang to complete this work.

References

[1] C. H. Wang, C. T. Doan, V. B. Nguyen, A. D. Nguyen, and S. L. Wang, "Reclamation of Fishery Processing Waste: A Mini-Review.," *Molecules*, vol. 24, no. 12, p. 2234, Jun. 2019. https://doi.org/10.3390/MOLECULES24122234

[2] X. Hu *et al.*, "Green, Simple, and Effective Process for the Comprehensive Utilization of Shrimp Shell Waste," *ACS Omega*, vol. 5, no. 30, pp. 19227–19235, Aug. 2020. https://doi.org/10.1021/ACSOMEGA.0C02705/ASSET/IMAGES/LARGE/AO0C02705_0007.JPEG

[3] S. Pawar, U. K. Togiti, A. Bhattacharya, and A. Nag, "Functionalized Chitosan-Carbon Dots: A Fluorescent Probe for Detecting Trace Amount of Water in Organic Solvents," *ACS Omega*, vol. 4, no. 6, pp. 11301–11311, Jun. 2019. https://doi.org/10.1021/acsomega.9b01208

[4] H. A. A. Cahyaningtyas, W. Suyotha, B. Cheirsilp, A. A. Prihanto, S. Yano, and M. Wakayama, "Optimization of protease production by Bacillus cereus HMRSC30 for simultaneous extraction of chitin from shrimp shell with value-added recovered products," *Environmental Science and Pollution Research*, vol. 29, no. 15, pp. 22163–22178, Mar. 2022. https://doi.org/10.1007/S11356-021-17279-8

[5] R. S. C. M. de Queiroz Antonino *et al.*, "Preparation and Characterization of Chitosan Obtained from Shells of Shrimp (Litopenaeus vannamei Boone).," *Mar Drugs*, vol. 15, no. 5, May 2017. https://doi.org/10.3390/md15050141

[6] A. Woźniak and M. Biernat, "Methods for crosslinking and stabilization of chitosan structures for potential medical applications," *J Bioact Compat Polym*, vol. 37, no. 3, pp. 151–167, May 2022. https://doi.org/10.1177/08839115221085738

[7] N. Alyani *et al.*, "Molecular Sciences The Potential of Insects as Alternative Sources of Chitin: An Overview on the Chemical Method of Extraction from Various Sources". https://doi.org/10.3390/ijms21144978

[8] K. Luo, Y. Wen, and X. Kang, "Halogen-Doped Carbon Dots: Synthesis, Application, and Prospects," *Molecules*, vol. 27, no. 14, Jul. 2022. https://doi.org/10.3390/MOLECULES27144620

[9] A. P. Francis and T. Devasena, "Toxicity of carbon nanotubes: A review," *Toxicol Ind Health*, vol. 34, no. 3, pp. 200–210, Mar. 2018. https://doi.org/10.1177/0748233717747472

[10] M. Kurian and A. Paul, "Recent trends in the use of green sources for carbon dot synthesis–A short review," *Carbon Trends*, vol. 3, Apr. 2021. https://doi.org/10.1016/J.CARTRE.2021.100032

[11] R. Tondwal and M. Singh, "Chitosan functionalization with a series of sulfur-containing α-amino acids for the development of drug-binding abilities," *J Appl Polym Sci*, vol. 135, no. 12, Mar. 2018. https://doi.org/10.1002/APP.46000

[12] J. Liu, R. Li, and B. Yang, "Carbon Dots: A New Type of Carbon-Based Nanomaterial with Wide Applications," *ACS Cent Sci*, vol. 6, no. 12, pp. 2179–2195, Dec. 2020. https://doi.org/10.1021/acscentsci.0c01306

[13] X. Zhou, K. Yi, Y. Yang, G. Xie, X. Ji, and Z. He, "A novel method for the synthesis of carbon dots assisted by free radicals," *Nano Res*, vol. 15, no. 10, pp. 9470–9478, Oct. 2022. https://doi.org/10.1007/S12274-022-4567-0

[14] H. Ababneh and B. H. Hameed, "Chitosan-derived hydrothermally carbonized materials and its applications: A review of recent literature," *International Journal of Biological Macromolecules*, vol. 186. Elsevier B.V., pp. 314–327, Sep. 01, 2021. doi: 10.1016/j.ijbiomac.2021.06.161

[15] N. Boudouaia, Z. Bengharez, and S. Jellali, "Preparation and characterization of chitosan extracted from shrimp shells waste and chitosan film: application for Eriochrome black T removal from aqueous solutions," *Appl Water Sci*, vol. 9, no. 4, Jun. 2019. https://doi.org/10.1007/S13201-019-0967-Z

[16] Y. Li, M. Hu, B. Qi, and X. Wang, "Preparation and Characterization of Biocompatible Quaternized Chitosan Nanoparticles Encapsulating CdS Quantum Dots," *J Biotechnol Biomater*, vol. 1, no. 4, Jul. 2011. https://doi.org/10.4172/2155-952X.1000108

[17] D. Zhao *et al.*, "Screening of Chitosan Derivatives-Carbon Dots Based on Antibacterial Activity and Application in Anti-Staphylococcus aureus Biofilm," *Int J Nanomedicine*, vol. 17, pp. 937–952, 2022. https://doi.org/10.2147/IJN.S350739

[18] J. Ebenezar, Ed., *Recent Trends in Materials Science and Applications*, vol. 189. Cham: Springer International Publishing, 2017. doi: 10.1007/978-3-319-44890-9

[19] Y. Hu *et al.*, "Synthesis of Carbon Dots by Varying Doped Elements and Application in Serine Detection," *J Fluoresc*, vol. 30, no. 6, pp. 1447–1456, Dec. 2020. https://doi.org/10.1007/S10895-020-02592-1

[20] N. M. Sarbon, S. Sandanamsamy, S. F. S. Kamaruzaman, and F. Ahmad, "Chitosan extracted from mud crab (Scylla olivicea) shells: physicochemical and antioxidant properties," *J Food Sci Technol*, vol. 52, no. 7, p. 4266, Jul. 2015. https://doi.org/10.1007/S13197-014-1522-4

[21] Y. Lv *et al.*, "A comprehensive study on KOH activation of ordered mesoporous carbons and their supercapacitor application," *J. Mater. Chem.*, vol. 22, no. 1, pp. 93–99, Dec. 2012. https://doi.org/10.1039/C1JM12742J

[22] Z. Tian, F. Lai, T. Heil, S. Cao, and M. Antonietti, "Synthesis of carbon frameworks with N, O and S-lined pores from gallic acid and thiourea for superior CO2 adsorption and supercapacitors," *Science China Materials 2020 63:5*, vol. 63, no. 5, pp. 748–757, Mar. 2020. https://doi.org/10.1007/S40843-019-1254-9

[23] S. Daniel, "Characterization of carbon dots," *Carbon Dots in Analytical Chemistry: Detection and Imaging*, pp. 43–58, Jan. 2022. https://doi.org/10.1016/B978-0-323-98350-1.00015-3

[24] J. B. Birks, "Fluorescence Quantum Yield Measurements," *J Res Natl Bur Stand A Phys Chem*, vol. 80A, no. 3, p. 389, May 1976. https://doi.org/10.6028/JRES.080A.038

[25] M. L. Tsaih and R. H. Chen, "The effect of reaction time and temperature during heterogenous alkali deacetylation on degree of deacetylation and molecular weight of resulting chitosan," *J Appl Polym Sci*, vol. 88, no. 13, pp. 2917–2923, Apr. 2003. https://doi.org/10.1002/APP.11986

[26] N. Yaghobi and F. Hormozi, "Multistage deacetylation of chitin: Kinetics study," *Carbohydr Polym*, vol. 81, no. 4, pp. 892–896, Jul. 2010. https://doi.org/10.1016/J.CARBPOL.2010.03.063

[27] A. A. Sarhan, D. M. Ayad, D. S. Badawy, and M. Monier, "Phase transfer catalyzed heterogeneous N-deacetylation of chitin in alkaline solution," *React Funct Polym*, vol. 69, no. 6, pp. 358–363, Jun. 2009. https://doi.org/10.1016/J.REACTFUNCTPOLYM.2009.02.009

[28] B. D. Mansuriya and Z. Altintas, "Carbon Dots: Classification, Properties, Synthesis, Characterization, and Applications in Health Care—An Updated Review (2018–2021)," *Nanomaterials*, vol. 11, no. 10, Oct. 2021. https://doi.org/10.3390/NANO11102525

[29] W. Dong, W. Li, Y. Chen, Y. Ye, and S. Jin, "Construction of a Fluorescent H2O2 Biosensor with Chitosan 6-OH Immobilized β-Cyclodextrin Derivatives," *Marine Drugs 2017, Vol. 15, Page 284*, vol. 15, no. 9, p. 284, Sep. 2017. https://doi.org/10.3390/MD15090284

[30] D. Chandra, M. T. H. Molla, M. A. Bashar, M. S. Islam, and M. S. Ahsan, "Chitosan-based nano-sorbents: synthesis, surface modification, characterisation and application in Cd (II), Co (II), Cu (II) and Pb (II) ions removal from wastewater," *Scientific Reports 2023 13:1*, vol. 13, no. 1, pp. 1–17, Apr. 2023. https://doi.org/10.1038/s41598-023-32847-3

[31] "(PDF) Removal of RR-23 dye from industrial textile wastewater by adsorption on cistus ladaniferus seeds and their biochar." Accessed: Mar. 30, 2024. [Online]. Available: https://www.researchgate.net/publication/322104015_Removal_of_RR-23_dye_from_industrial_textile_wastewater_by_adsorption_on_cistus_ladaniferus_seeds_and_their_biochar

[32] F. Kazeminava *et al.*, "Gentamicin-loaded chitosan/folic acid-based carbon quantum dots nanocomposite hydrogel films as potential antimicrobial wound dressing," *J Biol Eng*, vol. 16, no. 1, Dec. 2022. https://doi.org/10.1186/s13036-022-00318-4

[33] "IOP Conference Series: Earth and Environmental Science". https://doi.org/10.1088/1755-1315/1029/1/012023

Molecular dynamics simulation study on the effect of crosslinker concentration on polymer aerogel properties and its CO_2 adsorption

ABDUL Sami[1,2,a], KHAIRIRAIHANNA Johari[1,3,b], FARAH FADZEHAH Hilmi[4,c], SYED FARMAN Ali Shah[1,d], SITI MUSLIHA Mat Ghani[1,e], ALI SHAAN MANZOOR Ghumman[1,5,f], MEHTAB Ali Darban[1,g], Muhammad Haziq Mirza Bin Saidi[1,h], NURUL EKMI Rabat[1,i *]

[1]Chemical Engineering Department, Universiti Teknologi PETRONAS, 32610 Bandar Seri Iskandar, Perak, Malaysia

[2]Department of Chemical Engineering, NED University of Engineering and Technology, Karachi, 75270, Pakistan

[3]CO2 Research Centre (CO2RES), Universiti Teknologi PETRONAS, 32610 Bandar Seri Iskandar, Perak, Malaysia

[4]Radiation Processing Technology Division, Malaysian Nuclear Agency Bangi, 43000 Kajang Selangor, Malaysia

[5]Centre for Biofuel and Biochemical Research (CBBR), Institute for Sustainable Living, Universiti Teknologi PETRONAS, 32610, Bandar Seri Iskandar, Perak, Malaysia

[a]abdul_22000064@utp.edu.my, [b]khairiraihana.j@utp.edu.my, [c]farah@nuclearmalaysia.gov.my, [d]syed_22009368@utp.edu.my, [e]musliha.ghani@utp.edu.my, [f]alishaan.ghumman@utp.edu.my, [g]mehtab_22011506@utp.edu.my, [h]muhammad_19001323@utp.edu.my, [i]nurulekmi.rabat@utp.edu

Keywords: Aerogel, Molecular Dynamic Simulation, Hydrogel, CO_2 Capture

Abstract. Polyacrylamide-based aerogels represent a promising adsorbent for CO_2 capture due to their unique structural properties. However, optimizing these materials for enhanced CO_2 adsorption requires a comprehensive understanding of the relationship between crosslinker concentration and aerogel characteristics. The molecular simulation approach takes less time and resources to study the real adsorption performance of the polymer aerogel. Therefore, in this study, Grand Canonical Monte Carlo (GCMC) simulations investigated the crosslinker concentration (i.e. 1 and 7wt.%) impact on critical properties of polymer aerogels relevant to carbon capture. By examining the structural changes induced by differing crosslinker concentrations, we elucidate the relationship involved between molecular architecture and CO_2 adsorption performance. The computational results reveal that the polymer aerogel with 7wt% MBA has a higher degree of crosslinking compared to 1 wt.%. MBA. As a result, the glass transition temperature of polymer aerogel was increased from 385.5 to 495K. Furthermore, it was observed that the increase in crosslinker content significantly decreases the free volume and CO_2 adsorption of the polymer aerogel. The results of this research will help to understand the role of crosslinkers in designing the next-generation polyacrylamide-based aerogel for mitigating greenhouse gas emissions molecular simulation.

Introduction

Carbon dioxide (CO_2) emissions resulting from anthropogenic activities have led to exceptional levels of atmospheric CO_2 concentration, contributing significantly to global climate change [1] and its associated consequences[2], [3]. As a response to the urgent need for mitigating CO_2 emissions, researchers have been exploring various strategies to capture CO_2 from industrial flue gases and other emission sources [4]. Among these strategies, CO_2 adsorption using porous materials, particularly polymer aerogels, has garnered substantial attention due to their high surface area, tunable pore structure, and favorable adsorption properties [5]. Understanding the influence of different parameters, such as crosslinker concentration, on the performance of polymer aerogels in CO_2 adsorption is crucial for optimizing their design and enhancing their efficiency in CO2 capture processes [6].

The glass transition temperature (T_g) of a polymer aerogel plays a vital role in determining its extent of crosslinking, and CO_2 adsorption ability [7]. The concentration of crosslinkers within the polymer matrix has been identified as a critical factor influencing T_g which ultimately tailors the CO_2 capture capacity of polymer aerogel [8]. Molecular dynamics (MD) simulations offer a powerful tool for investigating the molecular-level interactions and dynamics governing the behavior of polymer aerogels with varying crosslinker concentrations [9] and glass transition temperature [10] upon CO_2 loading.. By employing MD simulations, it is possible to gain insights into the underlying mechanisms influencing the changes in T_g as a function of crosslinker concentration, thereby facilitating the design of aerogels with tailored morphology of the material [11].

In addition to Tg, the free volume within a polymer aerogel significantly impacts its gas permeability, and sorption capacity [12]. The presence of crosslinkers alters the polymer network's structure, thereby influencing the distribution and accessibility of free volume within the aerogel. Molecular dynamics simulations enable the characterization of free volume properties in polymer aerogels at the atomic scale, providing valuable information regarding the impact of crosslinker concentration on the spatial arrangement and mobility of polymer chains. Such insights are crucial for elucidating the structure-property relationships governing the performance of polymer in CO_2 capture applications [13].

X-ray diffraction (XRD) analysis serves as a powerful experimental technique for probing the structural properties of polymer aerogels, including crystallinity, chain orientation, and intermolecular packing [14]. The presence of crosslinkers within the polymer matrix can influence the XRD patterns by altering the arrangement of polymer chains and the degree of crystallinity within the aerogel structure [15]. Integrating MD simulations with XRD data offers a comprehensive understanding of the correlation between crosslinker concentration and structural characteristics in polymer aerogels, facilitating the interpretation of experimental observations and guiding the rational design of aerogel materials with tailored properties [16].

Moreover, the adsorption capacity of polymer aerogels for CO_2 capture is of significant interest for mitigating greenhouse gas emissions and addressing environmental challenges. The presence of crosslinkers affects the pore structure and surface chemistry of polymer aerogels, thereby influencing their CO_2 adsorption performance [17]. Molecular dynamics simulations provide valuable insights into the adsorption mechanisms and kinetics of CO_2 molecules within the aerogel matrix, offering predictive capabilities for optimizing crosslinker concentrations to enhance CO_2 capture efficiency [18]. By systematically investigating the effect of crosslinker concentration on CO_2 adsorption properties, this research contributes to the development of sustainable materials for environmental applications.

Herein, we investigated the effect of crosslinker concentration on polymer aerogel properties using molecular dynamic simulation on material studio software for the development of an efficient aerogels to mitigate environmental pollution through CO_2 capture. Firstly, a crosslinked polymer aerogel has been constructed and validated the crosslinking between MBA and polymer chain using radial distribution function analysis. Secondly, the aerogels obtained were examined in terms of their glass transition temperature property, XRD, and free volume analysis. Finally, the CO_2 adsorption ability of a crosslinked aerogel with an emphasis on the different crosslinker content investigated. This work aims to offer a new perspective on the polyacrylamide aerogel material using a Grand Canonical Monte Carlo (GCMC) simulation in the field of environmental sustainability.

Computational Methodology

The molecular dynamic simulation was done in BIOVIA material studio software to investigate the best polymeric aerogel for CO_2 capture. The details of the modules and process are given in *Table 1* and further stepwise explanation is discussed below.

Table 1. Modules, operating conditions, and parameters for molecular simulation work

Modules	Specification
Build	Homopolymer: Amide, Acrylates Random co-polymer: Amide-Acrylates Chain length:300 Number of Chain:01
Amorphous	Task: Construction Quality: Medium Temperature: 25°C Energy: COMPASS Density:0.7 g/cm^3
Dynamic simulation	Module: Forcite NPT process: 500 ps, time step of 1 fs at 298 K and 1 atm NVT process: 300 ps time step of 1fs at 500 K and 1 atm Thermostat and barostat: Berendsen
Sorption	Task: Adsorption Isotherm Method: Metropolis Quality: Medium Energy: COMPASS Pressure: 0.1 (kPa)–500 (kPa) Temperature: 298K

Construction of Crosslinked Aerogel. The initial obstacle encountered during the fabrication of the crosslinked copolymer aerogel involves the determination of the quantity of monomers present in the copolymer chain, as well as the quantity of copolymer chains within the vacuum slab or amorphous cell tool. The preference for a short chain length arises from the potential for increased intermolecular resistance during Material Studio MD simulation due to the increased likelihood of chain entanglement [19]. The size effect and unnecessary computations in molecular dynamic simulation can be reduced by a lesser number of polymer chains, as indicated by the studies conducted by Q. Wei et al. (2015) [20] and Mazlan et al. (2022) [19]. Therefore, the study employed the smallest number of copolymer chains. Based on the available data, it is evident that a quantity of 300 monomers is adequate for the representation of PAAm and PAAc in molecular dynamics simulation.

The second challenge in the construction of the copolymer structure is the fabrication of the polyacrylic acid-co-polyacrylamide structure. In order to obtain the reliable adsorption outcome that closely resembles the real-world situation, the structure was built using the experimental protocols [21].The structure of crosslinked PAA-co-PAAm can be illustrated in *Fig. 1*, taking into account the random and intricate crosslinking mechanisms. In this structure, each -CH2- unit in

Materials Research Proceedings 53 (2025) 322-333 https://doi.org/10.21741/9781644903575-31

MBA is crosslinked to a single copolymer chain. Smart Minimizer model is used to geometry optimize and energy minimize to have a real physical arrangement of atoms, improving the reliability of the simulation results.

Although some studies have represented crosslinking structures by manually constructing the structure. Arab eta l. (2013) claim that this approach is inaccurate. Several studies have been done on the crosslinking of polymers including crosslinking of epoxy based (Aghadavoudi et al., 2017; Shokuhfar & Arab, 2013; Tam & Lau, 2014; Yang & Qu, 2012), crosslinking of urea and formaldehyde (Arab & Shokuhfar, 2013), and crosslinking of acrylamide hydrogels with MBA crosslinker (An et al., 2019). These studies have used similar study where they employed cutoff distance to initiate the crosslinking of polymer chain and its crosslinking agent.

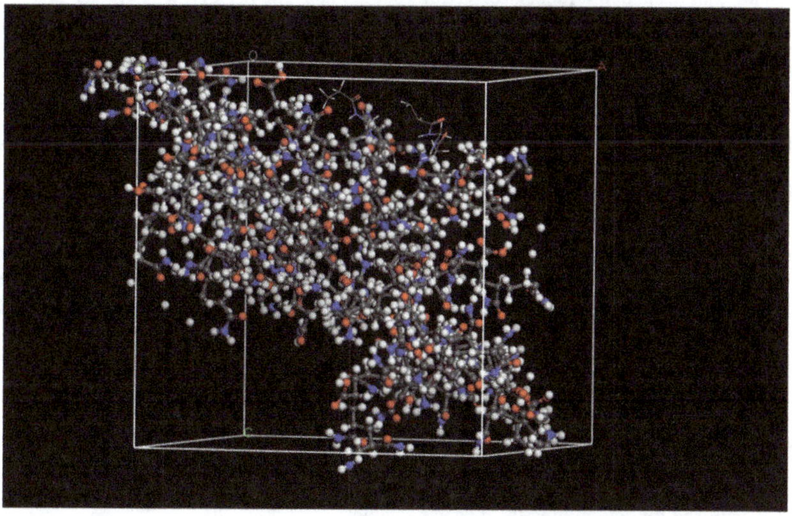

Fig. 1. Crosslinked polymer aerogel with MBA in amorphous cell

Afterwards, the copolymer's optimized structure was incorporated into a cell using the amorphous cell module. The process necessitates the input of a standard density of polymer aerogel, which is experimentally determined to be 1.0 g/cm^3. Nevertheless, in order to facilitate the ramping process and attain an equilibrium state of the actual polymeric aerogel density, a lower initial density has been introduced, deviating from the theoretical value.

Finally, an annealing procedure was executed within the Forcite module to facilitate the heat treatment process, which was guided by a temperature cycle protocol. The removal of materials internal energy of the constructed aerogel is a crucial stage in the construction of the whole framework, as it aims to promote greater relaxation of the simulated cells. Prior to commencing the cross-linking procedure, a 500 ps isothermal-isobaric (NPT) simulation was conducted at a temperature of 298 K and a pressure of 1 atm. This simulation aimed to determine the actual density and establish equilibrium in the system. The thermostat and barostat were manufactured by Berendsen. The ultimate density of the blend approached around 1 g/cm^3. Following the minimization process, the system underwent 300 ps of high temperature canonical (NVT) dynamics at 500 K and 1 atm. This was done to provide sufficient kinetic energy to the molecules and enhance the likelihood of curing sites being located inside the reaction cut-off distance. The aforementioned stages were iterated until the point at which the adoption of the cut-off distance rendered the formation of new covalent bonds nearly impossible.

Radial distribution function (RDF). RDF also known as the pair correlation function, is a fundamental analysis in molecular simulation. It describes the probability of finding a particle at a certain distance from another particle within the system. The crosslinked structure was validated with the radial distribution function (RDF) in Forcite module. In order to investigate the influence of the cross-linking on the structure RDF analysis was performed. RDFs on different pairs of atoms before and after crosslinking takes place for equilibrated structures were calculated at T = 298 K.

Glass Transition Temperature. The glass transition temperature (T_g) is a critical temperature in polymers, where it denotes the temperature at which an amorphous solid material transitions from a hard, glassy state to a rubbery or viscous state (Padhye, 2021). At temperatures below Tg, the polymer chains are locked into place, resembling a solid substance with high rigidity and brittleness, much like glass. However, as the temperature increases beyond Tg, the polymer chains start to gain mobility, allowing the material to exhibit more flexible and rubbery properties. The glass transition temperature is influenced by various factors, including molecular weight, chain flexibility, intermolecular forces, and polymer architecture (Gedde et al., 2021). It's a crucial parameter in the design and application of polymers since it determines their mechanical, thermal, and processing properties.

According to Arab & Shokuhfar (2013), polymers have different specific volume in glassy state and rubbery state. Generally, the specific volume of material will increase as temperature increases in linear manner. However, when the temperature exceeds T_g, the linear trend of specific volume will shift as it enters rubbery state. This result can be obtained by conducting annealing procedure using Forcite module where the material was heated and cooled from 300K to 600K repeatedly. The value of specific volume was obtained after the reading of specific volume is constant for several cycles.

Powder X-Ray Diffraction (XRD). X-ray diffraction (XRD), a very efficient analytical method for determining the phase of a crystalline substance, can also yield insights on the size of its unit cell. The average bulk composition of the material to be analyzed is calculated after homogenizing and finely powdering it. X-ray diffraction is predicated upon the phenomenon of X-ray scattering occurring within the atoms present in a crystal lattice. Equation (1) represents the greatest peak value utilised for determining the d-spacing value.

$$d = \frac{\lambda}{2\, sin\theta}$$

Where d is the highest peak, λ is x-ray radiation wavelength, θ is from the scattering angle value (2θ).

Free volume. The final aerogel structure consists of two separate phases: the solid phase, which is filled by the solid framework of the aerogel material, and the empty space phase, which is made up of voids between the solid particles within the structure. Pore volume, also known as free volume in material studio, refers to aerogels' low density and high porosity. It is static voids that were created by chain packing or transient gaps generated by thermally induced chain rearrangement (Z. Li, 2023). The FFV value has been calculated by using Equation (Amani et al., 2014), where observed specific volume, V, is calculated from the measured density, and the occupied volume, Vo

$$FFV = \frac{V - Vo}{V}$$

CO_2 Adsorption. The CO_2 adsorption isotherm study was conducted using sorption locator module at 298K. The CO_2 molecule structure was drawn and chosen as sorbate for the isotherm study. To study the effect of crosslinker content on CO_2 adsorption capacity, sorption study was conducted between 0.1kPa to 500 kPa and at 298K.

Results and Discussion

Effect of Crosslinker Concentration on Polymer Aerogel Properties. Radial Distribution Function. To investigate the covalent bond formation in the crosslinked aerogel due to varying crosslinker concentration, RDFs of poly (AAc-co-AAm) aerogel with varying crosslinking ratios were calculated for equilibrated structures at 298K. Fig. 3 displays the RDF of crosslinked structures. Peaks in RDF curves within 3.5 Å are primarily due to hydrogen and chemical bonds between atoms, whereas those over 3.5 Å are due to van der Waals and electrostatic interactions. The peaks at 1.1 Å correspond to direct chemical bonding between hydrogen as can be seen in Table *2*. The direct C-H and C-O bonds are represented by the peaks between 1.11 and 1.21 °C, whereas the direct C-N bond is represented by the peaks around 1.35 °C. Furthermore, the correlation between atoms two bonds apart, like hydrogen and carbon in H-C-C sequences and C-C-C sequences, is explained by the peak distances of 2.19 Å and 2.51 Å (Arab & Shokuhfar, 2013). Methylenebisacrylamide (MBA) crosslinking primarily involves the development of a covalent bond between the polymer chain's H-C-C sequences and the MBA double bond, resulting in the synthesis of C-C-C sequences. Fig. 2 show the hydrogen bonds (dash lines) in the crosslinked aerogel structure are the evidence of successful chemical bond formation and crosslinking between the polymer and MBA. Furthermore, Fig. 3 quantify the extent of crosslinking using RDF analysis.

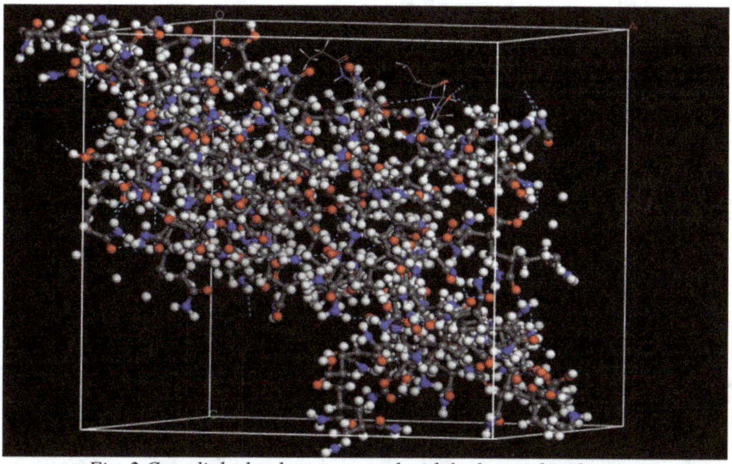

Fig. 2 Crosslinked polymer aerogel with hydrogen bonding

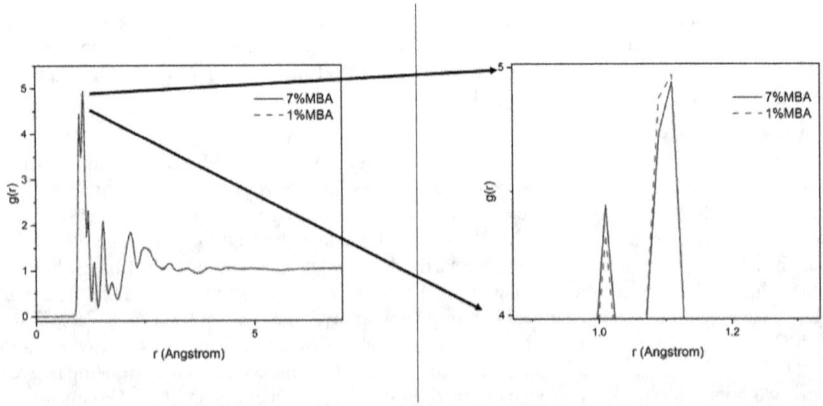

Fig. 3. RDF analysis of varying MBA concentration

By comparing the values of g(r)s between different crosslinker concentration (Fig. 3), we can easily find that the g(r)s between different crosslinker concentration has a following relationship of g(r)7% MBA > g(r)1% MBA. A large value of g(r) indicates a stronger intermolecular interaction, that's to say, intermolecular interaction in crosslinked structure containing 7% MBA is the strongest as compared to the crosslinked structure containing 1% MBA. The main reason behind this is that higher concentration of MBA as a crosslinker enhanced the probability of chain crosslinking with the aerogel structure. Which enables the polymer molecular chains in system closer to each other, and finally results a denser system.

Table 2.Chemical bonds represented by the peaks in RDF analysis

Peaks (Å)	Chemical bond
1.01	Hydrogen
1.11	C-H
1.21	C-O
1.35	C=O
1.55	C-N
2.19	H-C-C
2.51	C-C-C

Glass Transition Temperature Analysis. Fig. 3 shows specific volume of the aerogel increases with increasing temperature, and the trend does not remain linear. The shifting of linear trendline for specific volume indicates the transition from glass phase to rubbery phase. It is evident from the acquired data in Fig. 3 (b) that the linear trendline begins to shift at 464K, and a new linear trendline begins at 495K. Thus, the T_g for the produced aerogel is 495K, which is close to the simulated data (473K) with a 1.9% error. Fig. 3 (a) shows that lower degree of crosslinking lead to a lower T_g. Structure with 1% MBA has T_g around 385.5 K which is significantly lower than structure with 7% MBA. Similar research was done on epoxy by Chowdhury et al. (2020) using varying degrees of DETDA crosslinker, and a significant increase in T_g was noted as the degree of cure increased. This T_g number indicates that crosslinking has occurred and that the simulated material resembles to the aerogel that was synthesized and significantly effect on CO_2 adsorption capacity [10].

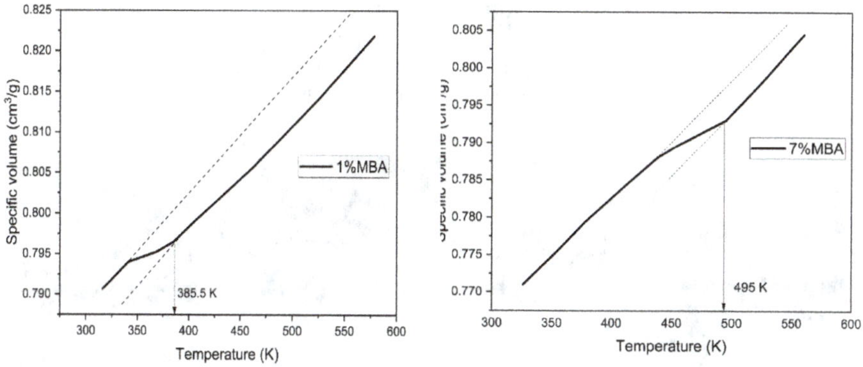

Fig. 4. The change of specific volume as a function of temperature of crosslinked poly (AAm-co-AAc) aerogel (a) 1%MBA and (b) 7%MBA.

Powder X-Ray Diffraction Analysis. X-ray powder diffraction (XRD) pattern for simulated materials were displayed in *Fig. 4*. The aerogel structures showed a peak at 2θ=19°, with the poly (AAm-co-AAm) with 1wt.% MBA having a wider peak than the the poly (AAm-co-AAm) with 7wt.% MBA. This finding indicates that an increased concentration of MBA, used as a crosslinking agent in the aerogel, will result in a higher level of crystallinity in the material. This changes in XRD also indicates that the crosslinking steps employed from Arab & Shokuhfar (2013) works in this study.

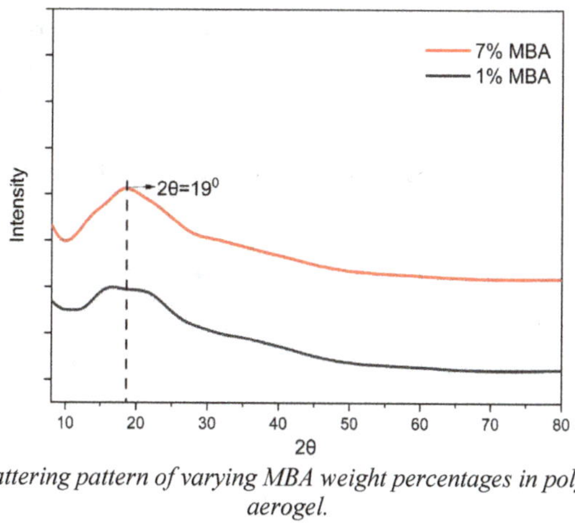

Fig. 5. X-ray scattering pattern of varying MBA weight percentages in poly (AAm-co-AAc) aerogel.

Materials Research Proceedings 53 (2025) 322-333 https://doi.org/10.21741/9781644903575-31

Effect of varying crosslinker concentration on free volume. In the simulation, an increase of MBA concentration resulted in a densification of the chain packing, leading to a reduction in the amount of free volume. The decrease in pore size within the structure of poly (AAm-co-AAc) aerogel can be attributed to the increased weight percentages of MBA, which leads to a higher degree of crosslinking of the polymer chain. The fractional free volume in the amorphous cells dropped as the weight percentage of MBA increased (specifically, for 1wt.% of 337094.39Å^3, 7wt%. of 224897.19 Å^3) for CO_2 gas. As illustrated in *Fig. 5*, the grey color represents the occupied volume, while the blue color represents the free volume. The observed reduction in free volume for the 7 wt.% MBA/poly (AAm-co-AAc) composite can be attributed to the increased crosslinking between the poly (AAm-co-AAc) chains.

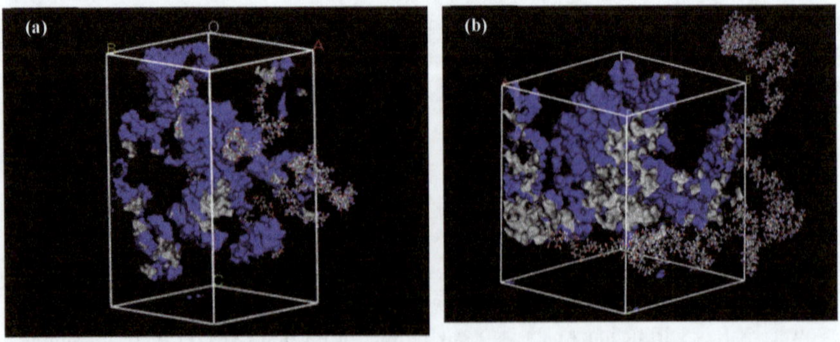

Fig. 6. Illustration of fractional free volume with MBA for (a) 1 wt.% and (b) 7wt.% within poly (AAm-co-AAc) aerogel.

Effect of Crosslinker Concentration CO_2 Adsorption Isotherm Analysis. This section examines the sorption behavior of the MBA crosslinked poly (AAm-co-AAc) aerogel by testing the amount of CO_2 gas adsorption. Fig. 6 presents the depiction of the CO_2 adsorption isotherms for various weight percentages of MBA. Fig. 6 illustrates that when the concentration of MBA increased, there was a corresponding drop in the adsorption value of carbon dioxide, with 1% MBA being more than 7% MBA. The addition of MBA resulted in a decrease in CO_2 adsorption, which can be attributed to the reduction in free volume inside the matrix. The simulation analysis demonstrated a direct correlation between adsorption and free volume. Specifically, it was observed that the free volume dropped as the weight percentage of MBA increased, up to a maximum of 7 wt.% in section 3.1.4. The computational findings indicate that the adsorption of CO_2 MBA crosslinked poly (AAm-co-AAc) is mostly influenced by the free volume, as evidenced by the CO_2 adsorption isotherm values.

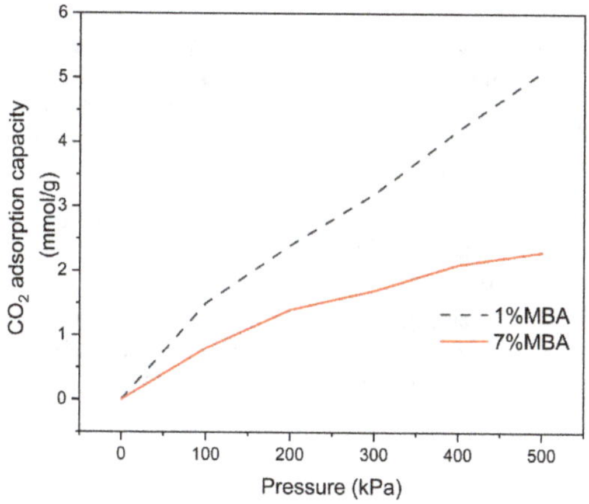

Fig. 7. CO_2 adsorption capacity of simulated poly (AAm-co-AAc) aerogels with varying MBA concentration at 298K.

Conclusion

In this study, the molecular dynamics method with COMPASS force-field was successfully utilized to simulate the cross linking of poly (AAm-co-AAc) with MBA crosslinking agent, yielding polymer networks with different extent of cross linking. The radial distribution Function (RDF) in Forcite module was performed to investigate the influence of MBA concentration on hydrogen bonding, which is a useful method to evaluate the cross-linking in the polymer structure. Then the annealing protocol, connolly surface and forcite scattering analysis method was applied to calculate the polymer aerogel properties of cross-linked structures. Finally, the sorption model was used to study the CO_2 adsorption capacity of the crosslinker polymer aerogel. Moreover:

- Based on the RDF analysis, the increase in g(r) intensity of the 7% MBA in the aerogel structure is a sign of the higher intermolecular attraction as compared to 1% MBA. Thus, it shows the polymer aerogel with 7wt% MBA has higher degree of crosslinking compared to 1 wt.%. MBA (Fig. 3).
- It was also noted that the increase in crosslinker concentration resulted in increased glass transition temperature from 385.5K to 495K (Fig. 4).
- The XRD results indicated that the polymer aerogel structure constructed in the molecular dynamics simulation with 1% MBA shows a highly amorphous structure in this work as compared to 7% MBA Fig. 4. A possible reason is that the polymer with a lower amount of crosslinker concentration formed fewer bonds between polymer chains. As a result, the network structure was less interconnected, leading to increased mobility of the polymer chains and a more disordered, amorphous arrangement.
- Furthermore, the varying of crosslinker content was shown that the polymer aerogel with 1% MBA concentration simulated the higher free volume, which is 337094.39 $Å^3$. The amorphous cell shown in Fig. 6 with blue colour demonstrate the voids in the polymer aerogel.
- The optimal efficacy of CO_2 adsorption up to 5mmol/g at 500kPa observed at the optimal crosslinker concentration of 1% (Fig. 7).

Acknowledgement
The author would like to thank the financial support from Yayasan Universiti Teknologi PETRONAS (YUTP) under cost centre 015LC0-439. The authors would like to thank the Chemical Engineering Department, Universiti Teknologi PETRONAS, Malaysia for providing technical services and facilities. The authors also would like to thank the Agensi Nuklear Malaysia, for providing technical services.

References

[1] S. Naeem, A. Ghaemi, and S. Shahhosseini, “Experimental investigation of CO_2 capture using sodium hydroxide particles in a fluidized bed,” *Korean J. Chem. Eng.*, vol. 33, no. 4, pp. 1278–1285, 2016. https://doi.org/10.1007/s11814-015-0237-1

[2] Y. Zhang, J. Zhuo, Y. Wu, and Q. Yao, “Molecular Simulation of the Adsorption Behaviors of CO_2/CH4 in Curvature, Planar, and Mixture Models,” *Energy and Fuels*, vol. 34, no. 4, pp. 4153–4161, 2020. https://doi.org/10.1021/acs.energyfuels.9b04418

[3] M. Nawaz, “Carbon Capture and Utilization : A Bibliometric Analysis from 2007–2021,” *energies*, 2022

[4] H. Liu, X. Wang, S. Chen, and S. Chen, “Sponge-like polyethyleneimine adsorbents with interconnected open channels enabling ultra-high CO_2 adsorption and separation capacity,” *Sep. Purif. Technol.*, vol. 330, no. PC, p. 125528, 2024. https://doi.org/10.1016/j.seppur.2023.125528

[5] A. Sami *et al.*, “Development of an amine-impregnated polymer aerogel for CO_2 capture,” *Colloids Surfaces A Physicochem. Eng. Asp.*, vol. 690, no. January, p. 133778, 2024. https://doi.org/10.1016/j.colsurfa.2024.133778

[6] Y. T. B. Farshid Aghadavoudi, Hossein Golestanian, “Investigating the Effects of Resin Cross-Linking Ratio on Mechanical Properties of Epoxy-Based Nanocomposites Using Molecular Dynamics,” *Polym. Compos.*, 2016

[7] Y. Chen, G. Shao, Y. Kong, X. Shen, and S. Cui, “Facile preparation of cross-linked polyimide aerogels with carboxylic functionalization for CO_2 capture,” *Chem. Eng. J.*, vol. 322, pp. 1–9, 2017. https://doi.org/10.1016/j.cej.2017.04.003

[8] S. Luo *et al.*, “Highly CO_2-selective gas separation membranes based on segmented copolymers of poly (ethylene oxide) reinforced with pentiptycene-containing polyimide hard segments,” *ACS Appl. Mater. Interfaces*, 2016. https://doi.org/10.1021/acsami.5b11355

[9] R. Sarangapani, S. T. Reddy, and A. K. Sikder, “Molecular dynamics simulations to calculate glass transition temperature and elastic constants of novel polyethers,” *J. Mol. Graph. Model.*, vol. 57, pp. 114–121, 2015. https://doi.org/10.1016/j.jmgm.2015.01.011

[10] H. Chitosan, J. Xu, M. Li, T. M. A. Elmaaty, A. M. Smink, and B. J. De Haan, “Molecular simulation and mathematical modelling of glass transition temperature depression induced by CO_2 plasticization in Polysulfone membranes,” 2017. https://doi.org/10.1088/1757-899X/226/1/012172

[11] K. Asif *et al.*, “A Molecular Simulation Study of Silica / Polysulfone Mixed Matrix Membrane for Mixed Gas Separation,” *Polymers (Basel).*, pp. 1–28, 2021

[12] S. S. M. Lock, K. K. Lau, A. M. Shariff, Y. F. Yeong, and M. A. Bustam, “Computational insights on the role of film thickness on the physical properties of ultrathin polysulfone membranes,” *RSC Adv.*, vol. 7, no. 70, pp. 44376–44393, 2017. https://doi.org/10.1039/c7ra07277e

[13] K. Asif, S. Sow, M. Lock, S. Ali, A. Taqvi, and N. Jusoh, “A molecular simulation study on amine-functionalized silica / polysulfone mixed matrix membrane for mixed gas separation,”

Chemosphere, vol. 311, no. P1, p. 136936, 2022. https://doi.org/10.1016/j.chemosphere.2022.136936

[14] C. Wei *et al.*, "Preparation and characterization of poly(acrylic acid-co-acrylamide)/montmorillonite composite and its application for methylene blue adsorption," *Colloid Polym. Sci.*, vol. 296, no. 4, pp. 653–667, 2018. https://doi.org/10.1007/s00396-018-4277-z

[15] S. Mishra, R. Bajpai, R. Katare, and A. K. Bajpai, "Radiation induced crosslinking effect on semiinterpenetrating polymer networks of poly(vinyl alcohol)," *Express Polym. Lett.*, vol. 1, no. 7, pp. 407–415, 2007. https://doi.org/10.3144/expresspolymlett.2007.58

[16] C. L. M. Camargo, N. S. Resende, C. A. C. Perez, C. R. A. Abreu, V. M. M. Salim, and F. W. Tavares, "Molecular dynamics simulation and experimental validation by X-ray data of hydroxyapatite crystalline structures," *Fluid Phase Equilib.*, vol. 470, pp. 60–67, 2018. https://doi.org/10.1016/j.fluid.2017.11.007

[17] A. Sami, K. Johari, F. F. Hilmi, and M. Rashid, "A new initiators-free technique for synthesizing stable amine-impregnated polymeric aerogel using electron beam radiation for CO2 capture," *Polymer (Guildf).*, p. 126987, 2024. https://doi.org/10.1016/j.polymer.2024.126987

[18] S. S. M. Lock *et al.*, "An atomistic simulation towards elucidation of operating temperature effect in CO_2 swelling of polysulfone polymeric membranes," *J. Nat. Gas Sci. Eng.*, vol. 57, no. July, pp. 135–154, 2018. https://doi.org/10.1016/j.jngse.2018.07.002

[19] N. Mazlan, N. Jusoh, S. Sow, and M. Lock, "Investigation of transport properties of 6FDA-durene polymeric membrane for landfill gas application using molecular simulation approach," *Chemosphere*, vol. 307, no. P3, p. 136019, 2022. https://doi.org/10.1016/j.chemosphere.2022.136019

[20] Q. Wei, Y. Wang, W. Chai, T. Wang, and Y. Zhang, "Effects of composition ratio on the properties of poly(vinyl alcohol)/poly(acrylic acid) blend membrane: A molecular dynamics simulation study," *Mater. Des.*, vol. 89, pp. 848–855, 2016. https://doi.org/10.1016/j.matdes.2015.10.048

[21] J. Xie, X. Liu, J. Liang, and Y. Luo, "Swelling Properties of Superabsorbent Poly (acrylic acid- co -acrylamide) with Different Crosslinkers," *J. Appl. Polym. Sci.*, 2009. https://doi.org/10.1002/app

Decarbonization Technology: ICDT2024 Materials Research Forum LLC
Materials Research Proceedings 53 (2025) 334-338 https://doi.org/10.21741/9781644903575-32

Designing a marine farm: An approach utilizing optimization model and algorithms

RUDI Nurdiansyah[1,a *], WILDANNA Indah Kusuma[1,b], and EFFENDI Mohamad[2,c]

[1]Department of Mechanical and Industrial Engineering, Universitas Negeri Malang, Indonesia

[2]Faculty of Industrial and Manufacturing Technology and Engineering, Universiti Teknikal Malaysia Melaka, Malaysia

[a]rudi.nurdiansyah.ft@um.ac.id, [b]wildanna.indah.2005166@students.um.ac.id, [c]effendi@utem.edu.my

Keywords: Marine Turbine, Marine Farm Design, Optimization Model, Optimization Algorithms

Abstract. This study investigates the design of a marine farm to optimize turbine placement while considering wake effects (turbine interactions) in the optimization model. The Toyapakeh Strait, Bali, Indonesia, serves as a potential location due to its strong tidal currents. Both ant colony optimization (ACO) and simulated annealing (SA) algorithms as optimization algorithms are implemented to approach the problem. Results show that the SA algorithm outperforms ACO in terms of objective function value (representing farm efficiency). The SA solution generates a less crowded turbine layout, potentially minimizing wake effects and maximizing power generation. This study highlights the effectiveness of optimization algorithms in designing efficient marine current farms.

Introduction

The increasing recognition of environmental sustainability's significance in the energy sector propels the advancement and adoption of renewable energy sources. Marine energy, known for its predictability, consistency, and energy density [1, 2], emerges prominently among these sources. This study focuses specifically on harnessing kinetic power derived from marine currents through marine current turbines (MCTs) in marine farms. Despite the potential economic and environmental benefits associated with tidal farm establishment, the commercialization of this technology remains limited to specific regions, primarily due to a lack of comprehensive understanding of the marine and its associated environmental challenges.

The deployment and design of MCT arrays are crucial to ensure the effectiveness and efficiency of the marine farm [2, 3]. This paper designs a marine farm using an optimization model approach that considers wake effects representing the interactions between MCTs. The wake effect is the disturbed fluid flow zone behind turbines, characterized by increased turbulence and decreased flow velocity as the turbine blades rotate to generate electricity [3].

We utilize ant colony optimization (ACO) algorithms and simulated annealing (SA) algorithms to solve the problem. [4] studies the effectiveness of SA algorithm in addressing similar problem to this study compared to tabu search algorithm. The effectiveness of ACO algorithm is investigated by [5], where ACO algorithm outperforms differential evolution algorithm to solve the complex optimization problem. The Toyapakeh Strait in Bali serves as a potential location to install the marine farm due to its strong tidal currents, with speeds ranging from 0.5 m/s to 3.2 m/s [6].

Model

We adapt a mathematical model proposed by [7] to design a marine farm considering interaction between MCTs. The model is as follows.

$$\min \frac{cost}{Power}. \tag{1}$$

Subject to

$$cost = N\left(\frac{2}{3} + \frac{1}{3}e^{(-0.00174N^2)}\right) \tag{2}$$

$$Power = \sum_{j=1}^{N} t_j \cdot P_j \tag{3}$$

$$P_j = \frac{1}{2} \cdot \rho \cdot C_p \cdot A \cdot v_j^3, \quad \forall j \in J. \tag{4}$$

$$A = \pi r^2. \tag{5}$$

$$v_j = \left(1 - \sqrt{\sum_{j=1}^{N} \frac{O_{ij}}{A}\left(1 - \frac{v_{ij}}{v_0}\right)^2}\right) v_0. \tag{6}$$

$$O_{ij} = \begin{cases} \dfrac{\sigma_{ij}^2\left(\theta_{ij} - \dfrac{\sin(2\theta_{ij})}{2}\right) + r^2\left(\beta_{ij} - \dfrac{\sin(2\beta_{ij})}{2}\right)}{A} \\ 0 \end{cases} \tag{7}$$

$$\theta_{ij} = \cos^{-1}\left(\frac{\sigma_{ij}^2 + |dy_{ij}|^2 - r^2}{2|dy_{ij}|\sigma_{ij}}\right) \tag{8}$$

$$\beta_{ij} = \cos^{-1}\left(\frac{\sigma_{ij}^2 - |dy_{ij}|^2 - r^2}{2|dy_{ij}|r}\right) \tag{9}$$

$$\sigma_{ij} = \frac{r}{2.59}\left(-15.542 \cdot I^2 + 21.361 \cdot I + 0.2184\right) \cdot \left[5.58\left(1 - e^{-0.051\frac{|dx_{ij}|}{2r}}\right) + 1.2\right] \tag{10}$$

$$\sqrt{{dx_j}^2 + {dy_j}^2} \geq D, \quad \forall j, j \in J; i \neq j. \tag{11}$$

$$\sum_{j \in J} t_j = N. \tag{12}$$

$$v_j \geq 0, \quad \forall j \in J. \tag{13}$$

$$P_j \geq 0, \quad \forall j \in J. \tag{14}$$

$$cost \geq 0. \tag{15}$$

$$Power \geq 0. \tag{16}$$

$$N \geq 0. \tag{17}$$

$$t_j \in \{0,1\}, \forall j \in J. \tag{18}$$

The objective function aims to minimize $cost$ per $Power$ as defined in Eq. (1). To calculate $cost$ of installing MCTs in a marine farm, which depends on the number of MCT, denoted as N, we utilize Eq. (2). Eq. (3) is used to determine $Power$, which is the sum of electricity generated by individual MCTs, represented as P_j, calculated in Eq. (4), where ρ stands for the density of marine water, C_p denotes the power coefficient of MCT, v_j represents the current velocity flow on MCT j, and r denotes the blade length of MCT. Interactions between MCT are represented by Eq. (6), where O_{ij} denotes overlapping area on MCT j caused by the wake expansion of MCT i (σ_{ij}) and v_{ij} indicates the current velocity flow on MCT j after flow on MCT i. The minimum distance between MCTs, D, is constrained by Eq. (11). Within the model, the initial current velocity, v_0, and turbulence intensity, I, serve as input parameters concerning the environment of the marine. The nonnegativity constraints and decision variables are represented in Eq. (13)-(18).

Methods

Both the ACO and SA algorithms are employed for solving the problem. First, we explain the ACO algorithm. The ACO is an optimization algorithm that mimics the collective intelligence of ant colonies by iteratively constructing solutions and updating pheromone trails to guide the search process. The algorithm consists of initialization, solution construction, local and global pheromone updates, and termination criteria. ACO has been widely applied to solve optimization problems, such as [8] and [9], leveraging both local and global information to find high-quality solutions in diverse problem domains. The input parameters of the ACO used in this study are as follows: the number of iterations is 1000, the number of ants is 50, the pheromone evaporation rate is 0.5, the pheromone importance is 2, and the pheromone update factor is 0.9.

The second algorithm used to solve the problem is the SA algorithm. The SA has been extensively used to approach optimization challenges across various domains, as demonstrated in [10]. This algorithm has three main steps: (1) initializing the solution; (2) searching for a new solution with a neighbor solution; (3) If the new solution is better than the initial solution, then decrease the temperature to a certain point using cooling schedule. If it has not reached the final temperature, then search for a new solution again. Steps (2) and (3) are repeated until a certain number of iterations is reached. The input parameters of the SA algorithm used in this study are as follows: initial temperature is 1000, final temperature is 1, cooling schedule is set to be 0.55, and the number of iterations is 100.

Results and Discussions

The effectiveness of the ACO and SA algorithms for designing a marine farm is assessed using the Toyapakeh Strait, located between Nusa Penida Island and Nusa Lembongan Island, Bali, as a case study. This strait offers favorable conditions for marine current energy resources, including suitable depth for turbine installation, dense population, and high electricity demand [11]. The Toyapakeh Strait has an average depth of around 30 meters, with turbines placed 15 meters above the seabed. Key environmental parameters include an average current velocity of 3.0 meters/second, marine water density of 1025 kg/m^3, and turbulence intensity of 10% [12]. Following the specifications outlined by [3] and [12], the horizontal axis turbine used in this study has a blade length of 9 meters, with power and thrust coefficients of 45% and 0.7, respectively, and a minimum turbine distance of 90 meters.

The marine farm layout consists of an 8×8 grid of cells, each 90 meters in length and width, totaling an area of 518,400 square meters. The current velocity is assumed to flow from the left of the marine farm layout similar to [3]. We also assume that all input parameters used in this study are constant. Fig. 1 depicts the marine farm design obtained through the ACO algorithm, resulting in an objective function value of $5.69 \cdot 10^{-7}$ and a power output of 40.64 Megawatts. Meanwhile, Fig. 2 illustrates the marine farm design obtained by the SA algorithm, yielding a power output of 44.01 Megawatts and an objective function value of $5.25 \cdot 10^{-7}$. We find that the SA algorithm outperforms the ACO algorithm. The directional arrows in Figs. 1 and 2 indicate the flow direction of the current, while 'X' symbols indicate that a turbine is placed within that cell. Notably, Fig. 2 illustrates minimized interactions among turbines, thereby highlighting its potential for enhanced power generation compared to the more crowded layout depicted in Fig. 1. Minimizing interactions among turbines generates more power output than overcrowding them in a marine farm [2, 3].

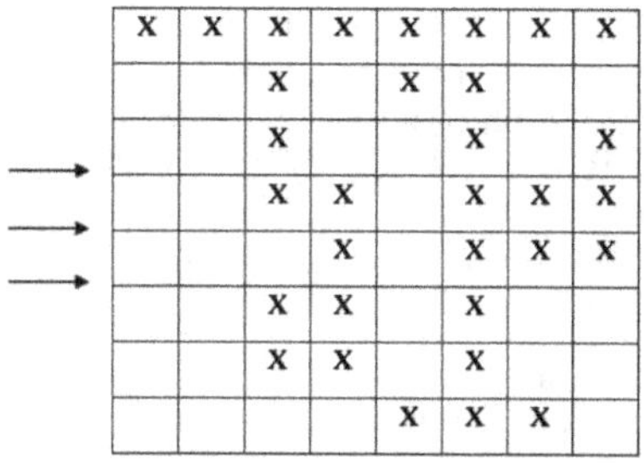

Fig. 1. *Marine Farm Design Obtained by ACO Algorithm*

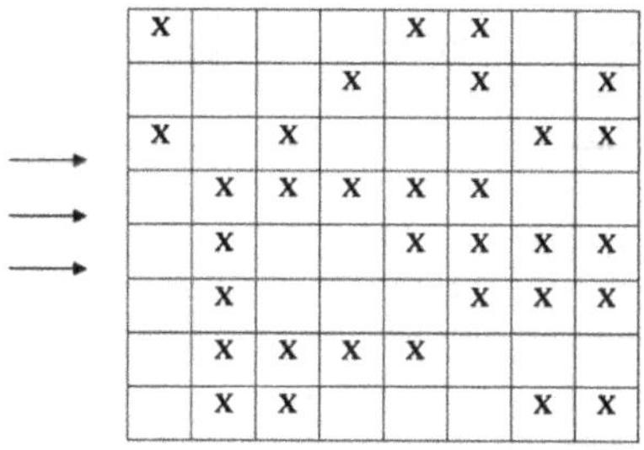

Fig. 2. *Marine Farm Design Obtained by SA Algorithm*

Conclusion

This paper employs an optimization modeling approach and algorithms (SA and ACO) to design a marine farm, with the Toyapakeh Strait in Bali identified as a potential installation site due to its strong tidal currents. The results indicate that the SA algorithm outperforms ACO, with reduced interactions among turbines leading to enhanced power generation. Future research will focus on the cable connections between the installed turbines. Additionally, the study will explore the dynamics of the marine environment and consider environmental constraints.

References

[1] A. S. Bahaj, "Generating electricity from the oceans,"Renewable and Sustainable Energy Reviews, vol. 15, no. 7, pp. 3399-3416, 2011. https://doi.org/10.1016/j.rser.2011.04.032

[2] Z. Ren, Y. Wang, H. Li, X. Liu, Y. Wen, and W. Li, "A coordinated planning method for micrositing of tidal current turbines and collector system optimization in tidal current farms," IEEE Transactions on Power Systems, vol. 34, no. 1, pp. 292-302, 2018. https://doi.org/10.1109/TPWRS.2018.2865310

[3] O. A. L. Brutto, J. Thiébot, S. S. Guillou, and H. Gualous, "A semi-analytic method to optimize tidal farm layouts-Application to the Alderney Race (Raz Blanchard), France," Applied Energy, vol. 183, pp. 1168-1180, 2016. https://doi.org/10.1016/j.apenergy.2016.09.059

[4] R. Nurdiansyah, A. Muid, and A. Dwiastuti, "Aplikasi algoritma simulated annealing untuk menentukan formasi tata letak turbin pada marine current turbine farm di selat Riau," in Prosiding Seminar Nasional Teknik Industri (SENASTI), vol. 1, pp. 50-57, October 2023.

[5] O. A. W. Riyanto, and B. Santosa, "ACO-LS algorithm for solving no-wait flow shop scheduling problem," in Proceedings Intelligence in the Era of Big Data: 4th International Conference on Soft Computing, Intelligent Systems, and Information Technology, vol. 4, pp. 89-97, March 2015. https://doi.org/10.1007/978-3-662-46742-8_8

[6] B. Rachmat and D. Ilahude, "Estimation of sea current energy potential by using calculation models of horizontal axis current turbine in Toyapakeh Strait, Nusa Penida, Bali," Bulletin of the Marine Geology, vol. 32, no. 2, 2018. https://doi.org/10.32693/bomg.32.2.2017.352

[7] R. Nurdiansyah, I. H. Hong, and J. C. P. Su, "The Optimization Model for Marine Current Turbines Installation Problem against Parameters Ambiguity," unpublished.

[8] A. Kumar, M. Thakur, and G. Mittal, "Planning optimal power dispatch schedule using constrained ant colony optimization," Applied Soft Computing, vol. 115, p. 108132, 2022. https://doi.org/10.1016/j.asoc.2021.108132

[9] Y. Wang and Q. Deng, "Optimization of maintenance scheme for offshore wind turbines considering time windows based on hybrid ant colony algorithm," Ocean Engineering, vol. 263, p. 112357, 2022. https://doi.org/10.1016/j.oceaneng.2022.112357

[10] K. Tole, R. Moqa, J. Zheng, and K. He, "A simulated annealing approach for the circle bin packing problem with rectangular items," Computers & Industrial Engineering, vol. 176, p. 109004, 2023. https://doi.org/10.1016/j.cie.2023.109004

[11] C. L. Dewangi, D. N. Sugianto, A. Rifai, and A. Yuningsih, "Kajian potensi energi arus laut di selat Toyapakeh, Nusa Penida, Bali," Journal of Oceanography, vol. 6, no. 1, pp. 22-29, 2017.

[12] Y. Dai, Z. Ren, K. Wang, W. Li, Z. Li, and W. Yan, "Optimal sizing and arrangement of tidal current farm," IEEE Transactions on Sustainable Energy, vol. 9, no. 1, pp. 168-177, 2017. https://doi.org/10.1109/TSTE.2017.2719042

Materials Research Proceedings 53 (2025) 339-346 https://doi.org/10.21741/9781644903575-33

Exploring CO_2/CH_4 separation potential of cellulose acetate-based membranes through molecular simulation

ZUNARA Bashir[1,2,a], SERENE Sow Mun lock[1,2,b*]

[1]Department of Chemical Engineering, Universiti Teknologi PETRONAS, 32610 Seri Iskandar, Perak Darul Ridzuan, Malaysia

[2]CO2 Research Center (CO2RES), Universiti Teknologi PETRONAS, 32610 Seri Iskandar, Perak Darul Ridzuan, Malaysia

[a]zunara_22009996@utp.edu.my, [b]sowmun.lock@utp.edu.my

Keywords: Cellulose Acetate (CA), CO_2/CH_4 Separation, Gas Separation, Molecular Dynamics (MD)

Abstract. The separation of CO2 from natural gas is a critical process for meeting methane product specifications and mitigating greenhouse gas emissions. While conventional separation methods are often complex, energy-intensive, and costly, membrane separation technology has emerged as a promising alternative due to its energy efficiency, simple design, scalability, and small footprint. This study focuses on cellulose acetate (CA)-based membranes for CO2/CH4 separation. The physical properties of CA membranes, including density, fractional free volume, and gas transport behaviors, are evaluated through molecular simulations using Material Studio software. The simulated CO2 affinity and separation performance are validated against experimental data, demonstrating the accuracy of the computational approach. The findings contribute to the development of efficient and sustainable membrane solutions for CO2 capture from natural gas, aligning with goals of environmental impact mitigation and energy- efficient processes. The insights from simulations and experimental validation provide a foundation for further optimization and scale-up of CA-based membranes for industrial gas separation applications.

Introduction

The rapid rise of global carbon dioxide (CO_2) emissions from burning fossil fuels for energy generation has emerged as a grave environmental issue, playing a significant role in driving global warming [1]. Projections suggest that by 2022, atmospheric CO_2 concentrations are anticipated to exceed a crucial level of 417 parts per million (ppm), marking a substantial increase from pre-industrial era levels. This alarming trend highlights the pressing need for effective strategies to curb CO_2 emissions, especially in sectors like natural gas purification, where efficient separation of CO_2 from methane (CH_4) is of utmost importance [2, 3] Membrane separation technologies have become viable solutions, owing to their numerous advantages. These include high energy efficiency, relatively low capital investment needs, reliable and consistent operation, and a straightforward design simplicity. The combination of these favorable traits has propelled membrane-based separation approaches as promising options to explore further [4].

Membrane-based separation methods present practical alternatives to traditional separation techniques, which frequently require complex equipment setups, higher energy demands, and substantial capital investments. Among the various membrane materials, cellulose acetate (CA) membranes have attracted significant interest and attention due to their potential advantages over conventional approaches.

In this context, owing to their exceptional film-forming properties, chemical and mechanical stability, hydrophilicity, and cost-effectiveness. Since the mid-1980s, CA-based membranes have been extensively utilized in natural gas processing, particularly for CO_2 separation from CH4, and have become industry standards for their reliability and performance [5].

CA as polymer have emerged to enhance membrane performance and mechanical strength. This hybrid approach capitalizes on the excellent selectivity of inorganic particles and the processability of polymer phases to address the trade-off issues often encountered with polymeric and inorganic membranes [6]. In recent decades, CA membranes have witnessed a significant surge in research interest. This heightened attention is fueled by growing environmental concerns and the pursuit of sustainable membrane solutions. CA membranes has paved the way for enhancing membrane performance while preserving their eco-friendly nature and cost-effectiveness. These advancements hold promise in addressing the pressing challenges of environmental sustainability and economic viability simultaneously [7].

The escalating demand for effective CO_2/CH_4 separation techniques in natural gas purification, coupled with the pressing need to reduce energy usage and environmental consequences, has spurred investigations into membrane-based separation technologies. CA membranes, owing to their notable properties and proven acceptance in industrial settings, play a crucial role in this field, offering efficient and environmentally sound solutions for gas separation applications [8].

In this research work, computational parameters and settings for molecular simulations using Material Studio software, covering geometric optimization, amorphous cell construction, annealing, molecular dynamics, structure analysis via X-ray scattering, and gas sorption studies. The cellulose acetate (CA) membrane is simulated to evaluate its performance in CO_2/CH_4 separation. It is observed that the physical properties of the simulated CA membrane, such as density and fractional free volume, as well as its gas transport properties, are in close agreement with experimental work. The percentage error between the simulated and experimental values is remarkably low. This validation of the simulation results with experimental data confirms the accuracy and reliability of the molecular modeling approach employed for studying the CA membrane system [9, 10]. Experimental investigations often face limitations in achieving the extreme temperatures and pressures that are relevant for real- world applications. Thus, computational simulations are crucial for exploring these conditions and providing insights that cannot be easily obtained through experiments. Understanding the gas sorption behavior and energy distribution of CO_2 and CH_4 within the CA membrane is also vital for effective design and performance prediction. The close correlation between simulation and experiment demonstrates the effectiveness of computational techniques in predicting the behavior and performance of CA membranes for gas separation applications, thereby providing a robust foundation for further research and optimization efforts.

Methodology

A computational framework to construct cellulose acetate can be used for the investigation of the gas transport behaviour of CO_2/CH_4. The built structure is subjected to a series of molecular treatment with the use of FORCITE Module. Firstly, the structure is refined through "Geometry Optimization" where the structure is optimized based on geometry consideration to generate a more stable and low energy structure.

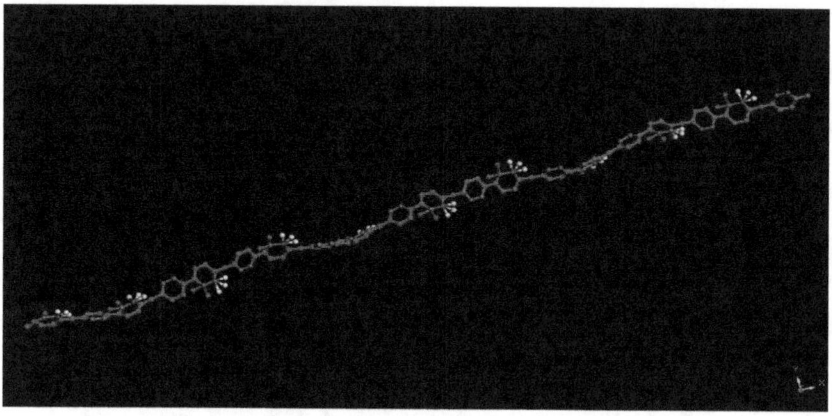

Fig. 1. Illustration of cellulose acetate polymer in Material Studio

Next, the material is packed with a periodic boundary condition (PBC) with the help of Amorphous Cell Module. Polymer is then subjected to annealing process to eliminate the undesirable configuration trapped during Amorphous Cell by relaxing the structure via continuous heating and cooling. The last part of molecular treatment is to run Molecular Dynamics (MD) where the simulation is further run for a prolonged time to reduce the energy and ensure then structure converges to a certain value that suggest thermodynamic equilibrium methodology. Table 1. displayed the operating modules and specifications for the process carried out in Materials Studio.

Table 1. Modules, operating conditions, and parameters for molecular simulation work [7].

Modules	Parameter
Geometric Optimization	Algorithm: Smart Quality: Medium Energy: 0.001 kcal/mol Force: 0.5 kcal/mol/A Max. iterations: 500 Energy: COMPASS
Amorphous Cell	Task: Construction Quality: Medium Temperature: 298.15 K Energy: COMPASS
FORCITE (Anneal)	Task: Anneal Quality: Medium Energy: COMPASS Annealing Cycles: 10 Initial Temperature: 298.15 K Final Temperature: 498.15 K Pressure: 1.013 × 10-4 kPa Time step: 1.0 fs
FORCITE (Dynamics)	Task: Dynamics Quality: Medium Energy: COMPASS

	Ensemble: Canonical (NVT) and Isothermal–Iso baric (NPT) Thermostat: Nose (0.01 Q ratio) Barostat: Berendsen (0.1 ps cell time constant) Time: 500 ps (NPT)
FORCITE Analysis (Scatter)	Radiation: X-ray Cutoff: 20 Å 2-Theta: 2.21◦ Wavelength: 1.54178 Å (5◦–50◦)
Sorption	Task: Adsorption Isotherm Method: Metropolis Quality: Medium Energy: COMPASS Pressure: 1.0 × 10-4 (kPa)–101.33 (kPa)

Once MD is completed, the polymer can be characterized based on density, glass transition temperature and Fractional Free Volume. These parameters are compared with the value obtained in the published literature to ensure the framework used is accurate and reliable. Finally, characterization is conducted to study its physical properties. The gas transport properties of CO_2/CH_4 and investigated through "Sorption" task [11].

Results and discussion

Physical Properties. Figure 2. depicts the density profile of a simulated system, likely a cellulose acetate membrane, over the course of a molecular dynamics simulation. The density exhibits fluctuations throughout the simulation, initially starting around 1.05 g/cm^3 and gradually decreasing to approximately 0.6 g/cm^3 towards the end of the 20 ps simulation time. These density fluctuations are expected as the system explores various configurations and energy states during the molecular dynamics simulation. The observed density variations provide insights into the structural dynamics and compactness of the simulated membrane system over time, which can contribute to a better understanding of its properties and behavior for gas separation applications.

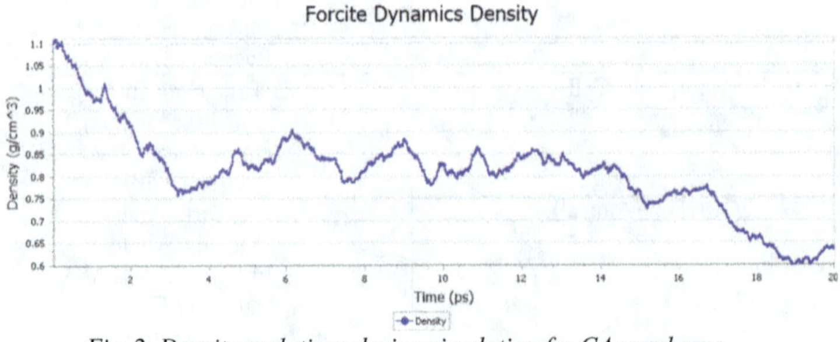

Fig. 2. Density evolutions during simulation for CA membrane.

Figure 3. presents the energy profiles from a molecular dynamics simulation, likely for a cellulose acetate membrane system. The different energy components, including potential energy, kinetic energy, non-bonded energy, and total energy, are plotted against the simulation time in picoseconds (ps). The potential energy (blue line) exhibits fluctuations around an average value of approximately 800 kcal/mol, indicating the variations in the potential energy of the system as it explores different conformational states during the simulation. The kinetic energy (red line) also shows fluctuations but maintains a relatively constant average value of around 200 kcal/mol, representing the kinetic energy associated with the motion of atoms and molecules.The non-bonded energy (green line) remains close to zero throughout the simulation, suggesting that the non-bonded interactions, such as van der Waals and electrostatic interactions, are well-balanced and do not contribute significantly to the overall energy of the system.

The total energy (green line) closely follows the potential energy profile, indicating that the potential energy is the dominant component of the total energy in this system. The fluctuations in the total energy reflect the exploration of different energy states as the system evolves during the simulation.These energy profiles provide insights into the energetic behavior and stability of the simulated membrane system.

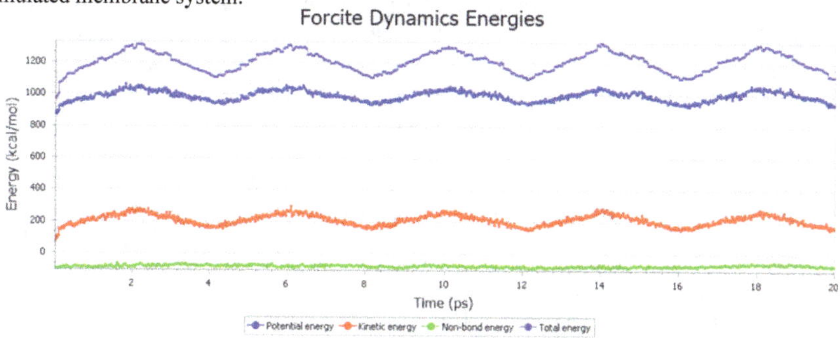

Fig. 3. Energy evolutions during simulation for CA membrane.

Table 2. Density data for simulated cellulose acetate membrane.

Simulated density of CA (g/cm^3)	Experimental Density (g/cm^3)	Reference
1.1	1.36	[12]

The density of the simulated cellulose acetate (CA) membrane was found to be 1.1 g/cm^3. This simulated value deviates from the experimental density reported in the literature, which is 1.36 g/cm^3 for CA membranes [12]. The discrepancy between the simulated and experimental densities could arise from various factors, such as differences in the simulation parameters, force field parameters, or the specific experimental conditions used in the reference study. Despite this difference, the simulated density value provides a reasonable estimate and serves as a starting point for further refinement and optimization of the computational models to better align with experimental observations.

The molecular simulations of CA membranes provide significant insights into their potential for CO_2/CH_4 separation, as reflected in the results. The density fluctuations observed during the simulation indicate how the membrane's structural dynamics might influence its performance in separating CO_2 from CH_4, with a lower density potentially affecting permeability. The energy profiles, particularly the variations in potential energy, offer a glimpse into the membrane's stability and interaction with gas molecules, which is crucial for optimizing separation efficiency. The simulated fractional free volume (FFV) of 0.13, which is close to the experimental value of 0.144, suggests a favorable environment for gas diffusion, directly impacting the membrane's ability to selectively separate CO_2 from CH_4. These results demonstrate the effectiveness of molecular simulations in predicting the performance of CA membranes for CO_2/CH_4 separation, highlighting their potential for developing efficient and sustainable gas separation technologies.

Free Volume Characteristics. The fractional free volume (FFV) of the simulated cellulose acetate (CA)-based membrane was determined to be 0.13. This value is reasonably close to the experimental FFV reported in the literature, which is 0.144 for CA membranes [13]. The fractional free volume represents the fraction of the total volume that is not occupied by the polymer chains and is available for gas molecules to diffuse through. The slight difference between the simulated and experimental FFV values could potentially be attributed to variations in the simulation parameters, force field parameters, or the specific experimental conditions employed in the reference study. Nonetheless, the simulated FFV value provides a reliable estimate and serves as a foundation for further investigations and optimizations of the CA-based membrane system.

Table 3. FFV data for simulated cellulose acetate-based membrane.

Fractional Free Volume (FFV) for CA	Experimental FFV	Reference
0.13	0.144	[13]

Sorption Isotherm. In Figure 4. The sorption isotherm exhibits a nonlinear trend, indicating that the CO_2 loading increases gradually with increasing fugacity (pressure) at lower fugacity values. However, at higher fugacity values, the slope of the curve becomes steeper, suggesting a more pronounced increase in CO_2 loading within the membrane material. This behavior is characteristic of many gas-polymer systems, where the gas molecules initially occupy the available free volume and adsorption sites within the polymer matrix. As the fugacity (pressure) increases, the polymer matrix becomes saturated, and additional gas molecules are forced to occupy higher-energy sorption sites, leading to a more rapid increase in loading.

The sorption isotherm data provides valuable information about the CO_2 sorption behavior in the cellulose acetate membrane, which is crucial for understanding and predicting the membrane's performance in gas separation applications. This information can be used to optimize membrane design, operating conditions, and to develop accurate models for describing the sorption and transport processes within the membrane.

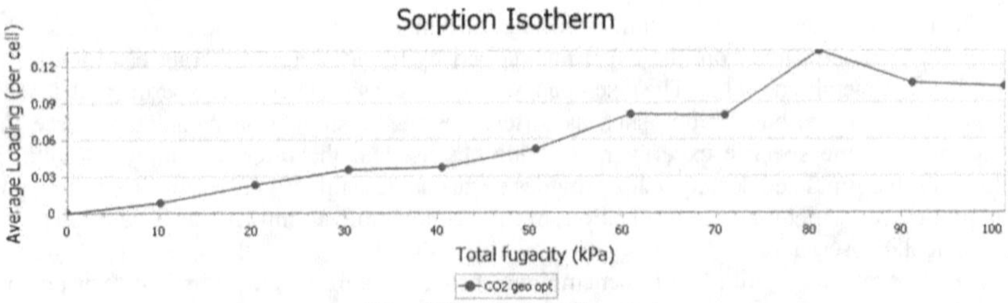

Fig. 4. Sorption isotherm

Figure 5 depicts the energy distribution profiles for two different gas species, likely CO2 and CH4, in a cellulose acetate membrane system. The blue curve represents the energy distribution for one gas species, while the red curve corresponds to the other gas species. Both distributions exhibit multiple peaks, indicating the presence of different energy states or configurations within the system. The energy distribution provides insights into the relative stabilities and populations of various conformational states occupied by the gas molecules within the membrane material. The heights and widths of the peaks are related to the probability of finding the gas molecules in specific energy states or configurations.

It is notable that the energy distributions for the two gas species differ, with distinct peak positions and intensities. This variation suggests that the gas molecules interact differently with the membrane material and experience different energy landscapes within the system. The energy distribution profiles can be further analyzed to understand the underlying interactions, sorption mechanisms, and diffusion processes that govern the separation performance of the cellulose acetate membrane for the specific gas mixture. This information is valuable for optimizing membrane design, operating conditions, and developing accurate models for predicting gas separation behavior.

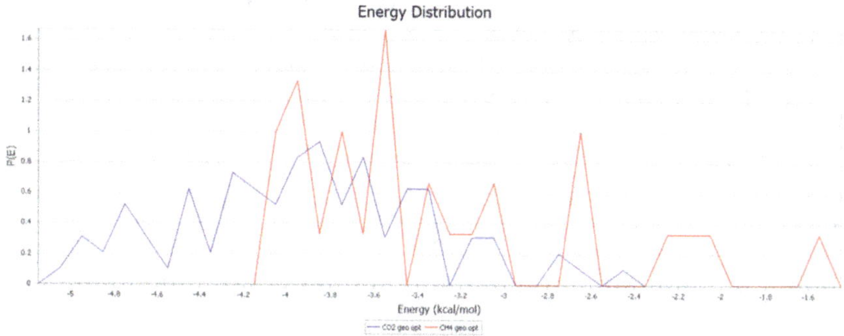

Fig. 5. Energy distribution

The nonlinear sorption isotherm for CO_2 indicates a significant increase in loading at higher pressures, highlighting the membrane's high affinity for CO_2, which is crucial for selective separation. The distinct energy distribution profiles for CO_2 and CH_4 demonstrate that these gases interact differently with the membrane, occupying unique energy states that facilitate their separation. These findings underscore the membrane's capability to effectively capture CO2 from CH4, thereby optimizing its design and operational parameters for enhanced gas separation performance.

Conclusion

The study focuses on evaluating the performance of CA membranes for CO2/CH4 separation through molecular simulations using Material Studio software. The simulated density, fractional free volume, and gas transport properties of the CA membrane are compared with experimental data, showing good agreement and validating the accuracy of the computational approach. The density profile from the molecular dynamics simulations exhibits fluctuations over time, reflecting the exploration of different configurations and energy states by the membrane system. The energy profiles, including potential, kinetic, and total energies, provide insights into the energetic behavior and stability of the simulated system. The sorption isotherm data reveals the nonlinear relationship between CO2 loading and fugacity (pressure) in the CA membrane, indicating the saturation of available sorption sites at higher pressures. This information is crucial for understanding and predicting the membrane's performance in gas separation. Overall, the findings from this study contribute to the development of efficient and sustainable membrane-based solutions for CO2 capture from natural gas streams. The insights gained from molecular simulations and experimental validation provide a solid foundation for further optimization and scale-up of CA-based membranes for industrial gas separation applications.

Summary

The authors acknowledge financial support from Universiti Teknologi PETRONAS, Malaysia. This work is done with the financial support from The Murata Science Foundation (Grant No.: 23MP07, Cost Centre: 015ME0-351)

References

[1] Naveed, U., N.E. Mohammad Rozali, and S. Mahadzir, Energy-Water-Carbon Nexus Study for the Optimal Design of Integrated Energy-Water Systems Considering Process Losses. Energies, 2022. 15(22): p. 8605. https://doi.org/10.3390/en15228605

[2] Iulianelli, A., et al., Novel bio-polymer based membranes for CO2/CH4 separation.

[3] International Journal of Greenhouse Gas Control, 2022. 117: p. 103657. https://doi.org/10.1016/j.ijggc.2022.103657

[4] Kim, W.-g., et al., Nanoporous layered silicate AMH-3/cellulose acetate nanocomposite membranes for gas separations. Journal of membrane science, 2013. 441: p. 129-136. https://doi.org/10.1016/j.memsci.2013.03.044

[5] Sridhar, S., S. Bee, and S. Bhargava, Membrane-based gas separation: Principle, applications and future potential. Chem. Eng. Dig, 2014. 1(1): p. 1-25.

[6] Scholes, C.A., G.W. Stevens, and S.E. Kentish, Membrane gas separation applications in natural gas processing. Fuel, 2012. 96: p. 15-28. https://doi.org/10.1016/j.fuel.2011.12.074

[7] Goh, P., et al., Recent advances of inorganic fillers in mixed matrix membrane for gas separation. Separation and Purification Technology, 2011. 81(3): p. 243-264. https://doi.org/10.1016/j.seppur.2011.07.042

[8] Golzar, K., et al., Molecular simulation study of penetrant gas transport properties into the pure and nanosized silica particles filled polysulfone membranes. Journal of membrane science, 2014. 451: p. 117-134. https://doi.org/10.1016/j.memsci.2013.09.056

[9] Vatanpour, V., et al., Cellulose acetate in fabrication of polymeric membranes: A review.

[10] Chemosphere, 2022. 295: p. 133914. https://doi.org/10.1016/j.chemosphere.2022.133914

[11] Shi, Y.-X., S.-H. Li, and Z.-P. Zhao, Molecular simulations of the effects of substitutions on the dissolution properties of amorphous cellulose acetate. Carbohydrate Polymers, 2022. 291: p. 119610. https://doi.org/10.1016/j.carbpol.2022.119610

[12] Bashir, Z., et al., A review on recent advances of cellulose acetate membranes for gas separation. RSC advances, 2024. 14(27): p. 19560-19580. https://doi.org/10.1039/D4RA01315H

[13] Asif, K., et al., A molecular simulation study on amine-functionalized silica/polysulfone mixed matrix membrane for mixed gas separation. Chemosphere, 2023. 311: p. 136936. https://doi.org/10.1016/j.chemosphere.2022.136936

[14] Matsuba, R., H. Kubota, and N. Matubayasi, All-atom molecular simulation study of cellulose acetate: amorphous structure and the dissolution of small molecule. Cellulose, 2022. 29(10): https://doi.org/10.1007/s10570-022-04616-4

[15] p. 5463-5478.

[16] Salahshoori, I., et al., Improvement efficiency of the of poly (ether-block-amide)-Cellulose acetate (Pebax-CA) blend by the addition of nanoparticles (MIL-53 and NH2-MIL-53): A molecular dynamics study. Journal of Polymer Research, 2021. 28(6): p. 223. https://doi.org/10.1007/s10965-021-02577-z

Materials Research Proceedings 53 (2025) 347-359 https://doi.org/10.21741/9781644903575-34

The TiO_2 catalytic combustion of microalgae *spirulina platensis*-synthetic waste blend: Thermal characteristics, kinetic and thermodynamic investigation

SUKARNI Sukarni[1,2,a*], ARDIANTO Prasetiyo[1,b], YAHYA Zakaria[1,c], POPPY Puspitasari[1,2,d], AVITA Ayu Permanasari[1,2,e], SAMSUDIN Anis[3,f], and ANWAR Johari[4,g]

[1]Center for Renewable Fuels Research (CRFR), Department of Mechanical and Industrial Engineering, Universitas Negeri Malang, Indonesia

[2]Center of Advanced Materials for Renewable Energy (CAMRY), Universitas Negeri Malang, Indonesia

[3]Agro Energy Research Center, Department of Mechanical Engineering, Universitas Negeri Semarang, Indonesia

[4]Centre of Hydrogen Energy, Institute of Future Energy, Universiti Teknologi Malaysia, Malaysia

[a]sukarni.ft@um.ac.id, [b]ardianto.1805168@students.um.ac.id, [c]yahya.zakaria.ft@um.ac.id, [d]poppy@um.ac.id, [e]avita.ayu.ft@um.ac.id, [f]samsudin_anis@mail.unnes.ac.id, [g]anwar@cheme.utm.my

Keywords: Combustion, Microalgae, Synthetic Waste, Kinetic, Thermodynamic

Abstract. This paper evaluates the catalytic impact of TiO_2 addition on the combustion behavior of the microalgae *Spirulina platensis* (SP) and synthetic waste (SW) blend using a thermogravimetric analyzer. The blend ratio of 30SP/70SW, 50SP/50SW, and 70SP/30SW (wt,%) with 5%TiO_2 were used for each experiment. The results show that the blended fuels with and without 5%TiO_2 decomposed in three stages, with the main combustion reaction taking place in the second stage. Assessment of the second stage revealed that the decomposition could be divided into two zones. Zone I started at 166-191 °C and finished at around 360-395 °C. Zone II started at the end temperature of Zone I and finished at around 490-507 °C. The presence of 5% TiO_2 forced the mass loss rate of the samples in Zone II toward a higher rate, except for the 30SP/70SW. The kinetic evaluation performed by using the Coats-Redfern method was implemented on the six selected models, i.e., three diffusion models (parabolic law (D1), valensi equation (D2), Ginstling-Broushtein equation (D3)), two-phase interfacial reaction model (shrinkage geometrical column (S1) and shrinkage geometrical spherical (S2)) and power-law model (P). The results showed that 5% of TiO_2 in the samples significantly reduced the activation energies.

Introduction

Considering the two main issues regarding future fuel security and the global warming associated with the CO_2 emission resulting from fossil fuel combustion, exploring renewable and environmentally friendly fuel sources has been drawing concern from researchers worldwide. Microalgae have been interesting among various future fuel candidates due to several advantages over the others [1]. First, microalgae cultivation does not need arable land; hence, it is not competing with land area for food production [2]. Second, microalgae have fast growth rates; their biomass production is 50 times higher than terrestrial plants [3]. Third, microalgae consume 2 g of CO_2 to produce 1 g of biomass. Assuming a growth rate of 50 g/m^2/d, one hectare of microalgae pond will absorb up to 1 tonne of CO_2 per day [4]. Thus, microalgae cultivation will neutralize CO_2 from biomass combustion [5].

A large amount of synthetic waste (SW) has been generated in Indonesia as a consequence of the fast-growing urban lifestyle [6]. The term SW refers to reference [6], including non-biodegradable materials that do not decompose naturally. It is mostly because of the plastic contained in the matter. Nationally, around 6.8 million tonnes of plastic waste have been produced per year, and it grows by around 5% annually [7]. It is thought that 22.54 kg/capita/year of plastic is used, and 60% of plastic products are used by the food and beverage industry [8]. A small amount of plastic waste was recycled, with a rate of just 7% of plastic waste produced [8]. A huge amount of plastic waste has been an environmental issue due to its accumulation in the land and ocean, interfering with soil fertility and polluting the water. A proper strategy for overcoming the SW problem has been an overwhelming challenge, and thermal conversion is considered the appropriate choice due to its significance in decreasing the volume and energy recovery [6].

From the importance of providing alternative energy and environmental mitigation, co-thermochemical conversion of microalga and SW are critically studied. Several thermochemical conversion techniques include pyrolysis, combustion, liquefaction, and gasification. Among them, combustion is the most favorable choice due to the flexibility of the fuel used, and high energy can be converted to mechanical work by means of combustion [9]. Moreover, combustion also significantly reduces the volume by around 90–95% [10,11] and mass by 70% [11]; therefore, it meets the context for overcoming the SW abundance.

Many works of literature have been presented related to the co-combustion of microalgae for energy utilization. For example, the co-combustion of coal and microalgae has been characterized [12–17], as well as between microalgae and textile dyeing sludge [18], and microalgae and lignocellulosic biomasses [19,20]. However, to the best of our knowledge, no literature has discussed the co-combustion of microalgae with solid waste, except Tang et al. [21] and Sukarni [6]. Tang et al. [21] have found that the combustion of MSW improved by microalgae addition, and in line with this investigation, Sukarni [5] has discovered that the presence of microalgae in SW has decreased the activation energy. However, the utilization of catalysts to improve the co-combustion of microalgae and waste has not been examined yet.

Previous literature has presented that the utilization of catalysts has a significant contribution to the improvement of solid fuel combustion. Prasetiyo et al. [22] have investigated the impact of TiO_2 on the thermal behavior and kinetics of *Tetraselmis chuii* combustion and noticed that TiO_2 forced the combustion process toward a lower temperature and therefore reduced the combustion time. The activation energy has significantly decreased by up to 33% because of TiO2 addition. In line with this previous result has also been presented in the literature of Sukarni et al. [23]. TiO2 nanoparticle addition has improved the coal-water fuel properties and its combustion parameters [24]. Coal combustion was significantly promoted by nano-TiO_2 additives [25]. However, no literature has discussed the effect of TiO2 on the thermal characteristics of microalgae-SW co-combustion.

This paper presented an in-depth investigation of the effect of TiO_2 on the combustion characteristic, kinetic, and thermodynamic parameters of microalgae *Spirulina platensis*-synthetic waste (SP-SW) blend. The combustion characteristics were evaluated by means of a thermogravimetric apparatus. The kinetic evaluation performed by using the Coats-Redfern method was implemented on the six selected models, i.e., three diffusion models (parabolic law (D1), valensi equation (D2), Ginstling-Broushtein equation (D3)), two-phase interfacial reaction model (shrinkage geometrical column (S1) and shrinkage geometrical spherical (S2)) and power-law model (P). The thermodynamic parameters were evaluated based on the most proper kinetic parameters resulting from kinetic evaluation.

Materials and Method

Materials. The Center for Development of Brackishwater Aquaculture, Jepara, Indonesia, has cultured the SP microalgae used in this experiment. After culturing for 8 days in the open pond, the SP was harvested. Finally, the sample was dried using an oven at 100 °C for 24 h to guarantee that the water had vaporized and the biomass was appropriate for combustion. Further SP preparation has been explained in the author's previous papers [6,26–28].

The SW samples used were collected from 5 dumpsters in Malang City, Indonesia. After washing SW with fresh water, they were sun-dried for two days. Next, the samples were individually powdered and sieved for less than 40 mesh size to obtain even powder. A certain weight of individual SW powder was then mixed together thoroughly. The kind of materials and their composition that arranged the final SW sample have been detailed in the author's earlier articles [6,27,28].

The SP and SW samples were mixed with the mass ratios of 30SP/70SW, 50SP/50SW, and 70SP/30SW (wt,%). The nanosize of TiO_2 powder was added to each blended fuel with the percentage of 5% mass ratios. Detailed nano TiO_2 used and how the mixing process was performed to obtain the final sample prior to testing have been presented in the author's previous work [22,23].

Combustion Method. The combustion was performed in the apparatus of thermogravimetric analysis Mettler Toledo TG-DSC1. Around 10 mg of the sample was mounted in the ceramic crucible and then placed in the combustion chamber. The air atmosphere continuously flows at a constant rate of 100 ml/min to ensure the sufficiency of oxidizer during combustion. The chamber was heated at a constant rate of 10 °C/min, starting from room temperature until 1000 °C. The degradation of the matter due to the thermal attack during the combustion process was recorded automatically using a computer as a mass loss parameter and known as thermogravimetric (TG) data. The rate of mass loss, known as derivative thermogravimetric (DTG), was obtained by deriving the TG data to temperature or time. From these results of TG and DTG data, the thermal characteristics could be analyzed, and the kinetic parameters could be examined. Subsequently, the thermodynamic parameter could be evaluated from the kinetic data resulted.

Kinetic Method. The fractional conversion of the material during the combustion process is presented as:

$$\alpha = 1 - \frac{m_t - m_f}{m_i - m_f} = \frac{m_i - m_t}{m_i - m_f} \quad (1)$$

where m_t is the sample's leftover at time t; m_f is the last sample remained at the completion of the reaction, and m_i is the initial sample's mass at the corresponding stage examined, overall in a unit of mg.

For the experiment with a non-isothermal setting, where the temperature changes are linear to the time changes multiplied by the heating rate programmed, denotes as $dT = dt\,\beta$, then, the fractional conversion rate that is generally expressed as an Arrhenius equation can be modified as follows:

$$\frac{d\alpha}{dt} = \frac{d\alpha}{dT/\beta} = k(T)f(\alpha) = A\exp\left(-\frac{E_a}{RT}\right)f(\alpha) \quad (2)$$

where T in Kelvin is the absolute reaction temperature, $k(T)$ is the rate constant, $f(\alpha)$ is the hypothetical model of the reaction mechanism, A in s^{-1} is the frequency factor, E_a in kJ/mol is the activation energy, R is the universal gas constant (8.314 J/mol K).

Eq. (2) can be rearrangement as follows:

$$\frac{d\alpha}{f(\alpha)} = \frac{A}{\beta}\exp\left(-\frac{E_a}{RT}\right)dT \quad (3)$$

The integration of Eq. (3) resulted in the following [29]:

$$g(\alpha) = \int_0^{\alpha} \frac{d\alpha}{f(\alpha)} = \frac{A}{\beta} \int_{T_0}^{T} exp\left(-\frac{E_a}{RT}\right) dT \quad (4)$$

where $g(\alpha)$ expressed the reaction model's integral form of the conversion function.

In this study, the integral Coats-Redfern was employed to approximately assess the kinetic parameters resulting from the combustion process of blended fuels with and without 5%TiO2. As a result, Eq. (4) can be rearrangement in the logarithmic form as follows [30–32]:

$$ln\left[\frac{g(\alpha)}{T^2}\right] = ln\frac{AR}{\beta E_a}\left(1 - \frac{2RT}{E_a}\right) - \frac{E_a}{RT} \quad (5)$$

The value of $2RT/E_a$ is very close to zero [30,33], therefore Eq. (5) can be approximated as follows:

$$ln\left[\frac{g(\alpha)}{T^2}\right] = ln\frac{AR}{\beta E_a} - \frac{E_a}{RT} \quad (6)$$

If the appropriate reaction model $g(\alpha)$ is selected, then the most straight line will be generated from the plot of ln $[g(\alpha)/T^2]$ versus $1/T$ with a coefficient correlation of R^2 very close to 1. As a consequence, the most proper model of reaction can be determined by the highest value of the coefficient correlation R^2. Subsequently, the slope of the regression line will be used to determine the activation energy, and the intercept of this line will be employed to estimate the frequency factor.

Table 1. Reaction mechanisms and model names with their $f(\alpha)$ and $g(\alpha)$

Symbol	Reaction mechanism Model name	Differential form $f(\alpha)$	Integral form $g(\alpha)$
Diffusion			
D1	Parabolic law	½α	α^2
D2	Valensi equation	$-[\ln(1-\alpha)]^{-1}$	$\alpha + (1-\alpha)\ln(1-\alpha)$
D3	Ginstling-Broushtein equation	$3/2[(1-\alpha)^{1/3} - 1]^{-1}$	$\left(1-\frac{2}{3}\alpha\right) - (1-\alpha)^{2/3}$
Phase of interfacial reaction			
S1	Shrinkage geometrical column	$2(1-\alpha)^{1/2}$	$1-(1-\alpha)^{1/2}$
S2	Shrinkage geometrical Spherical	$3(1-\alpha)^{2/3}$	$1-(1-\alpha)^{1/3}$
Power law			
P	Power law	1	α

The integral function $g(\alpha)$ is varied depending on the reaction mechanism and the model approximation developed. This research implements the integral Coats-Redfern on the six selected models, as listed in Table 1.

Thermodynamic Analysis. Considering the data from thermogravimetric analysis and kinetic parameters resulting from the previous step, then the thermodynamic parameters such as a change in Gibbs free energy (ΔG), the change in entropy (ΔS), and the change in enthalpy (ΔH) can be determined by following equation [34,35]:

$$\Delta G = E_a + R.Tm.\,ln\left(\frac{k_B.Tm}{h.A}\right) \quad (7)$$

$$\Delta H = E_a\alpha - R.T \quad (8)$$

$$\Delta S = \frac{\Delta H - \Delta G}{Tm} \quad (9)$$

where k_B is the Boltzmann constant (1.381×10^{-23} J/K), h is the Plank constant ($6{,}626 \times 10^{-34}$ J/s), and Tm is the peak temperature on the DTG graph.

Results and Discussion

Thermal Characteristics. Fig. 1 presents the TG and DTG curves of the blended fuels with and without 5%TiO_2 during the combustion process. In conformity with the DTG curve, it was obviously observed that overall samples decomposed in three stages. The different peaks appear along the combustion process in line with the release and burning of numerous components contained in the fuels: (1) moisture evaporation from the biomass, (2) main combustion stage that covers several different peaks, in which the most of organic ingredient of the blended fuels released and burned, (3) further degradation and combustion of the rest of carbonaceous materials in the char and decomposition of mineral components of the blended fuels, mostly from the microalgae biomass.

Table 2. Catalytic performance of the TiO_2 at the main combustion stage

Samples	Zone I				Zone II			
	Ti (°C)	Tmax (°C)	Tf (°C)	Mmax (%/s)	Ti (°C)	Tmax (°C)	Tf (°C)	Mmax (%/s)
30SP/70SW	191.04	272.60	363.82	-0.0396	363.82	475.19	501.12	-0.1539
30SP/70SW+5%TiO_2	190.10	268.12	362.02	-0.0425	362.02	467.93	496.78	-0.1318
50SP/50SW	172.93	269.32	366.51	-0.0483	366.51	470.72	506.87	-0.0977
50SP/50SW+5%TiO_2	173.55	266.28	360.11	-0.0463	360.11	461.16	490.50	-0.104
70SP/30SW	167.40	271.66	396.12	-0.0644	396.12	469.60	504.71	-0.0698
70SP/30SW+5%TiO_2	166.84	271.59	395.53	-0.0590	395.53	470.81	500.28	-0.0798

Ti: the initial temperature at the zone examined, Tmax: temperature of maximum mass loss rate at the zone examined, Tf: temperature of the final process at the zone examined, Mmax: maximum mass loss rate at the zone examined

Different peaks emerged at the main combustion stage, revealing the complexity of the material ingredient in blended fuel. This stage can be divided into two zones. Table 1 lists the temperature characteristics at the main combustion stage resulting from TG and DTG curves (Figure 1), from which the performance of TiO_2 during the thermal process could be identified. At Zone I, the initial temperature decomposition (Ti) shifted toward the lower temperature due to TiO_2 addition except for 50SP/50SW. The temperature of maximum mass loss rate (Tmax) and temperature of the final process (Tf) both also decreased, attributable to TiO_2 presence. In Zone II, almost overall Ti, Tmax, and Tf have the same tendency as in Zone I, where TiO_2 impacted these temperature characteristics toward the lower value. These results were in agreement with the study of Zhao et al. [25], in which case that adding nano-TiO_2 to coal led to a decrease in the burnout temperature by 32 °C. It was also found in Zone II that the mass-loss rate increased significantly due to TiO_2, except for the 30SP/70SW blend. Previous research [6] has discovered that increasing the SP ratio in the SP/SW blends leads to less mass loss rate due to increasing the low reactive combustible components from microalgal biomass. Therefore, it is more convincing that the addition of TiO_2 significantly advances the reactivity of char combustion at higher SP mass ratios.

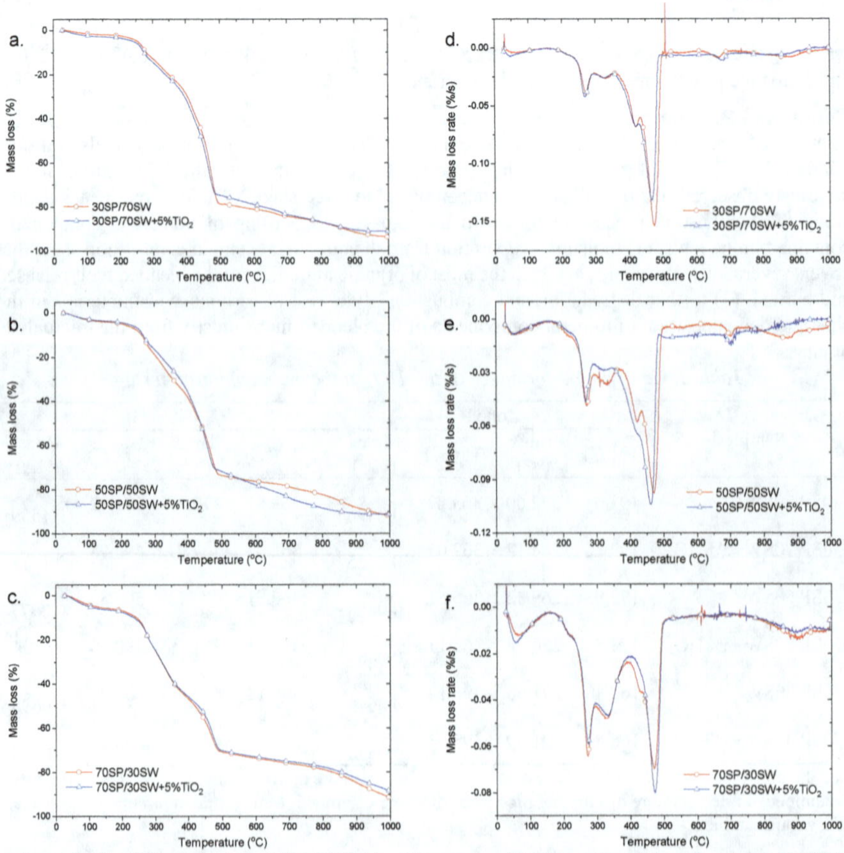

Figure 1. The TG (a, b, c) and the DTG (d, e, f) curves of the samples with and without 5%TiO_2 at 10 °C/min

Kinetic Parameters. By applying Eq. (6) to the selected model, as tabulated in Table 1, the most proper model can be achieved, as depicted in Table 3.

Table 3. The kinetic parameters of blended fuels with and without TiO_2 in different models

Symbol	Samples	Zone I			Zone II		
		Ea (kJ/mol)	R^2	Log A (min^{-1})	Ea (kJ/mol)	R^2	Log A (min^{-1})
D1	30SP/70SW	102.64	0.915	8.22	203.33	0.979	13.63
	30SP/70SW+5%TiO_2	91.24	0.895	7.19	187.81	0.986	12.67
	50SP/50SW	94.86	0.942	7.49	195.96	0.962	13.23
	50SP/50SW+5%TiO_2	88.55	0.909	6.95	179.51	0.981	12.21
	70SP/30SW	86.01	0.921	6.51	267.36	0.954	18.29
	70SP/30SW+5%TiO_2	85.34	0.933	6.45	258.44	0.993	17.55
D2	30SP/70SW	114.10	0.941	9.12	223.60	0.988	14.95
	30SP/70SW+5%TiO_2	104.07	0.926	8.23	211.04	0.993	14.23
	50SP/50SW	105.24	0.963	8.29	216.13	0.976	14.55
	50SP/50SW+5%TiO_2	100.90	0.938	7.95	202.03	0.991	13.73
	70SP/30SW	95.71	0.947	7.24	294.88	0.971	20.13
	70SP/30SW+5%TiO_2	94.88	0.957	7.17	289.62	0.998	19.64
D3	30SP/70SW	119.01	0.951	8.98	232.16	0.990	14.97
	30SP/70SW+5%TiO_2	109.67	0.937	8.15	221.01	0.995	14.37
	50SP/50SW	109.67	0.971	8.11	224.68	0.981	14.58
	50SP/50SW+5%TiO_2	106.27	0.948	7.86	211.72	0.993	13.85
	70SP/30SW	99.88	0.957	7.02	306.59	0.976	20.38
	70SP/30SW+5%TiO_2	98.97	0.966	6.94	303.01	0.998	20.01
S1	30SP/70SW	56.17	0.947	3.86	112.53	0.989	6.96
	30SP/70SW+5%TiO_2	51.64	0.932	3.45	107.37	0.994	6.68
	50SP/50SW	51.40	0.969	3.40	108.83	0.981	6.76
	50SP/50SW+5%TiO_2	49.90	0.943	3.29	102.71	0.993	6.42
	70SP/30SW	46.35	0.952	2.81	150.55	0.977	9.79
	70SP/30SW+5%TiO_2	45.89	0.962	2.77	149.19	0.998	9.63
S2	30SP/70SW	59.86	0.960	4.09	118.96	0.989	7.31
	30SP/70SW+5%TiO_2	55.85	0.947	3.72	114.87	0.993	7.12
	50SP/50SW	54.73	0.979	3.58	115.26	0.985	7.11
	50SP/50SW+5%TiO_2	53.95	0.957	3.55	110.00	0.994	6.84
	70SP/30SW	49.48	0.966	2.98	159.35	0.982	10.31
	70SP/30SW+5%TiO_2	48.96	0.974	2.94	159.25	0.996	10.24
P	30SP/70SW	46.63	0.897	3.12	95.77	0.976	5.90
	30SP/70SW+5%TiO_2	40.87	0.871	257	88.02	0.984	5.40
	50SP/50SW	42.78	0.928	2.74	92.12	0.956	5.69
	50SP/50SW+5%TiO_2	39.54	0.887	2.45	83.92	0.978	5.15
	70SP/30SW	38.26	0.899	2.22	127.71	0.950	8.29
	70SP/30SW+5%TiO_2	37.94	0.914	2.19	123.21	0.993	7.91

It can be understood from Table 3 that in Zone I, shrinkage geometrical spherical (S2) or contracting volume [36,37] has the highest coefficient correlation, which implies that this model was the most probable suitable model for degradation and combustion in Zone I. The S2 model took place in Zone I revealed that releasing and burning of volatile was controlled by the product of the reaction that formed a layer covered the surface of the materials [38,39], and reactions emerged in the boundary among the unreacted material surface and the ash layer toward the center

Materials Research Proceedings 53 (2025) 347-359 https://doi.org/10.21741/9781644903575-34

of the spherical form of sample. In Zone II, the diffusion model of Ginstling-Broushtein has the highest coefficient correlation. It means that the char resulted from the previous devolatilization reaction (in Zone I), subsequently reacted through diffusion mechanism in such ways reaction starting from the exterior of a spherical sample [40].

Table 3 also indicated that the activation energy of the blends with TiO_2 is lower than the blends without TiO_2 in the overall model in both Zone I and Zone II. It means that the amount of energy needed to initiate the reaction is reduced significantly by adding TiO_2. This result agrees with the earlier results of Prasetiyo et al. [22] and Zhao et al. [25]. The promotion of fuel blend combustion by TiO_2 can probably follow this mechanism: activated oxygen formed from the oxygen adsorbed on the surface of TiO_2 leads to an increased oxygen transfer rate from a gas phase to the surface of unreacted materials. In such ways, the reaction between materials and oxygen takes place more easily [25]. By this route, the blended fuel combustion may be advanced by TiO_2.

Thermodynamic Parameters. The thermodynamic parameters can be examined based on kinetic parameters that resulted previously. Table 4 presents the thermodynamic parameters of the various blended samples with and without TiO_2.

The parameter of enthalpy change (ΔH) reflects the energy barrier between reagents to form the activated complex. The positive ΔH demonstrates the endothermic property of the reaction; in this case, an external energy source was needed to provide a high energy level of reagents to convert into their activated complex in the transition state [30,32,41]. The values of ΔH have a similar trend to the activation energies, decreased due to the presence of TiO_2, justifying that TiO_2 addition reduced the energy needed by reagents to transform into the activated complex. The small difference between E_a and ΔH (around 4.7 kJ/mol in Zone I and around 5.9 kJ/mol in Zone II, both with and without TiO_2) explained that combustion reactions are feasible [42].

Gibbs free energy change (ΔG) indicates the increasing total energy of the system at the transition state, at which the reagents approach to their transformation becomes activated complex [43]. It presents a thorough evaluation of the disorder change of the system and its heat flow, and a higher ΔG implies a less favorable reaction. The ΔG had positive values thoroughly, in which case the reactions of activated complex formation occurred in an endergonic mode where the blended fuels reaction in the transition state is non-spontaneous, and the amount of energy is necessary to occur [44]. Table 4 shows that almost the whole ΔG with TiO_2 is less than ΔG without TiO_2, suggesting that the addition of TiO_2 leads to an increase in both the system disorder and reaction favorability.

The ΔS values resulting from this study are both negative and positive. The positive values of ΔS indicated that the reaction leads to an increase in the system disorder; therefore, it is favorable. Conversely, the negative values of ΔS demonstrated a low degree of system disorder, which means that activated complex formation in the system has less reactivity, and therefore, the reagents require more time to transform toward activated complex [43–45]. In Zone I, where the suitable model is S2, it has negative values of ΔS. This suggests that activated complex formation at the devolatilization phase has achieved into a condition close to a thermodynamic equilibrium [43,44], and therefore, bond disassociation tends to be more stable. In Zone II, in which the D3 model is the most probably appropriate, the values of ΔS were positive, in which case the blended fuel material turned disordered again after the volatile was released and burned [46].

Table 4. The thermodynamic parameters of blended fuels with and without TiO_2 in different models

Symbol	Samples	Stage I			Stage II		
		ΔH (kJ/mol)	ΔG (kJ/mol)	ΔS (kJ/mol.K)	ΔH (kJ/mol)	ΔG (kJ/mol)	ΔS (kJ/mol.K)
D1	30SP/70SW	97.91	153.08	-0.101	197.34	196.94	0.001
	30SP/70SW+5%TiO_2	86.49	151.93	-0.121	181.86	195.07	-0.017
	50SP/50SW	90.14	152.58	-0.115	190.03	195.39	-0.007
	50SP/50SW+5%TiO_2	83.81	151.46	-0.125	173.62	193.26	-0.027
	70SP/30SW	81.21	154.23	-0.134	261.34	194.72	0.089
	70SP/30SW+5%TiO_2	80.55	154.14	-0.135	252.37	196.24	0.075
D2	30SP/70SW	109.36	155.13	-0.084	217.61	198.39	0.025
	30SP/70SW+5%TiO_2	99.33	153.99	-0.101	205.10	195.25	0.012
	50SP/50SW	100.51	154.63	-0.099	210.20	196.73	0.018
	50SP/50SW+5%TiO_2	96.16	153.50	-0.106	196.15	194.38	0.002
	70SP/30SW	90.91	156.32	-0.120	288.87	196.21	0.124
	70SP/30SW+5%TiO_2	90.08	156.22	-0.121	283.55	197.71	0.112
D3	30SP/70SW	114.28	161.53	-0.086	226.17	206.95	0.026
	30SP/70SW+5%TiO_2	104.92	160.36	-0.103	215.07	204.26	0.014
	50SP/50SW	104.94	161.00	-0.103	218.75	204.85	0.018
	50SP/50SW+5%TiO_2	101.54	159.58	-0.108	205.85	202.31	0.005
	70SP/30SW	98.08	162.74	-0.124	300.57	204.37	0.129
	70SP/30SW+5%TiO_2	94.17	162.63	-0.125	296.94	205.87	0.122
S1	30SP/70SW	51.43	152.12	-0.184	106.54	201.67	-0.127
	30SP/70SW+5%TiO_2	46.89	151.11	-0.192	101.43	199.58	-0.132
	50SP/50SW	46.67	151.61	-0.193	102.90	200.25	-0.131
	50SP/50SW+5%TiO_2	45.17	150.66	-0.195	96.83	197.77	-0.137
	70SP/30SW	41.55	153.07	-0.205	144.53	198.83	-0.037
	70SP/30SW+5%TiO_2	41.09	153.03	-0.205	143.11	199.78	-0.076
S2	30SP/70SW	55.13	153.50	-0.180	112.97	203.16	-0.121
	30SP/70SW+5%TiO_2	51.10	152.46	-0.187	108.93	200.94	-0.124
	50SP/50SW	50.00	152.98	-0.189	109.33	201.70	-0.124
	50SP/50SW+5%TiO_2	49.21	152.01	-0.191	104.12	199.01	-0.129
	70SP/30SW	44.69	154.67	-0.202	153.33	200.32	-0.063
	70SP/30SW+5%TiO_2	44.17	154.42	-0.202	153.18	201.25	-0.065
P	30SP/70SW	41.89	150.31	-0.198	89.78	200.21	-0.147
	30SP/70SW+5%TiO_2	36.13	149.39	-0.209	82.07	198,47	-0.157
	50SP/50SW	38.05	149.79	-0.206	86.19	198.89	-0.152
	50SP/50SW+5%TiO_2	38.41	148.96	-0.212	78.04	196.71	-0.161
	70SP/30SW	33.46	151.23	-0.216	121.70	197.34	-0.102
	70SP/30SW+5%TiO_2	33.15	151.20	-0.216	117.13	198.40	-0.109

Conclusions

The impact of TiO_2 nanoparticles on combustion behavior, kinetic and thermodynamic parameters of the SP/SW blend have been evaluated. These results revealed that the initial temperature decomposition, the temperature of maximum mass loss rate, and the temperature of the final process decreased due to TiO_2 presence, both in Zone I and Zone II of the main decomposition and combustion stage. The activation energy of the blends was reduced significantly by adding TiO_2, which means that TiO_2 brings the sample toward a condition in such a way that the reaction between materials and oxygen occurs more easily. The thermodynamic parameters have validated that TiO_2 leads to reagents transformation into activated complex becomes at ease.

Acknowledgments
Acknowledgments are extended to the generous financial support received through the "Penelitian Dasar Unggulan Perguruan Tinggi 2023" program, provided under contract numbers 036/E5/PG.02.00.PL/2023 and 3.4.35/UN32.20.1/LT/2023.

References

[1] A.Y. Aminullah, S. Sukarni, R. Wulandari, M. Shahbaz, Pyrolysis Kinetics of Spirulina platensis and Non-condensable Gas Product Distribution in a Fixed-Bed Reactor, Journal of Mechanical Engineering Science and Technology (JMEST) 8 (2024) 151–167. https://doi.org/10.17977/UM016V8I12024P151

[2] L. Brennan, P. Owende, Biofuels from microalgae-A review of technologies for production, processing, and extractions of biofuels and co-products, Renewable and Sustainable Energy Reviews 14 (2010) 557–577. https://doi.org/10.1016/j.rser.2009.10.009

[3] Y. Li, M. Horsman, N. Wu, C.Q. Lan, N. Dubois-Calero, Biofuels from microalgae., Biotechnol Prog 24 (2008) 815–820

[4] P.M. Schenk, S.R. Thomas-Hall, E. Stephens, U.C. Marx, J.H. Mussgnug, C. Posten, O. Kruse, B. Hankamer, Second Generation Biofuels: High-Efficiency Microalgae for Biodiesel Production, Bioenergy Res 1 (2008) 20–43. https://doi.org/10.1007/s12155-008-9008-8

[5] S. Sukarni, Bahan bakar terbarukan dari Mikroalga: peluang dan tantangan untuk pasokan energi masa depan, Universitas Negeri Malang, 2021

[6] S. Sukarni, Thermogravimetric analysis of the combustion of marine microalgae Spirulina platensis and its blend with synthetic waste, Heliyon 6 (2020) e04902. https://doi.org/10.1016/j.heliyon.2020.e04902

[7] World Economic Forum, Radically Reducing Plastic Pollution in Indonesia: A Multistakeholder Action Plan, 2020

[8] Ministry of Environment and Forestry, National Plastic Waste Reduction Strategic Actions for Indonesia, Ministry of Environment and Forestry, Republic of Indonesia, 2020

[9] C. Gai, Z. Liu, G. Han, N. Peng, A. Fan, Combustion behavior and kinetics of low-lipid microalgae via thermogravimetric analysis, Bioresour Technol 181 (2015) 148–154. https://doi.org/10.1016/j.biortech.2015.01.045

[10] C. Montejo, C. Costa, P. Ramos, M.D.C. Márquez, Analysis and comparison of municipal solid waste and reject fraction as fuels for incineration plants, Appl Therm Eng 31 (2011) 2135–2140. https://doi.org/10.1016/j.applthermaleng.2011.03.041

[11] M. Li, J. Xiang, S. Hu, L.-S. Sun, S. Su, P.-S. Li, X.-X. Sun, Characterization of solid residues from municipal solid waste incinerator, Fuel 83 (2004) 1397–1405. https://doi.org/10.1016/j.fuel.2004.01.005

[12] A. Tahmasebi, M.A. Kassim, J. Yu, S. Bhattacharya, Thermogravimetric study of the combustion of Tetraselmis suecica microalgae and its blend with a Victorian brown coal in O2/N2 and O2/CO2 atmospheres, Bioresour Technol 150 (2013) 15–27. https://doi.org/10.1016/J.BIORTECH.2013.09.113

[13] H. Il Choi, J.S. Lee, J.W. Choi, Y.S. Shin, Y.J. Sung, M.E. Hong, H.S. Kwak, C.Y. Kim, S.J. Sim, Performance and potential appraisal of various microalgae as direct combustion fuel, Bioresour Technol 273 (2019) 341–349. https://doi.org/10.1016/J.BIORTECH.2018.11.030

[14] Y. Gao, A. Tahmasebi, J. Dou, J. Yu, Combustion characteristics and air pollutant formation during oxy-fuel co-combustion of microalgae and lignite, Bioresour Technol 207 (2016) 276–284. https://doi.org/10.1016/J.BIORTECH.2016.02.031

[15] B. Ye, R. Zhang, J. Cao, K. Lei, D. Liu, The study of co-combustion characteristics of coal and microalgae by single particle combustion and TGA methods, Journal of the Energy Institute 93 (2020) 508–517. https://doi.org/10.1016/J.JOEI.2019.07.001

[16] N.E. Magida, L.L. Bolo, S.P. Hlangothi, G. Dugmore, A.S. Ogunlaja, Co-combustion Characteristics of coal-Scenedesmus Microalgae Blends and Their Resulting Ash, Combustion Science and Technology 193 (2021) 419–436. https://doi.org/10.1080/00102202.2019.1658577

[17] C. Chen, Q.N. Chan, P.R. Medwell, G. Heng Yeoh, Co-Combustion Characteristics and Kinetics of Microalgae Chlorella Vulgaris and Coal through TGA, Combustion Science and Technology 192 (2020) 26–45. https://doi.org/10.1080/00102202.2018.1555820

[18] X. Peng, X. Ma, Z. Xu, Thermogravimetric analysis of co-combustion between microalgae and textile dyeing sludge, Bioresour Technol 180 (2015) 288–295. https://doi.org/10.1016/j.biortech.2015.01.023

[19] R. Lopez, C. Fernandez, X. Gomez, O. Marinez, M.E. Sanchez, Thermogravimetric analysis of lignocellulosic and microalgae biomasses and their blends during combustion, J Therm Anal Calorim 114 (2013) 295–305. https://doi.org/10.1007/s10973-012-2843-z

[20] Y. Tang, X. Ma, Z. Lai, Q. Yu, Oxy-fuel combustion characteristics and kinetics of microalgae and its mixture with rice straw using thermogravimetric analysis, Int J Energy Res 42 (2018) 532–541. https://doi.org/10.1002/er.3836

[21] Y. Tang, X. Ma, Z. Lai, Thermogravimetric analysis of the combustion of microalgae and microalgae blended with waste in N(2)/O(2) and CO(2)/O(2) atmospheres., Bioresour Technol 102 (2011) 1879–1885. https://doi.org/10.1016/j.biortech.2010.07.088

[22] A. Prasetiyo, S. Sukarni, R. Wulandari, P. Puspitasari, A Kinetic Study on Tetraselmis chuii Combustion: The Catalytic Impact of Nanoparticle Titanium Dioxide (TiO2) Additive, Journal of Advanced Research in Fluid Mechanics and Thermal Sciences 71 (2020) 39–49. https://doi.org/10.37934/arfmts.71.1.3949

[23] S. Sukarni, P. Ardianto, W. Retno, E.W. Aloon, P. Poppy, Thermal characteristic of tetraselmis chuii combustion influenced by titanium dioxide (Tio2) nanoparticle, Key Eng Mater 851 (2020) 149–155. https://doi.org/10.4028/www.scientific.net/KEM.851.149

[24] V.A. Pinchuk, A. V. Kuzmin, The effect of the addition of TiO2 nanoparticles to coal-water fuel on its thermophysical properties and combustion parameters, Fuel 267 (2020) 117220. https://doi.org/10.1016/J.FUEL.2020.117220

[25] Y. Zhao, S. Wang, Y. Shen, X. Lu, Effects of nano-TiO2 on combustion and desulfurization, Energy 56 (2013) 25–30

[26] S. Sukarni, S. Sumarli, I.M. Nauri, A. Prasetiyo, P. Puspitasari, Thermogravimetric analysis on combustion behavior of marine microalgae Spirulina platensis induced by MgCO3 and Al2O3 additives, International Journal of Technology 10 (2019) 1174–1183. https://doi.org/10.14716/ijtech.v10i6.3611

[27] S. Sukarni, A. Prasetiyo, S. Sumarli, I.M.I.M. Nauri, A.A.A.A. Permanasari, Kinetic analysis of co-combustion of microalgae Spirulina platensis and synthetic waste through the fitting model, in: MATEC Web of Conferences, EDP Sciences, 2018: p. 00009. https://doi.org/10.1051/matecconf/201820400009

[28] S. Sukarni, L.A.A. Saputra, The elemental composition changes of Spirulina platensis-synthetic waste mixture during the combustion analyzed by energy-dispersive X-ray, in: AIP Conf Proc, AIP Publishing, 2020: p. 030014. https://doi.org/10.1063/5.0000901

[29] C. Chen, X. Ma, Y. He, Co-pyrolysis characteristics of microalgae Chlorella vulgaris and coal through TGA, Bioresour Technol 117 (2012) 264–273. https://doi.org/10.1016/j.biortech.2012.04.077

[30] J. Chen, Y. Wang, X. Lang, X. Ren, S. Fan, Comparative evaluation of thermal oxidative decomposition for oil-plant residues via thermogravimetric analysis: Thermal conversion characteristics, kinetics, and thermodynamics, Bioresour Technol 243 (2017) 37–46. https://doi.org/10.1016/j.biortech.2017.06.033

[31] A. Prasetiyo, S. Sukarni, A.A. Wibowo, A.A. Permanasari, P. Puspitasari, Characteristics and modeling kinetics of oil palm frond petiole pyrolysis using thermogravimetric analyzer, in: IOP Conf Ser Earth Environ Sci, 2021. https://doi.org/10.1088/1755-1315/847/1/012018

[32] S.R. Naqvi, R. Tariq, Z. Hameed, I. Ali, M. Naqvi, W.H. Chen, S. Ceylan, H. Rashid, J. Ahmad, S.A. Taqvi, M. Shahbaz, Pyrolysis of high ash sewage sludge: Kinetics and thermodynamic analysis using Coats-Redfern method, Renew Energy 131 (2019) 854–860. https://doi.org/10.1016/j.renene.2018.07.094

[33] L. Vlaev, N. Nedelchev, K. Gyurova, M. Zagorcheva, A comparative study of non-isothermal kinetics of decomposition of calcium oxalate monohydrate, J Anal Appl Pyrolysis 81 (2008) 253–262. https://doi.org/10.1016/j.jaap.2007.12.003

[34] S. Sukarni, A. Prasetiyo, L. Fidiah, A.A. Permanasari, P. Puspitasari, Kinetics and thermodynamics study of organic waste combustion using thermogravimetric analysis, IOP Conf Ser Earth Environ Sci 847 (2021). https://doi.org/10.1088/1755-1315/847/1/012015

[35] S. Rasam, A. Moshfegh Haghighi, K. Azizi, A. Soria-Verdugo, M. Keshavarz Moraveji, Thermal behavior, thermodynamics and kinetics of co-pyrolysis of binary and ternary mixtures of biomass through thermogravimetric analysis, Fuel 280 (2020) 118665. https://doi.org/10.1016/j.fuel.2020.118665

[36] A. Khawam, D.R. Flanagan, Role of isoconversional methods in varying activation energies of solid-state kinetics: II. Nonisothermal kinetic studies, Thermochim Acta 436 (2005) 101–112. https://doi.org/10.1016/j.tca.2005.05.015

[37] A. Khawam, D.R. Flanagan, Role of isoconversional methods in varying activation energies of solid-state kinetics: I. isothermal kinetic studies, Thermochim Acta 429 (2005) 93–102. https://doi.org/10.1016/j.tca.2004.11.030

[38] A. Khawam, D.R. Flanagan, Solid-state kinetic models: Basics and mathematical fundamentals, Journal of Physical Chemistry B 110 (2006) 17315–17328. https://doi.org/10.1021/jp062746a

[39] S.Y. Yorulmaz, A.T. Atimtay, Investigation of combustion kinetics of treated and untreated waste wood samples with thermogravimetric analysis, Fuel Processing Technology 90 (2009) 939–946. https://doi.org/10.1016/j.fuproc.2009.02.010

[40] S.M. Alshehri, M.A.S. Monshi, N.M. Abd El-Salam, R.M. Mahfouz, Kinetics of the thermal decomposition of γ-irradiated cobaltous acetate, Thermochim Acta 363 (2000) 61–70. https://doi.org/10.1016/S0040-6031(00)00602-X

[41] S. Niu, H. Yu, S. Zhao, X. Zhang, X. Li, K. Han, C. Lu, Y. Wang, Apparent kinetic and thermodynamic calculation for thermal degradation of stearic acid and its esterification derivants through thermogravimetric analysis, Renew Energy 133 (2019) 373–381. https://doi.org/10.1016/j.renene.2018.10.045

[42] D.K.W. Gan, A.C.M. Loy, B.L.F. Chin, S. Yusup, P. Unrean, E. Rianawati, M.N. Acda, Kinetics and thermodynamic analysis in one-pot pyrolysis of rice hull using renewable calcium oxide based catalysts, Bioresour Technol 265 (2018) 180–190. https://doi.org/10.1016/j.biortech.2018.06.003

[43] L. Huang, J. Liu, Y. He, S. Sun, J. Chen, J. Sun, K.L. Chang, J. Kuo, X. Ning, Thermodynamics and kinetics parameters of co-combustion between sewage sludge and water hyacinth in CO2/O2atmosphere as biomass to solid biofuel, Bioresour Technol 218 (2016) 631–642. https://doi.org/10.1016/j.biortech.2016.06.133

[44] P. Parthasarathy, A. Fernandez, T. Al-Ansari, H.R. Mackey, R. Rodriguez, G. Mckay, Thermal degradation characteristics and gasification kinetics of camel manure using thermogravimetric analysis, J Environ Manage 287 (2021) 112345. https://doi.org/10.1016/j.jenvman.2021.112345

[45] Y.S. Kim, Y.S. Kim, S.H. Kim, Investigation of thermodynamic parameters in the thermal decomposition of plastic waste-waste lube oil compounds, Environ Sci Technol 44 (2010) 5313–5317. https://doi.org/10.1021/es101163e

[46] V. Vasudev, X. Ku, J. Lin, Pyrolysis of algal biomass: Determination of the kinetic triplet and thermodynamic analysis, Bioresour Technol 317 (2020) 124007. https://doi.org/10.1016/j.biortech.2020.124007

Materials Research Proceedings 53 (2025) 360-373 https://doi.org/10.21741/9781644903575-35

Fabrication of layered graphene-supported in-situ polymerized polyaniline (PANI) crystals as conductive electrode composite for hydrogen production

UMAIR Ishtiaq[1,4,a *], ALI SAMER Muhsan[2,4,b], BAWADI Abdullah[3,c], MUHAMMAD AFIF FIKRI bin Zambri[2,d]

[1]Petroleum Engineering Department, Universiti Teknologi PETRONAS, 32610 Bandar Seri Iskandar Perak, Malaysia

[2]Mechanical Engineering Department, Universiti Teknologi PETRONAS, 32610 Bandar Seri Iskandar Perak, Malaysia

[3]Chemical Engineering Department, Universiti Teknologi PETRONAS, 32610 Bandar Seri Iskandar Perak, Malaysia

[4]Center of Innovative Nanostructures and Nanodevices (COINN), Universiti Teknologi PETRONAS, 32610 Bandar Seri Iskandar Perak, Malaysia

[a]umair.ishtiaq_g02875@utp.edu.my, [b]ali.samer@utp.edu.my, [c]bawadi_abdullah@utp.edu.my [d]afif_17010458@utp.edu.my

Keywords: Water Electrolysis, Hydrogen Production, Polyaniline (PANI), Nano-Composite Electrode, Graphene Dispersion

Abstract. One of the potential routes for hydrogen production is through water electrolysis. Graphene (Gr) – Polyaniline (PANI) coated Fluorine-doped tin oxide (FTO) glass nano-composite electrode has potential to be a viable alternative to the commonly employed but rare and exorbitant platinum electrode. Well-dispersed graphene paste was synthesized and homogeneously tape-casted on FTO glass cuts in three separate coating layers onto which PANI crystals were grown by in-situ polymerization. The fabricated nano-composite electrode was characterized by Field Emission Scanning Electron Microscopy (FESEM) followed by Raman spectroscopy. It was well-observed that the triple coated electrode promoted the highest growth of PANI crystals onto it as densely coated graphene substrate rendered structural support and highest number of active sites that facilitated in-situ polymerization and subsequent growth of PANI crystals. This observation was supplemented by Raman spectroscopy results that clearly determined active presence of PANI crystals and multi-layered graphene by the analysis of D, G and 2D peaks. The conductivity of the electrode composite was quantified through resistance measurement by a multimeter. By increasing graphene layers from uncoated to triple coated, the average resistivity value reduced from 0.4Ω to 0.1Ω respectively, indicating presence of highest amount of PANI crystals on triple coated electrode thereby making it electro-conductive in order to be utilized for water electrolysis for hydrogen production.

Introduction

The primary energy consumption of the world is increasing by 1 to 2% annually [1, 2]. 85% of the energy sources are still dependent upon fossil fuels that has caused alarming situation for the environment due to rise in global warming. Hydrogen (H_2) is emerging as a clean and renewable fuel alternative to conventional fossil fuels which can be an ideal replacement for energy generation since it is high-density transportable sustainable energy carrier with environmental benefits and has the potential to play a crucial role in the transition to a low-carbon economy [3]. Hydrogen can be produced from different sources and methods to be use as a fuel such as Steam Methane Reforming (SMR) [4], electrolysis [5-7] and thermochemical water splitting [8, 9]. Recent studies have shown that hydrogen production through water splitting has proven to be an efficient method that has minimal energy consumption with high conversion rate as well as being environmentally friendly [10-12].

The most common electrode in the industry is made of Platinum (Pt) which has high conductivity and satisfactory catalytic activity especially for the HER process [13,14] but is exorbitant and scarce. In addition, the overpotential recorded from electrolysis process using platinum is high which results in higher energy consumption due to slow kinetics occurring at the electrodes [15]. Therefore, exploring low-cost, highly effective, and stable electrocatalysts for both hydrogen production process is highly desirable and currently under active research [16].

Modification of material matrix of electrodes by combining metallic elements in the electrocatalyst can be a highly effective way to increase the performance of HER process such as combination of existing electrodes with carbon structures, porous material, polymers as well as graphene [17]. The interaction of electrocatalyst with multi-metallic matrices shows a higher catalytic activity compared the monometallic matrices. This proves that the modification of electrodes with certain materials may improve the efficiency of the electrocatalyst. Some features of electrocatalyst are desirable to ensure high performance efficiency such as a high intrinsic activity with good binding potential, large active sites and high transfer ability for electrons for electric conductivity.

Recently, graphene-based materials have been employed as an electrocatalyst for both Hydrogen Evolution Reaction (HER) and Oxygen Evolution Reaction (OER) processes owing to their high surface area, excellent conductivity, stability, and low production cost [18]. Pristine graphene possesses poor catalytic activity. Graphene produced by chemical exfoliation method imparts considerable oxygen-related defects. The oxygenated graphene possesses both Lewis acidic and basic sites, which rises the unusual catalytic potential of graphene. Moreover, research has shown that addition of oxygen functional groups or hybridizing of graphene with other active electrolytic materials boost up overall capacity and electron transfer ability of graphene [19].

It was reported that novel composite materials having electrocatalytic characteristics for hydrogen production is comprised of conducting polymer poly(3,4-ethylenedioxythiophene) (PEDOT) with a non-conducting polymer of polyethylene glycol (PEG) family [20]. PEG composite was proven to have a higher electrocatalytic effect when compare with PEDOT-polymethylvinylether (PMVE) composite as well as pure PEDOT due to the capacity of PEG to bring and guide protons closer to PEDOT which allows the HER and OER process to occur comparable to Pt, neglecting all the possible downsides. Polyoxometalates are also one of the polymer composites used widely for HER due to its tiny nature, highly customizable components and has abundant molecular structures that includes sulphides, phosphides, nitrides, and carbides [21]. Polymer as alternative material was further proven where aniline and pyrrole were used as polymer through the process of carbonization and phosphorization where a polymer composite was created with increased electron density and conductivity [16]. Polyaniline (PANI) is a widely used polymer with unique properties such as high thermal, electrical, and environmental/chemical stability, as well as reasonable cost [22, 23]. It enjoys a conjugated structure that brings about conductivity in a doped state, where the dopants are usually acids endowing conductivity to PANI [24]. However, in high-technology applications, PANI should be armed with advanced materials due to its brittle structure and weak performance. Graphene (G) due to its high electron mobility, mechanical/thermal stability, high surface area volume, and capacitance has been utilized in various applications. G and PANI exhibit synergistic effects in G/PANI composites such as endow flexibility and electroactivity respectively [25]. In this regard, polyaniline is a new-found conductive polymer which has advantageous characteristics of low-cost conductive polymer, highly electrocatalytic and flexible in nature. In this study, polyaniline (PANI) graphene supported

FTO glass electrode is fabricated to be used as a conductive electrode which can be further explored for hydrogen production.

Material and methods

Materials

Chemicals namely absolute ethanol, Triton X-100, ethyl cellulose and terpineol were used as available in the laboratory. Graphene and FTO glass were purchased from Sigma Aldrich.

Preparation of Graphene Paste

Synthesis of graphene paste to coat the FTO glass was conducted in three stages described as follows:

Stage 1 - Synthetization of Graphene Solvent Exfoliant

50 ml of absolute ethanol was measured in a measuring cylinder and poured into a beaker. Two drops of Triton X100 were added using a dropper and then mixed by a magnetic stirrer at 500 rpm for about 5 mins. Then 2g of ethyl cellulose was weighted and added to the prepared solution. To avoid the solution from agglomerating, the mixture was mixed by a magnetic stirrer at the rate of 1000 rpm for 20 mins in an induced heat environment at 40 °C.

Stage 2 - Dissolving Graphene in Solvent Exfoliant

0.25g of graphene was measured and added to the solvent (prepared in Stage 1). Three-step sonication was performed via Probe Sonicator (model: Sonica Q500) to gradually dissolve the graphene without compromising yield of the paste. During sonication process, the graphene solution was immersed in an ice bath to avoid evaporation as the process was conducted at induced heated conditions. The specifications of sonicator were kept constant throughout the whole sonication process at 750 Watts and 20kHz. The first sonication was done at 60% power amplitude (PA), 900,000 J and pulse (50 s to 10 s) for about 30 minutes. The second sonication was done for about 45 minutes at increased power amplitude at 90% with similar energy and pulse. The third and last sonication was conducted with the exact same specification for 45 minutes. A total of 2 hours of sonication was performed in order to ensure that the graphene was well-dispersed in the solvent.

Stage 3 - Drying Graphene Paste

After performing three-step sonication in stage 2, the solution was set aside for 2 hours to ensure sediments to settle at the bottom of the beaker under the influence of gravity. 10 ml of terpineol as a thickener was added to the solution and was mixed by a magnetic stirrer at 300 rpm for about 10 mins. Further thickening of the solution was done by a bench-top centrifuge (Model: Eppendorf 5810/5810R) that contained 8 tubes in total. The solution was equally divided into three equal parts and poured into three centrifuge tubes. The remaining five empty tubes out of eight tubes were filled with water having the same mass so that weight can be evenly distributed. The tubes were then inserted in the centrifuge machine and centrifuging was conducted at 3000 rpm for durations of 10, 12 and 15 mins respectively. This centrifuge process separated any remaining sediments from the solution. The supernatant was carefully collected from the tubes yielding a highly dispersed graphene solution as shown in Figure 1(a). The solution was then transferred to a beaker and placed on a hot plate at 1000 rpm at 50°C for 1 hour. Further drying was done in the drying oven to remove the remaining ethanol in the solution. The oven was preheated at 40°C for 10 minutes and the solution was covered with a parafilm. To remove any remaining ethanol in the solution, drying was conducted for 2 hours in total in a stepwise manner consisting of four equal durations of 30 mins each. For the first three durations, the temperature was set constant at 40° but for the last duration of 30 mins, the temperature was ramped up to 60°C. This stepwise heating was performed to prevent rapid evaporation and promote thickening of the solution as shown in Figure 1(b).

a)

b)

***Fig.** 1(a): Thick graphene paste after centrifuge process (b) Reduced volume of graphene paste after step-wise heating*

Fabrication of graphene coated FTO coated glass electrode

FTO glass was coated by adopting tape-casting method [26]. Six pieces of FTO glass measuring 2.5cm × 3.5cm each were cut out from a FTO glass sheet. Tape was placed on the top and bottom of the glass pieces to create cavity in the middle part of the glass and the synthesized graphene paste was evenly roll coated onto the glass by the aid of a glass rod as shown in Figure 2(a). After coating the glasses with graphene paste, they were dried in the oven at 50°C for 15 mins as shown in Figure 2(b). This is considered as single coated glass electrode. Further, double and triple graphene-coated FTO glass were obtained in the same manner.

a) b)

***Fig.** 2(a): Schematic of cavity space creation and tape casting of graphene paste on FTO glass (b) Tape casted graphene coated FTO glass*

Sintering

Once all the graphene paste was roll-coated on the FTO glass cuts, these electrodes were sintered to fully evaporate the terpinol thickener so that the paste can fully adhere onto the glass without any contaminants. The translucency of the graphene turned opaque (from glossy to matte and smooth texture) once the specimen has undergone the sintering process. The higher the number of coatings of the graphene paste on the FTO glass, the darker the paste and vice versa. The sintering furnace had four temperature zones ranging maximum from 410°C to minimum at 190°C as shown in Figure 3(a). A total of 20 mins of sintering was performed for all six samples. Sintered graphene-coated FTO glasses were finally obtained for further in-situ polymerization process as shown in Figure 3(b).

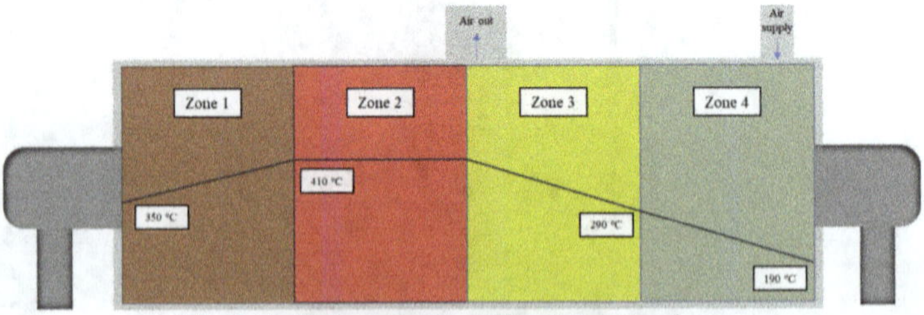

***Fig. 3(a)**: Temperature profile of sintering furnace*

Preparing PANI via in-situ polymerization

12M of fuming hydrochloric acid (HCl) was diluted to 500 ml of 0.4M and 1M concentration respectively and stored in a flat bottom flask. The amount of HCl needed to be diluted was calculated using the dilution equation as stated in Equation 1.

***Fig. 3(b)**: Sintered Graphene coated FTO Glass*

$$M_1 V_1 = M_2 V_2 \tag{1}$$

From the calculated volume of HCl needed using the dilution equation, 500 ml of 0.4M and 1.0M of HCl were obtained by diluting 16.7 ml and 41.7 ml of 12M HCl respectively followed by filling it with distilled water until it reaches the ring of the flat-bottomed flask. 1.2 moles of aniline monomer ($C_6H_5NH_2$) was dissolved in 50 ml of 0.4 HCl and the solution was stirred. Since the aniline is in liquid form, 1.2m moles were converted into ml. From the calculations done, 0.11 ml of aniline was measured (around two drops) and added into the HCl prepared in a beaker labelled as Solution 1. 0.41m moles of ammonium peroxydisulfate ($(NH_4)_2S_2O_8$) was mixed into another 50 ml of 0.4M HCl and stirred. Ammonium peroxydisulfate was in powder form thus 0.41m moles were converted into mass. From the calculation done, a total of 9.4 mg of ammonium peroxydisulfate was weighed and dissolved into the 50 ml of 0.4 M HCl prepared in the beaker and labelled as Solution 2. Both Solution 1 and Solution 2 were then vigorously mixed on a magnetic stirrer at 500 rpm for 5 mins as illustrated in Figure 4.

Materials Research Proceedings 53 (2025) 360-373 https://doi.org/10.21741/9781644903575-35

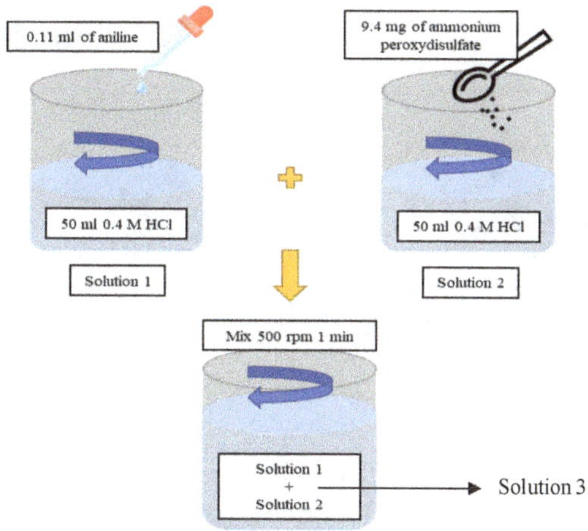

***Fig. 4**: Schematic of mixing of two solutions for PANI preparation*

Growing polyaniline (PANI) flake crystals on graphene-coated FTO glass

Solution 3 as shown in Figure 4 was prepared thrice in order to accommodate all three pairs of coated electrodes. The single, double and triple layered graphene coated FTO glasses were immersed respectively in three separate polyaniline mixtures. The polyaniline solution containing the electrode had to be immersed in a pre-cooled bath as heat was produced since the reaction of mixing both solutions was exothermic in nature. The solution was kept for 30 mins until it turned into blue precipitate (emeraldine base). The electrodes were taken out of the solution and treated with 1M of HCl until it turned green which indicated that the base turned to into salt (conductive) as shown in Figure 5(a). All the electrodes were washed with deionised water. The electrodes were dried in the oven at 40°C for about 5 mins for the moisture to dried out. The schematic of the above-mentioned process is shown in Figure 5(b).

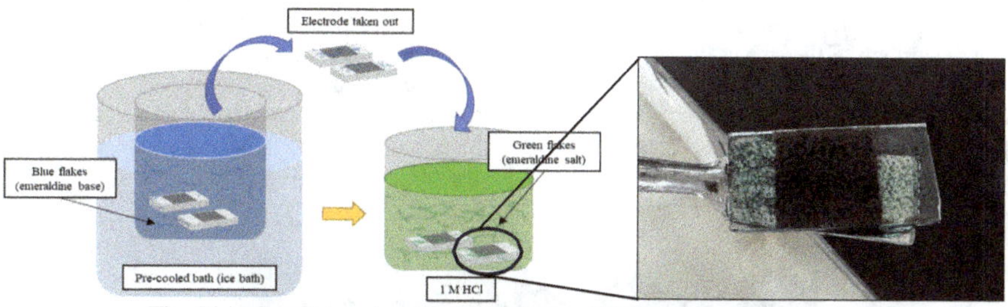

***Fig. 5**(a): Growing polyaniline and doping PANI to form conductive salt*

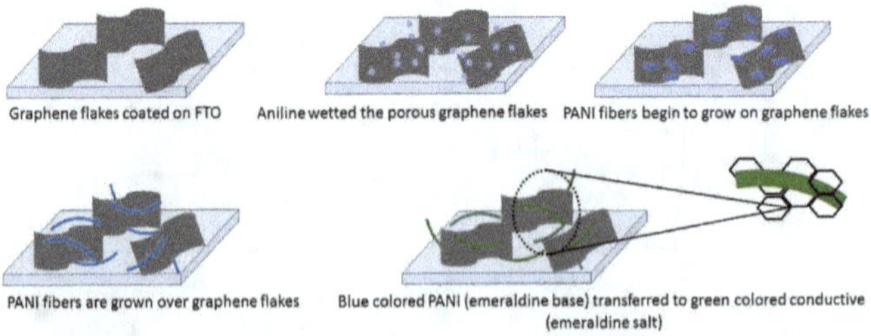

Fig. 5(b)*: Schematic illustration of PANI growth on graphene support*

Results and Discussion

FESEM Imaging

Single, double and triple coated FTO glass electrode with graphene matrix was visualized vertically through FESEM imaging to investigate thickness of the applied coatings in relation to the growth of polyaniline (PANI) as shown in Figure 5(b). The thickness of each coating was evaluated and listed in Table 1.

Table 1: *Thickness of graphene coatings*

Gr-Coating	Thickness
Single Coating	1.53 µm
Double Coating	3.41 µm
Triple Coating	4.40 µm

After the application of graphene coatings on the FTO glass substrate, glass rod was used to even- out the coated layer. While doing so, the major reason for non-uniformity of thickness was mainly due to human error of unsteady hands. Even after non-uniform coatings, the thickness of the coating layers well corresponded with the number of applied coatings. As shown in Figure 6(a), the thickness of the coating depicts single coating where miniscule presence of PANI flakes were witnessed. As the thickness of the coating was increased by double coating the electrode composite, PANI flakes are more distinctive and visibly clearer as shown in Figure 6(b). It can also be witnessed that the growth of flakes was distributed throughout the whole layer of the matrix in varying sizes. The growth of PANI flake crystals were more profound in triple coated electrode. As the thickness of the coated matrix increases, more active surface area along with structural support was available for the growth of flakes where crystals size was bigger and larger in number that were formed throughout the coating layer matrix as shown in Figure 6(c).

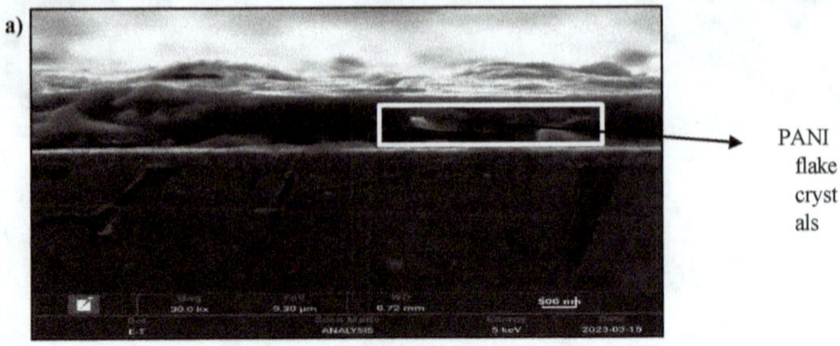

Materials Research Forum LLC
https://doi.org/10.21741/9781644903575-35

b)

PANI flake crysta ls

c)

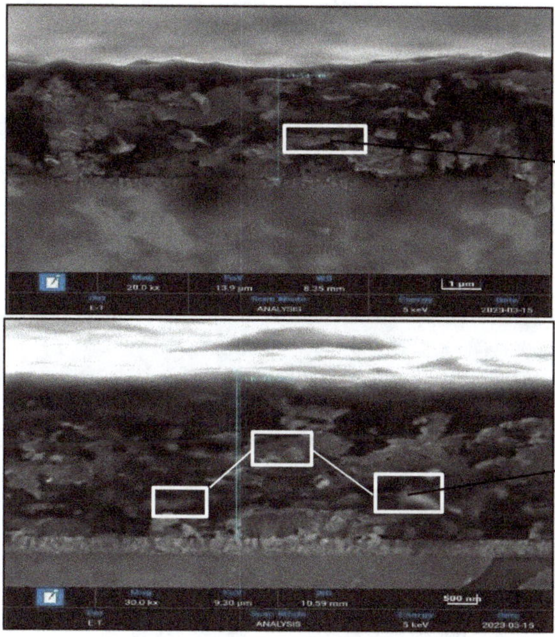

Large size PANI flake crystals

Fig. 6*(a): Field emission scanning electron microscopy images of single coated (thickness 1.53 μm) (b) double coated (thickness 3.41 μm) (c) triple coated (thickness 4.40 μm)*

Raman Spectroscopy
Gr-PANI electrode was characterized by Raman spectroscopy to verify the presence of PANI and graphene. The spectrums of single, double and triple coated PANI-Gr FTO glass are shown in Figure 7(a) – (c).

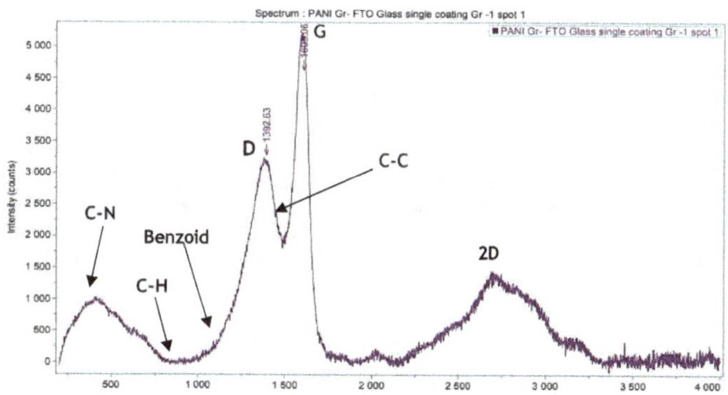

Fig. *7(a): Raman Spectrum Gr-PANI FTO Glass for single coating*

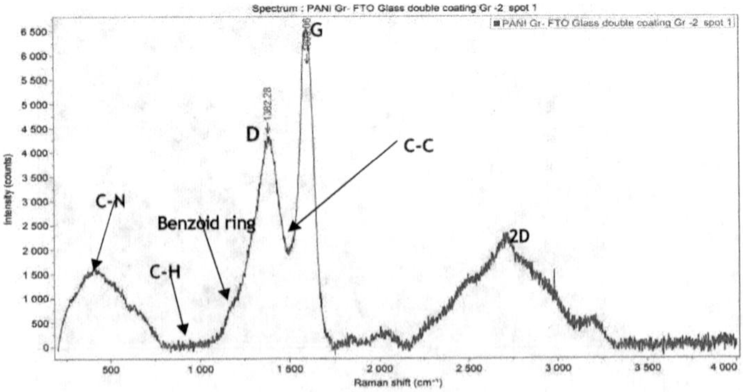

Fig. 7(b): Raman Spectrum Gr-PANI FTO Glass for double coating

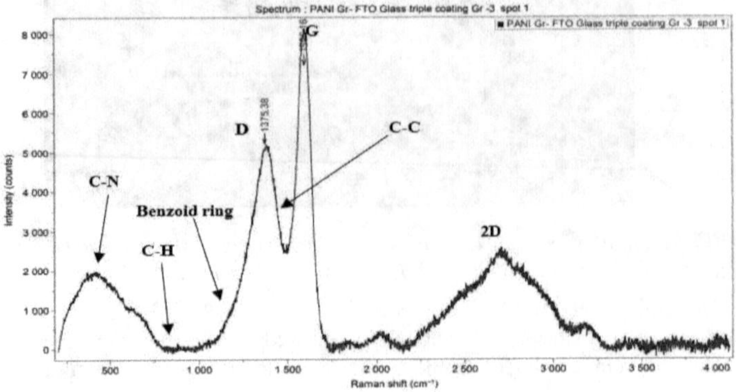

Fig. 7(c): Raman Spectrum Gr-PANI FTO Glass for triple coating

The peak at 400cm^{-1} indicates the presence of C-N bond bending whereas the stretch from 800-1100 cm^{-1} indicates C-H bending. The Raman shift from 1100 to 1400 cm^{-1} corresponds to C-C stretching of quinoid and bending of benzoid ring respectively [27]. Three prominent peaks of D, G, and 2D belongs to the graphene which confirms that both PANI and graphene are present in the fabricated Gr-PANI FTO glass electrode . First peak is a primary in-plane vibrational mode G peak which indicates detection of pure graphene. A different in-plane vibration was also being observed in the second order overtone which is the D and 2D peaks which are both dispersive and dependent on the excitation of electrons of the material [28]. The ratio of D/G peak intensities can be used to characterize the quality of graphene. The lower ratio of D/G will indicate that the graphene detected has low defect while if the ratio of D/G is higher, graphene is of low quality and has a higher defect density.

Since all the spectrums depict same pattern, minor discrepancies can be observed in the intensity of shift peaks that relate to the presence of graphene. As shown in Figure 7(a)-(c), the relationship of 2D and D peaks can be analyzed. According to [29], if the value of 2D is more than D, it indicates that there is only one layer of graphene being distributed on the sample. If the value of 2D is similar to D, it indicates that there are two layers of graphene and if the value of 2D is smaller than D, it indicates there are multilayers of graphene present.

Materials Research Proceedings 53 (2025) 360-373 https://doi.org/10.21741/9781644903575-35

For the Raman spectrum of all the coatings of the Gr-FTO glass electrode, the value of 2D is proven to be lesser than D which indicates that there is presence of multi-layered graphene on the FTO glass. Since the peak of 2D is active and noticeable, it shows that multi-layers of graphene were present in the coating. This shows that the layers of graphene on the FTO glass were proven to be at an optimized level. Conclusively, the presence of graphene in all single, double and triple coatings of electrode was proven to be in high purity due to low D/G ratio in multilayer due to the indication of 2D/D ratio and characteristics of 2D peaks. Too many layers of graphene may not be good for the performance of the electrode since it shows that the composition of the coating is fully graphene without the presence of PANI.

In addition, with the proven presence of graphene and PANI, the relationship between the number of coatings corresponds with the intensity of the compound present in the electrode. The relationship of the intensity differs with the number of coatings applied on the electrode is shown in Table 2.

Table 2. *Comparison of chemical composition intensity of PANI-Gr FTO Glass*

	Chemical Composition	Intensity (counts)		
		Single Coating	Double Coating	Triple Coating
PANI	C-N	1000	1550	2000
	C-H	100	200	300
	Benzoid ring	300	1000	1250
	C-C	2250	2400	3100
Gr	D	3250	4300	5100
	G	5200	6500	8000
	2D	1450	2250	2500

From Table 1. the intensity of PANI composition which are made of C-N, C-H, C-C and benzoid ring were recorded and compared among single, double, and triple coatings of graphene on fabricated PANI-Gr FTO glass electrode. For single coating, the count of C-N was recorded at 1000 for single coating, 1550 for double coating and 2000 for triple coating. As for C-H, the counts were 100, 200, and 300 for single, double, and triple coatings respectively. Additionally for the counts of benzoid ring, single coating was recorded at 300, while double and triple coatings were recorded at counts of 1000 and 1250 respectively.

The composition of Gr indication on the electrode that consists of D, G and 2D were also recorded and compared for its intensity. From the Raman spectrum of Gr-PANI FTO glass for all three coatings, the intensities of D which corresponds to graphene were 3250, 4300 and 5100 for single, double, and triple coatings respectively. As for value of G, the counts were recorded at 5200, 6500 and 8000 respectively for all three coatings. Finally, the value of intensity of 2D were recorded at 1450, 2250 and 2500 for single, double, and triple coatings respectively. The Raman spectrum intensity counts is graphically illustrated in Figure 8.

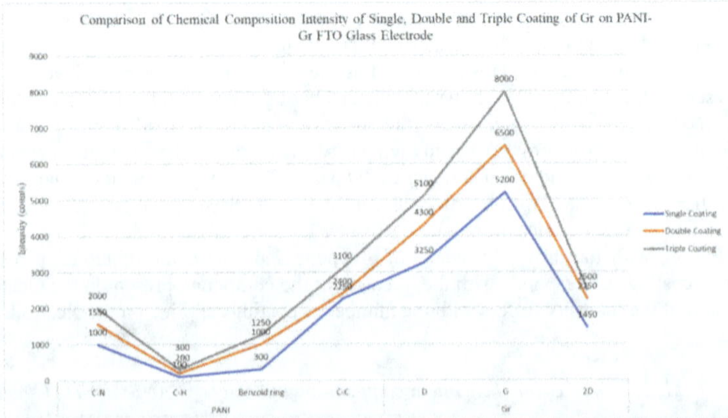

Fig. 8. *Raman spectrum intensity counts of different coatings on Gr-PANI FTO glass electrode*

Resistivity of Fabricated Electrode

The conductivity of Gr-PANI FTO glass electrode was quantified through the measurement of resistance using a multimeter. The probe of the multimeter was placed on the composite coatings. The results in terms of resistance values were compared among uncoated FTO glass and three different coatings on the electrode as shown in Table 3.

Table 3. *Comparison of Gr–PANI FTO Glass Resistivity*

	Resistivity Ω			
	Uncoated FTO glass	**Single Coating**	**Double Coating**	**Triple Coating**
Run 1	0.382	0.278	0.156	0.164
Run 2	0.423	0.343	0.172	0.132
Run 3	0.324	0.356	0.183	0.069
Average	0.4	0.3	0.2	0.1

Three runs per sample were conducted. It can be clearly observed that uncoated FTO glass has the highest resistivity value while the triple coated Gr-PANI FTO glass has the lowest resistivity value as shown in Table 3. These lower values of triple coated Gr-PANI FTO glass well correspond to high growth of PANI facilitated by graphene substrate. Denser graphene support surface renders higher active sites that encourage the growth of conductive PANI crystal flakes via in-situ polymerization making the coating electro-conductive in nature. The average resistivity values are graphically represented in Figure 9.

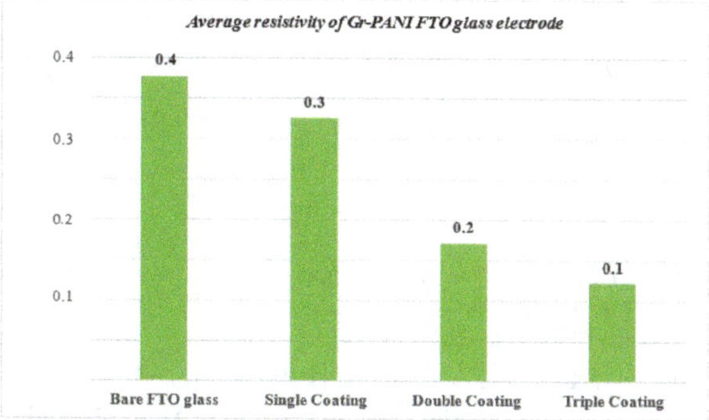

Fig. 9: Average resistivity values of different coatings on Gr-PANI FTO glass electrode

Conclusion

Effective water electrolysis provides a route for hydrogen generation of high purity. The initial results showed promising potential of the fabricated Graphene (Gr) – Polyaniline (PANI) Fluorine-doped tin oxide (FTO) glass nano-composite electrode to be a viable alternative to the commonly used but rare and exorbitant platinum electrode. Well-dispersed graphene paste was formulated that was homogenously coated on FTO glass cuts by means of tape-casting method in three separate coating layers. PANI crystals were grown onto graphene substrate by in-situ polymerization. It was well-observed through FESEM testing supplemented by Raman spectroscopy that the highest degree of PANI polymerization was observed on a triple coated graphene coated FTO glass. Densely coated graphene renders higher number of active sites along with structural support that promote physio-chemical adhesion and polymerization of PANI crystal flakes making the electrode electro-conductive having the lowest resistivity value of 0.1Ω.

References

[1] I. Arto, I. Capellán-Pérez, R. Lago, G. Bueno, and R. Bermejo, The energy requirements of a developed world, Energy for Sustainable Development, vol. 33, pp. 1–13. https://doi: 10.1016/J.ESD.2016.04.001.

[2] P. Nejat, F. Jomehzadeh, M. M. Taheri, M. Gohari, and M. Z. Muhd, A global review of energy consumption, CO2 emissions and policy in the residential sector (with an overview of the top ten CO2 emitting countries), Renewable and Sustainable Energy Reviews, vol. 43, pp. 843–862. https://doi: 10.1016/J.RSER.2014.11.066.

[3] S. E. Hosseini and M. A. Wahid, Hydrogen production from renewable and sustainable energy resources: Promising green energy carrier for clean development, Renewable and Sustainable Energy Reviews, vol. 57, pp. 850–866. https://doi: 10.1016/J.RSER.2015.12.112.

[4] L. Barelli, G. Bidini, F. Gallorini, and S. Servili, Hydrogen production through sorption- enhanced steam methane reforming and membrane technology: A review, Energy, vol. 33, no. 4, pp. 554–570. https://doi: 10.1016/J.ENERGY.2007.10.018.

[5] J. Mohammed-Ibrahim and H. Moussab, Recent advances on hydrogen production through seawater electrolysis, Mater Sci Energy Technol, vol. 3, pp. 780–807. https://doi: 10.1016/J.MSET.2020.09.005.

[6] S. Anwar, F. Khan, Y. Zhang, and A. Djire, Recent development in electrocatalysts for hydrogen production through water electrolysis, Int J Hydrogen Energy, vol. 46, no. 63,

pp. 32284–32317. https://doi: 10.1016/J.IJHYDENE.2021.06.191.

[7] Y. Zhang, L. Li, P. Xu, B. Liu, Y. Shuai, and B. Li, Hydrogen production through biomass gasification in supercritical water: A review from exergy aspect, Int J Hydrogen Energy, vol. 44, no. 30, pp. 15727–15736. https://doi: 10.1016/J.IJHYDENE.2019.01.151.

[8] M. Mehrpooya and R. Habibi, A review on hydrogen production thermochemical water- splitting cycles, J Clean Prod, vol. 275, p. 123836. https://doi: 10.1016/J.JCLEPRO.2020.123836.

[9] F. Safari and I. Dincer, A review and comparative evaluation of thermochemical water splitting cycles for hydrogen production, Energy Convers Manag, vol. 205, p. 112182. https://doi: 10.1016/J.ENCONMAN.2019.112182.

[10] C.-H. Liao, C.-W. Huang, and J. C. S. Wu, Hydrogen Production from Semiconductor-based Photocatalysis via Water Splitting, Catalysts, vol. 2, no. 4, pp. 490–516. https://doi: 10.3390/catal2040490.

[11] T. Hisatomi and K. Domen, Reaction systems for solar hydrogen production via water splitting with particulate semiconductor photocatalysts, Nat Catal, vol. 2, no. 5, pp. 387–399. https://doi: 10.1038/s41929-019-0242-6.

[12] R. D. Tentu and S. Basu, Photocatalytic water splitting for hydrogen production, Curr Opin Electrochem, vol. 5, no. 1, pp. 56–62. https://doi: 10.1016/J.COELEC.2017.10.019.

[13] A. Chen and P. Holt-Hindle, Platinum-Based Nanostructured Materials: Synthesis, Properties, and Applications, Chem Rev, vol. 110, no. 6, pp. 3767–3804. https://doi: 10.1021/cr9003902.

[14] M. Cao, D. Wu, and R. Cao, Recent Advances in the Stabilization of Platinum Electrocatalysts for Fuel-Cell Reactions, Chem Cat Chem, vol. 6, no. 1, pp. 26–45. https://doi: 10.1002/cctc.201300647.

[15] D. Navadeepthy, A. Rebekah, C. Viswanthan, and N. Ponpandian, Boosting the kinetics of oxygen and hydrogen evolution in alkaline water splitting using nickel ferrite /N-graphene nanocomposite as a bifunctional electrocatalyst, Int J Hydrogen Energy, vol. 46, no. 41, pp. 21512–21524. https://doi: 10.1016/J.IJHYDENE.2021.03.244.

[16] W. Li, C. Wang, and X. Lu, Conducting polymers-derived fascinating electrocatalysts for advanced hydrogen and oxygen electrocatalysis, Coord Chem Rev, vol. 464, p. 214555. https://doi: 10.1016/J.CCR.2022.214555.

[17] H. Khoshsang and A. Ghaffarinejad, Electrochemical synthesis of polyaniline silver-palladium composite coated nickel foam as an electrocatalyst for the hydrogen evolution reaction, Inorg Chem Commun, vol. 142, p. 109631. https://doi: 10.1016/J.INOCHE.2022.109631.

[18] H. Huang et al., Graphene Nanoarchitectonics: Recent Advances in Graphene-Based Electrocatalysts for Hydrogen Evolution Reaction, Advanced Materials, vol. 31, no. 48. https://doi: 10.1002/adma.201903415.

[19] B. Xia, Y. Yan, X. Wang, and X. W. (David) Lou, Recent progress on graphene-based hybrid electrocatalysts, Mater. Horiz., vol. 1, no. 4, pp. 379–399. https://doi: 10.1039/C4MH00040D.

[20] B. Winther-Jensen, K. Fraser, C. Ong, M. Forsyth, and D. R. MacFarlane, Conducting polymer composite materials for hydrogen generation, Advanced Materials, vol. 22, no. 15, pp. 1727–1730. https://doi: 10.1002/adma.200902934.

[21] D. Sun et al., Polymers and polyoxometalate induced Co/WC@NC for electrocatalytic hydrogen production, J Solid State Chem, vol. 308, p. 122879. https://doi: 10.1016/J.JSSC.2022.122879.

[22] S. Bhadra, D. Khastgir, N. K. Singha, and J. H. Lee, Progress in preparation, processing and applications of polyaniline, Prog Polym Sci, vol. 34, no. 8, pp. 783–810. https://doi: 10.1016/J.PROGPOLYMSCI.2009.04.003.

[23] H. Wang, J. Lin, and Z. X. Shen, Polyaniline (PANi) based electrode materials for energy storage and conversion, Journal of Science: Advanced Materials and Devices, vol. 1, no. 3, pp. 225–255. https://doi: 10.1016/J.JSAMD.2016.08.001.

[24] M. R. Saeb and P. Zarrintaj, Polyaniline/graphene-based nanocomposites, in Fundamentals and Emerging Applications of Polyaniline, Elsevier, 2019, pp. 165–175. https://doi: 10.1016/B978-0-12-817915-4.00010-5.

[25] H. Wang, Q. Hao, X. Yang, L. Lu, and X. Wang, Effect of Graphene Oxide on the Properties of Its Composite with Polyaniline, ACS Appl Mater Interfaces, vol. 2, no. 3, pp. 821–828. https://doi: 10.1021/am900815k.

[26] Z. Wang, J. Qian, J. Cao, S. Wang, and T. Wen, A study of multilayer tape casting method for anode-supported planar type solid oxide fuel cells (SOFCs), J Alloys Compd, vol. 437, no. 1–2, pp. 264–268. https://doi: 10.1016/J.JALLCOM.2006.07.110.

[27] S. A. Akbar, A. Rochliadi, V. Suendo, N. Saidi, Lelifajri, and A. Mardhiah, A raman spectroscopy study of the polyaniline electrode on ZN-polyaniline rechargeable batteries, Rasayan Journal of Chemistry, vol. 11, no. 4, pp. 1525–1531. https://doi: 10.31788/RJC.2018.1144064.

[28] L. M. Malard, M. A. Pimenta, G. Dresselhaus, and M. S. Dresselhaus, Raman spectroscopy in graphene, Phys Rep, vol. 473, no. 5–6, pp. 51–87. https://doi: 10.1016/J.PHYSREP.2009.02.003.

[29] A. Y. Lee et al., Raman study of D* band in graphene oxide and its correlation with reduction, Appl Surf Sci, vol. 536, p. 147990. https://doi: 10.1016/J.APSUSC.2020.147990.

In-situ epoxidation of palm kernel oil for chemical precursor: Catalysis and kinetic modelling

MOHD JUMAIN Jalil[1,a], SITI JUWAIRIYAH A Rahman[1,b] and INTAN SUHADA Azmi[1,c*]

[1]Chemical Engineering Studies, College of Engineering, Universiti Teknologi MARA, Cawangan Johor, Kampus Pasir Gudang, 81750 Masai, Johor, MALAYSIA

[a]mjumain0686@uitm.edu.my, [b]juwairiyah.rahman@gmail.com, [c]intansuhada@uitm.edu.my*

Keywords: Epoxidation, Palm Kernel Oil, Phosphoric Acid, Genetic Algorithm, Particle Swarm

Abstract. The increasing desire for eco-friendly epoxides made from vegetable oils has ignited curiosity in the epoxidation process of palm oil. However, limited research has explored the palm kernel oil with utilizing phosphoric acid as a catalyst during epoxidation. This study aims to investigate the influence of phosphoric acid as a homogeneous catalyst in the epoxidation reaction and examine the kinetics of the process employing particle swarm optimization (PSO) and genetic algorithm (GA). The in situ peracid mechanism, known for its cost-effectiveness and reduced environmental impact, was selected as the method for epoxide production. The experimental investigation involved epoxidizing palm kernel oil (PKO) to obtain epoxidized palm kernel oil (EPKO). Notably, a reaction time of 24 minutes resulted in a relative conversion of olefin (RCO%) reaching 51.4%. The results of this research enhance our comprehension of the phosphoric acid-catalyzed epoxidation of palm kernel oil and demonstrate the applicability of metaheuristic algorithms for kinetics modelling and optimization in this context. For the kinetics study, both GA and PSO methods were employed, and the sum of errors for both approaches was 0.28 for GA and 0.32 for PSO. These findings suggest that both GA and PSO methods can be effectively utilized for the purpose of kinetics study.

Introduction

Year after year, oil palm plantations expand in both size and output. The production of palm oil also increased at the same time as this surge. The palm seed is used to make palm kernel oil (PKO). In the manufacturing of epoxide for the synthesis of bio-polyol and polyurethane foam, PKO shows promise as a substitute for petroleum as a raw material [1]. The utilization and marketability of PKO globally has therefore demonstrated that it is a viable renewable resource. The key requirements for PKO are accessibility and affordability. The expansion of the oil palm industry in Malaysia has experienced remarkable growth and has gained significant recognition within the global oils and fats sector. [2]. Malaysia takes pride in its rapid ascent to becoming the globe's second-largest palm oil producer and exporter-based commodities within a relatively brief timeframe. Additionally, Malaysia's primary source of income is palm oil.

Epoxide is one of the basic chemicals utilized in the production of polyol, the primary raw material used in the synthesis of polyurethane (PU) [3]. PU must be manufactured in large quantities from sustainable sources in order to meet the product's high demand. High-molecular-weight polymer molecules called polyols have two or more hydroxyl groups and are highly cross-linked [4]. The first two atoms (carbon and hydrogen), which exist in large quantities in crude oil and natural gas, combine to make polyols, which are chemical compounds. Polyols are conventionally produced in large quantities using petrochemical feedstock. However, vegetable oil, due to its deficiency in hydroxyl groups, poses challenges in polyurethane (PU) chemistry [5]. Despite this limitation, polyols derived from vegetable oil have emerged as the predominant source

Materials Research Proceedings 53 (2025) 374-379 https://doi.org/10.21741/9781644903575-36

for bio-based PU [6]. Vegetable oil polyols can have hydroxyl groups inserted at double bond locations using a variety of methods.

Furthermore, in the context of the epoxidation process, hydrogen peroxide plays a vital role as an oxygen donor, while carboxylic acids, such as formic acid, act as oxygen carriers [7]. Unsaturated fatty acids' double bonds receive oxygen from hydrogen peroxide via the oxygen carrier, which then forms epoxide rings [8]. The epoxidation process relies heavily on catalysts. Catalysts are vital in chemical reactions as they help convert reactants into the desired products through mechanisms that lower activation energy barriers. The choice of catalyst in epoxidation process can significantly impact the yield of epoxide. Understanding the effect of catalysts on epoxide yield is essential for optimizing the synthesis process and achieving desired product outcomes. The rate of the epoxidation reaction is greatly sped up by catalysts. These compounds provide an alternative reaction pathway characterized by reduced activation energy, thereby facilitating the occurrence of the reaction. [9]. In this study, phosphoric acid was utilized as a catalyst instead of the commonly used sulfuric acid. The rate at which the epoxidation reaction occurs can be determined by kinetics research [10]. It offers useful details regarding the rate constants. The research utilized both genetic algorithm (GA) and particle swarm optimization (PSO) techniques, given their well-recognized acceptance in the field and frequently utilized by the research community in the context of epoxidation kinetics. Therefore, the goals of this work are to (1) investigate the phosphoric acid as catalyst in the formation of epoxide and (2) to determine the most suitable kinetic model for epoxide formation, either GA model or PSO model.

Research Methodology

Materials. Palm kernel oil procured from Orion Chemique Sdn. Bhd. was utilized in the study. Merck Sdn. Bhd. supplied hydrogen peroxide with a concentration of 30% and formic acid with a concentration of 85%. The titration procedure involved using hydrogen bromide at a concentration of 48%, crystal violet obtained from Merck Sdn. Bhd., and acetic acid at a concentration of 100% sourced from Qrec Sdn. Bhd.

Experimetal Setup. The experimental setup involved a precisely mixed 500 mL beaker where the reaction took place. The beaker was positioned on a hot plate functioning as a furnace. Initially, 50 g of palm kernel oil (PKO) was introduced into the 500 mL beaker. Subsequently, the beakers were placed in a water bath that had a thermostat for controlling the temperature. A constant swirling speed of 350 rpm was maintained for the PKO. The beaker was then filled with hydrogen peroxide (3.5 M) and formic acid (2 M). The solution temperature was monitored until it reached 90°C. At this point, phosphoric acid, serving as the catalyst for the experiment, was introduced into the mixture. The tracking of the reaction time commenced from this instant.

Subsequently, the solution underwent a titration procedure every 10 minutes for a duration of one hour. To obtain a 4 g sample for the analysis of oxirane content referring to the American Oil Chemists' Society (AOCS) Official Method Cd-957, the solution was extracted from the beaker using a 5 mL syringe and weighed. After transferring the mixture to a beaker, 10 mL of acetic acid and 2 drops of crystal violet were added to the sample. Next, the sample underwent titration with hydrogen bromide until the end point became visible, indicated by a bluish-green hue [11]. The relative conversion to oxirane (Eq.1) was calculated based on the Oxirane Oxygen Content (OOC) value obtained both theoretically (Eq. 2) and experimentally (Eq. 3)) [12]. The range of parameters was selected based on prior research [13].

$$RCO = \frac{OOC_{experimental}}{OOC_{theoretical}} \times 100\% \qquad \text{(Eq. 1)}$$

$$OOC_{theoretical} = \left\{\left(\frac{X_o}{A_i}\right) / \left[100 + \left(\frac{X_o}{2A_i}\right)(A_o)\right]\right\} \times A_o \times 100 \qquad \text{(Eq. 2)}$$

$$OOC_{experimental} = 1.6 \times N \times \frac{(V-B)}{W} \qquad \text{(Eq. 3)}$$

The variables X_O, A_i, A_O, N, V and W represent the initial iodine value, molar mass of iodine, molar mass of oxygen, normality of hydrogen bromide (HBr), volume of HBr solution, and weight of the sample, respectively.

Result and Discussion

Fig. 1 depicts the relationship between the relative conversion of olefin (RCO%) and reaction time for the epoxidation process catalyzed by phosphoric acid. A higher RCO% value indicates a greater extent of oxirane (epoxide) formation. As seen in Fig. 1, the maximum RCO (51.4%) was obtained during 24 minutes of reaction time. After that, ring degradation occurred where the RCO decreased to 19.7% at 35 minutes.

It can be seen that the RCO was quite low when using phosphoric acid as a catalyst. If compared to a work done by Azmi et al., [14] by utilizing sulfuric acid as a catalyst, the RCO was much higher, which is 82%. In addition, a work done by Jalil et al., 2014 [15] also shows that sulfuric acid generated high RCO value compared to phosphoric acid. The limited efficiency of phosphoric acid as a catalyst in epoxidation reactions can be attributed to several factors. Firstly, phosphoric acid exhibits relatively weaker acidity compared to sulfuric acid. The lower acidity level of phosphoric acid may result in reduced activation of the reactants, leading to slower reaction rates and lower yields of epoxide products [16]. The milder acid strength of phosphoric acid may not provide sufficient reactivity to promote the desired transformation efficiently. Furthermore, the reaction mechanism of epoxidation reactions involves complex processes, including the formation and stabilization of intermediates. Phosphoric acid may exhibit limitations in stabilizing key reaction intermediates, leading to the occurrence of side reactions that result in reduced yield of the desired epoxide product.

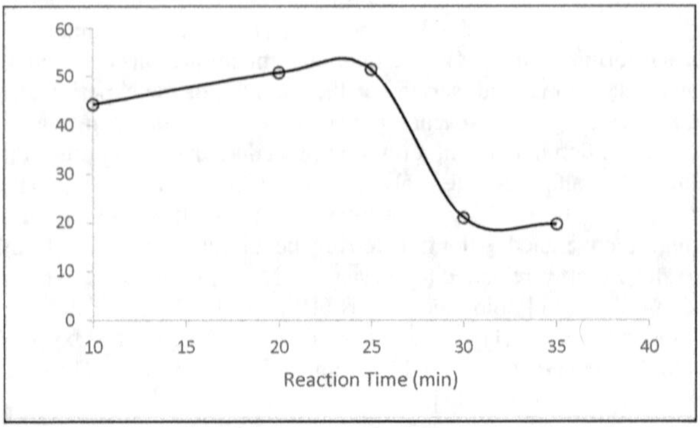

Fig. 1: Variation of Relative Conversion of Oxirane (RCO) with Reaction Time

Kinetics Study of Epoxidation Process by Using Genetic Algorithm (GA) and Particle Swarm Optimization (PSO). The utilization of genetic algorithm (GA) and particle swarm optimization

(PSO) in the investigation of epoxidation reaction kinetics enables efficient exploration of diverse reaction conditions and facilitates the identification of the optimal set of kinetic parameters. This capability allows for the identification of key factors influencing the reaction kinetics and assists in the design of reaction conditions that maximize epoxide yield. The rate constant (k) values are shown in Table 1. The coefficient k_{11} for the synthesis of performic acid was clearly larger than k_{12} by using PSO, as can be observed. This demonstrates that the quicker reply is the former. In the case of GA, the observed value of k_{11} was considerably smaller when compared to k_{12}. This difference could potentially be due to the presence of side reactions, which elevate the rate of reverse reactions to a greater extent. PSO produces a significantly greater reaction rate than GA when it comes to coefficient k2, which is the reaction that creates epoxide.

Table 1: Reaction Rate Constant by using GA and PS

Reaction Rate	Rate Constant, k (mol $L^{-1}min^{-1}$) by Genetic Algorithm	Rate Constant, k (mol $L^{-1}min^{-1}$) by Particle Swarm
k_{11}	0.81	20
k_{12}	10.64	1
k_2	0.20	20
Sum of error	0.28	0.32

Lastly for the sum of error, the error is much lower for PSO compared to GA which are 0.28 and 0.32, respectively. Overall, both methods can be applied to determine reaction rate for epoxidation process with minimum error between simulation and experiment as shown in Fig. 2.

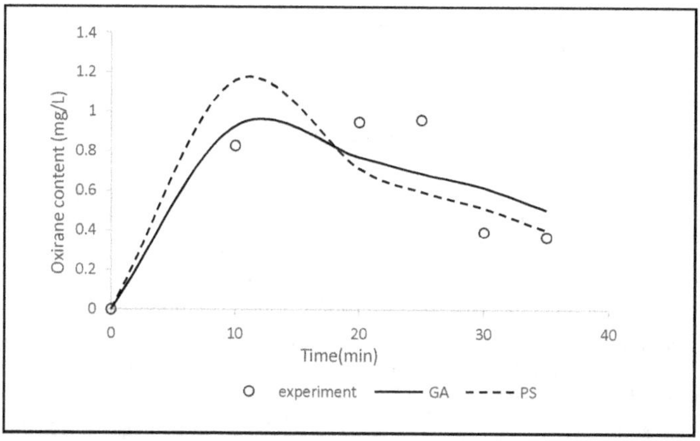

Figure 2: Genetic algorithm vs particle warm optimization method

Conclusion

The epoxidation of PKO offers an exciting opportunity to transform this renewable resource into a range of valuable products, including epoxidized palm kernel oil (EPKO). Researchers have explored the use of different catalysts and reaction parameters to improve the epoxidation yield of PKO. In conclusion, the use of phosphoric acid as a catalyst in the epoxidation reaction exhibits limitations in achieving high yields of epoxide products. While phosphoric acid is a widely used catalyst in various organic transformations, its performance in epoxidation reactions falls short when compared to alternative catalysts such as sulfuric acid. The application of computational intelligence techniques, which are GA and PSO, has gained attention as a powerful tool for optimizing reaction kinetics. In the kinetics study, both the GA and PSO methods were utilized, resulting in a lower sum of errors values. These findings suggest that both the GA and PSO methods exhibit effectiveness in conducting kinetics studies.

Acknowledgements
Authors acknowledge the Ministry of Higher Education (MOHE) for funding under the Fundamental Research Grant Scheme (FRGS) (Ref: 600-RMC/FRGS 5/3 (022/2023).

References

[1] W.N.W. Tajulruddin, A.R. Rahmat, Z.A.M. Yusof, A. Fakhari, Characterization of Polyol Synthesized from Epoxidized Palm Oil Using Fourier Transform Infra-Red and Nuclear Magnetic Resonance, J. Adv. Res. Appl. Mech. 10 (2015) 30–35. http://www.akademiabaru.com/doc/ARAMV10_N1_P30_35.pdf

[2] P. Kannan, N.H. Mansor, N.K. Rahman, T.S. Peng, S.M. Mazlan, A review on the malaysian sustainable palm oil certification process among independent oil palm smallholders, J. Oil Palm Res. 33 (2021) 171–180. https://doi.org/10.21894/jopr.2020.0056

[3] F.H. Yeoh, C.S. Lee, Y.B. Kang, S.F. Wong, S.F. Cheng, W.S. Ng, Production of biodegradable palm oil-based polyurethane as potential biomaterial for biomedical applications, Polymers (Basel). 12 (2020). https://doi.org/10.3390/POLYM12081842

[4] Sardari, A.A.S. Alvani, S.R. Ghaffarian, Synthesis and characterization of novel castor oil-based polyol for potential applications in coatings, J. Renew. Mater. 7 (2019) 31–40. https://doi.org/10.32604/jrm.2019.00045

[5] K. Polaczek, M. Kurańska, M. Auguścik-Królikowska, A. Prociak, J. Ryszkowska, Open-cell polyurethane foams of very low density modified with various palm oil-based bio- polyols in accordance with cleaner production, J. Clean. Prod. 290 (2021). https://doi.org/10.1016/j.jclepro.2021.125875

[6] M.A. Sawpan, Polyurethanes from vegetable oils and applications: a review, J. Polym. Res. 25 (2018). https://doi.org/10.1007/s10965-018-1578-3

[7] F. Marriam, A. Irshad, I. Umer, M. Arslan, Vegetable oils as bio-based precursors for epoxies, Sustain.Chem. Pharm.31 (2023)100935. https://doi.org/10.1016/j.scp.2022.100935

[8] W.F. Bohórquez, A. Orjuela, P.C.N. Rincón, J.G. Cadavid, J.A. García-Nunez, Experimental optimization during epoxidation of a high-oleic palm oil using a simplex algorithm, Ind. Crops Prod. 187 (2022) 1–8. https://doi.org/10.1016/j.indcrop.2022.115321

[9] E.R. Gunawan, D. Suhendra, P. Arimanda, D. Asnawati, Murniati, Epoxidation of Terminalia catappa L. Seed oil: Optimization reaction, South African J. Chem. Eng. 43 (2023) 128–134. https://doi.org/10.1016/j.sajce.2022.10.011

[10] Y. Meng, N. Kebir, X. Cai, S. Leveneur, In-Depth Kinetic Modeling and Chemical Analysis for the Epoxidation of Vegetable Oils in a Liquid–Liquid–Solid System, Catalysts. 13 (2023). https://doi.org/10.3390/catal13020274

[11] A.S.A. Hazmi, M.M. Aung, L.C. Abdullah, M.Z. Salleh, M.H. Mahmood, Producing Jatropha oil-based polyol via epoxidation and ring opening, Ind. Crops Prod. 50 (2013) 563–

[12] 567. https://doi.org/https://doi.org/10.1016/j.indcrop.2013.08.003

[13] N. Kamairudin, S.S. Hoong, L.C. Abdullah, H. Ariffin, D.R.A. Biak, Optimisation of epoxide ring-opening reaction for the synthesis of bio-polyol from palm oil derivative using

response surface methodology, Molecules. 26 (2021). https://doi.org/10.3390/molecules26030648

[14] E.R. Gunawan, D. Suhendra, P. Arimanda, D. Asnawati, Murniati, Epoxidation of Terminalia catappa L. Seed oil: Optimization reaction, South African J. Chem. Eng. 43 (2023) 128–134. https://doi.org/10.1016/j.sajce.2022.10.011

[15] I.S. Azmi, M.R. Mustafa, N.I.S. Mohd Zalman, M.D. Haqeem Norazlan, N. Mahdir, M.S. Md Zaini, M.J. Jalil, Synthesis of bio-polyol from epoxidized palm oleic acid by homogeneous catalyst, J. Elastomers Plast. 0 (2022) 1–13. https://doi.org/10.1177/00952443221147029

[16] M.J. Jalil, N. Mohamed, S.K. Jamaludin, A.M. Som, A.R. Mohamad Daud, Epoxidation of Palm Kernel Oil-Based Crude Oleic Acid, Adv. Mater. Res. 906 (2014) 125–130. https://doi.org/10.4028/www.scientific.net/AMR.906.125

[17] M. J. Jalil, I. S. Azmi, A. Farhan, M. Yamin, M. Zarqani, Y. Abdul, Catalytic Epoxidation of Oleic Acid and Subsequent Ring - Opening by In Situ Hydrolysis for Production Dihydroxystearic Acid, J. Polym. Environ. (2022). https://doi.org/10.1007/s10924-022- 02736-3

Materials Research Proceedings 53 (2025) 380-389 https://doi.org/10.21741/9781644903575-37

Effect of Cu loading to MCM-41 catalyst for propanediols production

NUR AKILA SYAKIDA IDAYU Khairul Anuar[1,a], ANITA Ramli[1,b*] and AFEEFAH Fakhruldin[1,c]

[1]Fundamental and Applied Sciences Department and Centre for Biofuel and Biochemical Research, Institute of Self-Sustainable Building, Universiti Teknologi PETRONAS, 32610 Seri Iskandar, Perak, Malaysia

[a]nurakila.khairul@utp.edu.my, [b]anita_ramli@utp.edu.my, [c]afeefah.fakhruldin@utp.edu.my

Keywords: Glycerol Hydrogenolysis, Propanediols, In-situ Hydrogen, Cu/MCM Catalyst, High-pressure Batch Reactor

Abstract. The catalytic activity of Cu/MCM for glycerol hydrogenolysis to 1,2-propanediol (1,2-PDO) and 1,3-propanediol (1,3-PDO) was tested in a high-pressure batch reactor at 180-200 °C for 3 h under 10 bars of N_2 with different solvent as in-situ hydrogen source. The commercial Al-MCM-41 was impregnated with 2.5, 5, 7.5 and 10 wt.% Cu using the incipient wetness impregnation method, dried at 100 °C for 24 h and calcined at 550 °C for 5 hours. The physicochemical properties of all catalysts were characterized using XRD, N_2 adsorption-desorption isotherms, and H_2-TPR. The synthesized Cu/MCM showed the presence of diffraction peaks, which could be assigned to a high-order two-dimensional hexagonal arrangement of MCM support at a low angle, while the CuO phase was detected at a wide angle diffraction peak. The surface area and pore volume showed a decreasing pattern as Cu content increased. The total hydrogen consumption of the Cu/MCM catalyst decreases at 2.5 wt.% of Cu content and gradually increases as the Cu content increases to 5, 7.5, and 10 wt.%. All catalysts were tested for glycerol hydrogenolysis without external hydrogen gas addition to determine the effect of different Cu loading on MCM support catalyst for 1,2-PDO and 1,3-PDO production.

Introduction

Due to the depletion of fossil fuel reserves and the ongoing rise of environmental concerns, research into renewable energy sources is of utmost importance for the creation of cleaner substitutes for liquid fuels and chemicals produced from petroleum [1]. One of these renewable, environmentally friendly fuels derived from biomass that may immediately replace conventional petroleum diesel is biodiesel. Glycerol production rises promptly as worldwide biodiesel production surges for blending with B10 diesel [2], which leads to an excess of glycerol on the market and a drop in glycerol pricing [3].

The hydrogenolysis of higher polyols of glycerol into high-value-added chemicals, like propanediols, has emerged as an attractive process for glycerol valorization. To reach an acceptable glycerol conversion and product selectivity, hydrogenolysis has to perform at high hydrogen pressure and high temperature, owing to the low solubility of hydrogen in aqueous glycerol solution. In addition, the molecular hydrogen is easily ignited upon contact with air, it certainly raises the safety concerns of the potential explosion due to its extreme ease of leakage [4]. Going forward, the development of more environmentally safe processes, such as catalytic transfer hydrogenolysis of glycerol in the presence of hydrogen donor from organic liquid, display much better atomic economics compared with conventional hydrogenolysis processes using excess H_2 [5]. Currently, 1,2-PDO derived from the hydrogenolysis of glycerol is employed in the production of unsaturated polyester resins and antifreeze [6] while the primary use of 1,3-PDO is the production of highly valuable polyester fibers, films, coatings and engineering plastics [7].

Glycerol hydrogenolysis has drawn much attention in recent years, and the majority of the studies on this subject have been developed by utilizing a variety of catalysts. The hydrogenolysis of glycerol employing a Cu combination on mordenite support as a catalyst has improved the selectivity of propanediols, with 90% conversion and 58.5% selectivity of 1,3-PDO using vapor phases at 210 °C. Azri et al. [8] reported that Cu-based catalyst produced 78.5% glycerol conversion and 79% selectivity of 1,2-PDO at 200 °C in 10 h reaction time. Samudrala et al. [9] demonstrated a high glycerol conversion (86%) and 1,3-propanediol selectivity (42%) are achieved using a continuous vertical fixed bed quartz reactor over the zeolite-based catalyst at 210 °C for 7 h. The acid sites of MCM-41 as support catalysts can perform the dehydration of glycerol to give acetol as the intermediate product, then proceeded by subsequent hydrogenation of acetol intermediate to propanediols which selectively occur on metal sites of Cu [10]. Therefore, in this research, different loadings of Cu were used on the high surface area of the MCM-41 catalyst to produce high-value-added propanediols.

Methodology

Commercial zeolite Al-MCM-41 was purchased from ACS Materials, LLC, and was impregnated with 2.5, 5, 7.5, and 10 wt.% copper (II) nitrate hexahydrate via the incipient wetness impregnation method. A sufficient amount of Cu was fully dissolved in deionized water before zeolite was added to the Cu aqueous solution. All catalysts prepared were stirred for 3 h at room temperature and then dried at 100 °C for 24 h and calcined at 550 °C for 5 h.

The crystallization state of prepared mesoporous materials of Al-MCM-41 supported Cu was determined using Bruker X–ray diffraction (XRD) model X'Pert^3 Powder and Empyrean, PANanalytical with Cu–Kα radiation. The diffractograms were recorded in the 2θ range of 1–10° (small angle) and 10–80° (wide angle) in the steps of 0.010° with a count time of 15 s at each point. The N_2 absorption-desorption was performed on Micromeritics ASAP 2020. Textural measurements such as specific surface areas, pore volume, and pore size distribution were obtained according to the Brunauer–Emmett–Teller theory and by the B.J.H method using N_2 at 78 K. Prior to the adsorption, the samples were pretreated under vacuum at 543 K for 3 h. Catalyst reduction properties was determined by a temperature program reduction using Thermo Scinetific TPDRO 1100 with a TCD detector. The catalysts were treated at 423 K under a argon flow (40 mL min^{-1}) prior to reduction. The TPR conditions were as follows: 125 mg of catalyst was reduced at atmospheric pressure under a stream of 5% v/v H_2 in argon (total flow rate of 25 mL min^{-1}), using a linear temperature program from room temperature to 1273 K, with a heating rate of 10 K/min.

The glycerol hydrogenolysis reaction was performed in a 100 mL autoclave stainless steel high-pressure batch reactor equipped with a thermocouple and a motor drive magnetically coupled to an internal stirrer shaft. Prior to the catalytic testing, the catalysts were reduced in a fixed-bed reactor under an H_2 flow of 25 mL min^{-1} at 300 °C for 3 h. The reduced catalyst (0.1-0.5 g) and 40 mL of the desired amount of purified dissolved organic solvent were introduced into the autoclave. The reactor was sealed and purged with pure N_2 more than three times and then pressurized up to 10 bars. The reaction conditions were tested at 180-200 °C, 40 mL purified glycerol dilute in different organic solvents (water, ethanol, and methanol), 0.2 g of catalyst, and 250 rpm stirring rate unless otherwise stated. After each reaction time, the system was cooled to room temperature, and the catalyst was separated by centrifugation at 2000 rpm for 10 min or filtrated using a vacuum pump. The separation of the compounds was performed on a Polaris Hi-Plex H column, 300 × 7.7 mm (Agilent Technologies, CA, USA), using the 5 mM H_2SO_4 mobile phase with a flow rate of 0.6 mL min^{-1}, column temperature T = 80 °C, and RID temperature T = 35 °C. The elution of the compounds was run for 25 minutes.

Result and Discussion

X-ray Diffractogram Analysis. The XRD patterns for all catalysts are shown in Figure 1. Figure 1a displays the low angle diffraction peak (1-10°) of Al-MCM-41 and different loading of Cu/MCM with a strong peak is assigned for (100) plane formed at 2θ = 2.1°. Meanwhile, there are other two diffraction peaks with low intensity centered at approximately 2θ = 3.8° and 4.4° represented the (110) and (200) planes, respectively according to ICSD 98-015-4906. Formation of (100), (110), and (200) planes attributed to the mesopores of a high-order two-dimensional hexagonal arrangement of MCM support [11]. In Figure 1a, the feature peaks of MCM-41 can be noticed showing that the framework of MCM-41 has no significant changes upon impregnation of Cu on mesoporous MCM-41 matrix. In addition, the intensity peak of MCM-41 was observed to decrease consistently as the loading of Cu was increased from 2.5 to 10 wt.%. The wide angle (20-80°) X-ray diffraction patterns of Al-MCM-41 and Cu/MCM catalysts with different loading are presented in Figure 1b. The XRD spectra of Cu/MCM exhibit two peaks formed at 2θ = 35.5° and 38.7° attributed to the CuO phase (ICSD 98-008-7122) that are assigned to miller indices of (002) and (111), respectively. The spectra signal for CuO formation shows an increased pattern as the loading of Cu increases. The intensity peak of MCM slightly reduced as metal loading increased in Figure 1b, indicating that the metal oxides are dispersed well due to the uniform distribution of the metal particles in the large surface area of the MCM-41 support [12].

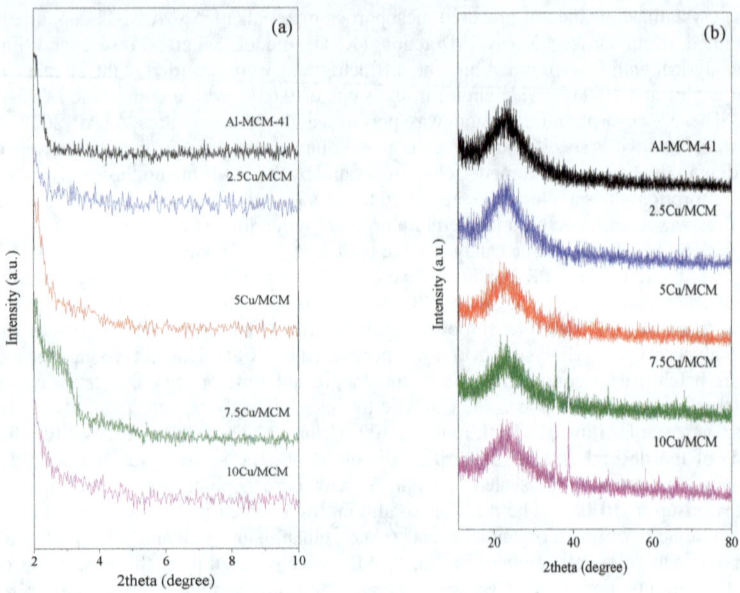

Figure 1. X-ray diffractogram of low angle (a) and wide angle (b) Al-MCM-41 and Cu/MCM catalysts with different loading.

N_2 Adsorption-Desorption Analysis. The surface area, pore size, pore volume, and crystallite size are summarized in Table 1. The surface area between Al-MCM-41 and 2.5Cu/MCM decreases when 2.5 wt.% of Cu is impregnated into MCM-41 catalyst. This indicated an interaction between

Cu species with the MCM framework. A study done by Rahman et al. [13] stated that there is a different decreasing pattern between MCM-41 and Ni/MCM-41 can be observed from the textural properties which indicates the blocking of pores by the deposition of NiO into MCM-41 support. The surface area was observed together with pore volume, which also demonstrated decreasing in value. However, as the loading of impregnated Cu increases from 7.5 to 10 wt.%, it can be observed that surface area and pore volume decrease significantly except for the 5Cu/MCM catalyst. The decreased nitrogen uptake indicated a limitation of mesopore volume in these samples due to the incorporation of copper content into MCM-41 [14]. For 5Cu/MCM cases, there is a contradiction in both surface area and pore volume compared to other catalysts. The high textural properties of 5Cu/MCM contributed by the Cu particles deposited much more prominent on the surface of support than deposited particles in the pores of MCM-41 [15].

Table 1. Textural properties of Al-MCM-41 and Cu/MCM catalysts

Catalyst	Surface area (m^2/g)	Pore size (nm)	Pore volume (cm^3/g)	Crystallite size of Cu (nm)
Al-MCM-41	745	3.48	0.78	-
2.5Cu/MCM	617	3.80	0.69	3.3
5Cu/MCM	692	3.89	0.82	3.1
7.5Cu/MCM	501	3.95	0.54	3.6
10Cu/MCM	492	3.68	0.56	4.1

There are slight differences that can be observed in pore size for all catalysts. The size of CuO particles was calculated from the XRD line broadening of the peak at the plane (111) using Scherrer's equation for all catalysts. The crystallite size of Al-MCM-41 support for all catalysts was not observed in this research due to the low signal-to-noise ratio and the inability to fit the dominant peak for Scherrer's calculation. The CuO particle size for all catalysts shows slight differences and exhibits almost identical crystal size as Cu content increases. This observation may be due to the uniform distribution of the metal particles in the large surface area of the MCM-41 support as described by Deshmane et al. [10]. The increase in XRD spectra of Cu/MCM is due to the inference of copper ions during the formation of ordered structures. This leads to increased order and pore blocking that could also be responsible for surface area and pore volume decrease as shown in Table 1.

Temperature Programmed Reduction of Hydrogen (H_2) Analysis. Figure 2 displays the H_2-TPR analysis of Al-MCM-41 and Cu/MCM catalyst with different loading of Cu impregnated on a support. As we can see, Al-MCM exhibits low-intensity peaks that appear at temperatures region around 56 to 791 °C with a total of four continuous broad peaks centered at 56, 353, 584, and 791 °C, respectively, while there are three broad shoulders formed around 272, 654, and 696 °C. Additionally, the intensity peak of Al-MCM shows a noticeable difference compared to MCM supported by different loadings of Cu. All Cu/MCM catalysts with different loading display a sharp and prominent reduction peak at substantially low-temperature regions. As Cu was introduced into support, we can observe that MCM reduction peak slowly vanished as it shifted to lower and higher temperature regions as Cu content increased. Both reduction of sharp peak centered at 277 and 305 °C associated with the reduction of Cu^{2+} to Cu^0 in aggregated copper oxide species and $Cu^{2+} \rightarrow Cu^+$ in monomeric copper cations, while the second peak assigned to the second step of monomeric cations reduction of $Cu^+ \rightarrow Cu^0$ [16]. Cu ions in 2.5Cu/MCM and 5Cu/MCM dispersed with a lower reduction of Al_2O_3 species as it shifted to a lower temperature and gave rise to the reduction peak around temperature 270 to 310 °C [17].

On the other hand, a broad peak detected around 870° C for 7.5Cu/MCM and 10Cu/MCM catalysts indicated that the SiO_2 species shifted to a higher temperature region of more than 800 °C as Cu loading increased. The reduction of copper silicate was not complete below 800 °C as copper species occluded in the silica walls of MCM-41 and hardly reducible copper silicate [18]. Table 2 summarizes the hydrogen consumption of Al-MCM-41 and Cu/MCM catalyst. The total hydrogen consumption of the Cu/MCM catalyst decreases at 2.5 wt.% of Cu content and gradually increases as the Cu content increases to 5, 7.5, and 10 wt.%. The increase of hydrogen uptake in the higher content of Cu is probably due to the contribution of Al_2O_3 and SiO_2 in MCM to promote the reduction of easily accessible surface of copper species at lower temperature regions [14].

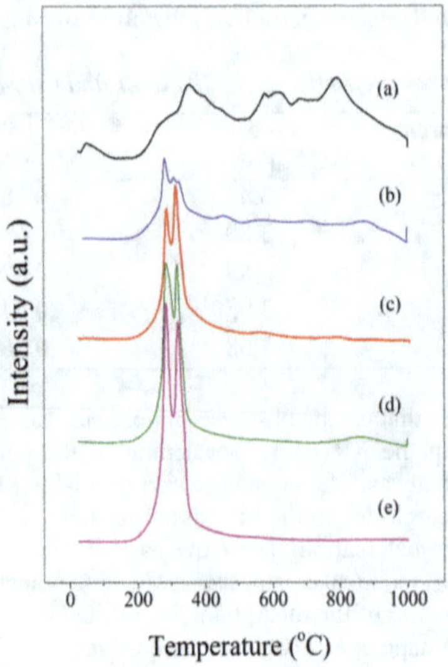

Figure 2. H_2-TPR analysis of Al-MCM-41 and Cu/MCM, (a) Al-MCM-41, (b) 2.5Cu/MCM, (c) 5Cu/MCM, (d) 7.5Cu/MCM and (e) 10Cu/MCM

Table 2. Reduction properties of Al-MCM-41 and Cu/MCM catalysts.

Catalyst	H_2 consumption (μmol H_2/g)				Total
	Peak 1	Peak 2	Peak 3	Peak 4	
Al-MCM-41	2707.9	2869.6	2845.0	2878.9	11301.4
2.5Cu/MCM	3978.7	3085.1	453.1	873.2	8390.1
5Cu/MCM	12813.4	11070.6	565.7	n.d	24449.7
7.5Cu/MCM	21516.0	20319.2	861.8	n.d	42697.0
10Cu/MCM	32749.1	30500.5	854.5	n.d	64104.1

Catalyst Testing. The catalytic performance of Al-MCM-41 and various Cu-MCM catalysts was evaluated using a high-pressure batch reactor, in terms of glycerol conversion and selectivity towards production of propanediols. Using the catalytic screening approach, only selected catalysts that give the highest yield of propanediols proceeded for further optimization on utilization of different organic solvents and reaction temperatures. Glycerol hydrogenolysis was carried out using ethanol as solvent and 0.2 g of catalyst with 10 wt.% of glycerol feedstocks under 10 bars of inert N_2 at 180 °C for 3 h. From the screening result demonstrated in Table 3, 7.5Cu/MCM and 10Cu/MCM catalysts give the highest 1,2-propanediols and 1,3-propanediols yield compared to other catalysts. In contrast, 2.5Cu/MCM and 5Cu/MCM only selectively produced 1,2-propanediols, while in the case of bare MCM-41, no propanediols were attained. The highest conversion of glycerol to 1,2-propanediols and 1,3-propanediols was 30.8% and 23.1% with selectivity of 18.25 and 14.6%, respectively by employing 10Cu/MCM as a catalyst. Whereas, utilizing of 7.5Cu/MCM catalyst recorded a conversion of 30.5% (1,2-PDO) and 20.4% (1,3-PDO), while selectivity of 17.7% (1,2-PDO) and 9.8% (1,3-PDO).

***Table 3**. Catalyst screening of all synthesized catalysts.*

Catalyst	Conversion (%)		Selectivity (%)	
	1,2-PDO	1,3-PDO	1,2-PDO	1,3-PDO
MCM-41	n.d	n.d	n.d	n.d
2.5Cu/MCM	24.1	n.d	8.5	n.d
5Cu/MCM	23.9	n.d	8.8	n.d
7.5Cu/MCM	30.5	20.4	17.7	9.8
10Cu/MCM	30.8	23.1	18.2	14.6

n.d = not detected

The combination of Cu as an alkaline metal into acid catalysts has been proven to enhance the catalytic performance for propanediol production [19]. The Cu-based exhibits excellent selectivity towards propanediols due to its properties, which are attributed to the tendency towards selectivity cleave of the C-O bond rather than the C-C bond. It has been described that alkaline metal, or specifically Cu, can promote the cleavage in the C-O bond of the primary hydroxyl group of glycerol to form acetol with simultaneously H_2 donor from liquid organic through hydrogenolysis without cleaved the C-C bond of glycerol molecules [20]. The TPR result showed that both catalysts were recorded with high H_2 uptake and were easily reducible at low temperatures. Thus, 7.5Cu.MCM and 10Cu/MCM catalyst fit to be used as an active site medium for conversion of glycerol to propanediols by improving the selectivity to cleavage the C-O bond where it can be proved by the highest conversion and selectivity towards both 1,2-propanediols and 1,3-propanediols production as shown in Table 3.

Effect of Solvent System as In-situ H_2. Figure 3 shows the effect of the different solvent systems by using ethanol, methanol, and water on propanediol production at 180 °C for 3 h and catalyzed by 10Cu/MCM under 10 bars of inert N_2. It was found that ethanol solvent resulted in the highest conversion and selectivity of propanediols, followed by methanol and water. The in-situ hydrogen production aided by ethanol solvent was consumed during the hydrogenation of intermediate acetol to propanediols over Cu metal. The hydrogen dehydrogenated from ethanol exists mainly in the form of the active hydrogen atom, which is adsorbed on the surface of the catalyst and enhanced to react with acetol quickly, which is the reason for the high conversion and selectivity of propanediols [21]. Furthermore, the high efficiency of ethanol as an in-situ hydrogen source could be attributed to active sites provided by Cu for the dehydrogenation of ethanol. This finding is agreed by Chang et al. [22] and Carotenuto et al. [23] stated that ethanol as competent in-situ hydrogen together with Cu metal improved the active site and selectivity for accelerating the acetol intermediate formation for further hydrogenate to form propanediols.

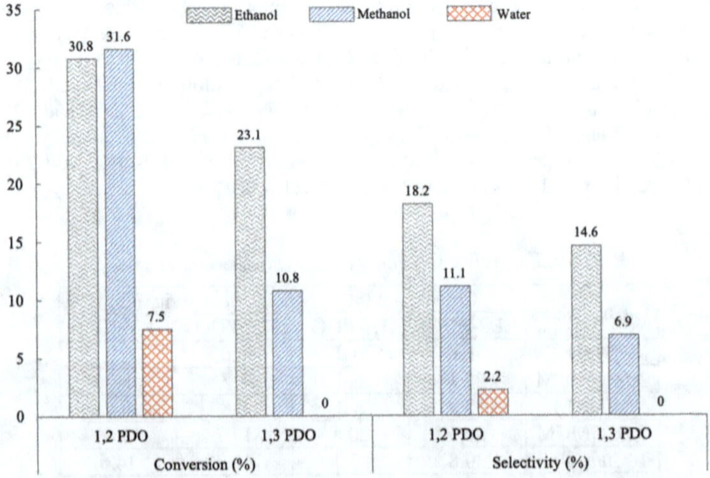

***Figure** 3. Effect of solvent as hydrogen donor catalyzed 10Cu/MCM.*

Effect of Reaction Temperature. Figure 4 demonstrated the three different reaction temperatures on glycerol hydrogenolysis over 10Cu/MCM using ethanol as in-situ hydrogen under 10 bars of nitrogen gas for 3 h with 10 wt.% of glycerol feedstocks. As expected, the conversion of glycerol to 1,2-propanediol and 1,3-propanediol shows a notable increase when the reaction temperature was raised from 180 °C to 200 °C. This was due to the improvement of reaction temperature, which favored glycerol conversion and led to high propanediol production. According to Mauriello et al. [24], the increased reaction temperature could be related to the improved activity of hydrogen production. Low glycerol conversion was noticed due to insufficient temperature at 180 °C to activate the reactant molecules and surface-active sites [25]. The selectivity to 1,2-propanediols was found to improve as temperature increased. However, selectivity to 1,3-propanediol was observed to slightly decline from 180 to 190 °C and then gradually increase when the temperature reached 200 °C. During the 190 °C reaction temperature, the production of side products such as acetol also increased, possibly due to it reaching the favored condition for the production of acetol. As the temperature rose, the catalyst active site went through a high activity for catalytic conversion by cleavage of the C-O bond, which led to the production of a large amount of desired products but also increased the production of side products such as acetol [26].

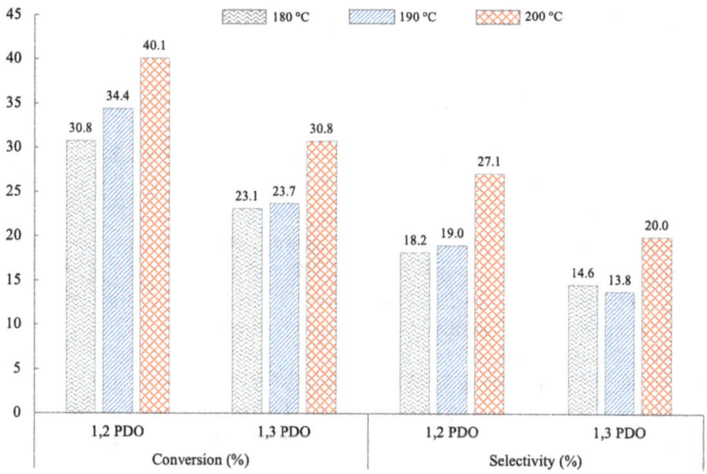

***Figure 4**. Effect of reaction temperature catalyzed 10Cu/MCM.*

Conclusion

Glycerol has been successfully produced 1,2-propanediol and 1,3-propanediol via catalytic transfer hydrogenolysis without an external hydrogen gas source. The screening and testing were determined under a high-pressure batch reactor using Al-MCM-41 and a different content of Cu on MCM support as the catalyst. The synthesized 10Cu/MCM was found to give considerable production of 1,2-propanediol (conversion 40.1%, selectivity 27.1%) and 1,3-propanediol (conversion 30.8%, selectivity 20.0%).

Acknowledgement

The research was funded by Yayasan Universiti Teknologi PETRONAS: Pre-Commercialization Research Grant (015PBC-029). The authors thank Universiti Teknologi PETRONAS for providing the necessary facilities to support this project.

References

[1] A. M. Ruppert, K. Weinberg, R. Palkovits, Hydrogenolysis Goes Bio: from Carbohydrates and Sugar Alcohols to Platform Chemicals. Angewandte Chemie International Edition. 51 (2012) 2564–2601. https://doi.org/10.1002/anie.201105125

[2] Zulqarnain, M. H.M. Yusoff, M. Ayoub, N. Jusoh, A. Z. Abdullah, The Challenges of a Biodiesel Implementation Program in Malaysia. Processes. 8 (2020) 1244. https://doi.org/10.3390/pr8101244

[3] T. Numpilai, C. K. Cheng, A. Seubsai, K. Faungnawakij, J. Limtrakul, T. Witoon, Sustainable Utilization of Waste Glycerol for 1,3-propanediol Production over $Pt/WO_x/Al_2O_3$ Catalysts: Effects of Catalyst Pore Sizes and Optimization of Synthesis Conditions. Environmental Pollution. 272 (2021) 116029. https://doi.org/10.1016/j.envpol.2020.116029

[4] Y. Liu, M. H. Wu, G. L. Rempel, F. T. T. Ng, Glycerol Hydrogenolysis to Produce 1,2-Propanediol in Absence of Molecular Hydrogen using a PD Promoted $Cu/MgO/Al_2O_3$ Catalyst. Catalysts. 11 (2021) 1299. https://doi.org/10.3390/catal11111299

[5] X. Liu, B. Yin, W. Zhang, X. Yu, Y. Du, S. Zhao, G. Zhang, M. Liu, H. Yan, M. Abbotsi-Dogbey, S. T. Al-Absi, S. Yeredil, C. Yang, J. Shen, W. Yan, X. Jin, Catalytic Transfer

Hydrogenolysis of Glycerol over Heterogeneous Catalysts: A Short Review on Mechanistic Studies. The Chemical Record. 21 (2021) 1792–1810. https://doi.org/10.1002/tcr.202100037

[6] N. D. Kim, J. R. Park, D. S. Park, B. K. Kwak, J. Yi, Promoter Effect of Pd in $CuCr_2O_4$ Catalysts on the Hydrogenolysis of Glycerol to 1,2-propanediol, Green Chemistry. 14 (2012) 2638. https://doi.org/10.1039/c2gc00009a

[7] A. Corma, S. Iborra, A. Velty, Chemical Routes for the Transformation of Biomass into Chemicals. Chemical Reviews. 107 (2007) 2411–2502. https://doi.org/10.1021/cr050989d

[8] N. Azri, I. Ramli, U. I. Nda-Umar, M. I. Saiman, Y. H. Taufiq-Yap, Effect of Different Supports for Copper as Catalysts on Glycerol Hydrogenolysis to 1,2-propanediol, Journal of King Saud University – Science. 33 (2021) 101417. https://doi.org/10.1016/j.jksus.2021.101417

[9] S. P. Samudrala, V. P. Kumar, M. L. Kantam, S. K. Bhargava, A. Srikanth, K. V. R. Chary, High Efficiency Conversion of Glycerol to 1,3-Propanediol using a Novel Platinum–Tungsten Catalyst Supported on SBA-15, Industrial & Engineering Chemistry Research. 54 (2015) 9104–9115. https://doi.org/10.1021/acs.iecr.5b01814

[10] B. Mallesham, P. Sudarsanam, B. V. S. Reddy, B. M. Reddy, Development of Cerium Promoted Copper–Magnesium Catalysts for Biomass Valorization: Selective Hydrogenolysis of Bioglycerol, Applied Catalysis B: Environmental. 18 (2016) 47–57. https://doi.org/10.1016/j.apcatb.2015.07.037

[11] B. Dou, Q. Hu, J. Li, S. Qiao, Z. Hao, Adsorption Performance of VOCs in Ordered Mesoporous Silicas with Different Pore Structures and Surface Chemistry, Journal of Hazardous Materials. 186 (2011) 1615–1624. https://doi.org/10.1016/j.jhazmat.2010.12.051

[12] V. G. Deshmane, R. Y. Abrokwah, D. Kuila, (2015). Synthesis of Stable Cu-MCM-41 Nanocatalysts for H_2 Production with High Selectivity *via* Steam Reforming of Methanol, International Journal of Hydrogen Energy. 40 (2015) 10439–10452. https://doi.org/10.1016/j.ijhydene.2015.06.084

[13] S. Rahman, C. Santra, R. Kumar, J. Bahadur, A. Sultana, R. Schweins, D. Sen, S. Maity, S. Mazumdar, B. Chowdhury, Highly Active Ga Promoted Co-HMS-X Catalyst towards Styrene Epoxidation Reaction Using Molecular O_2. Appl. Catal. A. 482 (2014) 61– 68. https://doi.org/10.1016/j.apcata.2014.05.024

[14] A. Jankowska, A. Chłopek, A. Kowalczyk, M. Rutkowska, M. Michalík, S. Liu, L. Chmielarz, (2020). Catalytic Performance of Spherical MCM-41 Modified with Copper and Iron as Catalysts of NH_3-SCR Process, Molecules. 25 (2020) 5651. https://doi.org/10.3390/molecules25235651

[15] A. S. Al-Fatesh, R. Kumar, S. O. Kasim, A. A. Ibrahim, A. H. Fakeeha, A. E. Abasaeed, H. Atia, U. Armbruster, C. Kreyenschulte, H. Lund, S. Bartling, Y. Mohammed, Y. A. Al-Baqmaa, M. S. Lanre, M. L. Chaudhary, F. S. Al-Mubaddel, B. Chowdhury, (2021). Effect of Cerium Promoters on an MCM-41-Supported Nickel Catalyst in Dry Reforming of Methane, Industrial & Engineering Chemistry Research. 61 (2021) 164–174. https://doi.org/10.1021/acs.iecr.1c03163

[16] A. Kowalczyk, A. Święs, B. Gil, M. Rutkowska, Z. Piwowarska, A. Borcuch, M. Michalik, L. Chmielarz, Effective Catalysts for the Low-Temperature NH_3-SCR Process Based on MCM-41 Modified with Copper by Template Ion-Exchange (TIE) Method. Appl. Catal. B Environ. 237 (2018) 927–937. https://doi.org/10.3390/molecules26061807

[17] G. Busca, L. Lietti, G. Ramis, F. Berti, Chemical and Mechanistic Aspects of the Selective Catalytic Reduction of NO_x by Ammonia over Oxide Catalysts: A review, Appl. Catal. B Environ. 18 (1998) 1–36. https://doi.org/10.1016/s0926-3373(98)00040-x

[18] A. Ungureanu, B. Drăgoi, A. Chirieac, C. Ciotonea, S. Royer, D. Duprez, A. Mamede, E. Dumitriu, Composition-Dependent Morphostructural Properties of Ni–Cu Oxide Nanoparticles Confined within the Channels of Ordered Mesoporous SBA-15 Silica. ACS Applied Materials & Interfaces. 5 (2013) 3010–3025. https://doi.org/10.1021/am302733m

[19] N. Azri, I. Ramli, U. I. Nda-Umar, M. R. Shamsuddin, M. I. Saiman, Y. H. Taufiq-Yap, Copper-Dolomite as Effective Catalyst for Glycerol Hydrogenolysis to 1,2-Propanediol. J. Taiwan Inst. Chem. Eng. 112 (2020) 112, 34–5. https://doi.org/10.1016/j.jtice.2020.07.011

[20] L. Zheng, S. Xia, Z. Hou, Hydrogenolysis of Glycerol over Cu-Substituted Hydrocalumite Mediated Catalysts, Appl. Clay Sci. 118 (2015) 68–73. https://doi.org/10.1016/s0926-3373(98)00040-x

[21] S. Xia, L. Zheng, L. Wang, P. Chen, Z. Hou, Hydrogen-Free Synthesis of 1,2-Propanediol from Glycerol over Cu–Mg–Al Catalysts. RSC Advances. 3 (2013) 16569. https://doi.org/10.1016/j.clay.2015.09.002

[22] F. Chang, H. Yang, L. S. Roselin, W. Kuo, Ethanol Dehydrogenation over Copper Catalysts on Rice Husk Ash Prepared by Ion Exchange, Applied Catalysis A: General. 304 (2006) 30–39. https://doi.org/10.1016/j.apcata.2006.02.017

[23] G. Carotenuto, R. Tesser, M. Di Serio, E. Santacesaria, (2013). Kinetic Study of Ethanol Dehydrogenation to Ethyl Acetate Promoted by a Copper/Copper-Chromite Based Catalyst, Catalysis Today. 203 (2013) 202–210. https://doi.org/10.1016/j.cattod.2012.02.054

[24] F. Mauriello, A. Vinci, C. Espro, B. Gumina, M. G. Musolino, R. Pietropaolo, Hydrogenolysis vs. Aqueous Phase Reforming (APR) of Glycerol Promoted by a Heterogeneous Pd/Fe Catalyst, Catalysis Science & Technology. 5 (2015) 4466–4473. https://doi.org/10.1039/c5cy00656b

[25] R. Arumugam, B. Ali, R. Manigandan, C. T. Da, M. Nguyen-Le, (2022). Hydrogenolysis of Glycerol to 1, 2-Propanediol on MgO/Ni_3C Catalysts Fabricated by a Solid-State Thermal Synthesis, Molecular Catalysis. 525 (2022) 112358. https://doi.org/10.1016/j.mcat.2022.112358

[26] F. Cai, D. Pan, J. J. Ibrahim, J. Zhang, G. Xiao, (2018). Hydrogenolysis of Glycerol over Supported Bimetallic Ni/Cu Catalysts with and without External Hydrogen Addition in a Fixed-Bed Flow Reactor, Applied Catalysis A: General. 564 (2018) 172–182. https://doi.org/10.1016/j.apcata.2018.07.029

Materials Research Proceedings 53 (2025) 390-398 https://doi.org/10.21741/9781644903575-38

Conversion of glucose to lactic acid using ordered mesoporous based supported ZrO_2 catalyst

FARINAA Md Jamil[1,a], ANITA Ramli[1,b*], NUR AKILA SYAKIDA IDAYU Khairul Anuar[1,c] and LIM Jun Wei[1,d]

[1]Fundamental and Applied Sciences Department and Centre for Biofuel and Biochemical Research, Institute of Self-Sustainable Building, Universiti Teknologi PETRONAS, 32610 Seri Iskandar, Perak, Malaysia

[a]farinaa.jamil@utp.edu.my, [b]anita_ramli@utp.edu.my, [c]nurakila.khairul@utp.edu.my, [d]junwei.lim@utp.edu.my

Keywords: Mesoporous Silica SBA-15, ZrO_2, Lactic Acid, Glucose

Abstract. Using the chemical catalytic approach, the conversion of glucose to lactic acid can be produced at high selectivity and most of the study focused on utilizing of catalyst. This present work discusses the conversion of glucose to lactic acid with the use of zeolite-based catalysts. For this purpose, synthetic SBA-15 was synthesized using the hydrothermal method at 100 °C for 2 days, dry at 100 °C for overnight, and calcined at 650 °C for 6h. The synthesized SBA undergo an alumination process with a Si/Al ratio of 25, dried at 60 °C for overnight and calcined at 550 °C for 6h followed by impregnation with 2 wt% of ZrO_2, dried at 100 °C overnight and calcined at 550 °C for 6h. The catalysts were characterized for their physicochemical properties using XRD and N_2 adsorption-desorption isotherm to evaluate the properties of catalyst to produce lactic acid. XRD diffraction shows the presence of the SBA-15 diffraction peak at a low angle. The N_2 adsorption-desorption analysis shows a decrease in surface area and pores size after alumination and impregnation process. All catalysts were screened for the production of lactic acid from glucose using a high-pressure batch reactor at 150 °C in inert condition of nitrogen gas for 3h. the impregnated Zr into the Al-SBA-15 demonstrated the highest conversion (94.9%) and selectivity of lactic acid (38.4%) while SBA-15 and Al-SBA-15 shows no conversion was detected at 200 °c temperature for 3h under 5 bars of nitrogen gas. It shows that Zr/Al-SBA-15 catalyst are suitable to convert lactic acid from glucose feedstock.

Introduction

Lactic acid is an important molecular compound that has been wide range of applications such as in the pharmaceutical, food industries, and chemical. It is harmless and can be categorized as Generally Recognized As Safe (GRAS) [1]. Traditionally it can be produced by fermentation method using a suitable microbial from a starch-rich crop feedstock. However, the process involves expensive methods such as membrane separation, purification, and extraction yet produces very low-purity lactic acid. To overcome this problem, chemo-catalytic process are the alternative ways and further study has been done to explore the process.

Lactic acid was first discovered by C. W. Scheele in 1781, who isolated the lactic acid from sour milk as an impure brown syrup. After 9 years, lactic acid was designated as acide lactique by Lavoisier1 around 1789. In 1981 CE Avery was the first person to commercially produce lactic acid in Littleton, MA, USA [2]. To this day, the utilization of microorganisms for the fermentation process remains prevalent in the realm of industrial production. Notably, Purac, Cargill, and Henan Jindan Lactic acid Technology Co. Ltd. are recognized as the foremost manufacturers of lactic acid on a global scale while the three largest consumer markets in the world are the United States followed by China and Western Europe.

There are two alternative routes for producing lactic acid namely fermentation and chemical catalysis. The fermentation of lactic acid involves the utilization of microorganism, specifically lactic acid bacteria to convert efficiently sugars into lactic acid through the fermentation of carbohydrates. Currently, the industrial production of lactic acid primarily relies on the fermentation of carbohydrates, such as glucose or sucrose with common microbes in this fermentation process include L. delbrueckii, L. amylophilus, and L. bulgaricus [3]. The selection of a suitable microorganism for the fermentation process is of utmost importance, as each microorganism possesses unique characteristics that can significantly influence the yield and selectivity of lactic acid [4].

Few types of catalysts have been study for the chemical catalytic process such as homogeneous catalyst and supported catalyst. The support catalyst can play an important role, increasing the surface area and the stability of the catalyst [3]. The support may also improve the catalytic activity by acting as a co-catalyst and its chemical properties can be changed by specific surface functional groups and physical properties tuned by controlling the pore structures [5]. In this study, ordered mesoporous silica, SBA had been choose as support to perform the conversion of lactic acid due to the high surface area, good stability to undergo the high temperature of the reaction and large pore that can allows bulky molecules to enter into the pores of the catalyst [6]. SBA-15 has a unique properties especially the higher surface area and pore structure to allow efficient interaction with glucose molecules, the Bronsted and Lewis acid sites that facilitate glucose activation in the initial steps of conversion, higher selectivity on lactic acid product stability under harsh reaction conditions and recyclability of the catalyst [7].

Employing metal oxide as an active species on support able to improving catalytic activity by narrow its selectivity for cleavage of bond to produce desired product. Zr based catalyst as supported material has been reported by Saleh & Abdullah [8] where oxide of Zr^{3+} was act as active component in conversion of lactic acid with glycerol conversion is 91.2% and lactic acid yield is 51.4% at 20 wt% of metal oxide loading into SBA support. Many research focus on the Ce and Zr supported zeolite for producing lactic acid. However this catalys mostly used glycerol as a feedstock for conversion to lactic acid. They studied with monometallic, bimetallic and sulfo-functionalized of SBA-15.

Zirconia-based acid catalysts have received much attention as both catalyst itself and catalyst support owing to their favorable physicochemical properties [9]. Doping with other oxoanions to zirconia can create additional electron deficiencies which may generate new acid sites or increase the strength of Brønsted acidity, thus enhancing its catalytic performance in reactions such as dehydration, hydrogenation, and hydrogen exchange, etc. It is shows that the catalytic performance of the zirconia-based catalysts for glucose conversions can be tuned by modifying its surface acidity properties by varying the crystalline structures, loading metals on zirconia, or supporting zirconia on other mesoporous materials. ZrO_2 catalyst can achieve the conversion of glucose up to 90%.

Another research has shown that bimetallic oxide catalysts containing Zr, such as CeZr/SBA-15 [10], and Zn-MMT [11], exhibit enhanced catalytic activity in promoting the selective oxidation of glucose to lactic acid. The presence of Zr in these catalysts contributes to the creation of active oxygen species, which facilitate the dehydrogenation of hydroxyl groups in glucose, leading to the production of lactic acid. Additionally, Zr-based catalysts have been found to possess high exchange affinity and Lewis acid sites, which are essential for catalytic conversion reactions, making Zr a key component in improving the efficiency and selectivity of the conversion process [3]. The Zr/SBA-15 catalyst has shown significant potential in the conversion of glucose to lactic acid. Studies have highlighted the effectiveness of ZrO_2-based catalysts in the hydrothermal conversion of multiple sugars to produce lactic acid, achieving a mass selectivity of 35.1% at 183°C [12].

This paper will report the preparation of Al/SBA-15 as the support using alumination process and ZrO_2 will be introduce into support using impregnation method. Only one weight percent of ZrO_2 will be evaluated for the conversion of glucose as a feedstock and selectivity of lactic acid using the batch reactor.

Methodology

Materials. All of the chemicals used for the preparation of the catalyst, reaction study, and product analysis were obtained from Sigma-Aldrich Chemical and used as received without any purification which is block co-polymer P123, TEOS, HCl, $Al(NO_3)_3.9H_2O$, sucrose and TMAOH in water solution. Distilled water was utilized in this experiment.

Catalyst Preparation. Synthesis of SBA-15 Support. SBA-15 are prepared by applying conventional procedure as described from Mazinani et al. [13]. A triblock copolymer P123 (6 g) was dissolved in distilled water for 3 hours. After that 240 g of 2 M HCl was added into the solution. The mixture was stirred at 40 °C for 2h. Then an amount of 18 g silica source, tetraethyl orthosilicate was dropwise to the mixture while stirring and let it age for 24h at 40 °C under an acidic environment. The mixture was put in a stainless-steel autoclave for the hydrothermal process at 100 °C for 2 days in static conditions. The white powder was filtered and washed a few times using distilled water before dry in oven at 100 °C overnight. The as-synthesized sample was calcined at 550 °C for 6h.

Alumination of SBA-15. Alumination process was referred to method from Gómez-Cazalilla et al. [14] with Si/Al ratio used in this experiment is 25. An appropriate amount of 1.2 M of $Al(NO_3)_3.9H_2O$ was mixed with 5.5 wt% TMAOH in a water solution. The mix solution was stirred at 60 °C for 2 hours. After that, the mixed solution was added to 1 g of pure silica SBA-15 under stirring conditions for 1 h at 60 °C. Then the powder product was washed and filtered using distilled water until the pH changed to neutral and continued to dry in an oven at 60 °C overnight. Lastly, the sample was calcined at 550 °C for 6h at 5 °C/min rate.

ZrO_2 metal loading into Al-SBA-15 catalyst. The incipient wetness impregnation technique was employed to load Zr metal into Al-SBA-15 samples using $Zr(NO_3)_4.5H_2O$ as a precursor to Zr. A suitable quantity of metal (2% loading) was dissolved in distilled water to generate impregnation solution before adding it into Al-SBA-15 catalyst. The mixture then was dried at 100 °C for 12h and calcined at 550 °C for 5h. The 2%Zr/Al-SBA-15 will be characterized and tested for glucose conversion.

Characterization. The characterization of the crystallinity of catalyst was performed using Bruker X-ray diffraction (XRD) model X'pert3 Powder and Empyrean (PAN Analytical), Billerica, MA, USA). All measurements were made at low angle in the 2θ range from 0.5-5 ° with step size of 0.01 and 10 s step duration.

The test of nitrogen adsorption-desorption were performed with a Micromeritics ASAP 2020, Norcoss, GA, USA). Samples were pre-treated at 110 °C for 10 hours. The specific surface area of catalyst were calculated by multiple-point Brunauer-Emmet-Teller (BET) method in the relative pressure (P/P_0) range from 0.03 to 0.30. The Barrett–Joyner–Halenda (BJH) technique, which is based on thermodynamics, was employed to estimate the distributions of the pore size from the isotherm's desorption branch. At a relative pressure of 0.998, the adsorbed quantity of liquid nitrogen indicated the total volume of the pores by inspecting the adsorption branch of the N_2 isotherm.

Conversion of sugars will be performed using a hydrothermal stainless autoclave with PTFE liner. 10 wt.% of glucose aqueous, and 0.1 g of catalyst will be added into the reactor vessel in turn under stirring. The autoclave was sealed by tightening all screws using an Allen key. The pressure at the gas regulator was adjusted to between 5 bars. Then the atmosphere in the reactor was replaced three times with N_2 and then 0.5 MPa N_2 will be charged. The reactor heat to temperature in the range 130 – 200 °C with reaction time varies at 3h. After the reaction was over,

Materials Research Proceedings 53 (2025) 390-398 https://doi.org/10.21741/9781644903575-38

the rector was left to cool down to 25 °C and the chiller was turned on and set at 10 °C. The obtained liquid was collected in glass vials. All catalysts will be tested using the same parameter. The separation of the compounds was performed on an ICSep ICE ORH 801 brand Transgenomic (Agilent Technologies, CA, USA) using the 5 mM H_2SO_4 mobile phase with a flow rate 0.6 mL min^{-1}, injection volume was 0.5 μL, column temperature 80 °C and RID temperature 35 °C. The evaluation of the compounds was run for 25 mins. The glucose conversion and selectivity of the lactic acid were calculated from the equation below:

$$Conversion\ (\%) = \frac{moles\ of\ glucose\ (in) - moles\ of\ glucose\ (out)}{moles\ of\ glucose\ (in)}\ x\ 100$$

$$Selectivity\ (\%) = \frac{moles\ of\ lactic\ acid}{moles\ of\ all\ product}\ x\ 100$$

Result and Discussion

Figure 1 displays the isotherms pattern for all catalysts. According to IUPAC, the nitrogen adsorption-desorption isotherms for SBA-15 support and supported catalysts show a type IV isotherm with an H1-type hysteresis loop. These isotherm features indicate the formation of mesoporous materials with 1-D cylindrical channels. There is sharp condensation detected at a relative pressure of 0.6-0.8 P/P_o which describes the uniformity of the mesoporous framework. As described by Pham et al. [15] the high degree of mesopore ordering leads to a sharp inflection at relative pressures between 0.6-0.8 which is consistent with well-defined 3 nm mesopores. The isotherm for SBA-15 shows a sharp reflection at relative pressure (P/P_o) indicating a higher N_2 adsorbing amount due to the multilayer adsorption in the formed mesopore. This suggested that the sample has a large pore size of 6.97 nm. The amount of adsorbed nitrogen for samples impregnated decreases as compared to pure silica due to the fact that the mesoporous of impregnated sample are filled by organic groups or metal salt which causes a decrease in the pore volume, size, surface, and total adsorption capacity [16].

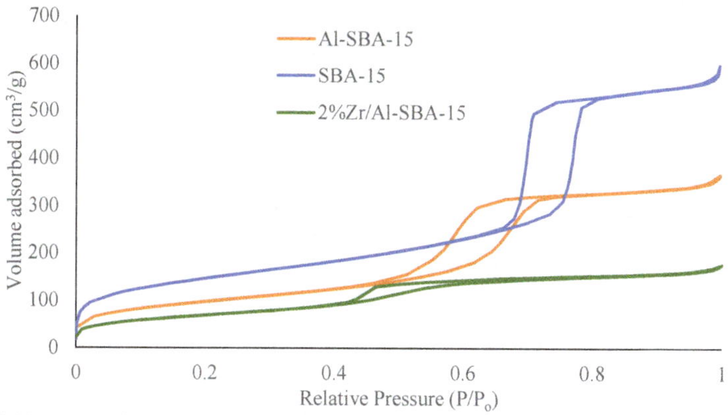

Fig. 1. Nitrogen adsorption-desorption isotherms of silica SBA-15 and impregnated samples

The textural of all catalysts such as surface area, pore size, pore volume, and particle size are summarized in Table 1. The table shows that SBA-15 catalyst exhibits the largest surface area which was 514 m^2/g compared to other catalysts. It is verified that the addition of Al_2O_3 and ZrO_2 into the SBA-15 framework resulted in a decrease of surface area which also can be related to the decrease of pore volume. As the incorporation of Al to mesoporous SBA-15 framework, the pores were filled by the impregnated metal and led to pore blockage [17]. This statement can be proved by observing the pore volume of SBA-15 and Al-SBA-15 which shows the decreasing pattern when Al_2O_3 was introduced as an impregnated metal. The same goes for 2%Zr/Al-SBA-15 which possesses the same pattern as Al-SBA-15 when ZrO_2 was impregnated into Al-SBA-15 catalyst. Meanwhile, the pore size of SBA-15 and Al-SBA-15 catalysts shows almost the same with slight differences in value which can be explained the hexagonal mesoporous of SBA-15 features was not altered when Al was impregnated [18]. As for 2%Zr/Al-SBA-15 cases the pore size shows significant declined compared to SBA-15 showing that there are some partial blockage of the pore in SBA-15 structure when Zr metal was introduce as impregnated metal which can be proved from isotherm plot in Figure 1 [19].

Table 1 Specific surface area, pore size, and total pore volume of sample catalyst

Sample	**Surface Area (m^2/g)**	**Pore Size (nm)**	**Pore Volume (cm^3/g)**
SBA-15	514	6.97	0.896
Al-SBA-15	344	6.12	0.529
2%Zr/Al-SBA-15	240	4.32	0.266

The formation of (100) was detected around 2θ=0.8° for all catalysts, while the formation of (110) and (200) was centered at 2θ=1.7°, 1.8° respectively as a reference by ICSD-98-009-6967. However, as Zr was added into the SBA-15 framework, (110) was decreased in intensity while (200) was shifted to the right and appeared as a broad peak. The result agreed by Wang et al. [20] that suggest the incorporation of Al and Zr into SBA-15 led to no damage in the mesostructured of the SBA-15 host. Also, no crystallized phase from the Al and Zr was detected in the diffractogram indicating that the Al and Zr were incorporated inside the pore of SBA-15. This observation is in agreement with the isotherms in Figure 1 where the amount of the absorbed N_2 decrease after incorporated Al and Zr into the pore of SBA-15.

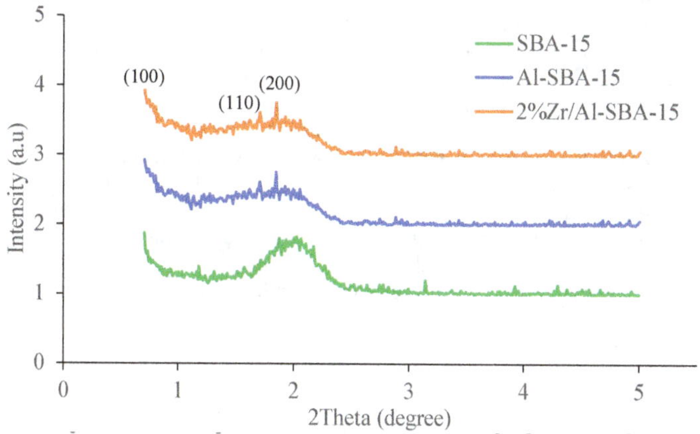

Catalytic Conversion of Glucose to Lactic Acid by using 2%Zr/Al-SBA-15. Using a catalytic screening approach, all synthesized catalysts were evaluated for their catalytic performance in producing lactic acid from glucose as a feedstock. Screening tests were performed at 150 °C in the inert condition of nitrogen gas at 5 bars for 3h using a high-pressure batch reactor. When the catalysts were tested for the production of lactic acid from glucose, the liquid produced in the presence of 2%Zr/Al-SBA-15 gave a conversion to lactic acid. In contrast, no conversion was recorded with SBA-15 and Al-SBA-15 catalysts. Figure 3 shows that the derived lactic acid was detected at the retention time of 10.09 minutes in the presence of 2%Zr/Al-SBA-15 catalyst. However, the chromatogram also shows the presence of three other derived compounds at the retention time of 4.62, 7.05 and 15.09 minutes. The other compounds are believed to be 5HMF, unreacted glucose residue, and levulinic acid, respectively, as compared to standard chromatogram. The Zr/Al-SBA-15 was chosen for further testing on reaction temperature ranging from 130 – 200 °C.

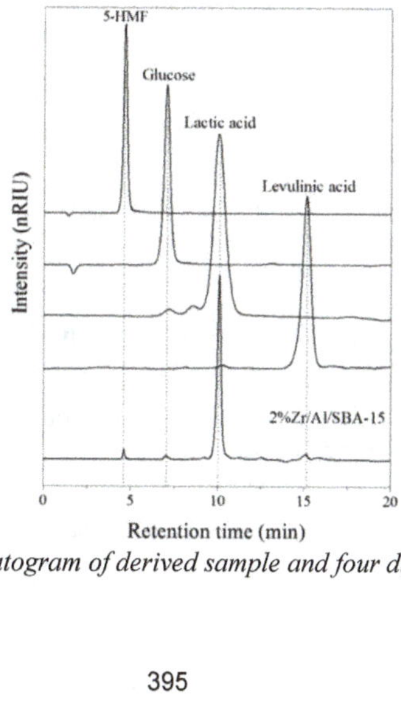

Figure 3 HPLC chromatogram of derived sample and four different standards

Figure 4 shows the effect of reaction temperature using the 2%Zr/Al-SBA-15 catalysts to study the lactic acid production in various ranges of reaction temperature for 3h at 0.1 g loading of the catalyst with 10 wt.% of glucose aqueous as feedstock under 5 bars of nitrogen pressure. As shown in Figure 4, the conversion and selectivity of lactic acid illustrate an increasing outline when the reaction temperature rose from 130 °C to 200 °C. The highest production of lactic acid was recorded at 200 °C reaction temperature with a conversion of 94.9% and selectivity of 38.43%. It could be explained that the catalytic rate of the catalyst increased as it reached the ideal temperature that favors the conversion of lactic acid [21]. The low conversion of lactic acid at a low temperature of 130 °C was due to the dihydroxyacetone or glyceraldehyde intermediate generated through retro aldol reaction that demand a higher energy to completely cleavage the C-C bond [22]. The low selectivity of lactic acid is due to the production of by-products such as gas and solid residue. As was mentioned by Xia et al, the presence of metal modifications in zeolites enhances catalytic activity but also contributes to the formation of insoluble residues due to complex interactions during the conversion that achieve a preferable condition for the production of other undesired products. [23].

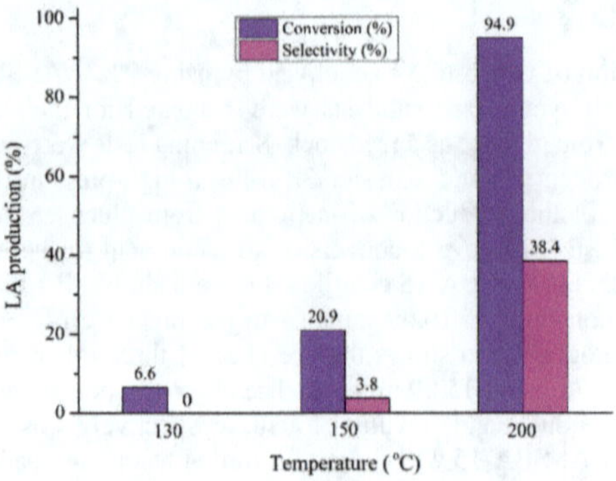

Figure 4 Effect of temperature reaction

Conclusion

Lactic acid was successfully produced from the glucose using high pressure batch reactor under 5 bars inert condition of N_2 at high 200 °C for 3h catalysed by 2%Zr/Al-SBA-15 with 10 wt% of glucose aqueous. The synthesized 2%Zr/Al-SBA-15 catalyst was found to produce high lactic acid production with 38.4% of selectivity and 94.9% of conversion.

Acknowledgement

The author would like to thank Yayasan Teknologi Petronas (cost center 015LC0-500) for research grant awarded to conduct the research.

References

[1] A. O. Ojo, O. de Smidt, Lactic Acid: A Comprehensive Review of Production to Purification, Processes. 11 (2023) 688. https://doi.org/10.3390/pr11030688.

[2] C. Miller, A Fosmer, B. Rush, T. McMullin, D. Beacom, and P. Suominen, "Industrial production of lactic acid, " Compr. Biotechnol., No. Jnuary, pp. 208-217. 2019. https://doi.org/10.1016/B978-0-12-809633-8.09142-1.

[3] F. A. Castillo Martinez, E. M. Balciunas, J. M. Salgado, J. M. Domínguez González, A. Converti, and R. P. de S. Oliveira, Lactic Acid Properties, Applications and Production: A

Review, Trends Food Sci. Technol.30 (2013) 70–83. https://doi.org/10.1016/j.tifs.2012.11.007.

[4] M. Hazim and B. I. N. Yaacob, "Selective conversion of glycerol to lactic acid using nickel supported on calcium oxide and magnesium oxide," 2018

[5] M. Shakeri, Z. K. Shal, and P. Van Der Voort, An Overview of the Challenges and Progress of Synthesis, Characterization and Applications of Plugged SBA-15 Materials for Heterogeneous Catalysis, Materials (Basel). 14 (2021). https://doi.org/10.3390/ma14175082.

[6] B. Dragoi, E. Dumitriu, C. Guimon, and A. Auroux, Acidic and Adsorptive Properties of SBA-15 Modified by Aluminum Incorporation, Microporous Mesoporous Mater., 121 (2009) 7–17. https://doi.org/10.1016/j.micromeso.2008.12.023.

[7] Zongling Li, Pengfei Wu, Jifeng Pang, Xianquan Li, Shangru Zhai, and Mingyuan Zheng, Catalytic Conversion of Sugars into Lactic Acid via a RuOx/MoS2 Catalyst, Catalysts, 2023, vol 13, pp 545. https://doi.org/https://doi.org/10.3390/catal13030545

[8] S. N. M. Saleh, A. Z. Abdullah, Zirconium–Cerium Oxides Supported on SBA-15 as Catalyst for Shape-Selective Synthesis of Lactic Acid from Glycerol, Waste and Biomass Valorization. 12 (2021) 2565–2578. https://doi.org/10.1007/s12649-020-01200-4.

[9] Hongxia Qin, Yu-Hang Zhang, Zhen Wang, Gui Yang, Photocatalytic Conversion of Fructose to Lactic Acid by BiOBr/Zn@SnO2 Material, Catalysts, vol 12, pp 719-719, 2022. https://doi.org/https://doi.org/10.3390/catal12070719.

[10] S. N. M. Saleh, M. Y. Ahmad, A. Ul-Hamid, A. Z. Abdullah, Elucidation of synergism in Ce-Zr/SBA-15 mesoporous catalysts and the effects on the activity in selective glycerol to lactic acid reaction, Powder Technology, vol 444, pp 120048, 2024. https://doi.org/https://doi.org/10.1016/j.powtec.2024.120048.

[11] S. M. Shahrul Nizan, N. F. I Che Sam, K. M. H. Elzaneen, M. S. Ideris, F. W. Harun, H. H. Azman, Understanding the cation exchange affinity in modified-MMT catalysts for the conversion of glucose to lactic acid, Royal Society of Chemistry Advances, vol 13, pp 31263, 2023. https://doi.org/https://doi.org/10.1039/d3ra05071h

[12] F. A. Piovano, A. S. Guadalupe, C. V. Bergamini, P Gimenez, Catalytic hydrothermal conversion of bio-derived sugars for lactic acid production using sol-gel synthesized zirconia,Latin American Applied Research. https://doi.org/10.52292/j.laar.2023.1145.

[13] B. Mazinani, A. Beitollahi, A. K. Masrom, S. Ibrahim, and F. Jamil, The Effect of Aging Temperature on the Pores of Mesoporous SBA-15 Silica, AIP Conf. Proc. 1502(2012). 272–279. https://doi.org/10.1063/1.4769150.

[14] M. Gómez-Cazalilla, J. M. Mérida-Robles, A. Gurbani, E. Rodríguez-Castellón, A. Jiménez-López, Characterization and Acidic Properties of Al-SBA-15 Materials Prepared by Post-Synthesis Alumination of a Low-Cost Ordered Mesoporous Silica, J. Solid State Chem. 180 (2007) 1130–1140. https://doi.org/10.1016/j.jssc.2006.12.038.

[15] S. T. Phạm, M. B. Nguyen, G. H. Le, T. Pham, T. T. T. Quan, T. D. Nguyen, T. L. Son, Cellulose Conversion to 5 Hydroxymethyl Furfural (5-HMF) Using Al-Incorporated SBA-15 as Highly Efficient Catalyst, J. Chem. 2019 (2019). https://doi.org/10.1155/2019/5785621.

[16] M. Kruk, M. Jaroniec, C. H. Ko, and R. Ryoo, Characterization of the porous structure of SBA-15, Chem. Mater. 12 (2000) 1961–1968. https://doi.org/10.1021/cm000164e.

[17] S. Kumaravel, S. Thiripuranthagan, E. Erusappan, M. Durai, T. Vembuli, and M. Durai,

Catalytic Conversion of Levulinic Acid under Noncorrosive Conditions using Ru/Zr/Al-SBA-15 Catalysts, Microporous Mesoporous Mater. 305 (2020). https://doi.org/10.1016/j.micromeso.2020.110298.

[18] A. Luengnaruemitchai, An Aluminum-Grafted SBA-15-Catalyzed Conversion of Glucose to Lactic Acid, 170 (2022). doi.org/10.1016/j.catcom.2022.106488

[19] C. Okutan, H. Arbag, N. Yasyerli, and S. Yasyerli, Catalytic Activity of SBA-15 Supported Ni Catalyst in CH_4 Dry Reforming: Effect of Al, Zr, and Ti Co-Impregnation and Al Incorporation to SBA-15, Int. J. Hydrogen Energy. 45 (2020) 13911–13928. https://doi.org/10.1016/j.ijhydene.2020.03.052.

[20] P. F. Wang, H. Jin, M. Chen, D. Jin, B. Hong, H. Ge, J. Gong, X. Peng, H. Yang, Z. Y. Liu, X. Q. Wang, Microstructure and Magnetic Properties of Highly Ordered SBA-15 Nanocomposites Modified with Fe_2O_3 and Co_3O_4 Nanoparticles, J. Nanomater. 2012 (2012) 21–23. https://doi.org/10.1155/2012/269861.

[21] Z. Shen, W. Chen, W. Zhang, M. Gu, D. Wu, X. Meng, H. Si, Y. Zhang, Efficient Catalytic Conversion of Glucose into Lactic Acid over Y-β and Yb-β Zeolites, ACS Omega. 7 (2022) 25200–25209. https://doi.org/10.1021/acsomega.2c02051.

[22] Y. Wang, W. Deng, B. Wang, Q. Zhang, X. Wan, Z. Tang, Y. Wang, C. Zhu, Z. Cao, G. Wang, H. Wan, Chemical Synthesis of Lactic Acid from Cellulose Catalysed by Lead(II) Ions in Water, Nat. Commun. 4 (2013). https://doi.org/10.1038/ncomms3141.

[23] M. Xia, W. Dong, M. Gu, C. Chang, Z. Shen and Y. Zhang, "Synergetic effects of bimetals in modified beta zeolite for lactic acid synthesis from biomass-derived carbohydrates", RSC Adv, vol 8, pp. 8965, 2018.

Decarbonization Technology: ICDT2024 Materials Research Forum LLC
Materials Research Proceedings 53 (2025) 399-403 https://doi.org/10.21741/9781644903575-39

Extraction of nickel from cathode material of lithium-ion batteries using pyroligneous acid

Cherry Nguyen[1,a], Philip Kwong[1,b,*]

[1]The University of Adelaid, Australia

[a]a1713113@adelaide.edu.au, [b]philip.kwong@adelaide.edu.au

Keywords: LIBs Recycling, Nickel, Pyroligneous Acid, Metal Extraction

Abstract. The study aims to investigate the potential use of pyroligneous acid (PA) in extracting nickel from the cathode active material of lithium-ion batteries (LIBs). The efficiency of nickel leaching was assessed by varying extraction temperature, time, and solid-to-liquid ratio. Optimal conditions were determined to be at 70 °C, 30 minutes, and a solid-to-liquid ratio of 20 g/L. Under these conditions, PA was able to extract up to 95% of nickel from the cathode active material. Kinetic studies revealed that nickel extraction by PA follows a chemical reaction control mechanism with an apparent activation energy of 31.8 kJ/kmol. Results from this study suggest that PA can be an cost-effective and efficient organic alternative to the harmful inorganic lixiviant commonly involved in the hydrometallurgical process of spent LIBs.

Introduction

Recent years have witnessed a significant surge in the LIBs demand typically for the electric vehicle industry, which is predicted to experience the growth from $7.8 billion in 2015 to $30.6 billion in 2024 [1]. The extensive use of LIBs in EVs has led to a substantial need for nickel, a critical element of high-energy-density cathodes for the next generation of batteries [2]. It is forecasted that a shortage of nickel may occur by 2025 [3]. Although LIBs have the potential to serve as a secondary resource of nickel, the lack of efficient technology for metal recovery has led to the disposal of over 11 million tons of spent LIBs through traditional methods, including stabilization, incineration, and landfilling, between 2017 and 2030 [4], [5]. The improper handling and disposal of spent LIBs have highlighted the need for more effective and sustainable recycling methods to mitigate the environmental impact associated with their disposal [5].

For extraction of nickel from spent LIBs, hydrometallurgy is considered as the most effective and feasible it requires less energy input, generates minimal toxic gas emissions, and importantly, achieves high metal extraction rates with excellent purity levels [6]. Potent acid leaching lixiviants such as sulfuric acid, nitric acid and hydrochloric acid are widely employed are commonly employed in this process. However, concerns have arisen recently due to the substantial emissions and the generation of large quantities of highly acidic and toxic sludges and tailings associated with their application [1], [7], [8]. As a result, there has been a shift in research focus towards exploring the viability of organic lixiviants as alternatives in the LIBs recycling process. However, the adoption of organic lixiviants is constrained by their higher costs and lower leaching efficiency when compared to their inorganic counterparts [7].

This study, therefore, focuses on proposing pyroligneous acid (PA) as a cost-effective yet efficient organic lixiviant for extraction of nickel from the cathode material. PA can be economically obtained as a condensed during the pyrolysis process of wood biomass. Due to presence of abundant organic acid contents, primarily acetic acid, PA can effectively remove undesirable metal contents such as copper, chromium, arsenic, cadmium and nickel from chromated copper arsenate and contaminated soil [9], [10]. However, to the best of our knowledge, no research has been done to demonstrate the potential application of PA in extraction of valuable metals from LIBs. This study will demonstrate the potential of PA to be applied in the LIBs recycling by investigating the optimal conditions (time, temperature and solid-to-liquid ratio) and conducting kinetic study for nickel extraction from cathode active material using PA.

Materials and Methods

Materials. Nickel – rich cathode active material (NMC 811, containing 7.40 ± 0.20 wt% of Li, 47.5 ± 1.50 wt% of Ni, 6.60 ± 0.60 wt% of Co and 5.50 ± 0.60 wt% of Mn) was supplied by Shandong Gelon LIB Co, Ltd. in the fine powder form (d_{50} = 13 ± 2 μm). PA was prepared by collecting condensable gas generated from the pyrolysis of wood biomass at 300 °C. The pH level and acidity of PA were determined to be 2.5 and 150 g of acetic acid equivalent/ L of PA, using pH meter and Titratable Acidity Titrator for Fruit Juice Analysis (Hanna Instrument, HI1131B), respectively.

Leaching Method. A 50 mL of pyroligneous acid was first heated up to the desired operating temperature (50 °C, 60 °C, 70 °C or 80 °C) by using a magnetic heater and stirrer (Heidolph Mr Hei – Standard). Afterward, the designated quantity of cathode active material was added in and stirred magnetically at 500 rpm for 50 minutes. During the metal extraction process, approximately 1 mL of leaching liquor was withdrawn and subsequently filtered every 10 minutes for characterization. The concentration of Ni ions in the leaching liquor were measured by Atomic absorption spectroscopy (AAS, Shimadzu 6300). The leaching efficiency was estimated calculated according to Equation 1:

$$\text{Leaching efficiency (\%)} = \frac{DC_{Ni}V_0}{m\text{wt\%}} \tag{1}$$

where D is a dilution factor, C_0 is the measured concentration of nickel in the diluted leaching liquor (ppm or 10^{-6} g/mL), V_0 is the volume of leaching liquor (mL), m is the weight of cathode active material (g) and wt % is the composition of nickel in the cathode active material (%).

Kinetic Study. Assume that the particles are spherical and shrinking due to occurrence of leaching reaction at the surface of particles, the shrinking core model was employed to study the kinetics of leaching process as shown in Equation (2): Chemical reaction control:

$$1 - (1 - X)^{\frac{1}{3}} = kt \tag{2}$$

where X is the leaching efficiency of metal, k (min^{-1}) and k′(min^{-1}) are reaction rate constants, and t (min) is the leaching time.

In addition, the apparent activation energies (E_a) associated to the extraction of Ni were calculated by plotting the apparent reaction rate constants k versus the temperature. The correlation between k and E_a could be described by the Arrhenius equation in the linear form:

$$\ln(k) = \ln A - \frac{E_a}{RT} \tag{3}$$

Results and Discussion

Optimization of Nickel Extraction by PA. Figure 1 illustrates the impact of critical process conditions involving time, temperature and solid-to-liquid ratio on the extraction of nickel from cathode active material. It was discovered that the extraction of nickel increase from 80 % to approximately 95 % as the prolonging of leaching time extends from 10 to 30 min. Afterward no significant improvement was when the time extended further to 30 min, suggesting that the equilibrium can be attained after 30 min (Figure 1a). Considering the operating temperature, the leaching efficiencies of nickel was found to drastically increase when the leaching temperature ranged between 50°C to 70°C. This trend can attribute to the fact that the leaching process of the cathode active material is endothermal [11], [12], [13]. Furthermore, higher leaching temperature can result into more frequent and more energetic collisions between molecules, which can help to accelerate reaction speed and consequently, enhance the efficiency of leaching process [13]. However, at the higher temperature than 70°C, the leaching efficiencies of all metals remained mostly unchanged or slightly decreased. On the other hand, the extraction of nickel typically decreases as the solid-to-liquid ratio increases from 10 g/L to 40 g/L, with the most significant reduction observed when the ratio surpasses 20 g/L. According to these findings, the optimal conditions for nickel extraction are determined to be reaction time of 30 min, operating temperature of 70°C, and a solid-to-liquid ratio of 20 g/L. Under these conditions, nickel extraction levels reach 95%. Figure 1d compares leaching performance of PA with commonly used lixiviants studied previously. Under identical operational conditions, the nickel extraction efficiency of PA was marginally inferior to that of the inorganic lixiviant (H_2SO_4/H_2O_2) but surpassed that of organic-based lixiviants. These findings underscore the considerable potential of PA to serve as a substitute for traditional lixiviants.

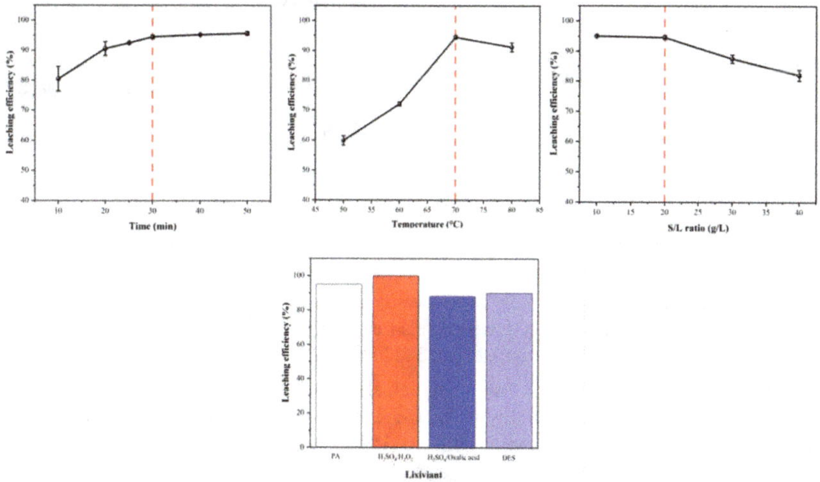

Fig. 1. Optimal conditions (a) Time, (b) Temperature and (c) Solid-to-liquid ratio for extraction of nickel from cathode material by PA (d) Leaching performance of PA in comparison to common lixiviants, involving H_2SO_4/H_2O_2 [14], H_2SO_4/Oxalic acid [15] and Deep eutectic solvent (DES) [16], considered similar operating conditions.

Kinetic Study. Figure 2a showed the plots of shrinking core model (chemical reaction control mechanism) applied for the extraction of nickel. A good linear relationship between the leaching efficiency of nickel and the chemical reaction kinetic could be apparently observed with the R^2 value in the range of 0.9869 – 0.991, indicating that the tested kinetic model can describe the nickel extraction. Based on the result described in Figure 2b, The E_a value for the extraction of nickel were found to be 31.8 kJ /kmol, which are comparable the results reported by Gao el at. and Meng el at. [17], [18].

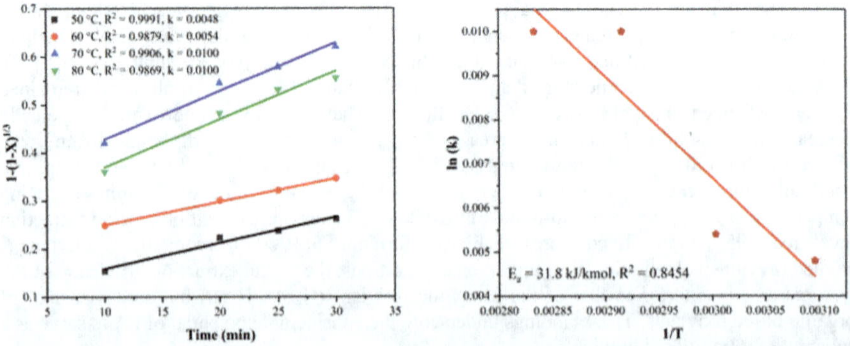

Fig. 2. Kinetic profile of nickel following the chemical reaction control mechanism for Shrinking Core Model

Conclusion

Under the optimal conditions of 70 °C, 30 min and solid-to-liquid ratio of 20 g/L, approximately 95 % of nickel was successfully extracted, following a chemical reaction control mechanism with an activation energy of 31.8 kJ/kmol. Future research should expand to explore the interaction between PA and nickel, as well as investigate the recovery of nickel from PA leaching liquor.

Acknowledgements

This study was supported by University of Adelaide Research Scholarship (UARS).

References

[1] Chaudhary, V., et al., Insights into the Eco-Friendly Recovery Process for Valuable Metals from Waste Lithium-ion Batteries by Organic Acids Leaching. Separation & Purification Reviews, 2023: p. 1-18. https://doi.org/10.1080/15422119.2022.2164650

[2] Nguyen, R.T., et al., Global Electrification of Vehicles and Intertwined Material Supply Chains of Cobalt, Copper and Nickel. Resources, Conservation and Recycling, 2021. 167: p. 105198. https://doi.org/10.1016/j.resconrec.2020.105198

[3] Jones, B., R.J.R. Elliott, and V. Nguyen-Tien, The EV revolution: The road ahead for critical raw materials demand. Applied Energy, 2020. 280: p. 115072. https://doi.org/10.1016/j.apenergy.2020.115072

[4] Natarajan, S. and V. Aravindan, Recycling Strategies for Spent Li-Ion Battery Mixed Cathodes. Acs Energy Letters, 2018. 3(9): p. 2101-2103. https://doi.org/10.1021/acsenergylett.8b01233

[5] Liu, Y.L., P.X. Xing, and J.G. Liu, Environmental performance evaluation of different municipal solid waste management scenarios in China. Resources Conservation and Recycling, 2017. 125: p. 98-106. https://doi.org/10.1016/j.resconrec.2017.06.005

[6] Jha, M.K., et al., Recovery of lithium and cobalt from waste lithium ion batteries of mobile phone. Waste Management, 2013. 33(9): p. 1890-1897. https://doi.org/10.1016/j.wasman.2013.05.008

[7] Golmohammadzadeh, R., F. Faraji, and F. Rashchi, Recovery of lithium and cobalt from spent lithium ion batteries (LIBs) using organic acids as leaching reagents: A review. Resources Conservation and Recycling, 2018. 136: p. 418-435. https://doi.org/10.1016/j.resconrec.2018.04.024

[8] Tesfaye, F., et al., Improving urban mining practices for optimal recovery of resources from e-waste. Minerals Engineering, 2017. 111: p. 209-221. https://doi.org/10.1016/j.mineng.2017.06.018

[9] Choi, Y.-S., B.J. Ahn, and G.-H. Kim, Extraction of chromium, copper, and arsenic from CCA-treated wood by using wood vinegar. Bioresource Technology, 2012. 120: p. 328-331. https://doi.org/10.1016/j.biortech.2012.06.045

[10] Li, Z.C., et al., Disinfection and removal performance for Escherichia coli, toxic heavy metals and arsenic by wood vinegar-modified zeolite. Ecotoxicology and Environmental Safety, 2019. 174: p. 129-136. https://doi.org/10.1016/j.ecoenv.2019.01.124

[11] Zhang, Y.J., et al., Use of grape seed as reductant for leaching of cobalt from spent lithium-ion batteries. Journal of Industrial and Engineering Chemistry, 2018. 66: p. 86-93. https://doi.org/10.1016/j.jiec.2018.05.004

[12] Yu, M., et al., A more simple and efficient process for recovery of cobalt and lithium from spent lithium-ion batteries with citric acid. Separation and Purification Technology, 2019. 215: p. 398-402. https://doi.org/10.1016/j.seppur.2019.01.027

[13] Zheng, Y., et al., Lithium fluoride recovery from cathode material of spent lithium-ion battery. Rsc Advances, 2018. 8(16): p. 8990-8998. https://doi.org/10.1039/C8RA00061A

[14] Vieceli, N., et al., Recycling of Lithium-Ion Batteries: Effect of Hydrogen Peroxide and a Dosing Method on the Leaching of LCO, NMC Oxides, and Industrial Black Mass. ACS Sustainable Chemistry & Engineering, 2023. 11(26): p. 9662-9673. https://doi.org/10.1021/acssuschemeng.3c01238

[15] Yang, C., et al., Recovery of Valuable Metals from Spent LiNi0.8Co0.1Mn0.1O2 Cathode Materials Using Compound Leaching Agents of Sulfuric Acid and Oxalic Acid. Sustainability, 2022. 14(21): p. 14169. https://doi.org/10.3390/su142114169

[16] Liao, Y., et al., A novel ternary deep eutectic solvent for efficient recovery of critical metals from spent lithium-ion batteries under mild conditions. Journal of Environmental Chemical Engineering, 2022. 10(6): p. 108627. https://doi.org/10.1016/j.jece.2022.108627

[17] Gao, W.F., et al., Selective recovery of valuable metals from spent lithium-ion batteries - Process development and kinetics evaluation. Journal of Cleaner Production, 2018. 178: p. 833-845. https://doi.org/10.1016/j.jclepro.2018.01.040

[18] Meng, F., et al., Selective recovery of valuable metals from industrial waste lithium-ion batteries using citric acid under reductive conditions: Leaching optimization and kinetic analysis. Hydrometallurgy, 2020. 191. https://doi.org/10.1016/j.hydromet.2019.105160

Reimagining space foods: Generation of O/W nanoemulsions via spontaneous emulsification in self-assembly, macrofluidics, and microfluidics to produce fortified beverages for astronauts

SVENJA Schmidt[1,2,3,4,a], ANH The Nguyen[1], HUY Quang Vu[1], NAM Nghiep Tran[1,b], NI Yang[4,c], MARIA Saarela[5,d], IAN Fisk[2,4,6,e], VOLKER Hessel[1,2,3,f*]

[1]School of Chemical Engineering, The University of Adelaide, Adelaide, Australia

[2]ARC Centre of Excellence in Plants for Space, The University of Adelaide, Adelaide, Australia

[3]Andy Thomas Centre for Space Resources, The University of Adelaide, Adelaide, Australia

[4]International Flavour Research Centre, University of Nottingham, Nottingham, United Kingdom

[5]South Australian Research and Development Institute (SARDI), The University of Adelaide, Adelaide, Australia

[6]International Flavour Research Centre (Adelaide), The University of Adelaide, Adelaide, Australia

[a]svenja.schmidt@adelaide.edu.au, [b]namnghiep.tran@adelaide.edu.au, [c]ni.yang@nottingham.ac.uk, [d]maria.saarela@adelaide.edu.au, [e]ian.fisk@nottingham.ac.uk, [f]volker.hessel@adelaide.edu.au

Keywords: Continuous Emulsification, Space Nutrition, Process Intensification

Abstract. Ensuring palatable, nutritionally balanced food options for long-term space missions remains a challenge. Our lab explores the concept of using fortified nanoemulsified beverages tailored to individual astronaut needs to improve life and well-being during space missions. Combining a self-assembly emulsification (solvent-free version of the spontaneous emulsification) with a continuous, microfluidic process has been demonstrated before to allow a resource-efficient production of beverage nanoemulsions. This study compares five different process setups (self-assembly only; self-assembly and batch process in small- and medium-scale; self-assembly and continuous process in single- and multi-contact) in regard to average droplet diameter of formed emulsions at constant surfactant concentration. Forced convection significantly decreases average droplet diameter in all cases, yet, the microfluidic continuous processes achieve the smallest average droplet size at 62-109 nm. Results indicated further that synthetic surfactant concentrations can be reduced by decreasing overall oil content, likely due to increased mass transfer of the surfactant between phases. This methodology not only supports astronaut nutrition but also provides a scalable framework for resource-efficient nanoemulsion-based production across the food, cosmetic, and pharmaceutical industries.

Introduction

Weight loss used to be a major concern in space exploration in low Earth orbit: during space missions of 6 months, astronauts were reported to lose on average 5-10 % of their body mass [1]. Applying different measures to ease food preparation and consumption allowed astronauts to maintain their body mass [2], however, these issues might resurface for long-term missions outside the low Earth orbit. As living in space is vastly different to living on Earth, many factors can contribute to the so-called 'space anorexia'. Next to e.g., the impact of microgravity on the digestive system, space foods are suspected to contribute to the loss in appetite [3]. Current options are almost entirely based on prepacked thermostabilized and dehydrated food items, placing food safety and shelf-life above palatability and flavor. Novel space food systems must therefore –

specifically for long-term space travel – (1) offer on-demand production to address limited storage, (2) offer healthy, flavorsome, diverse foods, while (3) ensure food quality and food safety.

Our lab has developed the concept of an on-demand fortified beverage production line that can adjust the beverage's composition to personal preferences (flavor) and nutritional needs (micronutrients). This way, both general space health risks as well as personal health concerns can be mitigated. The formulation of the beverages will be based on highly diluted oil-in-water nanoemulsions (= beverage nanoemulsions) to allow the incorporation of hydrophobic ingredients. Nanoemulsions are characterized by a mean droplet diameter of below 200 nm, which leads to high stability and promotes bioavailability of encapsulated bioactive molecules [4]. In the food and pharmaceutical industry, nanoemulsions are typically produced by high-energy emulsification processes, which rely on high mechanical forces to form droplets. As a side effect, these processes generate excessive heat, which in turn might degrade sensitive molecules. To avoid temperature peaks, we apply a solvent-free variant of the spontaneous emulsification: A surfactant of higher hydrophilicity than lipophilicity is initially dissolved in the oil phase. Once the oil phase touches the aqueous phase, the surfactant - due to its higher affinity to the aqueous phase – drives mass transfer between both phases, dragging small oil droplets along [5,6]. However, this method requires a high concentration of synthetic surfactant, which does not offer any additional nutritional benefit.

To reduce the surfactant concentration, we combine the spontaneous emulsification with a microfluidic/ flow-chemistry set-up while changing the processing mode from batch to continuous. Microfluidic technology is a space-proven technology [7] to control fluids very precisely by drastically decreasing the scale of channel diameters. Continuous processes are known to work more resource-efficiently (time, raw material, waste) than batch processes. Ultimately, we intend to obtain the fortified beverages via campaign manufacturing, i.e., using one production line for a range of products of similar characteristics. Therefore, our findings are not only relevant to beverage production in space, but rather apply in general to the sustainable production of emulsion-based products (medicine, cosmetics, foods) both on Earth and in space.

Materials and Methods

Materials. The oil phase consisted of ultra-filtered pharmaceutical grade medium-chain triglycerides (MCT; Vials Direct, NSW, Australia) and the non-ionic surfactant Tween® 80 (Sigma-Aldrich, VIC, Australia) at a surfactant-to-oil ratio of 1.0 (SOR, see equation 1, m: mass). Oil and surfactant were thoroughly mixed at 600 rpm for at least 30 min. For the aqueous phase, 1.0 M phosphate buffer solution (Sigma-Aldrich, VIC, Australia) was diluted to 5 mM using ultrapure water (Millipore Milli-Q®).

$$SOR = \frac{m_{Surfactant}}{m_{Oil}} \tag{1}$$

Self-assembly emulsification. Each experiment is prepared in a way that the final emulsion equals to 5 g, containing 0.1-10 wt% MCT. Oil phase is transferred in a glass vial (7 ml, Supelco®, Bellefonte, Pennsylvania, USA) and aqueous phase is added. Mixture is left undisturbed for 24 hours.

Self-assembly in combination with batch process. Beaker. Each experiment is prepared in a way that the final emulsion equals to 5 g, containing 0.1-10 wt% MCT. Oil phase is either transferred to the bottom of the beaker (0.1 wt%) or in a Pasteur pipette (1-10 wt%) to be added droplet by droplet to the aqueous phase which is stirred at 600 rpm. Once the oil phase is completely added, the mixture is stirred for an additional 10 min at 600 rpm. Burette. Each experiment is prepared in a way that the final emulsion equals to 100 g, containing 1-20 wt% MCT.

Oil phase is transferred to a burette (V = 50 ml, Purex, Corning, USA), which is positioned with the tip approx. 10 cm above a beaker containing the aqueous phase. The oil phase is slowly (droplet by droplet) added to the aqueous phase, while the mixture is stirred at 600 rpm. Once the oil phase is completely added, the mixture is stirred for an additional 10 min at 600 rpm.

Self-assembly in combination with continuous microfluidic process. Pump 1 (oil phase) and pump 2 (aqueous phase) (both Azura P 4.1S 10 ml, Knauer, Germany) are connected to either a T-mixer (single-contact microfluidic mixer, internal diameter: 0.5 mm, Tee LP PEEK 1/4-28 1/16" 0.040", Upchurch Scientific®, USA) or a low flow Corning reactor (multi-contact microfluidic mixer, 1 Module only, Corning® Inc., New York, USA). The oil phase is heated up to approx. 40°C to decrease its viscosity, and the temperature is hold during pumping with a heat mat placed around pump 1. Volume flows of pump 1 and pump 2 were set to 1.98 ml/min and 8.02 ml/min, respectively, to obtain an emulsion with 10 wt% MCT. Sampling of 50 ml of emulsion started after 7 min of start-up phase. During sample collection, the emulsion is stirred at 600 rpm, and, in case of the T-mixer, for an additional 10 min after sample collection ended.

Average droplet size analysis. The average droplet size was measured immediately after the sample collection using a static light scattering instrument (Mastersizer 2000 with a Hydro 2000MU Unit, Malvern, UK). Samples were diluted with phosphate buffer (5 mM) to avoid multi-scattering effects and stirred at 750 rpm during measurement. In case of smaller sample size, low oil content, and/or small average droplet size (i.e., all stirred beaker experiments, Corning reactor experiments, self-assembly at 0.1 wt% oil), average droplet size was determined with a dynamic light scattering device (Zetasizer, Malvern, UK) at 25°C. The refractive index of MCT was set to 1.445 for both methods according to literature data [6].

Statistical methods. Average droplet diameters were tested for outliers (Dixon test, two-sided, significance level 5%). Analysis of variance (ANOVA, 95% confidence interval) was conducted in combination with a Turkey pairwise comparison. Software: XLSTAT (Life Sciences version 2023.1.6, XLSTAT statistical and data analysis solution, Lumivero, Denver, US).

Results and Discussion

The spontaneous emulsification at 10 wt% oil content was implemented in 5 different process set-ups: (1) self-assembly without additional mixing; self-assembly in combination with a batch mixing process in (2) beaker (small-scale, approx. 5 ml) or (3) burette (medium-scale, approx. 100 ml); self-assembly in combination with a continuous, microfluidic process in (4) a single-contact T-mixer or (5) a multi-contact Corning® reactor. In a previous study, it was already demonstrated that a conventional emulsification (= surfactant is initially in the aqueous phase) leads to average droplet sizes in the micron range of the same model system, while the spontaneous emulsification leads consistently to smaller droplet dimeters regardless of the mixing process [8]. Furthermore, the set-ups (3-5) demonstrated different surfactant efficiencies (defined as decreasing average droplet size at constant surfactant concentration) at an SOR of 1.0: Continuous microfluidic processes demonstrated the highest surfactant efficiency (multi-contact > single-contact) for the spontaneous emulsification compared to batch processes [8]. This trend consisted when comparing the emulsion characteristics with set-up (1) and (2), as demonstrated by ANOVA testing (see Figure 1 (a)). Self-assembly without guided mixing (i.e., forced convection) leads to very large droplets outside the nanoemulsion range with considerable variation between single emulsification processes as indicated by the large standard deviation. Any form of forced convection significantly decreases droplet size, resulting in (approximately) nanoemulsions. Even though the two-sided Tukey test assigned the set-up (3) and (4) to the same group, removing setup (1) from an additional ANOVA test demonstrated that each set-up leads to emulsions of statistically significantly different average droplet diameters. The multi-contact Corning® reactor remains the setup with the greatest surfactant efficiency.

Materials Research Proceedings 53 (2025) 404-408 https://doi.org/10.21741/9781644903575-40

It has been previously established that surfactant content at constant oil content (10 wt%) strongly influences average droplet diameter, i.e., the droplet size decreases with increasing surfactant concentration [6,8]. Therefore, the influence of the oil content in the final emulsion on the average droplet diameter at constant relative surfactant content (SOR = 1.0) is investigated for set-ups (1-3). In all three cases, the average droplet diameter decreases with decreasing oil content. The greatest decrease (-85%) takes place in set-up (1), where nanoemulsions are formed when reducing the oil content to 0.1 wt% in the final emulsion. While for set-up (1), the total amount of oil phase is present in the vial at the beginning of the emulsification, for set-up (2-3) the oil phase is subsequently added via a pipette or a burette. Thus, the change in droplet size cannot be explained by a change in relative interfacial area, which remains constant ((1): vial ground area, (2-3): surface area of droplet). Instead, the decrease in particle size might be related to improved mass transfer of the surfactant from the oil phase into the aqueous phase due to a larger surfactant concentration difference between the oil phase and the aqueous phase. Since the absolute surfactant concentration decreases with decreasing oil content at constant SOR, less surfactant is present in the aqueous phase during and after emulsification. A greater surfactant concentration gradient is likely to drive surfactant diffusion from the oil phase into the aqueous phase, thereby supporting the formation of smaller droplets and increasing surfactant efficiency.

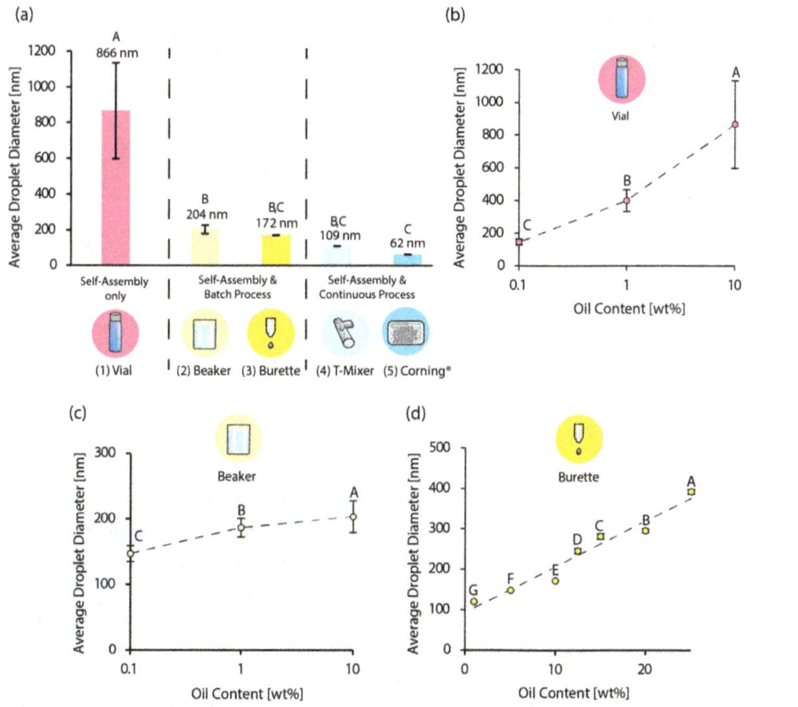

Fig.1. (a) average droplet diameter in relation to emulsification set-up (1-5) at 10 wt% oil content and SOR = 1.0; (b) average droplet diameter of set-up (1) in relation to oil content at SOR = 1.0; (c) average droplet diameter of set-up (2) in relation to oil content at SOR = 1.0; (d) average droplet diameter of set-up (3) in relation to oil content at SOR = 1.0.

Conclusion

Transferring the self-assembling spontaneous emulsification into a mixing process with forced convection improves surfactant efficiency, with the multi-contact Corning® reactor forming emulsions of the smallest particle size at constant surfactant concentration. Additionally, oil content has been identified as major factor in driving particle size at constant SOR in batch processes, likely due to increased surfactant concentration gradient between oil and aqueous phase. Therefore, decreasing the oil content at constant SOR is a promising approach for further increasing surfactant efficiency when conducting the spontaneous emulsification in continuous, microfluidic processes. Thereby, surfactant use in beverage emulsions can further be decreased, which has no nutritional benefit.

Acknowledgement

We would like to acknowledge the School of Chemical Engineering (University of Adelaide) to fund the travel to attend ESTCON 2024, the PhD scholarship from the Nottingham-Adelaide Alliance, the Hessel Research Group Fund which funded materials, Corning SAS (Paris, France), who donated the reactor investigated in this study, as well as the Engineering and Physical Sciences Research council [EP/R513283/1] and Biotechnology and Biological Sciences Research Council [BB/V018108/1].

Data Access Statement

The measurement data of the average droplet size measurements are achieved and publicly available in the Nottingham Research Repository (DOI: 10.17639/nott.7494).

References

[1] S.M. Smith, S.R. Zwart, G. Block, B.L. Rice, The nutritional status of astronauts is altered after long-term space flight aboard the international space station, The Journal of Nutrition, 135 (2005) 3 437-443. https://doi.org/10.1093/jn/135.3.437

[2] S.M. Smith, H. Lane, S. Zwart, Spaceflight Metabolism and Nutritional Support, Principles of Clinical Medicine for Space Flight (2019), 413-439. https://doi.org/10.1007/978-1-4939-9889-0_13

[3] H. Tang, H.H. Rising, M. Majji, R.D. Brown, Long-term space nutrition: A scoping review, Nutrients, 14 (2022) 1 194. https://doi.org/10.3390/nu14010194

[4] E. Acosta, Bioavailability of nanoparticles in nutrient and nutraceutical delivery, Current Opinion in Colloid & Interface Science, 14 (2009) 1 3-15. https://doi.org/10.1016/j.cocis.2008.01.002

[5] N. Anton, T.F. Vandamme, The universality of low-energy nano-emulsification, International Journal of Pharmacy, 377 (2009) 142-147. https://doi.org/10.1016/j.ijpharm.2009.05.014

[6] Komaiko, J., McClements, D.J. Low-energy formation of edible nanoemulsions by spontaneous emulsification: Factors influencing particle size, Journal of Food Engineering, 146 (2015) 122-128. https://doi.org/10.1016/j.jfoodeng.2014.09.003

[7] J. Nijhuis, S. Schmidt, N.N. Tran, V. Hessel, (2022). Microfluidics and macrofluidics in space: ISS-proven fluidic transport and handling concepts, Frontiers in Space Technology, 2 (2022) 779696. https://doi.org/10.3389/frspt.2021.779696

[8] S. Schmidt, A.T. Nguyen, H.Q. Vu, N.N. Tran, M. Saarela, I. Fisk, V. Hessel, Microfluidic spontaneous emulsification for generation of O/W nanoemulsions – opportunity for in-space manufacturing, Advanced Healthcare Materials, 12 (2023) 23, 2203363. https://doi.org/10.1002/adhm.202203363

Materials Research Proceedings 53 (2025) 409-416 https://doi.org/10.21741/9781644903575-41

Quantification and mitigation of carbon footprint in petroleum refinery process through simulation software

MUHAMMAD Nabil Fikri bin Shariffuddin[1,a], NABILA Farhana binti Jamaludin[1,2,b*], MUHAMMAD Fairis bin Hadipornama[3,c]

[1]Chemical Engineering Department, Universiti Teknologi PETRONAS, 32610, Seri Iskandar, Perak, Malaysia

[2]Institute of Smart and Sustainable Living, Universiti Teknologi PETRONAS, 32610, Seri Iskandar, Perak, Malaysia

[3]KINETICS Systems Malaysia Sdn. Bhd., 40300, Shah Alam, Selangor, Malaysia

[a]muhammad_19000556@utp.edu.my, [b]nabila.jamaludin@utp.edu.my, [c]Fairis@kinetics.net

Keywords: Petroleum Refinery, Carbon Quantification, Simulation Software, Mitigation Cases

Abstract. Petroleum refineries play a crucial role in producing various petroleum products through processes such as distillation, redistillation, cracking, and reforming. As the second-largest consumer of energy among industries, petroleum refineries significantly contribute to GHG emissions, particularly through combustion-related sources like process heaters and boilers. Therefore, it is crucial to adopt sustainable approaches to control and reduce these emissions. Simulation software like Aspen Plus/Symmetry can be utilized to quantify greenhouse gas emissions in petroleum refineries. This software allows for the selection of appropriate utility sources and energy strategies, ensuring compliance with sustainability regulations and facilitating the improvement of the refinery's carbon footprint in the future. Hence, this paper focuses to the quantification and mitigation of carbon footprints in petroleum refinery processes, with specific emphasis on the application of Aspen HYSYS. The results show 9.12 percentage differences in comparison with the manual calculation. A several mitigation options has been considered for different case and the differences in terms of GHG emissions has been analysed.

Introduction

Simulation and quantification methods play a crucial role in accurately calculating carbon emissions in most of the process industry. These methods provide valuable insights and benefits that aid in better decision-making, optimizing processes, and promoting sustainability efforts. Simulation software allows engineers to explore different process configurations and operational parameters. By identifying optimized operating conditions, the software helps industries reduce energy consumption and GHG emissions, leading to cost savings and improved environmental performance. On the other hand, traditional manual calculation was first being used in calculating total emission of carbon footprint from the process industry [1]. This method is currently most preferable since manual calculation does not require specialized software or advanced technical knowledge. It is accessible to a wider range of individuals, including smaller businesses, organizations with limited resources, or regions with limited technological infrastructure. Unfortunately, there is a claim that has been made to question the credibility between both methods in calculating carbon emission accurately [2]. This is because the accuracy factor of manual calculation when compared to simulation software depends on several factors, including the complexity of the system being analysed, the quality of data available, and the expertise of the person conducting the calculation [3].

In general, simulation software tends to offer a higher level of accuracy and precision for emission calculations compared to manual methods. According to research conducted by Bowen & Wittneben it was found that there are tensions between accuracy, consistency, and certainty in carbon accounting both within and across organizational fields [4]. Furthermore, scientific knowledge within the domain, and the organizational structures such as the IPCC usually provide different values of emission factor and thus create an inconsistency in the carbon accounting process [5]. It was also mentioned that there were some difficulties in measuring carbon emission from sources that are difficult to measure (e.g., land use) [6].

The literature that is closely related to this study is summarized in Table 1. Therefore, this study is required to compare the accuracy of the carbon emission quantifying method between manual calculation and simulation software.

Table 1 Review of carbon quantification with simulation software research studies

Title and Author	Objectives	Application	Simulation Software	Research Gaps
Petroleum Refinery Process Modelling: Integrated Optimization Tools and Applications [2]	To assess the environmental impact of a process or plant and identify optimization for reducing carbon dioxide (CO_2)	Petroleum Refinery Plant	Aspen HYSYS Simulation Software	No comparison with manual calculation method The methodology framework can be used as a reference.
Dynamic Simulation of Crude dynamic model Oil Distillation Plant using Aspen HYSYS. [11]	To simulate dynamic model of a complete distillation plant using Aspen HYSYS	Crude Oil Distillation	Aspen HYSYS Simulation Software	Results does not include CO_2 emission inventory The methodology framework can be used as reference.
The Determination of the Quantity of CO2 Produced in a Local Ethanol Plant using Aspen HYSYS Software [13]	To determine the quantity of CO_2 produced in a local ethanol plant using Aspen HYSYS software.	Ethanol Production Plant	Aspen HYSYS Simulation Software	Further research is required to develop comprehensive and reliable emission factor databases. Lack of standardized emission factors
Thermodynamic Analysis of Negative CO2 Emission Power Plant Using Aspen Plus, Aspen Hysys, and Ebsilon Software [15]	To compare the results of carbon emission obtained in three computing codes, namely Aspen Plus, Aspen HYSYS, and Ebsilon.	Power Generating Plant	Aspen Plus, Aspen HYSYS, Ebsilon.	No comparison with manual calculation method Parameter used based on assumption The comparison framework can be use as reference.

Materials Research Proceedings 53 (2025) 409-416 https://doi.org/10.21741/9781644903575-41

Methodology

This study consists of two main phases. The first phase involves collecting data on carbon footprint emissions from the petroleum refinery process. This data is obtained through a comprehensive analysis of documents, tutorial simulation textbook and journal articles related to the Petroleum Refinery Tutorial in Basrah, Syria [7]. The second phase focuses on compiling and analysing the collected data to establish a carbon accounting system for the refinery. A comparison between manual and simulation method will be calculated to determine the most reliable tools for current carbon emission quantification. Additionally, recommendations and strategies for managing greenhouse gas (GHG) emissions will be proposed based on the current emission trends. Aspen HYSYS software will be utilized to simulate the four major process units, which are crude oil feed preheating unit, atmospheric distillation unit, vacuum distillation unit and debutanizer unit, allowing for the determination of GHG emissions. An overall framework of the development of petroleum refinery carbon footprint accounting is presented in Figure 1.

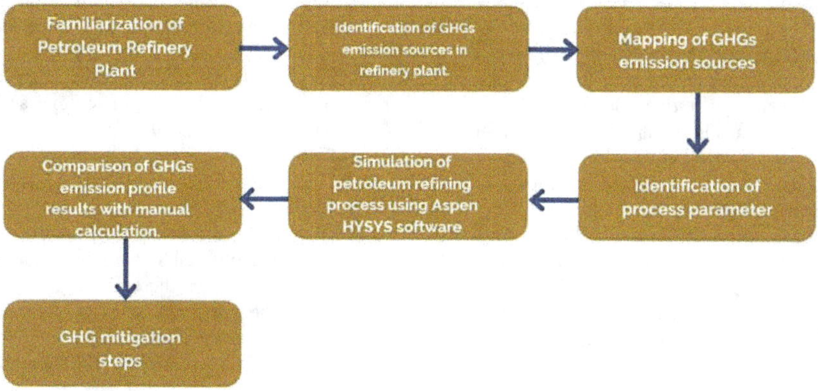

Fig.1: Overall methodology of quantification of carbon footprint in petroleum refinery using Aspen HYSYS simulation software.

Familiarization of Petroleum Refinery Plant. The process of plant familiarization plays a significant role in gaining a comprehensive understanding of plant operations, including production, capacity, and the various unit processes involved. By reviewing the process flow diagram or simplified block diagram, valuable information is obtained that helps identify the key indicators and parameters contributing to greenhouse gas (GHG) emissions during production. This step is essential as it provides a deeper insight into the plant's functioning and enables the identification of factors influencing GHG emissions throughout the production process.

Identification of GHGs Emission Sources in refinery plants. Recent case studies and research have indicated that the petroleum refining industry is a major source of GHG emissions, primarily due to the extensive use of fossil fuels for combustion purposes. The refinery's internal activities, such as the fluid catalytic cracking units (FCCU), hydrogen production units, and sulphur recovery plants, all significantly increase CO_2 emissions in addition to those from combustion-related sources.

Mapping of GHGs Emission Sources. In this step, the GHG mapping allows the identification of process units by streams that contributed to the GHG emission throughout the production.

Mapping of Greenhouse Gases (GHGs) will be conducted by determining which process unit within the refinery plant that produced high emission of GHGs and the units will be used in accounting of GHGS emissions.

Identification of Process Parameter. To ensure meaningful results, a sound understanding of chemical engineering principles is necessary to input appropriate values for parameters and evaluate the suitability of the obtained outcomes. In this study, relevant parameters such as pressure, temperature, flow rate, and composition will be acquired from reputable sources such as journal articles and reliable references. These parameters will then be utilized during the design phase of the refinery simulation.

Simulation of Petroleum Refining Process using Aspen HYSYS Software. Peng-Robinson equation of state (EOS) is the preferred fluid package for the process under investigation since it can manage the complex components (pseudo-components). Coded variables called pseudo-components make model fitting and design creation easier. The crude assay, or chemical examination of crude oil feedstock by petroleum testing laboratories, is what is used to provide the crude oil feed qualities. Using HYSYS V12, a complete simulation for the Petroleum Refinery including Crude Oil Preheat Train, Pre-flashing unit, Heating unit, Atmospheric Crude Tower and Vacuum Crude Tower will be designed.

Manual Calculation of GHG. The first step in manual GHG calculation is defining the scope of assessment and setting boundaries for the system under consideration. This step is crucial as it helps in focusing the analysis on relevant activities and clarifies the extent of responsibility for emissions. Scope 1 emissions encompass direct emissions from sources that the organization owns or controls, such as combustion of fossil fuels. Scope 2 emissions arise from the indirect emissions associated with purchased energy. Scope 3 emissions include all other indirect emissions in the organization's value chain, including transportation, waste generation, and employee commuting.

Emission factors are key parameters used to convert activity data into GHG emissions. These factors represent the amount of GHGs emitted per unit of activity. Once the activity data and emission factors are in hand, the next step involves performing the actual emission calculations. The basis used for the emission factor is based on activity or process, the corresponding activity data (flow rate) will be multiplied by the relevant emission factor. The result will yield the GHG emissions value for each of the desired specific activities. After calculating the GHG emissions for the desired activity, the values will be summed up to determine the total emissions for the four main process unit systems. This step provides a comprehensive overview of the overall carbon footprint when compared with the carbon emission value obtained from the simulation.

Comparison of GHGs Emission Profile Results between Manual Calculation. Once the values are calculated, the accounting of GHG emissions can be performed, allowing for a reliable comparison between the results obtained through simulation and manual calculations. This comparison will help to evaluate the accuracy of both simulation and manual calculation methods used in a refinery plant. To ensure easy reference and documentation, the calculated values will be organized and presented in tabular form. A detailed methodology steps of the simulation development is outlined separately as shown in Figure 2.

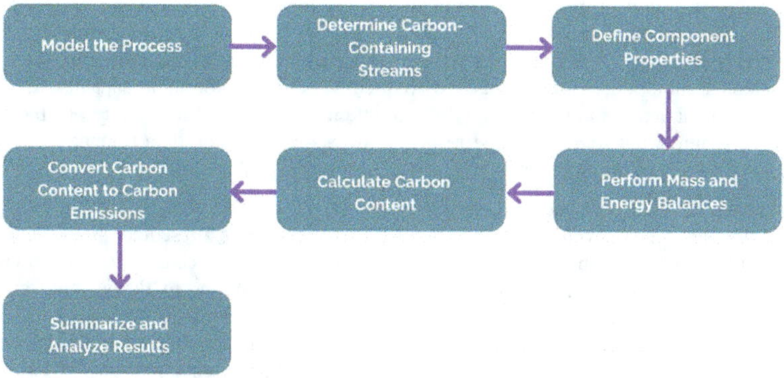

Figure 2 Simulation methodology of quantification of carbon footprint in petroleum refinery using Aspen HYSYS simulation software.

GHG Mitigation Steps. Relevant GHG mitigation measures will be considered and analysed. The performance of each optimization case was compared to the base case, which represented the original simulation without any optimization.

Results and Discussion

The final overall simulation of selected units in a typical refinery process was simulated as shown in Appendix 1. Aspen HYSYS software has been utilized to simulate the four major process units, which are crude oil feed preheating unit, atmospheric distillation unit, vacuum distillation unit and debutanizer unit.

A comparison of the total CO_2 emission values obtained through simulation and manual calculation is shown in Table 2 below. The result shows quite a significant difference in terms of the amount of total carbon emitted per year. The value from simulation and manual differs from about 9.12 % by taking the assumption of simulation value to be the true value while the value from manual method as the observed value.

Table 2: Comparison of carbon emission value for simulation and manual method

Method	Carbon emission value (kg/hr)
Simulation Software (Aspen HYSYS)	700 kg/hr
Manual Calculation	770.22 kg/hr

This is due to the primarily contingent on the complexity of the system being analysed and the quality of input data. The simulation method utilizes an older version of the EPA emission factor table when estimating carbon emissions. In comparison, manual calculations rely on the most up to-date EPA emission factor values, resulting in a slight difference in the total carbon emission figures. Hence, it can be concluded that manual calculation provides a more reliable and trustworthy overall total carbon emission value than the simulation method. Nevertheless, the simulation method remains valuable due to its ability to automatically identify and analyse the streams that have the potential to contribute to carbon emissions. These features help to ease the calculation process and, despite the use of outdated data, makes it a practical and efficient tool for assessing emissions much faster.

The precision of both methods relies heavily on the data's reliability and the robustness of the fundamental assumptions. Simulations are frequently preferred for their capacity to manage complex mathematical computations and difficult interrelationships. However, despite the method selected, it is important to validate the results using actual emission data (manual method), to ensure the accuracy of the findings.

Optimization Simulation Cases

Optimized cases were developed following the two approaches presented in the simulation modification flowchart as shown in Figure 2 and Figure 3, three new cases have been developed. The optimization framework can change all the process conditions of the existing crude oil distillation unit for minimizing either energy consumption or emissions. The results of the optimization will be different for these cases. The operating variables considered for optimization are the temperature of the feed and reflux ratio of the distillation column as shown in **Table 3.**

Table 3 Comparison between base case, case 1 and case 2.

Case	GHG Emission (kg/hr)	Optimization Parameter
Base	350.3	N/A
1	331.9	Increase Temp
2	296.3	Reduced Reflux

In the comparison of the three cases involving greenhouse gas (GHG) emissions, the base case which represent the original simulation without optimization recorded emissions of 350.3 kg/hr. In Case 1, where the temperature feed was increased, a notable reduction in GHG emissions to 331.9 kg/hr was observed. This decrease can be attributed to the improved efficiency associated with elevated temperatures, as higher temperatures often enhance reaction kinetics and reduce energy consumption, resulting in a more environmentally friendly operation. In Case 2, where the reflux ratio of the distillation column was reduced, GHG emissions further decreased to 296.3 kg/hr. This reduction is linked to the optimized operation of the distillation column, as a lower reflux ratio can enhance the energy efficiency of the separation process. Both optimization strategies, whether through temperature adjustments or reflux ratio reduction, demonstrate their effectiveness in minimizing environmental impact by promoting more energy-efficient and sustainable industrial processes.

Conclusion

This study has showed the potential of simulation tools to comprehensively model and analyse complex refinery processes, offering valuable insights into carbon emissions. By accurately quantifying emissions, these software solutions empower refineries to make informed decisions and provide wide areas and option for improvement in a short period of time. Additionally, the use of simulation tools significantly speeds up the process, saving time and resources that can be utilized in other areas within the organization. Unfortunately, the role of simulation software in mitigation strategies cannot be overstated. Both simulation and manual calculation have its own capabilities and offer reliable with trustworthy value of carbon emission for each of the organization resources and capabilities. The ability to simulate various scenarios and evaluate the impact of process modifications, energy efficiency measures, and emissions reduction technologies provides refineries with a roadmap for achieving substantial carbon footprint reductions. This not only aligns with regulatory compliance but also reflects the industry's commitment to sustainability and environmental survivability.

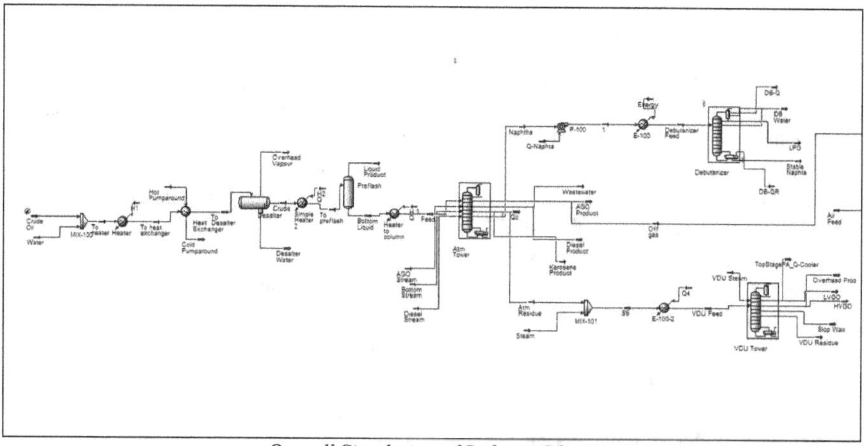

Overall Simulation of Refinery Plant

Acknowledgment
This work was supported by Universiti Teknologi PETRONAS and The Murata Science Foundation Grant, 015ME0-346. This assistance has been instrumental in the successful completion of our work, and we appreciate the contribution to our research endeavours.

References

[1] Bowen, F., & Wittneben, B. (2011). Carbon accounting. Accounting, Auditing & Accountability Journal, 24(8), 1022–1036. https://doi.org/10.1108/09513571111184742.

[2] Khalaf, A. (2018). Steady State Simulation of Basrah crude oil refinery distillation unit using Aspen Hysys. *Thi-Qar University Journal for Engineering Sciences.* https://doi.org/10.31663/tqujes.9.2.312(2018)

[3] Leontief, W.W. (1972) "Air Pollution and the Economic Structure Empirical Results of Input-Output Computations," in Input Output Economics 2nd ed., Oxford University Press, New York, 1986, pp.273-293.

[4] Leontief, W.W. (1970) "Environmental Repercussions and the Economic Structure: An Input-Output Approach," Review of Economics and Statistics, Vol.52, No.3, pp.262-271.

[5] Madugula, A. C., Sachde, D., Hovorka, S. D., Meckel, T. A., & Benson, T. J. (2021). Estimation of CO_2 emissions from petroleum refineries based on the total operable capacity for carbon capture applications. Chemical Engineering Journal Advances, 8, 100162. https://doi.org/10.1016/j.ceja.2021.100162

[6] Moro, L.F.L., Zanin, A.C. and Pinto, J.M. (1998) 'A planning model for refinery diesel production', Computers and Chemical Engineering, Vol. 22, pp.1039–1042.

[7] M. Wilkinson, J. Boden, T. Gilmartin, C. Ward, D. Cross, R. Allam, N. Ivens, CO2 capture from oil refinery process heaters through oxyfuel combustion, in: Proceedings of the Greenhouse Gas Control Technologies 6th International Conference, Elsevier, 2003, pp. 69–74, https://doi.org/10.1016/b978-008044276- 1/50012-x.

[8] Neiro, S.M.S. and Pinto, J.M. (2005) 'Multiperiod optimization for production planning of petroleum refineries', Chemical Engineering Communications, Vol. 192, pp.62–88.

[9] Pinto, J.M. and Moro, L.F.L. (2000) 'A planning model for petroleum refineries', Brazilian Journal of Chemical Engineering, Vol. 17, Nos. 4–7, pp.575–586.

[10] Smith, J. A., Johnson, M. B., Brown, C. D., & White, E. F. (2010). Simulation-Based Carbon Emission Quantification in Petroleum Refinery Processes: A Review of Methodologies and Challenges. Journal of Environmental Engineering, 42(3), 123-145.

[11] Sotelo, D., Favela-Contreras, A., Lozoya, C., Beltran-Carbajal, F., Dieck-Assad, G., & Sotelo, C. (2019). Dynamic simulation of a crude oil distillation plant using AspenHYSYS®. International Journal of Simulation Modelling, 18(2), 229–241. https://doi.org/10.2507/ijsimm18(2)465

[12] U.S. DOE (Department of Energy). 2007. Energy and Environmental Profile of the U.S. Petroleum Refining Industry. Prepared by Energetics, Inc., Columbia, MD. November 2007.

[13] Yousuo, Digieneni & Obuebite, Ann Amalate. (2023). The Determination of the Quantity of CO 2 Produced in a Local Ethanol Plant using Aspen HYSYS Software. 7. 41.

[14] Zhang, N., and Zhu, X.X. (2000) 'A novel modelling and decomposition strategy for overall refinery optimisation', Computers and Chemical Engineering, Vol. 24, pp.1543–1548.

[15] Ziółkowski, P., Madejski, P., Amiri, M., Kuś, T., Stasiak, K., Subramanian, N., PawlakKruczek, H., & Mikielewicz, D. (2021). Thermodynamic analysis of negative CO2 emission power plant using Aspen Plus, Aspen Hysys, and Ebsilon Software. Energies, 14(19), 6304. https://doi.org/10.3390/en14196304

Utilising the intrinsic characteristics of blind-hole nanoporous anodic alumina for blue energy

Khanh Nhien Vu[1,2,a], Cheryl Suwen Law[1,2,b,*], and Abel Santos[1,2,c*]

[1]School of Chemical Engineering, The University of Adelaide, Adelaide, SA 5005, Australia

[2]Institute for Photonics and Advanced Sensing (IPAS), The University of Adelaide, Adelaide, South Australia 5005, Australia

[a]khanhnhien.vu@adelaide.edu.au, [b]suwen.law@adelaide.edu.au, [c]abel.santos@adelaide.edu.au

Keywords: Nanoporous Anodic Alumina, Osmotic Energy Generation, Iontronic Membrane, Ionic Current Rectification, Barrier Oxide Layer

Abstract. Osmotic energy generation is a promising renewable energy technology that converts the osmotic potential between two saline solutions of different concentration into electricity. Ion exchange membranes (IEMs) have gained attention as promising materials, but the high ionic resistance and low power output have prevented IEMs from achieving satisfactory performances for real-life applications. Nanoporous anodic alumina (NAA) is a promising platform material to develop IEMs with precisely tailored properties because of its intrinsic ionic current rectification (ICR) properties. NAA provides new venues to control the transport of ions through its barrier oxide layer, to generate current associated with the selective flow of ions across its nanochannels. This study explores the capability of blind-hole NAA membranes for osmotic energy generation. Featuring a surface area of 13.4 mm², NAA membranes demonstrate the ability to produce a power density output of approximately 2.8×10^{-4} W m^{-2} under conditions of a 10^5-fold salinity ratio at pH 3 in a $CaCl_2$ electrolyte solution. This achievement is coupled with a notable efficiency rate of 7%.

Introduction

The rising energy demand, along with climate change concerns from fossil fuel emissions and global geopolitical instability, has made a significant redirection of research efforts and investments towards cleaner and more sustainable energy solutions [1]. One such promising resource is blue energy, derived from osmotic power extracted from the difference in salt concentration between two solutions separated by an ion exchange membrane. Blue energy has long been regarded as a highly attractive renewable energy source, boasting a substantial theoretical power yield (i.e., 0.8 kW h m^{-3}) [2]. But in reality, a low power output with high maintenance cost has hindered the exploitation of this resource. There are numeral factors for this obstacle, but one main reason is claimed to be high membrane resistance, beside pore dimension, internal resistance, or charge polarisation. Because of that, a novel ion-exchange membrane (IEM) with better ion-transportation is expected to raise the power harvested.

Nanoporous anodic alumina (NAA)-based heterogeneous membranes are often used as it helps increase transmembrane ion flux while having high mechanical stability, also it is industrially scalable, economically competitive, and highly versatile, using a membrane that has functions of a solid-state diode such as rectification could also increase the amount of power converted. In a previous study of Kim et al had indicated that one unique intrinsic property of NAA is ion-current rectification (ICR) through its barrier oxide layer (BOL) [3–7]. This barrier layer is influenced on the electrolyte, the pH condition of the electrolyte, and the electric field bias. Ionic rectification in ion exchange membranes could control and allow ions flow in one desirable direction but not the opposite. Osmotic energy is released followed by Gibb's free energy conversion when two

Materials Research Proceedings 53 (2025) 417-422 https://doi.org/10.21741/9781644903575-42

solutions with different salinity concentration are mixed [8]. Making use of ionic rectification phenomena to slow down the energy conversion could result in higher power harvested. Hybrid membranes with MOF-modified layer, or chemical composition has been attached to NAA layer as an active layer, to create ion rectification through the membrane and turn into a rectifying diode. However, no research has been made on NAA itself as a platform for osmotic energy conversion. Inspired by this, we produce blind-hole NAA by a two–step anodisation and investigate its capability as an ion–exchange membrane for osmotic energy harvesting.

Results and Discussion

Fabrication of NAA Membranes. Aluminum sheets were cut into square chips measuring 1.5 × 1.5 cm². These chips were then subjected to a cleaning process, first in ethanol and then in Milli-Q water, both with sonication for 5 minutes each to eliminate surface impurities. Once cleaned, the aluminum chips were air-dried and stored in a dry environment. Subsequently, the chips underwent electropolishing at approximately 6°C using a potential of 20 V for 3 minutes, with continuous stirring. The electropolishing was carried out in an electrolyte mixture comprising 4 parts ethanol to 1 part HClO4, aimed at smoothing the surface roughness and preventing undesirable effects related to localized electric fields, thereby promoting uniform, self-organized pore growth. The resulting electropolished aluminum substrates, with a reflective appearance, were then subjected to anodization in two stages using three different water-based acid electrolytes with a concentration of 0.3 M. This process was conducted to engineer the desired geometric characteristics of the barrier oxide layer (BOL) in nanoporous anodic alumina (NAA) membranes [9].

Blue Energy Production in Blind-Hole NAA Membranes. The equation (1) below can be used to determine the output power density of the blind-hole NAA membrane system when it is exposed to a specific salinity gradient [10]. The electrical power is extracted using an external circuit that is connected to an electrical load resistor (Fig.1a).

$$P_D = \frac{I^2 \times R}{A} \quad (1)$$

where P_D (W m^{-2}) is the output power density calculated, I (A) is the current passing through the membrane, R (W) is the external load resistance, and A (m^2) is the area of the membrane. Power and current density profiles from a blind-hole NAA membrane exposed to a salinity gradient of *[KCl]* = 4×10^6 are displayed in Fig. 2. From $J = 1\times10^{-2}$ A m^{-2} at $R = 1$ W to $J = 5\times10^{-4}$ A m^{-2} at R = 10 MW, it is evident that the current density flowing via the external electrical circuit drops following a sigmoidal pattern as R increases. It is important to remember that when $R = R_{Max} = Rm$, P_D becomes P_{DMax} [11]. Equation (1) then can be converted so that the greatest theoretical power that may be extracted from a membrane system is determined by the balance between ionic transport selectivity, membrane resistance, and salinity gradient concentration:

$$P_{DMax} = \frac{\left[(t_+ + t_-)\times\frac{RT}{F}Ln\left(\frac{a_H}{a_L}\right)\right]^2}{4R_m} \quad (2)$$

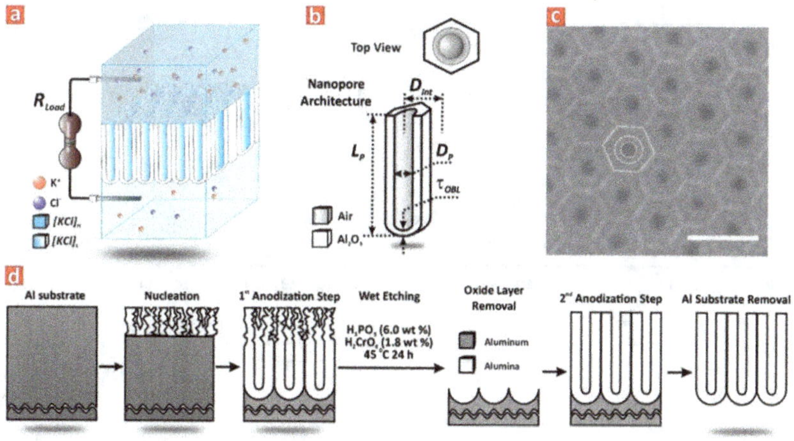

Fig.1. Blue energy conversion under a salinity gradient and fabrication and characterization of blind-hole NAA membranes produced by the two-step anodization process. (a) Illustration of electrolyte arrangement in the experimental setup used to convert salinity gradients into blue energy in blind-hole NAA membranes through an external electrical load (RLoad) and asymmetric electrolyte concentration. (b) Schematic of the main geometric features of the nanopore architecture in blind-hole NAA membranes (c) FEG-SEM bottom view image showing the BOL of a blind-hole NAA membrane. (d) Diagram illustrating the various phases of the two-step anodization process: electropolished aluminum substrate; first anodization step; nanopore nucleation and growth; removal of anodic oxide sacrifical layer by wet chemical etching; second anodization step; formation of self-organized nanoporous structure; and removal of aluminum substrate.

Therefore, membranes must be extremely selective to ionic transport of anions or cations and offer lower barrier to ionic transport in order to maximize a system's capacity to capture blue energy at a given concentration gradient. The current density evolution in a blind-hole NAA membrane subjected to different salinity gradients (i.e., *[KCl]* = 4×10^0, 4×10^1, 4×10^3, 4×10^5, and 4×10^6) is depicted in Fig. 2a. As R increases, J drops until it reaches its lowest value at $R = 10$ MΩ. Except for the trend observed at *[KCl]* = 4×10^0, which reaches its minimum J value at $R \sim 1\times10^4$ Ω, this is true for all salinity gradients. With a membrane area of around 13.4 mm^2 at pH of 6, blind-hole NAA membranes manufactured in oxalic acid electrolyte under mild anodization conditions extract a maximum blue power of approximately 1.5×10^{-4} W m^{-2} at *[KCl]* of 4×10^5.

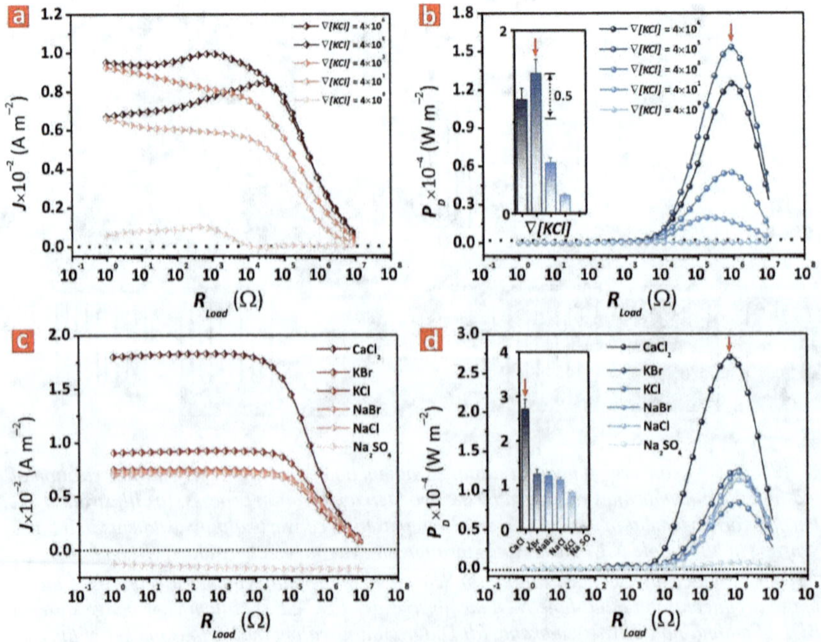

Fig. 2. Blue energy conversion in blind-hole NAA membranes produced by the two-step anodization process. (a) Current density (J) profile generated from a blind-hole NAA membrane subjected to distinct salinity gradients ([KCl] = [KCl]H/[KCl]L = 4×10⁰–4×10⁶) under a varying external electrical load and pH 6. (b) Power density (PD) profile generated from a blind-hole NAA membrane subjected to distinct salinity ratio ([KCl] = [KCl]H/[KCl]L = 4×10⁰–4×10⁶) with inset showing a bar chart summarizing the maximum power density (PDMax) obtained at the optimal RLoad (NB: red arrows indicate the best performing salinity gradient conditions in terms of PDMax). (c) Current density (J) profiles generated from a blind-hole NAA membrane subjected to a salinity gradient ([KCl] = 1×10⁶) under varying external electrical load in distinct electrolyte solutions (i.e., CaCl2, KBr, NaBr, NaCl, KCl, and Na2SO4). (d) Power density (PD) profiles generated under a salinity gradient ([KCl] = 1×10⁶) and varying external electrical load in distinct electrolyte solutions, and inset showing a bar chart summarizing the maximum power density (PDMax) obtained at the optimal RLoad (NB: red arrows indicate the best performing electrolyte in terms of PDMax).

In blind-hole NAA membranes, osmotic power conversion is mostly dependent on ion transport by hopping from vacancy to vacancy across the BOL. This motivates us to investigate if employing different electrolytes can boost the system's output power. As a result, it is conceivable that when ions move through the BOL under the influence of a salinity gradient, ions with varying sizes and charges ought to interact with space charges at various levels. To do this, we examined the formation of osmotic power in blind-hole NAA membranes when they were subjected to electrolytes containing six different salts: KCl, NaCl, KBr, NaBr, $CaCl_2$, and Na_2SO_4. It is apparent that the current density of the blind-hole NAA membrane is about one order of magnitude bigger in the $CaCl_2$ electrolyte than that recorded in KCl, NaCl, KBr, NaBr, and Na_2SO_4. The $CaCl_2$ electrolyte had the greatest output of osmotic power density, with a value of $P_{DMax} = \sim 2.8\times10^{-4}$ W m^{-2} at $R \sim 6.4\times10^5$ Ω over a working area of ~13.4 mm^2.

Several studies have demonstrated the high power density achieved by such membranes under various conditions. For example, Zhou et al. reported a mesoporous carbon-silica/anode alumina oxide (MCS/AAO) hybrid membrane generating 5.04 W m-2 with KCl. Other variations, including photo-switchable membranes, ionic diode system or heterogeneous membranes combining MOF (Metal-Organic Framework) modifications with NAA, have shown power densities ranging from 1.47 to 27.8 W m^2, depending on concentration gradients and electrolyte types [12]. It is noteworthy that the working areas employed in these studies in the field are four orders of magnitude smaller than that used in our and even further from the commercialised membranes and increasing the membrane's area significantly reduce the power density harvested. However, when a comparable working area is employed, the native BOL of NAA membranes may produce as much blue energy as that documented for other cutting-edge systems including hybrid heterogenous membranes [13]. Therefore, considering the scalability and ease of anodization technique, it is feasible to infer that as-produced NAA membranes provide a basic yet adaptable platform to create blue energy in a simple and cost-effective manner.

In conclusion, NAA membranes offer a straightforward and versatile platform for generating blue energy in a cost-effective manner, thanks to the scalability and simplicity of anodization technology. However, several fundamental questions remain, including how to tune the system by engineering the barrier oxide layer (BOL) through anodization, optimizing the saline electrolyte, and understanding the impact of the working area on energy generation efficiency.

Conclusion

In summary, we present the initial discovery of osmotic energy generation in blind-hole nanoporous anodic alumina membranes. The maximum power density output recorded was approximately 2.8×10^{-4} W m^{-2} for a working area of 13.4 mm^2, with a conversion efficiency of 7%. The utilization of a saline electrolyte with significantly different cation and anion sizes enhances the barrier oxide layer's selectivity towards smaller ions over larger ones. Consequently, through careful design considerations regarding pH and saline ratio composition, these membranes hold potential for state-of-the-art blue power energy generation.

Acknowledgements

We would like to acknowledge and pay our respects to the Kaurna-Adelaide people, the Traditional Custodians of the land on which the work was performed. The authors thank the support provided by the Australian Research Council through the grant DP220102857, the School of Chemical Engineering, The University of Adelaide, and the Institute for Photonics and Advanced Sensing (IPAS). C.S.L. thanks the support provided by The University of Adelaide and IPAS through her Future Making Fellowship.

References

[1] R. Vakulchuk, I. Overland, D. Scholten, Renewable Sustainable Energy Rev. 2020, 122, 109547. https://doi.org/10.1016/j.rser.2019.109547

[2] N. Y. Yip, M. Elimelech, Environ. Sci. Technol. 2012, 46, 5230, https://doi.org/10.1021/es300060m

[3] Y. D. Kim, S. Choi, A. Kim, W. Lee, ACS Nano 2020, 14, 13727 https://doi.org/10.1021/acsnano.0c05954

[4] M. Mibus, C. Jensen, X. Hu, C. Knospe, M. L. Reed, G. Zangari, Appl. Phys. Lett. 2014, 104, 244103 https://doi.org/10.1063/1.4882656

[5] H. Strathmann, Ion-Exchange Membrane Separation Processes, Elsevier Science, 1st edn, 2004, vol. 9. https://doi.org/10.1016/S0927-5193(04)80032-9

[6] A. Campione, L. Gurreri, M. Ciofalo, G. Micale, A. Tamburini, A. Cipollina, Desalination 2018, 434, 121. https://doi.org/10.1016/j.desal.2017.12.044

[7] J. Cao, X. -P. Zhao, M. R. Younis, Z. -Q. Li, X. -H. Xia, C. Wang, Anal. Chem. 2017, 89, 10957 https://doi.org/10.1021/acs.analchem.7b02765

[8] B. E. Logan, M. Elimelech, Nature 2012, 488, 313 https://doi.org/10.1038/nature11477

[9] A. Santos, J. Mater. Chem. C, 2017, 5, 5581, c) M. Norek, J. Electrochem. Soc. 2022, 169, 123503. https://doi.org/10.1149/1945-7111/aca937

[10] J. Su, D. Ji, J. Tang, H. Li, Y. Feng, L. Cao, L. Jiang, W. Guo, Chin. J. Chem. 2018, 36, 417. https://doi.org/10.1002/cjoc.201800067

[11] G. Laucirica, A. G. Albesa, M. E. Toimil-Molares, C. Trautmann, W. A. Marmisollé, O. Azzaroni, Nano Energy 2020, 71, 104612 https://doi.org/10.1016/j.nanoen.2020.104612

[12] T. Xiao, J. Ma, Z. Liu, B. Lu, J. Jiang, Z. Nie, R. Luo, J. Jin, Q. Liu W. Li, J. Zhai, J. Mater. Chem. A 2020, 8, 11275 https://doi.org/10.1039/D0TA02162H

[13] C. -Y. Lin, S. -F. Chang, K. -T. Kuo, S. Garner, S. C. Pollard, S. -H. Chen, J. -P. Hsu, ACS Appl. Mater. Interfaces 2023, 15, 43094. https://doi.org/10.1021/acsami.3c05831

Materials Research Proceedings 53 (2025) 423-432 https://doi.org/10.21741/9781644903575-43

Predicted binary isotherms of CO_2/N_2 gas mixture for post-combustion capture and direct air capture using gallate-based metal-organic frameworks

MARHAINA Ismail[1,a], NIK ABDUL AZEEM Nik Saleduddin[2,a] and MOHAMAD AZMI Bustam[3,a *]

[1]Carbon Dioxide Research Centre (CO2RES), Universiti Teknologi PETRONAS, 32610 Bandar Seri Iskandar, Perak, Malaysia

[2]Centre of Research in Ionic Liquid (CORIL), Universiti Teknologi PETRONAS, 32610 Bandar Seri Iskandar, Perak, Malaysia

[a]marhainaismail@gmail.com, [b]nikazeem26@gmail.com, [c]azmibustam@utp.edu.my

Keywords: IAST, Binary Isotherm, Post-Combustion, Direct Air Capture

Abstract. Post-combustion capture and direct air capture are two important processes for reducing CO_2 content in the atmosphere and mitigating climate change. Post-combustion capture involves capturing CO_2 emissions produced from power plants, while direct air capture removes CO_2 directly from the ambient air. Gallate-based MOFs can offer remarkable CO_2 adsorption capacity. However, there is scarcity of reported works related to N_2 adsorption, limiting the scope of research on binary isotherms of CO_2/N_2 mixture. Therefore, this work aims to investigate the adsorption of pure N_2 gas using gallate-based MOFs. Additionally, prediction of binary isotherms of CO_2/N_2 mixtures was carried out to assess the efficiency of these frameworks for post-combustion capture and direct air capture. The predicted binary isotherms of CO_2/N_2 mixture revealed that gallate-based MOFs hold potential as adsorbents for both post-combustion capture and direct air capture. The IAST selectivity illustrated that gallate-based MOFs possess the capability to selectively capture CO_2 from CO_2/N_2 mixture, with Mg-gallate exhibiting the highest values, followed by Ni-gallate and Co-gallate. Therefore, Mg-gallate, Co-gallate and Ni-gallate can be suggested as promising adsorbents for post-combustion capture and direct air capture.

Introduction

The global concern over climate change has intensified due to the swift increase in atmospheric carbon dioxide (CO_2) levels. This will lead to global warming that can cause the future consequences such as extreme weather changes, rising of sea level, food shortages and disruption, disturbing ecosystem and biodiversity. In order to keep global temperature increase to less than 2°C above pre-industrial levels by the end of this century, the European Union (EU) has established a goal to reduce CO_2 emissions by at least 40% by 2030 compared to 1990, and by 30% compared to 2005 [1]. Efforts to reduce CO_2 emissions have led to substantial growth in research and development focused on capturing and separating CO_2, both from stationary sources (industrial processes and power generation) with high CO_2 concentrations and directly from the atmosphere. The United Nations Framework Convention on Climate Change (UNFCCC), the Intergovernmental Panel on Climate Change (IPCC), the Kyoto Protocol, and the Paris Agreement all acknowledge carbon capture and storage (CCS) as a viable technology in the combat against global warming [2]. In CCS, CO2 is captured from the sources, liquefied prior to transportation and stored for future purposes.

The CO_2 capture can be classified into four main methods which are post-combustion capture, direct air capture (DAC), pre-combustion capture and oxy-fuel combustion [3]. Post-combustion capture refers to the process of capturing CO_2 from the flue gas of a combustion system [4]. The

Materials Research Proceedings 53 (2025) 423-432 https://doi.org/10.21741/9781644903575-43

flue gas predominantly comprises 68–75% N_2, 10–15% CO_2, 5–10% H_2O, 2–5% O_2 and trace amounts of NO_x and SO_x [1]. Direct air capture is a process of capturing CO_2 directly from the surrounding atmosphere, specifically aimed at decreasing atmospheric concentrations [3]. If CO_2 emissions are not properly managed, the current atmospheric CO_2 concentration of 400 ppm is projected to rise to 500 ppm by 2030 [5]. Pre-combustion entails the conversion of fossil fuels into syngas, a mixture of hydrogen (H_2) and carbon monoxide (CO), which is then subjected to water-gas shift (WGS) reactions [5]. This process transforms the syngas into a mixture of H_2 and CO_2 and CO_2 is captured prior to combustion. Oxyfuel-combustion is the combustion of fuel feedstocks in an environment containing nearly pure oxygen (95–99%) or a mixture of O_2 and CO_2, leading to a flue gas with a significantly high concentration of CO_2 [6]. This flue gas mainly consists of CO_2 and water vapor. The CO_2 capture is simplified as shown in Fig. 1 below:

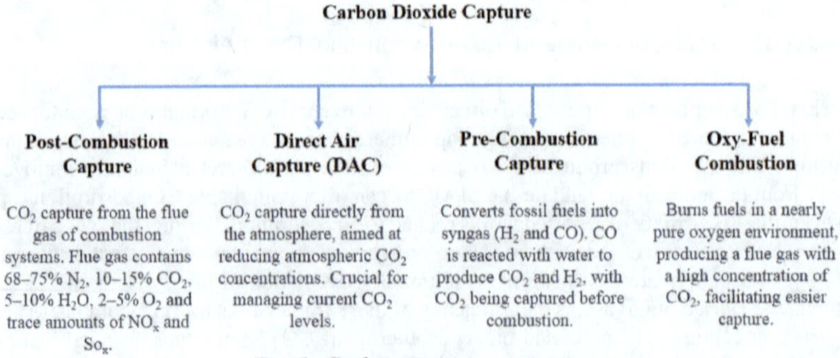

Fig. 1: Carbon capture processes

There are four primary CO_2 capture technologies including chemical absorption, adsorption, membrane separation and cryogenic distillation [2]. However, membrane separation faces challenges in industrial applications primarily due to its high cost [7]. Meanwhile, chemical absorption and cryogenic distillation require high energy consumptions [8, 9]. While CO_2 capture through physical adsorption boasts relatively low energy consumption due to its simple regeneration process, significant efforts are still needed to enhance the development of advanced adsorbents that offer both high selectivity and capacity [10]. Recently, metal-organic frameworks (MOFs) stand out as valuable candidates for fulfilling critical roles as selective adsorbents. This is attributed to their tunable binding strengths influencing adsorption selectivity, high surface areas contributing to substantial capacity and large pore sizes facilitating rapid diffusion kinetics [11].

According to our past studies, gallate-based MOFs (Mg-gallate, Co-gallate and Ni-gallate) have exhibited exceptional efficiency in the adsorption of CO_2 [12-14]. However, there is a lack of study on adsorption of N_2 using gallate-based MOFs. As a result, there are very limited binary isotherms available for CO_2/N_2 mixture adsorption. Therefore, this work will report on the adsorption of pure N_2 gas using gallate-based MOFs. Subsequently, prediction on binary isotherms of CO_2/N_2 mixture was conducted to evaluate the performances of gallate-based MOFs for post-combustion capture and direct air capture. The IAST selectivity was calculated to assess their efficiency in adsorbing CO_2 from a CO_2/N_2 mixture. It will help to evaluate the potential of these frameworks for further experimental works based on their performances. This paper aims to suggest promising adsorbents in capturing CO_2 for post-combustion capture and direct air capture, with the goal of significantly reducing CO_2 emissions to address global warming and achieve environmental sustainability.

Methodology

The synthesis and characterization analysis of Mg-gallate, Co-gallate and Ni-gallate were reported in our previous works, including their pure equilibrium isotherms for CO_2 adsorption [12- 14].

Pure N_2 Adsorption. Prior to the adsorption measurement, Mg-gallate, Co-gallate and Ni-gallate with a mass of 0.2-0.25 g were first degassed at 393 K for 24 hours under vacuum. The pure equilibrium isotherms for N_2 adsorption were measured at ambient condition (298 K and 1 bar) by employing volumetric gas adsorption using BET BELSORP-mini II (BEL Japan,Inc). The purity of N_2 gas is 99.9995 %, supplied by Air Products Malaysia.

Prediction of Binary Isotherms of CO_2/N_2 Gas Mixture. The binary isotherms of CO_2/N_2 mixture were predicted based on experimental CO_2 and N_2 pure equilibrium isotherms using Ideal Adsorbed Solution Theory (IAST) calculations by the Python package (pyIAST) [15]. For post-combustion capture, the compositions of CO_2/N_2 were varied at 10:68, 10:75, 15:68, and 15:75, based on the flue gas content [1]. Meanwhile, the compositions of CO_2/N_2 for direct air capture was fixed at 0.04:78, following the atmospheric content. The calculation flow consists of experimental data loading, model fitting and generation of binary isotherms, as simplified in Fig. 2 below:

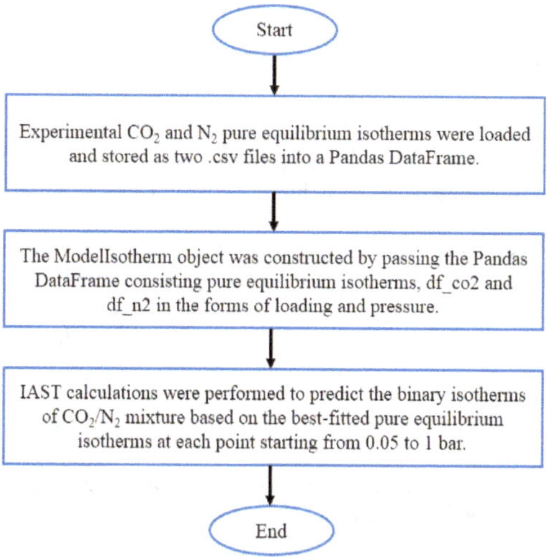

Fig.2: pyIAST calculation

Based on the predicted binary isotherms of CO_2/N_2 mixture, the IAST selectivity was calculated using Eq. 1 [16].

$$S_{ij} = \frac{x_i/x_j}{y_i/y_j} \tag{1}$$

where x_i and x_j denote mole fractions of component i and j in the adsorbed phase, while y_i and y_j signify mole fractions of component i and j in the gas phase (dimensionless).

Result and Discussions

Pure Equilibrium Isotherms. The pure equilibrium isotherm is measured as a plot of fractional coverage of adsorbate or adsorbed amount against the relative pressure of the gas. It explains the surface affinity of the adsorbent towards the adsorbate, evaluating the feasibility of the adsorption process as a unit operation. The experimental pure equilibrium isotherms of CO_2 and N_2 for Mg-gallate, Co-gallate and Ni-gallate at 298 K are plotted as shown in Fig. 3 below:

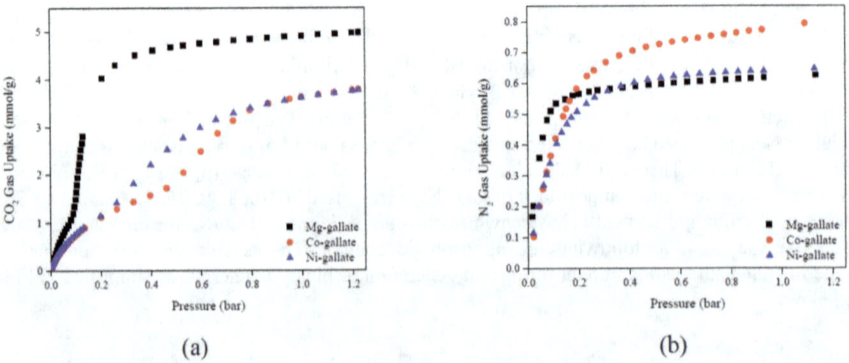

Fig. 3: Pure equilibrium isotherms of (a) CO_2; (b) N_2 for Mg-gallate, Co-gallate and Ni-gallate

The pure equilibrium isotherms of CO_2 are re-plotted and detailed explanation can be found in our previous works [12-14]. Generally, the CO_2 uptakes increased with pressure in which Mg-gallate demonstrated the highest uptake compared to Co-gallate and Ni-gallate.

The pure equilibrium isotherms of N_2 exhibit that N_2 uptakes increased with pressure in which Co-gallate revealed the highest uptake compared to Ni-gallate and Mg-gallate, following Type I isotherm. The Type I isotherm corresponds to Langmuir isotherm [17]. This type of isotherm is a concave to the relative pressure axis and the uptake approaches a limiting point [18]. It represents the N_2 uptakes increased rapidly at the low pressure region and gradually occupied until reaching saturation (horizontal asymptote) once the monolayer capacity is achieved.

The CO_2 uptakes were found to be higher than N_2 due to thermodynamic equilibrium effect, reflecting differences in affinity between the frameworks and gas molecules. Since the interactions of gallate-based MOFs and CO_2 are stronger, resulting in selectivity values greater than unity. Selectivity is calculated by dividing the capacity values with pressure at constant temperature, which is the most straightforward approach. Table 1 tabulates the capacity and selectivity values of Mg-gallate, Co-gallate and Ni-gallate at 298 K and 1 bar.

Table 1: Capacity and Selectivity of Mg-gallate, Co-gallate and Ni-gallate at 298 K and 1 bar.

Gallate-based MOF	CO_2 Gas Uptake [mmol/g]	N_2 Gas Uptake [mmol/g]	Selectivity
Mg-gallate	4.92	0.62	7.9
Co-gallate	3.63	0.78	4.7
Ni-gallate	3.62	0.64	5.7

Mg-gallate, Co-gallate and Ni-gallate can act as promising adsorbents in capturing CO_2 from CO_2/N_2 gas mixture since they offered remarkable capacity and selectivity. Capacity and selectivity serve as preliminary factors to evaluate the gas adsorption behaviors and to assess the performance of various adsorbents.

Analytical Model Fitting. Predicting binary isotherms prior to conducting experiments provides a valuable roadmap for researchers, guiding decision-making by identifying the most promising candidates for further experimental investigation in the development of adsorption processes and materials. Therefore, the predication of binary isotherms of CO_2/N_2 via IAST calculation using Python Package comes into picture. Firstly, the analytical models were employed for the experimental isotherm fittings. The available analytical models in Python are Henry, Langmuir, approximated Temkin, Dual-site Langmuir, quadratic, and BET models. The model parameters are tabulated in Table 2.Table 2: Analytical Model Parameter.

Model	Parameter	CO_2			N_2		
		Mg-gallate	Co-gallate	Ni-gallate	Mg-gallate	Co-gallate	Ni-gallate
Henry	K_H	5.7421	3.6559	3.8227	0.8992	1.1051	0.9189
	RMSE	1.3555	0.3335	0.4621	0.2817	0.2545	0.2349
Langmuir	M	6.2675	8.0105	6.0054	0.6431	0.8882	0.7164
	K_L	4.5690	0.7915	1.5162	30.7560	8.5625	11.3287
	RMSE	0.3882	0.1786	0.1096	0.0198	0.0210	0.0142
Approximated Temkin	M	2.6270	9.6645	5.2978	0.6053	0.7622	0.6461
	K_T	3.74	0.6837	1.5744	19.9321	6.4725	8.3168
	θ	−8.1530	0.4130	−0.4899	−1.1373	−1.4232	−1.1597
	RMSE	0.2330	0.1776	0.1073	0.0103	7.97×10^{-3}	2.94×10^{-3}
Dual-site Langmuir	M_1	5.25×10^{-4}	5.25×10^{-4}	2.6383	5.25×10^{-4}	5.25×10^{-4}	0.4896
	K_1	5.0×10^{-4}	5.0×10^{-4}	1.5162	5.0×10^{-4}	5.0×10^{-4}	11.3287
	M_2	5.0×10^{-4}	5.0×10^{-4}	3.3671	5.0×10^{-4}	5.0×10^{-4}	0.2267
	K_2	5.0×10^{-4}	5.0×10^{-4}	1.5163	5.0×10^{-4}	5.0×10^{-4}	11.3287
	RMSE	3.2372	2.1577	0.1096	0.5490	0.6238	0.0142
Quadratic	M	2.5036	4.7543	2.7635	0.3087	0.4084	0.3390
	K_A	−0.4963	1.3728	3.0598	21.9514	9.0862	13.3144
	K_B	60.6852	0.3118	3.2617	888.71	97.1651	147.39
	RMSE	0.1953	0.1778	0.1077	6.98×10^{-3}	5.40×10^{-3}	4.51×10^{-3}
BET	M	5.25×10^{-4}	6.2152	5.25×10^{-4}	0.6845	1.1151	0.8508
	K_A	5.0×10^{-4}	1.0521	5.0×10^{-4}	26.2164	5.9361	8.2827
	K_B	5.0×10^{-4}	0.0676	5.0×10^{-4}	−0.0854	−0.2103	−0.1732
	RMSE	3.2372	0.1784	2.2070	0.0176	0.0149	7.37×10^{-3}

Based on Table 2, Langmuir model was observed to offer the highest goodness-of-fit, since it possessed the lowest RMSE values with the most appropriate parameter values. The Langmuir isotherm assumes that: (1) monolayer adsorption; (2) uniform distribution of adsorption sites; (3) constant adsorption energy; and (4) negligible interaction among adsorbate molecules [19]. Fig. 4 illustrated the experimental data with their respective model fittings.

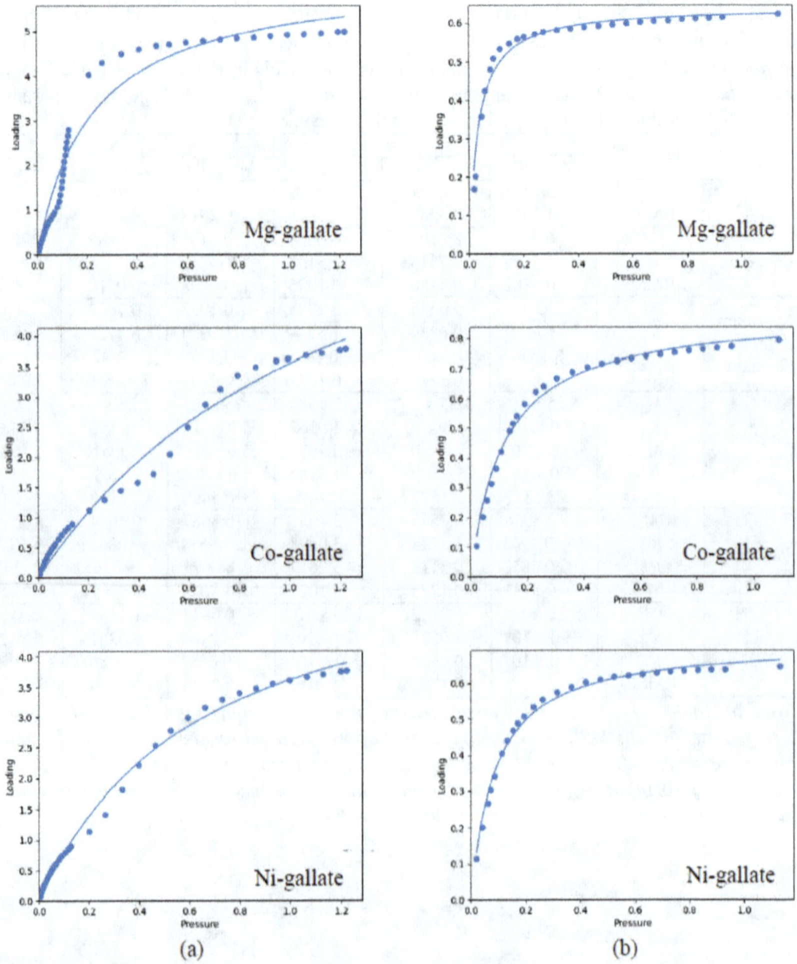

Fig. 4: Experimental data (dot) and model fitting (line) of (a) CO_2; (b) N_2 for Mg-gallate, Co-gallate and Ni-gallate

Subsequently, the IAST calculation was performed based of Langmuir-fitted isotherms to generate the predicted binary isotherms of CO_2/N_2 mixture.

Predicted Binary Isotherms of CO_2/N_2 Mixture. The predicted binary isotherms of CO_2/N_2 mixture for Mg-gallate, Co-gallate and Ni-gallate for post-combustion capture are shown in Fig. 5

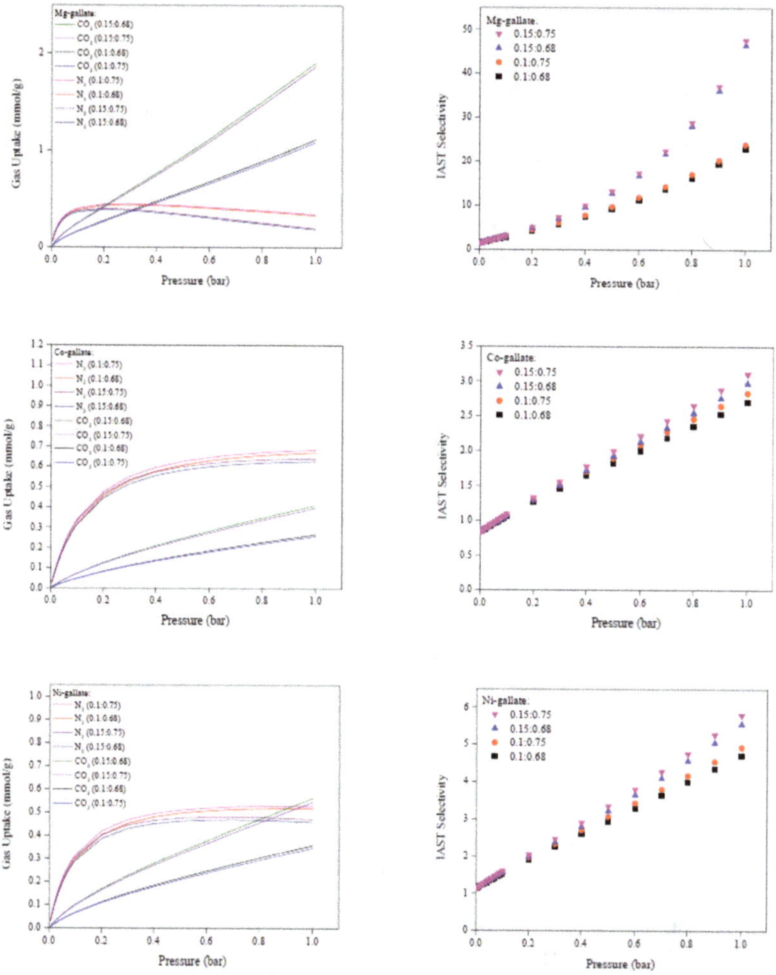

(a) (b)

Fig. 5: (a) Predicted binary isotherms of CO_2/N_2 mixture; (b) IAST selectivity for Mg-gallate, Co-gallate and Ni-gallate for post-combustion capture

The predicted CO_2 uptakes were observed to increase with pressure and gas phase composition. Meanwhile, for predicted N_2 uptakes exhibited inconsistent patterns; Co-gallate increased with pressure, while Mg-gallate and Ni-gallate started to show a decrement at certain points (although it was not too significant for Ni-gallate). This phenomenon may occur due to more N_2 molecules compete for the limited number of adsorption sites as pressure increases. As a result, the adsorption sites become saturated, leading to a decrease in the rate of adsorption and, consequently, a decrease in the adsorbed N_2 amount.

Despite, the predicted binary isotherms demonstrated promising IAST selectivity in which Mg- gallate presented the highest values, followed by Ni-gallate and Co-gallate, as displayed in Fig. 5

(b). In this case, selectivity denotes the capability of gallate-based MOFs to preferentially adsorb CO_2 over N_2 in a CO_2/N_2 gas mixture. Selectivity plays an essential role in gas separation processes by facilitating the effective capture of particular gases from a mixture. The adsorbent with high selectivity will adsorb the desired gas more strongly while allowing other gases to pass through or adsorb to a lesser extent. This selectivity facilitates the effective separation of gases, resulting in enhanced purity of the targeted end product.

Fig. 6 (a) illustrates the predicted binary isotherms of CO_2/N_2 mixture for Mg-gallate, Co-gallate and Ni-gallate for direct air capture.

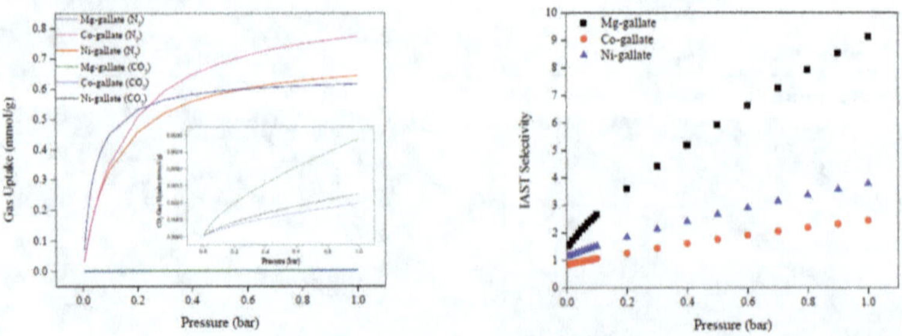

(a) (b)

Fig. 6: (a) Predicted binary isotherms of CO_2/N_2 mixture; (b) IAST selectivity for Mg-gallate, Co-gallate and Ni-gallate for direct air capture

The predicted CO_2 and N_2 uptakes for Mg-gallate, Co-gallate and Ni-gallate exhibited an increasing pattern as pressure increased. Since the content of CO_2 in atmospheric air is very low (0.04 %), the predicted CO_2 uptakes were found to be correspondingly low. However, IAST selectivity also suggested that Mg-gallate, Co-gallate and Ni-gallate can capture CO_2, with Mg-gallate demonstrated the highest values.

Conclusion

The pure equilibrium isotherms of N_2 demonstrated that the uptakes of N_2 increased with increasing pressure, with Co-gallate displaying the highest uptake in comparison to Ni-gallate and Mg-gallate. The predicted binary isotherms of CO_2/N_2 mixture suggested that gallate-based MOFs can be promising adsorbents for post-combustion capture and direct air capture. IAST selectivity demonstrated that gallate-based MOFs can selectively capture CO_2 from CO_2/N_2 mixture in which Mg-gallate offered the greatest values, followed by Ni-gallate and Co-gallate. Therefore, Mg-gallate, Co-gallate, and Ni-gallate are candidates worthy of further experimental investigation under real-practice conditions.

Acknowledgement

Funding for this research was provided by a grant from Yayasan Universiti Teknologi PETRONAS (YUTP Grant 015LC0-364). Additional support was received from the Carbon Dioxide Research Centre (CO_2RES) and the Centre of Research in Ionic Liquids (CORIL) at Universiti Teknologi PETRONAS.

References

[1] C. Chao, Y. Deng, R. Dewil, J. Baeyens, and X. Fan, "Post-combustion carbon capture," Renewable and Sustainable Energy Reviews, vol. 138, p. 110490, 2021. https://doi.org/10.1016/j.rser.2020.110490

[2] S. Y. W. Chai, L. H. Ngu, and B. S. How, "Review of carbon capture absorbents for CO2 utilization," Greenhouse Gases: Science and Technology, vol. 12, pp. 394-427, 2022. https://doi.org/10.1002/ghg.2151

[3] T. M. Gür, "Carbon dioxide emissions, capture, storage and utilization: Review of materials, processes and technologies," Progress in Energy and Combustion Science, vol. 89, p. 100965, 2022. https://doi.org/10.1016/j.pecs.2021.100965

[4] A. I. Osman, M. Hefny, M. Abdel Maksoud, A. M. Elgarahy, and D. W. Rooney, "Recent advances in carbon capture storage and utilisation technologies: a review," Environmental Chemistry Letters, vol. 19, pp. 797-849, 2021. https://doi.org/10.1007/s10311-020-01133-3

[5] Z. Hu, Y. Wang, B. B. Shah, and D. Zhao, "CO2 Capture in Metal-Organic Framework Adsorbents: An Engineering Perspective," Advanced Sustainable Systems, vol. 3, p. 1800080, 2019. https://doi.org/10.1002/adsu.201800080

[6] X. Wang and C. Song, "Carbon capture from flue gas and the atmosphere: a perspective," Frontiers in Energy Research, vol. 8, p. 560849, 2020. https://doi.org/10.3389/fenrg.2020.560849

[7] C. Chen, X. Feng, Q. Zhu, R. Dong, R. Yang, Y. Cheng, et al., "Microwave-assisted rapid synthesis of well-shaped MOF-74 (Ni) for CO2 efficient capture," Inorganic chemistry, vol. 58, pp. 2717-2728, 2019. https://doi.org/10.1021/acs.inorgchem.8b03271

[8] E. E. Ünveren, B. Ö. Monkul, Ş. Sarıoğlan, N. Karademir, and E. Alper, "Solid amine sorbents for CO2 capture by chemical adsorption: a review," Petroleum, vol. 3, pp. 37-50, 2017. https://doi.org/10.1016/j.petlm.2016.11.001

[9] M. Songolzadeh, M. Soleimani, M. Takht Ravanchi, and R. Songolzadeh, "Carbon dioxide separation from flue gases: a technological review emphasizing reduction in greenhouse gas emissions," The Scientific World Journal, vol. 2014, 2014. https://doi.org/10.1155/2014/828131

[10] R. S. Liu, X. D. Shi, C. T. Wang, Y. Z. Gao, S. Xu, G. P. Hao, et al., "Advances in post-combustion CO2 capture by physical adsorption: from materials innovation to separation practice," ChemSusChem, vol. 14, pp. 1428-1471, 2021. https://doi.org/10.1002/cssc.202002677

[11] R. Krishna and J. R. Long, "Screening metal-organic frameworks by analysis of transient breakthrough of gas mixtures in a fixed bed adsorber," The Journal of Physical Chemistry C, vol. 115, pp. 12941-12950, 2011. https://doi.org/10.1021/jp202203c

[12] M. Ismail, M. A. Bustam, N. E. F. Kari, and Y. F. Yeong, "Ideal Adsorbed Solution Theory (IAST) of Carbon Dioxide and Methane Adsorption Using Magnesium Gallate Metal-Organic Framework (Mg-gallate)," Molecules, vol. 28, p. 3016, 2023. https://doi.org/10.3390/molecules28073016

[13] M. Ismail, M. A. Bustam, and Y. F. Yeong, "Adsorption of Carbon Dioxide and Methane on Cobalt Gallate-Based Metal-Organic Framework (Co-Gallate): Equilibrium Isotherm, Thermodynamic and Kinetic Studies," Journal of Advanced Research in Fluid Mechanics and Thermal Sciences, vol. 108, pp. 151-163, 2023. https://doi.org/10.37934/arfmts.108.2.151163

[14] M. Ismail, M. A. Bustam, Y. Y. Fong, and A. Ahmad, "Unary Adsorption Isotherm of Carbon Dioxide and Methane of Nickel Gallate Metal-organic Framework," Journal of Advanced Research in Micro and Nano Engineering, vol. 19, pp. 51-62, 2024. https://doi.org/10.37934/armne.19.1.5162

[15] C. M. Simon, B. Smit, and M. Haranczyk, "pyIAST: Ideal adsorbed solution theory (IAST) Python package," Computer Physics Communications, vol. 200, pp. 364-380, 2016. https://doi.org/10.1016/j.cpc.2015.11.016

[16] N. F. Cessford, N. A. Seaton, and T. Düren, "Evaluation of ideal adsorbed solution theory as a tool for the design of metal-organic framework materials," Industrial & engineering chemistry research, vol. 51, pp. 4911-4921, 2012. https://doi.org/10.1021/ie202219w

[17] S. Alafnan, A. Awotunde, G. Glatz, S. Adjei, I. Alrumaih, and A. Gowida, "Langmuir adsorption isotherm in unconventional resources: Applicability and limitations," Journal of Petroleum Science and Engineering, vol. 207, p. 109172, 2021. https://doi.org/10.1016/j.petrol.2021.109172

[18] M. Thommes, K. Kaneko, A. V. Neimark, J. P. Olivier, F. Rodriguez-Reinoso, J. Rouquerol, et al., "Physisorption of gases, with special reference to the evaluation of surface area and pore size distribution (IUPAC Technical Report)," Pure and applied chemistry, vol. 87, pp. 1051-1069, 2015. https://doi.org/10.1515/pac-2014-1117

[19] J. Wang and X. Guo, "Adsorption isotherm models: Classification, physical meaning, application and solving method," Chemosphere, vol. 258, p. 127279, 2020. https://doi.org/10.1016/j.chemosphere.2020.127279

Decarbonization Technology: ICDT2024 Materials Research Forum LLC
Materials Research Proceedings 53 (2025) 433-443 https://doi.org/10.21741/9781644903575-44

Tailoring superparamagnetic nanoparticle for adsorption of carbon dioxide

NUR KAMALIYAH Dzil Razman[1,a] and OH Pei Ching[1,b*]

[1]Chemical Engineering Dept, Universiti Teknologi Petronas, Malaysia

[a]nur_22007862@utp.edu.my, [b]peiching.oh@utp.edu.my

Keywords: Magnetic, Superparamagnetic, Solvothermal Method, Nanoparticles, Adsorption

Abstract. In recent years, green technologies have been extensively developed, including adsorption technology. In industrial installations, swing adsorptions are favoured for its continuous process. It also posed a large energy demands of almost 40% of overall energy supply. In a temperature swing adsorption (TSA) process, the purging or regeneration phase of the bed requires external heat, whether it to be from steam or hot gas, which largely contributed to the high demand. Hence, an extension of temperature swing adsorption was introduced to overcome the issue by using magnetize particles that can generate in-situ heat upon exposure to induction field. This study attempted to synthesize a magnetic nanoparticle that is superparamagnetic, to be embedded in metal organic framework (MOF). Superparamagnetic iron oxide (Fe_3O_4) was synthesised through one-pot solvothermal method using iron (III) chloride as its precursor at 250°C, for 10 hours. Synthesized Fe_3O_4 reported to have spherical shape from field emission scanning electron microscopy (FESEM), with crystalline size of 0.4 nm. Highest value of magnetic saturation was 50 emu/g, as obtained from vibrating sample magnetometry (VSM) test. From the VSM result, it was also firmed that superparamagnetic nanoparticles were achieved due to the absence of hysteresis loop and near zero retentivity of 0.11 emu/g. Parametric study on the reaction parameters and reactants concentration was also conducted. It was found that reaction temperature and concentration of sodium acetate able to reduce the particle size. Thus, with the result obtained, magnetite with high magnetic saturation and particle stability was tailored to be embedded with MOF to produce a magnetic framework composite.

Introduction

Magnetic nanomaterials are functionalized nanosized materials that exhibit magnetic characteristics such as magnetoresistance [1], superparamagnetic [2], opto-magnetic and magnetic anisotropy [3]. Due to their unique properties, this material has potential in various application in many industries, ranging from information storage, nanoelectronics, sensors [4], biomedical [5], wastewater treatment [6] and bioremediation. Recently, researchers have been included carbon capture and storage into it list of application as magnetic nanoparticles have large surface area and great adsorption capacity [7]. Naturally, magnetic materials can be found almost everywhere, from interstellar materials to fishes and other insects that stores them as iron storage protein, ferritin [8]. Magnetic oxides, specifically iron oxides, have several forms to it, which are hematite (α-Fe_2O_3), maghemite (γ-Fe_2O_3), ferrites, wurtzite (FeO), goethite (α-FeOOH) and magnetite (Fe_3O_4).

Among all forms of iron oxides, magnetite possesses interesting properties due to the presence of iron cations in two valence states, Fe^{2+} and Fe^{3+} [9]. Generally, magnetite is ferrimagnetic, or attracted to magnets, at temperature below 858 K [10], which is the Curie temperature. However, it can become superparamagnetic if the particle diameter is below critical range [11]. There are many advantages of being superparamagnetic, one of it is the absence of remnant magnetization.

Thus, making it a profitable study for many researchers. Additionally, its characteristics and physical properties can be tailored according to the desired application. Hence, another reason for its desirability.

Currently, the most prominent application for magnetite is in the medical field, for the use of cancer therapy by magnetic hyperthermia and drug carrying agent. For hyperthermia treatment of, it utilizes the mechanism of remote heating of the iron oxide, to increase the surrounding cells temperature, which maximizes therapeutic effect of the region [12], [13]. This thermal effect can be activated through exposure of alternating magnetic fields, which aligns the particle domains through Néel and Brownian Relaxation [14]. The high-speed spinning motion of the domains produced friction, and in turn, increased the temperature of the particle, hence, the term, remote heating.

The same concept is applied to the current study of carbon capture using temperature swing adsorption (TSA), as conventional temperature swing adsorption process is energy intensive. In fact, its thermal energy consumption for a TSA installation can reach up to 24.45 – 48.84 MJ/kg for a 500 MWe coal-fired power plant [15]. Hence, one of the proposed methods is by employing the self-heating mechanism of iron oxide nanoparticles in the adsorbent used in the column [16]. Consequently, external heat sources such as hot gas or steam are no longer required during purging stage. The use external heat sources can be replaced with an induction system that will activate the magnetite particles, which in turn, generate heat through its Brownian-Néel movement. Consequently, the heat generated will instigate the desorption process and efficiently take over the role of hot gas or steam.

Numerous methods were developed to synthesize magnetite, but it all started with an iron source, such as iron salts, ferrous chloride ($FeCl_2$) or ferric chloride ($FeCl_3$). There are the base-up techniques like biosynthesis, chemical vapour deposition (CVD), pyrolysis, sol-gel, and hydrothermal heating [17], [18]. There are also hierarchical methods such as ball milling, thermal decay, laser removal, sputtering and nanolithography. Each method comes with its own set of complexities and boundaries. Most mechanical methods listed under hierarchical techniques lacks accuracy and consistency [19], [20]. Meanwhile, base-up techniques provide high crystallinity and consistency, but requires extended reaction time [21].

The aim for this study was to synthesize a magnetic material that is suitable to be embedded in metal organic framework (MOF) adsorbent. To optimize the material, several goals were set, which were (1) high magnetization saturation, (2) uniform spherical shape, (3) small size dispersion and (4) superparamagnetic. These goals were achieved via careful method selection and controlling reaction variables such as reactants' concentrations and reaction parameters. Method selection was done by considering the resulting product size distribution and shape uniformity, which can be achieved by having control over crystallization steps, from nucleation and growth, until coarsening and agglomeration. Careful precursor selection and reagent ratios may help to control its elemental composition and stoichiometry, while reaction parameters such as temperature and time can help with crystal uniformity. To achieve control over these factors, a solvothermal method was selected for synthesis, with manipulation of amount of precursor and reductant, reaction time and temperature.

At the end of the study, several conclusions can be made with regards to the optimization of the synthesis process of spherical magnetite. Firstly, magnetization saturation of nanoparticles and amount of iron precursor does not have a linear relationship. Similarly, increment of sodium acetate does not cause the same effect to particle size as well. As for reaction parameters, increment of reaction temperature able to synthesize a more stable nanoparticles as observed from their surface charge that was closer to the integer 30. Additionally, it also produced smaller crystalline size as well. On the other hand, increment of reaction time has no significant effect on the yielded nanoparticles except for a more uniform particle size and bigger crystalline size too.

Materials Research Proceedings 53 (2025) 433-443 https://doi.org/10.21741/9781644903575-44

Experimental Section

Materials. The materials used in this study are iron (III) chloride hexahydrate, ($FeCl_3.6H_2O$), ≥99% reagent grade from Sigma-Aldrich, as the iron precursor, trisodium citrate dihydrate ($Na_3Cit.2H_2O$), ≥99% analysis grade, from Merck, as electrostatic stabilizer, ethylene glycol (EG), ≥99% reagent grade from Merck, as solvent and reductant, sodium acetate anhydrous (NaAc), from Merck, as its alkali source, ethanol and deionized water was used during the washing process.

Experimental Procedure. The experiment was adapted from Jia Liu and colleagues, which had synthesized magnetite (Fe_3O_4), aiming for a material that will be bio-compatible and water dispersible [22]. The process started by dissolving 0.003 mol, or 0.75g of the iron precursor, $FeCl_3$, 0.001 mol, or 0.2 g of trisodium citrate in 20 mL of ethylene glycol. The mixture was stirred until all the $FeCl_3$ powder dissolved, and a reddish solution was formed. Then, 0.015 mol, or 1.2 g of sodium acetate was added, which turned the mixture to a foamy light brown color. After the mixture was homogeneous, which was around 30 minutes, the mixture was transferred to a hydrothermal flask and heated in the oven for 250°C for 10 hours. After 10 hours, the blackish product was washed, using ethanol and deionized water consecutively, for 3 cycles. This step is important to remove any unreacted materials and reduce undesired maghemite (γ-Fe_2O_3). To hasten the process, neodymium magnets were applied for magnetic separation. Lastly, the product was dried in a vacuum oven for an overnight at 70°C. The boiling point of ethanol is 70°C, and to ensure that the product had completely dried, the weight was measured for every hour until there was no fluctuation.

Characterization. A list of characterizations was applied. Morphology analysis was conducted using Zeiss Supra 55VP, field emission scanning electron microscopy (FESEM), while Panalytical Xpert3 powder x-ray diffraction (XRD) using CuKα emission with 2θ diffraction angle was deployed to analyze the structure and phase composition. The particle size and uniformity were analyzed using Malvern Zetaziser Nano-ZSP, particle size analyzer. Finally, magnetization curve was generated by using vibrating-sample magnetometer.

Vibrating sample magnetometry (VSM) was applied to sense the magnetization of the sample. This method of measurement involves the vibration of the sample within a uniform magnetic field, G, thus forming an electric current in sensing coils [23]. The produced voltage is in proportion to the magnetic moment of the sample. The result of this analysis will explain the shape and size-dependent properties such as coercivity, saturation magnetization, and remnant magnetization [24]. It can also confirm whether a ferromagnetic material is also superparamagnetic based on its hysteresis curve. For superparamagnetic materials, there should be no hysteresis loop observed with close to zero remnant magnetization [25]. There are several components that make up the magnetization curve, which are the magnetic field (G) axis, flux density (B) or Oe, retentivity (M_R) and coercivity (H_C).

Results and Discussions

The main objective of this experiment is to determine the optimized process of synthesizing Fe_3O_4 that is suitable for a core-shell structure of Mg-MOF-74. Critical characteristics are particle shape and size, magnetization saturation, and particle uniformity. The parameters in sight are the amount of magnetic precursor, $FeCl_3$, in the range of 0.65 g to 0.85 g, amount of sodium acetate, from 0.9 g to 1.2 g, reaction time of 8 hours until 12 hours and lastly, reaction temperature from 180°C to 250°C. From the experiment, it was determined that the combination of 0.75 g of $FeCl_3$ with 1.2 g of NaAc, reacted in 250°C for 12 hours is the optimum method to produce magnetite particles that are uniformly nanosized, with high magnetization and low retentivity.

Figure 1: Spherical magnetite nanoparticle as observe under 100kx magnification using FESEM.

Crystal Morphology. Characteristically, Fig.1 shows that spherical nanoparticles were obtained. Notably, the spheres are not truly smooth as it was a result of agglomerated nanoclusters like Liu et al obtained and this can contribute to its ability for adsorption due to its large surface area [22] Formation of magnetite, Fe_3O_4, was also affirmed using XRD pattern by identifying their peaks. All samples, as shown in Fig. 3 – 5, have peaks located at 30°, 35°, 43°, 53°, 57° and 63°, in line with JCPDS No. 98-003-5001, which is the reference for Fe_3O_4.

Effect of Amount of Iron (III) Chloride. To produce a ferrimagnetic material, a high anisotropic magnetic nanoparticle must be introduced [26]. For this reaction, the source of magnetic material will be the precursor, iron (III) chloride, that is initially paramagnetic. As mentioned, a strict control over the reaction kinetics enables control over the particle size and crystallinity. Additionally, solvothermal method relies on the stoichiometric interaction between iron precursors and its reducing agent, thus, tuning the concentration of the mixture may enable control over the nucleation rate [27]. There are several routes to change the mixture's concentration, but by changing the amount of Fe^{3+} through manipulation of amount of $FeCl_3$, the amount of nucleation sites also varies accordingly. Apart from particle size, reactant concentrations also affect the magnetization saturation, M_S, as well.

To affirm this, the weight of iron (III) chloride was manipulated, from 0.65 g to 0.85 g. As observed from the magnetization curve in Figure 2, hysteresis loop is almost nonexistent in all three curves, parallel with the near zero coercive force values of minimum 0.11167 emu/g and maximum of 0.56964 emu/g. The absence of hysteresis loop and small coercive force values showed that all sets produced superparamagnetic nanoparticles, with different M_S. The whole summary of magnetic properties is in Table 1. Based on the result obtained, it shows the ideal amount of iron precursor is 0.75 g as it can produce particles with highest M_s and lowest coercivity and retentivity.

Apart from their magnetic properties, the precursor's concentration also affects the crystallinity of the nanoparticles as shown in Fig 3. and set with 0.75 g of $FeCl_3$ had the steepest peaks. Steep peaks in XRD graph indicate good crystallinity within the crystal lattice. Additionally, by using Scherrer's equation, the crystallite diameter for each sample are 0.11 nm, 0.09 nm, and 0.06 nm, respectively. Considering all the results obtained, it shows that although set with 0.85g of $FeCl_3$ produced the smallest crystal, it does not have the highest M_S and has the highest value of coercivity as well. Hence, the optimum amount of iron precursor is 0.75 g.

$$D = \frac{k\lambda}{\beta \cos\theta} \quad (1)$$

Equation 1 is Scherrer's equation, where D is the average crystallite size in nm, k is the Scherrer's constant, λ is the x-ray wavelength for CuKα in Å, β is the peak broadening in radian and θ is Bragg's angle.

Effect of Sodium Acetate. Another method of controlling the particle diameter is by controlling sodium acetate's concentration in the reaction mixture. Inferentially, more iron (III) hydroxide was formed if the amount of sodium acetate increases [27]. When reaction in high temperature occurs, iron (II) hydroxide was produced from the reduction with ethylene glycol, which will react with iron (III) hydroxides to produce magnetite clusters. With increment of sodium acetate, more iron (III) hydroxide was produced, thus increasing the nucleation and growth rate [28]. However, there is a critical point where particle size reduces, which is the point where Fe^{3+} ions are in excess and generate a dielectric double-layer around the particle, thus restricts further growth [29]. Further excess in iron (III) hydroxides will increase the alkalinity of the reaction that promotes hydrolysis and thus, increases the crystal growth rate. Finally, the particle size takes a turn, and expands.

Table 1: Magnetic properties of samples containing different amount of $FeCl_3$.

Amount of $FeCl_3$ (g)	Magnetization (emu/g)	Coercivity (G)	Retentivity (emu/g)
0.65	47.83	2.5757	0.2581
0.75	50.413	0.98734	0.11167
0.85	47.085	5.4158	0.56964

It was assumed that the crystal formation consists of clusters of agglomerated nanosized primary particles that is caused by surface charge [30], [31], [32]. Surface charge is indicated by zeta potential, which is the electrical potential in the interfacial double layer. It also indicates particle stability or their tendency to agglomerate. Hence, particle size and zeta potential go relatively linear as shown in Table 2. Based on Table 2, there was a reduction in particle size as the amount of NaAc increases. This shows that 1.2 g of NaAc can limit crystal growth. However, it seems to increase afterwards when NaAc is in excess, which also causes a rise in its size dispersity. Overall, 1.2 g of sodium acetate is sufficient to control the growth and stability of nanoparticles.

Effect of Reaction Temperature. In this experiment, three reaction temperatures were compared, 180°C, 200°C and 250°C. Complexation reaction is a temperature-dependent process, where temperature differences able to affect several properties such as particle size, crystallinity, and stability [4]. Based on diffraction analysis, it was known that higher temperature able to produce better crystals from narrow peaks at 250°C, and this observation is supported by the calculated crystalline size of 0.44 nm, which is the smallest compared to the other temperatures. In the opposing side, reaction produced at 180°C did not produce any magnetic material at all, as there are no peaks observed from XRD data.

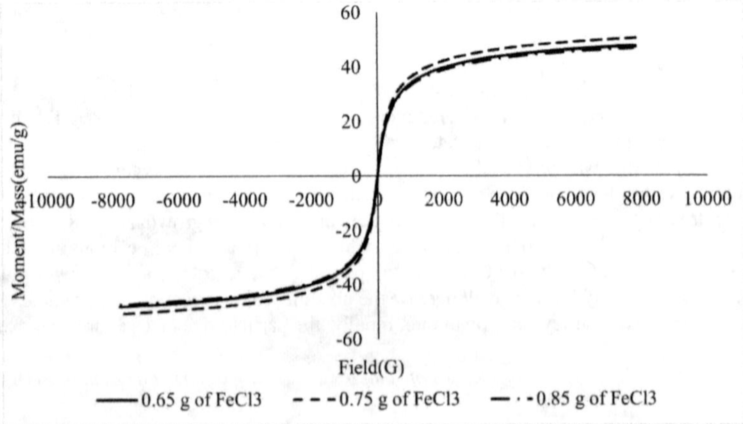

Figure 2: Magnetization curves from three samples with varying amount of

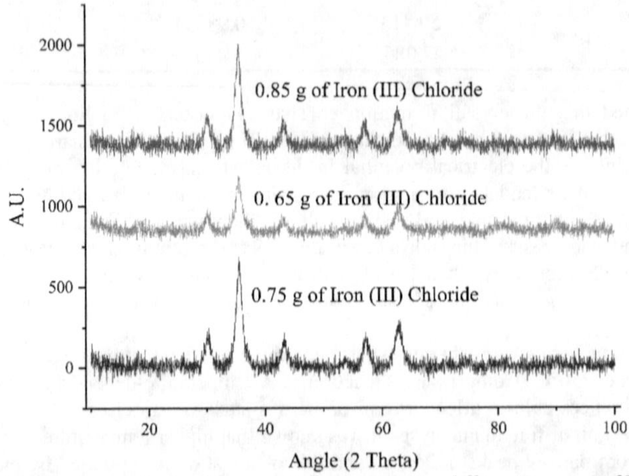

Figure 3: XRD patterns obtained from samples containing different amount of $FeCl_3$. 0.75 g of iron (III) chloride gave the most intense peaks compared to the others. The peaks are located at 30.2°, 35.5°, 43.2°, 53.6°, 57.1° and 63°, aligning with JCPDS No. 98-003-5001.

Table 2: Particle size analysis and surface energy from samples containing different amount of sodium acetate.

Amount of Sodium Acetate (g)	Particle Size (nm)	Zeta Potential (mV)	Polydispersity Index
0.9	289.7	-39.4	0.142
1.2	284.0	-32.0	0.178
1.5	335.7	-33.2	0.208

On the other hand, the particle size analysis shows fluctuations across the range. However, by comparing the stability of the particle through zeta analysis, reaction temperature of 250°C produced the most stable synthesis. Hence, it can be concluded that reaction temperature of 250°C able to promote good crystal formation with great stability. These findings are supported by a study made by Ajibade and his colleagues, where the magnetite quantum dots prepared in 250°C had the smallest mean diameter of 3.25 nm, compared to the particle yielded when the temperature was 165°C that had a diameter of 3.90 nm [33].

Effect of Reaction Time. As previously elaborated, magnetite is a cluster of agglomerated nanosized primary particles and with extended time, the agglomeration period also continues, hence forming a larger cluster of nanoparticles. These can be seen from the increment of particle diameter from 8 hour to 10 hour, when ~50% increment was observed as it went from 597.7 nm to 1140.0 nm. In fact, the dispersity is also close to 1.0, which indicates active agglomeration is taking place during the 10th hour. However, the dispersity reduces during the 12th hour from 0.895 to 0.423, because of other forces affecting the surface charges, such as the dipole-dipole interaction and Van der Waals force, that constricts the size of nanoclusters [34]. The effect from these forces had changed the particle size from 1140 nm to 1059 nm.

Table 3: Summary of effects of reaction parameters towards particle size and surface charge.

No	Temperature (°C)	Time (hour)	Particle Size (nm)	Crystalline Size (nm)
1	180	10	13030	4.0
2	200	10	2612	0.9
3	250	10	5764	0.4
4	200	8	597.7	1.3
5	200	12	1059	1.3

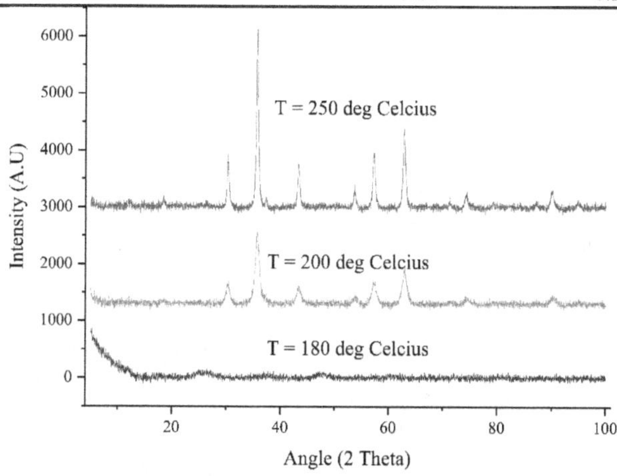

Figure 4: XRD patterns from samples synthesized in varying temperature. Reactions conducted above the boiling point of ethylene glycol able to produce peaks, while reaction temperature of 180°C did not displays any peak. 30.2°, 35.5°, 43.2°, 53.6°, 57.1° and 63°, aligning with JCPDS No. 98-007-7592.

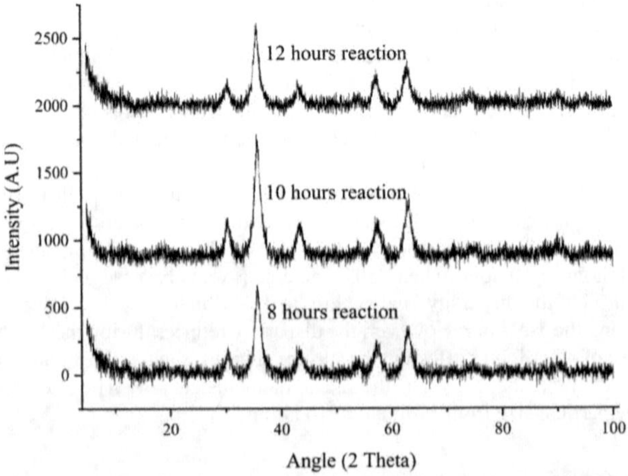

Figure 5: XRD patterns from samples synthesized in varying reaction time, which shows similar peaks for all three sets. The peaks' locations are at, 30.2°, 35.5°, 43.2°, 53.6°, 57.1° and 63°, aligning with JCPDS No. 98-015-8745.

Conclusion

By using iron (III) chloride as the precursor, and sodium acetate as alkali source, spherical magnetic nanoparticles was synthesized. Several conclusions were established at the end of the parametric study. It was found that magnetic saturation of the yield did not have a linear relationship with the amount of iron (III) chloride in the reaction mixture, and the amount that achieved the highest magnetic saturation at 50.412 emu/g was 0.003 mol of $FeCl_3$. Additionally, sodium acetate can be used to control the particle diameter as it controls the formation of iron (III) hydroxide during complexation reaction. The optimum amount of sodium acetate was found to be 0.015 mol. As for the reaction process parameters, a more stable particles were produced at 250°C, as compared to their surfaces charges from particles synthesized at 200°. Meanwhile, reaction temperature of 180°C was identified as threshold temperature as it does not yield any desired product. Contrastingly, extended reaction time did not pose any significant effect on the yield, except for a more uniform particle size but bigger crystalline size. In the final analysis, a tailored superparamagnetic spherical nanoparticle is synthesized through one-pot solvothermal method, ready for the next step of embedment in a core-shell magnetic framework composite structure.

References

[1] A. Behera, S. S. Mohapatra, and D. K. Verma, "Nanomaterials: Fundamental Principle and Applications," in Nanotechnology and Nanomaterial Applications in Food, Health, and Biomedical Sciences, Apple Academic Press, 2019, pp. 163–194. https://doi.org/10.1201/9780429425660-4.

[2] N. Barroso et al., "Magnetic sustentation as an adsorption characterization technique for paramagnetic metal-organic frameworks," Commun Chem, vol. 6, no. 1, Dec. 2023. https://doi.org/10.1038/s42004-022-00799-w.

[3] X. M. Liu and J. K. Kim, "Solvothermal synthesis and magnetic properties of magnetite nanoplatelets," Mater Lett, vol. 63, no. 3–4, pp. 428–430, Feb. 2009. https://doi.org/10.1016/j.matlet.2008.11.001.

[4] E. Oddo, R. M. Pesce, M. Derudi, and L. Magagnin, "Amino-functionalized magnetic nanoparticles for CO2 capture," Int J Smart Nano Mater, vol. 12, no. 4, pp. 472–490, 2021. https://doi.org/10.1080/19475411.2021.1987350.

[5] L. S. Ganapathe, M. A. Mohamed, R. M. Yunus, and D. D. Berhanuddin, "Magnetite (Fe3O4) nanoparticles in biomedical application: From synthesis to surface functionalisation," Magnetochemistry, vol. 6, no. 4. MDPI AG, pp. 1–35, Dec. 01, 2020. https://doi.org/10.3390/magnetochemistry6040068.

[6] J. S. T. Hernandez et al., "Characterization of Fe3O4 Nanoparticles for Applications in Catalytic Activity in the Adsorption/Degradation of Methylene Blue and Esterification," Molecules, vol. 27, no. 24, Dec. 2022. https://doi.org/10.3390/molecules27248976.

[7] H. ; Pashaei, Ghaemi, and Ahad, "Review of CO2 Capture Using Absorption and Adsorption Technologies," 2022.

[8] I. Castellanos-Rubio et al., "A Milestone in the Chemical Synthesis of Fe3O4 Nanoparticles: Unreported Bulklike Properties Lead to a Remarkable Magnetic Hyperthermia," Chemistry of Materials, vol. 33, no. 22, pp. 8693–8704, Nov. 2021. https://doi.org/10.1021/acs.chemmater.1c02654.

[9] M. Karimi, M. Shirzad, J. A. C. Silva, and A. E. Rodrigues, "Carbon dioxide separation and capture by adsorption: a review," Environmental Chemistry Letters, vol. 21, no. 4. Springer Science and Business Media Deutschland GmbH, pp. 2041–2084, Aug. 01, 2023. https://doi.org/10.1007/s10311-023-01589-z.

[10] D. Digitalcommons@umaine and N. Carter, "Physical Properties of Iron Oxide Nanoparticles," 2015. [Online]. Available: https://digitalcommons.library.umaine.edu/honors

[11] M. Rukhsar, Z. Ahmad, A. Rauf, H. Zeb, M. Ur-Rehman, and H. A. Hemeg, "An Overview of Iron Oxide (Fe3O4) Nanoparticles: From Synthetic Strategies, Characterization to Antibacterial and Anticancer Applications," Crystals, vol. 12, no. 12. MDPI, Dec. 01, 2022. https://doi.org/10.3390/cryst12121809.

[12] A. Włodarczyk, S. Gorgoń, A. Radoń, and K. Bajdak-Rusinek, "Magnetite Nanoparticles in Magnetic Hyperthermia and Cancer Therapies: Challenges and Perspectives," Nanomaterials, vol. 12, no. 11. MDPI, Jun. 01, 2022. https://doi.org/10.3390/nano12111807.

[13] M. R. Ghazanfari, M. Kashefi, S. F. Shams, and M. R. Jaafari, "Perspective of Fe3O4 Nanoparticles Role in Biomedical Applications," Biochemistry Research International, vol. 2016. Hindawi Limited, 2016. https://doi.org/10.1155/2016/7840161.

[14] A. E. Deatsch and B. A. Evans, "Heating efficiency in magnetic nanoparticle hyperthermia," Journal of Magnetism and Magnetic Materials, vol. 354. pp. 163–172, 2014. https://doi.org/10.1016/j.jmmm.2013.11.006.

[15] S. Li et al., "Energy dissipation evaluation of temperature swing adsorption (TSA) cycle based on thermodynamic entropy insights," Sci Rep, vol. 9, no. 1, Dec. 2019. https://doi.org/10.1038/s41598-019-53398-6.

[16] M. Gholami, B. Verougstraete, R. Vanoudenhoven, G. V. Baron, T. Van Assche, and J. F. M. Denayer, "Induction heating as an alternative electrified heating method for carbon capture process," Chemical Engineering Journal, vol. 431, Mar. 2022. https://doi.org/10.1016/j.cej.2021.133380.

[17] A. Ali et al., "Synthesis, characterization, applications, and challenges of iron oxide nanoparticles," Nanotechnology, Science and Applications, vol. 9. Dove Medical Press Ltd, pp. 49–67, Aug. 19, 2016. https://doi.org/10.2147/NSA.S99986.

[18] M. A. Mikhailova et al., "SUPPLEMENTARY INFORMATION Revisiting Syntheses of Fe 3 O 4 Nanoparticles in Water and Lower Alcohols for Memristive Application," 2021.

[19] M. Pashangpour and A. A. Peyghan, "Adsorption of carbon monoxide on the pristine, B- and Al-doped C3N nanosheets," J Mol Model, vol. 21, no. 5, Apr. 2015. https://doi.org/10.1007/s00894-015-2648-7.

[20] V. Phouthavong et al., "Magnetic Adsorbents for Wastewater Treatment: Advancements in Their Synthesis Methods," Materials, vol. 15, no. 3. MDPI, Feb. 01, 2022. https://doi.org/10.3390/ma15031053.

[21] W. Lei et al., "Synthesis and magnetic properties of octahedral Fe3O4 via a one-pot hydrothermal route," Physics Letters, Section A: General, Atomic and Solid State Physics, vol. 381, no. 4. Elsevier B.V., pp. 314–318, Jan. 30, 2017. https://doi.org/10.1016/j.physleta.2016.09.018.

[22] J. Liu et al., "Highly water-dispersible biocompatible magnetite particles with low cytotoxicity stabilized by citrate groups," Angewandte Chemie - International Edition, vol. 48, no. 32, pp. 5875–5879, Jul. 2009. https://doi.org/10.1002/anie.200901566.

[23] J. Ren, Z. Zhu, Y. Qiu, F. Yu, J. Ma, and J. Zhao, "Magnetic field assisted adsorption of pollutants from an aqueous solution: A review," Journal of Hazardous Materials, vol. 408. Elsevier B.V., Apr. 15, 2021. https://doi.org/10.1016/j.jhazmat.2020.124846.

[24] R. Azimikia, H. Mazaheri, A. H. Joshaghani, and E. Joudaki, "An experimental study of acid gas absorption by method hybrid nanofluid spraying in the presence of continuous and alternating magnetic fields," Journal of Materials Research and Technology, vol. 13, pp. 919–932, Jul. 2021. https://doi.org/10.1016/j.jmrt.2021.05.013.

[25] P. Nanta, K. Kasemwong, and W. Skolpap, "Isotherm and kinetic modeling on superparamagnetic nanoparticles adsorption of polysaccharide," J Environ Chem Eng, vol. 6, no. 1, pp. 794–802, Feb. 2018. https://doi.org/10.1016/j.jece.2017.12.063.

[26] M. D. Nguyen, H. V. Tran, S. Xu, and T. R. Lee, "Fe3O4 nanoparticles: Structures, synthesis, magnetic properties, surface functionalization, and emerging applications," Applied Sciences (Switzerland), vol. 11, no. 23. MDPI, Dec. 01, 2021. https://doi.org/10.3390/app112311301.

[27] S. I. U. Madrid, U. Pal, and F. S.-D. Jesus, "Controlling size and magnetic properties of Fe3O4 clusters in solvothermal process," Adv Nano Res, vol. 2, no. 4, pp. 187–198, Dec. 2014. https://doi.org/10.12989/anr.2014.2.4.187.

[28] M. Kalubowilage, K. Janik, and S. H. Bossmann, "Magnetic nanomaterials for magnetically-aided drug delivery and hyperthermia," Applied Sciences (Switzerland), vol. 9, no. 14. MDPI AG, Jul. 01, 2019. https://doi.org/10.3390/app9142927.

[29] S. E. Favela-Camacho, E. J. Samaniego-Benítez, A. Godínez-García, L. M. Avilés-Arellano, and J. F. Pérez-Robles, "How to decrease the agglomeration of magnetite nanoparticles and increase their stability using surface properties," Colloids Surf A Physicochem Eng Asp, vol. 574, pp. 29–35, Aug. 2019. https://doi.org/10.1016/j.colsurfa.2019.04.016.

[30] Y. V. Kolen'Ko et al., "Large-scale synthesis of colloidal Fe3O4 nanoparticles exhibiting high heating efficiency in magnetic hyperthermia," Journal of Physical Chemistry C, vol. 118, no. 16, pp. 8691–8701, Apr. 2014. https://doi.org/10.1021/jp500816u.

[31] W. Li, J. Wu, S. S. Lee, and J. D. Fortner, "Surface tunable magnetic nano-sorbents for carbon dioxide sorption and separation," Chemical Engineering Journal, vol. 313, pp. 1160–1167, 2017. https://doi.org/10.1016/j.cej.2016.11.013.

[32] Y. Wei, B. Han, X. Hu, Y. Lin, X. Wang, and X. Deng, "Synthesis of Fe3O4 nanoparticles and their magnetic properties," in Procedia Engineering, Elsevier Ltd, 2012, pp. 632–637. https://doi.org/10.1016/j.proeng.2011.12.498.

[33] P. A. Ajibade, L. P. Mona, and S. P. Songca, "Effects of Reaction Temperature and Time on Structural, Morphological and Optical Properties of Magnetite Quantum Dots Prepared by Solvothermal Method," J Inorg Organomet Polym Mater, 2023. https://doi.org/10.1007/s10904-023-02841-3.

[34] G. C. Hermosa et al., "Investigations of the effective parameters on the synthesis of monodispersed magnetic Fe3O4 by solvothermal method for biomedical applications," AIP Adv, vol. 10, no. 1, Jan. 2020. https://doi.org/10.1063/1.5130063.

Decarbonization Technology: ICDT2024 Materials Research Forum LLC
Materials Research Proceedings 53 (2025) 444-451 https://doi.org/10.21741/9781644903575-45

Carbon capture and storage techniques for CO_2 reduction from energy sector

CHUNG Hong Tan[1,a,*], SIEH Kiong Tiong[1,b,*], ODAY A Ahmed[2], JOHNNY Siaw Paw Koh[1], CHONG Tak Yaw[1], TALAL Yusaf[3,4], KUMARAN Kadirgama[5,6,7]

[1]Institute of Sustainable Energy (ISE), Universiti Tenaga Nasional (UNITEN), Jalan IKRAM-UNITEN, 43000 Kajang, Selangor, Malaysia

[2]Department of Electrical Engineering, University of Technology, Baghdad, Iraq

[3]School of Engineering and Technology, Central Queensland University, Brisbane, QLD 4009, Australia

[4]College of Engineering, Almaaqal University, Basra 61003, Iraq

[5]Advance Nano Coolant-Lubricant (ANCL), College of Engineering, Universiti Malaysia Pahang, Pekan 26600, Malaysia

[6]Faculty of Mechanical and Automotive Engineering Technology, Universiti Malaysia Pahang, Gambang 26300, Malaysia

[7]Centre for Research in Advanced Fluid and Processes, Universiti Malaysia Pahang, Pekan 26600, Malaysia

[a]chung.hong@uniten.edu.my, [b]siehkiong@uniten.edu.my

Keywords: Absorption, Adsorption, Carbon Capture and Storage, CO_2, Membrane

Abstract. With the growing global population comes an ever-increasing need for energy. The current global energy supply is still being satisfied primarily by fossil fuels. The worldwide combustion of fossil fuels has released huge amounts of greenhouse gases (GHGs), particularly carbon dioxide (CO_2), into the atmosphere. To keep global mean temperature rise below 1.5°C, GHG emissions must be minimized and become net zero by 2050. One of the best methods of reducing GHG emissions is by capturing CO_2 using carbon capture and storage (CCS) technology from industries and power plants. Three commonly utilized CCS processes are absorption, adsorption, and membrane techniques. Absorption is the most established technique, but suffers from degradation of equipment due to corrosive absorbents and costly absorbent regeneration. Meanwhile, adsorption does not perform well in large-scale applications because the adsorbents have relatively lesser capacity and capability to seperate CO_2 than absorption technique. Compared to absorption and adsorption, membrane technique requires lower energy to operate, but becomes inefficient with low partial pressures of CO_2. In this review, the mechanisms of CO_2 capture using absorption, adsorption, and membrane techniques as well as their pros and cons are discussed in detail.

Introduction

Since the Industrial Revolution, human civilization and technology have expanded rapidly. With the swift advances in technological capabilities comes the demand for massive amounts of power. Currently, the global demand for power is still being satisfied primarily by burning fossil fuels. The burning of fossil fuels on a global scale has released huge amounts of greenhouse gases (GHGs), particularly carbon dioxide (CO_2), into the atmosphere. The volume of CO_2 emitted by the energy sector and all other industries that utilize fossil fuels account for roughly 65% of the global GHG emissions [1]. As high concentrations of atmospheric CO_2 have led to global warming and undesirable weather conditions across the world, research to reduce CO_2 emissions has become necessary and critical. If the worldwide mean temperature rise is to be kept below 1.5°C as determined by the outcome of COP-21 in Paris in 2015, a significant decrease in total CO_2 emission will need to be achieved, and paving the way for net zero emissions by 2050 [2].

Huge efforts have been expended to research and develop renewable energy as replacements for fossil fuels in the hopes of decreasing greenhouse gas emissions. One promosing pathway for renewable energy is the conversion of captured atmospheric CO_2 into synthetic hydrocarbon fuels such as methane, light gases, and gasoline [3]. In a study by Holechek et al. (2022), the authors demonstrated that it was possible to achieve a complete energy transition from fossil fuels to renewable energy by 2050, if all countries could work together to implement national and international policies such as research on renewables, energy conservation steps, carbon taxes, as well as carbon capture, utilization, and storage (CCUS) techniques [4]. However, in order to develop renewable energy, fossil fuels are still necessary to power the research institutions and manufacturing industries. One of the most effective method to minimize GHG emissions is by capturing CO_2 gas via carbon capture and storage (CCS) techniques. Absorption, adsorption, membrane technique are among the most widely researched CCS techniques [5]. Therefore, in this review, the mechanisms of these three CCS techniques as well as their pros and cons are discussed in detail.

Absorption

The technique of absorbing CO_2 can be broadly categorized into two types, namely chemical absorption and physical absorption. Chemical absorption occurs when CO_2 reacts with solvents to create chemical compounds, whereby the absorbed CO_2 can be separated from the solvent in subsequent processes. Meanwhile, physical absorption occurs when CO_2 does not react with the solvents and is instead contained within the solvents via physical soaking [6]. Chemical absorption of CO_2 is performed in two phases. Firstly, the solvent flows countercurrent against the treated gas stream and absorbs the CO_2 portion of the gas stream. Then, the solvent passes into a stripping column where the CO_2 is separated from the solvent via heating. The regenerated solvent is recycled back into the absorber, while the pure CO_2 accumulated at the top of the stripping column is compressed and stored [7, 8]. The first phase of chemical absorption shows optimum performance at low temperatures and high pressures, whereas the second phase performs better at high temperatures and low pressures [8]. Since chemical absorption of CO_2 occurs at lower pressures, it is a suitable carbon capture technique for post-combustion capture. The commonly used chemical solvents are amine and carbonate solutions [9]. Lu et al. (2023) designed a demonstration project for the installation of a 150 kt/a carbon capture and storage (CCS) system for the Shaanxi Guohua Jinjie Power Plant in China. Coal was the primary fuel source for the power plant, and the CCS system combined the usage of three different CO_2 absorbers, which were compound amine absorber, ionic liquid, and phase change absorber. After introducing three energy-conserving components in the CCS system, which were split flow desorption, interstage cooling, and mechanical vapor recompression heat pump, the final energy requirement for absorbers regeneration was reduced to 2.32 GJ/t CO_2. The proposed CCS system could achieve an adsorbent loss of ≤ 1.0 kg/t CO_2, overall steam requirement of ≤ 1.27 t/t CO_2, and positive internal rate of return when commercial price of CO_2 was above 203.43 Yuan/t. The authors concluded that this proposed project was both technologically and financially viable [10].

Physical absorption of CO_2 is carried out using either organic or inorganic solvents that are inert to CO_2 gas. This technique operates according to Henry's law which states that, at constant temperature, the concentration of dissolved gas in a liquid is directly proportional to the partial pressure of the gas in contact with the liquid [9]. Since physical absorption of CO_2 is dependent on gas pressure, it is a more suitable carbon capture technique when encountering high pressures of CO_2 gas stream [11]. Physical absorption technique is appropriate for pre-combustion capture in an integrated gasification combined cycle (IGCC) power plant. This is because the syngas feed stream in an IGCC is produced at higher pressures, resulting in higher partial pressures of CO_2 gas. The physical solvents are then regenerated and recycled back into the stripping column. One advantage of physical solvents over chemical solvents is that physical solvents require less energy to be regenerated. On the other hand, the main drawback of physical absorption is the need to cool down the gas stream before passing through the stripping column. This is due to the increase in solvent performance at lower temperatures. The cooling of the gas stream results in an overall decrease in plant efficiency [12]. The commonly used and commercially available physical solvents include Purisol, Rectisol, and Selexol [11]. When comparing the different commercial physical solvents, Selexol was observed to have lesser energy requirement and simpler process integration, thereby making it a better option for capturing CO_2 gas [6]. Berstad et al. (2013) simulated the integration of low-temperature carbon capture technique in an IGCC, and compared the performance of two carbon capture methods, which were low-temperature syngas separation and Selexol. It was found that both methods could achieve 80 – 83% CO_2 capture. However, the low-temperature method could also output liquid CO_2 which required lower energy to pressurised compared to the gaseous CO_2 generated from Selexol absorption. This simulation indicated the potential for low-temperature CO_2 capture method to be employed in industrial settings [13]. The main research focus of chemical and physical absorption techniques is the discovery and testing of different solvents for efficient capture at different partial pressures of CO_2 in the gas stream [5]. The advantages and disadvantages of three CCS techniques (absorption, adsorption, and membrane) are compared in Table 1.

Table 1: Advantages and disadvantages of different CO_2 capture techniques [14]

Technique	Mechanism	Advantages	Disadvantages
Absorption (Chemical)	Chemical reaction between CO_2 gas and solvent	• Well-established process • High efficiency with low CO_2 concentration • Suitable for pre- and post-combustion capture	• Expensive cost of operation • Relatively low rates of absorption and desorption • Elevated temperatures needed to regenerate absorbent • Corrosion of equipment by absorbent • Deterioration of absorbent
Absorption (Physical)	CO_2 gas dissolves in solvent	• Well-established process • High efficiency at high pressure and low temperature • Lower cost of solvent • Suitable for post-combustion capture	• Low efficiency at low pressure and high temperature • Elevated temperatures needed to regenerate absorbent • Relatively low selectivity to CO_2 • Loss of absorbent
Adsorption (Chemical)	Reaction between CO_2 gas and adsorbent to create carbonates or bicarbonates	• Can function at elevated temperatures • Large adsorption capacity • Suitable for pre-combustion capture	• Reduced capacity after each cycle • Larger energy requirement than physical adsorbent • Deterioration of adsorbent by thermal and oxidative processes • Sensitive to impurities like NO_x and SO_x
Adsorption (Physical)	Non-reactive bonding between CO_2 gas and adsorbent	• High efficiency at high pressure and low temperature • Lesser production of waste • Suitable for pre-combustion capture	• Needs high pressure for better efficiency • Efficiency reduces with increasing temperature • Relatively low selectivity to CO_2

Materials Research Proceedings 53 (2025) 444-451 https://doi.org/10.21741/9781644903575-45

			• Water vapor competes with CO_2 for adsorption • Elevated temperatures needed to regenerate adsorbent
Membrane	Permeability of CO_2 gas from feed stream into permeate stream	• Ease of process operation and management • Cheaper cost of operation • Suitable for pre- and post-combustion capture	• Expensive cost of membrane fabrication • Liquid retention in membrane decreases CO_2 mass transfer • Relatively low selectivity to CO_2

Adsorption

Adsorption refers to process of adhering a compound onto a solid surface and extracting the compound from its original mixture. Unlike absorption method where CO_2 bonds with the whole solvent, in adsorption, the CO_2 adheres to the surface of the adsorbent via the creation of chemical or physical bonds. The intermolecular forces between the CO_2 molecules and the surface molecules of the adsorbent allow the extraction of CO_2 from the incoming gas streams [15]. Key parameters of the adsorption process, which include type of adsorbent material, adsorbent pore size, operating temperature and pressure of the separation unit, as well as the surface force of adsorbent, strongly determines the single layer or multiple layer adsorption of CO_2 gas [16]. An adsorption process generally goes through the following steps: (1) adsorbent is loaded into stripping column; (2) CO_2-rich gas stream flows through the column; (3) CO_2 molecules attach to surface of adsorbent until the adsorbent becomes saturated; (4) the saturated adsorbent is extracted and regenerated, and the desorbed CO_2 is pressurised and stored; and (5) the regenerated adsorbent is recycled back into the column and the adsorption cycle repeats [15].

There are four common adsorbent regeneration methods utilized for CO_2 adsorption with single bed configuration, namely temperature swing adsorption (TSA), pressure swing adsorption (PSA), vacuum swing adsorption (VSA), and electrical swing adsorption (ESA). In TSA, the saturated adsorbent is regenerated via heating to a higher temperature where the chemical and physical bonds between the CO_2 molecules and adsorbent surface molecules are broken. The freed CO_2 can then be accumulated and stored. However, the heating requirement for desorption and cooldown phase before reuse of the adsorbent increase both the cost and time consumption for the TSA process [9, 16]. The extended time needed for adsorbent regeneration in TSA can be improved by using the ESA method. In ESA, the saturated adsorbent is heated up by passing electric current with low voltage across the adsorbent. Although this method can regenerate the adsorbent more swiftly, it also uses up high-grade energy (in this case, electricity) rather than low-grade energy (like heat) as utilized in TSA [17].

In PSA, a high pressure CO_2-rich gas stream is pumped into the stripping column. The high pressure promotes adsorption of CO_2 and the resulting CO_2-lean gas stream exits the top of the column. When the adsorbent becomes saturated, the column pressure is lowered to encourage desorption of CO_2 from the adsorbent. The desorbed CO_2 is then purged from the column and undergo subsequent collection and storage steps [18, 19]. On the other hand, the VSA method is used instead when the feed gas stream is near atmospheric pressure or slightly pressurized (up to a maximum of 1.5 atm). The adsorbed CO_2 is later collected by applying vaccum condition in the column [9, 20].

When the concentration of CO_2 in the feed gas stream is high, PSA is more preferable. Meanwhile, when CO_2 concentration in the feed stream is low, TSA is more favorable. This is because PSA requires longer operating time when dealing with feed stream containing lower partial pressure of CO_2 [21]. Compared to absorption technique, the advantages of adsorption technique include high adsorption of CO_2 at normal temperature and pressure, high stability of adsorbents, lower cost of adsorbent regeneration, and lesser energy consumption [22]. One of the most critical parameter that determines the performance and configuration of adsorption process is the choice and characteristics of the adsorbent. A variety of materials have received intense research due to their porous structures and favorable interactions with CO_2 molecules, such as activated carbons, zeolites, molecular sieves, metal-organic frameworks, and hydrotalcites. Therefore, the primary research focus in adsorption technique is the development of novel, efficient, and more selective adsorbents for capturing CO_2 [23].

Membrane technique

When compared to absorption and adsorption techniques, membrane technique for CO_2 capture typically requires lower energy input and more environmentally friendly as less harsh chemicals are needed to construct the membranes. These benefits has attracted huge efforts into researching novel and viable membranes for CO_2 capture [24, 25]. Membranes are selective physical barrier that can facilitate the transfer or blockage of materials between mediums, as well as separation of compounds in gas and liquid streams [15]. Membranes can be created from inorganic or organic materials. When a CO_2-rich feed gas stream enters the membrane stripping column, the CO_2 molecules diffuse through the membrane into a capturing stream known as a permeate stream. The permeate stream, initially low in CO_2 concentration, flows countercurrent to the feed gas stream and absorbs the CO_2 molecules that diffused through the membrane. The permeate stream exits the column with high CO2 concentration, whereas the feed stream exits with low CO_2 concentration and also known as the retentate stream [26].

Three common configurations for using membranes are flat sheet, spiral wound, and hollow fiber arrangment. The selectivity of membranes can be altered to better suit the types of gases present in the feed stream [27]. For carbon capture, membrane techniques are broadly categorized into two types, gas separation membrane (GSM) method and gas absorption membrane (GAM) method. The GSM method normally employs cylindrical membranes to segregate CO_2. The feed stream is passed through the inner region of the membrane at high pressure, where CO_2 diffuses out of the membrane and into the surrounding permeate stream with lower pressure. On the other hand, the GAM method employs a microporous membrane to isolate CO_2 from the feed stream. One advantage of GAM over GSM is that the equipment size of GAM is more compact [5].

When compared to traditional gas absorption technique using packed column, the benefits of CO_2 capture using membrane technique include: (1) minimization of operation issues such as flooding, channeling, foaming, and entrainment; (2) high rates of CO_2 separation due to larger surface area of membranes; and (3) ease of controlling gas and liquid flow rates [28, 29]. However, the primary drawback of membranes is that their efficiency decreases when CO_2 concentration decreases. In this regard, membranes are highly inflexible and become uneconomical when CO_2 concentration dips below 20% [30]. This drawback indicates that the membrane technique is not suited for post-combustion CCS system. Other drawbacks of membranes include limited lifespan which necessitates periodic membrane replacements, and increased resistance in mass transfer of membranes during long continuous operations. Ideally, all the pores in the membrane ought to be fully occupied by the gas stream so that CO_2 diffusion into the permeate stream can be maximized. But in real life operations, this condition is not always feasible. The liquid stream tends to occupy the membrane pores and reduces CO_2 diffusion rates across the membrane. This phenomenon is termed membrane wetting. During extended operations, the membrane risk becoming partially or fully wetted, which significantly impacts the membrane efficiency [27, 30].

Membranes are broadly categorized into two types, namely organic (polymer) and inorganic (ceramic) membranes. CO_2 capture via polymer membranes are more commonly used in industries, but the low CO_2 selectivity of polymer membranes hinders their large-scale applications. Meanwhile, ceramic membranes display better CO_2 separation, higher stability, and greater resistance to corrosion, resulting in more robust membranes that are suitable for large-scale utilization. However, ceramic membranes are not appealing to industries due to the high cost of materials and fabrication of the membranes [31, 32]. Thus, the main research area for membrane technique in capturing CO_2 is to improve the selectivity of membrane materials toward CO_2 and increase the efficiency of CO_2 separation [32].

Conclusion

Fossil fuels still remain as the main power source in the current world even though combustion of fossil fuels is harming the environment. For existing power plants that utilize fossil fuels, post-combustion CCS techniques (such as absorption and membrane) are more beneficial as these techniques can be installed without expensive modifications to existing plants. For new and/or existing power plants that have pressurized CO_2 feed streams (like an integrated gasification combined cycle power plant), pre-combustion techniques (which include absorption, adsorption, and membrane) are suitable for capturing CO_2. The main disadvantage of CCS techniques is the rising cost of electricity production due to the need for additional energy to capture CO_2 as well as extra energy and cost to regenerate or replace the medium for CO_2 capture (absorbent, adsorbent, and membrane). When comparing the three CCS techniques (absorption, adsorption, and membrane), the absorption technique is the most established process, but its drawbacks include degradation of equipment due to corrosive nature of absorbents and expensive cost of absorbent regeneration. The main research focus of absorption technique is to develop different and efficient solvents for various concentrations of CO_2 in the feed stream. Meanwhile, the adsorption technique suffers at large-scale application because the adsorbents have relatively lesser capacity and ability to remove CO_2 than absorption technique. This leads to numerous research on developing novel, efficient, and more selective adsorbents toward CO_2 capture. Compared to absorption and adsorption, membrane technique requires lower energy input to function, but becomes more inefficient with decreasing partial pressures of CO_2. When CO_2 concentration dips below 20%, membrane technique become uneconomical. Thus, the primary research area for membrane technique is to enhance the selectivity of membrane materials toward CO_2 and improve the efficiency of CO_2 separation. Despite the drawbacks of each CCS technique, CCS technology still remains as one of the best method of minimizing anthropogenic CO_2 emissions into the atmosphere. Numerous research should continue to examine and develop more efficient CCS technologies in order to curb global warming and achieve net zero emissions by 2050.

Acknowledgements

The authors would like to acknowledge financial support from AAIBE Chair of Renewable Energy (ChRE), grant number 201801 KETTHA, Universiti Tenaga Nasional, Kajang, Selangor, Malaysia.

References

[1] US Environmental Protection Agency (EPA). Global Greenhouse Gas Emissions Data. 2024 [29 March 2024]; Available from: https://www.epa.gov/ghgemissions/global-greenhouse-gas-emissions-data.

[2] J. Rogelj, et al., Paris Agreement climate proposals need a boost to keep warming well below 2 C. Nature. 534 (2016) 631-639. https://doi.org/10.1038/nature18307.

[3] C.G. Okoye-Chine, et al., Conversion of carbon dioxide into fuels—A review. Journal of CO_2 Utilization. 62 (2022) 102099. https://doi.org/10.1016/j.jcou.2022.102099.

[4] J.L. Holechek, et al., A global assessment: can renewable energy replace fossil fuels by 2050? Sustainability. 14 (2022) 4792. https://doi.org/10.3390/su14084792.

[5] N.S. Sifat, Y. Haseli, A critical review of CO_2 capture technologies and prospects for clean power generation. Energies. 12 (2019) 4143. https://doi.org/10.3390/en12214143.

[6] A. Padurean, C.-C. Cormos, P.-S. Agachi, Pre-combustion carbon dioxide capture by gas–liquid absorption for Integrated Gasification Combined Cycle power plants. International Journal of Greenhouse Gas Control. 7 (2012) 1-11. https://doi.org/10.1016/j.ijggc.2011.12.007.

[7] C.-H. Yu, C.-H. Huang, C.-S. Tan, A Review of CO_2 Capture by Absorption and Adsorption. Aerosol and Air Quality Research. 12 (2012) 745-769. https://doi.org/10.4209/aaqr.2012.05.0132.

[8] W.L. Theo, et al., Review of pre-combustion capture and ionic liquid in carbon capture and storage. Applied Energy. 183 (2016) 1633-1663. https://doi.org/10.1016/j.apenergy.2016.09.103.

[9] S. Rackley, Absorption capture systems, Carbon Capture and Storage, Elsevier: Amsterdam, The Netherlands, 2017, pp. 115-149.

[10] S. Lu, et al., Research and design experience of a 150 kt/a CO_2 capture and purification project in the Shaanxi Guohua Jinjie power plant. Separation and Purification Technology. 320 (2023) 124089. https://doi.org/10.1016/j.seppur.2023.124089.

[11] D. Jansen, et al., Pre-combustion CO2 capture. International Journal of Greenhouse Gas Control. 40 (2015) 167-187. https://doi.org/10.1016/j.ijggc.2015.05.028.

[12] J.D. Figueroa, et al., Advances in CO_2 capture technology—The U.S. Department of Energy's Carbon Sequestration Program. International Journal of Greenhouse Gas Control. 2 (2008) 9-20. https://doi.org/10.1016/S1750-5836(07)00094-1.

[13] D. Berstad, R. Anantharaman, P. Nekså, Low-temperature CCS from an IGCC Power Plant and Comparison with Physical Solvents. Energy Procedia. 37 (2013) 2204-2211. https://doi.org/10.1016/j.egypro.2013.06.100.

[14] J. Singh, D.W. Dhar, Overview of Carbon Capture Technology: Microalgal Biorefinery Concept and State-of-the-Art. Frontiers in Marine Science. 6 (2019) 29. https://doi.org/10.3389/fmars.2019.00029.

[15] M.K. Mondal, H.K. Balsora, P. Varshney, Progress and trends in CO_2 capture/separation technologies: A review. Energy. 46 (2012) 431-441. https://doi.org/10.1016/j.energy.2012.08.006.

[16] A. Meisen, X. Shuai, Research and development issues in CO_2 capture. Energy Conversion and Management. 38 (1997) S37-S42. https://doi.org/10.1016/S0196-8904(96)00242-7.

[17] S. Lillia, et al., A comprehensive modeling of the hybrid temperature electric swing adsorption process for CO_2 capture. International Journal of Greenhouse Gas Control. 74 (2018) 155-173. https://doi.org/10.1016/j.ijggc.2018.04.012.

[18] S. Sircar, Pressure Swing Adsorption. Industrial & Engineering Chemistry Research. 41 (2002) 1389-1392. https://doi.org/10.1021/ie0109758.

[19] C. Voss, Applications of Pressure Swing Adsorption Technology. Adsorption. 11 (2005) 527-529. https://doi.org/10.1007/s10450-005-5979-3.

[20] J. Zhang, P.A. Webley, P. Xiao, Effect of process parameters on power requirements of vacuum swing adsorption technology for CO2 capture from flue gas. Energy Conversion and Management. 49 (2008) 346-356. https://doi.org/10.1016/j.enconman.2007.06.007.

[21] C.A. Grande, Advances in Pressure Swing Adsorption for Gas Separation. ISRN Chemical Engineering. 2012 (2012) 982934. https://doi.org/10.5402/2012/982934.

[22] N.A. Rashidi, S. Yusup, An overview of activated carbons utilization for the post-combustion carbon dioxide capture. Journal of CO_2 Utilization. 13 (2016) 1-16. https://doi.org/10.1016/j.jcou.2015.11.002.

[23] I. Hinkov, et al., Carbon dioxide capture by adsorption. Journal of Chemical Technology and Metallurgy. 51 (2016) 609-626. https://hal.science/hal-03355471/.

[24] T.C. Merkel, et al., Power plant post-combustion carbon dioxide capture: An opportunity for membranes. Journal of Membrane Science. 359 (2010) 126-139. https://doi.org/10.1016/j.memsci.2009.10.041.

[25] S. Japip, et al., Highly permeable zeolitic imidazolate framework (ZIF)-71 nano-particles enhanced polyimide membranes for gas separation. Journal of Membrane Science. 467 (2014) 162-174. https://doi.org/10.1016/j.memsci.2014.05.025.

[26] R. Khalilpour, et al., Membrane-based carbon capture from flue gas: a review. Journal of Cleaner Production. 103 (2015) 286-300. https://doi.org/10.1016/j.jclepro.2014.10.050.

[27] M.H. Ibrahim, et al., CO_2 Capture Using Hollow Fiber Membranes: A Review of Membrane Wetting. Energy & Fuels. 32 (2018) 963-978. https://doi.org/10.1021/acs.energyfuels.7b03493.

[28] M.H. El-Naas, et al., Evaluation of the removal of CO_2 using membrane contactors: Membrane wettability. Journal of Membrane Science. 350 (2010) 410-416. https://doi.org/10.1016/j.memsci.2010.01.018.

[29] S. Rajabzadeh, et al., Effect of membrane structure on gas absorption performance and long-term stability of membrane contactors. Separation and Purification Technology. 108 (2013) 65-73. https://doi.org/10.1016/j.seppur.2013.01.049.

[30] S. Rostami, P. Keshavarz, S. Raeissi, Experimental study on the effects of an ionic liquid for CO_2 capture using hollow fiber membrane contactors. International Journal of Greenhouse Gas Control. 69 (2018) 1-7. https://doi.org/10.1016/j.ijggc.2017.12.002.

[31] A.A. Olajire, CO_2 capture and separation technologies for end-of-pipe applications – A review. Energy. 35 (2010) 2610-2628. https://doi.org/10.1016/j.energy.2010.02.030.

[32] M. Kárászová, et al., Post-combustion carbon capture by membrane separation, Review. Separation and Purification Technology. 238 (2020) 116448. https://doi.org/10.1016/j.seppur.2019.116448.

Materials Research Proceedings 53 (2025) 452-457 https://doi.org/10.21741/9781644903575-46

Utilization of carbon dioxide in chemical and materials treatment sector

CHUNG Hong Tan[1,a,*], SIEH Kiong Tiong[1,b,*], ODAY A Ahmed[2], JOHNNY Siaw Paw Koh[1], CHONG Tak Yaw[1], TALAL Yusaf[3,4], KUMARAN Kadirgama[5,6,7]

[1]Institute of Sustainable Energy (ISE), Universiti Tenaga Nasional (UNITEN), Jalan IKRAM-UNITEN, 43000 Kajang, Selangor, Malaysia

[2]Department of Electrical Engineering, University of Technology, Baghdad, Iraq

[3]School of Engineering and Technology, Central Queensland University, Brisbane, QLD 4009, Australia

[4]College of Engineering, Almaaqal University, Basra 61003, Iraq

[5]Advance Nano Coolant-Lubricant (ANCL), College of Engineering, Universiti Malaysia Pahang, Pekan 26600, Malaysia

[6]Faculty of Mechanical and Automotive Engineering Technology, Universiti Malaysia Pahang, Gambang 26300, Malaysia

[7]Centre for Research in Advanced Fluid and Processes, Universiti Malaysia Pahang, Pekan 26600, Malaysia

[a]chung.hong@uniten.edu.my, [b]siehkiong@uniten.edu.my

Keywords: Acid Mine Drainage, Alumina, CO_2 Sequestration, CO_2 Utilization, Iron and Steel

Abstract. CO_2 capture and utilization is necessary to reduce the increasing atmospheric CO_2 concentration from the rapid burning of fossil fuels for energy. The high CO_2 concentration in the atmosphere is largely responsible for global warming, which is negatively affecting the environment and health of all living beings. Pathways to utilize CO_2 need to be explored in order to minimize the cost of CO_2 capture. One way to utilize CO_2 is in three sectors involved in the chemical and materials treatment industry, namely iron and steel sector, alumina sector, and acid mine drainage treatment. In these three sectors, CO_2 is less commonly applied but can be employed to potentially enhance their industrial processes. In the three sectors, CO_2 can be used as a neutralizing agent to lower the pH of greatly alkaline products and by-products, or an additive for temperature control and dust reduction of blast furnaces in steel production. This review discusses in detail the roles and mechanisms of CO_2 in these three sectors.

Introduction

With the worldwide increase in greenhouse gas emissions, especially carbon dioxide (CO_2), carbon capture and utilization (CCU) techniques have attracted the attention of many institutions, industries, and governments around the globe. CCU techniques are currently regarded as a necessity to curb the rising atmospheric CO_2 concentration and prevent global warming. Many climate models have predicted that if CO_2 emissions are not drastically reduced, the resulting global warming will have severe effects on the environment and all living beings [1]. If worldwide CO_2 emissions can be reduced to below 5 gigatons of CO_2 per year by 2050, then there exists roughly a 2 in 3 chance of restricting the mean global temperature rise to below 2°C by 2100. The aforementioned CO_2 emissions value is a major reduction compared to current worldwide emissions of above 48 gigatons of CO_2 per year [2, 3].

Since CO_2 capture process consumes a notable amount of capital and energy to operate in industrial settings, it is necessary to implement CO_2 utilization pathways to minimize the cost of capture. Although CO_2 has been utilized in various industrial applications for decades, such as food and beverage sector, urea synthesis, and enhanced oil recovery, it is still important to find other avenues to utilize CO_2 [4]. One area where CO_2 is less commonly applied but can be employed is in the chemical and materials treatment sector, which includes iron and steel manufacturing, alumina production, and rare earth elements recovery from acid mine drainage [1]. In these three sectors, CO_2 can be used directly in the processing steps to improve the output of the products and by-products. In this review, the roles and mechanisms of CO_2 in the three sectors (iron and steel, alumina, and acid mine drainage) are discussed in detail.

Employment of CO_2 in iron and steel sector

The production of steel typically follows these steps: (1) extraction of iron ores from mines; (2) recovery of iron metal from the ores using flotation or magnetic separation; (3) pelletization of iron metal for easy transportation and utilization; (4) reduction of iron pellets into pig iron using a blast furnace; and (5) upgrading of pig iron into steel via a basic oxygen furnace or an electric arc furnace. During the steel production processes, it is important to monitor and control the presence of CO_2. In the iron recovery step, CO_2 is produced when iron ores are reduced to iron metal via Equation 1. CO_2 is also emitted as a byproduct when reducing materials such as coke are burned to obtain carbon monoxide (CO) for the reaction in Equation 1, as shown in Equation 2 [5].

$$Fe_2O_3 + 3CO \rightarrow 2Fe + 3CO_2 \quad (1)$$

$$C + O_2 \rightarrow CO_2 \quad (2)$$

In the pig iron conversion step, external CO_2 can be injected and recirculated inside the blast furnace to regulate the temperature of the tuyere region. This forms a mixing effect on the gas phase within the furnace and promotes the synthesis of CO via appropriate oxidation of coke. This method utilizes about 50 kg of CO_2 per ton of iron. Although the injected CO_2 can be recirculated multiple times in the furnace, but all the CO_2 will still be purged to the atmosphere at the end of the process [5].

In the iron upgrading step, CO_2 can be used to prevent excessive oxidation and control the temperature of molten steel in the basic oxygen furnace. These two functions of CO_2 aid in reducing the formation of dust in the basic oxygen furnace. Dust formation is a critical issue in steel production because this phenomenon decreases the overall yield of steel. In a study by Yi et al. (2009), the effects of CO_2 injection into a basic oxygen furnace for steel production were examined using the facilities at Fujian Sanming Iron and Steel Plant in China. It was observed that the presence of CO_2 in the furnace decreased the formation of dust by 12.5% [6]. Being the current largest country in steel production, China has been researching the applications of CO_2 in many different metallurgical operations since the 1980s [7].

Other than CO_2 recirculation in the furnaces, another method to utilize CO_2 in the iron and steel industry is via steel slag carbonation. Carbonated slag is a practical building material that can be mixed with asphalt and concrete, or used as a coarse aggregate. Meanwhile, fine carbonated slag may find usage as an additive in cement mortar or a conditioner for soil. The applications of carbonated slag contribute to CO_2 sequestration, and based on a study by Myers et al. (2019), this method can potentially sequester up to 268 million tons of CO_2 annually. After considering the industrial conditions for slow solidification of carbonated slag and usage of renewable energy to power the grinding equipment, the total amount of CO_2 sequestered from 2020 to 2100 was estimated to be 26 – 42 gigatons of CO_2. This estimated range would be influenced by the future global appetite for iron and steel [8].

In this first phase of steel production, the iron metal needs to be extracted from the iron ores. The majority of iron ores, in particular hematite ores, are processed using reverse flotation method. During reverse flotation, finely ground iron ores are added into strongly alkaline water. The targeted iron-bearing compounds sink to the bottom, whereas the gangue silica floats to the surface with the help of collector reagents and aeration. This process outputs a slurry of iron ore concentrate that requires dewatering step before further treatment or transportation. The slurry is usually dewatered via filtration [9]. Although it is common practice to perform floation and filtration sequentially, the conditions for both processes are completely different. Flotation requires mild dispersion under strongly alkaline state, whereas filtration requires mild aggregation under acidic state [10]. In an investigation by Kawatra and Claremboux (2019), CO_2 gas was demonstrated to be effective in reducing the pH of iron ore concentrate slurry down to 6.7, and increasing the filtration rates of the treated slurry by up to 23.7%. The CO_2 requirement for this step is roughly 0.54 kg of CO_2 per ton of iron ore. This showed that CO_2 can play an important role in enhancing the filtration step and improving the economics of iron production [9].

Employment of CO_2 in alumina sector

The most frequently utilized ore to create alumina [also known as aluminium oxide (Al_2O_3)] is bauxite. The alumina is then processed to become aluminium metal. The Bayer process is the most established industrial method to process bauxite into alumina. In the Bayer process, bauxite ores are digested in highly alkaline sodium hydroxide solution and the resulting aluminium oxides are isolated from the mixture. The main by-product of the digestion step is solid wastes which are very caustic, usually with pH above 13. The caustic solid wastes is also termed as red mud. Proper disposal of red mud is a very challenging task, but must be carried out to avoid harming the environment and health of surrounding living beings. An incident that highlighted the dangers of red mud is the 2010 Ajka red mud disaster in Hungary, where red mud spillage from a disposal pond caused chemical burns to hundreds of people, several deaths, and huge losses in damage to properties [11].

The highly caustic property of red mud is responsible for most of its risks and dangers. To alleviate this issue, CO_2 could be used to neutralize the high pH of red mud. The main benefit of using CO_2 is its cheap cost and abundance. The CO_2 neutralization method would cause most of the caustic species to be altered to carbonates, and the resulting slurry would still be mildly alkaline due to some caustic species being buried deep within the slurry. Sahu et al. (2010) demonstrated that CO_2 could drop the pH of red mud to less than 7. However, the pH of red mud would jump back up to 9 after a few hours since the alkaline species would leach out from within the slurry. The CO_2 treatment plan for red mud would be to gas the slurry with CO_2 periodically with intervals between 5 – 24 h to ensure the slurry was adequately neutralized [12]. External calcium source like gypsum could also be added to red mud to encourage higher uptake of CO_2. This phenomenon was attributed to the precipitation of $CaCO_3$ which promoted the reaction between CO_2 and red mud [11, 13].

The carbonates produced from CO_2 neutralization of red mud can become a permanent sink for sequestration of CO_2, provided that the carbonates are not disturbed. Considering that global output of red mud is more than 140 million tons and an estimated 83 kg of CO_2 per ton of red mud is needed for the neutralization method, more than 11.6 million tons of CO_2 can be sequestered annually. From existing inventory of red mud, an additional 0.33 gigatons of CO_2 can also be captured [14]. The neutralized red mud can then be further processed or utilized, such as the recovery of rare earth elements often found in large concentrations within red mud slurries [13, 15].

Employment of CO_2 in acid mine drainage

After the ceasing of mining activities, water that has been polluted through contact with the mines may flow into the surrounding water bodies and soil. This polluted water is called abandoned mine drainage, and usually leaches out from coal mines. In regions with histories of mining activities, abandoned mine drainage is a common source of water pollution. There are three types of abandoned mine drainage: (1) acid mine drainage, which is the most frequently encountered water pollution; (2) alkaline mine drainage, which usually follows the presence of calcite or dolomite; and (3) metal mine drainage, which forms when the polluted water contains very high concentration of lead and other heavy metals from the mines. The most common type, acid mine drainage, refers to the creation and leaching of greatly acidified water that contain high concentration of heavy metals. The acidified water is a result of surface water (rainwater, pond water, or snowmelt) chemically reacting with shallow subsurface water that is in contact with rocks comprising sulfur-bearing minerals. This chemical reaction produces sulphuric acid. When the acid touches rocks, heavy metals can be leached from the rocks via chemical reactions with the acid. This leaching event may be significantly boosted by bacterial activity. The final liquid mixture can become extremely toxic, and upon seeping into surface water, groundwater, or soil, the polluted water may cause detrimental effects on the health of humans, animals, and plants [16]. Other than causing health problems, the leached heavy metals often also comprise large amounts of rare earth elements, which are very valuable and utilized in many modern devices such as computers, batteries, as well as screens of smartphones and televisions. This indicates that unresolved acid mine drainage are losses on two fronts, deteriorating health of surrounding living beings and missed opportunity to recover valuable heavy metals and rare earth elements [17]. In particular, based on a 2019 report by the United States International Trade Commission, the U.S. imported over 90% of its rare earth elements in 2019, and China contributed 78% of the imports [18]. Within acid mine drainage, the content of each element in the group of rare earth elements can range from 30 – 1200 mg/L. It was approximated that 700 – 3400 tons of rare earth elements could be reclaimed from regions suffering from acid mine drainage annually. This highlights the importance and value in solving acid mine drainage issues [19].

There are two common ways to treat acid mine drainage to reclaim the rare earth elements. Sodium hydroxide is usually used to neutralize the acid mine drainage where the heavy metals can be recovered via the formation of metal hydroxides. However, the sodium hydroxide method is not very effective at recovering rare earth elements. Another option is to use carbonates to better isolate the rare earth elements compared to hydroxide or oxide treatment. This is because carbonates possess larger negative Gibbs free energy values which promote the chemical reactions between carbonates and rare earth elements, thereby isolating the valuable elements in the form of carbonate compounds. In the second method, CO_2 could be utilized as the carbonate source to precipitate the rare earth elements. The usage of CO_2 for acid mine drainage presents an underexplored field that is a candidate for future research and development [20].

Conclusion

The rapid combustion of fossil fuels for energy has increased atmospheric CO_2 concentration by a large amount compared to pre-industrial era. The high CO_2 concentration in the atmosphere has led to the onset of global warming. There is a 2 in 3 chance of restricting the mean global temperature rise to below 2°C, if worldwide CO_2 emissions can be greatly reduced from present time of above 48 gigatons of CO_2 per year to below 5 gigatons of CO_2 per year by 2050. To achieve this reduction in CO_2 emissions as well as curb rising global temperatures and global warming, CCU techniques are necessary. Since CO_2 capture negatively impacts the economics of any industry, CO_2 utilization pathways need to be developed to minimize the cost of capture. In this review, the utilization of CO_2 in three chemical and materials treatment sectors are discussed, which include iron and steel sector, alumina sector, and acid mine drainage treatment. In the iron and steel sector, CO_2 can be used to: (1) neutralize highly alkaline iron-rich slurry from flotation step, which can enhance filtration rates of the slurry; (2) promote synthesis of CO in blast furnace to convert iron into pig iron, and reduce the amount of dust generation; and (3) form carbonated steel slag where the CO_2 can be sequestered. In the alumina sector, CO_2 can be employed to neutralize red mud, which is the by-product of aluminium production. CO_2 can be permanently sequestered in the form of carbonates after reaction with red mud, whereas the neutralized red mud presents opportunities for further utilization like reclaimation of rare earth elements. In the treatment of acid mine drainage, CO_2 can act as a carbonate source to isolate rare earth elements from acid mine drainage via the precipitation of carbonates. Since CO_2 is a low cost and abundant resource, more research should be conducted to improve the utilization and efficiency of CO_2 in these three sectors.

Acknowledgements
The authors would like to acknowledge financial support from AAIBE Chair of Renewable Energy (ChRE), grant number 201801 KETTHA, Universiti Tenaga Nasional, Kajang, Selangor, Malaysia.

References

[1] S. Valluri, V. Claremboux, S. Kawatra, Opportunities and challenges in CO_2 utilization. Journal of Environmental Sciences. 113 (2022) 322-344. https://doi.org/10.1016/j.jes.2021.05.043

[2] J. Rogelj, et al., Paris Agreement climate proposals need a boost to keep warming well below 2 C. Nature. 534 (2016) 631-639. https://doi.org/10.1038/nature18307

[3] A. Olhoff, J.M. Christensen, Emissions Gap Report 2020. UNEP DTU Partnership. 2020. Available from: https://www.unep.org/emissions-gap-report-2020

[4] Q. Zhu, Developments on CO_2-utilization technologies. Clean Energy. 3 (2019) 85-100. https://doi.org/10.1093/ce/zkz008

[5] Y. Lin, et al., Low-Carbon Development for the Iron and Steel Industry in China and the World: Status Quo, Future Vision, and Key Actions. Sustainability. 13 (2021) 12548. https://doi.org/10.3390/su132212548

[6] C. Yi, et al., Experimental research on reducing the dust of BOF in CO2 and O2 mixed blowing steelmaking process. ISIJ International. 49 (2009) 1694-1699. https://doi.org/10.2355/isijinternational.49.1694

[7] L. Ren, et al., A review of CO_2 emissions reduction technologies and low-carbon development in the iron and steel industry focusing on China. Renewable and Sustainable Energy Reviews. 143 (2021) 110846. https://doi.org/10.1016/j.rser.2021.110846

[8] C.A. Myers, T. Nakagaki, K. Akutsu, Quantification of the CO_2 mineralization potential of ironmaking and steelmaking slags under direct gas-solid reactions in flue gas. International Journal of Greenhouse Gas Control. 87 (2019) 100-111. https://doi.org/10.1016/j.ijggc.2019.05.021

[9] S.K. Kawatra, V. Claremboux, Application of Surface Chemical Fundamentals to Improving Industrial Filtration Rates. Mineral Processing and Extractive Metallurgy Review. 40 (2019) 292-297. https://doi.org/10.1080/08827508.2019.1598404

[10] J.A. Halt, S.K. Kawatra, Does the Zeta Potential of an Iron Ore Concentrate Affect the Strength and Dustiness of Unfired and Fired Pellets? Mineral Processing and Extractive Metallurgy Review. 38 (2017) 132-141. https://doi.org/10.1080/08827508.2017.1288114

[11] P. Renforth, et al., Contaminant mobility and carbon sequestration downstream of the Ajka (Hungary) red mud spill: The effects of gypsum dosing. Science of The Total Environment. 421-422 (2012) 253-259. https://doi.org/10.1016/j.scitotenv.2012.01.046

[12] R.C. Sahu, R.K. Patel, B.C. Ray, Neutralization of red mud using CO_2 sequestration cycle. Journal of Hazardous Materials. 179 (2010) 28-34. https://doi.org/10.1016/j.jhazmat.2010.02.052

[13] Y.-S. Han, et al., Bauxite residue neutralization with simultaneous mineral carbonation using atmospheric CO_2. Journal of Hazardous Materials. 326 (2017) 87-93. https://doi.org/10.1016/j.jhazmat.2016.12.020

[14] K. Evans, The History, Challenges, and New Developments in the Management and Use of Bauxite Residue. Journal of Sustainable Metallurgy. 2 (2016) 316-331. https://doi.org/10.1007/s40831-016-0060-x

[15] R.M. Rivera, et al., Extraction of rare earths from bauxite residue (red mud) by dry digestion followed by water leaching. Minerals Engineering. 119 (2018) 82-92. https://doi.org/10.1016/j.mineng.2018.01.023

[16] U.S. Environmental Protection Agency. Abandoned Mine Drainage. 2023 [28 April 2024]; Available from: https://www.epa.gov/nps/abandoned-mine-drainage.

[17] R.D. Kiggins, The Strategic and Security Implications of Rare Earths, in R.D. Kiggins, Editor, The Political Economy of Rare Earth Elements: Rising Powers and Technological Change, Palgrave Macmillan UK: London, 2015, pp. 1-19. https://doi.org/10.1057/9781137364241_1

[18] K. Van Veen, A. Melton, Rare Earth Elements Supply Chains, Part 1: An Update on Global Production and Trade. 2020. Available from: https://www.usitc.gov/publications/332/executive_briefings/ebot_rare_earths_part_1.pdf

[19] C.R. Vass, A. Noble, P.F. Ziemkiewicz, The Occurrence and Concentration of Rare Earth Elements in Acid Mine Drainage and Treatment By-products: Part 1—Initial Survey of the Northern Appalachian Coal Basin. Mining, Metallurgy & Exploration. 36 (2019) 903-916. https://doi.org/10.1007/s42461-019-0097-z

[20] B. Vaziri Hassas, M. Rezaee, S.V. Pisupati, Precipitation of rare earth elements from acid mine drainage by CO_2 mineralization process. Chemical Engineering Journal. 399 (2020) 125716. https://doi.org/10.1016/j.cej.2020.125716

Enhancing recursive-PCA-based process monitoring using t-SNE visualization

OKTORIFO Gardiola[1,a], ABDULHALIM SHAH Maulud[1,2,b*], MUHAMMAD Nawaz[1,c] and NABILA FARHANA Jamaludin[1,3,d]

[1]Department of Chemical Engineering, Universiti Teknologi PETRONAS, Bandar Seri Iskandar, Perak 32610, Malaysia

[2]Centre for Carbon Capture, Utilisation and Storage, Institute of Sustainable Energy and Resources, Universiti Teknologi PETRONAS, Bandar Seri Iskandar, Perak 32610, Malaysia

[3]Centre of Advance Process Safety (CAPS), Institute of Smart and Sustainable Living (ISSL), Universiti Teknologi PETRONAS, Bandar Seri Iskandar, Perak 32610, Malaysia

[a]oktorifo_22009993@utp.edu.my, [b]halims@utp.edu.my, [c]muhammad.nawaz@utp.edu.my, [d]nabila.jamaludin@utp.edu.my

Keywords: Adaptive Process Monitoring, Process Monitoring, PCA, Recursive PCA, t-SNE

Abstract. Certain industrial chemical processes are known to change over time but are still regarded as normal; this can lead to false alarms when monitored using a non-adaptive process monitoring technique. Adaptive process monitoring techniques can improve product quality and minimize downtime by reducing false alarms. Since most industrial chemical processes are time-variant, standard principal component analysis (PCA) is not the best method for monitoring a slowly changing process. It is crucial to predetermine the data's characteristics to choose the best process monitoring technique. It has been discovered that t-distributed stochastic neighbour embedding (t-SNE) is a valuable data visualisation method. This paper uses t-SNE to visualise simulated data from a continuously stirred tank reactor (CSTR) process. It has been found that the visualized data using t-SNE is worth monitoring with recursive PCA due to its non-linear data characteristics. Consequently, sample-wise recursive PCA is utilized to identify slow changes caused by variations in ambient temperature during the day and night. Compared to standard PCA, the recursive PCA model performs noticeably better in reducing false alarms due to its effective adaptation to slow changes.

Introduction

Process monitoring is essential in industrial chemical processes to produce high-quality products and guarantee safe working conditions. Because of recent technological developments, modern industry's manufacturing processes have risen in size, complexity, and intelligence [1, 2]. Therefore, early fault detection and diagnosis (FDD) is required to improve plant safety, enhance product quality, and reduce downtime [1, 3, 4]. Researchers have developed several effective techniques to enhance process monitoring for various industrial chemical process techniques. Chemical process monitoring increasingly relies on effective multivariate statistical process monitoring (MSPM) techniques [5-8]. MSPM's fundamental objective is to efficiently extract different process features using a particular multivariate analysis procedure. The high-dimensional data is then projected into a less-dimensional space to assess the statistics. Various MSPM techniques are available to be implemented; however, principal component analysis (PCA) [9-12] and partial least squares (PLS) [13, 14] are the leading MSPM techniques that are frequently implemented to monitor chemical processes. PCA is frequently used to monitor industrial chemical processes statistically, especially in a continuous chemical process, due to its non-linearity in minimum processing [15]. Although PCA has been implemented over the years as an effective process monitoring tool, several drawbacks exist. One of the drawbacks is the limitation to implementing PCA to parameter-varying processes, where it interprets the natural changes that occur in the process as faults, resulting in innumerable false alarms which led to credibility decrement of the monitoring system [16]. Furthermore, another major limitation of PCA-based monitoring is that the PCA model that is developed from data is time-invariant, while most industrial processes are time-varying [15, 17]. Over the years, researchers have determined alternative methods to monitor time-varying processes. In addition, time-varying processes including slow changes might occur in continuous stirred tank reactor (CSTR) [8].

Recursive PCA provides recursive determination of relevant parameters for time-varying processes, increasing the ability to implement PCA. To adapt to the time variation, in recursive PCA, the mean for covariance-based PCA and correlation-based PCA is recursively updated, algorithms for PCA calculation including sample-wise and block-wise updates are constructed, the number of principal components as the significant number of component changes are recursively determined, and the confidence limits for Squared Prediction Error (SPE) and Hotelling's T^2 statistics in real-time are recursively determined to support adaptive process monitoring [15]. Thus, it can be deduced that recursive PCA will be an effective process monitoring method to detect slow changes in a process, and not classify it as process errors.

Nevertheless, before implementing recursive PCA to process data, the process data must be initially visualized to determine its characteristics. One widely used data visualization technique is t-distributed stochastic neighbor embedding (t-SNE). t-SNE effectively captures a significant portion of the local patterns in high-dimensional data while uncovering broader global structures, including clusters across various scales [18].

This paper will focus on the recursive PCA method and its ability to detect slow changes in temperature in a process that occurred in the CSTR. More specifically, recursive PCA [15] will be utilized within an online monitoring framework (updating models on a sample-by-sample basis) for simulated data. The effectiveness of recursive PCA will be evaluated in comparison to conventional PCA.

Methods

Simulated Data Generation. The data was generated from a process that occurred in CSTR with a total time of 14 days (20,160 minutes). In the generated data, the dynamic ambient temperature was considered to simulate realistic process behaviour. The change in ambient temperature affected the inlet coolant temperature of the CSTR's cooling jacket, with an interval of 10°C between the lowest and highest points. Moreover, the noise was also applied to simulate realistic data (Fig. 1). In addition, there are a total of 10 variables in the data, which include liquid holdup (h), concentration inside the reactor (C), temperature inside the reactor (T), temperature of coolant (T_c), flow inside the reactor (F), flow of coolant (F_c)flow of reactor inlet (F_i), concentration of reactor inlet (C_i), temperature of reactor inlet (T_i), and coolant temperature inlet (T_{ci}). Consequently, the data series was scaled to zero mean and unit variance corresponding to normal process operation without any disturbance from the dynamic ambient temperature behaviour, which was then considered as the scaled reference data.

From (Fig. 1), it can be observed that the ambient temperature is changing between day and night conditions. During the daytime, the Tci can reach up to 305°C; at nighttime, the Tci can reach as low as 295°C.

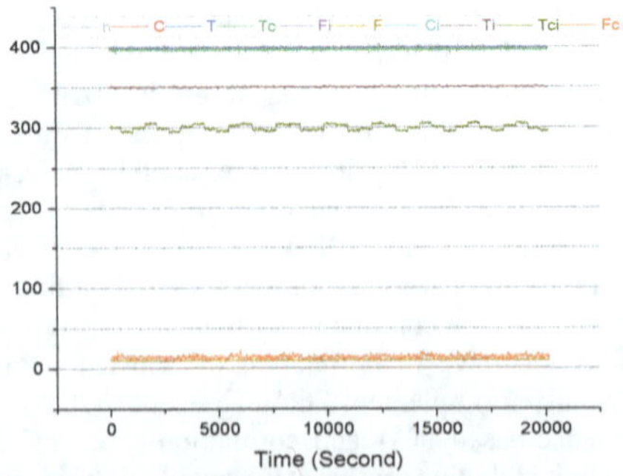

Fig. 1. Disturbed data

Conventional PCA and Recursive PCA Model. Initially, conventional PCA [16] was applied to the simulated data and was built using the scaled reference data. SPE and T^2 statistic control limits of 95% and 99% were calculated (Fig. 3). Consequently, the recursive PCA model was built for the disturbed data scaled to zero mean and unit variance corresponding to the reference data. In the recursive PCA, the sample means, standard, deviation, correlation matrix, SPE and T^2 statistics limits, and number of principal components were recursively calculated, which will be further discussed in the next section of this paper (Fig. 4).

Results and Discussion

t-SNE Data Visualization. To begin with, the scaled disturbed data is visualized with t-SNE. According to (Fig. 2), the red cross represents the data in daytime, and the blue circle represents the data in nighttime. Therefore, according to (Fig. 2), it can be observed that the data is not uniform and not well distributed, as there is a difference in data characteristics between daytime and nighttime data. In contrast, the daytime data contributed to the left side of the graph, and the nighttime data mainly contributed to the middle and right sides of the graph. Hence, it can be deduced that the process data in this paper is worth monitoring using an effective adaptive process monitoring method of recursive PCA.

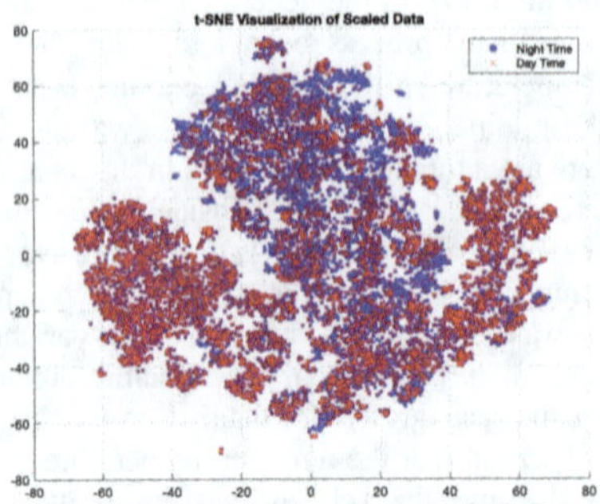

Fig. 2. t-SNE data visualization from scaled disturbed data

Conventional PCA and Recursive Pca Results. Consequently, scaled disturbed data is utilized to develop the conventional PCA model. In this conventional PCA, the number of principal components and confidence limits for T^2 statistic and SPE is only calculated once throughout the entire data series.

Materials Research Proceedings 53 (2025) 458-463 https://doi.org/10.21741/9781644903575-47

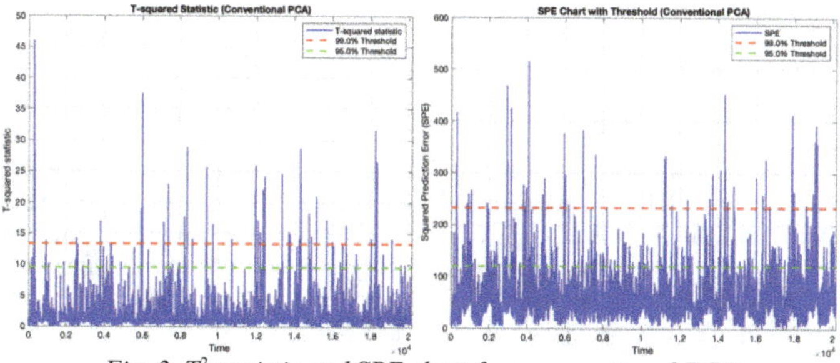

Fig. 3. T^2 statistic and SPE chart from conventional PCA

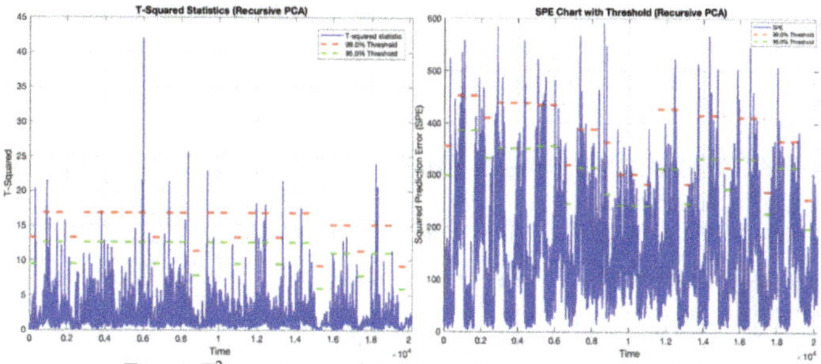

Fig. 4. T^2 statistic and SPE chart from recursive PCA

According to the T^2 statistic and SPE chart from conventional PCA plotted in (Fig 3), it can be deduced that the 95 and 99% confidence limits are stationary at certain points. This is due to the single calculation of the number of principal components and confidence limits for T^2 statistics and SPE throughout the data series. As a result, certain data points in T^2 statistic and SPE chart (Fig. 3) are beyond the confidence limits, although the process is still considered normal, resulting in false alarms. In addition, a drawback of utilizing the PCA model for online monitoring lies in its inability to adapt to changes over time, particularly in the mean, variance, and correlation structure among variables [16]. Furthermore, it can be observed from (Fig. 1) that before time point 3000, the T_{ci} started to increase as it approached daytime. From (Fig. 3), it can be observed that the T^2 statistic and SPE from conventional PCA are beyond the confidence limits. Subsequently, the same phenomena apply to the rest of the data series. Thus, in case of slow changes in the temperature of the coolant, the conventional PCA cannot adapt to such changes. Therefore, these phenomena show a need to apply recursive PCA to the data series to enhance the accuracy of the process monitoring by decreasing the possibility of generating false alarms.

In recursive PCA, the number of principal components and the confidence limits for T^2 statistic and SPE were recursively calculated when the new data became available. In contrast, the day and night data were predetermined and segmented into different data segments in this case. Hence, the number of principal components and the confidence limits for T^2 statistic and SPE were recursively calculated throughout the segments, creating a recursive algorithm that can adapt to slow change of coolant temperature inlet, significantly reducing false alarms. According to (Fig. 4), it can be observed that the confidence limits for both T^2 and SPE charts recursively changed for each segment of the data. This is due to the recursive PCA algorithm that adapts to the data in the respective segment by recursively calculating its number of principal components and confidence limits. Hence, creating an adaptive process monitoring technique that can significantly reduce the possibility of false alarm generation. In addition, a similar effective performance of recursive PCA was also found when it was used as the process monitoring method of the Rapid Thermal Annealing (RTA) process [15].

Conclusion

Proper data visualization must be performed to determine whether process data needs to be monitored adaptively. t-SNE is an efficient and effective data visualisation method that determines data characteristics. The visualized data using t-SNE shows the importance of monitoring by recursive PCA, as it was observed that the data is non-linear between daytime and nighttime. Recursive PCA is effective as an adaptive process monitoring technique due to its recursive calculation of the number of principal components and confidence limits for Hotelling's T^2 statistic and SPE, reducing the possibility of false alarm generation.

Acknowledgements

The authors acknowledge the funding support provided by the Ministry of Higher Education Malaysia through the Fundamental Research Grant Scheme (FRGS).

References

[1] M. Nawaz, A. S. Maulud, H. Zabiri, and H. Suleman, Review of Multiscale Methods for Process Monitoring, With an Emphasis on Applications in Chemical Process Systems, IEEE Access, vol. 10. (2022) 49708-49724. https://doi.org/10.1109/ACCESS.2022.3171907

[2] Q. Jiang, X. Yan, and W. Zhao, Fault Detection and Diagnosis in Chemical Processes Using Sensitive Principal Component Analysis, Industrial & Engineering Chemistry Research, vol. 52, no. 4. (2013) 1635-1644. https://doi.org/10.1021/ie3017016

[3] M. Ayoubi and R. Isermann, Neuro-fuzzy systems for diagnosis, Fuzzy sets and Systems, vol. 89, no. 3. (1997) 289-307.

[4] K. Severson, P. Chaiwatanodom, and R. D. Braatz, Perspectives on process monitoring of industrial systems, Annual Reviews in Control, vol. 42. (2016) 190-200. https://doi.org/https://doi.org/10.1016/j.arcontrol.2016.09.001

[5] E. Naderi and K. Khorasani, A data-driven approach to actuator and sensor fault detection, isolation and estimation in discrete-time linear systems, Automatica, vol. 85. (2017) 165-178. https://doi.org/https://doi.org/10.1016/j.automatica.2017.07.040

[6] M. Kano *et al.*, Comparison of multivariate statistical process monitoring methods with applications to the Eastman challenge problem, Computers & Chemical Engineering, vol. 26, no. 2. (2002) 161-174. https://doi.org/https://doi.org/10.1016/S0098-1354(01)00738-4

[7] Q. Jiang and B. Huang, Distributed monitoring for large-scale processes based on multivariate statistical analysis and Bayesian method, Journal of Process Control, vol. 46. (2016) 75-83. https://doi.org/https://doi.org/10.1016/j.jprocont.2016.08.006

[8] M. Nawaz, A. S. Maulud, H. Zabiri, S. A. A. Taqvi, and A. Idris, Improved process monitoring using the CUSUM and EWMA-based multiscale PCA fault detection framework, Chinese Journal of Chemical Engineering, vol. 29. (2021) 253-265. https://doi.org/https://doi.org/10.1016/j.cjche.2020.08.035

[9] M. Z. Sheriff, M. Mansouri, M. N. Karim, H. Nounou, and M. Nounou, Fault detection using multiscale PCA-based moving window GLRT, Journal of Process Control, vol. 54. (2017) 47-64. https://doi.org/https://doi.org/10.1016/j.jprocont.2017.03.004

[10] T. Ait-Izem, M. F. Harkat, M. Djeghaba, and F. Kratz, On the application of interval PCA to process monitoring: A robust strategy for sensor FDI with new efficient control statistics, Journal

of Process Control, vol. 63. (2018) 29-46. https://doi.org/https://doi.org/10.1016/j.jprocont.2018.01.006

[11] N. Basha, M. Nounou, and H. Nounou, Multivariate fault detection and classification using interval principal component analysis, Journal of Computational Science, vol. 27. (2018) 1-9. https://doi.org/https://doi.org/10.1016/j.jocs.2018.04.017

[12] M. Nawaz, A. S. Maulud, H. Zabiri, H. Suleman, and L. D. Tufa, Multiscale Framework for Real-Time Process Monitoring of Nonlinear Chemical Process Systems, Industrial & Engineering Chemistry Research, vol. 59, no. 41. (2020) 18595-18606. https://doi.org/10.1021/acs.iecr.0c02288

[13] G. Li, S. J. Qin, and D. Zhou, Geometric properties of partial least squares for process monitoring, Automatica, vol. 46, no. 1. (2010) 204-210. https://doi.org/https://doi.org/10.1016/j.automatica.2009.10.030

[14] G. Wang and S. Yin, Enhanced quality-related fault detection approach based on OSC and M-PLS, IEEE Trans. Ind. Inf, vol. 1. (2015).

[15] W. Li, H. H. Yue, S. Valle-Cervantes, and S. J. Qin, Recursive PCA for adaptive process monitoring, Journal of Process Control, vol. 10, no. 5. (2000) 471-486. https://doi.org/https://doi.org/10.1016/S0959-1524(00)00022-6

[16] L. M. Elshenawy, S. Yin, A. S. Naik, and S. X. Ding, Efficient Recursive Principal Component Analysis Algorithms for Process Monitoring, Industrial & Engineering Chemistry Research, vol. 49, no. 1. (2010) 252-259. https://doi.org/10.1021/ie900720w

[17] B. M. Wise, N. B. Gallagher, S. Watts Butler, D. D. White, and G. G. Barna, Development and Benchmarking of Multivariate Statistical Process Control Tools for a Semiconductor ETCH Process: Impact of Measurement Selection and Data Treatment on Sensitivity, IFAC Proceedings Volumes, vol. 30, no. 18. (1997) 35-42. https://doi.org/https://doi.org/10.1016/S1474-6670(17)42377-9

[18] L. Van der Maaten and G. Hinton, Visualizing data using t-SNE, Journal of machine learning research, vol. 9, no. 11. (2008).

Potential electrolyte additives for hydrogen economy: Screening of amino-based ionic liquids through COSMO-RS

Muhammad Usman[1,a], Mohd Dzul Hakim Wirzal[1,2b *], Junaid Afridi[1c], Sohail Ahmad Shahani[1,2d], Noor e Hira[1,e] and Shafiq M Hizam[1,f]

[1]Chemical Engineering Department, Universiti Teknologi PETRONAS, 32610 Seri Iskandar, Malaysia

[2]Center of Research in Ionic Liquids (CORIL), Universiti Teknologi PETRONAS, 32610 Bandar Seri Iskandar, Perak, Malaysia

[a]muhammad_22011476@utp.edu.my, [b]mdzulhakim.wirzal@utp.edu.my, [c]junaid_22009546@utp.edu.my, [d]sohail_21000836@utp.edu.my, [e]noor_22000794@utp.edu.my, [f]shafiq.hizam@utp.edu.my

Keywords: Lionic liquids, Hydrogen Energy, COSMO-RS, Electrolyte Additives, Ionic Conductivity

Abstract. The identification of hydrogen as a prospective energy source for the substitution of non-renewable energy sources such as fossil fuels and natural gas has been acknowledged. Nevertheless, the existing methodologies used in the production of hydrogen exhibit significant drawbacks, mostly focused on the substantial release of greenhouse gas (GHG) emissions linked to the process. One potential approach to produce environmentally friendly hydrogen is alkaline water electrolysis, whereby potassium hydroxide (KOH) is used as the electrolyte. However, alkaline water electrolyzers, which use potassium hydroxide (KOH) as an electrolyte, have significant technological challenges in their performance. Commonly used ionic liquids as an electrolyte additive in this process are non-biodegradable. Previously amino-based ionic liquids had not been used as an electrolyte additive. Therefore, it is necessary to explore some biocompatible ionic liquids to be used as an electrolyte additive. Amino-based ionic liquids include a class of chemicals that exhibit biocompatibility and provide the possibility of customization, making them a promising option for improving electrolyte properties. This study will use the Cosmo-RS quantum methodology to evaluate amino-based ionic liquids with favorable physicochemical characteristics. The anions, cations, and water structures were created using TURBOMOLE version 21.0.0 (TmoleX) and the triple valance polarised basis set (TZVP) program package. Choline glycine and Choline Serine were found to have high ionic conductivity of 0.9578 mS/cm and 0.6945 mS/cm respectively. It was also observed that proline, glycine, serine, and asparagine anions can interact with water molecules better as compared to other anions as they show high peaks in non-polar regions.

Introduction

Hydrogen is a carbon-free alternative energy source that might be used in future energy systems. It has the benefits of being environmentally friendly and having high energy density. It has sparked considerable interest as a clean and renewable energy source because of its potential to reduce dependency on fossil fuels and greenhouse gas emissions [1-4]. Future energy systems might employ hydrogen, a carbon-free energy source with the advantages of being harmless to the biological system and having a high energy density [5, 6]. Hydrogen's energy yield is around 2.75 times that of hydrocarbon fuels [7]. Hydrogen produced by alkaline water electrolysis is considered green hydrogen as there is no carbon emission in this process if the source of electricity is renewable [1]. By 2050, the demand for hydrogen is expected to increase by about eight times compared to 2020. Therefore, a variety of issues might make it challenging to properly deploy hydrogen technologies in the energy sector [7]. One well-established and extremely promising technique to produce hydrogen is alkaline water electrolysis [8]. Using an alkaline electrolyte as a conducting medium, this electrochemical process splits water molecules (H_2O) into hydrogen (H_2) and oxygen (O_2) gasses.

The use of potassium hydroxide (KOH) as an electrolyte can cause electrode deterioration and reduced efficiency during long periods of operation. The need for regular maintenance and electrolyte quality in KOH-based systems adds to the operating cost and undermines the efficiency of hydrogen production [9]. According to a study conducted by K. Chen et al, room-temperature ionic liquids as an electrolyte additive can modify the adsorption and charge transfer processes at the metal–electrolyte interface as shown in Fig.1 [10]. Ionic Liquid facilitates the mass transfer process and enhances its catalytic performance for hydrogen evaluation reaction [11].

The most widely used IL, represented by electrolyte additives derived from imidazolium cation, represents environmental and toxicological issues [12]. ILs with choline chloride as the basis of the cation are gaining popularity as a substitute for the imidazole-based alternative due to their non-toxic nature, biocompatibility, and cost-effectiveness [13].

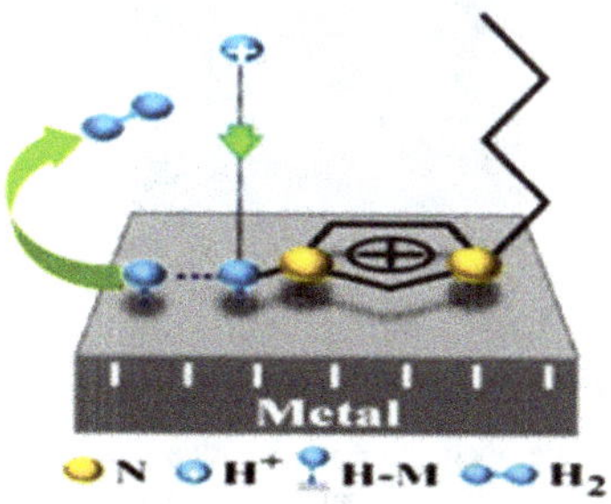

Fig. 1: IL interaction at electrode electrolyte interface [10]

The motivation for this work is the requirement for advanced electrolyte compositions that may enhance the efficiency and feasibility of electrolytically produced hydrogen in KOH-based alkaline electrolyzers and an understanding of the mechanism of particular amino-based ionic liquids. ILs can be prepared by mixing various cations and anions, and experimenting to find the best ones for specific applications is time-consuming and costly [14, 15]. According to wlazło et al, it is very difficult to select potential ionic liquid by experimental method[16]. The COSMO-RS technique reduces the cost of experimental trials by providing valuable information about the physiochemical properties of ionic liquids for pre-screening before experimental use. By using this quantum chemical approach, we can get predictions about molecular-level interaction among different ions.

Methodology

A quantum chemical technique was used in this study to screen potential amino acid ionic liquids (AAILs) as an electrolyte additive according to previous studies [17-19]. The assessment is carried out via COSMO-RS software to evaluate vital physicochemical properties, including molecular geometry, solute ability to interact with water, and sigma profiles. For this research work, choline hydroxide was selected as the cation $[(C_5H_{14}NO - OH)]$ and L-proline [Pr], L-serine [Sr], L-glycine [Gly], Asparagine [Asp], Leucine [Lu], Lysine [Ly] and L-alanine [Ala] were among the

Materials Research Proceedings 53 (2025) 464-472 https://doi.org/10.21741/9781644903575-48

amino acid anions that were chosen to go through a screening procedure using COSMO-RS(Conductor-like Screening Model for Real Solvents), Structures of these ions are illustrated in Figure 2. COSMO-RS software was used to estimate the thermodynamic characteristics of AAILs as it is not practically possible to choose possible ILs by experimental methods [14].

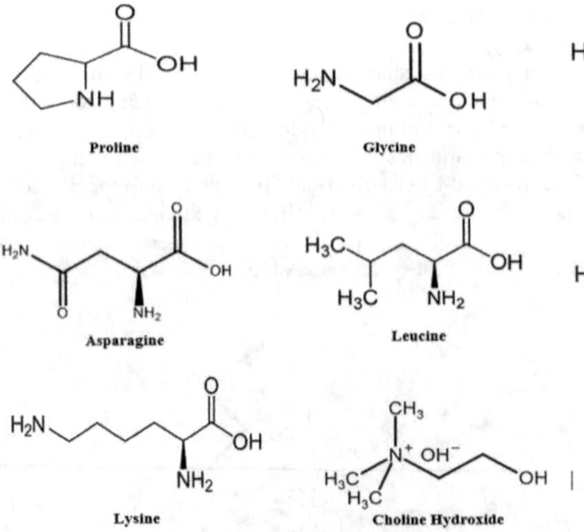

Fig. 2. Different ionic structures of ILs were used in this study.

COSMO-RS serves as a developed screening model, adept at forecasting the physicochemical attributes of ionic liquids, such as viscosity and ionic conductivity. Utilizing the program package of triple valance polarized basis set (TZVP) and TURBOMOLE version 4.4.1 (T-Molex), the structure of the anions, cations, as well as water was built.

A 3D molecular structure of the target molecule was created initially. After that by employing a triple -ξ polarised basis set (TZVP) and B3LYP parameterization, DFT computing was used to optimize the geometry. After exporting the optimized structures to COSMOthermX version 19.0.4, the σ-profiles and σ-surfaces of the structures were generated. The high-effectiveness AAILs as an electrolyte additive were identified by using the sigma profiles, conductivity, viscosity, and activity coefficient obtained from COSMO-RS.

Sigma Profiles. Sigma profiles were calculated for ionic liquids. Every σ-profile is the total of all the σ-profiles of a segment that the molecule is divided into [20]. The equation (1) describes σ-profiles:

$$p_i(\sigma) = \frac{\Sigma_i x_i n_i p_i(\sigma)}{\Sigma_i x_i n_i p_i} = \Sigma_i^N x_i\, p_i \quad (1)$$

Whereas

$p_i(\sigma)$ = sigma profile of component i

x_i = Mole fraction of component i in the mixture.

n_i = Number of segments of component i

p_i = Probability distribution of the surface charge density for pure component i

Materials Research Proceedings 53 (2025) 464-472 https://doi.org/10.21741/9781644903575-48

Ionic conductivity prediction COSMO-RS. The ionic conductivity (x=κ) is determined by equation (2):

$$\ln(x(T)x_o) = d^* + e^*(\ln(r*_m)) + f^*(\ln/\sigma)) + E_{die1}/E_{die10} + g^*(E/(RT)) + i^*(E_{die1}(T_o))/(R^*T^2) + d \quad (2)$$

where $T_0 = 1$ K is the starting temperature and $E_{die10} = 1$ kJ/mol is the field without the dielectric. Molecular radius is denoted by the phrase r*m, symmetry number by, and gas constant by R. The regression parameter values for ionic conductivity are d = 8.784179, e = -15.668627, f = -1.141751, g = 0.025274, h = - 0.054461, and i = 19.107936. To create the model, the experimental data were correlated to determine the parameter values.

Viscosities calculation in COSMO-RS. Ionic liquid viscosity plays an important role during the charge transfer process in the electrolyte system. The temperature-dependent viscosity is calculated using equation (3).

$$\ln(\eta_i(T)) = \ln(\eta_i(T_{Room})) + \max[0, (c_{T_0} + c_T \ln(\eta_i(T_{Room})))] \left(\frac{1}{T} - \frac{1}{T_{Room}}\right) \quad (3)$$

Where η is the viscosity, T is temperature and c is the correlation parameter.

Activity Coefficient. The activity coefficient of a component at infinite dilution is determined by using the chemical potential values in both mixed and pure states. This can be illustrated by equation (4).

$$\ln\gamma = \exp\left(\frac{\mu-\mu^0}{RT}\right) \quad (4)$$

Where μ is the chemical potential of the solvent and μ0 is the chemical potential of the pure compound. The lower value of ln Υ for the solute in the solvents indicates that the solvent has a strong ability to interact with and dissolve the solute effectively.

Results and Discussion

The COSMO-RS (Conductor-like Screening Model for Real Solvents) was used for advanced screening of amino ionic liquids by measuring physicochemical characteristics including ionic conductivity, viscosity, activity coefficient, and sigma profiles. In this study amino amino-based ionic liquids were treated as individual molecules with equimolar fractions to facilitate COSMO-RS calculations.

Sigma Profile. Incorporating the σ-profile probability distribution into this study facilitates the understanding of the cation and anion compatibility with water. There are three parts in the σ-profiles graph: H-bond donor (σ < -1 e/A2), non-polar (-1 e/A2 < σ < 1 e/A2), and the H-bond acceptor (σ > 1 e/A2) [21]. The formation of a peak by the choline cation is observed at the H-bond donor region as shown in Figure 3. It showed that choline has the ability to interact with water at its hydrogen bond acceptor region. On the other hand glycine, proline, serine, and alanine anions showed their peaks in non-polar and hydrogen bond acceptor regions. This trend depicted that these anions tend to interact with molecules that are hydrogen bond donors. As water molecules showed a peak in the hydrogen bond donor region, these anions tend to interact with water resulting in good solubility in water.

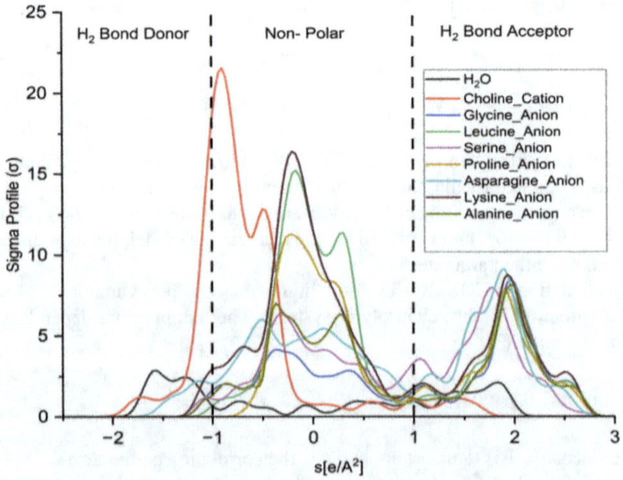

Fig. 3. Sigma profiles of different ions

Ionic Liquid Conductivities. The ionic conductivity of electrolytes plays an important role during electrolytic reactions. According to de Fátima Palhares et al. (2018), the electrolytic system's efficiency may be increased by raising the electrolyte's conductivity and lowering the dominating resistances [22].

The COSMO-RS software was used in this study to evaluate the ionic conductivity of several choline-based ionic liquids, building upon previous research findings [23]. The order of conductivities of these ionic liquids is [Ch][Gly] > [Ch][Sr] > [Ch][Ala] > [Ch][Pr] > [Ch][Asp] > [Ch][Lu] > [Ch][Ly]. The findings reveal varying levels of conductivity across the several substitutes of choline, with Choline glycine exhibiting the highest ionic conductivity at 0.9578 mS/cm. Choline glycine has superior molecular transportability and serves as an efficient electrolyte for applications requiring high conductivity. On the other hand, it was shown that Choline lysine and Choline leucine had significantly lower conductivities of 0.02556 mS/cm and 0.06698343 mS/cm, respectively, this following the previous studies which stated that longer alkyl chains provide poorer ionic conductivity. This is so because longer chains provide stronger intermolecular interactions, therefore hindering ion mobility [24-26]. Another study conducted by F. Fiegenbaum et al. reveals that high ionic conductivity of ionic liquid electrolytes leads to higher efficiency and low activation energy of reaction that facilitates proton transfer in the system [27]. L Amaral et al. experimentally found that the addition of 1Vol% of IL in KOH electrolyte increased the current density of the system from $8.08 \times 10^{-3} mAcm^2$ to 2.1×10^{-2} mAcm^2. This study also implies that ILs have the potential to enhance the efficiency of the electrolyte solution towards hydrogen production, most likely by promoting superior ionic conductivity and stabilizing the reaction environment [11].

The aforementioned results not only demonstrate the potential of COSMO-RS as an innovative method for studying ionic liquids but also provide a basis for the development of ionic liquids with enhanced conductivity suitable for specific applications and current demands.

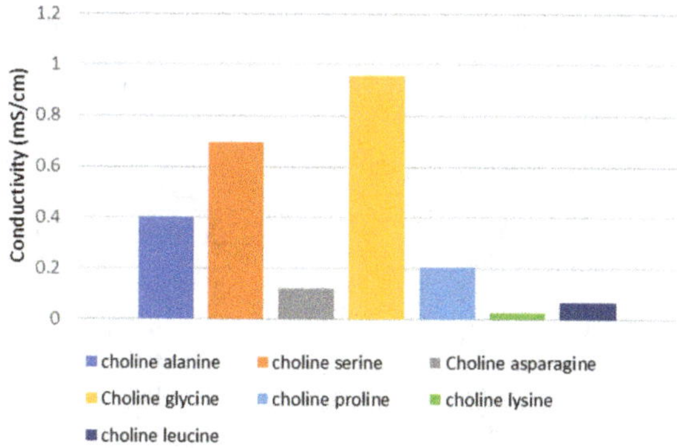

Fig.4. Predicted conductivities of ionic liquids

Viscosities. Viscosities of different anions and choline cation was also studied using the COSMO-RS approach. Ionic liquids having high viscosities in solution slow down the charge transfer process.[26, 28, 29]. Therefore, to be considered as a better electrolyte additive for hydrogen production, ionic liquids with less viscosities are more favorable. The order of amino-based ionic liquids according to their viscosities was [Ch][Ly] > [Ch][Lu] > [Ch][Asp] > [Ch][Pr] > [Ch][Ala] > [Ch][Gly] > [Ch][Sr]. Choline lysine, Choline leucine, and choline asparagine show higher viscosities (2787.49 mPaS, 1379.15mPaS, and 936.45mPaS respectively) shown in Fig 5. as compared to other AAILs (Amino acid ionic liquids) making them less effective as high viscosity hinders ionic mobility in electrolytic system.

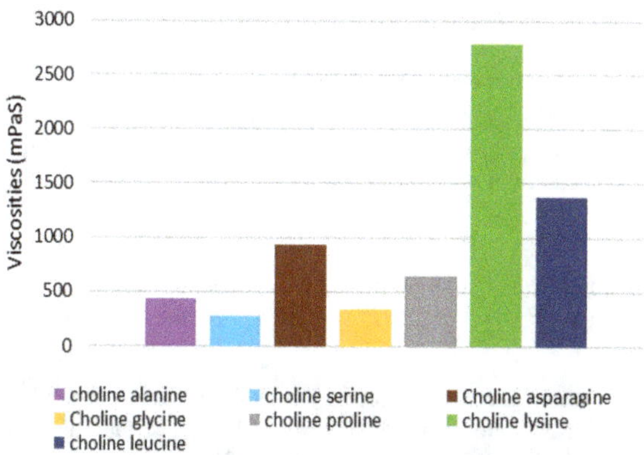

Fig. 5. Viscosities of different anions with choline hydroxide

Activity Coefficients. The activity coefficients of a range of choline-based ionic liquids, which include various anions such as proline, lysine, glycine, leucine, alanine, asparagine, and serine, have also been studied. The activity coefficients help to calculate valuable quantitative measurements of the intermolecular interactions among ions in solutions. Choline serine and choline asparagine show higher activity coefficient results as compared to other anions (0.1188 and 0.0752 respectively) as shown in Table 1. This implies that these anions are less favourable to be added in water because they exhibit strong self-ionic interactions as compared to interaction

Materials Research Proceedings 53 (2025) 464-472 https://doi.org/10.21741/9781644903575-48

with water because of large molecule shape or polarity difference with water. On the other hand, choline glycine show less activity coefficient result (0.03) indicating its suitability to be solvated by water molecules. Ionic liquids with less activity coefficient can interact with water molecules in a better way due to their small size and hydrogen bonding.

Table 1: Predicted activity coefficient results

Ionic Liquid	Activity Coefficient
Choline Serine	0.118837
Choline Asparagine	0.075283
Choline Proline	0.048195
Choline Leucine	0.042379
Choline Lysine	0.033682
Choline Alanine	0.034376
Choline Glycine	0.030029

Conclusion

This research work aimed to study 7 amino-based ionic liquids using the COSMO-RS technique and screening these by measuring their physiochemical properties. COSMO-RS predictions serve as a basis for the screening of ionic liquids to be incorporated as electrolyte additives. Sigma profile results revealed that choline cation can interact with water molecules having hydrogen ions at the hydrogen bond acceptor region. Choline proline showed a peak only in non-polar and hydrogen bond acceptor regions predicting that it can interact with water molecules at the hydrogen bond donor region. Ionic liquids according to conductivity value were in order of: [Ch][Gly] > [Ch][Sr] > [Ch][Ala] > [Ch][Pr] > [Ch][Asp] > [Ch][Lu] > [Ch][Ly] and it showed that of that choline glycine has the highest ionic conductivity value of 0.9578 mS/cm indicating its ability to increase the charge transfer process. Furthermore, activity coefficient results predicted that ionic liquids with less activity coefficient value can interact with water molecules due to polarity and hydrogen bonding. Based on the above-mentioned results choline proline $[(C_5H_{14}NO\text{-}C_5H_9NO_2)]$ and choline glycine $[(C_5H_{14}NO\text{-}C_2H_5NO_2)]$ are screened out to be used as electrolytes for further research due to their suitable physiochemical properties. Choline-based ionic liquids should be explored as electrolyte additives in future experimental work to evaluate their performance in hydrogen production.

Acknowledgment

The authors are grateful to Universiti Teknologi PETRONAS Malaysia for its financial assistance. This work is done with the support of (FRGS) grant with number 015MA0-152.

References

[1] L. Amaral et al., "Toward tailoring of electrolyte additives for efficient alkaline water electrolysis: salicylate-based ionic liquids," ACS Applied Energy Materials, vol. 1, no. 9, pp. 4731-4742, 2018. https://doi.org/10.1021/acsaem.8b00858

[2] C. Tarhan and M. A. Çil, "A study on hydrogen, the clean energy of the future: Hydrogen storage methods," Journal of Energy Storage, vol. 40, p. 102676, 2021. https://doi.org/10.1016/j.est.2021.102676

[3] M. Rasul, M. Hazrat, M. Sattar, M. Jahirul, and M. Shearer, "The future of hydrogen: Challenges on production, storage and applications," Energy Conversion and Management, vol. 272, p. 116326, 2022. https://doi.org/10.1016/j.enconman.2022.116326

[4] A. Pareek, R. Dom, J. Gupta, J. Chandran, V. Adepu, and P. H. Borse, "Insights into renewable hydrogen energy: Recent advances and prospects," Materials Science for Energy Technologies, vol. 3, pp. 319-327, 2020. https://doi.org/10.1016/j.mset.2019.12.002

[5] M. Yue, H. Lambert, E. Pahon, R. Roche, S. Jemei, and D. Hissel, "Hydrogen energy systems: A critical review of technologies, applications, trends and challenges," Renewable and Sustainable Energy Reviews, vol. 146, p. 111180, 2021. https://doi.org/10.1016/j.rser.2021.111180

[6] Z. Abdin, A. Zafaranloo, A. Rafiee, W. Mérida, W. Lipiński, and K. R. Khalilpour, "Hydrogen as an energy vector," Renewable and sustainable energy reviews, vol. 120, p. 109620, 2020. https://doi.org/10.1016/j.rser.2019.109620

[7] Y. Gong, J. Yao, P. Wang, Z. Li, H. Zhou, and C. Xu, "Perspective of hydrogen energy and recent progress in electrocatalytic water splitting," Chinese Journal of Chemical Engineering, vol. 43, pp. 282-296, 2022. https://doi.org/10.1016/j.cjche.2022.02.010

[8] J. Brauns and T. Turek, "Alkaline water electrolysis powered by renewable energy: A review," Processes, vol. 8, no. 2, p. 248, 2020. https://doi.org/10.3390/pr8020248

[9] S. Hu et al., "A comprehensive review of alkaline water electrolysis mathematical modeling," Applied Energy, vol. 327, p. 120099, 2022. https://doi.org/10.1016/j.apenergy.2022.120099

[10] K. Chen, B. Xu, L. Shen, D. Shen, M. Li, and L.-H. Guo, "Functions and performance of ionic liquids in enhancing electrocatalytic hydrogen evolution reactions: a comprehensive review," RSC advances, vol. 12, no. 30, pp. 19452-19469, 2022. https://doi.org/10.1039/D2RA02547G

[11] L. Amaral, D. Cardoso, B. Šljukić, D. Santos, and C. Sequeira, "Room temperature ionic liquids as electrolyte additives for the HER in alkaline media," Journal of The Electrochemical Society, vol. 164, no. 4, p. F427, 2017. https://doi.org/10.1149/2.0011706jes

[12] A. Nikpour, M. Moosavi, and M. Torkzadeh, "Structural, spectral, and dynamical behaviors of biodegradable choline-based ionic liquids (CBILs) confined inside neutral and charged mxene nanopores for supercapacitor applications," Colloids and Surfaces A: Physicochemical and Engineering Aspects, vol. 674, p. 131968, 2023. https://doi.org/10.1016/j.colsurfa.2023.131968

[13] A. Le Donne and E. Bodo, "Cholinium amino acid-based ionic liquids," Biophysical reviews, vol. 13, no. 1, pp. 147-160, 2021. https://doi.org/10.1007/s12551-021-00782-0

[14] Y. Cao, Z. Wu, Y. Zhang, Y. Liu, and H. Wang, "Screening of alternative solvent ionic liquids for artemisinin: COSMO-RS prediction and experimental verification," Journal of Molecular Liquids, vol. 338, p. 116778, 2021. https://doi.org/10.1016/j.molliq.2021.116778

[15] Z. Song, J. Wang, and K. Sundmacher, "Evaluation of COSMO-RS for solid-liquid equilibria prediction of binary eutectic solvent systems," Green Energy & Environment, vol. 6, no. 3, pp. 371-379, 2021. https://doi.org/10.1016/j.gee.2020.11.020

[16] M. Wlazło, E. I. Alevizou, E. C. Voutsas, and U. Domańska, "Prediction of ionic liquids phase equilibrium with the COSMO-RS model," Fluid Phase Equilibria, vol. 424, pp. 16-31, 2016. https://doi.org/10.1016/j.fluid.2015.08.032

[17] N. Islam, H. W. Khan, A. A. Gari, M. Yusuf, and K. Irshad, "Screening of ionic liquids as sustainable greener solvents for the capture of greenhouse gases using COSMO-RS approach: Computational study," Fuel, vol. 330, p. 125540, 2022. https://doi.org/10.1016/j.fuel.2022.125540

[18] K. Yu et al., "Ionic liquids screening for lignin dissolution: COSMO-RS simulations and experimental characterization," Journal of Molecular Liquids, vol. 348, p. 118007, 2022. https://doi.org/10.1016/j.molliq.2021.118007

[19] C. Y. Foong et al., "COSMO-RS prediction and experimental investigation of amino acid ionic liquid-based deep eutectic solvents for copper removal," Journal of Molecular Liquids, vol. 333, p. 115884, 2021. https://doi.org/10.1016/j.molliq.2021.115884

[20] H. W. Khan et al., "Evaluating ionic liquids for its potential as eco-friendly solvents for naproxen removal from water sources using COSMO-RS: computational and experimental validation," Environmental Research, vol. 231, p. 116058, 2023. https://doi.org/10.1016/j.envres.2023.116058

[21] S. A. Shahani, H. W. Khan, A. M. Shaiff, M. Goto, and M. Moniruzzaman, "Screening of ionic liquids for the solubility enhancement of quinine using COSMO-RS," Journal of Molecular Liquids, p. 125388, 2024. https://doi.org/10.1016/j.molliq.2024.125388

[22] D. D. A. de Fátima Palhares, L. G. M. Vieira, and J. J. R. Damasceno, "Hydrogen production by a low-cost electrolyzer developed through the combination of alkaline water electrolysis and solar energy use," International Journal of Hydrogen Energy, vol. 43, no. 9, pp. 4265-4275, 2018. https://doi.org/10.1016/j.ijhydene.2018.01.051

[23] P. Nancarrow, A. Al-Othman, D. K. Mital, and S. Döpking, "Comprehensive analysis and correlation of ionic liquid conductivity data for energy applications," Energy, vol. 220, p. 119761, 2021. https://doi.org/10.1016/j.energy.2021.119761

[24] R. A. A. Talip, W. Z. N. Yahya, and M. A. Bustam, "Understanding the physicochemical and transport properties of pyrazolium based ionic liquids bearing iodide and triiodide anions," Journal of Molecular Liquids, vol. 346, p. 118270, 2022. https://doi.org/10.1016/j.molliq.2021.118270

[25] R. A. A. Talip, W. Z. N. Yahya, and M. A. Bustam, "Viscosity and ionic conductivity of imidazolium based ionic liquids bearing triiodide anion," in E3S Web of Conferences, 2021, vol. 287: EDP Sciences, p. 02015. https://doi.org/10.1051/e3sconf/202128702015

[26] W.-L. Yuan, X. Yang, L. He, Y. Xue, S. Qin, and G.-H. Tao, "Viscosity, conductivity, and electrochemical property of dicyanamide ionic liquids," Frontiers in chemistry, vol. 6, p. 59, 2018. https://doi.org/10.3389/fchem.2018.00059

[27] F. Fiegenbaum, E. M. Martini, M. O. de Souza, M. R. Becker, and R. F. de Souza, "Hydrogen production by water electrolysis using tetra-alkyl-ammonium-sulfonic acid ionic liquid electrolytes," Journal of power sources, vol. 243, pp. 822-825, 2013. https://doi.org/10.1016/j.jpowsour.2013.06.077

[28] O. Hollóczki, F. Malberg, T. Welton, and B. Kirchner, "On the origin of ionicity in ionic liquids. Ion pairing versus charge transfer," Physical Chemistry Chemical Physics, vol. 16, no. 32, pp. 16880-16890, 2014. https://doi.org/10.1039/C4CP01177E

[29] Y. Zheng, Y. Zheng, Q. Wang, and Z. Wang, "Density, viscosity, and electrical conductivity of 1-alkyl-3-methylimidazolium dicyanamide ionic liquids," Journal of Chemical & Engineering Data, vol. 66, no. 1, pp. 480-493, 2020. https://doi.org/10.1021/acs.jced.0c00754

Comparative study of MCF-supported Ni-Zr and Ni-CaO catalysts for acetylation in biodiesel by-product utilization

NADZIRAH Azmi[1,a], MOHAMMAD FAUZI Ahmad Termezi[1,b], JAD HARIZZ Harun[1,c], AHMAD IMRAN AFIF Muhammad Izzuddin[1,d], and MOHD HIZAMI Mohd Yusoff[1,e*]

[1]HICoE—Centre for Biofuel and Biochemical Research, Institute of Sustainable Energy and Resources, Department of Chemical Engineering, Universiti Teknologi PETRONAS, Seri Iskandar 32610, Perak, Malaysia

[a]nadzirah_21000025@utp.edu.my, [b]mohammad_18002364@utp.edu.my, [c]jad_20000609@utp.edu.my, [d]ahmad_19000602@utp.edu.my, [e]hizami.yusoff@utp.edu.my

Keywords: Glycerol, Acetylation, Catalyst, MCF, Acetin

Abstract. Malaysia's biodiesel production is expanding quickly, and the main by-product produced is surplus crude glycerol, which has limited application in local industries. Based on the current situation, the conversion of glycerol to higher-value products like acetin through acetylation is an attractive alternative. It has been demonstrated by previous research that the optimal temperature, catalyst type, and reaction time for acetylating glycerol are well-known. However, the effect of catalyst synthesis factors like metal loading still has limited research. Therefore, in this study mesostructured cellular foam (MCF) supported nickel (Ni) to zirconium (Zr) and nickel (Ni) to calcium oxide (CaO) catalysts were synthesised to investigate their performance on the glycerol conversion to acetins. Scanning electron microscopy-energy dispersive x-ray (SEM-EDX) analysis and Fourier transform infrared spectroscopy (FTIR) were done to study the physiochemical and morphological properties of the catalysts. The properties of the catalyst were then correlated with the catalytic activity in the glycerol acetylation with acetic acid reaction, where the impact of various metal loadings (0.5% Ni, 1.0% CaO and Zr) on the glycerol conversion and monoacetin concentration was examined. The reaction was conducted at a reaction duration of 60 minutes, reaction temperature of 115 °C, 5 wt.% catalyst loading and glycerol/acetic acid molar ratio of 1:9. It was found that 0.5:1.0 Ni-CaO/MCF catalyst has the highest conversion of glycerol compared to other catalysts which is 82.70% while Ni-Zr/MCF catalyst has potential to produce a higher concentration of monoacetin compared to other catalysts which is around 69%. These findings show the significance of catalyst composition and surface properties in affecting product selectivity and conversion efficiency in glycerol conversion reactions.

Introduction

The major byproduct of the transesterification process that results in the manufacture of biodiesel is 10 wt% of glycerol [1]. Glycerol is a colourless, odourless, and non-toxic compound [2]. Since glycerol is a waste byproduct that is produced in huge quantities during the production of biodiesel, its usage in the synthesis of value-added chemicals has gained interest in great industries. Chemical purity achieved by the purification of crude glycerol yields a valuable industrial chemical used in personal care, cosmetic, and pharmaceutical products [3]. Nevertheless, the glycerol industry is already well-developed, and purification is expensive. Thus, the cost of producing biodiesel is directly impacted by the ongoing drop in the price of crude glycerol. The successful application of this crude glycerol will be crucial in advancing the commercialization and advancement of biodiesel.

Acetylation reaction involving glycerol with acetic acid or acetic anhydride will produce three distinct acetin compounds which are monoacetin (MA), diacetin (DA), and triacetin (TA). Acetic acid provides a safer, less poisonous, and environmentally friendly option compared to acetic anhydride [4]. Acetylation of glycerol has been gaining interest due to the potential production of acetin for several applications such as cosmetics, pharmaceuticals, food, polymer, and fuel industries as a humectant, emulsifier, plasticizer, and fuel additive [5]. Diacetin and triacetin can be employed as antiknock additives in gasoline and can enhance the cold flow and viscosity qualities of fuels, including biodiesel [6]

Mufrodi et al. [7] revealed that silica-alumina and modified zeolite performed better at 100 °C with 88.65% and 89.22% conversion of glycerol and 47.69% and 59.42% selectivity of TA, respectively. Meanwhile, amberlyst-15 acid resin was able to achieve a 98% conversion of glycerol and 90% selectivity of TA after 60 minutes at 110 °C. Nonetheless, the primary disadvantage of Amberlyst-15 is its low thermal stability, which prevents regeneration by causing corrosion issues due to its breakdown. Kulkarni et al. [8] discovered that, under optimum conditions, the glycerol conversion for CeO_2, ZrO_2, and $CeO_2 - ZrO_2$ mixed metal oxide catalyst was 75.6%, 81.8%, and 62% after 30 minutes of reaction. Compared to pure metal oxides to mixed metal oxides, the pure metal oxides exhibit more catalytic activity.

The glycerol acetylation reaction is carried out with the assistance of homogenous catalysts such as mineral acids (orthophosphoric, hydrochloric, or sulfuric acid) [9]. Nevertheless, these acids have several disadvantages, such as dangerous operating conditions, equipment damage, separation challenges, and poor catalyst reusability [10]. Since heterogeneous solid catalysts are less hazardous, numerous attempts have been made to make this process more environmentally beneficial such as mixed metal oxides [8], mesoporous silica [11], and zeolites [12]. However, these catalysts continue to have poor glycerol conversion.

In this study, mesostructured cellular foam (MCF) is selected as a catalyst support due to its unique characteristics, particularly in the acetylation of the glycerol process. The high surface area which is around 600 -1200 m^2/g [13] and well-defined pore structure provided by MCFs provide numerous active sites for catalytic reactions. In terms of functionalizing the acetylation reaction effectively, the presence of Lewis and Brønsted acidic sites on the catalyst surface is significant; it ensures that glycerol conversion rates are very high while targeting products [14]. Furthermore, MCF-based catalysts are heterogeneous which makes them easily separable and reusable, two critical aspects to improve catalyst utilization and efficiency in industrial processes [15]. Overall, MCFs have the desired characteristics of structural properties and catalytic functionality which makes them an appropriate support for catalysts used in glycerol acetylation reactions thereby enhancing reaction efficiency as well as product yield.

Nickel-based catalysts are some of the best options for glycerol acetylation because they have high catalytic activity, economical value, and stability properties. They increase reaction rates, provide cost benefits compared to other metals such as palladium or platinum, and remain active under difficult situations [16]. Furthermore, nickel catalysts enhance selectivity towards required products while reducing side products and increasing the overall yield. Moreover, their reusability helps to reduce costs as well as waste generation. Nickel-based catalysts are adaptable to different reaction circumstances contributing hugely to making the glycerol acetylation process more efficient, affordable, and sustainable [17].

Ni-Zr and Ni-Cao were chosen because it is a metal oxide-type catalysts that are also proven to be a catalyst for other reactions [18], [19]. Due to the limited research on this catalyst for this project's esterification reaction, it will be used for analysis. Therefore, the objective of this study is to evaluate the effectiveness of MCF-supported Ni-Zr and Ni-CaO catalysts in converting glycerol into valuable acetin.

Materials Research Proceedings 53 (2025) 473-481 https://doi.org/10.21741/9781644903575-49

Experimental

Synthesis of Catalysts. For this study, MCF was used as support for the catalyst. MCF was created with some modifications according to the method suggested by Hermida et al. [20]. Firstly, 4g of Pluronics 123 (P123) was dissolved in 70 ml of 1.6 M HCl and heated to 40°C on a hot plate. Next, to synthesise the microemulsion, 6.8 mL of 1,3,5-trimethylbenzene (TMB), which functions as a swelling agent for the polymer, is added and agitated for two hours at 40°C. For fifteen minutes, the liquid was agitated further after adding 9.2 mL of tetraethyl orthosilicate (TEOS). In terms of information, TEOS is used as a source of ethanol, a co-surfactant, and silica. After that, the mixture was poured into a Teflon cup and placed in the oven for 20 hours at 40°C to produce precondensed silica foam. The material was further aged in a static oven at 80°C for 24 hours. After filtering and washing the solid product with deionized water, it was dried for 7 hours at 100°C in an oven to further eliminate moisture. After that, the template was removed and mesocellular foam (MCF) was created by calcination, which was done in ambient air at 500 °C for 6 hours at a heating rate of 2°C/min.

To prepare the catalyst, calcium nitrate hexahydrate ($Ca(NO_3)_2 \cdot 6H_2O$) with purity > 99.9%, and nickel nitrate hexahydrate ($Ni(NO_3)_2 \cdot 6H_2O$) (purity > 99.9%), were dissolved in distilled water to form a metal precursor solution with a Ni-to-Ca molar ratio of 0.5:1. After that, the MCF foam was immersed in the precursor solution while stirred for 90 minutes at 60°C and 400 rpm to ensure sufficient contact time. The impregnated MCF foam was dried at 105°C for about 12 hours after impregnation. After that, the dried material was calcined for six hours at 900°C in a furnace to turn the precursors into oxides and encourage metal-support interactions. Lastly, the catalyst was gradually cooled before storing it in sealed jars.

For Ni-Zr/MCF catalyst preparation, nickel nitrate hexahydrate ($Ni(NO_3)_2 \cdot 6H_2O$) with purity > 99.9%, and zirconium(iv) oxychloride octahydrate ($ZrOCl_2.8H_2O$) (purity > 99.5%), were dissolved in distilled water to create a metal precursor solution with a Ni-to-Zr molar ratio of 0.5:1. Then, the MCF foam was immersed in the precursor solution stirring for 90 minutes at 60°C with speed of 400 rpm ensuring sufficient contact time. After impregnation, the impregnated MCF foam was dried for 12 hours at 110°C to remove excess solvent. Next, the dried material was calcined in a furnace at temperatures 500°C for 6 hours to convert the precursors into oxides and promote metal-support interactions. Finally, the catalyst was allowed to cool slowly and stored in airtight containers.

Characterization of Catalysts. The morphological structure of the catalysts with varying metal loadings was observed using scanning electron microscopy (SEM) examination. The ZEISS Model EVOLS15 with Energy-dispersive X-ray (EDX) was used to capture the SEM pictures. EDX was used to determine the elemental composition present on the surface of the produced catalysts.

Fourier transform infrared spectroscopy (FTIR) was used to examine the specific chemical bonds and catalysts present in the Ni-Zr, MCF-supported Ni-Zr, and Ni-Cao/MCF. A Perkin-Elmer spectrum will be used to record the samples' FTIR spectra. The spectrum was recorded between 4000 and 400cm^{-1} wave number range, and the applied method was KBr.

Acetylation of Glycerol. The acetylation process began by mixing 10 mL of glycerol and 5.0 wt.% of the synthesized catalysts with acetic acid on a stirring hot plate. The molar ratio of glycerol to acetic acid was 1:9. Next, the mixture was transferred into a reaction flask, which was then placed in a microwave oven. The reactor setup included a microwave, a 250 mL three-neck flask for the reaction mixture, a condenser, a thermocouple, and a magnetic stirrer. The irradiation temperature was set to 115°C for 90 minutes while maintaining stirring with the magnetic stirrer. Once the reaction was completed, the reaction flask was removed. All samples were then filtered to separate the catalysts from the reaction solution. High-performance liquid chromatography (HPLC), and gas chromatography with mass spectrometry (GC-MS) were used to further evaluate the liquid samples.

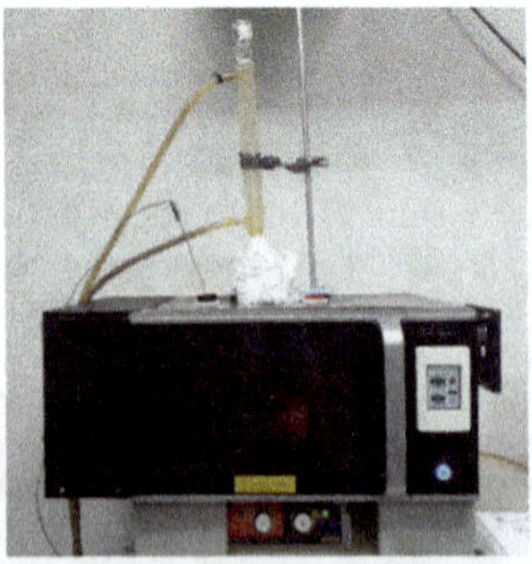

Fig. 1 Setup of Microwave Reactor

Results and Discussion

Characterization of Catalyst. Scanning electron microscopy-energy dispersive x-ray (SEM-EDX) and Fourier transform infrared spectroscopy (FTIR) techniques were used to investigate and assess the chemical and physical characteristics of the Ni-CaO/MCF, Ni-Zr/MCF, and Ni-Zr catalysts.

Scanning Electron Microscopy-Energy Dispersive X-ray (SEM-EDX). Figure 2 shows a close-packed arrangement of catalysts with large grains and a restricted number of macro-pores. The remarkable pore size of MCF materials—up to 50 nm in diameter—is well-known for its advantageous effects on reactants and products [20]. In a range of reactions, this facilitates their ability to overcome mass transfer limitations. The unique form of Zr/MCF particles is most likely caused by the lack of 1,3,5-trimethylbenzene (TMB) during catalyst production. It was considered that TMB's role as a swelling agent was visible and that it might have affected the change in morphology from a rod-like shape to a full and discontinuous spherical shape. Moreover, it was evident that the hole had widened due to the inadequate amount of TMB, which may have suggested the typical rod-like materials or hexagonal structure for MCF.

The elemental composition of 0.5:1.0 Ni-CaO/MCF, 0.5:1.0 Ni-Zr/MCF, and Ni-Zr catalysts were determined by EDX analysis. By this analysis, it was confirmed that there was the presence of Ni, Ca, and Zr on each catalyst. However, the experimental value of Ni, Ca, and Zr elements is quite different from the theoretical value. There could be a few reasons why the theoretical and experimental values of these elements differ. The elements distribution can be affected by the synthesis techniques, temperatures, and compositions used for catalyst formation [21].

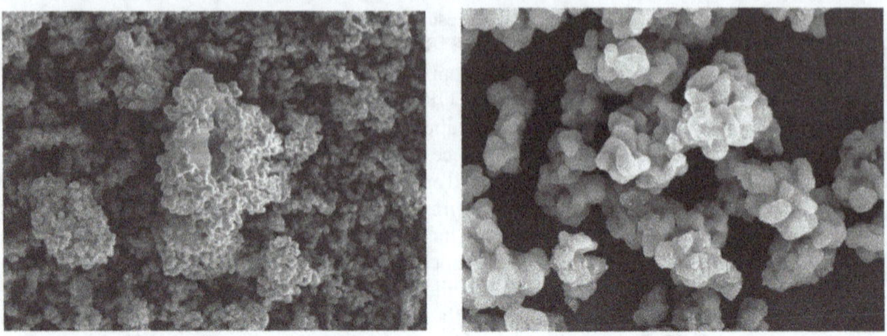

(a) (b)

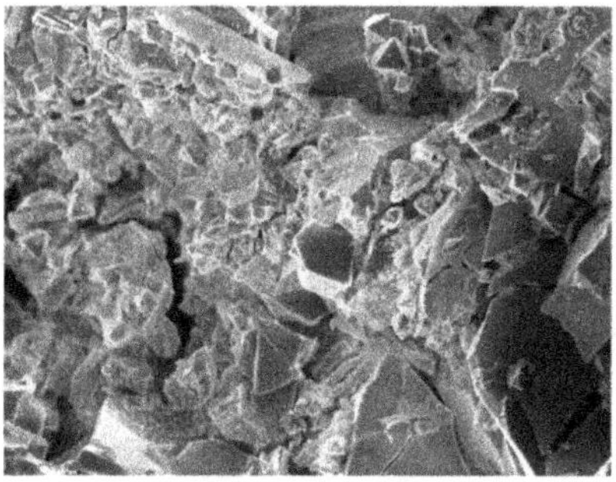

(c)

Fig. 2 SEM image of catalysts (a) 0.5:1.0 Ni-CaO/MCF (b) 0.5:1.0 Ni-Zr/MCF (c) Ni-Zr

Table 1: Chemical composition of the catalysts

Element	Weight (%)		
	0.5:1.0 Ni-CaO/MCF	0.5:1.0 Ni-Zr/MCF	Ni-Zr
Silicon (Si)	50.73	66.71	-
Oxygen (O)	49.68	-	-
Carbon (C)	8.54	-	-
Nickel (Ni)	0.70	2.30	17.35
Calcium (Ca)	0.34	-	-
Zirconium (Zr)	-	30.98	82.65

Fourier Transform Infrared Spectroscopy (FT-IR). The consistent features observed in 0.5:1.0 Ni-CaO/MCF and 0.5:1.0 Ni-Zr/MCF catalysts, including O-H stretching vibrations (3448.40-3459.38 cm^{-1}), C=O stretching vibrations (1638.96-1642.18 cm^{-1}), Si-O stretching vibrations (1076.90-1082.32 cm^{-1}), and C-H bending vibrations (801.49-802.16 cm^{-1}), suggest the presence of water/hydroxyl groups, carbonyl groups, silica/silicate components, and possibly C-H groups [22] [23]. These vibrational characteristics are indicative of the chemical composition and structure of the catalysts, providing insights into the functional groups and components present in the catalyst materials. The consistency in these vibrational peaks across the catalysts indicates similarities in their molecular compositions and supports the understanding of their reactivity and catalytic properties based on the identified functional groups and components.

The absorption bands at 2060 and 1617 cm^{-1} correspond to Ni-Zr, while 1634 cm^{-1} and 1079 cm^{-1} relate to Ni-Zr/MCF, indicating the presence of hydroxyl (-OH) groups and O-H bonds in the spectra [24] [25]. The absorptions at 3372 and 3454 cm-1 signify the stretching vibration of CH-OH groups from the precursor, and the peaks at 630 and 740 cm^{-1} represent the zirconia stretching vibration band [26]. Additionally, the FT-IR peaks at 462 and 459 cm^{-1} in the 460–600 cm^{-1} range are attributed to the stretching vibration modes of υ(Ni–O) [27]. These findings collectively elucidate the distinct vibrational modes and bonds present in the analyzed spectra, providing valuable insights into the molecular composition and structural characteristics of the materials.

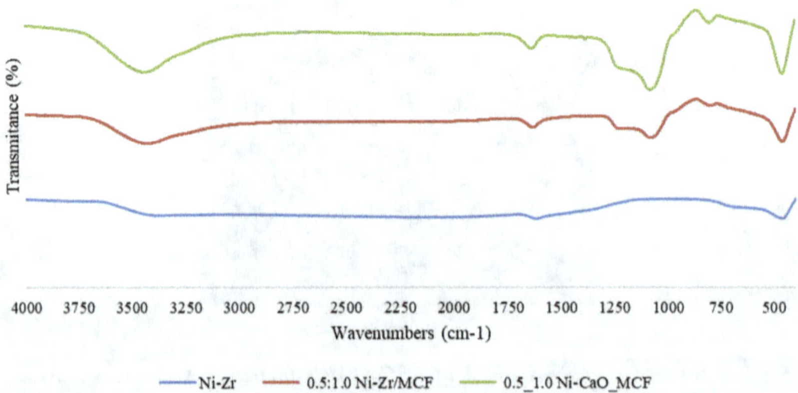

Fig. 3 FTIR spectrum of catalyst Fig. 3 FTIR spectrum of Ni-Zr, 0.5:1.0 Ni-Zr/MCF and 0.5:1.0 Ni-CaO/MCF catalysts

Performance of 0.5:1.0 Ni-CaO/MCF, 0.5:1.0 Ni-Zr/MCF, and Ni-Zr Catalysts on the Acetylation Reaction. 0.5:1.0 Ni-CaO/MCF, 0.5:1.0 Ni-Zr/MCF, and Ni-Zr catalysts were used for the acetylation of glycerol. The catalysts' performance was assessed by examining the conversion of glycerol and their capacity to facilitate selective reactions leading to acetin. Table 1 shows the effect of bimetallic loading in MCF on glycerol conversion. 0.5:1.0 Ni-CaO/MCF catalyst has the highest conversion of glycerol compared to other catalysts which is 82.70% while 76.54% for Ni-Zr catalyst and 71.78% for 0.5:1.0 Ni-Zr/MCF catalyst. The 0.5:1.0 Ni-CaO/MCF catalyst exhibits the highest glycerol conversion of 82.70% compared to other catalysts due to its unique characteristics [28]. The Ni-CaO/MCF catalyst benefits from the synergistic effect of Ni and CaO, providing suitable basicity and stability for efficient glycerol acetylation. The Ni-CaO/MCF catalyst shows high catalytic activity, with Ni-CaO interactions enhancing the catalytic performance [29]. Additionally, the CaO component contributes to the stability and basic properties of the catalyst, leading to enhanced glycerol conversion [30]. The utilization of Ni and CaO in the catalyst structure enhances the conversion and selectivity rates, making the 0.5:1.0 Ni-CaO/MCF catalyst a superior choice for glycerol acetylation compared to Ni-Zr and Ni-Zr/MCF catalysts.

Besides, when comparing the glycerol conversion between Ni-Zr and Ni-Zr/MCF catalysts, the Ni-Zr catalyst has a higher value compared to Ni-Zr/MCF catalyst. The EDX spectra of the Ni-Zr catalyst show that high levels of nickel give a higher value for glycerol conversion because the amount of nickel loading has a significant impact on catalyst performance [28].

Table 2. Results for performance of the catalysts on the glycerol acetylation reaction

Catalyst	Glycerol Conversion (%)
0.5:1.0 Ni-CaO/MCF	82.70
0.5:1.0 Ni-Zr/MCF	71.78
Ni-Zr	76.54

The results shown in Table 3 prove that there is production of monoacetin for the acetylation process. Based on the results, 0.5:1.0 Ni-Zr/MCF catalysts have a high potential to produce a high concentration of monoacetin compared to 0.5:1.0 Ni-CaO/MCF and Ni-Zr catalysts. The specific composition of the 0.5:1.0 Ni-Zr/MCF catalyst, with a combination of nickel and zirconium on the

MCF support, may create active sites that favor the formation of monoacetin. Zirconium could play a role in promoting selective reactions leading to monoacetin due to its unique chemical properties. In order to ensure the best results, it is crucial to employ a suitable metal loading and support, such as the MCF, in the catalyst.

Table 3. GC-MS analysis of 0.5:1.0 Ni-CaO/MCF, 0.5:1.0 Ni-Zr/MCF and Ni-Zr catalysts

Catalyst	Compound	Retention time (min)	Area (%)
0.5:1.0 Ni-CaO/MCF	Monoacetin	10.160	59.49
0.5:1.0 Ni-Zr/MCF	Monoacetin	10.344	69.07
Ni-Zr	Monoacetin	10.297	54.17

Conclusion

In conclusion, the 0.5:1.0 Ni-Zr/MCF catalyst demonstrates a higher potential for generating a concentrated monoacetin product, attributed to its unique catalyst composition and surface properties. Conversely, the 0.5:1.0 Ni-CaO/MCF catalyst exhibits superior glycerol conversion efficiency. These results highlight the importance of surface characteristics and catalyst composition in determining conversion efficiency and product selectivity in glycerol conversion processes. Optimizing these parameters through tailored catalyst design methodologies has the potential to improve the overall selectivity and efficiency of glycerol valorization processes.

Acknowledgment

The authors would like to acknowledge the Ministry of Higher Education for the monetary support via Fundamental Research Grant Schemes (FRGS/1/2020/TK0/UTP/03/9), and for the conferment of the Higher Institution Centre of Excellence (HICoE) award (015MA0-052) to the Centre for Biofuel and Biochemical Research (HICoE-CBBR). Special thanks are extended to Universiti Teknologi PETRONAS for generously providing research facilities essential for conducting this research.

References

[1] M. J. P. Calderon *et al.*, “Producing polyglycerol polyester polyol for thermoplastic polyurethane application: A novel valorization of glycerol, a by-product of biodiesel production,” *Heliyon*, vol. 9, no. 9, p. e19491, Sep. 2023. https://doi.org/10.1016/j.heliyon.2023.e19491

[2] K. Abida and A. Ali, “A review on catalytic role of heterogeneous acidic catalysts during glycerol acetylation to yield acetins,” *Journal of the Indian Chemical Society*, vol. 99, no. 6, p. 100459, Jun. 2022. https://doi.org/10.1016/j.jics.2022.100459

[3] M. Ayoub and A. Z. Abdullah, “Critical review on the current scenario and significance of crude glycerol resulting from biodiesel industry towards more sustainable renewable energy industry,” *Renewable and Sustainable Energy Reviews*, vol. 16, no. 5, pp. 2671–2686, Jun. 2012. https://doi.org/10.1016/j.rser.2012.01.054

[4] F. M. Perez, M. N. Gatti, N. N. Nichio, and F. Pompeo, “Bio-additives from glycerol acetylation with acetic acid: Chemical equilibrium model,” *Results in Engineering*, vol. 15, p. 100502, Sep. 2022. https://doi.org/10.1016/j.rineng.2022.100502

[5] B. Zada, M. Kwon, and S.-W. Kim, “Current Trends in Acetins Production: Green versus Non-Green Synthesis,” *Molecules*, vol. 27, no. 7, p. 2255, Mar. 2022. https://doi.org/10.3390/molecules27072255

[6] A. Malaika, D. Mesjasz, and M. Kozłowski, "Maximizing the selectivity to triacetin in glycerol acetylation through a plastic waste-derived carbon catalyst development and selection of a reaction unit," *Fuel*, vol. 333, p. 126271, Feb. 2023. https://doi.org/10.1016/j.fuel.2022.126271

[7] Z. Mufrodi, E. Astuti, M. Syamsiro, Sutiman, and S. Purwono, "Triacetin Synthesis as Bio-Additive from Glycerol Using Homogeneous and Heterogeneous Catalysts," *Key Eng Mater*, vol. 849, pp. 90–95, Jun. 2020. https://doi.org/10.4028/www.scientific.net/KEM.849.90

[8] R. M. Kulkarni, J. K. Dhanyashree, E. Varma, and S. P. Sirivibha, "Microwave assisted glycerol esterification using CeO2, ZrO2, mixed metal oxide catalyst," *Mater Today Proc*, Oct. 2023. https://doi.org/10.1016/j.matpr.2023.10.130

[9] J. N. Appaturi *et al.*, "Selective synthesis of triacetyl glyceride biofuel additive via acetylation of glycerol over NiO-supported TiO2 catalyst enhanced by non-microwave instant heating," *Appl Surf Sci*, vol. 545, p. 149017, Apr. 2021. https://doi.org/10.1016/j.apsusc.2021.149017

[10] M. A. Betiha, H. M. A. Hassan, E. A. El-Sharkawy, A. M. Al-Sabagh, M. F. Menoufy, and H.-E. M. Abdelmoniem, "A new approach to polymer-supported phosphotungstic acid: Application for glycerol acetylation using robust sustainable acidic heterogeneous–homogenous catalyst," *Appl Catal B*, vol. 182, pp. 15–25, Mar. 2016. https://doi.org/10.1016/j.apcatb.2015.09.010

[11] R. Jothi Ramalingam, J. N. Appaturi, T. Pulingam, H. A. Al-Lohedan, and D. M. Al-dhayan, "In-situ incorporation of ruthenium/copper nanoparticles in mesoporous silica derived from rice husk ash for catalytic acetylation of glycerol," *Renew Energy*, vol. 160, pp. 564–574, Nov. 2020. https://doi.org/10.1016/j.renene.2020.06.095

[12] L. G. Tonutti, H. P. Decolatti, C. A. Querini, and B. O. Dalla Costa, "Hierarchical H-ZSM-5 zeolite and sulfonic SBA-15: The properties of acidic H and behavior in acetylation and alkylation reactions," *Microporous and Mesoporous Materials*, vol. 305, p. 110284, Oct. 2020. https://doi.org/10.1016/j.micromeso.2020.110284

[13] K.-H. Lee, S. Lee, D. Shin, and H.-S. Hahm, "Dimethyl carbonate synthesis via transesterification of ethylene carbonate and methanol using ionic liquid catalysts immobilized on mesoporous cellular foams," *Research on Chemical Intermediates*, vol. 42, no. 1, pp. 109–121, Jan. 2016. https://doi.org/10.1007/s11164-015-2396-4

[14] F. R. Tentor, D. B. Dias, M. R. Gomes, J. G. P. Vicente, L. Cardozo-Filho, and M. E. Berezuk, "Glycerol Acetylation with Propionic Acid Using Iron and Cobalt Oxides in Al-MCM-41 Catalysts," *Bulletin of Chemical Reaction Engineering & Catalysis*, vol. 15, no. 3, pp. 829–844, Dec. 2020. https://doi.org/10.9767/bcrec.15.3.9020.829-844

[15] N. Hidayati, R. P. Sari, and H. Purnama, "Catalysis of glycerol acetylation on solid acid catalyst: a review," *Jurnal Kimia Sains dan Aplikasi*, vol. 23, no. 12, pp. 414–423, Jan. 2021. https://doi.org/10.14710/jksa.23.12.414-423

[16] J. N. Appaturi *et al.*, "Highly Active Nickel (II) Oxide-Supported Cerium Oxide Catalysts for Valorization of Glycerol into Oxygenated Fuel Additives," *Materials*, vol. 16, no. 13, p. 4713, Jun. 2023. https://doi.org/10.3390/ma16134713

[17] Y. Wang *et al.*, "Highly active nickel-loaded β-SiC nanowire catalysts for photocatalytic H2 production by water splitting," *AIP Adv*, vol. 13, no. 12, Dec. 2023. https://doi.org/10.1063/5.0173850

[18] Q. Zhang *et al.*, "Synthesis of Silicotungstic Acid/Ni-Zr-O Composite Nanoparticle by Using Bimetallic Ni-Zr MOF for Fatty Acid Esterification," *Catalysts*, vol. 13, no. 1, p. 40, Dec. 2022. https://doi.org/10.3390/catal13010040

[19] H. Liu, Y. Tang, X. Ma, and W. Yue, "Catalytic pyrolysis of corncob with Ni/CaO catalysts for hydrogen-rich gas: Synthesis modes and catalyst/biomass ratios," *Journal of Industrial and Engineering Chemistry*, vol. 123, pp. 51–61, Jul. 2023. https://doi.org/10.1016/j.jiec.2023.03.021

[20] L. Hermida, A. Z. Abdullah, and A. R. Mohamed, "Synthesis of monoglyceride through glycerol esterification with lauric acid over propyl sulfonic acid post-synthesis functionalized SBA-15 mesoporous catalyst," *Chemical Engineering Journal*, vol. 174, no. 2–3, pp. 668–676, Nov. 2011. https://doi.org/10.1016/j.cej.2011.09.072

[21] T. N. Afonasenko *et al.*, "The Formation of Mn-Ce-Zr Oxide Catalysts for CO and Propane Oxidation: The Role of Element Content Ratio," *Catalysts*, vol. 13, no. 1, p. 211, Jan. 2023. https://doi.org/10.3390/catal13010211

[22] Z. J. Berkson *et al.*, "Active Site Descriptors from 95Mo NMR Signatures of Silica-Supported Mo-Based Olefin Metathesis Catalysts," *J Am Chem Soc*, vol. 145, no. 23, pp. 12651–12662, Jun. 2023. https://doi.org/10.1021/jacs.3c02201

[23] H. Windeck, F. Berger, and J. Sauer, "Spectroscopic Signatures of Internal Hydrogen Bonds of Brønsted-Acid Sites in the Zeolite H-MFI," *Angewandte Chemie*, vol. 135, no. 25, Jun. 2023. https://doi.org/10.1002/ange.202303204

[24] M. Aamir, W. Aleem, M. N. Akhtar, A. A. Din, G. Yasmeen, and M. N. Ashiq, "Synthesis and characterizations of Co–Zr doped Ni ferrite/PANI nanocomposites for photocatalytic methyl orange dye degradation," *Physica B Condens Matter*, vol. 624, p. 413392, Jan. 2022. https://doi.org/10.1016/j.physb.2021.413392

[25] H. A. Dawoud, L. Abu Ouda, and S. K. K. Shaat, "FT-IR Studies of Nickel Substituted Polycrystalline Zinc Spinel Ferrites for Structural and Vibrational Investigations," *Chem Sci Trans*, vol. 6, no. 2, Apr. 2017. https://doi.org/10.7598/cst2017.1315

[26] S. Wang, L. Chen, B. Liu, J. Song, and C. Li, "First-principles studies on electronic structure and optical properties of two-dimensional ZrO_2," *Modern Physics Letters B*, vol. 36, no. 12, Apr. 2022. https://doi.org/10.1142/S021798492250049X

[27] D. Kumar, A. Singh, B. S. Saini, B. C. Choudhary, V. Shinde, and R. Kaur, "Effect of Ni Doping on the Structural and Optical Properties of ZrO2 Thin Films," *J Electron Mater*, vol. 50, no. 1, pp. 65–74, Jan. 2021. https://doi.org/10.1007/s11664-020-08558-0

[28] J. N. Appaturi *et al.*, "Highly Active Nickel (II) Oxide-Supported Cerium Oxide Catalysts for Valorization of Glycerol into Oxygenated Fuel Additives," *Materials*, vol. 16, no. 13, p. 4713, Jun. 2023. https://doi.org/10.3390/ma16134713

[29] Ma. E. Manríquez-Ramírez, I. Elizalde, R. Ramírez-López, M. Trejo-Valdez, and M. Estrada-Flores, "Acetylation of glycerol using MgO–CaO catalysts with different CaO loadings," *Reaction Kinetics, Mechanisms and Catalysis*, vol. 130, no. 1, pp. 417–431, Jun. 2020. https://doi.org/10.1007/s11144-020-01774-z

[30] X. Zhang *et al.*, "Preparation of mesoporous CaO-ZrO2 catalysts without template for the continuous synthesis of glycerol carbonate in a fixed-bed reactor," *Appl Catal A Gen*, vol. 590, p. 117313, Jan. 2020. https://doi.org/10.1016/j.apcata.2019.117313

Materials Research Proceedings 53 (2025) 482-497 https://doi.org/10.21741/9781644903575-50

Optimization of alkaline electrolysis system for hydrogen production: A simulation study

MUNIRAH NURFATIHAH Mohd Nasir[1,a]* and SITI NUR AZELLA Zaine[1,b]*

[1]Chemical Engineering Department, Universiti Teknologi PETRONAS, Seri Iskandar, 32610 Perak, Malaysia

[a]munirah_19001255@utp.edu.my, [b]nurazella.zaine@utp.edu.my

Keywords: Alkaline Electrolysis, Hydrogen Production, Simulation, Aspen Plus

Abstract. Global economic development and population growth have increased the energy demand, necessitating a reduction in the dependence on non-renewable sources. Rising energy demand necessitates a shift from fossil fuels to renewable hydrogen production. While nickel-based electrodes are common in alkaline electrolysis, their performance can be further enhanced. Previous studies indicate Ni/TiO_2 electrodes as promising electrode materials; however, these findings are based on experimental studies and a preliminary simulation study is needed to further be utilised in industrial-scale applications. This study aims to develop and validate a simulation model to predict the performance of an alkaline electrolyser using the new Ni/TiO_2 coated electrode system. The base case model of the alkaline electrolysis system was developed using Aspen Plus and validated by comparing the simulation result with the reference data, with the maximum acceptable overall percentage difference of 5%. Based on the result obtained, changing electrode material from conventional Ni to Ni/TiO_2 coated electrode reduces the activation overpotential by 4.13% and it improves the hydrogen production rate by 4.3% due to its enhanced electrocatalytic properties. TiO_2 coating enhanced the electrocatalyst efficiency, leading to higher hydrogen evolution rates which is mainly attributed to the increased active surface area and improved charge transfer which has eventually increased the amount of hydrogen produced.

Introduction

The need for energy has grown significantly due to the development of the global economy as well as the increase in the number of populations. Fossil fuels which include coal, crude oil and natural gas are the current primary energy sources. In moving towards Net-Zero Carbon Emissions and sustainability agenda, current dependence on fossil fuels must be minimized since it is categorized as non-renewable energy and will eventually be depleted [1]. Therefore, renewable energies must be explored extensively to replace the current demand for fossil fuel consumption. Renewable energy will continuously meet the energy demand since it will remain viable throughout the lifetime. Furthermore, it does not exhibit harmful significant impacts on the environment, especially in the aspect of climate change. Common types of renewable energies include wind, solar and water (hydropower). In the aspect of hydrogen production, both renewable and non-renewable energy can be utilised, where various industrial methods and technological advancements are available, whereby each method requires feed and energy such as non-renewable fossil fuels as well as renewable water and biomass. However, this study will mainly focus on renewable energy, specifically water utilization for the electrolysis process.

There are four methods classified under the electrolysis process which are alkaline electrolysis, proton exchange membrane (PEM) electrolysis, anion exchange membrane (AEM) electrolysis and solid oxide electrolysis (SOEL) [2]. Electrolysis is a potential option for carbon-free hydrogen production since it is a process of separating water into hydrogen and oxygen. The separating process occurs in a unit known as an electrolyser, which varies in size, depending on the size and requirements of hydrogen production plants. In electrolyser, anode and cathode are separated by a separator and both electrodes are immersed in electrolyte [1]. Since the type of electrode materials and type of electrolyte used in the electrolysis process can vary, hence different electrolyser functions in different manners with different degrees of performance. From the industrial point of view, alkaline electrolysis is considered the most practised method and a mature technology. The alkaline electrolyser system operates by the movement of hydroxide ions through the electrolyte from cathode to anode. At the cathode, hydrogen will be produced meanwhile oxygen will be produced at the anode [1, 2]. In industry, common electrolytes used in alkaline electrolyser are potassium hydroxide (KOH) and sodium hydroxide (NaOH) and the type of electrode material used is nickel (Ni) coated stainless steel as well as nickel alloys [1]. Hence, concerning hydrogen production through an alkaline electrolysis system, a simulation study was conducted by developing the simulation model to optimize the production process as well as maximize the amount of hydrogen produced.

Alkaline Electrolysis System

The alkaline electrolysis system includes two electrodes, an anode and a cathode which are immersed in an aqueous electrolyte and separated by a separator, in which the function of this separator is to physically separate these two electrodes and simultaneously separate the product gases produced at respective anode and cathode [1]. Product gas intermixing can also be avoided with the aid of this separator. At each respective electrode, hydrogen can be collected at the cathode meanwhile oxygen can be collected at the anode. An alkaline electrolyser enables the use of cheaper and more readily available electrode metals such as Nickel and Cobalt [1, 2]. The most used strong base in electrolysis is KOH, which produces potassium cations and hydroxide anions. The concentration of KOH in the water is typically between 20wt% - 40wt%, preferably within 25wt% - 30wt% [3]. NaOH or sodium chloride (NaCl) can be used as an alternative option, producing sodium cation and hydroxide or chloride anion. However, KOH is preferred due to its higher ionic conductivity and lower solubility of CO_2 in a KOH-water solution. When using NaOH, CO_2 from the air will easily dissolve in the electrolyte, forming carbonate ions and further reducing the ionic conductivity [3].

Electrode and Electrocatalyst

Porous electrodes are now commonly used in industrial applications of alkaline electrolysers instead of nonporous solid electrodes. This is because the porous electrodes offer a much larger surface area for contact between the reactants, the electronic conductor, and the electrolyte that fills the empty spaces within the porous structure [4]. The performance of alkaline electrolysis cells is greatly influenced by various stages of electrochemical reactions at the electrode-electrolyte boundary such as the transfer of reagents to the electrode surface, diffusion of reactive substances within the electrode pores, electrochemical transformation on the catalytic sites, ion and charge transfer, as well as diffusion of species within the electrode pores and towards the external surface [5]. The electrode itself acts as a catalyst to some extent by affecting the activation energy of the electrochemical reaction. However, in electrode design, it is common to use doping or coating techniques to create a more stable and active layer [6]. Similar to catalysts, these electrocatalysts facilitate charge transfer and reduce the activation energy of the reaction, leading to a decrease in overpotential. The choice of electrocatalyst is influenced by the electronic structure of the electrodes. For example, metals like Ni, Pd, and Pt with specific electronic configurations exhibit lower overvoltage values, while metals like Zn, Cd, and Hg with different electronic configurations show higher values [6]. Non-noble metals, including those doped with Li, also exhibit electrocatalytic activities. The physical properties of electrode materials, such as increasing the active area of a nanostructured electrode can reduce operating current density and improve efficiency [1]. Current research on electrode material and electrocatalysts are summarised in Table 1.

Table 1: Comparison of electrode material and electrocatalyst

Electrode Materials	Operating conditions and procedure	Tafel Slope [mV/dec]	$V_{hydrogen}$ [mL]	Reference
Ni Foam (NF)/ NiCo-2/Ag	30 min of electrolysis with constant potential of 3V with current density of 5mA/cm^2 and at 1.5V	0.48	108.1	[7]
C felt (CF) /NiGa		Not provided	28.9	[8]
NiCuF-2		0.515	104.1	[9]
AgNiCu		0.362	105.7	[9]
NiFe Nanocone Array (NA)	Reach current density of 200mA/cm^2 and 500mA/cm^2 at 1.684V and 1.841V respectively	44.4	Not provided	[10]
IrO_2-NF		81.5	Not provided	[10]
Ni/TiO_2	Hydrogen production at a voltage level of 6V	110	5815	[11]

Based on Table 1, as per a current study conducted among different materials, it is concluded that Ni/TiO_2 is the most promising electrode material to be used in this study since the experimental studies done are conducted at large-scale production, where this electrode can produce the highest amount of hydrogen as compared to another electrode. Although the Ni/TiO_2 experiment was conducted at a higher voltage level, which eventually led to the high amount of hydrogen produced, however when doing an indirect comparison with other materials through comparison at similar voltage levels, Ni/TiO_2 is still the most promising electrode to be chosen in this study to further verify the experimental results shown in Table 1 by using Aspen Plus simulation

Electrolyte

Alkaline electrolysis systems utilise electrolytes consisting of aqueous solutions of KOH or NaOH. These salts are dissolved in demineralized water to enhance the conductivity of the solution, hence minimizing ohmic losses. However, these electrolyte solutions can be corrosive to the electrodes, which can result in a decrease in their catalytic activity and overall lifespan [2]. To mitigate the corrosive effects of the electrolytes, a molecular electrocatalyst can be added. The energy requirements for water splitting can be reduced by employing these measures.

Separator

The separator in an alkaline electrolysis cell serves one main purpose which is separating the generated gases. Its ability to conduct electricity is influenced by its tortuosity and porosity, which affects the liquid electrolyte's conductivity within the separator's pores [2]. To fulfil its functions, the separator needs to be stable in highly alkaline conditions and have high conductivity for the transport of hydroxide ions. In an alkaline electrolyser, the movement of ions through the separator can be described using Fick's law, which relates the diffusion of ions to the separator's area, thickness, porosity, and the concentration gradient of the dissolved component between the cathode and anode compartments. These mathematical models accurately represent how ions diffuse through the separator and highlight the significance of factors in determining ion transport such as area, thickness, tortuosity, and porosity.

Electrochemical Model

To assess the efficiency of the alkaline electrolysis system through overall cell voltage, electrochemical models are employed. A commonly used method to view the electrochemical behaviour is through a polarization curve. This model serves as the basis for the system's modelling. Researchers often describe the polarization curve using the cell voltage, V_{cell} approach which includes the total of overpotentials, and reversible voltage and it is represented by Eq. 1 [12-16].

$$V_{cell} = V_{rev} + V_{act} + V_{ohmic} + V_{conc} \tag{1}$$

The reversible voltage, V_{rev} represents the minimum voltage required to produce hydrogen through water electrolysis. V_{act} represents the activation overvoltage for the electrode, V_{ohmic} represents the ohmic overvoltage and V_{conc} represents the concentration overvoltage. However, concentration overvoltage is neglected in most of the studies since significant concentration loss occurs when the current density is very high, which is significantly higher than the typical operating range of current density in alkaline electrolysers [17]. Eq. 1 also serves as the fundamental equation for constructing electrochemical models that describe the performance of cells or stacks in alkaline electrolysis. To estimate the cell voltage, each respective voltage must be first determined. For reversible cell voltage, is defined as the minimum voltage needed to begin the electrolysis process, which is represented by Eq. 2 [17],

$$V_{rev} = E^{\circ}_{rev} + (T - T_{ref}) * \frac{\Delta S^{\circ}}{nF} + \frac{RT}{2F} [\ln\left(\frac{(P-P_{H2O})^{\frac{3}{2}}}{\frac{P_{H2O}}{P^{\circ}_{H2O}}}\right)] \tag{2}$$

E°_{rev} is the minimum reversible voltage for electrolysis to occur at standard conditions, T is the operating temperature, T_{ref} is the reference temperature at which E°_{rev} and ΔS° are estimated, $\frac{\Delta S^{\circ}}{nF}$ represents entropy change of standard state, R is the universal gas constant, F is the Faraday constant, P is the operational pressure, P_{H2O} is the vapour pressure of water over the electrolyte and $P^{\circ}{}_{H2O}$ is the vapour pressure of pure water.

Next, the activation overvoltage, V_{act} quantifies the energy required to break chemical bonds before an electrochemical reaction occurs. It is influenced by the kinetics of electronic charge transfer reactions taking place at the electrode's reaction site [17]. The activation overpotential depends on the current density and exchange current density. Current density refers to the flow of electrons or ions per unit surface area, while exchange current density represents the current in the absence of net electrolysis and zero overpotential. The activation voltage arises from an overvoltage across the electrode [17] and is represented by Eq. 3,

$$V_{act} = b_{anode} \ln\left(\frac{j_{anode}}{j_{o-anode}}\right) + b_{cathode} \ln\left(\frac{j_{cathode}}{j_{o-cathode}}\right) \tag{3}$$

The Tafel equation can be used to determine the activation overpotential, which is expressed by Eq. 3 for anode and cathode activation, respectively. The Tafel slope determines the slope of the activation overpotential in the linear region of the polarization curve, which can be obtained from previous literature papers. From Eq. 3, b_{anode} and $b_{cathode}$ represent the Tafel slope, j_{anode} and $j_{cathode}$ is the current density meanwhile $j_{o-anode}$ and $j_{o-cathode}$ is the effective exchange current density [17].

Next, ohmic overpotential, V_{ohmic} arises from the presence of electrical resistance in electrodes, electrolytes, and membranes. The total ohmic resistance is the combined resistance caused by the electrode, electrolyte, and separator. Multiplying the total ohmic resistance by the current yields the ohmic overpotential, which can be represented by Eq. 4,

$$V_{ohmic} = I(R_{electrode} + R_{electrolyte} + R_{separator}) \tag{4}$$

To estimate the resistance caused by electrode, electrolyte and separator, Eq. 5, Eq. 6 and Eq. 7 listed are used respectively,

$$R_{electrode} = \rho_{eff-a}\frac{t_a}{A_a}[1 + k_a(T - T_{ref})] + \rho_{eff-c}\frac{t_c}{A_c}[1 + k_c(T - T_{ref})] \quad (5)$$

$$R_{electrolyte} = \frac{1}{\sigma_{KOH,free}}\left(\frac{d_{am}}{A_a} + \frac{d_{cm}}{A_c}\right) \quad (6)$$

$$R_{separator} = \frac{1}{\sigma_{KOH}}\left(\frac{\tau_m^2 t_m}{\omega_m \epsilon_m A_m}\right) \quad (7)$$

From Eq. 5, $\rho_{eff-a,c}$ represents the effective resistance of anode and cathode respectively, $t_{a,c}$ is the thickness of anode and cathode, $A_{a,c}$ is the cross-sectional area of electrode and $k_{a,c}$ represents the temperature coefficient of resistivity [17]. From Eq. 6, $\sigma_{KOH,free}$ represents bubble-free ionic conductivity of KOH solution, d_{am} is the distance between anode and separator meanwhile d_{cm} is the distance between cathode and separator. From Eq. 7, τ_m represents the tortuosity of the separator, t_m is the separator thickness, ω_m is the separator wettability meanwhile ϵ_m is the porosity of the separator.

Before estimating the overall cell voltage, V_{rev} , V_{act} and V_{ohmic} must be first determined by using all the equations shown above. In summary, these mathematical models accurately depict the behaviour of electrolysers by considering several parameters. The models have been obtained and validated from previous literature by [17].

Development and Validation of Simulation Model

In developing the simulation model of the alkaline electrolysis system, Aspen Plus software was utilised since this software includes a comprehensive thermodynamic and property database which facilitates accurate modelling of the chemical reactions involved in alkaline electrolysis. In addition, Aspen Plus is also chosen as this software provides better visualization of streams flowing in and out of each equipment. Firstly, several assumptions were considered [18] before developing this simulation model which are;

1. Overall alkaline electrolysis system operates at steady-state conditions.
2. All involved gases behave similarly to ideal gas behaviour.
3. Electrolytic cells are uniform in performance.
4. The alkaline electrolysis stack operates at a balanced pressure for both electrodes.

Then, the development of a simulation model was performed starting from the stack component of the alkaline electrolysis system. Specifically for the electrolysis stack, based on previous studies, two main techniques were used to develop the electrolysis stack which is: (i) stack model will be developed in Aspen Custom Modeler, then the developed model will be integrated into Aspen Plus, and (ii) direct usage of available equipment in Aspen Plus. For the first technique, Aspen Custom Modeler (ACM) tools will be utilised since there are no electrolyser stack models that are readily available in the Aspen Plus software. To develop a stack operation unit, the Aspen Properties file will be created and transferred to the ACM tool to specify components and choose a suitable property method. In ACM, an electrochemical model for alkaline electrolysis cells as well as equations involving mass and energy balances occurring in the stack will be included in the coding. All mathematical equations in the electrochemical model stated in the previous section will be calculated automatically after completing the development of the stack model. The developed stack model can be further utilised in Aspen Plus by integrating the model into Aspen Plus.

On the other hand, the second technique involves the manual calculation using Microsoft Excel for all equations shown in the electrochemical model section previously. From the calculations, the overall cell voltage can be estimated as well as the fractional conversion of water that is required to be input in the Aspen Plus later. Then, an overall electrolysis simulation model can be developed in Aspen Plus to further obtain the amount of hydrogen produced. From both techniques, the second technique is utilised in this study since it is less complex as compared to the first technique. However, since this second technique is infrequently used in previous research, the validation of this technique must be done by comparing the simulation result with the past literature that utilised the complex first technique.

In developing the overall simulation model, the property method used is Electrolyte-Non-Random-Two-Liquid (ELECNRTL) since it is recommended to use this property method for simulations involving electrolytes. ELECNRTL can represent both aqueous and organic electrolyte systems across the entire range of electrolyte concentrations. For the stack operation unit, data for nickel-coated stainless-steel electrodes was used and several required parameters from the literature are displayed in Table 2, Table 3, Table 4, Table 5, Table 6 and Table 7.

Table 2: Parameters required for alkaline electrolysis stack

Parameter	Value	Unit	Sources
Stack Temperature, T_{stack}	75	°C	
Stack Operating Pressure, P_{stack}	7	bar	[18]
Electrolyte (KOH) Concentration	35	wt%	
Cell Number, N	12	cells	
Electrical Power Input, W_{stack}	10	kW	
Cell Current Density, j	0.42	$A.cm^{-2}$	

Table 3: Feedwater condition in alkaline electrolysis system

Parameter	Value	Unit	Sources
Temperature	25	°C	[18, 19]
Pressure	1	bar	
Mass Flowrate	900	kg/hr	

Table 4: Electrochemical model constant at a reference temperature of 25°C

Parameter	Value	Unit	Sources
Standard Entropy Change, ΔS°	-162.86	J/mol.K	
Faraday Constant, F	96485	C/mol	
Universal Gas Constant, R	8.314	J/mol.K	[17, 20]
Standard Enthalpy Change, ΔH°	-286020	J/mol	

Table 5: Anode electrode (Nickel) material parameter input

Parameter	Value	Unit	Sources
Electrode Roughness Factor, γ_a	1.25		
Free Activation Energy, ΔG_a	41500	J/mol.K	
Reference Exchange Current Density, $j_{o-ref,a}$	1.35×10^{-5}	A/cm^2	
Surface Area, A_a	300	cm^2	
Thickness of Electrode, t_a	0.2	cm	[17, 20, 21]
Resistivity of 100% Dense Ni Electrode, ρ_a	6.4×10^{-6}	Ω.cm	
Temperature Coefficient of Resistivity, k_a	0.00586	/°C	
Distance between Anode and Membrane, d_{am}	0.125	m	
Porosity, $\epsilon_{a,c}$	0.3		
Tafel Slope, b	0.117	V	
Current Density, j_a	0.0738800	A/cm^2	[22]
Exchange Current Density, $j_{o,a}$	0.0000720	mA/cm^2	

Materials Research Proceedings 53 (2025) 482-497 https://doi.org/10.21741/9781644903575-50

Table 6: Cathode electrode material parameter input

Parameter	Value	Unit	Sources
Electrode Roughness Factor, γ_c	1.05		[17, 20, 21]
Free Activation Energy, ΔG_c	23450	J/mol.K	
Reference Exchange Current Density, $j_{o-ref,c}$	1.85×10^{-3}	A/cm²	
Surface Area, A_c	300	cm²	
Thickness of Electrode, t_c	0.2	cm	
The resistivity of 100% Dense Ni Electrode, ρ_c	6.4×10^{-6}	Ω.cm	
Temperature Coefficient of Resistivity, k_c	0.00586	/°C	
Distance between Cathode and Membrane, d_{cm}	0.125	m	
Porosity, $\epsilon_{a,c}$	0.3		
Nickel Electrode			
Tafel Slope, b	0.117	V	[22]
Current Density, j_c	0.0738800	A/cm²	
Exchange Current Density, $j_{o,c}$	0.0000720	A/cm²	
Ni/TiO₂ Electrode			
Tafel Slope, b	0.123	V	[22]
Current Density, j_c	0.0550600	A/cm²	
Exchange Current Density, $j_{o,c}$	0.0003470	A/cm²	

Table 7: Separator material parameter input

Parameter	Value	Unit	Sources
Surface Area, A_m	300	cm²	[17, 20]
Thickness, t_m	0.05	cm	
Porosity, ϵ_m	0.42		
Tortuosity, τ_m	2.18		
Wettability Factor	0.85		

Since the second technique is used, all required data was utilised in the manual calculation using Microsoft Excel. From the calculations, it is desired to calculate the fractional conversion of water to hydrogen and this value was further used in Aspen Plus simulation to further estimate the amount of hydrogen produced. Then, the simulation results obtained were validated by comparing them with the data obtained by [18] by calculating the percentage difference between both data, as represented by Eq.8. Then, the percentage difference for each stream was summed up and divided by the total streams to obtain the overall percentage difference as shown by Eq. 9.

$$\text{Percentage Difference (\%)} = \frac{\text{Simulation Value} - \text{Reference Value}}{\text{Reference Value}} \times 100\% \quad (8)$$

$$\text{Overall Percentage Difference (\%)} = \frac{\text{Total Percentage Difference}}{\text{Number of Streams}} \quad (9)$$

The simulation results obtained were compared with the data obtained from the base case developed before, in terms of the amount of hydrogen produced as well as the overall cell voltage. The percentage increment of hydrogen production between the conventional Ni electrode and Ni/TiO_2 coated electrode was calculated by using Eq. 10.

$$\text{Increment Percentage} = \frac{H_2\,(Ni/TiO_2) - H_2\,(Ni)}{H_2\,(Ni)} \times 100\% \quad (10)$$

Result and Discussion

To develop an alkaline electrolysis system simulation model values such as the fractional conversion of water must be estimated since it is needed to be input in the Aspen Plus simulation.

Therefore, calculation steps involving Eq.1 to Eq. 7 were performed, and few assumptions were made before estimating overall cell voltage and fractional conversion which include,

1. Voltage loss due to the concentration effect is negligible since concentration losses happen at a very high current density of approximately 800,000 A/cm^2, which is relatively higher than the normal operating range of current density in the alkaline electrolyser [17].
2. Overall cell voltage is influenced by reversible voltage, activation overpotential as well as ohmic overpotential.
3. Both anode and cathode are made up of conventional Nickel electrodes.
4. Fractional bubble coverage on the electrode surface is constant and does not influence the activation overpotential.

The calculation results for cell voltage and fractional conversion are displayed in Table 8.

Table 8: Cell voltage and fractional conversion for hydrogen production

Electrode	Cell Voltage			Fractional Conversion
	Reversible Voltage (V_{rev})	Activation Overpotential (V_{act})	Ohmic Overpotential (V_{ohmic})	
Ni	1.2210 V	0.7046 V	0.0515 V	0.002893
Ni/TiO_2	1.2210 V	0.6230 V	0.0515 V	0.003018

From Table 8, the respective voltage is similar for both type of electrode utilised since it is assumed that each respective voltage is independent of the number of cells. To elaborate more, for reversible voltage, parameters that will affect this voltage are the concentration of KOH and stack temperature [17], in which these two parameters are kept constant. However, stack pressure slightly affects the reversible voltage, which can be neglected [17]. Hence, the reversible voltage is similar for both cases. On the other hand, the activation overpotential is dependent on the type of electrode material used. Since both cases utilised different type of electrodes, therefore the activation overpotential differs, in which Ni/TiO_2 have lower activation overpotential. Meanwhile, ohmic overpotential is dependent on the electrode, electrolyte and separator resistance. Since ohmic overpotential is considered insignificant in this study, hence it is assumed that the ohmic overpotential is constant for both cases. For fractional conversion of water, the value is different for different types of electrodes since it is influenced by the overall cell voltage of electrolyser stack. Fractional conversion of water is affected by the cell voltage, current density, active electrode area and number of cells. Since current density, as well as active electrode area, is assumed to be constant, hence only the number of cells will affect the fractional conversion of water. It is shown that the overall cell voltage is inversely proportional to the fractional conversion, therefore lower overall cell voltage will increase the fractional conversion of water.

In hydrogen production, the overall process includes the conversion of water to hydrogen gas and oxygen gas, then the produced oxygen and hydrogen are further purified using a separator. The developed simulation model is displayed in Figure 1. The data input and unit operation utilised in developing the simulation model are described in Table 9.

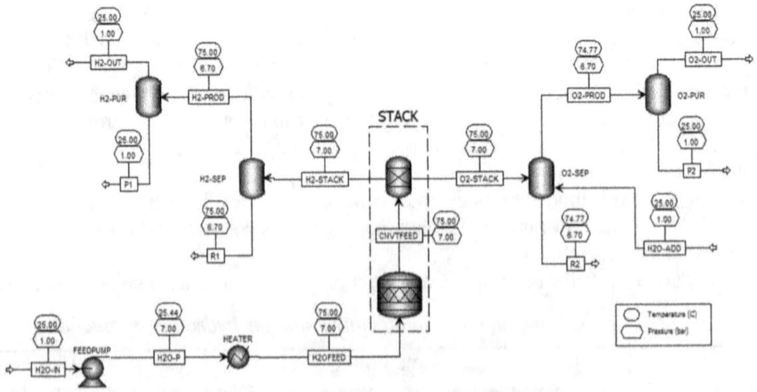

Fig. 1: Hydrogen production simulation model development

Table 9: Justification on data input and equipment used

Equipment	Process Descriptions and Specifications
Pump (FEEDPUMP)	At the beginning of the electrolysis process, the KOH-water inlet was fed into the feed pump to increase the pressure of the feed stream for atmospheric pressure to approximately a pressure of 7 bar. Parameters specified in this simulation include, Feed Inlet: • T = 25°C, P = 1 bar • Flowrate: 900 kg/hr • KOH concentration: 35wt% Pump Specification: • Discharge pressure, P = 7 bar
Heat Exchanger (HEATER)	The pressurised KOH-water inlet flows into the heater to preheat the feed to 75°C, before entering the electrolyser stack. Parameters specified in this simulation include, Inlet: • P = 7 bar, T = 25°C Heater Specifications: • P = 7 bar, T = 75°C
Electrolyser (STACK)	The main purpose of this stack is to separate hydrogen and oxygen components in water through an electrolysis process. Since electrolyser unit operation is unavailable in Aspen Plus V12.1, therefore electrolyser unit operation was simulated using two different equipment which are reactor (RStoic) and component separator (Sep), [23]. Data for conventional nickel electrodes for both anode and cathode were used in this electrolyser. For the reactor equipment (RStoic), it is used when the reaction kinetics are unknown, however, the stoichiometry and fractional conversion are known. It was utilised in this study since the reaction stoichiometry

Equipment	Process Descriptions and Specifications
	is known and reaction kinetics is negligible. This RStoic was used for water decomposition reaction to hydrogen and oxygen, in which data input specified were, Reactor Specifications [18]: • P = 7 bar • Duty: 10kW Reaction: • $H_2O \rightarrow H_2 + \frac{1}{2}O_2$ • Estimated fractional conversion of water: 0.002893 For the separator equipment (Sep), the function of this equipment is to separate the inlet stream into any number of outlet streams. Hence, in this study, Sep was utilised to separate two components which are hydrogen gas and oxygen gas. Hydrogen gas will be separated from oxygen gas, meanwhile, unreacted water and KOH are separated equally to both outlets, anode and cathode. Separator Specifications and Assumptions: • P = 7 bar
Flash Separator (H_2-SEP and H_2-PUR)	Produced hydrogen and oxygen gas flow into respective flash separators to separate both hydrogen and oxygen from water and KOH for further product purification. For H_2 purification, the produced gas flows into the first flash separator (H_2-SEP) to separate produced gas from unconverted water and KOH, then the produced gas flows into the second flash separator (H_2-PUR) to further purify the produced gases to approximately >75wt% [18]. H_2-SEP Specifications: • P = 6.7 bar, no duty H_2-PUR Specifications: • T = 25°C, P = 1 bar
Flash Separator (O_2-SEP and O_2-PUR)	For O_2 purification, the produced gas flows into the first flash separator (O_2-SEP) to separate produced gas from unconverted water and KOH, then the produced gas flows into the second flash separator (O_2-PUR) to further purify the produced gases to approximately >90wt% [18]. O_2-SEP Specifications: • P = 6.7 bar, no duty O_2-PUR Specifications: • T = 25°C, P = 1 bar

To validate the developed simulation model, simulation results were compared with data from previous literature by [18]. Firstly, the validation of the developed simulation model was conducted using small-scale production by calculating the percentage difference of the most significant parameters, which are hydrogen production and oxygen production as shown in Table 11.

Table 11: Simulation model validation through the main product produced

Parameter		Literature Data [18]	Simulation Result	Percentage Difference (%)
Product yield (kg/hr)	Hydrogen	0.173	0.186	7.60
	Oxygen	1.329	1.446	8.78
Product purity (wt%)	Hydrogen	78.7	77.2	1.91
	Oxygen	98.2	98.2	0.00

Based on the data presented in Table 11, both products from the simulation show close proximity to the previous literature data. Both hydrogen and oxygen product yields obtained from simulation are slightly higher than the literature data. This is because, in developing simulation models, it is assumed that all equipment operates ideally by default, where heat losses in the overall system are minimized, which eventually leads to higher hydrogen production in the simulation model. Moreover, the simulation model was developed by using a second technique, which is more simplified compared to the technique used in the literature data. The technique used is simpler since the main purpose of the simulation model is to provide a general process flow and the products produced are mainly under ideal conditions. From the calculated percentage difference, the value is 7.60% and 8.78% for hydrogen and oxygen respectively and it is considered within the acceptable range. On the other hand, for product purity, it is observed that the simulation model predicts slightly lower product purity of hydrogen, and similar product purity was obtained for oxygen. The percentage difference for hydrogen and oxygen product purity is 1.91% and 0% respectively, which is relatively low and within the acceptable range. In addition, the overall percentage difference between these two data was calculated and the overall simulation model accuracy was estimated as shown in Table 12. From the overall percentage difference of the main flow stream, it is observed that the obtained percentage difference is minimal, which is only 2.46%. To further validate the developed simulation model, the overall simulation accuracy was estimated. It is observed that overall simulation accuracy is approximately 97.54%, which is considered a relatively high value of accuracy. Therefore, from this simulation model validation, the developed simulation model can be further utilised for Ni/TiO_2 electrode hydrogen production.

Table 12: Percentage difference calculations

Stream Name	Units	H_2-OUT	H_2-PROD	H_2-STACK	H_2O-ADD	H_2O FEED	O_2-OUT	O_2-PROD	O_2-STACK
Temperature	C	0.000	0.012	0.000	0.000	3.093	0.000	0.052	0.000
Pressure	bar	0.000	0.000	0.000	0.000	0.000	0.000	0.000	0.000
Mass Flows	kg/hr	9.549	5.072	0.378	0.000	0.378	8.844	8.359	0.398
H2O	kg/hr	9.730	8.375	0.376	0.000	0.387	12.414	4.068	0.376
H2	kg/hr	7.597	7.597	6.392	0.000	0.000	0.000	0.000	0.000
O2	kg/hr	0.000	0.000	0.000	0.000	0.000	8.780	8.780	7.053
Percentage Difference		4.48	3.51	1.19	0.00	0.64	5.01	3.54	1.30

Overall Simulation Percentage Difference	2.46	%
Overall Simulation Accuracy	97.54	%

For better visualisation, the relationship between hydrogen production for both type of electrodes was summarised in Table 13,

Table 13: Hydrogen production for Ni and Ni/TiO₂ electrode

Number of Cells	Input Parameter	Electrode	Cell Voltage (V)	Fractional Conversion	Hydrogen Production (kg/hr)
12	P_{stack}: 7 bar T_{stack}: 75°C 35wt% KOH	Anode: Ni Cathode: Ni	23.725	0.002893	0.186
		Anode: Ni Cathode: Ni/TiO_2	22.746	0.003018	0.194

Based on data presented in Table 13, it is observed that hydrogen production increases when a Ni/TiO_2 coated electrode is used as the cathode. This might be due to the overall cell voltage reduction. From the calculated cell voltage, it is observed that the overall cell voltage is lower for Ni/TiO_2 coated electrodes compared to conventional Ni electrodes. As stated before, reduction in cell voltage is contributed solely by activation overpotential. Assuming the constant value of reversible voltage and ohmic overpotential, changing the type of electrode material from conventional Ni to Ni/TiO_2 coated electrode has contributed to the reduction of activation overpotential. The percentage decrement of overall cell voltage was estimated as shown below, in which the overall cell voltage decreases significantly by 4.13%.

$$\text{Decrement Percentage} = \frac{V_{cell}\,(Ni) - V_{cell}\,(Ni/TiO_2)}{V_{cell}\,(Ni)} \times 100\% = 4.13\%$$

Moreover, Ni/TiO_2 coated electrodes exhibit higher fractional conversion of water due to the reduction in activation overpotential. The fractional conversion of water is dependent on the moles of water converted, which will eventually increase the amount of hydrogen produced. This can be explained, since overall cell voltage is inversely proportional to the fractional conversion of water, hence lower cell voltage contributes to the increased fractional conversion of water, which further leads to an increment in hydrogen production. To compare the hydrogen production of conventional nickel electrodes with the hydrogen production of the Ni/TiO₂ system as presented in Table 13, the percentage increment of hydrogen produced was estimated,

$$\text{Increment Percentage} = \frac{H_2\,(Ni/TiO_2) - H_2\,(Ni)}{H_2\,(Ni)} \times 100\% = 4.30\%$$

The findings in this study were compared with other electrode materials from previous literature. Table 14 shows the hydrogen production rate for experimental scale hydrogen production. It is observed that the hydrogen production rate is higher when utilising Ni/TiO_2 as the electrode material.

Table 14: Hydrogen production comparison with previous literature

Cathode Electrode	Hydrogen Gas Volume Produced (mL)	Hydrogen Production (mL/min)	Hydrogen Production (mL/min.cm$^{2)}$	Reference
NF/NiCo-2/Ag	108.1	3.60	1.8	[7]
C@FeZn/Ru	85.6	1.43	0.715	[24]
CF/NiGa	28.9	0.96	0.24	[8]
NiCu-F	104.1	3.47	1.735	[9]
AgNiCu	105.7	3.52	1.76	
Cu	7.7	0.77	-	[25]
Graphite	5.2	0.52	-	
Ni/TiO_2	-	4487.5	14.96	This study

Based on the data presented, it is observed that the hydrogen production rate is significantly higher when utilising Ni/TiO_2 as the electrode material compared to other electrode materials. However, the value obtained for Ni/TiO_2 is considerably higher compared to the experimental data since it is a simulation value which considers ideal gases and does not include losses that may occur throughout the electrolysis process. From the comparison of electrode materials for the cathode, Ni/TiO2 electrode material significantly affects the hydrogen production rate in an alkaline electrolysis system. To summarise, changing electrode material from conventional Ni to Ni/TiO_2 coated electrode improves hydrogen production through its production rate by 4.30%, which is considered a significant increment due to its enhanced electrocatalytic properties. TiO_2 coating enhanced the electrocatalyst efficiency, leading to higher hydrogen evolution rates. This phenomenon is mainly attributed to the increased active surface area and improved charge transfer which has eventually increased the amount of hydrogen produced [22, 26].

Conclusion

Growing energy demand necessitates a transition to renewable sources. This study focuses on hydrogen production from water using alkaline electrolysis, a mature, low-cost technology. Aspen Plus was used to model the system due to its comprehensive capabilities. AEM and SOEL electrolyser are currently in development, and it has not been made widely available. On the other hand, PEM is currently considered as one of the suitable options that is ready for energy transition, however, since this system utilises Pt catalyst which makes this technology slightly expensive and challenging to expand its application. While nickel-based electrodes are common in alkaline electrolysis system, Ni/TiO_2-coated electrodes were chosen due to its superior performance reported in literature. The developed model accurately predicted system behaviour, with a 2.46% overall percentage difference. Switching to Ni/TiO_2 electrodes reduced activation overpotential by 4.3% and increased hydrogen production rate by 4.3%, aligning with findings by Gomez et al. and Uzal et al. This improvement is attributed to enhanced electrocatalytic properties, specifically increased surface area and charge transfer.

Future Work Recommendation

In this study, the development of an alkaline electrolysis system simulation model utilised Aspen Plus software Version 12.1, where the electrolysis stack was modelled using two equipment which are a stochiometric reactor and a separator. However, the electrolysis stack equipment has been upgraded to Aspen Plus Version 14. For future work improvement, the electrolysis stack system can be modelled using this feature to observe the distinct differences between these two techniques in developing electrolysis stack equipment. In addition to that, economic perspective of new Ni/TiO_2 system can also be evaluated to further observe the technical and economic feasibility of the new system. On the other hand, studies involving other types of alkaline electrolysis system optimization can be conducted. As stated in the problem statement section, the optimization of alkaline electrolysis systems for hydrogen production includes parametric optimization and overpotential reduction. Specifically for optimization involving overpotential reductions, which are classified into three distinct categories: (i) electrode materials enhancement, (ii) electrode fabrication method, and (iii) separator materials enhancement. TiO_2 material can be further studied in terms of its suitability to be used as a separator in alkaline electrolysis systems as per preliminary experimental studies conducted by [27].

References

[1] Ji, M. and J. Wang, Review and comparison of various hydrogen production methods based on costs and life cycle impact assessment indicators. International Journal of Hydrogen Energy, 2021. 46(78): p. 38612-38635. https://doi.org/10.1016/j.ijhydene.2021.09.142

[2] Vincent, K. Water Electrolysis Alkaline Technology. 2022.

[3] Kuleshov, V.N., et al., Electrode-diaphragm assembly for alkaline water electrolysis. Journal of Physics: Conference Series, 2020. 1683(5): p. 052011. https://doi.org/10.1088/1742-6596/1683/5/052011

[4] Young, G.J. and R.B. Rozelle, Fuel cells. Journal of Chemical Education, 1959. 36(2): p. 68. https://doi.org/10.1021/ed036p68

[5] Banoth, P., C. Kandula, and P. Kollu, Introduction to Electrocatalysts, in Noble Metal-Free Electrocatalysts: New Trends in Electrocatalysts for Energy Applications. Volume 2. 2022, American Chemical Society. p. 1-37. https://doi.org/10.1021/bk-2022-1432.ch001

[6] Mert, M.E., Production of NF/NiCo/Ag catalyst for hydrogen generation in alkaline medium. International Journal of Energy Research, 2022. 46(4): p. 4130-4141. https://doi.org/10.1002/er.7415

[7] Koca, M.B., et al., NiGa modified carbon-felt cathode for hydrogen production. International Journal of Hydrogen Energy, 2019. 44(27): p. 14157-14163. https://doi.org/10.1016/j.ijhydene.2018.09.031

[8] Mert, M.E., et al., Design and performance analysis of a PV-assisted alkaline electrolysis for hydrogen production: An experimental and theoretical study. Fuel, 2024. 355: p. 129497. https://doi.org/10.1016/j.fuel.2023.129497

[9] Liu, P., et al., A superhydrophilic NiFe electrode for industrial alkaline water electrolysis. International Journal of Hydrogen Energy, 2023. https://doi.org/10.1016/j.ijhydene.2023.07.253

[10] Lattieff, F.A., et al., Modeling of electrocatalytic hydrogen evolution via high voltage alkaline electrolyzer with different nano-electrocatalysts. International Journal of Hydrogen Energy, 2023. https://doi.org/10.1016/j.ijhydene.2023.08.062

[11] PANCOTTO, A., Characterization and modeling of alkaline electrolyzers for hydrogen production. 2021.

[12] Amores, E., J. Rodríguez, and C. Carreras, Influence of operation parameters in the modeling of alkaline water electrolyzers for hydrogen production. International Journal of Hydrogen Energy, 2014. 39(25): p. 13063-13078. https://doi.org/10.1016/j.ijhydene.2014.07.001

[13] Ulleberg, Ø., Modeling of advanced alkaline electrolyzers: a system simulation approach. International Journal of Hydrogen Energy, 2003. 28(1): p. 21-33. https://doi.org/10.1016/S0360-3199(02)00033-2

[14] Abdin, Z., C.J. Webb, and E.M. Gray, Modelling and simulation of an alkaline electrolyser cell. Energy, 2017. 138: p. 316-331. https://doi.org/10.1016/j.energy.2017.07.053

[15] Jang, D., H.-S. Cho, and S. Kang, Numerical modeling and analysis of the effect of pressure on the performance of an alkaline water electrolysis system. Applied Energy, 2021. 287: p. 116554. https://doi.org/10.1016/j.apenergy.2021.116554

[16] Jang, D., et al., Numerical modeling and analysis of the temperature effect on the performance of an alkaline water electrolysis system. Journal of Power Sources, 2021. 506: p. 230106. https://doi.org/10.1016/j.jpowsour.2021.230106

[17] Niroula, S., et al., Parametric Modelling and Optimization of Alkaline Electrolyzer for the Production of Green Hydrogen. IOP Conference Series: Materials Science and Engineering, 2023. 1279(1): p. 012005. https://doi.org/10.1088/1757-899X/1279/1/012005

[18] Sánchez, M., et al., Aspen Plus model of an alkaline electrolysis system for hydrogen production. International Journal of Hydrogen Energy, 2020. 45(7): p. 3916-3929. https://doi.org/10.1016/j.ijhydene.2019.12.027

[19] Sakas, G., et al., Dynamic energy and mass balance model for an industrial alkaline water electrolyzer plant process. International Journal of Hydrogen Energy, 2022. 47(7): p. 4328-4345. https://doi.org/10.1016/j.ijhydene.2021.11.126

[20] Rodríguez, J. and E. Amores CFD Modeling and Experimental Validation of an Alkaline Water Electrolysis Cell for Hydrogen Production. Processes, 2020. 8, DOI: 10.3390/pr8121634. https://doi.org/10.3390/pr8121634

[21] Henao, C., et al., Simulation tool based on a physics model and an electrical analogy for an alkaline electrolyser. Journal of Power Sources, 2014. 250: p. 58-67. https://doi.org/10.1016/j.jpowsour.2013.10.086

[22] Gómez, M.J., et al., Effect of TiO2 content on Ni/TiO2 composites electrodeposited on SS316L for hydrogen evolution reaction. Electrochimica Acta, 2021. 378: p. 138136. https://doi.org/10.1016/j.electacta.2021.138136

[23] Muazzam, Y., et al., Thermo-Economic Analysis of Integrated Hydrogen, Methanol and Dimethyl Ether Production Using Water Electrolyzed Hydrogen. Resources, 2022. 11: p. 85. https://doi.org/10.3390/resources11100085

[24] Farsak, M., et al., The noble metal loading binary iron-zinc electrode for hydrogen production. International Journal of Hydrogen Energy, 2017. 42(10): p. 6455-6461. https://doi.org/10.1016/j.ijhydene.2016.11.078

[25] Rustana, C.E., et al., The Effect of Voltage and Electrode Types on Hydrogen Production from The Seawater Electrolysis Process. Journal of Physics: Conference Series, 2021. 2019(1): p. 012096. https://doi.org/10.1088/1742-6596/2019/1/012096

[26] Uzal, H., A. Döner, and H. Bayrakçeken, Hydrogen evolution behavior of nickel coated TiO2. International Journal of Hydrogen Energy, 2020. 45(60): p. 34879-34887. https://doi.org/10.1016/j.ijhydene.2019.12.138

[27] Ali, M.F., et al., A study on the effect of TiO2 nanoparticle size on the performance of composite separators in alkaline water electrolysis. Journal of Membrane Science, 2023. 678: p. 121671 https://doi.org/10.1016/j.memsci.2023.121671

Decarbonization Technology: ICDT2024 Materials Research Forum LLC
Materials Research Proceedings 53 (2025) 498-511 https://doi.org/10.21741/9781644903575-51

Assessing risks in hydrogen production facilities: A quantitative approach

NURSYAWALEENA Hani[1,a], AFIQAH SHAHIRAH Shariman[1,b], SYAZA IZYANNI Ahmad[1,c,*]

[1]Chemical Engineering Department, Universiti Teknologi PETRONAS, Seri Iskandar, 32610 Perak, Malaysia

[a]nursyawalee_18000789@utp.edu.my, [b]afiqah_19000545@utp.edu.my, [c]syaza.ahmad@utp.edu.my

Keywords: Hydrogen, Energy, Steam Methane Reforming, Quantitative Risk Assessment, Simulation

Abstract. Hydrogen, also known as H_2, has emerged as a promising sustainable energy source, with increasing interest in its potential to replace our reliance on fossil fuels. The global demand for hydrogen is expected to increase from 70 million tonnes in 2019 to 120 million tonnes by 2024. The safety of hydrogen producing facilities is a major concern during the process operation. Hydrogen production carries a considerable operational risk, and catastrophic consequences are possible because hydrogen is easily inflammable and explosive. Thus, understanding the risks associated with hydrogen production is crucial. This paper aims to identify key hazards, safety, and evaluating the risks in the hydrogen production process, focusing on health, safety, and environmental (HSE) properties. Steam Methane Reforming (SMR) is one of the most cost-effective and widely used methods of creating hydrogen. Hence, the SMR process was chosen to be analyzed for risk scenarios using a comprehensive Quantitative Risk Assessment (QRA) integrating PHAST (Process Hazard Analysis Software Tool) and SAFETI (Software for the Assessment of Flammable, Explosive, and Toxic Impact) simulations. Results from the PHAST (Process Hazard Analysis Software Tool) simulation conducted highlighted jet fires and flash fires as the primary hazards, with relatively small threat areas in leak and rupture scenarios. For instance, the pressure swing adsorption reactor recorded significant flame lengths and proximity to the lower flammability limit (LFL), indicating potential flash fire risks. SAFETI (Software for the Assessment of Flammable, Explosive, and Toxic Impact) analysis reveals that fatality rates are within acceptable standards. Ultimately, this study aims to inform safety measures and regulatory standards in the hydrogen production industry, supporting the advancement of sustainable energy alternatives.

Introduction

Human activity and population growth have led to a huge increase in global energy demand. Fossil fuels are the global primary source of energy today, yet they pollute and degrade the environment by emitting greenhouse gases [1].

The greenhouse effect is closely related to energy production, which is the sector of the economy that emits the most carbon dioxide (CO_2), accounting for 80% of global emissions. Fossil fuels account for the biggest share of the global energy matrix. These fuels are depletable and carry dangers associated with extraction and consumption, as well as the production of gases that contribute to the greenhouse impact [2]. The continuous development and fast urbanisation of the world population and economy have resulted in an immense rise in energy demand, driving the transfer from fossil-based fuels to alternative clean renewables [3].

In terms of carbon neutrality, the global energy structure is changing from fossil fuels to renewable energy in order to reduce carbon dioxide (CO_2) emissions. Hydrogen is a significant alternative to help renewable energy development because of its widespread availability, clean and carbon-neutral nature, and numerous application possibilities [4].

Applications for hydrogen, a renewable and non-polluting energy source, may be found in many industries. With a little environmental impact, hydrogen is a clean energy source. Since the hydrogen atom (H) is so reactive chemically, it is not frequently seen in nature in its free elemental form. Unless it is activated in some way, hydrogen is not extremely reactive at normal temperature; nonetheless, in order for a hydrogen molecule to break into hydrogen atoms, very high temperatures are required. In the natural world, hydrogen is mostly bonded to oxygen or carbon atoms [2]. Hydrogen is found in nature in compounds rather than as a single element. A process is used for producing hydrogen, separating it from its individual components [5]. Current methods of hydrogen production, including natural gas reforming, pose fire and explosion risks [6]. SMR, or steam methane reforming, is a common industrial technique [7]. Steam methane reforming (SMR) works by turning the hydrocarbons in methane to hydrogen, H_2 and carbon dioxide, CO_2. This process typically occurs in two stages: a reformer stage and a water-gas shift reactor. Following these basic procedures, the product produced may be purified to remove any contaminants and ensure that the hydrogen meets industry standards [8]. Globally, each year approximately 50 million tonnes of hydrogen is produced [7].

Despite being the most abundant element in the universe, hydrogen is only found on Earth in chemicals like water and living organisms due to its reactivity. It is a colourless, flammable, and odourless gas that poses a risk to safety, especially if a leak goes unnoticed and builds up in a small space where it could eventually catch fire and explode [3]. Common hydrogen production methods, including steam methane reforming, involve working with dangerous chemicals and high-pressure gases. This can make accidents more likely if safety precautions are not taken [9]. For instance, in 1996 a pipe burst in a steam methane reformer (SMR). A 24-inch-diameter stainless steel pipe ruptured at startup, causing the process gas flow to skip the high-temperature shift converter (HTS). The SMR plant yard was filled with process gas from equipment in the process that was both upstream and downstream of the broken line. The process gas leaving the equipment was a mixture of hydrogen, carbon monoxide, carbon dioxide, steam, and methane at a pressure of 550 psig and 650 °F [9].

Therefore, the highly flammable nature of hydrogen necessitates thorough risk assessments for safe production [3]. PHAST (Process Hazard Analysis Software Tool) and SAFETI (Software for the Assessment of Flammable, Explosive, and Toxic Impact) are key software tools in process safety management. PHAST (Process Hazard Analysis Software Tool) assesses risks in industrial processes involving hazardous materials, simulating potential hazard scenarios like fires and toxic releases. SAFETI (Software for the Assessment of Flammable, Explosive, and Toxic Impact) performs quantitative risk assessments for facilities, combining process data, equipment reliability, and failure scenarios to determine risk levels. The majority of quantitative and semi-quantitative risk evaluations carried out in hydrogen-related industries have focused on hydrogen fuel stations and transportation. Not many quantitative risk assessments (QRAs) have been used in the production of hydrogen [7].

A study conducted by Zarei et al. (2013) identified that the desulphurization reactor posed the highest risk in case of a full-bore rupture. The desulphurization reactor, heat exchanger, reformer (furnace), hydrogen purification absorbers, and purge gas buffer are examples of hydrogen units where consequence modelling for vapour cloud explosions via quantitative risk assessment (QRA) was carried out using PHAST (Process Hazard Analysis Software Tool) software [10]. Results from a more recent study by Huser et al. (2021) showed that in case of an explosion, the surrounding people will be affected. The SAFETI (Software for the Assessment of Flammable,

Materials Research Proceedings 53 (2025) 498-511 https://doi.org/10.21741/9781644903575-51

Explosive, and Toxic Impact) simulation was used for calculating risk in situations where safety distances and risk contours are employed to evaluate the risk of hydrogen stations [11]. These studies done regarding safety of hydrogen production, mainly focuses on assessments of potential accidents, selective equipment, and scenario-based with specific conditions. Furthermore, both PHAST (Process Hazard Analysis Software Tool) and SAFETI (Software for the Assessment of Flammable, Explosive, and Toxic Impact) are often utilized separately. Advanced computer models and simulation approaches can predict different accident scenarios and assess their potential consequences. However, unless integrated into a single quantitative risk assessment (QRA) framework, these techniques fall short of providing the actionable information required for effective risk mitigation [12].

This study aims to systematically assess risks associated with hydrogen production through quantitative risk assessment (QRA) with the help of consequence modelling simulations like PHAST (Process Hazard Analysis Software Tool) and frequency analysis software such as SAFETI (Software for the Assessment of Flammable, Explosive, and Toxic Impact). By evaluating potential hazards, such as fire and explosion, this approach ensures safety measures are in place, promoting hydrogen as a clean fuel alternative while mitigating environmental impacts.

Chemicals and Methods

For safety and risk management in the hydrogen production project, a systematic quantitative risk assessment (QRA) approach was utilized.

The first step in the quantitative risk assessment (QRA) process was to clearly define the objectives. This involved understanding why the risk assessment was being conducted and what it aimed to achieve. The objectives included ensuring regulatory compliance, strengthening safety measures, preventing accidents, and reducing possible dangers to people, property, and the environment. Setting clear objectives helped guide the entire risk assessment process and ensured that all relevant aspects were considered. After defining the objectives, the next step was to describe the process in detail. This involved documenting the operations, systems, and activities that were part of the process being assessed. A comprehensive process description helped identify all possible sources of hazards by including information about the equipment used, operational procedures, chemicals handled, the overall workflow, and initial research into facility design, operations, and safety regulations. This step was crucial for understanding the context in which the hazards existed. Potential scenarios and risks were outlined, with data collection gathering relevant operational details. Consequence modeling was the process of evaluating the potential impacts of the selected scenarios. This involved analyzing the outcomes of risk events, such as the extent of damage, the number of people affected, and environmental impacts.

PHAST (Process Hazard Analysis Software Tool) software simulated accident scenarios for consequence modeling, while SAFETI (Software for the Assessment of Flammable, Explosive, and Toxic Impact) software conducted frequency analysis. Frequency estimation helped in understanding the probability of various incidents and was essential for quantifying the overall risk. It provided a basis for prioritizing risk mitigation measures based on the likelihood of occurrence. The final step was risk estimation, which combined the results of frequency estimation and consequence modeling. Risk estimation involved calculating the overall risk by considering both the likelihood and severity of the identified scenarios. This provided a quantitative measure of risk, which was used to make informed decisions about risk management. Risk estimation helped prioritize risks, allocate resources for mitigation, and develop strategies to reduce or eliminate potential hazards. Results were summarized, identifying key risks, and aiding in prioritizing risk management efforts. This approach helps understand facility weaknesses and plan mitigation strategies. Figure 1 illustrates the systematic steps of the quantitative risk assessment process.

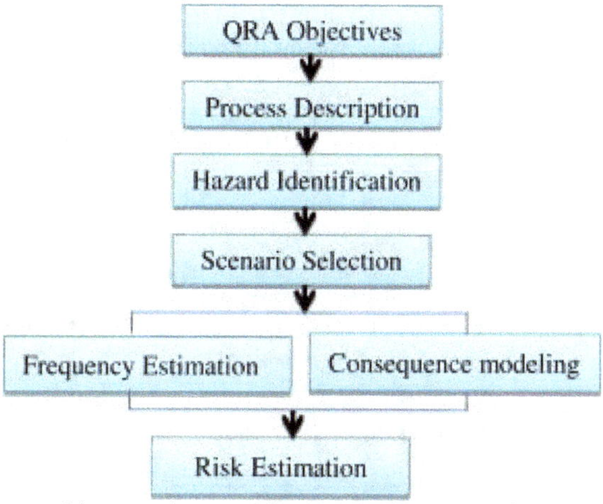

Fig. 1. Flow diagram of the procedure used for QRA (Reference: (Jafari, Zarei, & Badri, 2012))

Results and Discussion

Hazard Identification and Scenario Selection. Prior to doing a quantitative risk assessment, process hazards were identified. All relevant information for the risk assessment was gathered from the hydrogen production process. The information typically includes the geographical location of the unit, atmospheric data of the location, P&IDs (process instrumentation diagrams), PFDs (process flow diagrams), plant layout, production procedures, physical and chemical characteristics of chemical, and other applicable details [7]. Kelly Ibsen's concept (2006) served as the basis for the design of the hydrogen production plant. Kerteh, Terengganu, is the favoured site for the proposed plant.

The dispersion and dilution of chemicals released from a source are affected differently depending on different air conditions, and this data is necessary to understanding this occurrence. Wind speed significantly influences the dispersion of chemicals. Higher wind speeds generally enhance the horizontal spread and dilution of the chemical released, reducing chemical concentrations near the source. Conversely, lower wind speeds may result in higher concentrations close to the source due to limited dispersion. The ability of the atmosphere to disperse chemicals is measured by the atmospheric stability class. This approach uses surface wind speeds, insolation during the day, and cloudiness during the night to classify stability [13]. Figure 2 shows the estimation for each atmospheric stability class.

In this study, stability classs C and F were chosen for the consequence modelling in PHAST (Process Hazard Analysis Software Tool) and SAFETI (Software for the Assessment of Flammable, Explosive, and Toxic Impact). Class F represents very stable atmospheric conditions, typically occurring during nighttime with clear skies and low wind speeds. The atmosphere is characterized by minimal turbulence and weak vertical mixing. Due to constrained vertical dispersion, the chemical released under class F conditions stays concentrated close to the ground. This raises the concentration of chemicals around the source, increasing the danger of exposure in the nearby area. Meanwhile, class C indicates slightly unstable atmospheric conditions, often occurring during the daytime with moderate sunlight and higher wind speeds. The atmosphere exhibits more turbulence and better vertical mixing compared to stable conditions. The chemical spreads more evenly in both vertical and horizontal directions under class C conditions. As a result, the chemical becomes wider and less concentrated, harmful elements dispersing across a greater surface area and the concentrations close to the source decreasing.

Daytime			
Wind speed	Fine (0 - 3 oktas)	Sunny periods / intervals (4 - 6 oktas)	Cloudy to overcast (7 - 8 oktas)
0 - 3m/s	A	B	C
3 - 5m/s	B	B	C
> 5m/s	C	C	D
Nighttime			
Wind speed	Clear (0 - 3 oktas)	Cloud periods / intervals (4 - 6 oktas)	Cloudy to overcast (7 - 8 oktas)
0 - 3m/s	G	F	E
3 - 5m/s	F	E	D
> 5m/s	D	D	D

Fig. 2. Atmospheric stability class (Reference: (Hung, 2020))

Six atmospheric stabilities are recognised: A, which is highly unstable or convective; B, which is moderately unstable; C, which is somewhat unstable; D, which is neutral; E, which is moderately stable; and F, which is extremely stable. Table 1 below shows the atmospheric data used, while Table 2 below summarizes the parameters and assumptions applied to simulate leak and catastrophic rupture scenarios for the major units in hydrogen production.

Table 1. Atmospheric data for consequence modelling

Period	Average temperature (°C)	Relative humidity (%)	Atmospheric stability class	Wind velocity (m/s)
Day	28	83	C	2.5
Night	3.4	70	F	1.5

Table 2. Parameters and assumptions for leak and catastrophic rupture scenarios

Equipment	Leak diameter (mm)	Tank head (m)	Elevation (m)	Concentration of interest (ppm)	Averaging time (s)
Hydro-desulphurization reactor	152	1	1	500	1000
Reforming bed reactor				4100	
High temperature shift reactor					
Pressure swing adsorption reactor					

Consequence Modelling Results for Leak and Catastrophic Rupture Scenarios via PHAST (Process Hazard Analysis Software Tool) Simulation. The results of leakage (methane and hydrogen release) from the reactor units and catastrophic rupture from the reactor units were analyzed through Figure 3 and Figure 4. Based on Figure 3 and Figure 4 above, it was observed that all four reactors show relatively small threat zones with a 152 mm leakage occurrence and catastrophic rupture scenario. When compared, it was revealed that the affected area in weather conditions at 2.5/C are notably larger than that of in weather condition at 1.5/F. Table 3 below shows the comparison data.

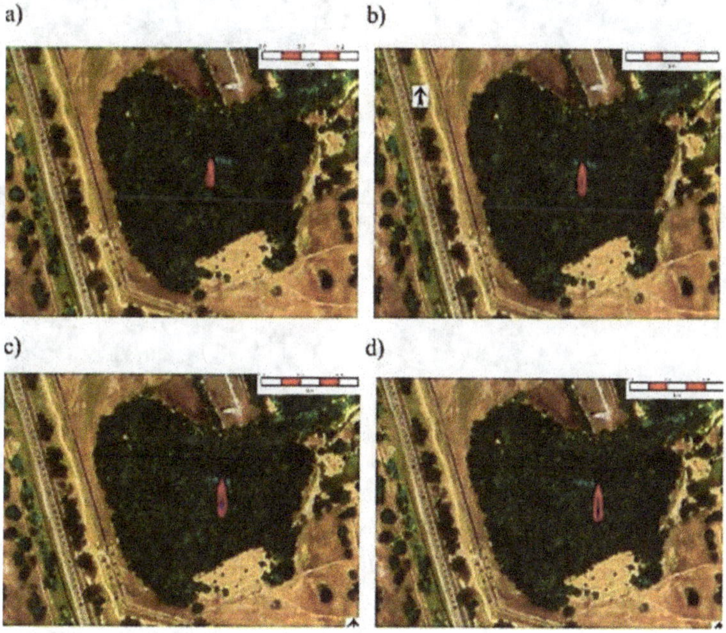

Fig. 3. GIS and contour of 152 mm leak: (a) Hydro-desulfurization reactor, (b) Reforming bed reactor, (c) High-temperature shift reactor, and (d) Pressure swing adsorption reactor

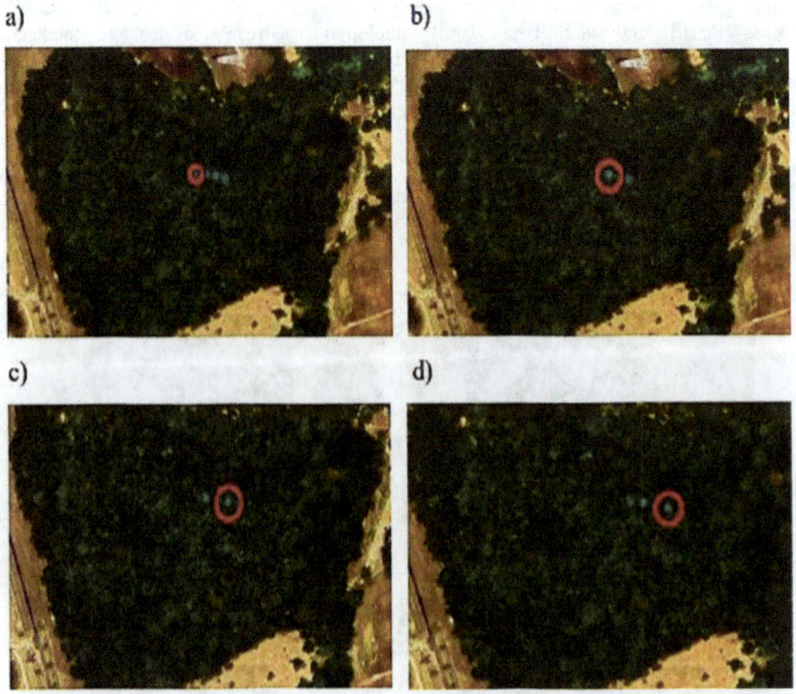

Fig. 4. GIS and Contour of Rupture: (a) Hydro-desulfurization reactor, (b) Reforming bed reactor, (c) High-temperature shift reactor, and (d) Pressure swing adsorption reactor

Table 3. Comparison between weather conditions at 2.5/C and 1.5/F for leak and rupture scenario

Equipment	Scenario			
	Leak		Rupture	
	Affected area size (m)		Affected area size (m)	
	2.5/C	1.5/F	2.5/C	1.5/F
Hydro-desulphurization reactor	156	136	31	24
Reforming bed reactor	122	106	66	58
High temperature shift reactor	150	106	78	68
Pressure swing adsorption reactor	198	160	108	92

In general, the catastrophic rupture scenario has a significantly shorter maximum affected distance than the leak scenario. These might result from the intensity and rate of release in each case. In most catastrophic rupture scenarios, a significant amount of hazardous chemicals are released suddenly and violently. Compared to a continuous leak scenario, the intensity tends to decrease more quickly even when the initial release rate is higher. Because of the quick drop in intensity, the downwind distances to specified intensity levels may be shortened in the event of a catastrophic failure. Weather conditions, like atmospheric stability class (C) and wind speed (2.5

m/s), affect dispersion rates, with unstable atmospheres enhancing turbulence and influencing toxic release dispersion.

Large leaks will spread hydrogen far from the leak source, into flammable levels. If the leak ignites right away, it will respond as a jet fire and put people and the surrounding region at risk of thermal shocks. A late ignition of a release allows time for reactive hydrogen/air clouds to build, potentially leading to explosions. High congestion accelerates flames and may result in high explosive pressures. Leaks in closed or semi-confined locations, such as inside a container, result in hydrogen accumulation and a significant risk of devastating explosion pressures. This process also boosts explosion pressures due to a lack of gas escaping during the combustion process. The release rate is determined by the leak size and hydrogen pressure, with smaller rates producing shorter flame lengths or smaller explosive clouds. However, in confined places, even relatively small leak rates can produce explosive clouds throughout the space [14]. For example, a jet fire could be triggered by ignited leaks in high-pressure pipework. The length of the flame varies depending on the fuel release rate, which is determined by the pressure and size of the hole. The momentum of the emission contributes to the exceptional orientation of the flames [15].

Consequence Modelling Results for Jet Fire and Flash Fire Scenarios Using PHAST (Process Hazard Analysis Software Tool) Simulation. Based on the results of the PHAST (Process Hazard Analysis Software Tool) simulation for the jet fire scenario (Table 4), it was discovered that the pressure swing adsorption (PSA) reactor has the longest flame length (59.1395 at weather conditions 2.5/C) and the greatest distance downwind to the intensity levels when compared to the other three other major reactor units involved in the hydrogen production process. A longer downwind distance indicates a wider radius from the source that is exposed to potentially dangerous heat and thermal radiation levels. This increases the potential for injuries, fatalities, and damage. People and equipment located further downwind might be exposed to hazardous conditions for longer periods, increasing the risk of severe consequences. In order to safeguard individuals and property against the thermal consequences of the jet fire, it is necessary to create wider safety zones surrounding the event.

When a hydrogen leak is detected and separated by safety measures, it is shut down, and no accidents occur in general. When hydrogen leaking is not identified and isolated, and there is no ignition source, a non-ignition leakage accident can occur, resulting in frostbite or asphyxia. If an ignition source ignites immediately, a jet flame will ensue, resulting in a personal burn mishap. If the ignition is delayed, hydrogen will collect; if there is obvious overpressure, a vapour cloud explosion accident will occur, causing shock injuries to personnel; if there is hydrogen accumulation but no overpressure, flash fire accidents will occur [4].

Flash fires are predicted to erupt as the final event. In the absence of fire, hydrogen diffuses into the atmosphere with no impact. If all safety equipment fails, the hydrogen will continue to leak, resulting in jet fires with immediate ignition and flash fires with delayed ignition. A flash fire is an abrupt and brief igniting of a combustible chemical that occurs when it is rapidly disseminated in the air and ignited. Flash fires are characterised by their high temperature and rapid dispersion, yet they normally have a short duration [6].

For a flash fire, having a longer distance downwind from the Lower Flammability Limit (LFL) is usually preferred. The longer distance shows that the concentration of combustible gases needed for burning occurs further away from the source. Flash fire flames will cause severe damage to equipment as well as serious harm to people. In the worst-case scenario, especially during the maintenance period, it may claim lives. People within the flash fire envelope, lower flammability limit (LFL) will experience fatalities due to severe radiation dosages [6]. PHAST (Process Hazard Analysis Software Tool) simulation results (Table 5) revealed that leak scenarios have larger downwind distances to lower flammability limit (LFL) than catastrophic rupture scenarios, influenced by multiple factors.

Table 4. Jet Fire Results from Consequence Modelling Simulation

Equipment	Scenario	Weather	Flame length (m)	Distance downwind to intensity level 1 (m)	Distance downwind to intensity level 2 (m)	Distance downwind to intensity level 3 (m)
Hydro-desulphurization reactor	Leak	Category 1.5/F	57.88	99.59	78.90	65.59
		Category 2.5/C	58.27	99.35	79.01	65.99
Reforming bed reactor		Category 1.5/F	39.51	58.69	49.78	45.22
		Category 2.5/C	39.51	58.66	49.77	45.19
High temperature shift reactor		Category 1.5/F	46.02	72.63	59.87	52.10
		Category 2.5/C	46.01	72.26	59.67	52.00
		Category 1.5/F	58.71	101.61	80.31	66.58
		Category 2.5/C	59.14	101.39	80.45	67.02

Table 5. Flash Fire Results from Consequence Modelling Simulation

Equipment	Scenario	Weather	Distance downwind to LFL (m)
Hydro-desulphurization reactor	Leak	Category 1.5/F	45.11
		Category 2.5/C	47.03
	Rupture	Category 1.5/F	10.13
		Category 2.5/C	10.13
Reforming bed reactor	Leak	Category 1.5/F	56.52
		Category 2.5/C	64.91
	Rupture	Category 1.5/F	21.81
		Category 2.5/C	21.97
High temperature shift reactor	Leak	Category 1.5/F	69.64
		Category 2.5/C	79.62
	Rupture	Category 1.5/F	21.24
		Category 2.5/C	21.76
Pressure swing adsorption reactor	Leak	Category 1.5/F	92.12
		Category 2.5/C	105.38
		Category 1.5/F	21.29
		Category 2.5/C	22.68

Risk Modelling Results via SAFETI (Software for the Assessment of Flammable, Explosive, and Toxic Impact) Analysis. Results for individual flammable, toxic, and total risks are shown in Figure 5 below.

These display risk contours for catastrophic rupture and leak scenarios. Purple indicates a 1 fatality in 1,000 years area, smallest and around one equipment. Red and grey circles signify 1 fatality in 10,000 and 100,000-years risks, limited to areas near process equipment. Pink denotes 1

Materials Research Proceedings 53 (2025) 498-511 https://doi.org/10.21741/9781644903575-51

fatality in 1,000,000 years, including nearby equipment. Yellow and orange lines denote 1 fatality in 10,000,000 and 100,000,000 years. The study confirms all risk contour lines within the plant area, ensuring operational safety with risks confined near process equipment.

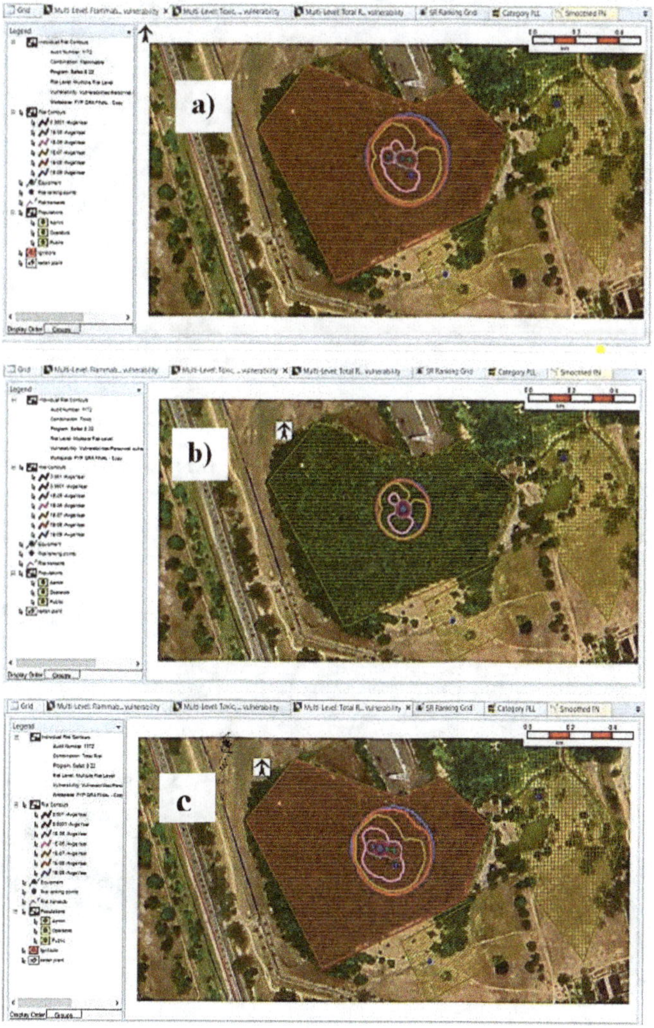

Fig. 5. Multi-Level Risks Contour for Flammable (a), Toxicity (b), Total Risk (c)

Fig. 5 (a). Multi-Level Risks Contour for Flammable

Fig. 5 (b). Multi-Level Risks Contour for Toxicity

Materials Research Proceedings 53 (2025) 498-511 https://doi.org/10.21741/9781644903575-51

Fig. 5 (c). Multi-Level Risks Contour for Total Risk

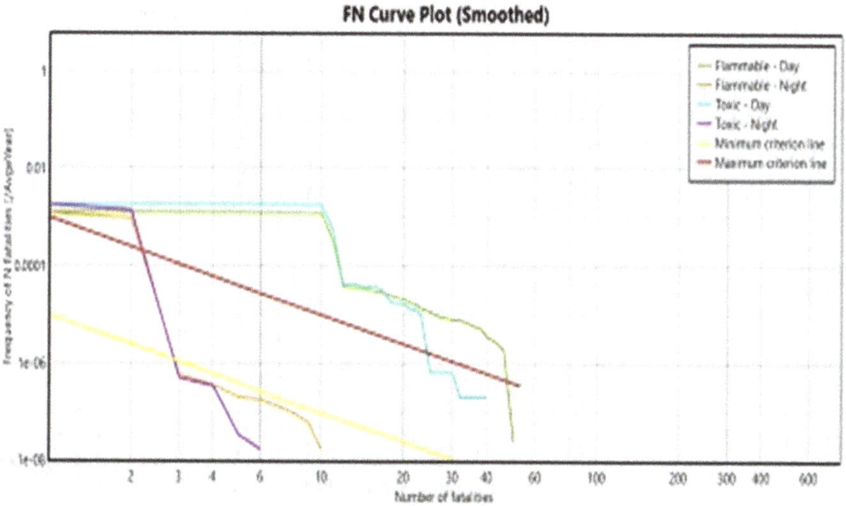

Fig. 6. F-N Curve (Smoothed)

F-N curves in Figure 6 above illustrate risk scenarios, showing fatalities decrease as frequency increases for both flammable and toxic incidents day and night. Flammable incidents peak at about 50 and 10 fatalities, exceeding minimum criteria, with frequencies of $1.5\text{x}10^{-8}$/year and $1.1\text{x}10^{-8}$/year. Toxic incidents peak at around 40 and 6 fatalities, with frequencies below $1.3\text{x}10^{-7}$/year and $1.2\text{x}10^{-8}$/year, meeting safety standards. Yearly fatality rate, combined into Potential Loss of Life (PLL), contrasts risk reduction approaches, aiding cost-effectiveness evaluation. Table 6 below shows PLL: one flammable incident every 112 years, one toxic incident every 104 years, and roughly one fatal incident every 54 years.

Table 6. *Potential Loss of Life (PLL)*

Group Name	Group Type	Total (Average/Year)
Flammable	Combination	0.008951711
Toxic	Combination	0.009599984
Total Risk	Combination	0.018551694

Summary

In summary, hydrogen offers a promising path towards sustainable energy and fossil fuel substitution. This study emphasizes the necessity of comprehending risks inherent in hydrogen production. Through a detailed quantitative risk assessment (QRA) using PHAST (Process Hazard Analysis Software Tool) and SAFETI (Software for the Assessment of Flammable, Explosive, and Toxic Impact) simulations, primary hazards like jet and flash fires were identified, with leak and rupture scenarios posing minimal threats. For instance, the pressure swing adsorption reactor showed potential for jet fire and flash fire due to significant flame lengths and proximity to the lower flammability limit (LFL). SAFETI (Software for the Assessment of Flammable, Explosive, and Toxic Impact) analysis confirms fatality rates for each risk scenario are within acceptable limits. These findings contribute valuable insights to enhance safety measures and regulatory standards, ultimately advancing the implementation of sustainable energy alternatives in hydrogen production. Additionally, it is recommended for future studies to look into the environmental implications of the production of hydrogen and investigate appropriate control strategies. Moreover, the effectiveness of existing safety protocols and emergency response strategies in dealing with hydrogen-related hazards should be further studied. Additionally, studies into novel techniques and resources for safer hydrogen storage and transportation would aid in the wider adoption of hydrogen as a source of energy.

Acknowledgement

We extend our sincerest appreciation and gratitude to our esteemed organization, Universiti Teknologi PETRONAS for their support and provision of resources in achieving our objectives. (015LC0-469).

References

[1] E. B. Agyekum, C. Nutakor, A. M. Agwa and S. Kamel, "A Critical Review of Renewable Hydrogen Production Methods: Factors Affecting Their Scale-Up and Its Role in Future Energy Generation," Modelling and Simulation of Industrial and Environmental Processes with Membranes , vol. 12, no. 2, 2022. https://doi.org/10.3390/membranes12020173

[2] C. B. B. Farias, R. C. S. Barreiros, M. F. d. Silva, A. A. Casazza, A. Converti and L. A. Sarubbo, "Use of Hydrogen as Fuel: A Trend of the 21st Century," Environmental Biotechnology for Green Energy, vol. 15, no. 1, p. 311, 2022. https://doi.org/10.3390/en15010311

[3] O. I. Ahmed, N. Mehta, E. M. Ahmed, M. Hefny, A. Al-Hinai, A. H. Ala'a and R. W. David, "Hydrogen production, storage, utilisation and environmental impacet: a review," Environmental Chemistry Letters, vol. 20, pp. 153-188, 2022. https://doi.org/10.1007/s10311-021-01322-8

[4] X. Zhang, G. Qiu, S. Wang, J. Wu and Y. Peng, "Hydrogen Leakage Simulation and Risk Analysis of Hydrogen Fueling Station in China," Sustainability , vol. 14, no. 19, p. 12420, 2022. https://doi.org/10.3390/su141912420

[5] S. F. Ardabili, B. Najafi and S. Shamshirband, "Computational intelligience approach for modeling hydrogen production: a review," Engineering Applications of Computational Fluid Mechanics, vol. 12, no. 1, 2018. https://doi.org/10.1080/19942060.2018.1452296

[6] I. Mohammadfam and E. Zarei, "Safety risk modelling and major accidents analysis of hydrogen and natural gas release: A comprehensive risk analysis framework," International Journal of Hydrogen Energy, vol. 40, no. 39, pp. 13653-13663, 2015. https://doi.org/10.1016/j.ijhydene.2015.07.117

[7] M. J. Jafari, E. Zarei and N. Badri, "The quantitavie risk assessment of a hydrogen generation unit," International Journal of Hydrogen Energy, vol. 37, no. 24, pp. 19241-19249, 2012. https://doi.org/10.1016/j.ijhydene.2012.09.082

[8] A. Bahadori, "Modeling and analysis of hydrogen production in steam methane reforming (SMR) process," Petroleum Science and Technology, vol. 37, no. 12, 2019. https://doi.org/10.1080/10916466.2019.1587466

[9] M. Calabrese, M. Portarapillo, A. D. Nardo, V. Venezia, M. Turco, G. Luciani and A. D. Benedetto, "Hydrogen Safety Challenges: A Comprehensive Review on Production, Storage, Transport, Utilization, and CFD-Based Consequence and Risk Assessment," Energies, vol. 17, no. 6, p. 1350, 2024. https://doi.org/10.3390/en17061350

[10] E. Zarei, M. J. Jafari and N. Badri, "Risk Assessment of Vapor Cloud Explosions in a Hydrogen Production Facility with Consequence Modeling," Journal of Research in Health Sciences, vol. 13, no. 2, pp. 181-187, 2013.

[11] A. Huser, M. Bucelli, K. Zakariyya, M. M. Skinnemone and P. Nadim, "Quantitative Risk Analysis of Scaled-Up Hydrogen Facilities," International Conference for Hydrogen Safety , 2021.

[12] R. Rezaei, M. Arjmand and M. Heidari, "A novel risk assessment framework for hydrogen production, storage, and utilization systems using integrated fuzzy logic and bow-tie approach," Journal of Loss Prevention in the Process Industries, vol. 70, p. 104393, 2021.

[13] H. Fan-yiu, "Effect of Atmospheric Stability on the Dispersion of Radioactive Materials," Hong Kong Observatory, 2020.

[14] S. Lee, M. V. Ottermo, K. Vaagsaether and H. Henriksen, "Hazard Identification for Gaseous Hydrogen in Storage, Transportation, and Use," in ESREL 2022, Dublin, 2022. https://doi.org/10.3850/978-981-18-5183-4_R13-04-240

[15] X. Zhang, L. Hu, Q. Wang, X. Zhang and P. Gao, "A mathematical model for flame volume estimation based on flame height of turbulent gaseous fuel jet," Energy Conversion and Management , vol. 103, pp. 276-283, 2015. https://doi.org/10.1016/j.enconman.2015.06.061

Optimizing hydrogen production through steam gasification of biodiesel-derived waste glycerol: A parametric approach

KENNETH Chong Yih Haur[1,a] , MUHAMMAD Ayoub[1,2,b,*]

[1]Chemical Engineering Department, Universiti Teknologi PETRONAS, Bandar Seri Iskandar, Perak, Malaysia

[2]HICoE-Centre for Biofuel and Biochemical Research (CBBR), Institute of Self-Sustainable, Universiti Teknologi PETRONAS, Bandar Seri Iskandar, Perak, Malaysia

[a]kenneth_19001147@utp.edu.my, [b]muhammad.ayoub@utp.edu.my

Keywords: Glycerol, Hydrogen Production, Parametric Studies, Steam-to-Glycerol (S/G) Ratio, Waste-to-Hydrogen

Abstract. The high dependence on fuels including biodiesel, which garnered attention as an alternative for sustainable fuel sources. Despite these benefits, biodiesel production generates glycerol as a by-product, leading to oversupply challenges and exacerbating greenhouse emissions, and this phenomenon results in an eager movement for cleaner energy transition from waste. The focus on mitigating climate change has led to global initiatives such as the transition of waste-to-hydrogen for alternative clean fuel, in achieving cleaner energy and sustainability objectives. In these studies, the utilization of excess waste glycerol for maximum hydrogen production potential through steam gasification has been emphasized, by altering the core gasification parameters. The experimental data obtained from previous experimental studies forms the basis for parametric studies by using Aspen HYSYS. To achieve optimum H2 yield, several parameters including steam-to-glycerol (S/G) ratio, temperature, and pressure were investigated and identified as key parameters influencing gas yield. Based on the simulation, the impact of the S/G ratio on hydrogen and carbon monoxide (CO) yields is explored, with an optimum ratio of 0.074 identified. Temperature studies indicate 680˚C as the optimal condition for the highest hydrogen production when (S/G) ratio is 0.5, while pressure parametric studies underscore the significance of maintaining low pressure for optimal gas yields. To maximize the overall hydrogen yield, the gasification process is optimized by setting the steam-to-biomass (S/G) ratio at 0.072 and maintaining a temperature of 800 ˚C under atmospheric conditions. This configuration results in the highest hydrogen production, reaching 0.57. Recommendations for experimental improvements are provided to align experimental and simulation data. The utilization of simulation methods enhances accuracy, and findings contribute to the broader goal of sustainable and efficient clean energy production from biomass gasification.

Introduction

The growth of the global population, which results in the high reliability of fossil fuels, has led to increased greenhouse gas emissions, posing a significant threat to the current environmental quality. Fossil fuel has become one of the main energy resources that is widely used for daily energy usage in most of the countries [1]. With that, the approach to producing cleaner energy is now becoming more crucial in addressing current climate change and plays a vital role in the global transition toward a more sustainable and resilient energy ecosystem. Since the late 20th century, there has been a strong emphasis on the reduction of greenhouse gases (GHGs). As a result, a global initiative such as Sustainable Development Goals (SDG), and The Net Zero Emissions by 2050 Scenario (NZE Scenario) has been started to decrease greenhouse emissions and limit the increase in the average global temperature to 3°C above pre-industrial levels [2]. In recent years, hydrogen (H_2)

gas as a clean energy source has been prioritized and also has been established as a viable solution for mitigating greenhouse gas emissions and aligning with the United Nations (UN) Sustainable Development Goals (SDG) in providing better environment and sustainability to the world [3]. Dincer discovered that the use of H_2 fuel was still in the traditional method, with a market value of approximately \$50 billion for an annual production of 40 million tonnes, hydrogen gas is primarily supplied as a chemical substance rather than a fuel [4]. Its predominant applications include serving as a processing agent in oil refineries and in various chemical production processes, including methanol, ammonia, and pharmaceuticals. However, this situation will be altered when hydrogen production methods start to transform as direct fuel supply, or as synthesis fuel to utilize fertilizers or heavy oils in the future, to meet the current nation goals, where one of it is about the initiative to achieve zero carbon emission [5]. This condition has driven policymakers around the globe to opt for alternative energy opportunities.

One of the alternative fuels that has garnered attention in recent decades is biodiesel, contributing to the sustainable production of energy [6]. In the production of biodiesel, emissions of carbon monoxide and carbon dioxide are reduced compared to conventional diesel production. Biodiesel is derived from animal fats and vegetables, and various methods can be employed for its production, including transesterification, thermal degradation, and emulsion [7]. However, the process of biodiesel formation will produce glycerol as by- products which alters the current global market supply where the supply exceeded the demand. Based on the research, 10 kg of glycerol will be formed from every 100 kg of biodiesel produced [8]. In the past few decades, there has been a notable rise in the demand for biodiesel, as depicted in Figure 1. This trend has concurrently impacted the glycerol production. Failure to properly utilize the excess glycerol may lead to waste issues and pose challenges to sustainable production [9].

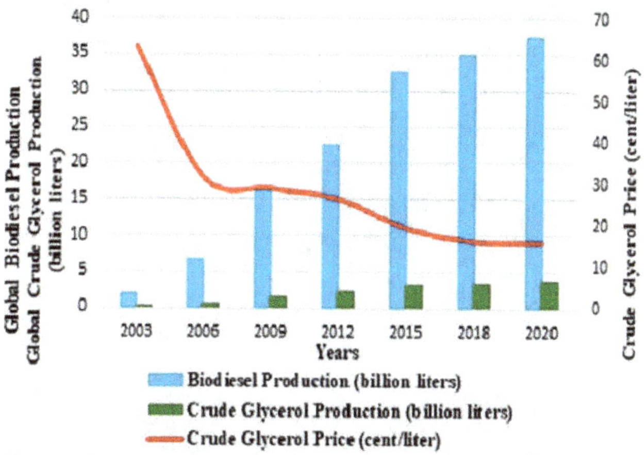

Fig. 1. Global crude glycerol price vs. crude glycerol production [10]

To tackle the increasing of glycerol waste issue, this waste can be used to produce hydrogen in the current waste-to- hydrogen industries [11]. The production can be done by subjecting crude glycerol to high-temperature cracking (>600°C) via glycerol-steam reforming, where significant gas products such as hydrogen and carbon monoxide can be obtained [12]. Utilizing glycerol derived from biodiesel production for renewable resource-based hydrogen production presents a sustainable alternative for cleaner fuel production. According to a study, 1 mole of glycerol has the potential to yield up to 4 moles of hydrogen, establishing it as a relatively suitable feedstock for hydrogen production [13]. Ayoub et al. [14] has proposed a route for hydrogen production from glycerol via steam-enhanced gasification and tested in a small-scale experimental system to study the effect of steam-to-glycerol (S/G) ratio, temperature, and the presence of catalyst towards the rate of H_2 and CO production. However, there are very limited studies that have been conducted on the modelling and simulation of hydrogen synthesis from glycerol via steam- enhanced gasification.

In these studies, the simulation of H_2 production process from crude glycerol is focused by Aspen HYSYS software. After that, a parametric study was performed on the process to explore the operation limits for H_2 production process using single step process. The effect of attributes such as temperature, steam-to-glycerol (S/G) ratio, and pressure, are studied to determine the impacts on the formation of H_2 and CO and the hydrogen yield in this process.

Methodology

The experimental data from the research done by Ayoub et al. [14] is applied to conduct the parametric studies. In the experiment, steam gasification of pure and crude glycerol was conducted in a continuous fixed-bed reactor in a tubular furnace. The specification is shown as follows:

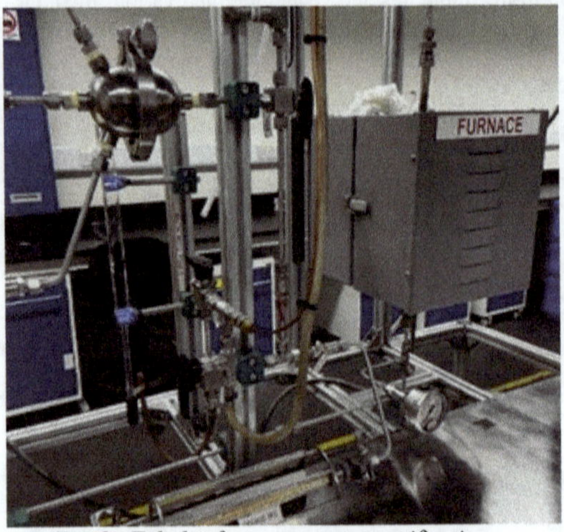

Fig. 2. Tubular furnace reactor-gasification

Table 1: Steam Gasification Specification For Parametric Studies

Parameter	Value
Temperature (˚C)	650-800
Pressure (atm)	1
Flow rate (g min^{-1})	5.4
Catalyst	Ni/Al2O3
Gasifying Agent	Steam

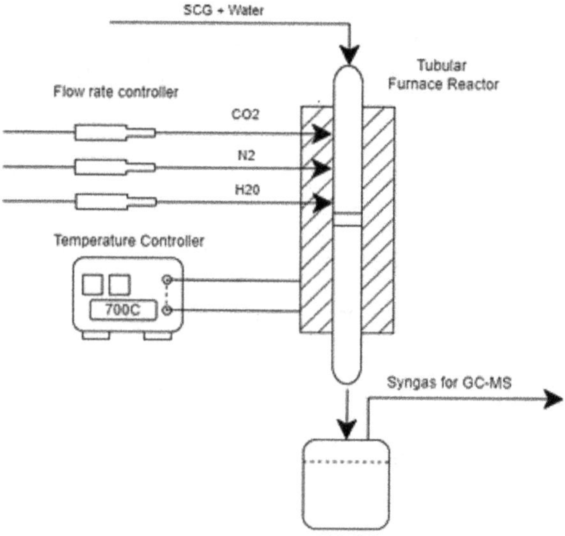

Fig. 3. Schematic diagram of biomass gasification

For the feedstock, crude glycerol was obtained from the by- product of transesterification of waste cooking oil for biodiesel production. The TGA analysis was conducted to determine the composition of crude glycerol.

Table 2: Composition of Crude Glycerol

	Glycerol (wt%)	Methanol (wt%)	KOH (wt%)	H2O (wt%)
Crude Glycerol	60	31	1.05	7.5

Table 2 shows the composition of crude glycerol used in the conducted experimental studies. The composition of this crude glycerol will be further applied into the modelling and simulation studies.

The overall process route for this gasification for organic waste involves gasification process with several major reaction schemes including Steam Reforming of Glycerol (1), Water Gas Shift Reaction (WGSR) (2), Steam Reforming of Methanol (3), and Steam Reforming of Methane (4) [14]-[15].

$$C_3H_8O_3 \xrightarrow{H_2O} 3CO + 4H_2 \tag{1}$$

$$CO + H_2O \leftrightarrow CO_2 + H_2 \tag{2}$$

$$5CH_3OH{+}H_2O \rightarrow 3CO_2 + 7H_2 + 2CH_4 \tag{3}$$

$$CH_4 + H_2O \leftrightarrow CO + 3H_2 \tag{4}$$

The conversion and the equilibrium constant are stated as follow, as on the component yield obtained [14]:

Table 3: Reaction Conversion and Equilibrium Constant

Reaction	**Type of Reaction**	**Conversion/ Equilibrium ($ln(K_{eq})$)**
Steam Reforming by Glycerol	Conversion	$X_{C_3H_8O_3} = 1$
Water-Gas Shift Reaction	Equilibrium	$\ln(K_{eq}) = -12.11 + \frac{5318.69}{T} + 1.012ln(T) - 0.0001T$
Steam Reforming of Methanol	Conversion	$X_{CH_3OH} = 1$
Steam Reforming of Methane	Equilibrium	$\ln(K_{eq}) = -20.55 - \frac{22920.58}{T} + 7.195ln(T) - 0.0029T$

In the steam reforming of glycerol and methanol, the reaction performance will be based on the conversion of glycerol and methanol since they are more expensive than water. The gas produced from the reaction was injected into GC-MS with TCD detector to analyze the gas composition. N_2 was used as carrier gas in this analysis.

The overall process route for this biomass gasification is illustrated in Figure 4:

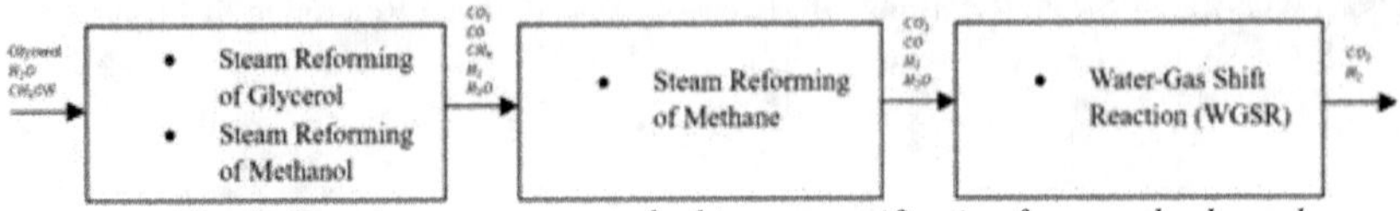

Fig. 4. Reaction process route for biomass gasification from crude glycerol

In the parametric study of hydrogen production processes, several core assumptions underlie the development and modelling efforts was assumed:

- The reactions are in ideal conditions.
- Glycerol and methanol are used up in the gasification reactions (x=1).
- Isothermal reactions with constant volume.

Materials Research Proceedings 53 (2025) 512-525 https://doi.org/10.21741/9781644903575-52

- Tar and char formation is negligible in the simulation.
- Perfect mixing and uniform temperature and pressure distribution inside the reactor.

To further study about the effects of temperature, steam-to- glycerol (S/G), and pressure towards the CO and H_2 production, Aspen HYSYS is used to study the relationships and the parameter conditions to achieve optimum CO and H_2 production in this glycerol-to-hydrogen production via gasification.

First, all the components were assigned for the simulation including glycerol. For the fluid properties package selection, Wilson and Peng-Robinson are used for this simulation. Glycerol was assigned into the Wilson property package whereas the rest of the gases were assigned into Peng- Robinson. This is because Wilson can effectively capture the behaviour of binary systems and demonstrated superior fitting accuracy when applied to polar component such as glycerol [16]. For Peng-Robinson, it is suitable for gas-processing- related processes, and well fit for gases simulation [15].

The constructed simulation model, using Aspen HYSYS, as shown in Figure 5 is derived from the reactions stated in Table 3. The configuration of this simulation model consists of 3 major sections, and the tool models used for the simulation, and the specification based on the experimental setup [14] are listed as follows:

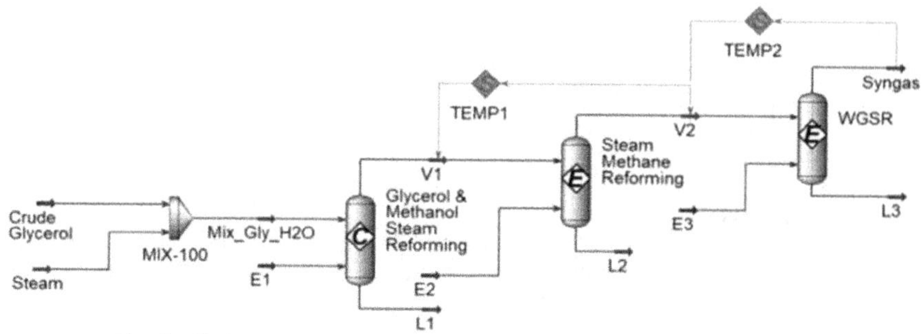

Fig. 5. Hydrogen production via biomass gasification process simulation

Materials Research Proceedings 53 (2025) 512-525 https://doi.org/10.21741/9781644903575-52

Table 4: Simulation Model Specification

Model	Function	Reaction Involved	Specification
Mixer (MIX-100)	To mix the crude glycerol and steam	-	Flow rate=5.4 g/min
Conversion Reactor	To module steam reforming reaction, occur in the tubular furnace reactor.	Steam Glycerol Reforming Steam Methanol Reforming	Temperature = 800˚C Pressure = 101.3kPa
Equilibrium Reactor	To module steam methane reforming	Steam Methane Reforming	Temperature = 800˚C Pressure = 101.3kPa
	reaction, occur in the tubular furnace reactor		
	To module water- gas shift reaction occurs in the tubular furnace reactor.	Water-Gas shift Reaction	Temperature = 800˚C Pressure = 101.3kPa
Set (TEMP1 &TEMP2)	To module the same temperature for all reactor models used when conducting case studies.	-	-

Result and Discussion

Comparison Between Experimental and Simulation Results. The data shows a big difference between the real and simulated amounts of methane (CH_4), with a huge error of 118900%. This difference likely happened because the chemical reactions during the experiment studies, especially with methane, might not have finished. This incomplete reaction probably led to not enough methane converting and, as a result, less carbon dioxide (CO_2) being produced in the experiment in which indicates low accuracy results obtained from the experimental studies. Other than that, based on the studies conducted by Berend and Robin, the negligence of tar and char formation in the simulation studies ignored the mechanism of tar decomposition via radical reaction and the formation of methane [17]. This difference suggests that the reaction, especially with the formation of methane, might not have simulated properly during the experiment and cause lower methane yield obtained from the simulation studies. Several factors also could be causing this, including low space-time for the reaction, changes in temperature, change in phases for water, or the reactants not completely converting. Therefore, parametric studies using simulation method can overcome the issues stated as above if the assumptions can be simulated into it, to obtain high-accuracy data. However, the hydrogen yield remains relatively unchanged and constant in both experimental and simulation studies with low error.

Table 5: Experimental and Simulation Results for Product and Component Distribution of Steam Gasification of Crude Glycerol at Steam to Glycerol Ratio 50:50 at 800°C

Product	Experimental Results [14]	Simulation result	Percentage Error
	mol %	mol %	%
H2	61.4	66.7	7.95
CO	20.5	18.4	11.41
CO2	6.2	14.9	58.39
CH4	11.9	0.01	118900

Steam-to-Glycerol (S/G) Ratio. The steam-to-glycerol (S/G) ratio plays a crucial role in the production of hydrogen fuel through biomass gasification when steam serves as the gasifying agent. This is attributed to the direct influence of the reactant quantity on the formation of products. Glycerol is an essential component in the gasification reaction as it is the pioneering carbon source for gasification to occur and steam further enhances the gasification reaction, where they react to generate CO_2, CO, and H_2. The ratio between reactants indicates how high the hydrogen yield can be achieved.

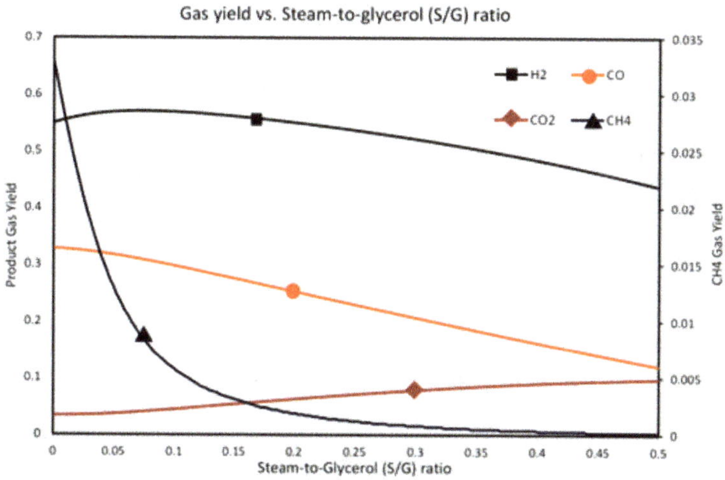

Fig. 6. Gas yield vs. Steam-to-glycerol (S/G) ratio

Based on the result illustrated in Figure 6, the H_2 and CO yield increase when (S/G) ratio increases. However, a peak of 0.57 is determined for H_2 yield when (S/G) ratio is at 0.074. This indicates that the optimum (S/G) ratio is at 0.074 (the ratio of 7.4% steam to 92.6% glycerol) to achieve the highest potential H_2 yield in this process. Based on the related studies [18], which state that the H_2 yield decreases when feedstock is increased with its flow rate, it only applies when the ratio of steam-to- glycerol is above 0.074. For methane, it starts to increase drastically from 0.0025 to 0.0325 after (S/G) ratio = 0.2. This is due to the presence of large quantities of steam favoring the glycerol reforming reaction [19], and sufficiently supplied as reactants to form syngas.

Temperature. Temperature is an influential factor that can affect the overall product yield in biomass gasification, given that all reactions involved are highly temperature dependent. This is particularly evident in the Steam Methane Reform and Water-Gas Shift reactions, where temperature changes can influence the formation of products on either side of the equilibrium reaction. Based on the experimental studies, the temperature of 650˚C to 800˚C was set for the gasification process of glycerol [14].

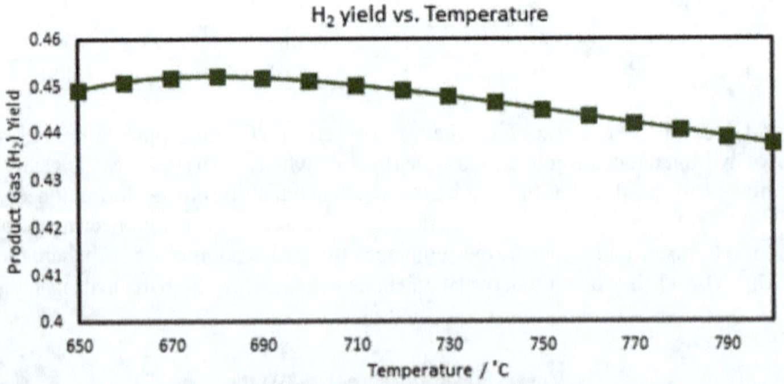

Fig. 7. H2 yield vs. Temperature

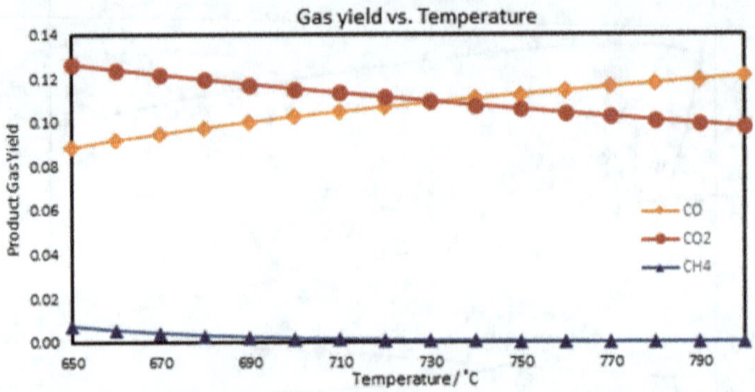

Fig. 8. CO, CO2 and CH4 yield vs. Temperature

By referring to both plotted graphs as shown in Figure 7 and Figure 8, the optimum temperature is determined to be 680˚C. The yield of H_2 continues to decrease, whereas the yield of CO decreases as the temperature increases from the optimum at 680˚C. This phenomenon may be attributed to the temperature- dependent nature of the reactions, wherein the temperature increase inhibits the formation of H_2 and CO gases but favors CO formation. Besides that, most of the reactions that occur in the reactor excluding Steam Methane Reforming (SMR), are moderate exothermic in the overall reaction, which is mostly favored by low-temperature conditions [19, 20]. Given that previous studies have conclusively determined the optimum temperature to be 658˚C, the temperature findings in our study align significantly with and strongly reference these established results [21].

Pressure. Pressure has also been studied parametrically to determine the effect of pressure on the production yield, especially for H_2. Different pressure conditions were assigned to study the significant effect on the gas yield.

By initiating with the lowest possible pressure for the gasification, which is at 50 kPa, the H_2 gas yield decreased significantly when pressure increased, same goes for CO yield. However. The CO_2 and CH_4 yield rise when pressure increases. According to Le Chatelier's Principle, an increase in pressure generally favors the reaction that produces fewer total moles of gas [22]. In this context, nearly every reaction generates a greater total number of gas moles compared to the respective reactant in the reactions, except for the Water-Gas Shift Reaction (WGSR). The rise in pressure inhibits the formation of H_2 and CO, positioned on the right side of the reaction, and favors the formation of CH_4 from the Steam Methane Reforming reaction, in which it is located on the left side of the reaction. However, atmospheric pressure will be taken as the best condition to trade-off between optimal gas yields and cost considerations in the gasification process.

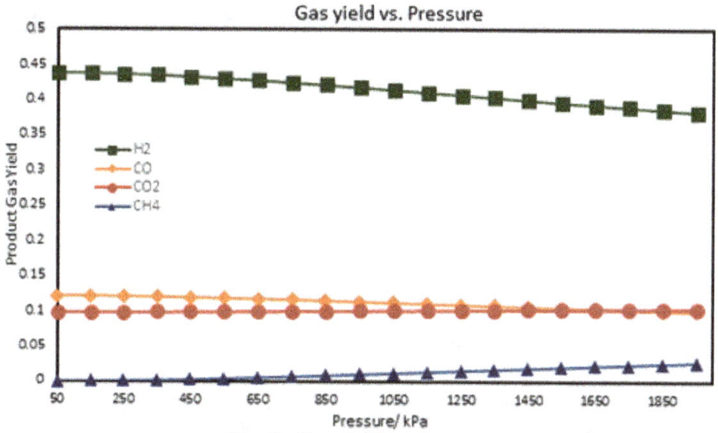

Fig. 9. Gas yield vs. Pressure

Optimum H_2 Yield. To enhance the H_2 yield, the investigation focuses on optimizing conditions by introducing 7.4 wt% steam alongside 92.6 wt% crude glycerol. This composition is examined to determine the optimal temperature. The emphasis on the reactant ratio stems from its significant attribute in altering the composition of the syngas products. This statement is supported by Figure 4.6, which illustrates that as the steam-to-glycerol (S/G) ratio decreases, the optimum H_2 yield shifts towards lower values as temperature increases. This observation leads to the conclusion that the optimal yield of 0.57 is achieved precisely at an S/G ratio of 0.074 and a temperature of 800°C. Overall, the reduction in the ratio between steam and biomass results in higher temperatures to enhance H_2 yield.

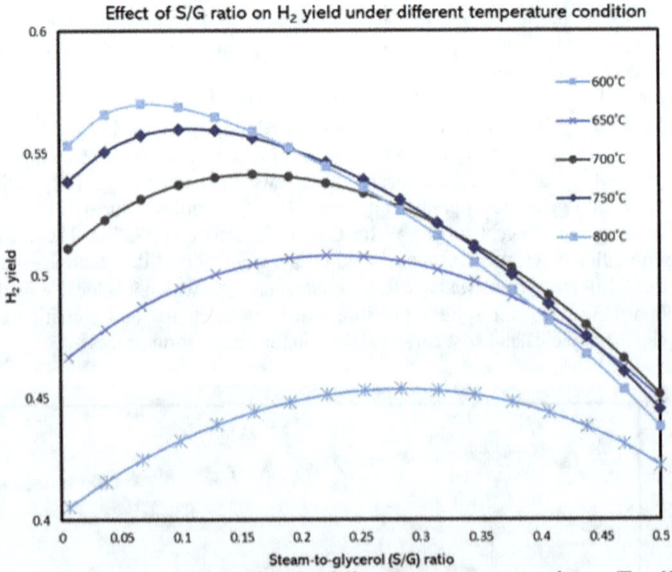

Fig. 10. Effect of S/G ratio on H2 yield under different temperature condition (T = 600 °C, 650 °C, 700 °C, 750 °C, 800 °C)

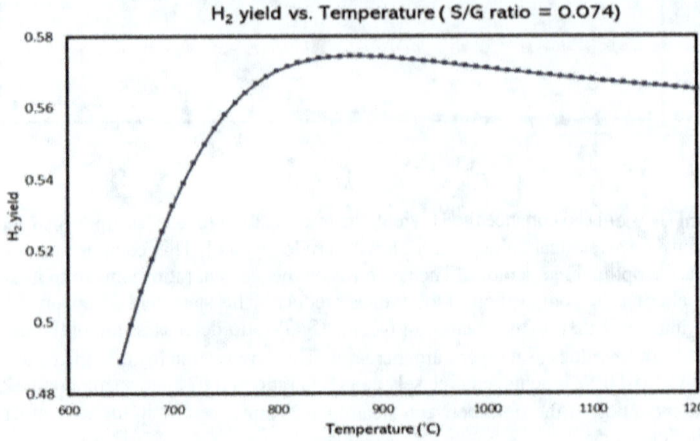

Fig. 11. H2 yield vs. temperature when (S/G) ratio = 0.074

In a comprehensive view, maintaining atmospheric pressure proves to be highly conducive to achieving optimal conditions. Specifically, injecting reactants with a composition of 7.4 wt% steam and 92.6 wt% crude glycerol at a temperature of 870 °C (as indicated in Figure 11.) results in the highest achievable H_2 yield, reaching 0.5743. This shows the effectiveness of steam gasification using waste crude glycerol under these specified conditions.

Conclusion

In conclusion, the comparison between experimental and simulated highlighted a significant difference in CH_4, which emphasizing the limitation of experimental methods to capture the intricacies of gasification reactions due to errors occur throughout the experiment. Several recommendations can be made to eliminate the offset between the data obtained from the experiment and simulation is that the reaction can be taken when the water is pre-heated into steam (gaseous phase) to reduce the reaction time for the phase change in the continuous tubular furnace reactors. The injection time can be reduced for a longer reaction time, ensuring complete conversion.

The investigation into the (S/G) ratio shows a significant role in producing the highest possible H_2 yield. The observed increase in hydrogen (H_2) and carbon monoxide (CO) yield with an increasing S/G ratio reached an optimum at 0.074 (the ratio of 7.4% steam to 92.6% glycerol), indicating the importance of sufficient glycerol and steam supply for the desirable gasification reactions.

For temperature, the optimal temperature determined from the experimental studies was found to be 680˚C for the highest hydrogen production at a 50:50 ratio condition. The low pressure also highlighted its impact on the production of H_2 and CO. However, maintaining atmospheric pressure (101.3 kPa) was recommended to balance the trade-off between optimal gas yields and cost considerations.

Overall, maintaining an S/G ratio of 0.074, operating at a temperature of 800˚C, and considering atmospheric pressure to ensure the highest hydrogen production yield from the biomass gasification method via crude glycerol at 0.57. These findings contribute valuable insights to the advancement of glycerol gasification processes, aligning with the overarching goal of achieving sustainable and efficient waste-to-hydrogen fuel production in the future.

Acknowledgements

The authors wish to thank the Chemical Engineering Department of Universiti Teknologi PETRONAS for the facilities for the completion of this project.

References

[1] R. Zhang, G. Guo, Y. Gao, L. Wang, N. Zhang, and J. Cai, "Compared study on the bentonite supported nickel-based catalysts materials," Materials Letters, vol. 113, pp. 146-148, 2013. https://doi.org/10.1016/j.matlet.2013.09.019

[2] B. Hare, N. Höhne, J. Gütschow, M. Schaeffer, and M. Vieweg- Mersmann, "Climate Action Tracker," 2015.

[3] T. M. Mata, A. A. Martins, and N. S. Caetano, "Microalgae for biodiesel production and other applications: a review," Renewable and Sustainable Energy Reviews, vol. 14, no. 1, pp. 217-232, 2010. https://doi.org/10.1016/j.rser.2009.07.020

[4] I. Dincer, "Green methods for hydrogen production," International Journal of Hydrogen Energy, vol. 37, no. 2, pp. 1954-1971, 2012. https://doi.org/10.1016/j.ijhydene.2011.03.173

[5] D. B. Levin and R. Chahine, "Challenges for renewable hydrogen production from biomass," International Journal of Hydrogen Energy, vol. 35, no. 10, pp. 4962-4969, 2010. https://doi.org/10.1016/j.ijhydene.2009.08.067

[6] A. Tamošiūnas, P. Valatkevičius, D. Gimžauskaitė, V. Valinčius, and M. Jeguirim, "Glycerol steam reforming for hydrogen and synthesis gas production," International Journal of Hydrogen Energy, vol. 42, no. 17, pp. 12896-12904, 2017. https://doi.org/10.1016/j.ijhydene.2016.12.071

[7] J. Iyyappan, G. Baskar, B. Bharathiraja, and R. Saravanathamizhan, "Malic acid production from biodiesel derived crude glycerol using morphologically controlled Aspergillus niger in batch fermentation," Bioresource Technology, vol. 269, pp. 393-399, 2018. https://doi.org/10.1016/j.biortech.2018.09.002

[8] J. Chen, S. Yan, X. Zhang, R. D. Tyagi, R. Y. Surampalli, and J. R. Valéro, "Chemical and biological conversion of crude glycerol derived from waste cooking oil to biodiesel," Waste Management, vol. 71, pp. 164-175, 2018. https://doi.org/10.1016/j.wasman.2017.10.044

[9] L. R. Kumar, S. K. Yellapu, R. D. Tyagi, and X. Zhang, "A review on variation in crude glycerol composition, bio-valorization of crude and purified glycerol as carbon source for lipid production," Bioresource Technology, vol. 293, p. 122155, 2019. https://doi.org/10.1016/j.biortech.2019.122155

[10] N. A. Roslan, S. Z. Abidin, A. Ideris, and D. V. N. Vo, "A review on glycerol reforming processes over Ni-based catalyst for hydrogen and syngas productions," International Journal of Hydrogen Energy, vol. 45, no. 36, pp. 18466-18489, 2020. https://doi.org/10.1016/j.ijhydene.2019.08.211

[11] J. Lui, W. H. Chen, D. C. Tsang, and S. You, "A critical review on the principles, applications, and challenges of waste-to-hydrogen technologies," Renewable and Sustainable Energy Reviews, vol. 134, p. 110365, 2020. https://doi.org/10.1016/j.rser.2020.110365

[12] N. H. Zamzuri, R. Mat, N. A. S. Amin, and A. Talebian-Kiakalaieh, "Hydrogen production from catalytic steam reforming of glycerol over various supported nickel catalysts," International Journal of Hydrogen Energy, vol. 42, no. 14, pp. 9087-9098, 2017. https://doi.org/10.1016/j.ijhydene.2016.05.084

[13] S. J. Yoon, Y. M. Yun, M. W. Seo, Y. K. Kim, H. W. Ra, and J. G. Lee, "Hydrogen and syngas production from glycerol through microwave plasma gasification," International Journal of Hydrogen Energy, vol. 38, no. 34, pp. 14559-14567, 2013. https://doi.org/10.1016/j.ijhydene.2013.09.001

[14] M. Ayoub, S. Sathiaseelan, and S. A. M. Johari, "Experimental Studies of Gasification of Glycerol for The Production of Hydrogen," International Journal of Biomass and Renewables, vol. 12, no. 2, pp. 6-12, 2023. https://doi.org/10.61762/ijbrvol12iss2art24516

[15] M. N. Hussin, "Parametric Study for Production of Dimethyl Ether (DME) as a Fuel from Palm Wastes."

[16] J. Soujanya, C. A. Reddy, B. Satyavathi, and T. Sankarshana, "Experimental vapour-liquid equilibrium data of the quaternary system Methanol (1) + Isopropyl alcohol (2) + Water (3) + Glycerol (4) along with Isopropyl alcohol (2) + Glycerol (4) and Isopropyl alcohol (2) + Water (3) binary data at atmospheric and sub-atmospheric pressures," Fluid Phase Equilibria, vol. 409, pp. 327-333, 2016. https://doi.org/10.1016/j.fluid.2015.09.052

[17] B. J. Vreugdenhil, R. Zwart, and J. P. A. Neeft, "Tar formation in pyrolysis and gasification," 2009.

[18] R. Maceiras, J. Feijoo, V. Alfonsín, L. Pérez, M. A. Álvarez-Feijoo, P. Falcón, and J. P. Vallejo, "Influence of alumina fixed bed in steam reforming of glycerol for hydrogen production," Energy Reports, vol. 9, pp. 309-315, 2023,n doi: 10.1016/j.egyr.2023.08.092. https://doi.org/10.1016/j.egyr.2023.08.092

[19] Twigg MV. Catalyst handbook. 2nd ed. Frome, England: Wolfe Publishing Ltd.; 1989.

[20] V. Hacker and S. Mitsushima, Eds., Fuel Cells and Hydrogen: From Fundamentals to Applied Research. Elsevier, 2018.

[21] M. A. Haider and N. D. Chaturvedi, "An energy-efficient and cleaner production of hydrogen by steam reforming of glycerol using Aspen Plus," Int. J. Hydrogen Energy, 2023, doi: 10.1016/j. ijhydene.2023.09.089.

[22] S. Adhikari, S. D. Fernando, and A. Haryanto, "Hydrogen production from glycerol: An update," Energy Conversion and Management, vol. 50, no. 10, pp. 2600-2604, 2009. https://doi.org/10.1016/j.enconman.2009.06.011

[23] Y. Zhang, G. Zhang, B. Zhang, F. Guo, and Y. Sun, "Effects of pressure on CO2 reforming of CH4 over carbonaceous catalyst," Chemical Engineering Journal, vol. 173, no. 2, pp. 592-597, 2011. https://doi.org/10.1016/j.cej.2011.08.008

Materials Research Proceedings 53 (2025) 526-541 https://doi.org/10.21741/9781644903575-53

Carbon accounting tool for greenhouse gas emission sources in hydrogen production facilities

Farah Anissa Sukri[1,a*], Syaza Izyanni Ahmad[1,b*]

[1]Department of Chemical Engineering, Universiti Teknologi PETRONAS, Seri Iskandar, Malaysia

[a]farahanissasukri60@gmail.com, [b]syaza.ahmad@utp.edu.my

Keywords: Carbon Footprint, Steam-Methane Reforming, Hydrogen, Carbon Capture

Abstract. Hydrogen production facility stands out as one of the major CO_2 contributors in oil and gas industry. As hydrogen has emerged as a clean energy source in the future, this research is conducted to focus on carbon footprint assessment in hydrogen production facilities. This study aims to quantify amount of carbon dioxide through direct and indirect emissions which comprised of stationary combustion, process emissions and energy consumption. A comparative analysis has been done between SMR facility in Hydrogen production plant in company A, partial oxidation (POX) and alkaline electrolysis (AEL). The study revealed that gridline-based electrolysis shows the highest CO_2 emissions per production of 18 tCO_2e/tH_2, whereas hydropower-based electrolysis had the lowest emissions of 0 tCO_2e/tH_2. The study concluded that SMR exhibited the best efficiency among POX and AEL due to its overall low emissions per production of 8.6 tCO_2e/tH_2, high product throughput, and relatively low investment cost. Reforming technologies emit high amount of CO_2 emissions due to large fuel consumption as feedstock, but greater hydrogen throughput compared to electrolysis. As SMR emerged as the most efficient technology, the study recommended implementing CO_2 mitigation at process emissions, which accounted for the largest CO_2 emissions.

Introduction

Climate change is a serious issue that represents a persistent alteration in weather patterns across tropics to polls. The Intergovernmental Panel on Climate Change (IPCC) emphasized the acceleration of global warming comes from agriculture and industrial processes [1,2]. Most critical GHG emission is attributed to carbon dioxide (CO_2) due to its highest daily emission rate. 76% of the world GHG emission comes from CO_2, 16% methane, 6% nitrous oxide and 2% from others [3]. Hydrogen energy has emerged as high potential source of electric power generation apart from natural gas. Production of hydrogen comes from various sources such as reforming, pyrolysis, gasification, and electrolysis. Some of these methods however may not be a sustainable approach and can impact the environment.

Carbon emissions produced from process whether in direct or indirect form. Carbon accounting is a method used to evaluate carbon emissions in a business organization, together identifying the critical greenhouse gas emissions sources in hydrogen production facility. Examples of commercial hydrogen production technology include SMR, POX and AEL. The efficiency of SMR takes up about 65% to 75% among other commercial production methods such as gasification, electrolysis, and biological method as listed in Table 1.

Table 1. List of hydrogen production from fossil fuels [4,5,6]

Production Method	Feedstock	Efficiency (%)	Maturity	Cost ($/kg H_2)
Steam Reforming	Light hydrocarbons	74-85	Commercial	2.27
Partial Oxidation	Hydrocarbons, heavy fuel oil, coal	60-75	Commercial	1.48
Autothermal Reforming	Light hydrocarbons	60-75	Early Commercial	1.48
Pyrolysis	Hydrocarbons	35-40	Commercial	1.70
Gasification	Coal	30-40	Commercial	2.00
Electrolysis	Electricity	60-80	Commercial	10.30

Steam-Methane Reforming (SMR). SMR uses hydrocarbon of natural gas as feedstock. Preparation of steam and natural gas with high composition of methane must be done to achieve their feed specification prior to reformer. If the hydrocarbon feed contains heavier hydrocarbon such as ethane, a pre-reformer can be used upstream to break the hydrocarbon chain. As the methane gas is very low at the temperature of -30°C, preheating of natural gas must be done to elevate the temperature up to 20°C [7]. The concern of this process is to avoid condensation from occurring when mixing natural gas with high temperature of steam. Mixture of steam and natural gas will be fed into packed bed reactors with heat supplied reaching a temperature of 500°C to 950°C [8]. To avoid catalyst poisoning, desulphurization of natural gas must be done prior to reforming process. Syngas output from the reformer consists of carbon monoxide that could not be released directly to atmosphere. Hence, water-shift gas reaction is carried out at 300°C-450°C, and with the aid of a catalyst, which can enhance the reaction rate of reaction [9]. Hydrogen product obtained from WGS usually will be purified using pressure-swing adsorber (PSA).

Partial Oxidation. Partial oxidation serves as an alternative to SMR and is typically used for heavier hydrocarbons or when there is access to pure oxygen. This process offers rapid dynamic response, but as a trade-off, it has lower efficiency which is resulting in lower concentration of hydrogen product [10]. This process is prone to coke formation and requires operation at elevated temperature with range temperature of 1150°C to 1500°C to ensure complete conversion [10]. When catalyst is used, the temperature will be lowered to 780°C to 900°C [10]. Increasing oxygen consumption leads to higher air flow rates and causes greater fuel combustion. To prevent large of fuel consumption from combustion, reduction of start-up times is required by increasing the temperature rapidly. Generally, catalyst is used to obtain short start-up time due to fast exothermic nature of reaction. Nevertheless, sintering of catalyst may occur due to unable localize the hot spots and cause overheating. Partial oxidation does not need external heating to maintain the reaction which results in less energy required for fuel vaporization [7].

Alkaline Electrolysis. Alkaline water electrolysis is a process of transporting hydroxide ions through the electrolyte from the cathode to the anode while hydrogen is being generated at the cathode [6]. At the cathode side, two molecules of alkaline solution were reduced to one molecule of hydrogen, and two hydroxyl ions produced as exhibited in Eq. 1. H_2 exits the cathode surface to recombine in a gaseous state, while hydroxyl ions (OH) transfer through the porous diaphragm under the influence of the electrical circuit between anode and cathode [6]. Upon reaching the anode, these ions undergo discharge, resulting in the formation of half molecule of oxygen (O_2) and one molecule of water (H_2O). At the electrode surface, oxygen recombines discharged as hydrogen as demonstrated in Eq. 2 and Fig. 1. Alkaline electrolysis operates at lower temperatures ranging from 30°C to 80°C using aqueous solution of KOH or NaOH as the electrolyte, with concentrations typically around 20% to 30% [6]. From an economic standpoint, alkaline electrolysis is often viewed as a more economical option compared to other electrolysis method for large-scale hydrogen gas production. The reasoning behind this lies in alkaline electrolysis's ability to operate at higher current densities, leading to increased hydrogen production efficiency per unit of electrical energy [11].

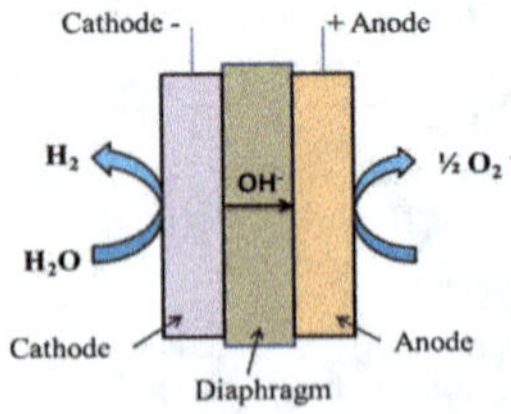

Fig. 1. Schematic illustration alkaline electrolysis [6]

Reaction at anode: $2OH^- \rightarrow H_2O + \frac{1}{2}O_2 + 2e^-$ Eq. 1

Reaction at cathode: $2H_2O + 2e^- \rightarrow H_2 + 2OH^-$ Eq. 2

Several companies including largest oil and gas company in Malaysia has set a vision of clean hydrogen production by 2050 as part of decarbonization roadmap. To initiate clean hydrogen production, carbon assessment must be done in hydrogen production facility to evaluate the annual carbon emission. However, most industrial processes company in Malaysia conducted carbon assessment according to their in-progress project and specific facility only. Effective GHG reporting is based on scope classification, scope 1, 2 or 3, to assess direct and indirect GHG emission in a specific organization and boundaries [12]. Even if there are carbon assessments conducted according to scope classification, there are still no precision of carbon accounting developed for hydrogen production facility, specifically in Malaysia. Table 2 exhibits past research regarding carbon accounting for industrial process and very few research conducted for hydrogen production method. Upon to that, this research of study is conducted to develop a carbon accounting tool for carbon footprint evaluation in hydrogen production facility and together propose carbon reduction strategy from critical greenhouse gas emissions path.

Table 2. Past research on carbon accounting for industrial processes

Author	Objective	Focus Area	Research Gaps
[13]	To evaluate carbon footprint of coal-based hydrogen production without carbon capture and effect of carbon footprint towards cost of hydrogen production.	Coal-based hydrogen production	Focus on cost imposed towards carbon emissions and coal-based hydrogen production method.
[14]	To calculate carbon footprint for offshore survey project of oil and gas industry.	Offshore transportation	Carbon assessment for mobile combustion only.
[15]	To quantify carbon footprint of steel production based on different type of production method	Steel production	Focus on steel production industry.
[16]	To quantify direct and indirect carbon emissions in steel production process.	Steel production	Focus on steel production industry.
[17]	To quantify greenhouse gas emissions, water consumption and solid waste between different method and material of wood chemical treatment.	Wood chemical treatment	Focus on wood chemical treatment facility.
[18]	To build carbon footprint model for Life Cycle Assessment (LCA) of oil and gas pipelines.	Oil and gas pipeline construction phase	Focus on carbon footprint of oil and gas pipeline and covers all scope from cradle to grave.
[19]	To quantify water and carbon footprint on dimethyl ether production.	Coal gasification	Focus on water footprint instead of carbon footprint.

Research Method

Project Description and Boundaries. Evaluation of steam-methane reforming for carbon accounting has been selected as the main facility in this project, followed by comparative analysis with partial oxidation and electrolysis technology. This process emits large amounts of carbon that will be captured through absorption process at PSA adsorber. Boundary of this study has been limited to operational only in which it is defined as the organization's greenhouse gas control and its ownership. Evaluation of scope 1 and 2 have been selected. In scope 1, all direct emissions were assessed around this process unit except fugitive and mobile combustion and scope 2, the consumption of steam and purchased electricity were evaluated. The boundaries have been set the same for partial oxidation and electrolysis technologies.

Process Flowsheet Determination. Process flowsheet for SMR has been taken from a hydrogen production plant in company A as in Fig. 2. Meanwhile, flowsheet for partial oxidation has been taken based on Fluor Daniel Inc data as in Fig. 3 [20]. Flowsheet for electrolysis has been referred from ELB company in Germany as in Fig. 4 [21], where the process flow for AEL is standard for most electrolysis plant.

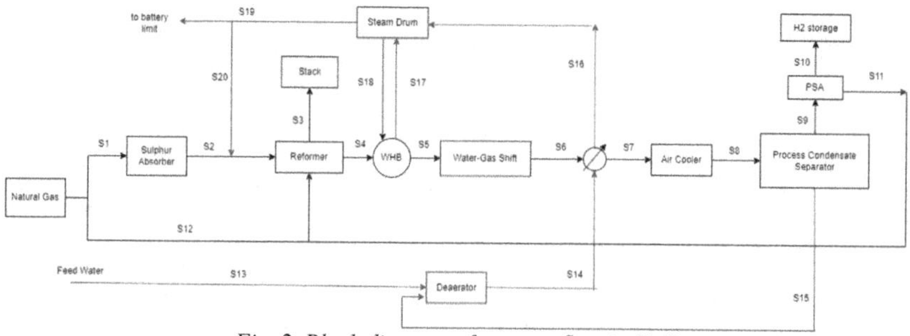

Fig. 2. Block diagram of process flow in SMR

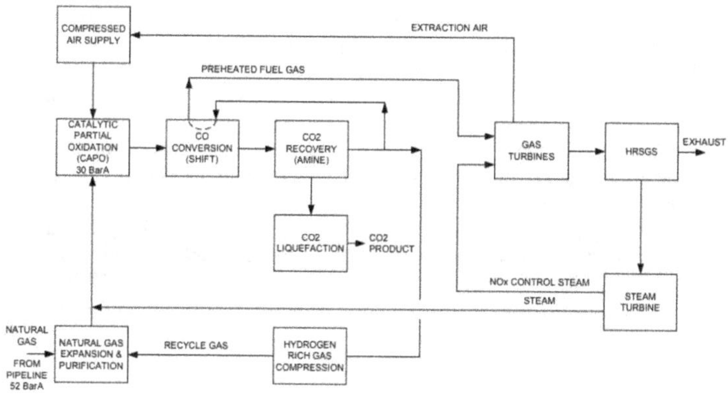

Fig. 3. Block diagram of process flow in POX

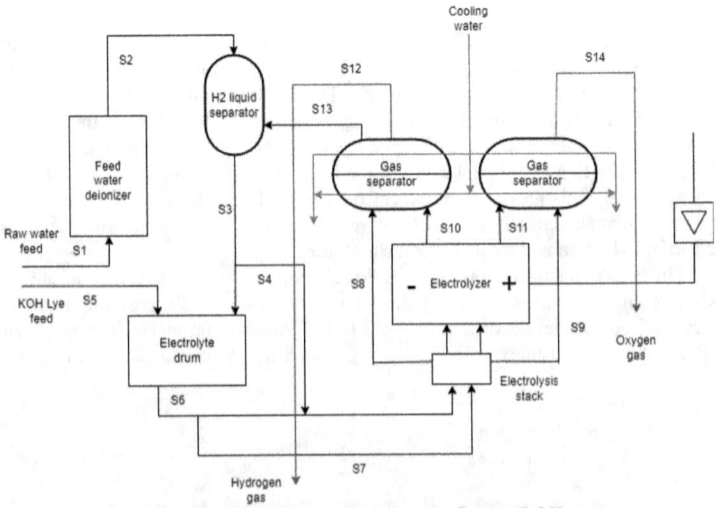

Fig. 4. Block diagram of process flow in POX

Fig. 2 shows flowsheet of steam-methane reforming unit in company A. The process includes pretreatment of sulphur removal, reforming of steam and methane, water-gas shift reaction, heat recovery process and finally purification through pressure-swing adsorption process (PSA) before obtaining the final hydrogen product. Fig. 3 shows base case flow diagram of catalytic partial oxidation process from Daniel Fluor Inc. In this process, oxygen is used instead of steam to combust together with natural gas. Other than that, similar process of reforming applied to POX such as water-gas shift process and hydrogen purification. Fig. 4 shows general process flow diagram of AEL to represent electrolysis facility from ELB. The process includes deionization of water and electrolysis only. However, this process is the most expensive compared to SMR and POX due to high power requirements with limited product throughput.

Data Collection and Processing. Data needed for carbon accounting such as mass input, synthesis gas generation, fuel, and energy consumption, as well as properties of raw materials were taken from respective company A, Fluor Daniel Inc and ELB [20,21].

Determination of carbon Emissions Sources. Carbon emissions sources have been determined through the analyzed stream from process flowsheet. SMR process consists of three major carbon emissions sources comprised of stationary combustion, process emissions and energy consumption. Partial oxidation comprised of process emissions and energy consumption while electrolysis only consists of energy consumption only. The summary of carbon emissions is listed as in Table 3.

Table 3. Type of emissions from each facility

Facility	Scope of emission	Type of emission	Elaboration
SMR	1	Stationary combustion	Natural gas fed to combustion chamber
		Process emission	Conversion of CO to CO_2
	2	Energy consumption	Production of steam as feed, operation of pump
POX	1	Process emission	Conversion of CO to CO_2
	2	Energy consumption	Production of steam, operation of pump and compressor
AEL	2	Energy consumption	Power consumption for electrolysis

Carbon Accounting Tool Development. Calculation for carbon emissions was evaluated in period of 8000 hours standard of operation and reported as annual emissions. American Petroleum Institute (API) Compendium methodology, revision 2021 was used as formula reference as shown in Table 4 [22]. Tier 3 is used for SMR carbon accounting since each facility data provided by company A to be assessed accurately. Tier 1 has been selected for POX and AEL using default emission factor provided by API Compendium [22,23].

Table 4. Summary equation for carbon accounting [22]

Type of Emission	Formula	Exhibit
Stationary combustion	$\mathbf{E_{CO2} = FC \times MW_{fuel} \times wt\%\ C \times (\frac{44}{12})/23.685}$	4.4b
Process emission	$\mathbf{E_{CO2} = FC \times wt\%\ C \times 44/12}$	6-42
Energy consumption	Steam consumption (unknown source):	
	$\mathbf{E_{CO2} = m_{steam}(\Delta H_g - \Delta H_{sat}) \times EF}$	8.3
	Electricity consumption (known source):	
	$\mathbf{E_{CO2} = (NG \times EF) - E_{CO2,steam} \times \frac{Power\ rate}{\Delta H_{electric}}}$	8.1
	Electricity consumption (unknown source):	
	$\mathbf{E_{CO2} = Power\ rate \times EF}$	8.1

All emissions value calculated in unit of tonne CO_2 per year. To standardize carbon accounting unit, all GHG emissions shall be converted to carbon equivalent. For carbon dioxide, carbon equivalent is calculated using formula in Eq. 3 [22].

$$E_{CO2e} = E_x \times \frac{1\ \text{ton}\ CO_2 2}{\text{ton}\ CO_2} \qquad \text{Eq. 3}$$

Results And Discussion

Steam-Methane Reforming (SMR).

Table 5. Carbon accounting for Steam-Methane Reforming

SECTION A - Stationary combustion

No	Type of substance combusted	Amount of combusted substances	Unit	Emission factor (tCO_2e/tonne)	Tonnes CO_2 Emissions per year
1	SMR 1 - Natural Gas	5938.4	tonne/year	3.22	19096.47
2	SMR 2 - Natural Gas	3111.2	tonne/year	2.97	9236.51
3	SMR 3 - Natural Gas	27347.2	tonne/year	3.04	83241.89
		Total			111574.88

SECTION B - Process Emission

No	Type of substance processed	Amount of combusted substances	Unit	Emission factor (tCO_2e/tonne)	Tonnes CO_2 Emissions per year
1	SMR 1 - Natural Gas	38160	tonne/year	2.67	101837.29
2	SMR 2 - Natural Gas	117544	tonne/year	2.63	308599.79
3	SMR 3 - Natural Gas	181512	tonne/year	2.64	478559.50
		Total			888996.58

SECTION C - Purchased Energy Consumption

Type	Facility of Purchased Energy	Amount of Consumption	Unit	Emission factor (tCO_2e/kWh)	Tonnes CO_2 Emissions per year
Electricity	SMR 1	1760	MWh/year	0.035	61.54
	SMR 2	9600	MWh/year	0.035	335.66
	SMR 3	19200	MWh/year	0.035	671.33
		Total			1068.53

Table 5 shows parameters for carbon emissions calculation. Parameters required are such amount of fuel combusted, energy consumption and calculated emission factor. The calculation divided into three sections, section A, B and C, to distinguish carbon emissions from stationary combustion, process emissions and purchased energy consumption. Table 6 only consists of section B and C while table 7 and 8 only consist of section C.

Partial Oxidation (POX).

Table 6. Carbon accounting for Partial Oxidation

SECTION B - Process Emission

No	Type of substance processed	Amount of combusted substances	Unit	Emission factor (tCO_2e/tonne)	Tonnes CO_2 Emissions per year
1	Natural Gas	1070784	tonne/year	2.72	2911788.82
		Total			2911788.82

SECTION C - Purchased Energy Consumption

Type	Source of Purchased Energy	Amount of Consumption per year	Unit	Emission factor	Unit	Tonnes CO_2 Emissions per year
Electricity	Combined Cycle Plant	745	MWh/year	0.0727	tCO_2e/ MWh	54.16
Steam	Combined Cycle Plant	2382336	tonne/year	0.2195	tCO_2e/t	522958.42
		Total				523012.58

Alkaline Electrolysis AEL-1.

Table 7. Carbon accounting for Alkaline Electrolysis AEL-1

SECTION C - Purchased Energy Consumption

Type	Source of Purchased Energy	Amount of Consumption	Unit	Emission factor (tCO2e/MWh)	Tonnes CO_2 Emissions per year
Electricity	Electricity Gridline	19073.20	MWh/year	0.37764	7202.80
		Total			7202.80

Alkaline Electrolysis AEL-2

Table 8. Carbon accounting for Alkaline Electrolysis AEL-2

SECTION C - Purchased Energy Consumption

Type	Source of Purchased Energy	Amount of Consumption	Unit	Emission factor (tCO2e/MWh)	Tonnes CO_2 Emissions per year
Electricity	Hydropower	19073.20	MWh/year	0.00	0.00
		Total			0.00

Comparison of CO_2 Emissions between Each Steam-Methane Reforming Unit

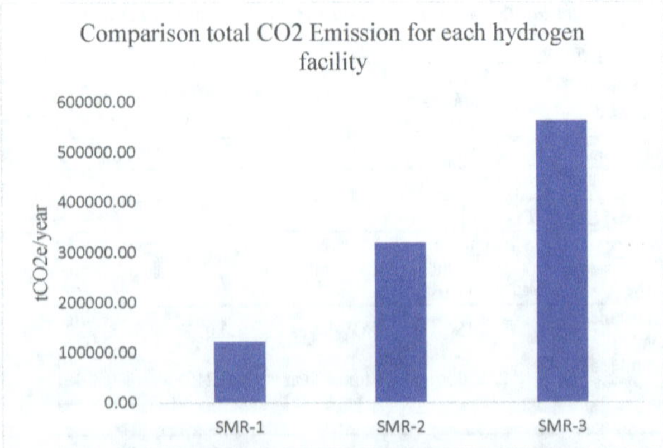

Fig. 5. CO_2 emissions from each SMR facility

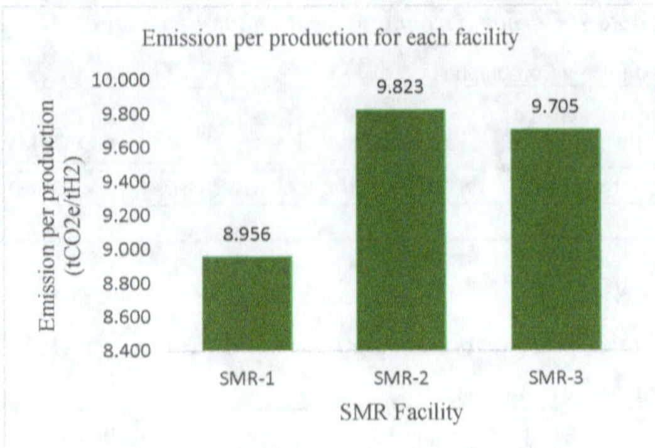

Fig. 6. CO_2 emissions per production for each facility

Table 8. CO_2 emission for each SMR facility

Hydrogen Facility	SMR-1	SMR-2	SMR-3
Total Direct CO_2 Emissions (tCO_2e/y)	121019.16	317846.80	561814.05
Total Indirect CO_2 Emissions (tCO_2e/y)	61.54	335.66	671.33
Total CO_2 Emissions (tCO_2e/y)	121080.70	318182.47	562485.37
H_2 Production (tonne/year)	13520	32392.8	57956
Emission per production (tCO_2e/tH_2)	8.956	9.823	9.705

Fig. 5 and Fig. 6 shows results of carbon accounting for steam-methane reforming facilities in company A. Carbon emissions evaluated excluding all carbon equivalent (CH_4, N_2O) due to minor contribution to the main objective of this research. Assuming 8000 hours of operation, SMR-3 shows the largest number of total CO_2 emissions in span of a year, amounting to 561,814.05 tonne CO_2 emissions per year. The carbon emission calculated is based on the direct and indirect emission within the boundary of hydrogen production unit. This substantial amount is mainly influenced by the amount of fuel feedstock fed to the reformer which is 181,512 tonne/year.

Consequently, SMR-3 exhibits the highest number of direct emissions compared to SMR-1 and SMR-2.

From Table 8, direct emission exhibits the largest CO_2 emissions from SMR facility, mainly contributed from combustion of substance to heat up the reformer and feedstock used for hydrogen production. Stationary combustion of natural gas with sufficient oxygen converts all carbon component into carbon dioxide, assumed 100% carbon oxidation as shown in Eq. 4. The equation provided is valid since major amount of CH_4 component in natural gas content is used for combustion which are 89 mol% for SMR-2 and SMR-3, and 95 mol% for SMR-1. However, SMR-1 consists of 39 mol% of methane, 15 mol% of ethane, 8 mol% propane, 4 mol% butane and 31 mol% of hydrogen as fuel feedstock. The amount of CO_2 produced in 1 kmol basis can be deduced from the chemical equation Eq. 5.

$$CH_4 + O_2 \rightarrow CO_2 + H_2O \qquad \text{Eq. 4}$$

$$0.39CH_4 + 0.15C_2H_6 + 0.08C_3H_8 + 0.04C_4H_{10} + 0.31H_2 + 1.88O_2 + 7.0688N_2 \rightarrow 0.85CO_2 + 7.0688N_2 + 2.06H_2O \qquad \text{Eq. 5}$$

CO_2 produced for combustion in SMR-1 is relatively lower due to consumption of recycled blended hydrogen for combustion. This makes formation of carbon oxidized less in the combustion process, where 1 kmol of fuel produces 0.85 kmol of CO_2.

Main feedstocks used for reforming are hydrocarbon and steam that produces carbon monoxide in the syngas. Water-gas shift reaction shifts the reaction equilibrium to favor formation of hydrogen and carbon dioxide at lower temperature. Huge amount of CO_2 formed with limited H_2 formed from WGS process. From the overall stoichiometric Eq. 6, 1 mol of CH_4 produces 1 mol of CO_2. Major amount of CH_4 feedstock is used in all SMR units which are 89 mol% for SMR-2 and SMR-3, and 95 mol% for SMR-1. Major amount of CO_2 produced due to its larger molecular weight compared to hydrocarbon which causes 16 kg of methane to produce 44 kg of carbon dioxide, equivalent to 4 times CO_2 production from its feedstock.

$$CH_4 + 2H_2O \rightarrow CO_2 + 4H_2 \qquad \text{Eq. 6}$$

Table 8 shows indirect CO_2 emission from cogeneration contributes minor value towards the total emissions. This emission origin from electricity consumption for each SMR facility, mainly used for pump and compressor driver. Boiler efficiency for steam and electricity of cogeneration plant was considered in calculating emission factor for electricity consumption. Higher efficiency results in lower emission factor due to less combustion required to produce steam and electricity. Steam generation for the SMR facility is achieved through a waste heat boiler, which operates without combustion, thereby eliminating another source of CO_2 emissions.

Evaluation of the emissions according to the annual production was done to conclude the assessment of facility efficiency regarding CO_2 emissions. Based on Fig. 6, SMR-2 emits the largest CO_2 emission per production of 9.823 tCO_2e/tH_2. Based on the calculation made, factor that affects different CO_2 emissions for each facility depends on the mass flowrate of fuel and carbon content from the fuel used. Comparison of emission factor for each fuel feedstock has been done as shown in Table 9 for process emission only since it is the largest CO_2 contributor compared to stationary combustion.

Table 9. Fuel feedstock properties for each SMR facility

Type of substance	Fuel consumption (t/y)	wt% C	Emission factor (tCO_2/t)	Hydrogen production (t/y)	Product-to-feed ratio
SMR 1 - Natural Gas	38192.00	0.728	2.669	13520.00	0.354
SMR 2 - Natural Gas	117548.00	0.716	2.625	32392.80	0.276
SMR 3 - Natural Gas	181516.80	0.719	2.637	57956.00	0.319

Despite low emission factor of SMR-2, product-to-feed ratio for shows the lowest value, indicating lowest hydrogen conversion of feedstock to product. This concludes that large carbon emissions per production for SMR-2 is caused by low conversion of feedstock to product.

From the justification provided, most efficient unit to produce hydrogen is SMR-1 since it emits the least CO_2 emissions per production in total of 8.956 tCO_2e/tH_2 and has higher product-to-feed ratio compared to SMR-2 and SMR-3. SMR-3 stands as the second optimal SMR facility that obtains low emission value per production and high product-to-feed ratio. Despite the high efficiency of SMR-1 show, utilization of SMR-1 is the lowest due to small capacity and high operating cost imposed on the facility. This inference justifies maximization of SMR-3 utilization is advantageous compared to SMR-2.

Comparison of CO_2 emissions between each steam-methane reforming with partial oxidation and electrolysis technology

Table 10. CO_2 emissions for each hydrogen production technology

Hydrogen Facility	SMR	POX	AEL-1	AEL-2
Total Direct CO_2 Emissions (tCO_2e/y)	561814.1	2911788.82	0.0	0.0
Total Indirect CO_2 Emissions (tCO_2e/y)	671.33	246326.56	7202.8	0.00
Total CO_2 Emissions (tCO_2e/y)	562485.4	3158115.37	7202.8	0.0
H_2 Production (tonne/year)	57956	293299	400	400
Emission per production (tCO_2e/tH_2)	9.705	10.768	18.01	0.0

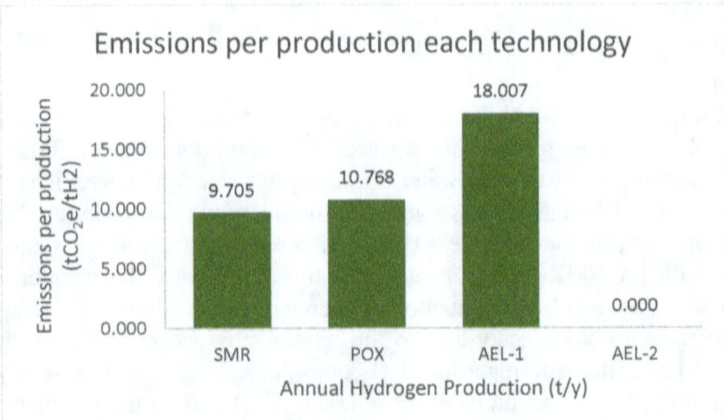

Fig. 7. Emission per production for each technology

A comparison of the overall CO_2 emissions among various hydrogen production technologies has been conducted. SMR-3 has been selected for technology comparison due to its optimal efficiency. Two alkaline electrolysis has been compared due to different type of power supply that causes different value of emissions. The findings reveal that partial oxidation exhibits the highest total CO_2 emissions of 561814.1 tCO_2e/year. This result mainly affected from the significant

Materials Research Proceedings 53 (2025) 526-541 https://doi.org/10.21741/9781644903575-53

difference in amount of feedstock used for both SMR and POX. AEL-1 and AEL-2 stand out as the lowest total CO_2 emissions at 7,202.8 tCO_2e/year and 0 tCO_2e/year respectively. Global demand for green hydrogen production is 400 tonne per year due to its high price compared to fossil fuel-based hydrogen [24]. According to Fig. 7, electrolysis using AEL technology generates no direct emissions since it relies on carbon-free reactants. The hydrogen production process through electrolysis involves the separation of hydrogen and oxygen gases from water, as illustrated in Eq. 7.

$$H_2O \rightarrow \frac{1}{2}O_2 + H_2 \qquad \text{Eq. 7}$$

However, the operation of the plant demands a high voltage of electricity. The electricity requirement for the entire process, including both the system and stack, is 4.2 to 4.8 kW, producing 1400 Nm^3 of H_2 per hour [21]. Assuming 400 tH_2/year of production, the amount of electricity required is 19,073.2 MWh per year. Emission factor used was 0.37764 tCO_2e/MWh sourced from Environment Protection Agency's (EPA) eGrid database [25]. AEL-2 consists of zero emissions due to renewable energy consumption of hydropower. Emission factor for renewable energy such as solar, hydropower and wind are considered 0 tonne CO_2 per kJ for scope 1,2 and 3 [22]. However, in life cycle process of generating electricity through hydropower, 4.5g CO2/kWh shall be generated from biochemical activities for capacity between 100 kW to 10 MW [26]. This emission does not include in scope 1 or 2 and shall be reported separately [22].

The ultimate findings regarding total emissions per unit of production reveal high CO_2 emissions from AEL-1 amounting to 18.01 tCO_2 emissions for each tonne of hydrogen generated. Electrolysis technology based on gridlines leads to significant CO_2 emissions per unit of hydrogen produced because it involves large combustion of fuel to generate a relatively small quantity of hydrogen [27]. This statement can be justified from Fig. 8, where the value of CO_2 emissions per production for gridline-based electrolysis ranges from 13 tCO_2e/tH_2 to 25 tCO_2e/tH_2.

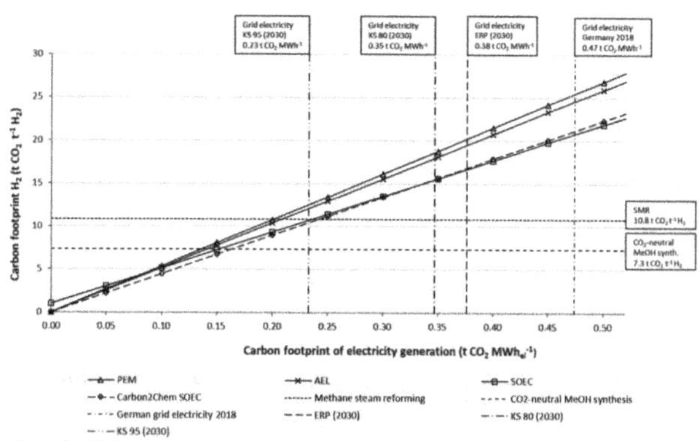

Fig. 8. CO_2 emissions of electrolysis from gridline electricity [27]

Partial oxidation (POX) stands as the second highest CO_2 emissions per production technology with a total value of 10.768 tCO_2e/tH_2. In contrast, SMR exhibits lower CO_2 emission per production in amount of 9.705 tCO_2e/tH_2. From Figure 4.9, 92% of POX and 99% of SMR emissions are origin from direct emission. Higher direct carbon dioxide emission of SMR is due to the combustion required for heating the reformer. Additional of indirect POX's CO_2 emissions due to production of HP steam from combined cycle plant. This steam is used to preheat the natural gas used prior to reformer for faster reaction. These factors result in a significant disparity in direct and indirect emissions between POX and SMR. Significant comparison that can be made to deduce the difference in the amount of emissions factor between POX and SMR is the amount of hydrogen produced per feedstock fed as shown in Table 11.

Table 11. Product-to-feed ratio for each technology

Type of substance	Fuel consumption (t/y)	wt% C	Emission factor (tCO_2/t)	Hydrogen production (t/y)	Product-to-feed ratio
SMR	181516.80	0.719	3.04	57956.00	0.319
POX	1070784	0.742	2.72	293299.000	0.274

Due to different amount of feedstock used for each technology, product-to-feed ratio was evaluated to identify the technology with best efficiency. From Table 11, SMR has higher product-to-feed ratio of 0.319 compared to POX with value of 0.274 in terms of H_2 production. This process can be justified from Eq. 6 where 4 mol of hydrogen is produced from every mol of fuel feedstock. However, from overall equation of POX, 1 mol of hydrocarbon produces 2 mol of hydrogen. This discrepancy shows that SMR utilizes a lower amount of feedstock and achieves a higher hydrogen production, signifying that SMR technology attains a better throughput compared to POX with lower process emissions.

Grey hydrogen emits about 8 to 10 $kgCO_2$ per kgH_2, equivalent to 8 to10 tCO_2e/tH_2 [28]. It indicates utilization of hydrocarbon in producing hydrogen without carbon capture facility. Values obtained from the carbon assessment are from the theoretical emissions excluding all types of carbon capture such as pressure-swing adsorption for SMR and carbon capture and storage (CCS) for POX. Actual number of direct CO_2 emissions can be obtained by calculating the venting rate at the stack, where most of the CO_2 has been absorbed at the PSA column and CCS facility.

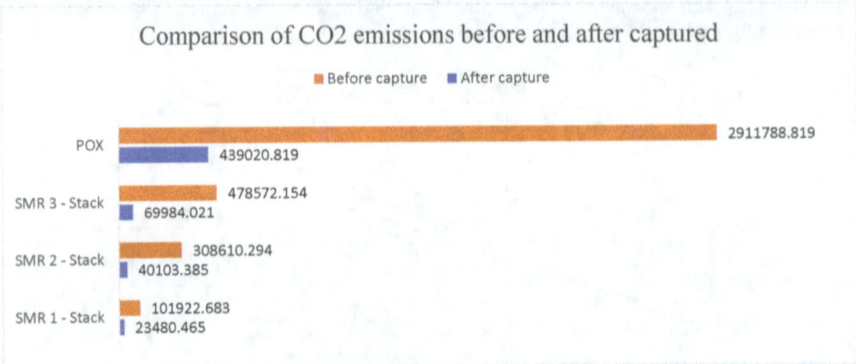

Fig. 9. Comparison of CO_2 emissions before and after captured

Process emission has shown its major factor to contributing the total emissions in a reforming facility. Fig. 9 shows installation of carbon capture for process emissions in both SMR and POX facilities. The actual CO_2 emissions was calculated based on the amount of flue gas venting rate to the atmosphere. CO_2 byproduct produced during the reaction was absorbed by both PSA for SMR, and CCS for POX before released to the stack. Note that amount of CO_2 released solely from process emissions and emissions from stationary combustion is not counted. Results from Figure 4.12 shows a significant decrement in CO_2 emissions before and after captured wherein around 85% to 87% from the calculated emissions has been captured and adsorbed. Implementation of carbon capture is essential as one of the strategies that can be made to mitigate critical source of CO_2 emissions in the plant and this practice has been complied with many companies. Captured CO_2 from PSA and CCS can be extracted in form of gas or liquid to be utilized for many uses such as enhancement of oil recovery and insulation material.

While the advantages of utilizing electrolysis as a green technology for hydrogen production are acknowledged, certain drawbacks have raised concerns. The limited quantity of hydrogen that can be generated, along with the high capital expenditure (CAPEX) and operational expenditure (OPEX) compared to SMR and POX processes, are reasons for the lower utilization of this method. The amount of investment cost for electrolyzing operation is 800 to 1,500 Euro per kWh [21]. Electrolyzer cell degrades faster after several usages in which replacement of the cells is required after 60,000 hours [21]. This makes the process less efficient compared to reforming technology. As a result, further research is needed to address the challenge of integrating electrolysis into mainstream hydrogen production processes, thereby replacing conventional reforming technologies.

Conclusion

In conclusion, this study successfully developed a carbon accounting tool for hydrogen production facility using an Excel spreadsheet, to trace annual carbon emissions across different scopes. By quantifying carbon emissions for scope 1 and scope 2 in two case studies involving SMR and SMR-POX-AEL from company A, Fluor Daniel Inc., and ELB, the results showed that SMR-2 has the highest CO_2 emissions per production at 9.823 tCO_2e/tH_2, while SMR-1 has the lowest at 8.956 tCO_2e/tH_2. SMR-1 demonstrated the highest SMR efficiency due to its lower CO_2 emissions and higher product-to-feed ratio. Comparative analysis revealed that green hydrogen production from alkaline electrolysis AEL-2 has the lowest carbon emissions, approaching 0 $tCO_2e/tH2$, while AEL-1 has the highest at 18 tCO_2e/tH_2 due to non-renewable power supply. Although AEL-2 is most efficient in terms of emissions, SMR is more attractive due to lower investment costs and emissions per production. Mitigation of CO_2 emissions from SMR can be achieved by implementing carbon capture technologies such as PSA or CCS, which currently capture 85% to 87% of emissions. Overall, this study achieved all objectives, providing insights into reducing greenhouse gas emissions and identifying areas for improvement and innovation in clean energy technologies, contributing to the goal of decarbonization by 2050.

Acknowledgement

This research study was supervised and supported by Universiti Teknologi PETRONAS under Research University Grant 015LC0-469.

References

[1] Intergovernmental Panel on Climate Change. (2023). Synthesis Report of the IPCC Sixth Assessment Report (AR6). AR6 Synthesis Report Climate Change 2023.

[2] Gale (2021). Scholarly articles on global warming and climate change | gale. Global Warming.

[3] United States Environment Protection Agency (EPA). (2023). Global Emissions by Gas. Global Greenhouse Gas Emission Data | US EPA.

[4] Megía, P. J., Vizcaíno, A. J., Calles, J. A., & Carrero, A. (2021). Hydrogen Production Technologies: From Fossil Fuels Toward Renewable Sources. A Mini Review. Energy & Fuels, 35(20), 16403-16415. https://doi.org/10.1021/acs.energyfuels.1c02501

[5] Rathee, D. (2023, July 31). Types of electrolysers: CEF explains. CEEW.

[6] Shiva Kumar, S., & Himabindu, V. (2019). Hydrogen production by PEM water electrolysis - a review. Materials Science for Energy Technologies, 2(3), 442- 454. https://doi.org/10.1016/j.mset.2019.03.002

[7] Erickson, P. A., Tang, H.-Y. (Ray), & Vernon, D. R. (2014, July 29). Reformation of Hydrocarbon Fuels. Routledge Handbooks Online.

[8] Rapier, R. (2022, October 12). Estimating The Carbon Footprint Of Hydrogen Production.Forbes. production.

[9] Roy, S., & Tiwari, A. K. (2021). Efficient water-gas shift catalysts for H2O and co dissociation using Cu-Ni step alloy surfaces. The Journal of Physical Chemistry, 125(25), 13819-13835. https://doi.org/10.1021/acs.jpcc.1c02363

[10] Voldsund, M., Jordal, K., & Anantharaman, R. (2016). Hydrogen production with CO2 Capture. International Journal of Hydrogen Energy, 41(9), 4969-4992. https://doi.org/10.1016/j.ijhydene.2016.01.009

[11] Zeng, K., & Zhang, D. (2010). Recent progress in alkaline water electrolysis for hydrogen production and applications. Progress in Energy and Combustion Science, 36(3), 307-326. https://doi.org/10.1016/j.pecs.2009.11.002

[12] MyCarbon (2014). National Corporate GHG Reporting Programme for Malaysia. Ministry of Natural Resources and Environment (NRE) Malaysia

[13] Li, J., Wei, Y.-M., Liu, L., Li, X., & Yan, R. (2022). The carbon footprint and cost of coal-based hydrogen production with and without carbon capture and storage technology in China. Journal of Cleaner Production, 362, 132514. https://doi.org/10.1016/j.jclepro.2022.132514

[14] Lam, C. H., Hassim, M. H., & Ho, W. S. (2023). Carbon footprint of an offshore survey by oil and Gas Service Provider Company. Journal of Energy and Safety Technology (JEST), 5(2), 87-96. https://doi.org/10.11113/jest.v5n2.116

[15] Suer, J., Traverso, M., & Jäger, N. (2022). Carbon footprint assessment of hydrogen and steel. Energies, 15(24), 9468. https://doi.org/10.3390/en15249468

[16] Tian, W., An, H., Li, X., Li, H., Quan, K., Lu, X., & Bai, H. (2022). CO2 accounting model and Carbon Reduction Analysis of iron and steel plants based on intra- and inter-process carbon metabolism. Journal of Cleaner Production, 360, 132190. https://doi.org/10.1016/j.jclepro.2022.132190

[17] Lewis, G. M., Eastin, I., & Keoleian, G. A. (2022). Life Cycle Carbon Accounting of thermally modified wood in the US inland northwest. ACS Sustainable Chemistry & Engineering, 10(31), 10252-10259. https://doi.org/10.1021/acssuschemeng.2c02023

[18] Huang, L., Liao, Q., Yan, J., Liang, Y., & Zhang, H. (2021). Carbon footprint of Oil Products Pipeline Transportation. Science of The Total Environment, 783, 146906. https://doi.org/10.1016/j.scitotenv.2021.146906

[19] Liu, Y., Xu, H., Wang, Y., Cui, P., Sun, C., Zhu, Z., & Wang, Y. (2023). Life cycle water footprint and carbon footprint analysis of coal gasification to clean fuel dimethyl ether. Fuel, 357, 129884. https://doi.org/10.1016/j.fuel.2023.129884

[20] Rao, A., Francuz, D., Scherffius, J., & West, E. (2000). CO2 capture by partial oxidation - IEAGHG. IEA Greenhouse Gas R&D Programme.

[21] Buttler, A., & Spliethoff, H. (2018). Current status of water electrolysis for energy storage, grid balancing and sector coupling via power-to-gas and power-to-liquids: A Review. Renewable and Sustainable Energy Reviews, 82, 2440-2454. https://doi.org/10.1016/j.rser.2017.09.003

[22] Campbell, L., Toolen, J., Grubert, D., & Napp, G. (2021a). Compendium of Greenhouse Gas Emissions Methodologies For The Natural Gas And Oil Industry. Compendium of Greenhouse Gas Emissions Methodologies.

[23] Darío, G. R., & John, W. D. (2006). 2006 IPCC guidelines for National Greenhouse Gas Inventories. IPCC.

[24] Tonelli, D., Rosa, L., Gabrielli, P., Caldeira, K., Parente, A., & Contino, F. (2023). Global land and water limits to electrolytic hydrogen production using wind and solar resources. Nature Communications, 14(1). https://doi.org/10.1038/s41467-023-41107-x

[25] Grid Electricity Emissions Factors. (2023). 2022 grid electricity emissions factors v0.1 - February 2023.

[26] Smoot, G. (2023, July 20). What is the carbon footprint of Hydropower Energy? A life-cycle assessment. Impactful Ninja.

[27] Tenhumberg, N., & Büker, K. (2020). Ecological and economic evaluation of hydrogen production by different water electrolysis technologies. Chemie Ingenieur Technik, 92(10), 1586-1595. https://doi.org/10.1002/cite.202000090

[28] Editor. (2023, January 17). Gh2 facts: CO2 emissions per kg of hydrogen depending on the method of production. Hydrogen Newsletter.

Materials Research Proceedings 53 (2025) 542-549 https://doi.org/10.21741/9781644903575-54

Development of amino acid deep eutectic solvent (AADES) as a crosslinker for carboxylated nitrile butadiene rubber (XNBR)

NURFATHIAH IZZATY Mohd Faridz Hilmy[2,a], AIMAN HANIF Hanafi[1,b] and WAN ZAIREEN NISA Yahya[1,2,c*]

[1]Department of Chemical Engineering, Universiti Teknologi PETRONAS, 32610 Seri Iskandar, Perak, Malaysia

[2]Centre of Research in Ionic Liquid (CORIL), Universiti Teknologi PETRONAS, 32610 Seri Iskandar, Perak, Malaysia

[a]nurfathiah.faridz@utp.edu.my, [b]aiman_17001040@utp.edu.my, [c]zaireen.yahya@utp.edu.my

Keywords: Amino Acid-Based Deep Eutectic Solvents (AADES), Carboxylated Acrylonitrile Butadiene Rubber (XNBR), Crosslinker

Abstract. Carboxylated acrylonitrile butadiene rubber (XNBR) a high-performance synthetic nitrile rubber is gaining much demand due to enhance mechanical properties in many applications including rubber gloves. The common crosslinker for XNBR, namely sulfur and zinc oxide with presence of accelerator which pose risks of Type IV Allergic Contact Dermatitis (ACD) and environmental concerns. To overcome this problem, a new amino acid-based deep eutectic solvent (AADES) is proposed as a green and sustainable ionic crosslinking agent. The ionic crosslinking between AADES and XNBR demonstrates superior machanical and thermal properties, offering a viable alternative to conventional methods. Fourier Transform Infrared Spectroscopy (FTIR) and Nuclear Magnetic Resonance were used to identify key structures in both cured XNBR and AADES. Mechanical and thermal analyses were performed using Universal Tensile Test, Thermal Gravimetric Analysis, and Differential Scanning Calorimetry on cured XNBR and AADES samples. Notably, XNBR crosslinked with a 1:3 DES molar ratio showed the highest tensile strength of 4.84 MPa, while a 1:2 DES molar ratio resulted in the highest elongation at break of 120%. Cured XNBR exhibited increased thermal stability, indicating successful ionic crosslinking between L-lysine: Glycerol and XNBR matrices. This study underscores AADES's potential as a promising crosslinking agent in polymer development, offering enhanced performance and reduced environmental impact.

Introduction

The development of rubber materials has led various modifications over time, for both industrial and domestic settings, owing to their unique elasticity and flexibility. In rubber industries, there are several types of synthetic nitrile rubber namely Buna-N nitrile rubber (NBR), hydrogenated acrylonitrile butadiene rubber (HNBR), and carboxylated acrylonitirle butadiene rubber (XNBR) each offer distinct advantages and find varied applications across the rubber industry. Among these options, carboxylated acrylonirile butadiene rubber (XNBR) stands out as the most superior modified synthetic rubber, boasting exceptional tensile strength and remarkable resistance to abrasion. Carboxylated acrylonitrile butadiene rubber (XNBR) is a synthetic rubber that has carboxylic acid group added to the polymer chain made from acrylonitrile 1,3-butadiene.

To improve the elasticity and mechanical strength of rubber, the cross-linking process has been widely used, typically using sulfur and peroxide. However, this conventional crosslinking method leaves environmental challenges during the rubber`s end-of-life cycle, primarily due to resource wastage [1]. The inclusion of cross-linking additives in this procedure results in the release of volatile organic compounds (VOCs), which produce undesirable odors. Furthermore, the permanent nature of covalent crosslinks formed during sulfur vulcanization renders the material non-recyclable, as stated by Das et al. [2]. Efforts to recycle sulfur from cross-linked polymers through desulfurization processes face challenges, as breaking the cross-linking bonds requires prior disintegration of the polymer chain.

Moreover, sulfur vulcanization is a time-consuming process that necessitates accelerators such as thiuram for immediate vulcanization effects. However, these accelerators are associated with occupational dermatitis, particularly Allergic Contact Dermatitis (ACD), leading to skin inflammation. Cure activators, like zinc-oxide , play a pivotal role in efficient rubber cross-linking but pose environmental risks through zinc leaching during the rubber's lifespan and in the wastewater discharge. In response to this concern, researchers have explored the modification of zinc oxide as a cross-linking agent to facilitate ionic cross-linking within the rubber matrix. The presence of carboxyl group in the XNBR allows it to be crosslinked by various agents including ionic crosslinkers.

Consequently, numerous studies have delved into utilizing alternative ionic crosslinkers. In a study by Das et al., they reported the usage of zwitterionic compounds on XNBR, namely two amino acids, L-lysine and tryptophane as a cross-linking agent to enhance elastomer mechanical properties in addition to zinc oxide [2]. It is found that, the presence of L-lysine significantly influences the mechanical properties of cross-linked XNBR, enhancing the tensile strength and elongation at break, as well as reduces the optimum cure time and scorch time with increasing L-lysine concentration [2]. Apart from that, L-lysine can be found in several other applications such as in water management in the oil and gas industry through hydrogel cross-linking [3], and possess an excellent antimicrobial activity as a carbon source in food and pharmaceutical indutries [4] and animal feeds [5].

On the other hand, a deeep eutectic solvent (DES) is a mixture comprising a hydrogen bond donor (HBD) and a hydrogen bond acceptor (HBA), forming through hydrogen bonding. It can be easily synthesized by stirring two components together under constant heat. Recently, there has been exploration into the application of DES in rubber products due to their cost-effectiveness and simple chemical synthesis, which contrasts with the complexity of regular ionic liquids. Glycerol, which is the main byproduct of biodiesel production has been extensively used in many industrial applications namely food and pharmaceutical industries. Glycerol-based DESs have been reported in different applications as well as lysine-based DESs have been investigated as biomass pretreatment as it could enhance the efficacy of the subsequent enzymatic hydrolysis [6].

Given its establised safety profile in medical contexts, L-lysine and glycerol emerge as a viable alternative for rubber crosslinker. Therefore, this paper aims to develop the deep eutectic mixture based on L-lysine and glycerol as a green crosslinker and its crosslinking towards XNBR with regards to both chemical and mechanical properties are presented.

Experimental Section

Materials. Carboxylated Nitrile Rubber (XNBR) were comprised of 25-35 phr nitrile content, 60-70 phr of butadiene and 4-6 phr carboxylic acid content. Compounding ingredients namely antifoam agent, whitening agent, antioxidant as well as coagulants were commercial grades and used without further modifications. L-lysine (99wt% purity; CAS Number: 56-87-1) and glycerol (99.5wt% purity; CAS Number: 56-81-5) were purchased from Sigma Aldrich.

Preparation of Amino Acid Deep Eutectic Solvent (DES). The AADES mixtures were prepared at standard room temperature and pressure by mixing two pure components with predefined weight. The binary mixtures which were L-lysine and glycerol were prepared based on the mole ratio of lysine to glycerol 1:1, 1:2, 1:3 respectively. The mixtures were stirred and heated to 80°C for 30 minutes until they fully melted and became homogeneous. Afterward, they were cooled to room temperature and stored under a vacuum.

Preparation of Rubber Composite and Rubber Film. The rubber composites were comprised of AADES, XNBR, antifoam agent, whitening agent as well as antioxidant in the stirred mixture. The mixture was stirred in a beaker at 500-1000 rpm for 24 hours at room temperature. Prior to the curing process, the ceramic plate was dipped in the coagulant at 51°C and dried in the oven at 95°C. Following this, the plate was placed into the rubber composite at an appropriate depth. The composite-coated plate was then cured in the oven at 105°C for 20 minutes to form a film of rubber. Once the film had dried and cooled to room temperature, it was gently peeled off from the ceramic plate.

Characterization. ^{1}HNMR spectroscopy was employed to analyze the spectra of 1:2 L-lysine: Glycerol DES and 1:3 L-lysine: Glycerol DES using the Bruker 500 MHz Avance III spectrometer. The samples were set up in NMR tubes, with deuterated methanol (CD_3OD) serving as the reference solvent. Chemical shifts for both samples were measured in ppm relative to CD_3OD.

Fourier transform infrared spectroscopy (FT-IR) absorbance spectra were recorded in the range of wavenumbers from 400 to 4000 cm^{-1}. The measurements were performed with the use of a Thermo Scientific Nicolet iS5 FT-IR spectrometer with OMNIC 3.2 software. Attenuated Total Reflectance (ATR) equipped with a single reflection diamond ATR crystal was used for all investigations.

The decomposition temperature (T_g) was determined by thermal gravimetric analysis (TGA) using Perkin Elmer STA6000. The samples were tested at the temperature range of 30°C to 800°C at a heating rate of 10°C/min.

The melting temperature (T_{fus}) was measured by differential scanning calorimetry technique (DSC) using Mettler Toledo DSC 1 STAR System equipped with liquid nitrogen. The calibration experiment for rubber films were carried out with a temperature range of -80°C to 270°C at heating rate of 5°C/min while DES samples were analyzed at the temperature of -80°C to 180°C at the same heating rate of 5°C/min with heating, cooling and heating cycle for both type of samples.

The elongation at break (EOB) and tensile strength (TS) measurements were conducted using Universal Tensile Machine (UTM) with a speed of 500 mm/min at ambient temperature. The samples were cut according to ASTM D412 standard.

Result and Discussion

The formation of deep eutectic solvent using binary mixtures composed of L-lysine and glycerol at 1:1, 1:2, and 1:3 was evaluated. Only L-lysine: glycerol of mole ratio 1:2 and 1:3 successfully formed homogeneous deep eutectic solvent whereas 1:1 ratio shows a highly viscous gel-like mixture. Hence, only the two samples proceeded for structural and thermal analysis while for the development of AADES: XNBR composites, all molar ratios were used.

Chemical Structure Analysis. NMR Characterization. ^{1}HNMR spectroscopy was carried out to confirm the structure of both hydrogen bond donor (HBD), glycerol and hydrogen bond acceptor (HBA), l-lysine relative to CD_3OD. Based on Fig.1, the NMR spectra of l-lysine for both 1:2 AADES and 1:3 AADES were denoted as 1-5 while the glycerol for both samples were labelled as a, b, c. The chemical shift of l-lysine for samples 1:2 AADES were observed at δ(ppm) = 1.503 (2H, m) at position 2, 1.669 (3H, q) at position 3, 1.810 (3H, m) at position 1, 2.864 (2H, t) at position 4, and 3.410 (1H, t) at position 5 while there was a slight shift of signal for l-lysine in 1:3 AADES at δ(ppm) = 1.550 (2H, m) at position 2, 1.693 (3H, q) at position 3, 1.793 (3H, m) at position 1, 2.869 (2H, t) at position 4, and 3.490 (1H, t) at position 5.

The ^{1}H signal of glycerol for 1:2 AADES, can be seen at δ(ppm) = 3.560 (4H, t) at position a, 3.685 (4H, m) at position b, and 3.706 (2H, m) at position c. In addition to the subtle change chemical shift, there was also an increased in the intensity of peaks observed in 1:3 AADES, attributed to the higher dosage of glycerol. Fig. 1 (b) showed the signals at δ(ppm) = 3.560 (6H, t) at position a, 3.685 (6H, m) at position b, and 3.706 (3H, m) at position c.

Materials Research Proceedings 53 (2025) 542-549 https://doi.org/10.21741/9781644903575-54

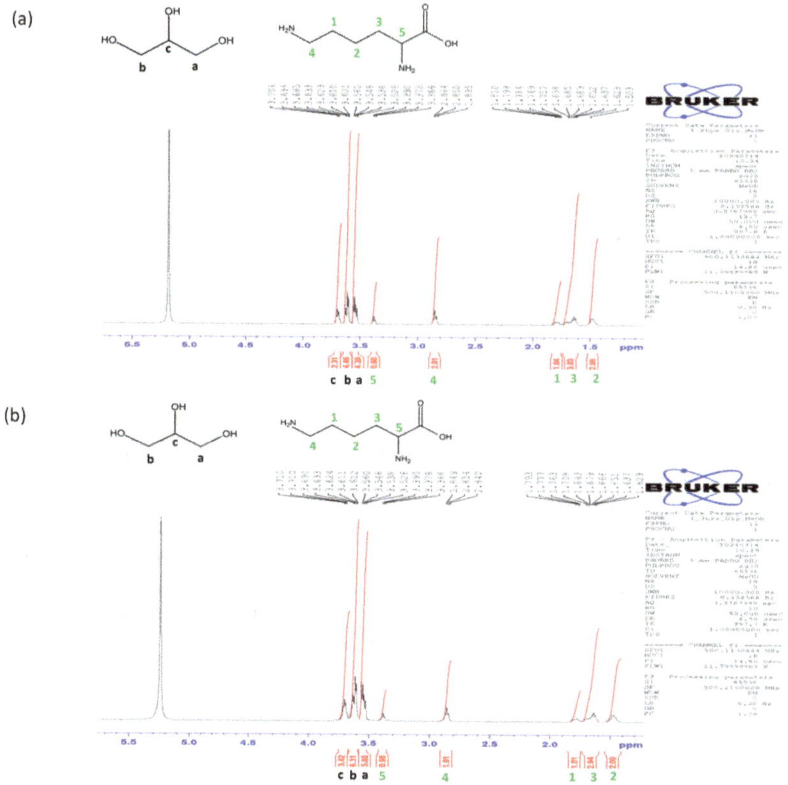

Fig.1 NMR spectra of (a) 1:2 AADES and (b) 1:3 AADES

FTIR Analysis. FT-IR was conducted to determine the functional groups structure of DES crosslinking towards XNBR. The FTIR spectra of pure rubber and both [1:2 DES]: XNBR and [1:3 DES]: XNBR films on Fig. 2 showed absorption peaks at 3439 cm^{-1}, 3307 cm^{-1} and 3288 cm^{-1} which corresponded to the O-H stretching bands of XNBR. Moreover, consistent stretching vibrations were observed across all samples at 2919 cm^{-1}, 2844 cm^{-1}, and 2235 cm^{-1}. These bands correspond to the asymmetric and symmetric stretching of the methylene group ($-CH_2$), and the stretching vibration of nitrile triple bonds of XNBR. At 1603 cm^{-1} and 1541 cm^{-1} carbonyl stretching indicated the presence of carboxylic acid of XNBR, while 1437 cm^{-1} and 966 cm^{-1} displayed stretching and bending of C=C bonds.

For both [1:2 DES]: XNBR and [1:3 DES]: the peaks of the crosslinking spectrum were observed within the region of 3200 – 3500 cm^{-1}, 1438 -1578 cm^{-1}. These peaks most likely correspond to stretching vibration of COO^-, C=C bonds or N-H bending vibrations of lysine, indicated its contribution to the composition. Besides, the peaks between 1150 cm^{-1} and 1023 cm^{-1} may indicated the stretching vibrations of C-O-C bonds in glycerol in the solvent mixture.

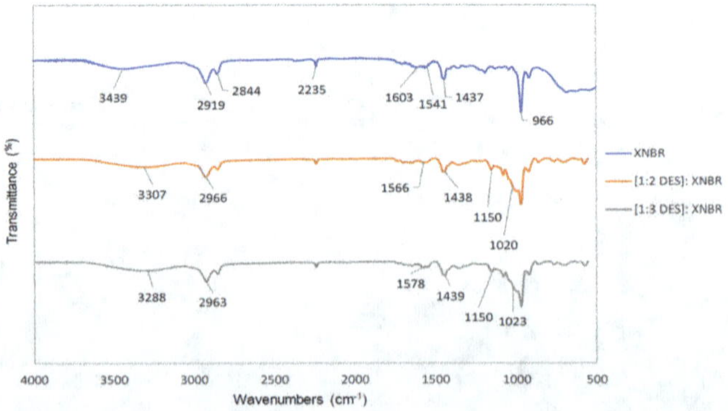

Fig. 2 FTIR spectra of samples obtained within the range of 500 cm^{-1} to 4000 cm^{-1}

Thermal Analysis. Thermogravimetric Analysis (TGA). TGA was executed to determine thermal stability of the AADES and rubber films by analyzing the decomposition temperature of the samples. Fig. 3 illustrates the onset decomposition temperature of AADES, comprising HBA (L-lysine) and HBD (Glycerol) at different molar ratios (1:2 and 1:3), exhibited two decomposition stages, which began with glycerol decomposition followed by L-lysine. For instance, in the [1:2 DES]: XNBR, glycerol vaporization occurred at 175°C followed by L-lysine decomposition at 410°C. The thermal stability decreased with increasing AADES molar ratio of HBD content, as evidenced by lower onset decomposition temperatures. According to Wenjun et al., by nature, HBD has poor thermal stability and low boiling point, causing it to vaporize and decompose first during thermogravimetric analysis. Consequently, an increase in the quantity of HBDs results in the system comprising a mixture of AADES and excess glycerol [7]

The AADES: XNBR films showed enhanced thermal stability compared to pure AADES, possibly due to hydrogen bonding interactions between AADES and XNBR matrices, proving the presence of crosslinking within them. Based on Table 1., both molar ratios exhibited similar initial onset decomposition temperatures, with two major decomposition stages involving AADES followed by XNBR decomposition. Overall, the findings suggest that AADES composition significantly influences thermal behaviour, impacting the stability of rubber composites.

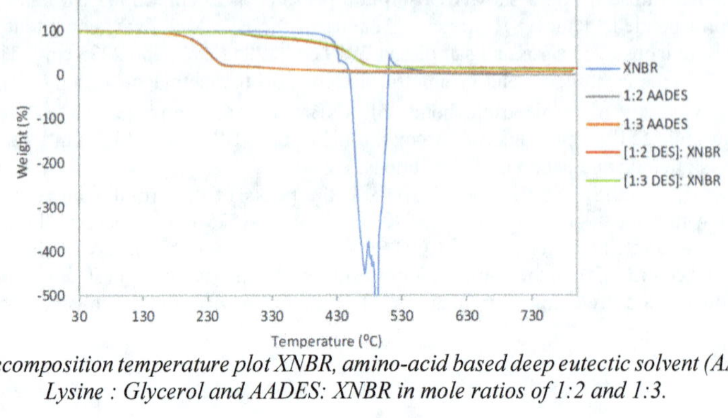

Fig. 3 Decomposition temperature plot XNBR, amino-acid based deep eutectic solvent (AADES), Lysine : Glycerol and AADES: XNBR in mole ratios of 1:2 and 1:3.

Table 1. Decomposition temperature for each AADES and rubber films at different molar ratio

Component	L-lysine:Glycerol (molar ratio)	Heating rate (°C/min)	Onset decomposition temperature (°C)		
			L-lysine	Glycerol	XNBR
AADES	1:2		410	190	-
	1:3	10	370	187	-
AADES: XNBR	1:2		288		402
films	1:3		306		429

Differential Scanning Calorimetry (DSC). The glass phase transition (T_g) and melting point (T_{fus}) temperatures of the pure AADES at a mole ratio of (1:2 and 1:3) and rubber films are presented in Fig. 4. The T_g of the 1:2 AADES was found to be -59.43°C. However, increasing the dosage of glycerol lowers this temperature to -64.39°C. Notably, the AADES mixture remains in a liquid state at room temperature, resulting in the absence of a melting peak in the DSC thermogram within the range of -80°C to 140°C. This may be attributed to glycerol`s hygroscopic nature that allows the AADES to be liquid stable within the respective temperature range.

Based on Fig. 4, the T_g rises from -21.27°C to -16.18°C with an increase in AADES mole ratio from 1:2 to 1:3. Additionally, [1:3 DES]: XNBR demonstrated a significant increase in its second T_g at 154.55°C. This increase is attributed to the dispersion of L-Lysine throughout the XNBR matrices via the plasticization effect of glycerol, indicating ionic bonding between L-lysine and the COO^- group (Lysine-carboxylate) of XNBR, creating a 3D network structure in the XNBR rubber chain [8]. Moreover, for [1:2 DES]: XNBR film, multiple T_g and T_m peaks ranging from 124.06°C to 144.44°C were observed. At lower concentrations, there may be fewer available crosslinking sites, leading to the formation of a less uniform network structure with varying degrees of crosslinking. This non-uniformity in crosslinking density could contribute to the observation of multiple peaks in thermal analysis [9].

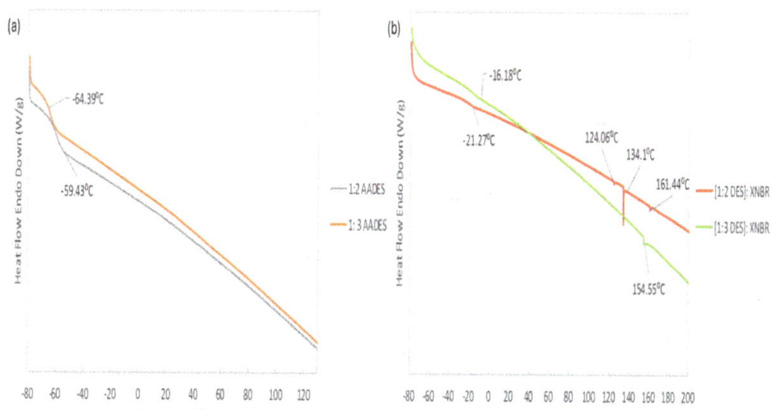

Fig. 4 The DSC curves of AADES mixtures, Lysine : Glycerol in mole ratios of (a) 1:2 and 1:3 and AADES: XNBR in mole ratios of (b) 1:2 and 1:3

Mechanical Properties (Tensile Strength & Elongation at Break). Tensile strength and elongation at break analyses were conducted to determine the effect of crosslinking AADES towards the mechanical strength of the rubber. Based on Table 2. above, AADES which were 1:1, 1:2 and 1:3 mole ratio of CholLys : Glycerol with rubber films were tested. The examined samples were shaped into dumbbell forms following ASTM D412 type C specifications, with dimensions measuring 11.5 cm in length and 0.6 cm in width. The tensile strength of rubber films increases with the amount of glycerol in AADES, progrssing from 3.46 MPa at 1:1 ratio to 3.93 MPa at 1:2 ratio and peaking at 4.84 MPa at 1:3 ratio. Increasing the amount of hydrogen bond donors (HBDs) in a AADES can enhance its ability to form stronger hydrogen bonds with other components in the system, such as polymers like (XNBR). These hydrogen bonds contribute to better intermolecular interactions between AADES and the polymer matrix. This stronger hydrogen bonding interaction between the AADES and the polymer can lead to improved dispersion and distribution of the AADES within the polymer matrix which promote more effective reinforcement effects resulting increased tensile strength of the material [10].

According to Thongpin et al., increasing glycerol content enhances the plasticization effect, thereby improving the rubber films elongation. This finding is supported by Kruzelak at al., who explained how glycerol`s plasticization effect increases the rubber films mobility and elacticity along the rubber chain segments, leading to increased elongation at break [11]. However, as indicated in Table 2, the highest elongation at break can be shown in [1:2 DES]: XNBR, reaching 120% compared to [1:3 DES]: XNBR, which stands at 101%. This difference may be attributed to the increased presence of HBDs that can lead to stronger interactions between AADES and the polymer chains. While this can enhance tensile strength, it can also restrict the movement of polymer chains and causing stiffness due to the excessive plasticization of HBDs, making them less mobile and more resistant to deformation. As a result, the material may exhibit reduced elongation at break [12].

Table 2. Mechanical properties for each AADES and rubber films at different molar ratio

Component	L-lysine:Glycerol (molar ratio)	Tensile Strength (MPa)	EOB (%)
AADES : XNBR	1:1	3.46	110
	1:2	3.93	120
	1:3	4.84	101

Conclusion

This study investigated the effect of L-lysine-Glycerol based Deep Eutectic Solvent (DES) on the crosslinking properties of cured XNBR rubber. Results indicated that the new DES L-lysine:Glycerol AADES, at mole ratios of 1:2 and 1:3 have successfully resulted in the production of ionic-crosslinked XNBR samples exhibiting the highest tensile strength, notably at 4.84 MPa for the 1:3 DES molar ratio. However, n increase in glycerol content within DES led to a decline in elongation at break attributed to glycerol`s excessive plasticization effect. The optimal molar ratio, identified as 1:2 AADES, facilitated the formation of an ionic-crosslinking network between L-lysine with glycerol and XNBR. Thermal analysis demonstrated enhanced stability in the cured rubber, with lower glass transition temperatures indicating improved flexibility and durability. Hence, it can be concluded that L-lysine-Glycerol deep eutectic solvent (DES) could be potential green rubber crosslinker which is way more sustainable and safer to used in the next generation.

References

[1] G. Zhang., H. Feng., K. Liang., Z. Wang., X. Li., X. Zhou., B. Guo and L. Zhang "Design of next-generation cross-linking structure for elastomers toward green process and a real recycling loop," Science Bulletin, vol. 65, no. 11, pp. 889–898, Jun. 2020, doi: https://doi.org/10.1016/j.scib.2020.03.008.

[2] M. Das, S. Pal, and K. Naskar, "A new route of cross-linking of carboxylated acrylonitrile-butadiene rubber via zinc oxide -amino acid network formation," Journal of Applied Polymer Science, vol. 138, no. 11, p. 49996, Oct. 2020, doi: https://doi.org/10.1002/app.49996.

[3] Y. Zhang, W. Yu, D. Han, J. Meng, H. Wang, and G. Cao, "L–lysine ameliorates sepsis-induced acute lung injury in a lipopolysaccharide-induced mouse model," Biomedicine & pharmacotherapy, vol. 118, pp. 109307–109307, Oct. 2019, doi: https://doi.org/10.1016/j.biopha.2019.109307.

[4] L. Wang., C. Zhang, J. Zhang, Z. Rao, X.Xu, Z. Mao, X. Chen "Epsilon-poly-L-lysine: Recent Advances in Biomanufacturing and Applications," Frontiers in Bioengineering and Biotechnology, vol. 9, Sep. 2021, doi: https://doi.org/10.3389/fbioe.2021.748976.

[5] L. A. Letti Junior, G. V. M. Letti, and C. R. Soccol, "Development of an L-Lysine Enriched Bran for Animal Nutrition via Submerged Fermentation by Corynebacterium glutamicum using Agroindustrial Substrates," Brazilian Archives of Biology and Technology, vol. 59, p. e16150519, May 2016, doi: https://doi.org/10.1590/1678-4324-2016150519 .

[6] Y. Liang, W. Duan, X. An, Y. Qiao, Y. Tian, and H. Zhou, "Novel betaine-amino acid based natural deep eutectic solvents for enhancing the enzymatic hydrolysis of corncob," Bioresource technology, vol. 310, pp. 123389–123389, Aug. 2020, doi: https://doi.org/10.1016/j.biortech.2020.123389.

[7] W. CHEN, Z. XUE, J. WANG, J. JIANG, X. ZHAO, and T. MU, "Investigation on the Thermal Stability of Deep Eutectic Solvents," Acta Physico-Chimica Sinica, vol. 34, no. 8, pp. 904–911, 2018, doi: https://doi.org/10.3866/pku.whxb201712281.

[8] M. Das, T. R. Aswathy, S. Pal, and K. Naskar, "Effect of ionic liquid modified graphene oxide on mechanical and self-healing application of an ionic elastomer," European Polymer Journal, vol. 158, p. 110691, Sep. 2021, doi: https://doi.org/10.1016/j.eurpolymj.2021.110691 .

[9] Z. Hasan, Md Zillur Rahman, B. Saha, and Sazzad Hossain Ador, "Progress in recovery, recycling and reuse of polymers, biopolymers and their composites," Elsevier eBooks, Jan. 2023, doi: https://doi.org/10.1016/b978-0-323-96020-5.00185-0 .

[10] C Thongpin, C Wongprat, P Srithavorn, and P Muenchee, "Effect of Glycerol on Gluten filled in Natural Rubber/WG Blend," IOP conference series. Materials science and engineering, vol. 965, no. 1, pp. 012033–012033, Nov. 2020, doi: https://doi.org/10.1088/1757-899x/965/1/012033.

[11] Ján Kruželák, Klaudia Hložeková, A. Kvasničáková, M. Džuganová, I. Chodák, and I. Hudeč, "Application of Plasticizer Glycerol in Lignosulfonate-Filled Rubber Compounds Based on SBR and NBR," Materials, vol. 16, no. 2, pp. 635–635, Jan. 2023, doi: https://doi.org/10.3390/ma16020635.

[12] N. A. Ramli, F. Rosli, M. A. Hamdan, and F. A"Synthesishesis Of Carrageenan-Based Biocomposite Plasticized With Deep Eutectic Solvent And Characterization Of Its Mechanical Properties," Malaysian Journal of Analytical Sciences, Feb. 28, 2024. https://mjas.analis.com.my/mjas/v28_n1/html/28_1_5.html

Materials Research Proceedings 53 (2025) 550-554 https://doi.org/10.21741/9781644903575-55

Nuclear fusion: Future clean energy source

Mustakimah Mohamed[1,a *], Serene Lock Sow Mun[1,2,b] and Azmi Mohd Shariff[1,2,c]

[1]CO2 Research Centre (CO2RES) Universiti Teknologi PETRONAS, 32610 Seri Iskandar, Perak, Malaysia

[2]Chemical Engineering Department, Universiti Teknologi PETRONAS, 32610 Seri Iskandar, Perak, Malaysia

[a]mustakimah.md@utp.edu.my, [b]sowmun.lock@utp.edu.my, [c]azmish@utp.edu.my

Keywords: Clean Energy, Fusion, Nuclear, Electricity

Abstract. Nuclear fusion is an alternative technology for low-carbon energy generation that able to create a greener economy and energy grid on a large scale. It also offers a pathway to resolve energy security and unequal distribution of energy resources since seawater is its ultimate fuel source. Development and testing of new materials and technologies are unceasing to achieve net fusion energy through national and international collaboration as well as private partnerships. The ever-growing number of research works reported various designs and magnet-based fusion devices, such as stellarators, lasers, and tokamaks. This article provides an overview on the utilization of nuclear energy as a clean energy source, as well as the strategies and progress towards establishing successful commercial fusion energy to the grid and transition to a reliable clean energy source. The International Thermonuclear Experimental Reactor (ITER) has been an international reference program for fusion energy development and most developed countries with nuclear development capacity are aiming to complete their in-house fusion energy facilities in parallel to ITER.

Global Outlook of Nuclear Energy for Power Generation

Nuclear energy is one of the alternative innovative technologies that offers stable and clean energy generation. It offers minor GHG emissions and has the potential to be a cost-competitive technology in the long-run operation [1]. The Intergovernmental Panel on Climate Change (IPCC) explained that nuclear power can provide stable low-carbon electricity. For instance, the UN's Economic Commission for Europe (UNECE) reported the range of CO2 released from nuclear energy in 2022 was 5.1–6.4 g CO2/kWh, which is the lowest CO2 emitted among all power generation technologies. Fig. 1 compares the number of CO2-equivalent emissions per unit of electricity generated by major energy sources based on the lifecycle analysis reported by the United Nations (UN) IPCC [2]. The report indicates the median value for CO2 emitted from nuclear power plants is 12 g CO2/kWh, which is equivalent to the amount of CO2 emitted from wind and lower compared to solar and other sources [3]. Coal and biomass co-firing are recognized as the energy sources with the largest estimated CO2 emissions, which are more than 60-fold higher than nuclear power. The United States (US) was reported as the largest supplier of global nuclear electricity generation by the end of 2022. The country generated 30% of the total nuclear energy generation, which is equivalent to 772 TWh, whereas China contributed half of that amount, at 16% of the total [4]. This capacity has also been contributed to by the initiation of nuclear reactors in Japan since the Fukushima Daiichi incident in 2011. However, the US's dependency on nuclear electricity was considered minor compared to France, Russia, and South Korea. Nuclear electricity is the dominant source power in France, where 63% of the electricity in the country is generated by nuclear sources, followed by Slovakia and Ukraine, at 59% and 58%, respectively [5]. The details of nuclear electricity supplied by major countries in the world and share of the total nuclear energy supply in the country are further illustrated in Fig. 2.

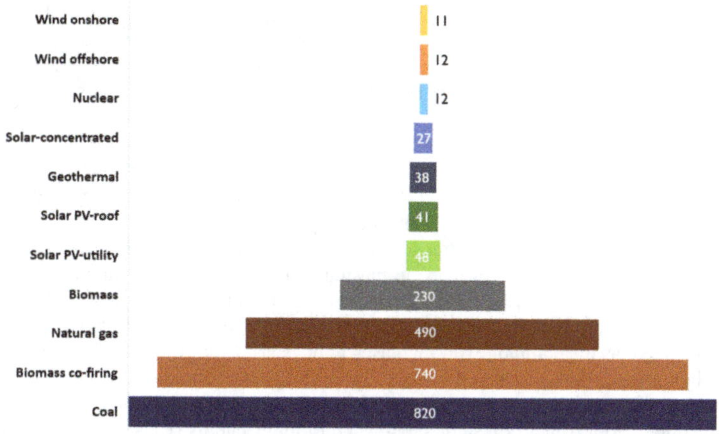

Fig. 1. CO2 emissions (gCO2 equivalent/kW) of different electricity sources based on a life cycle analysis

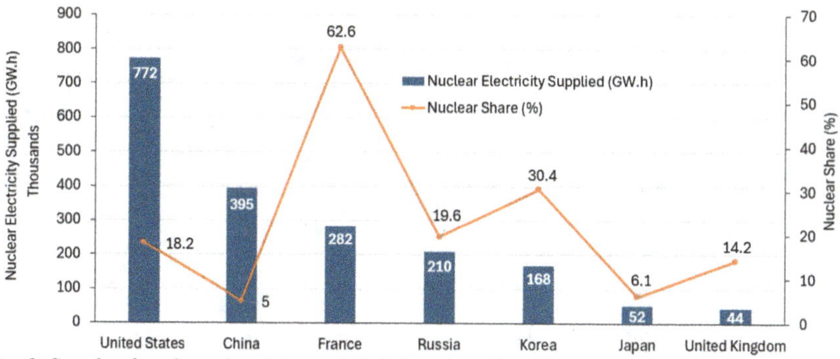

Fig. 2. Supply of nuclear electricity and global nuclear share by major countries in the world as of 2022

Introduction To Nuclear Fission and Fusion Reaction

Nuclear energy is produced from the core of an atom known as the nucleus, which contains protons and neutrons. Neutrons are surrounded by electrons that carry negative electrical charges and protons that carry positive electrical charges. The energy that holds the nucleus together is enormous and nuclear energy can be released by breaking those bonds, splitting the nucleus into several parts, which is known as the fission reaction [6]. The currently deployed nuclear power plants are employing nuclear fission reactions using uranium atoms [6]. The uranium atom (U-235) has an unstable particle arrangement and any excitement to the atom can disintegrate the nucleus. Figure 3 (a) illustrates the splitting of U-235, which comprises 92 protons and 143 neutrons. Neutron bombardment into the nucleus of U-235 will split the atom and two or three neutrons are released each time the splitting happens, which creates the potential occurrence of chain reactions [7]. The atom splitting releases tremendous amounts of heat and radiation that can be harvested for energy generation.

The fission reaction poses a great risk for reactor meltdown due to the uncontrolled runaway chain reaction during the atom splitting. Therefore, comprehensive water treatment plants and radioactive waste management systems are crucial elements to the nuclear power plant for the safe disposal of radioactive wastes and all released materials. It is a national-scale disaster occurrence should an incident occur, which may lead to a nuclear explosion and the release of radiation. The history of nuclear incidents shows that radioactive contamination poses fatal threats to health and the surrounding areas, in both the short and long term. The known risks and challenges of performing fission reactions have promoted the transition to another means of producing nuclear energy, known as the fusion reaction. The fusion reaction is fundamentally safe, has no runaway chain reaction, utilizes fuels with shorter half-lives, and eliminates highly radioactive and long-life products [8]. The reaction takes place in plasma, which is a hot gas containing positive ions and electrons that move freely [9]. Fig. 3 (b) demonstrates the fusion reaction using deuterium (D) and tritium (T), which release the neutron, helium, and energy. Deuterium and tritium are hydrogen isotopes that serve as the common working gases in nuclear fusion reactions. The reaction yields a heavier helium nucleus and emits a neutron. Interestingly, the combined mass of the resulting helium nucleus and the neutron is slightly less than the combined mass of the original deuterium and tritium nuclei. Referring to Einstein's renowned formula, energy (E) is equal to the product of mass (m) and speed of light (c2), the mass difference during the reaction is converted into a substantial amount of energy, which is introduced as the fusion energy [10].

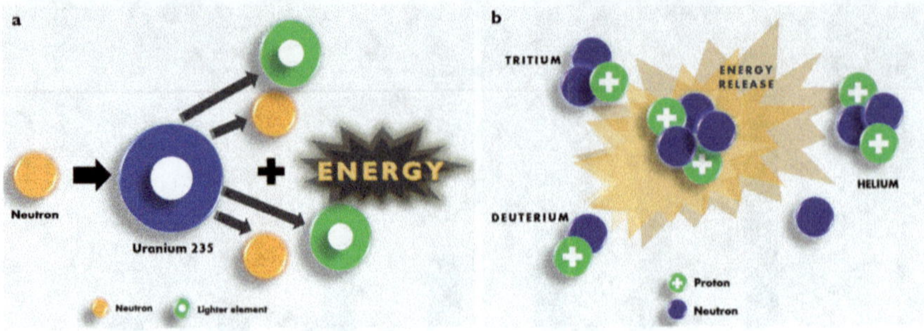

Fig. 3. (a) Splitting of uranium atom during the fission reaction (b) Fusion reaction using deuterium and tritium

Commercialization Of Fusion Technology

Nuclear fusion was first introduced into international dialogue during the 26th Conference of Parties (COP26) in 2021 by the United Nations Framework Convention on Climate Change (UNFCC) [11]. This technology has diverse market sectors and can significantly impact the national and global economy. For instance, electricity generation is the most interesting market sector aimed by the private sector, whereas industrial heat and hydrogen/clean fuels are considered the most potential spin-off markets to invest in [12]. However, there are other market sectors with the potential for booming, such as off-grid energy, medical uses, as well as space and marine propulsion. The findings of a market survey of the private sector on the market potential of fusion technology are illustrated in Fig. 4. The progress and enhancements in the design and construction of future reactors demand sufficient funding to realize cost-effective, safe, and dependable fusion energy production. Globally, there are an estimated of 50 private companies operating in fusion industries as of 2023. Rapid expansion was observed within the years of 2017 and 2021, that refers to an increament from 15 to 45 companies in 4 years [12]. Up to 2023, the fund for the global fusion program had gained a total of USD 6.2 billion, where USD 5.9 billion is contributed by private companies [12]. The US experienced a noticeably increased transition of participation by private sectors into fusion R&D. The private funding in the past years has been increasingly channeled into fusion companies located in Canada, the UK, the EU, China, Japan, and Australia [13].

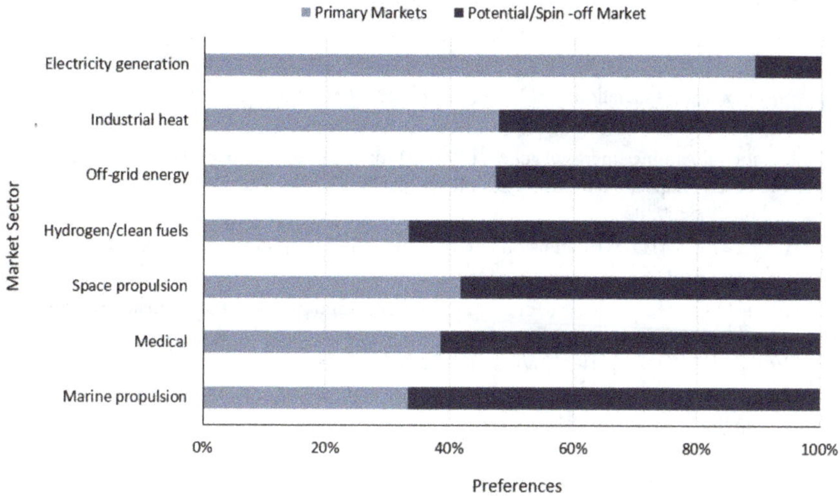

Fig. 4. Preferences of market sector aims by private sectors for fusion technology, as surveyed in 2023

Summary

Nuclear power plants release minor tracing of greenhouse gases during operation and have minimal emissions throughout their lifespan. The technology provides 11% of global electricity, a third of which is carbon-free. Fusion reactors are considered as promising future of green energy, offering limitless potential of stable energy security and combating climate change. Given the urgency of climate change, expanding nuclear energy is crucial. By pairing with smart grids system, nuclear fusion power will offer huge aid to shift to cleaner electricity sources, ensuring reliable and sustainable energy security in the future.

References

[1] Abdullah, W.S.W., Osman, M., Kadir, M.Z.A.A., Verayiah, R., (2019). The potential and status of renewable energy development in Malaysia, Energies, 12, 2437 https://doi.org/10.3390/en12122437

[2] Mathew, M.D., (2022). Nuclear energy: A pathway towards mitigation of global warming. Prog. Nucl. Energy, 143, 1040880 https://doi.org/10.1016/j.pnucene.2021.104080

[3] Schlömer, S., Bruckner, T., Fulton, L., Hertwich, et al., (2014). Annex III: Fifth Assessment Report on the Intergovernmental Panel on Climate Change; Cambridge University Press: Cambridge, UK; New York, NY, USA. Available online: https://www.ipcc.ch/site/assets/uploads/2018/02/ipcc_wg3_ar5_annex-iii.pdf (accessed on 4 Jan 2024)

[4] International Atomic Energy Agency (IAEA). Nuclear Share of Electricity Generation in 2022. IAEA-Power Reactor Information System (PRIS). Available online:

https://pris.iaea.org/PRIS/WorldStatistics/NuclearShareofElectricityGeneration.aspx (accessed on 15 Apr 2024)

[5] Wiatros-Motyka, M., (2023). Global Electricity Review 2023. In Oliver Then; Ember: Atlanta, GA, USA

[6] International Energy Agency, (2024). Electricity-Analysis and forecast to 2026. January 2024. Available online: https://www.iea.org/reports/electricity-2024 (accessed on 15 Apr 2024)

[7] World Nuclear Association (2023). World Nuclear Performance Report 2023; World Nuclear Association: London, UK

[8] International Atomic Energy Agency, (2022). World Survey of Fusion Devices 2022. Vienna. Available online: https://www.iaea.org/publications/15253/world-survey-of-fusion-devices-2022 (accessed on 29 Sept 2023)

[9] IAEA, (2023). IAEA World Fusion Outlook 2023. In Fusion Energy: Present and Future; IAEA: Vienna, Austria

[10] Zheng, J.; Qin, J.; Lu, K.; Xu, et al., (2022). Recent progress in Chinese fusion research based on superconducting tokamak configuration. In Innovation; Cell Press: Cambridge, MA, USA, Vol. 3 https://doi.org/10.1016/j.xinn.2022.100269

[11] White, S., (2021). Taking Fusion to COP26; EUROfusion Consortium Research Institute:Garching, Germany, 1 Dec 2021. Available online: https://euro-fusion.org/member-news/taking-fusion-to-cop26/ (accessed on 4 Jan 2024)

[12] Fusion Industry Association, (2023). The Global Fusion Industry in 2023. United States. Available online: https://www.fusionindustryassociation.org/wp-content/uploads/2023/07/FIA%E2%80%932023-FINAL.pdf (accessed on 3 January 2024).

[13] Hsu, S.C., (2023). U.S. Fusion Energy Development via Public-Private Partnerships. In Journal of Fusion Energy; Springer: Berlin/Heidelberg, Germany, Vol. 42 https://doi.org/10.1007/s10894-023-00357-9

Effect of temperature and ZSM-5 catalyst dosage on carbon char yield from catalytic pyrolysis of waste tire

AMANI HUSSEIN[1,2,a*], RAIHAN Mahirah Biniti Ramli[1,2,b*], SURIATI Binti Sufian[1,2,c], and ABID Salam Farooqi[1,2,d]

[1]Centre of Innovative Nanostructures & Nanodevices (COINN), Universiti Teknologi PETRONAS, 32610 Seri Iskandar, Perak, Malaysia

[2]Department of Chemical Engineering, Universiti Teknologi PETRONAS, 32610 Seri Iskandar, Perak, Malaysia

[a]amani_22011516@utp.edu.my, [b]raihan.ramli@utp.edu.my, [c]suriati@utp.edu.my, [d]abid_20001276@utp.edu.my

Keywords: Waste Tires, ZSM-5, Solid Char, Slow Pyrolysis, Carbonous Materials, Fast Pyrolysis

Abstract. The disposal of waste tires has become a substantial environmental challenge attributed to their accumulation and potential hazards. Pyrolysis emerges as a viable approach for the valorization of waste tires, into valuable products such as pyrolytic char. This carbon-rich char can be effectively used for pollutant removal or further processed into activated carbon, which is widely utilized in purification and catalytic applications. This study explores the impact of pyrolysis temperature and zeolite catalyst dosage on the yield of carbonous materials from the catalytic degradation of pyrolytic char derived from waste tires pyrolysis. Firstly, slow pyrolysis of waste tires at 500°C for 3 hours yielded 60% solid char. Subsequently, The pyrolytic char was subjected to catalytic degradation at temperatures varying from 600°C to 800°C and catalyst dosage ranging from 0 to 1.5. The findings discovered diverse patterns in carbon char yield with increasing catalyst dosage across varying temperatures. At 800°C, a linear rise in char yield from 80% to 71% was noted with increasing catalyst dosage, attributed to coke formation on the surface of the catalyst. Conversely, at 700°C and 600°C, the carbon char yield reduced from 88% to 68% and 85% to 65%, respectively, with increasing catalyst dosage, potentially due to the promotion of liquid product formation. The present research demonstrates the significant impact of pyrolysis temperature and catalyst dosage on the carbon char yield derived from waste tires, providing valuable insight for optimizing the catalytic decomposition, for the valorization of waste tire-derived carbonaceous materials.

Introduction

The waste tires accumulation has evolved into a formidable environmental challenge, with an estimated 1.5 billion rubber tires produced annually, of which 50% are discarded without any treatment [1]. Like any other type of waste, the objective behind waste tire elimination is to minimize its environmental effect. In general, used rubber tires are rich in carbon sources, hence appropriate disposal is important to prevent environmental issues and lengthen their valuation[2]. At the end of their life cycle, used rubber tires can be transformed into valuable products including carbon black, gaseous products, and pyrolytic oil [3].

Among the various pathways to convert waste into valuable products, thermochemical conversion approaches (including pyrolysis and gasification) have received considerable attention in research and development activities [4], [5]. Pyrolysis, in particular, has been identified as a favorable technique for waste tire disposal because of its negligible release of toxic gases and

Materials Research Proceedings 53 (2025) 555-563 https://doi.org/10.21741/9781644903575-56

recover of valuable products such as char, gas, and pyrolytic oil, rendering the process economically sustainable[5], [6], [7], [8].

The pyrolysis process is influenced by several factors, including reactor design, feedstock composition, temperature, pressure, heating rate, residence time, catalyst etc. Notably, the temperature at which pyrolysis occurs and catalyst type, exert a significant influence on the properties of the pyrolytic products [9]. The present study examines the impact of these factors on the pyrolytic char of waste tires The incorporation of catalysts can improve the quality and yield of the products [10]. Catalysts frequently used in the pyrolysis of tires are ZSM-5, Zeolite and Y-type and zeolite (USY). [11], [12], [13], [14], [15].

Zeolites, as acidic solid catalysts, possess an alumina-silicate framework with well-defined pores structures and high surface area, impacting their cracking capabilities [12]. While several investigations have explored the effect impact of zeolite catalysts on the gas and oil yields from the pyrolysis of waste tires, the effect of zeolite catalysts on the char product has not been thoroughly investigated. [11], [13], [14], [15], [16], [17], [18], [19].

Therefore, the present research aims to fill this gap by examining the impacts of pyrolysis temperature and zeolite catalyst dosage on the yield of carbon char obtained from waste tires. Specifically, the objectives are: (1) to investigate the slow pyrolysis condition for maximizing the solid char yield from waste tires, and (2) to examine the distinct temperature range (600- 800°C) and catalyst dosage (0 to 1.5) influence on the carbon char yield during the catalytic pyrolysis.

Materials and Methods

Materials. Waste tires crumbled were obtained from Universiti Teknologi Petronas (UTP), Malaysia. ZSM-5 zeolite was provided by Sigma-Aldrich Malaysia

Pretreatment of Waste Tires. The crumbled waste tires were sieved to a size between 1-5mm. The rubber crumbles were Subsequently, rinsed with distilled water to eliminate impurities, followed by being dried in an oven overnight set at a temperature of 80°C.

Synthesis of Solid Char from Waste Tires. A batch of 30g of the treated rubber was pyrolyzed in a horizontal ceramic tube furnace for 3 hours at a temperature of 500°C, using a heating rate of 10°C/min. The furnace was purged with a continuous nitrogen flow to ensure an oxygen-free environment during the pyrolysis process.

Catalyst Preparation. Before the experiment, the ZSM-5 zeolite catalyst was calcined at a temperature of 500°C for two hours, with the heating rate maintained at 10°C/min in a muffle furnace in the presence of air. This process aimed to convert them into their acidic form and remove impurities before being used in catalytic pyrolysis.

Catalytic Pyrolysis of Pyrolytic Char for Carbon Char Production. The impact of temperature and catalyst dosage on the yield of carbonous materials from the catalytic pyrolysis of pyrolytic char was investigated using an LT quartz tube furnace. For each experimental trial, 1g of solid char, derived from step (2.3), was mixed with the desired catalyst dosage (varying between 0 and 1.5 by weight) and positioned in a quartz boat. Subsequently, the quartz boat was placed into a quartz tube furnace and exposed to the following conditions: pyrolysis temperatures ranging from 600°C to 800°C,heating rate of 30°C/min, and a consistent duration of 20 minutes under a continuous nitrogen flow to keep an inert atmosphere.

After the catalytic degradation process, the carbon char samples were carefully removed from the LT quartz tube furnace. The carbon char was first separated manually from the catalyst particles. Subsequently, the samples were rinsed multiple times with distilled water to eliminate any residual catalyst particles. Post-rinsing, the samples were exposed to desiccation in an oven at 80°C for a duration of 12 hours. to ensure complete moisture removal and facilitate consistent conditions for subsequent analysis. The carbonous materials yield was determined using the following formula:

Materials Research Proceedings 53 (2025) 555-563 https://doi.org/10.21741/9781644903575-56

Yield (%) = (Weight of carbonous materials / Weight of solid char) x 100

where "Weight of carbonous materials" refers to the weight of the carbon char obtained after the catalytic pyrolysis process, and "Weight of solid char" refers to the initial weight of the solid char used in the catalytic pyrolysis process.

Results and Discussion

Yield of solid char from Waste Tire Pyrolysis. Solid char yield is a crucial parameter that reflects the efficiency of converting rubber from waste tires into valuable solid char during slow pyrolysis. In the present study, we observed a high 60% solid char yield, achieved under pyrolysis conditions shown in Figure 1.

Temperature plays a critical role; As reported by numerous researchers, a temperature of 500°C under atmospheric pressure is considered optimal for achieving total tire conversion during waste tire pyrolysis [20], [21], [22], [23] [24]. Lower temperatures may not fully convert the rubber, leaving it sticky [9]. Conversely, increasing the temperature leads to a reduction in char yield [7]. Our findings concur, demonstrating that maintaining a temperature of 500°C leads to achieving a high char yield of approximately 60%. This observed 60% solid char yield is aligned with the findings of Wang et al.,[25] who studied a temperature ranging from 400- 600°C, determining 500°C as the ideal temperature for efficient decomposition of organic components while maximizing char yield from waste tire pyrolysis.

The extended residence time of 3 hours employed in this study facilitated the complete decomposition of tire components and allowed for secondary conversion reactions of the primary pyrolysis products. These secondary reactions include carbonization, dealkylation, and dehydrogenation mechanisms, promoting the formation of condensed coke and solid char deposits from vapors and liquids, thereby enhancing the overall char yield[26]. Similarly, Caputo et al.[27] reported that longer residence times favored the formation of carbon residue through secondary thermal degradation reactions, ultimately increasing char production.

Furthermore, the increased solid char yield was attributed to the slow heating rate of 10°C/min utilized in this investigation. Slow pyrolysis, characterized by lower heating rates, facilitates gradual thermal decomposition[28]. This allows more complete carbonization to occur, forming stable graphitic solid char. This observation aligns with the findings of Wang et al. [25] where they also applied a heating rate of 10°C/min and reported an increase in carbonaceous deposits as a result of the slower decomposition of the organic components in the feedstock.

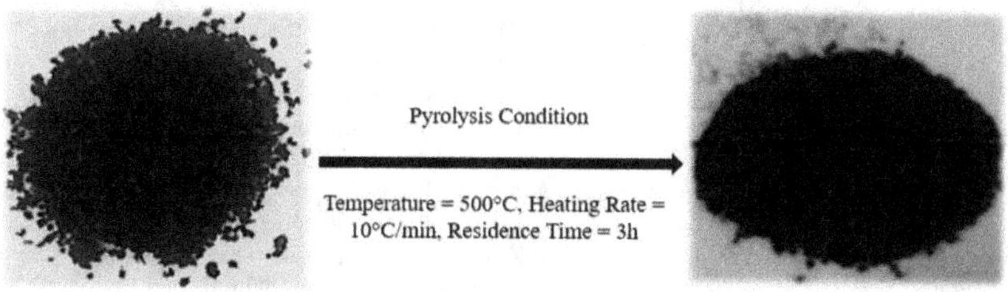

Fig. 1: Char Formation from waste tires Precursor via slow Pyrolysis.

Influence of Temperature and Zeolite Ratio on Carbonous Materials Yield. The catalytic degradation of pyrolytic char derived from waste tire pyrolysis was investigated using a ZSM-5 zeolite catalyst. The product yield was evaluated under three distinct temperatures, 600°C, 700°C, and 800°C. Concurrently, the catalyst dosage was varied (ranging from 0 to 1.5). The results

Materials Research Proceedings 53 (2025) 555-563 https://doi.org/10.21741/9781644903575-56

revealed a contrasting trend in the product yield with increasing catalyst dosage at the three temperatures investigated.

Temperature had a significant influence on the catalytic activity of ZSM-5 and the product yield. As shown in Figure 2, when the temperature was maintained at 800°C, the product yield exhibited a linear increase from approximately 71% to 80% as the catalyst dosage increased from 0 to 1.5, representing a 12.7% increase. This positive correlation can be ascribed to the formation of coke deposits on the surface of the zeolite catalyst, consequently increasing the overall solid residue or char yield [29]. The formation of this carbonaceous coke is attributed to the polycondensation of volatiles, which is facilitated by the acidic nature of the ZSM-5 catalyst [19]. Additionally, the observed raised in solid residue across the whole range of char- to-catalyst ratios studied can be ascribed to the corresponding augmentation in catalyst weight, comprised of reaction residues homogeneously dispersed within the feedstock before the pyrolysis process. [14].

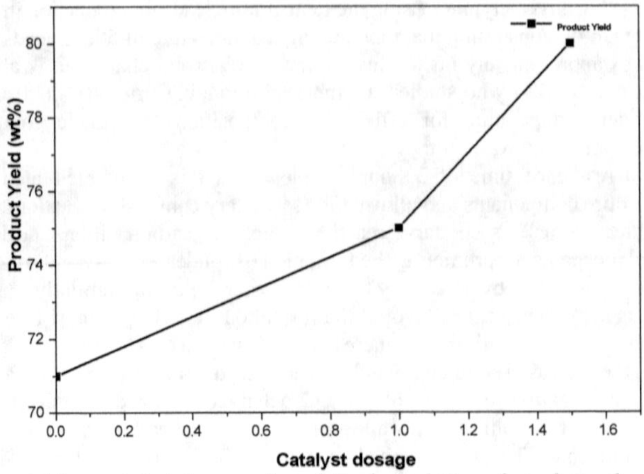

Fig.2: Product yield vs. catalyst dosage of catalytic degradation of pyrolytic char over ZSM-5 at 800°C.

However, at lower temperatures of 700°C and 600°C, an increasing catalyst dosage led to a decrease in product yield. At 700°C (Figure 3), the product yield showed a decreasing trend, dropping from around 85% to 68% with an increasing catalyst dosage from 0 to 1.5 a substantial decrease of 20%. At 600°C (figure 5), the yield reduced from 88% to 65% over the same range of catalyst dosages, representing a 26% decrease. This behavior suggests that temperature significantly influences the performance of zeolite catalysts in pyrolysis by impacting surface reactions, adsorption-desorption processes, and overall catalyst performance [30]. Furthermore, increasing the catalyst dosage led to a reduction in char yield, possibly due to the improved promotion of liquid product formation by the ZSM-5 zeolite catalyst. This was supported by the detection of pyrolytic oil in the furnace during experiments conducted at high catalyst dosage. This aligns with Qu et al. [13]who suggest that zeolite catalysts can facilitate the cracking of pyrolytic vapors into liquid products[13]. These observations are visually supported by the char morphology in Figure 4.

Materials Research Proceedings 53 (2025) 555-563 https://doi.org/10.21741/9781644903575-56

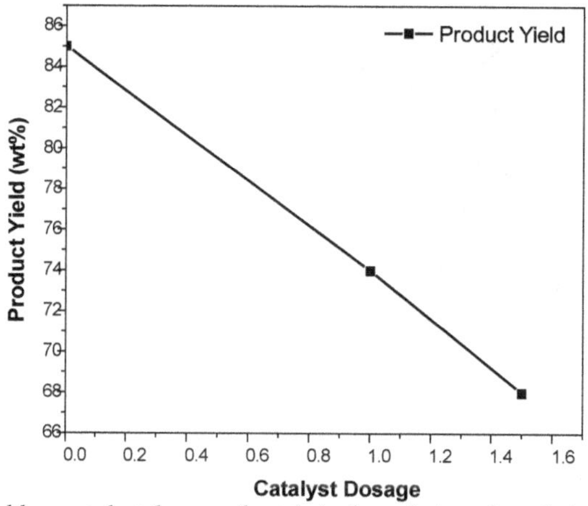

Fig. 3: Product yield vs. catalyst dosage of catalytic degradation of pyrolytic char over ZSM-5 at 700 °C.

ZSM-5 plays a crucial role in pyrolysis processes due to their unique structure and properties. The shape selectivity of ZSM-5's crystal structure, with a pore diameter of approximately 5.6 Å, promotes the cracking of larger hydrocarbons into smaller, more volatile compounds[31]. The catalyst's acidity, influenced by its Si/Al ratio, provides active sites for various reactions, including isomerization, alkylation, and dealkylation[32]. These properties affect the formation and composition of pyrolysis products, such as tars and coke [19].

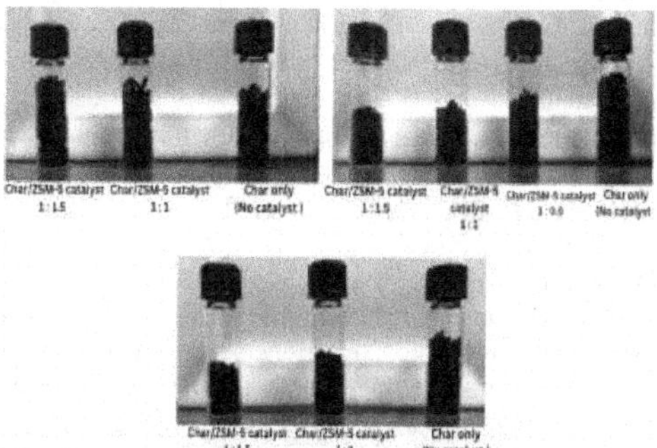

Fig. 4: Impact of Zeolite Catalyst Dosage and Temperature on Catalytic Pyrolysis of Tire-Derived Char. (a) At 800°C, (b) At 700°C and (c) At 600°C.

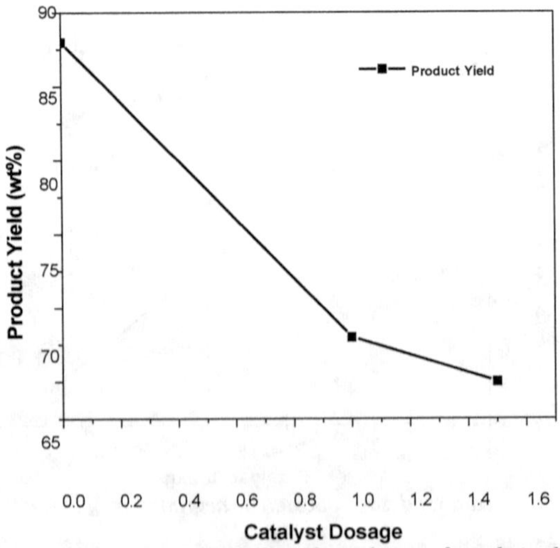

Fig. 5: Product yield vs. catalyst ratio of catalytic degradation of pyrolytic char over ZSM-5 at 600 °C.

Comparing the yields at different temperatures with no catalyst (dosage 0), we observe that the yield decreases from 88% at 700°C to 85% at 600°C (a 3.5% decrease), and then further drops to 71% at 800°C (a 19.3% decrease from 700°C). This linear decreasing trend aligns with the expected behavior of pyrolysis processes, where increasing temperature leads to lower char yield and higher oil and gas yields. As the temperature rises further, the char yield continues to decrease while the gas yield increases, likely due to the secondary cracking of oil vapors [9], [24], [33], [34].

At the highest catalyst dosage (1.5), the yield increases with temperature: 65% at 600°C, 68% at 700°C (a 4.6% increase), and 80% at 800°C (a 17.6% further increase). This indicates that higher temperatures enhance the catalytic effect on product yield, with the most dramatic impact observed between 700°C and 800°C, emphasizing the critical role of temperature in determining the catalyst's effect on product yield [30].

The yield of carbon char under various catalytic and temperature conditions is notable for its industrial and environmental applications. It is used as an adsorbent to remove pollutants from wastewater and acts as a support material for catalysts in industrial processes [35], [36].

Influence of Reaction Conditions on the Carbonous Materials Yield. Fast pyrolysis is typically defined by brief durations of material residence, elevated temperature conditions, and rapid thermal transfer rates. [37]. The relatively short residence time of 20 minutes employed in these experiments may have played a role in the high char yield. A shorter residence time can minimize the occurrence of secondary reactions resulting in high char yield [33]. Additionally, the rapid heating rate of 30°C/min used likely hinders the release of volatiles, promoting solid char formation and ultimately leading to a higher char yield [38]. Moreover, the low nitrogen flow rate may have further contributed to the increased char yield by extending the residence duration of volatiles in the reactor. This prolonged residence time, facilitated by the low flow rate, favors the formation of heavier products in the process [39].

Conclusion

The study on the influence of pyrolysis temperature and zeolite catalyst on the carbon char yield obtained from the catalytic degradation of solid char derived from waste tires. The important findings can be recapitulated as follows:

1. The slow pyrolysis of waste tire at 500°C for 3 hours with a heating rate of 10°C/min resulted in a high solid char yield of approximately 60%, demonstrating the efficacy of these conditions in maximizing solid char production.
2. The catalytic degradation of pyrolytic char over ZSM-5 zeolite showed significant temperature-dependent effects: at 800°C, increasing the catalyst dosage led to a linear increase in carbon char yield from 80% to 71%, indicating enhanced coke formation; however, at 700°C and 600°C, higher catalyst dosage resulted in decreased carbon char yields, likely due to promotion of liquid product formation.
3. The combination of fast pyrolysis characterized by a short residence time of 20 minutes, high temperature, and rapid heating rates of 30°C/min likely facilitated the high char yield through the promotion of solid char formation.

Overall, this study demonstrates the significant impact of pyrolysis temperature and zeolite catalyst dosage on the char yield. Further investigations into the characterization and possible applications of the produced carbon char are recommended to fully exploit the value of this waste-derived resource.

Acknowledgement

The authors acknowledge financial support from Universiti Teknologi PETRONAS, Malaysia. This work is done with the financial support from Urif Incentive for Centre of Innovative Nanostructure And Nanodevices (COINN). Moreover, the author would like to express appreciation to Dr. Adel Mohammed Al-dhahebi for his invaluable assistance and guidance in the operation of the LT furnace, which contributed significantly to the successful completion of this work.

References

[1] Y. Qiao, Z. Chen, X. Wu, and Z. Li, "Effect of demineralization on waste tire pyrolysis char physical, chemical characteristics and combustion characteristics," *Energy*, vol. 284, p. 129219, Dec. 2023. https://doi.org/10.1016/j.energy.2023.129219.

[2] S. Maroufi, M. Mayyas, and V. Sahajwalla, "Nano-carbons from waste tyre rubber: An insight into structure and morphology," *Waste Management*, vol. 69, 2017, doi:10.1016/j.wasman.2017.08.020.

[3] J. Nisar *et al.*, "Pyrolysis of waste tire rubber: Influence of temperature on pyrolysates yield," *J Environ Chem Eng*, vol. 6, no. 2, 2018. https://doi.org/10.1016/j.jece.2018.05.021.

[4] A. enkosi Mavukwana and C. Sempuga, "Recent developments in waste tyre pyrolysis and gasification processes," 2022. https://doi.org/10.1080/00986445.2020.1864624.

[5] P. Parthasarathy *et al.*, "Influence of process conditions on product yield of waste tyre pyrolysis- A review," Aug. 01, 2016, *Springer New York LLC*. https://doi.org/10.1007/s11814-016-0126-2.

[6] A. Alsaleh and M. L. Sattler, "Waste Tire Pyrolysis: Influential Parameters and Product Properties," Dec. 01, 2014, *Springer Nature*. https://doi.org/10.1007/s40518-014-0019-0.

[7] M. M. Rahman, Y. Yu, and H. Wu, "Valorisation of Waste Tyre via Pyrolysis: Advances and Perspectives," Oct. 20, 2022, *American Chemical Society*. doi: 10.1021/acs.energyfuels.2c02053.

[8] H. Afash, B. Ozarisoy, H. Altan, and C. Budayan, "Recycling of Tire Waste Using Pyrolysis: An Environmental Perspective," *Sustainability*, vol. 15, no. 19, p. 14178, Sep. 2023, doi:10.3390/su151914178.

[9] W. Han, D. Han, and H. Chen, "Pyrolysis of Waste Tires: A Review," Apr. 01, 2023, *MDPI*.

https://doi.org/10.3390/polym15071604.
[10] B. Rijo, A. P. Soares Dias, and Ł. Wojnicki, "Catalyzed pyrolysis of scrap tires rubber," *J Environ Chem Eng*, vol. 10, no. 1, Feb. 2022. https://doi.org/10.1016/j.jece.2021.107037.
[11] A. Abedeen, M. S. Hossain, U. Som, and M. D. Moniruzzaman, "Catalytic cracking of scrap tire- generated fuel oil from pyrolysis of waste tires with zeolite ZSM-5," *International Journal of Sustainable Engineering*, vol. 14, no. 6, pp. 2025–2040, 2021, doi:10.1080/19397038.2021.1951883.
[12] G. H. A. Razzaq and N. S. Majeed, "Pyrolysis of scrap tire by utilizing zeolite as catalyst," in *Materials Today: Proceedings*, Elsevier Ltd, 2021, pp. 4606–4611. https://doi.org/10.1016/j.matpr.2020.12.1232.
[13] W. Qu *et al.*, "Pyrolysis of waste tire on ZSM-5 zeolite with enhanced catalytic activities,"*Polym Degrad Stab*, vol. 91, no. 10, pp. 2389–2395, Oct. 2006. https://doi.org/10.1016/j.polymdegradstab.2006.03.014.
[14] J. I. Osayi and P. Osifo, "Utilization of Synthesized Zeolite for Improved Properties of Pyrolytic Oil Derived from Used Tire," *International Journal of Chemical Engineering*, vol. 2019, 2019. https://doi.org/10.1155/2019/6149189.
[15] A. Abedeen, M. S. Hossain, U. Som, and M. D. Moniruzzaman, "Catalytic cracking of scrap tire- generated fuel oil from pyrolysis of waste tires with zeolite ZSM-5," *International Journal of Sustainable Engineering*, vol. 14, no. 6, pp. 2025–2040, 2021, doi:10.1080/19397038.2021.1951883.
[16] S. Muenpol, R. Yuwapornpanit, and S. Jitkarnka, "Valuable petrochemicals, petroleum fractions, and sulfur compounds in oils derived from waste tyre pyrolysis using five commercial zeolites as catalysts: Impact of zeolite properties," *Clean Technol Environ Policy*, vol. 17, no. 5, pp. 1149–1159, Jun. 2015. https://doi.org/10.1007/s10098-015-0935-8.
[17] Suhartono, M. Harsanti, W. Septiyanti, Suharto, and F. Achmad, "Zeolite catalytic pyrolysis of waste tire into fuel in gasoline hydrocarbon range," in *IOP Conference Series: Earth and Environmental Science*, 2022. https://doi.org/10.1088/1755-1315/969/1/012030.
[18] W. Qu *et al.*, "Pyrolysis of waste tire on ZSM-5 zeolite with enhanced catalytic activities,"*Polym Degrad Stab*, vol. 91, no. 10, pp. 2389–2395, Oct. 2006. https://doi.org/10.1016/j.polymdegradstab.2006.03.014.
[19] M. Arabiourrutia, M. Olazar, R. Aguado, G. López, A. Baron, and J. Bilbao, "HZSM-5 and HY zeolite catalyst performance in the pyrolysis of tires in a conical spouted bed reactor," *Ind Eng Chem Res*, vol. 47, no. 20, pp. 7600–7609, Oct. 2008. https://doi.org/10.1021/ie800376d.
[20] S. Ucar, S. Karagoz, A. R. Ozkan, and J. Yanik, "Evaluation of two different scrap tires as hydrocarbon source by pyrolysis," *Fuel*, vol. 84, no. 14–15, 2005, doi: 10.1016/j.fuel.2005.04.002.
[21] C. Berrueco, E. Esperanza, F. J. Mastral, J. Ceamanos, and P. García-Bacaicoa, "Pyrolysis of waste tyres in an atmospheric static-bed batch reactor: Analysis of the gases obtained," in *Journal of Analytical and Applied Pyrolysis*, 2005. https://doi.org/10.1016/j.jaap.2004.10.007.
[22] A. M. Mastral, R. Murillo, M. S. Callén, T. García, and C. E. Snape, "Influence of process variables on oils from tire pyrolysis and hydropyrolysis in a swept fixed bed reactor," *Energy and Fuels*, vol. 14, no. 4, 2000. https://doi.org/10.1021/ef990183e.
[23] R. Murillo, E. Aylón, M. V. Navarro, M. S. Callén, A. Aranda, and A. M. Mastral, "The application of thermal processes to valorise waste tyre," in *Fuel Processing Technology*, 2006. https://doi.org/10.1016/j.fuproc.2005.07.005.
[24] C. Sathiskumar and S. Karthikeyan, "Recycling of waste tires and its energy storage application of by-products –a review," Dec. 01, 2019, *Elsevier B.V. https://doi.org/*10.1016/j.susmat.2019.e00125.
[25] M. Wang, L. Zhang, A. Li, M. Irfan, Y. Du, and W. Di, "Comparative pyrolysis behaviors of

tire tread and side wall from waste tire and characterization of the resulting chars," *J Environ Manage*, vol. 232, pp. 364–371, Feb. 2019. https://doi.org/10.1016/j.jenvman.2018.10.091.
[26] J. D. Martínez, N. Puy, R. Murillo, T. García, M. V. Navarro, and A. M. Mastral, "Waste tyre pyrolysis - A review," 2013, *Elsevier Ltd.* https://doi.org/10.1016/j.rser.2013.02.038.
[27] P. Caputo *et al.*, "Char from Pyrolysis of Waste Tires to Increase Bitumen Performances,"*Applied Sciences*, vol. 14, no. 1, p. 30, Dec. 2023. https://doi.org/10.3390/app14010030.
[28] S. M. R. Costa, D. Fowler, G. A. Carreira, I. Portugal, and C. M. Silva, "Production and Upgrading of Recovered Carbon Black from the Pyrolysis of End-of-Life Tires," Mar. 01, 2022, *MDPI.* https://doi.org/10.3390/ma15062030.
[29] X. Jin *et al.*, "Coke formation on activated carbon during catalytic upgrading of coal pyrolysis volatiles," *Ranliao Huaxue Xuebao/Journal of Fuel Chemistry and Technology*, vol. 49, no. 5, 2021. https://doi.org/10.1016/S1872-5813(21)60047-6.
[30] V. Daligaux, R. Richard, and M. H. Manero, "Deactivation and regeneration of zeolite catalysts used in pyrolysis of plastic wastes—a process and analytical review," 2021. doi:10.3390/catal11070770.
[31] M. Olazar, M. Arabiourrutia, G. Lôpez, R. Aguado, and J. Bilbao, "Effect of acid catalysts on scrap tyre pyrolysis under fast heating conditions," *J Anal Appl Pyrolysis*, vol. 82, no. 2, pp. 199–204, 2008. https://doi.org/10.1016/j.jaap.2008.03.006.
[32] S. M. Pai, R. K. Das, S. A. K. Kumar, L. Kumar, A. L. Karemore, and B. L. Newalkar, "Emerging trends in solid acid catalyst alkylation processes," in *Catalysis for Clean Energy and Environmental Sustainability: Petrochemicals and Refining Processes - Volume 2*, 2021. https://doi.org/10.1007/978-3-030-65021-6_4.
[33] J. Xu *et al.*, "High-value utilization of waste tires: A review with focus on modified carbon black from pyrolysis," Nov. 10, 2020, *Elsevier B.V. https://doi.org/*10.1016/j.scitotenv.2020.140235.
[34] A. Hijazi, A. H. Al-Muhtaseb, S. Aouad, M. N. Ahmad, and J. Zeaiter, "Pyrolysis of waste rubber tires with palladium doped zeolite," *J Environ Chem Eng*, vol. 7, no. 6, 2019, doi:10.1016/j.jece.2019.103451.
[35] V. Benedetti, S. S. Ail, F. Patuzzi, D. Cristofori, R. Rauch, and M. Baratieri, "Investigating the feasibility of valorizing residual char from biomass gasification as catalyst support in Fischer-Tropsch synthesis," *Renew Energy*, vol. 147, 2020. https://doi.org/10.1016/j.renene.2019.09.050.
[36] N. Jagadeesh and B. Sundaram, "Adsorption of Pollutants from Wastewater by Biochar: A Review," *Journal of Hazardous Materials Advances*, vol. 9, 2023, doi:10.1016/j.hazadv.2022.100226.
[37] N. Gao, F. Wang, C. Quan, L. Santamaria, G. Lopez, and P. T. Williams, "Tire pyrolysis char: Processes, properties, upgrading and applications," Nov. 01, 2022, *Elsevier Ltd.* doi:10.1016/j.pecs.2022.101022.
[38] F. Wang, N. Gao, C. Quan, and G. López, "Investigation of hot char catalytic role in the pyrolysis of waste tires in a two-step process," *J Anal Appl Pyrolysis*, vol. 146, Mar. 2020. https://doi.org/10.1016/j.jaap.2019.104770.
[39] Z. Wang *et al.*, "Co-pyrolysis of waste tire and pine bark for syngas and char production," *Fuel*, vol. 274, Aug. 2020. https://doi.org/10.1016/j.fuel.2020.117878.

Prediction of densities of novel deep eutectic solvent: Step towards carbon dioxide (CO_2) mitigation using green solvent

GHULAM Murshid[1, a,*], Shaima Said Al-Salmani[1], Khashayar Nasrifar[1]

[1]Department of Petroleum & Chemical Engineering, Sultan Qaboos University, Muscat. Oman

[a]g.murshid@squ.edu.om

Keywords: CO_2, Deep Eutectic Solvent, Critical Properties, Density, Rackett Equation

Abstract. The density is one of the important physical property essential in effective design and development of new processes to capture CO_2 from various industrial processes such as post combustion (power plants). The usual practice is to measure the densities of the solvent but sometime it is not practically possible to measure the density due to limited resources. Therefore, it is important to develop estimation methods to predict the important physical properties of solvents such deep eutectic solvents. In this work, the densities of deep eutectic solvent consist of Ethanolamine hydrochloride (EAHC) and Diethylenetriamine (DETA) at ratio 1:9 with 30 % water by volume was predicted at various temperature ranging from 290 K to 350 K which is usually the operating temperature range of absorber and desorber. The critical properties of salt DETA and hydrogen bond donor (EAHC) were estimated using the modified Joback-Reid method, while for aqueous DES, Lee-Kesler method was employed. The Rackett equation was employed to predict the densities of the aqueous DES (EAHC:DETA, 1:9) at various temperatures. The estimated densities were compared with the measured density data and a good agreement was found which is evident from the R^2 values and anova analysis.

Introduction

The earth's climate contains a various greenhouse gases which include carbon dioxide (CO_2), methane (CH_4), and nitrous oxide (N_2O) [1, 2]. Global greenhouse gases emissions had been on the rise since the beginning of the 21st century. This increase is primarily due to the usage of fossil fuel, deforestation, and other human activities [2]. As a result, the concentration of GHG in the earth's atmosphere is immensely increased, boosting the natural greenhouse effect, which may have a significant impact on life on the earth [3]. Relaying on fossil fuel as a source of energy, the global environmental and socio-economic conditions are becoming unpredictable. Fossil fuel burning tends to generate significant volumes of greenhouse gases such as, CO_2, CH_4, NO_2 and others, which have severe impacts on the climate [4]. There has been an increase in the earth temperature by a significant degree of magnitude in past few decades [5]. GHG emissions must be drastically reduced to avoid catastrophic climate changes as 50 billion tons of GHG are emitted each year. Effective reduction of GHG emissions required understanding the source of emissions, where the emissions come from. Below pie chart, (Fig. 1), shows that 73.2% of total greenhouse gases (GHG) emission is coming from energy production sources. However, agriculture, forestry and land use contribute to 18.4% of GHG emissions [6].

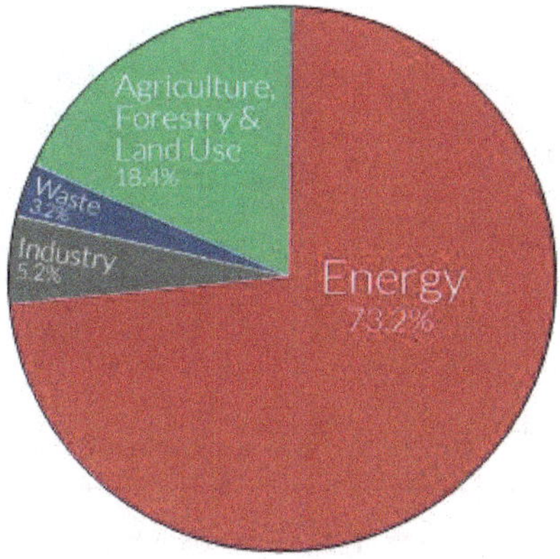

Fig. 1. Global GHG Emissions by Sectors [5].

Carbon dioxide (CO_2) is considered to be the most significant source of GHG because of the dependency of the world's economies on fossil fuels causing an increase in the emission of CO_2 into the atmosphere. By using renewable energy sources such as biomass, solar and wind energy, GHG concentration in the atmosphere can be stabilized. As fossil fuels are plentiful and readily available, the high cost of renewable energies delays their introduction to the industry. Additionally, there are some barriers that prevent the technological systems from being changed to one that is dependent on fossil fuels. Therefore, fossil fuels will continue to provide most of the energy in the coming years [7-9]. For stabilizing atmospheric CO_2 concentrations, carbon capture and storage (CCS) is the preferred technology. There are three steps in CCS: capturing CO_2, transporting CO_2, and storing CO_2. Using CCS, CO_2 is captured from exhaust gas and injected into deep reservoirs containing fluids (oil and gas) for thousands of years. In addition to maintaining existing carbon-based infrastructure, CCS minimizes the effects of CO_2 on earth's climate system [7]. This particular research focuses on the capture step which is essential and key step towards the effective storage or utilization.

CO_2 emissions in Oman during 2021 increased by 9 megatons (Mt) or 10.49% compared to 2020. Oman placed 141st out of 184 nations in the list of countries for CO_2 emissions in 2021 with 94.62 megatons of emissions. It will be beneficial to examine the behaviour of its CO_2 emissions per resident for future policy making. Fig. 2 demonstrates how Oman's per-capita CO_2 emissions climbed in 2021, when they were 17.95 tons per capita.

CO_2 removal is critical process and different technologies have been suggested to minimize CO_2 emission, including chemical/ physical solvent absorption, solid adsorption, membrane separation and cryogenic separation which are discussed below in details [8]. However, each of these technologies has its own benefits and disadvantages in terms of capital and operational expenses, purifying capacity, dependability, ease of operation and maintenance, weight and footprint of the units [9]. Different types of liquid solvents are used as absorbents to capture CO_2 from gas stream. There are typically two types of absorption processes. Chemical absorption processes are those in which the solvent and the dissolved gas interact chemically. Absorption into an amine solvent like MEA (MonoEthanolAmine) is the most industrially utilized method of capturing CO_2 from flue gas [10]

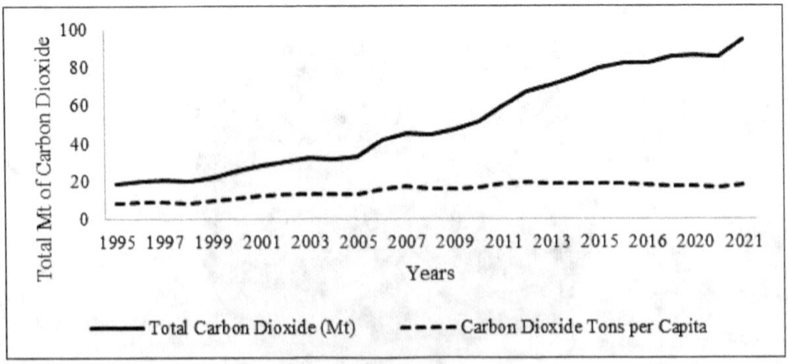

Fig. 2. CO_2 Emission in Oman per Year and per Capita [5].

The physiochemical properties of amines are crucial for the design of the processes. There is a wide range of amine solvents available in industry, like primary amines (MEA, DGA), secondary amines (DEA), tertiary amines (MDEA, TEA), hindered amines (AMP) and cyclic amines (Piperazine). Monoethanolamine (MEA), diethanolamine (DEA), di-2-propanolamine (DIPA) and N-methyldiethanolamine (MDEA) are considered as the most important alkanolamine for industrial applications. Solvent MEA is widely recognized as a benchmark solvent for efficient CO_2 capture because alkanolamines solvents have high chemical potential to selectively absorb CO_2 from flue gas with low CO_2 partial pressure, low solvent cost, and ease of solvent regeneration [10, 11]. In addition, CO_2 is acidic in nature while other flue gases are inert in nature and alkanolamines solvents have at least one hydroxyl group, which helps to reduce the vapor pressure and increase the CO_2 solubility in the solution [11]. However, the amine absorption process still has some technical challenges that hinder its widespread application. The regeneration of treated amines requires high temperature and therefore very energy intensive. Moreover, decomposition of amine-based chemicals causes equipment corrosion, formation of harmful compounds and release of volatile compounds [10].

Ionic liquids have received scientists and researchers' attention since the mid-19th century [12]. Blanchard et al.[13] was the initial researchers who studied how soluble is CO_2 in imidazolium-based ILs. ILs gained additional consideration in CO_2 capture area because of the attractive features as non-volatility, its stability with temperature change, and acid gases are highly soluble in these solvents when compared with conventional solvents. These special characteristics of ILs offer the possibility to decrease the thermal energy requirements for solvent regeneration [14]. ILs are also called designer solvents because of their properties like, viscosity, melting point, density, and hydrophobicity can be adjusted to meet the needs of a given process by only modifying the structure of the ions. However, even though ILs offer such a high potential for usage as "green" solvents for industrial processes, their widespread industrial application still faces difficulties. This is because of challenging synthetic processes and costly chemical raw materials. As a result, eutectic mixtures, also known as deep eutectic solvents (DESs), have been recognized as affordable substitute for ILs [12].

In recent years, the scientific community has paid close attention to deep eutectic solvents (DES). Deep eutectic solvents (DESs) have demonstrated the capacity to absorb CO_2 from flue gases. A DES is a mixture of a salt and a hydrogen-bond donor (HBD), which make a clear liquid when heated gently at moderate temperature and mixed in a certain molar ratio [15]. Similar to ILs, DES mostly consists of ionic species, making it an attractive solvent with significant CO_2 dissolving capabilities [12]. DESs have several key benefits and advantages over ILs, including

Materials Research Proceedings 53 (2025) 564-575 https://doi.org/10.21741/9781644903575-57

the potential to be biodegradable, nontoxic, and nonflammable as well as the potential for very low vapor pressure [12]. Another key feature of DES is that DES production is less complicated and cheaper than ILs. Deep eutectic solvents are so termed because an eutectic point will develop when the two constituent parts are combined in the right proportion so that its melting point will be lower than that of an ideal liquid mixture as it is described by Fig. 3 [12].

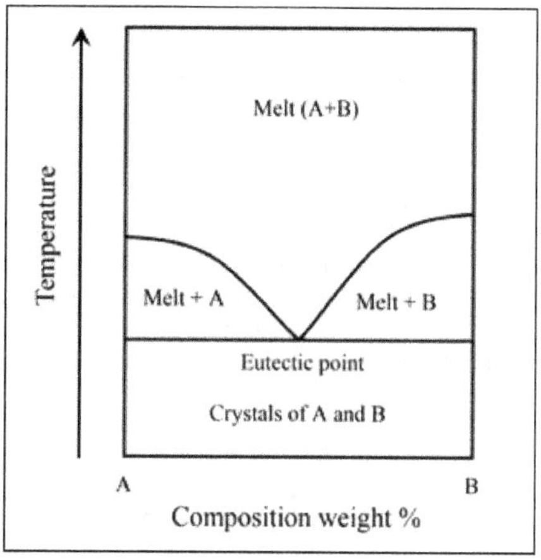

Fig. 3. Eutectic point in phase diagram of Eutectic solvent.

The exact measurement of physical properties of novel DES is very important for gas-liquid mass transfer calculations as well as for absorber design. These properties such as density which shall be measured accurately over a wide operating range of temperature for effective mass transfer calculations and consequently applying the in various design calculations. However, DES is a tuneable solvent and various complex combinations of HBD and Salt can be used to produce large number of DES for screening purpose. However, it is not possible economically to synthesis large number of solvents for measuring their properties prior to screening and choosing the best solvent in terms of CO_2 capture. It is important to use empirical correlations and thermodynamic model to predict the essential properties. Therefore, in this work, a novel deep eutectic solvent (DES), ethanolamine hydrochloride (EAHC) and diethylenetriamine (DETA) (1:9) with 33 vol% water was synthesized and its properties such as density, and molar volume were estimated.

Materials and methods

Chemicals. All the chemicals which include DETA and EAHC were purchased from sigma Aldrich with the highest purity and were used without further purification. Their chemical structures are summarized in the following Table 1. The high purity distilled water was used to prepare aqueous DES solution with ratio (1:9) to 33 % by volume water. The preparation method details can be found in our previously published work [25]

Table. 1. Chemical structure of HBD and Salt used to synthesize DES

Name	Formula	Mw	Structure
Diethylenetriamine	$C_4H_{13}N_3$	103.17	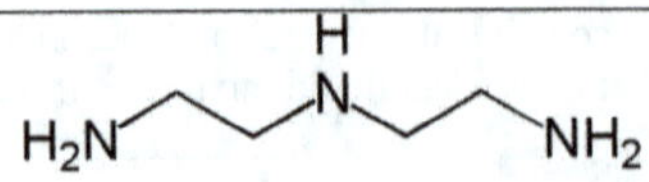
Ethanolamine hydrochloride	C_2H_8ClNO	97.54	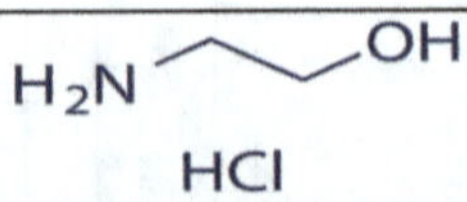

Defining the novel DES in Aspen-plus. The user defined function was used to define the novel DES in the aspen plus since this solvent was not available in aspen data base and UNIFAC group distribution method was employed to define the DES in the aspen environment. The important steps are presented in the following Fig. 4 for better illustration and understanding.

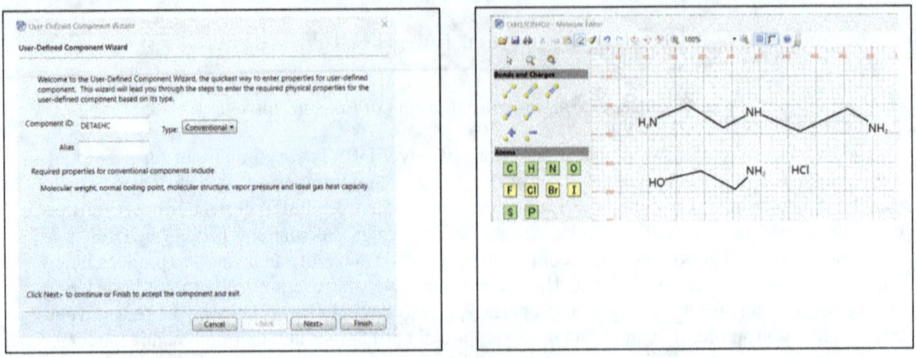

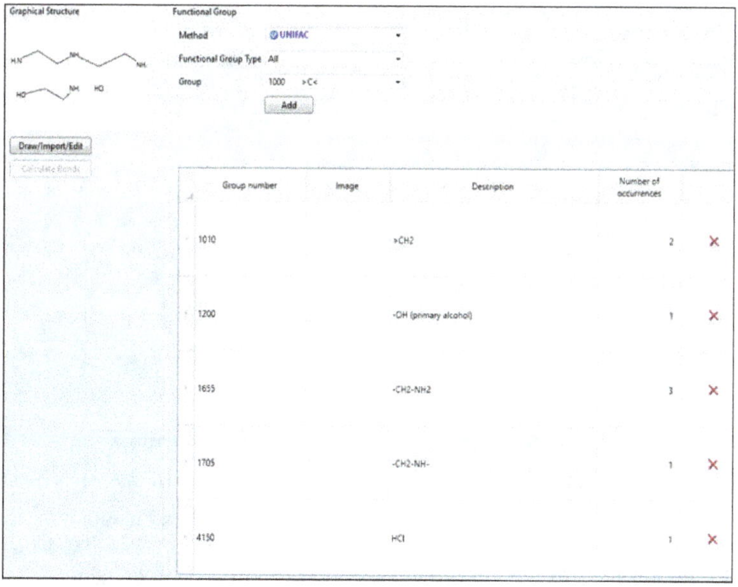

Fig. 4. Pictorial description of defining the functional groups in aspen plus

Density Measurement. The density of aqueous DES (1:9) with 30 % water by volume was measured using DMA 4500 Anton Paar density meter capable of measuring the density ranging from 0 $g{\cdot}cm^{-3}$ to 3 $g{\cdot}cm^{-3}$ up to a temperature of 100 °C with a measuring accuracy of 0.00005 $g{\cdot}cm^{-3}$. The density meter was calibrated using distilled water before every run and the data was compared with literature to make sure the integrity of the measured and reported data.

Results and discussion

The density meter was calibrated before performing any measurements with 30% solution of monodiethanolamine (MDEA) and the data was compared with literature. The details of the calibration data can be found in our previous work [25]. The density of novel DES (1:9) (DEAT:EAHC) and 30% water by volume was measured over a wide range of temperature and the data is presented in the Fig. 5 as a function of temperature. There is a linear decrease in density values with increasing temperature from 280 K to 350 K. This decrease could be associated to the volume expansion of the aqueous DES.

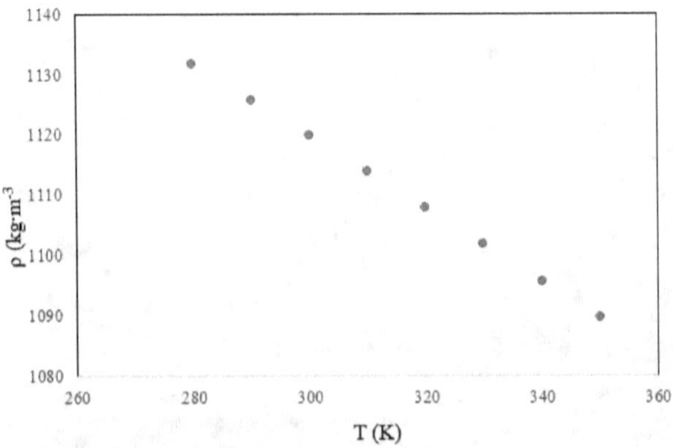

Fig. 5. Experimentally measured data of (1:9) (EAHC:DETA) at various temperatures

However, In order to estimate the density of the novel DES via Aspen plus, it is mandatory to calculate the critical properties of the novel DES such as temperature, pressure, molar volume and molecular weight and normal boiling temperature. This input data will further facilitate the physical properties estimation. Therefore, , Joback group contribution method was employed to estimate the normal boiling point Tb, critical pressure P_c, critical temperature T_c and critical compressibility factor Z_c. Those properties were predicted using the modified Lyderson – Joback-Reid method because this method shows good accuracy. In this technique, the Lyderson method is used to calculate the critical volume and critical pressure whereas Joback-Reid method was used to calculate the normal boiling point and critical temperature. Molecular weight and normal boiling temperature are the most fundamental inputs for group-contribution property estimation [22-26]. The equations which are proposed by Lydersen and Joback – Reid are the following [22]:

$$T_b = 198.2 + \sum n_i \Delta T_{bMi} \quad \text{Eq. 1}$$

$$T_c = \frac{T_b}{0.5703+1.0121\sum n_i \Delta T_{Mi} - (\sum n_i \Delta T_{Mi})^2} \quad \text{Eq. 2}$$

$$V_c = 6.75 + \sum n_i + \Delta V_m \quad \text{Eq. 3}$$

$$P_c = \frac{M}{(0.2573+\sum n_i \Delta P_{Mi})^2} \quad \text{Eq. 4}$$

$$Z_c = \frac{P_c V_c}{R T_c} \quad \text{Eq. 5}$$

Where, n_i represents the frequency of the ith group of atoms in the molecule, ΔTbMi is the contribution to the normal boiling temperature (K) of ith group of atoms in the molecule, ΔTM is their contribution to the critical temperature (K), ΔPM is their contribution to the critical pressure (bar), ΔVM is their contribution to the critical molar volume (cm3/mol) and M is the molecular mass of the molecule (g/mol). ΔTbM, ΔTM, ΔPM, and ΔVM are constants for the modified Lydersen–Joback–Reid method. Zc is critical compressibility factor and R is an ideal gas constant = 8.314 J/K.mol [22].

Materials Research Proceedings 53 (2025) 564-575 https://doi.org/10.21741/9781644903575-57

The acentric factor can be calculated for each component using the equation which has described by Valderrama and Robles [22]

$$\omega_i = \frac{(T_b-43\text{ K})(T_c-43\text{ K})}{(T_c-T_b)(0.7T_c-43\text{K})}\log\left(\frac{P_c}{P_b}\right) - \frac{(T_c-43\text{ K})}{(T_c-T_b)}\log\left(\frac{P_c}{P_b}\right) + \log\left(\frac{P_c}{P_b}\right) - 1 \qquad \text{Eq. 6.}$$

where is Tc and Pc are the estimated critical properties and Tb is the normal boiling temperature at the normal boiling pressure (Pb = 110132.5 Pa). ω_i is the acentric factor of pure component in DES [22].

After estimating the critical properties of HBD and salt separately using modified Lydersen–Joback–Reid method, Lee-Kesler mixing rule equations are recommended by Plocker et al. [27] to calculate the critical properties of binary DES mixtures because of its applicability to those constituents which have different molecular size [22, 27].

$$T_{c_m} = \frac{1}{V_{c_m}{}^{1/4}} \Sigma_i \Sigma_j y_i y_j V_{c_{ij}}{}^{1/4} T_{c_{ij}} \qquad \text{Eq. .7}$$

$$P_{c_m} = 0.2905 - 0.085\, \omega_m \frac{RT_{c_m}}{V_{c_m}} \qquad \text{Eq. 8.}$$

$$V_{c_m} = \sum_i \sum_j y_i y_j V_{c_{ij}} \qquad \text{Eq.9}$$

$$\omega_m = \Sigma_i y_i \omega_i \qquad \text{Eq. 10}$$

$$T_{c_{ij}} = \left(T_{c_i} T_{c_j}\right)^{\frac{1}{2}k_{ij}} \qquad \text{Eq.11}$$

$$V_{c_{ij}} = \frac{1}{8}\left(V_{c_i}^{\frac{1}{3}} V_{c_j}^{\frac{1}{3}}\right)^3 \qquad \text{Eq.12}$$

In these equations, m refers to the DES mixture, i and j refer to the pure components. Kij is binary parameter which is simplified to be a unit value because of no published data are available [27]. The estimated critical properties of the DES are summarized in the Table 2.

Table. 2. Estimated Properties of Studied DES

Property	Value	Unit
T_{CM}	939.65	[K]
V_{CM}	0.5885	[cm^3/kmol]
P_{CM}	35.14	[bar]
ω_M	0.5631	-
MW	200.7	[g/mol]
T_b	736.35	[K]

The molar volume data was required to estimate the densities of the novel DES. The density was calculated using Gunn-Yamada. The calculated values were used to determine the Rackett parameter as shown by the equation.

$$V_m^l = \frac{RT_c\left(Z_m^{RA}\right)^{1+(1-T_r)^{2/7}}}{P_c} \qquad \text{Eq. 13}$$

where R is ideal gas constant 8.314 J/K.mol, T_c is the DES critical temperature and T_r=T/T_c is the reduced temperature.

Then, the density (ρ) can be calculated by the following equation:

$$\rho = \frac{MW\,(\frac{kg}{kmol})}{V_m\,(\frac{m^3}{kmol})} \qquad \text{Eq. 14}$$

The estimated molar volume is presented in the Fig. 6 as a function of temperature. It is evident from the data that the molar volume increase by increasing the temperature which could be due to the expansion of the solution as the higher temperatures. The estimated density data was compared with the measured data and presented in the Fig. 7 along with R^2 value. The R^2 value shows that a good agreement between measured and the calculated values.

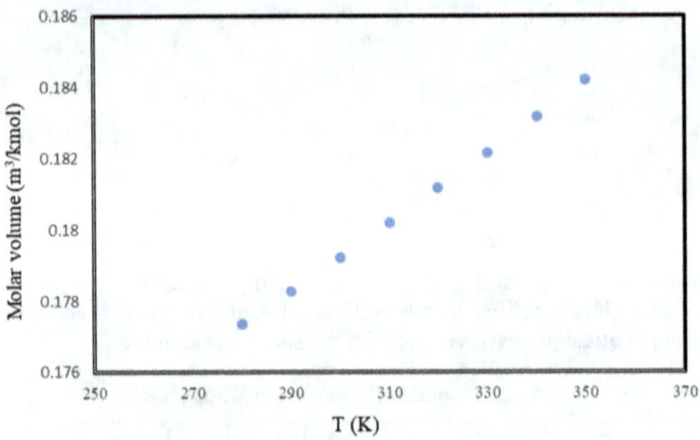

Fig. 6. Effect of Temperature on Molar Volume (m^3/kmol)

Materials Research Proceedings 53 (2025) 564-575 https://doi.org/10.21741/9781644903575-57

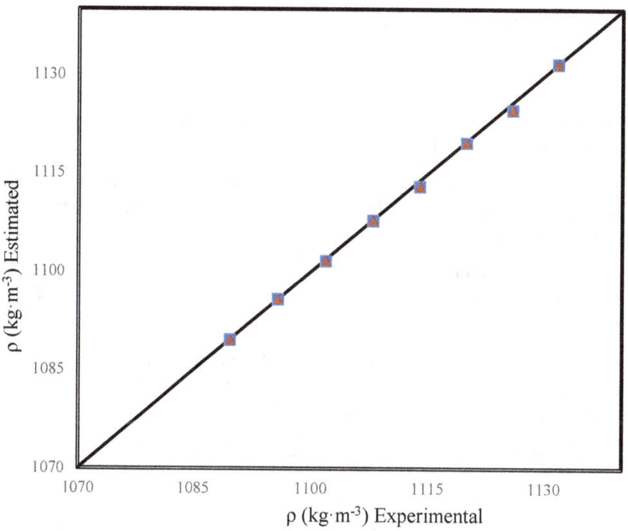

Fig. 7. Estimated and Measured density data of novel DES (DETA:EAHC) 30% water

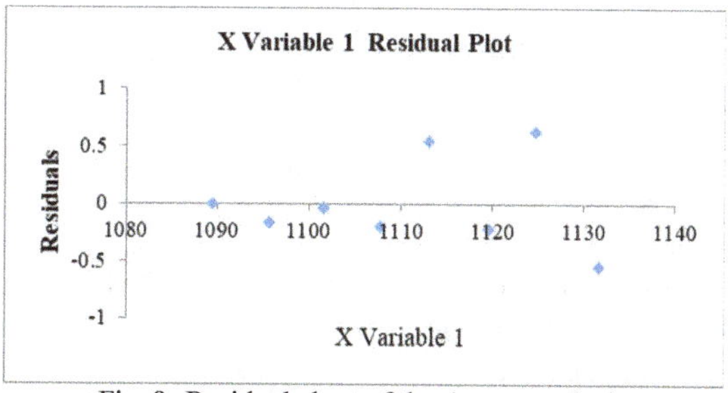

Fig. 8. Residual chart of the Anova analysis

The anova analysis was also performed which is a unique statistical tool to validate the regressed data. The R square value (0.999) and the F-significant smaller value (1.18586E-10) shows that the little deviation of the estimated densities from the measured values. The similar trend has been depicted in the residual chart (Fig. 8) which shows a uniform distribution of the data points across the horizontal line.

Conclusion

The current work shows that important design data can also be estimated using empirical correlations as a function of temperature. The critical properties of hydrogen bond donor (HBD) and salt were estimated and these properties were further used to calculate the critical properties of the novel DES using Lee-Kesler equation. The densities of the DES (DETA:EAHC) were then estimated and a good agreement was found between measured and the estimated density values which were supported by the Anova analysis. The prediction of the design properties can expedite the screening and selection of the solvent process for effective CO_2 capture without going through extensive experimental work.

References

[1] V. Fischer, Properties and applications of deep eutectic solvents and low-melting mixtures. (2015) https://d-nb.info/1070996343/34.

[2] J.C.M. Pires, F.G. Martins, M.C.M. Alvim-Ferraz, M. Simões, Recent developments on carbon capture and storage: An overview. Chemical engineering research and design, 89(2011) 1446-1460. https://doi.org/10.1016/j.cherd.2011.01.028

[3] S.A. Aromada, L.E. Øi, Simulation of improved absorption configurations for CO2 capture. in Proceedings of the 56th Conference on Simulation and Modelling (SIMS 56), (2015). 10.3384/ecp1511921

[4] A.D.Ebner, J.A. Ritter, State-of-the-art Adsorption and Membrane Separation Processes for Carbon Dioxide Production from Carbon Dioxide Emitting Industries. Sep Sci & Tech. 44 (2009) 1273-1421. https://doi.org/10.1080/01496390902733314

[5] A.W. Sabah, Y. Charabi., R. Al-Maamari, G.A. Al-Rawas, A, Gastli, K. Chan, CO2 greenhouse emissions in Oman over the last forty-two years: Review. Renew. Sustain. Energy Rev. (2015) 52 1702-1712. 10.1016/j.rser.2015.07.193

[6] M. Crippa, D. Guizzardi, M. Banja, CO2 emissions of all world countries. Publications Office of the European Union. (2022) 10.2760/07904

[7] Countryeconomy.com. Oman - CO2 Emission. 2021 [cited 2022 07/Nov/2022]; Available from: https://countryeconomy.com/energy-and-environment/co2-emissions/oman.

[8] Kumar,P.Singh, P.Raizada, C.M.Hussain, Impact of COVID-19 on greenhouse gases emissions: A critical review. Sci. Total Environ. 806. (2022) 150349 https://doi.org/10.1016/j.scitotenv.2021.150349.

[9] H.Ritchie, P.Rosado, M.R.Rosado, CO_2 and Greenhouse Gas Emissions. (2020) https://ourworldindata.org/greenhouse-gas-emissions

[10] G. Hu, K.H. Smith, Y. Wu, K.A. Mumford, S.E. Kentish, Carbon dioxide capture by solvent absorption using amino acids: A review, Chin. J. Chem. Eng. 26 (2018) 2229-2237. https://doi.org/10.1016/j.cjche.2018.08.003

[11] M. Hoorfar,Y. Alcheikhhamdon, B. Chen, A novel tool for the modeling, simulation and costing of membrane based gas separation processes using Aspen HYSYS: Optimization of the CO2/CH4 separation process, Comput. Chem. Eng. 117 (2018) 11-24. https://doi.org/10.1016/j.compchemeng.2018.05.013

[12] O. Norouzi, M. Heidari, A.D. Norouzi, Technology Overview of Renewable Natural Gas Production from Organic Wastes Based on Canadian Incentives and Standards for Gas Injection into Existing Pipelines. Fuel Commun. 11 (2022) 100056. https://doi.org/10.1016/j.jfueco.2022.100056

[13] A.A. Khan, G.N. Halder, A.K. Saha, Carbon dioxide capture characteristics from flue gas using aqueous 2-amino-2-methyl-1-propanol (AMP) and monoethanolamine (MEA) solutions in packed bed absorption and regeneration columns, Int. J. Greenhouse Gas Control. 32 (2015) 15-23. https://doi.org/10.1016/j.ijggc.2014.10.009

[14] G. Mondino, C.A. Grande, R. Blom, L.O. Nord, Evaluation of MBTSA technology for CO2 capture from waste-to-energy plants, Int. J. Greenhouse Gas Control. 118 (2022) 103685. https://doi.org/10.1016/j.ijggc.2022.103685

[15] M. Younas, M. Sohail, L. K. Leong, M.J.K. Bashir, S.Smathi, Feasibility of CO2 adsorption by solid adsorbents: a review on low-temperature systems. Int. J. Env. Sci. & Tech. 13 (2016) 1839-1860. 10.1007/s13762-016-1008-1

[16] J. Xu, H. Wu, Z. Wang, Z. Qiao, S. Zhao, Recent advances on the membrane processes for CO2 separation. Chin. J. Chem. Eng. 26 (2018) 2280-2291. https://doi.org/10.1016/j.cjche.2018.08.020

[17] U. Arachchige, Viscosities of pure and aqueous solutions of monoethanolamine (MEA), diethanolamine (DEA) and N-methyldiethanolamine (MDEA). Annual Transactions of the Nordic Rheology Society. 21 (2013) 299.

[18] M.K. Hadj-Kali, S. Mulyono, I. Alnashef, A. Fakeeha, F. Mjalli, A. Hayyan, Solubility of CO2 in deep eutectic solvents: Experiments and modelling using the Peng–Robinson equation of state. Chem. Eng. Res Des. 92 (2014) 1898-1906. https://doi.org/10.1016/j.cherd.2014.02.004

[19] L.A. Blanchard, D. Hancu, E.J. Beckman, J.F. Brennecke, Green processing using ionic liquids and CO2. Nature. 399 (1999) 28-29. https://doi.org/10.1038/19887

[20] S. Zhang, X. Lu, Q. Zhou, X. Li, X. Zhang, S. Li, Ionic liquids: physicochemical properties. 1st Edition, Elsevier, 2009. https://doi.org/10.1016/B978-0-444-53427-9.X0001-2

[21] N.R. Mirza, N.J. Nicholas, Y.W. Sandra, K.W.Geoffrey, Estimation of normal boiling temperatures, critical properties, and acentric factors of deep eutectic solvents. J.Chem. Eng.Data. 60 (2015) 1844-1854. 10.1021/acs.jced.5b00046

[22] M.K Aroua, M.Z. Haji-Sulaiman, K. Ramasamy, Modelling of carbon dioxide absorption in aqueous solutions of AMP and MDEA and their blends using Aspenplus, Sep. Purif. Technol. 29 (2002) 153-162. https://doi.org/10.1016/S1383-5866(02)00071-0

[23] J. Hack, N. Maeda, D.M. Meier, Review on CO2 Capture Using Amine-Functionalized Materials. ACS Omega, 44 (2022) 39520-39530. https://doi.org/10.1021/acsomega.2c03385

[24] I.M.Saeed, P. Alaba, S.A. Mazari, W.J. Basirun,V.S. Lee, N. SabzoiI, Opportunities and challenges in the development of monoethanolamine and its blends for post-combustion CO2 capture, Int. J. Greenhouse Gas Control. 79 (2018) 212-233. https://doi.org/10.1016/j.ijggc.2018.11.002

[25] S.M. Farouq, S.F. Mehdi, K. Ganesh, G. Murshid, N. Jamil, A.M. Suhaib, Experimental and theoretical study of the physicochemical properties of the novel imidazole-based eutectic solvent, J. Mol. Graph. 118 (2023) 108319. 10.1016/j.jmgm.2022.108319

[26] K. Shahbaz, F.S. Mjalli, M.A. Hashim,I.M. AlNashef, Prediction of deep eutectic solvents densities at different temperatures, Thermochim. Acta. 515 (2011) 67-72. https://doi.org/10.1016/j.tca.2010.12.022

[27] U.Plocker, H. Knapp, J. Prausnitz, Calculation of High-Pressure Vapor-Liquid Equilibria from a Corresponding-States Correlation with Emphasis on Asymmetric Mixtures. Ind Eng. Chem. Proc. DD. 17 (1978) 324-332. https://doi.org/10.1021/i260067a020

Decarbonization Technology: ICDT2024 Materials Research Forum LLC
Materials Research Proceedings 53 (2025) 576-583 https://doi.org/10.21741/9781644903575-58

Solvent aided crystallization for purification of biodiesel synthesized from coconut oil

ACHOL Aguek Tengteng[1,a], WAN Nur Aisyah Wan Osman[2,b] and SHAFIRAH Samsuri[3,b*]

[1]Department of Chemical Engineering, Universiti Teknologi PETRONAS, 32610 Bandar Seri Iskandar, Perak, Malaysia

[2]HICoE-Centre for Biofuels and Biochemical Research (CBBR), Institute of Engineering, Universiti Teknologi PETRONAS, 32610 Bandar Seri Iskandar, Perak, Malaysia

[a]achol_20000146@utp.edu.my, [b]wan_22000197@utp.edu.my, [c]shafirah.samsuri@utp.edu.my

Keywords: Biodiesel, Coconut Oil, Coolant Temperature, Crystallization Time, Stirring Speed, Solvent Aided Crystallization

Abstract. According to recent studies, employing refined cooking oils can increase biodiesel purity to as high as 98.86% to 100%. The use of solvent-aided crystallization (SAC) to purify biodiesel made from unrefined coconut oil (CO), which has better qualities like reduced density and ideal viscosity that improve fuel efficiency, is, nonetheless, mostly lacking in scientific literature. By examining the efficacy of SAC in purifying coconut oil-derived biodiesel utilizing 1-butanol as the solvent, this study seeks to bridge that gap.This paper assessed the effects of several SAC parameters, including coolant temperature, crystallization duration, and stirring rate, on the purity and yield of biodiesel using a One-Factor-at-a-duration (OFAT) methodology. The outcomes showed purities ranging from 89.83% to 100%, highlighting the possibility of high purity levels being attained by coconut oil following SAC treatment in contrast to other oils such as palm oil. Significantly, crude biodiesel made from coconut oil already showed a 90% purity, which is significantly higher than the 30.63% purity usually seen in palm oil. The study produced 100% pure biodiesel by adjusting SAC parameters, such as a coolant temperature of 9°C, crystallization period of 20 minutes, and stirring rate of 201 rpm. Additional analysis of the biodiesel's physicochemical characteristics, such as density, viscosity, pour point, cloud point, total acid number, and oxidative stability, demonstrated that SAC has the potential to be an efficient and sustainable way to meet the high purity requirements needed for industrial biodiesel applications. These results imply that SAC employing 1-butanol might greatly improve the purity of biodiesel derived from coconut oil, assisting in the shift to renewable energy sources and promoting global sustainability objectives. It is advised that more study be done to improve SAC conditions and see whether it can be used with different feed stocks.

Introduction

Biodiesel can be defined as a fuel composed of mono-alkyl esters of long-chain fatty acids derived from vegetable oils or animal fats. Currently the easiest and most practical method for the production of Biodiesel is using transesterification method of vegetable oil and alcohol to produce Ethyl Esters in this case Biodiesel. Transesterification can be defined as a chemical reaction to achieve high-quality fuel with hardly any contaminants that could sabotage its excellence, and it is crucial to be performed to meet this project's main objective. There are several purification methods such as Water washing/Wet washing, ion-exchange adsorbents, membrane-based adsorbents, and SAC. Biodiesel and glycerol are separated by gravitational settling and centrifugation due to the differences in their densities. This project aims for the production of biodiesel which is highly viscous liquids hence in this case we deploy the SAC separation

Materials Research Proceedings 53 (2025) 576-583 https://doi.org/10.21741/9781644903575-58

technique where sufficient Solvent is added to reduce the viscosity of the sample and control the crystallization kinetics in a favorable way since rapid Crystallization occurs in less Viscous Samples [1].in which one mole of Triglycerides of vegetable oil/fat react with 3 moles of alcohol to produce 3 moles of alkyl esters (Biodiesel) and 1 mole of glycerol as a by-product. The excess alcohol that was not used in the reaction separates partly from the fuel and partly from the glycerol at the end of the process and can recover and be reused through a process of distillation [1].

Typically, a strong base such sodium hydroxide or Potassium hydroxide can be used as the catalyst and is normally not consumed during the reaction hence ends up in the glycerol layer. A triglyceride is a molecule made up of 3 fatty acids attached to a glycerol backbone. The 3 fatty acids on each molecule could be same or they could be different. The fatty acid makeup of the oil. or fat is responsible for some of the properties found in the resulting biodiesel. The transesterification reaction will proceed quickly at first and eventually slow down as it reaches equilibrium. At this point there will be some unreacted glycerides left in the biodiesel. It is possible to improve the quality by removing the glycerin that has settled out and to run another reaction, thus pushing the equilibrium point further to the products side [1].

After the production of biodiesel, it will still need to be purified because it contains many contaminants such as glycerol, soap, Triglycerides, water and the catalyst used that need to be separated and treated. Biodiesel purification is an important stage in biodiesel production to SAC is carried out at low temperatures; this is supported by other studies mentioning that the biodiesel would be volatile at high temperatures in the range of 340-375 $°C$ which were obtained from thermal analyses of thermogravimetric analysis (TGA) and Differential Scanning Calorimetry (DSC). The fundamental idea behind SAC is to introduce sufficient amounts of an aiding agent into the solvents to selectively lower the viscosity of melts and change the kinetics of crystallization [2]. In a low-viscosity sample solution, fast crystallization happens when the assistant solvents are injected. This process is able to get beyond the main obstacle in the difficult-to-separate process of biodiesel–glycerol. Purified biodiesel was the result of successfully removing undesirable glycerol, methanol, and soap components using SAC.

In this experiment, coconut oil is used as the biodiesel feedstock to produce crude biodiesel, in which the crude biodiesel undergoes purification via SAC to remove all undesirable contaminants as listed above. The lack of research on SAC for the purification of biodiesel produced from coconut oil represents a critical research gap in the current scientific literature. Despite the growing importance of biodiesel as a renewable energy source and the abundance of coconut oil feedstock in tropical regions, the absence of studies investigating SAC's efficiency in purifying coconut oil-derived biodiesel impedes the development of efficient and sustainable purification methods. Consequently, there is a pressing need for research projects to address this gap and explore the potential of SAC for enhancing the purity and quality of coconut oil-based biodiesel, thus contributing to the advancement of both biodiesel production and renewable energy technologies [1]. This study explores SAC for purifying coconut oil biodiesel, analyzing its effectiveness, impact on efficiency, and quality, aiming to determine its potential as a cost- effective processing method.

Methodology

Materials. Unrefined Coconut oil is used to produce purified biodiesel. The coconut oil is bought from a store located nearby. On the other hand, the chemicals used for transesterification reactions include potassium hydroxide (KOH) and methanol (CH3OH), coolant used in refrigerated bath is ethylene glycol (C2H6O2) while the solvent used for SAC process is 1-butanol (C4H10O).

Production of biodiesel via catalyzed transesterification reaction method. The method for catalyzed transesterification reaction is adapted from [3]. 5.1g of KOH was dissolved in 90ml of CH3OH. Consequently, 500ml of coconut oil was heated up to 60°C and the solution of KOH and CH3OH was poured into the heated cooking oil upon reaching the desired temperature. The mixture was stirred rapidly for about 20 minutes at 500 rpm in order to allow complete mixing. Upon completing the stirring process, the mixture needed to undergo a gravity settling process for overnight to remove the majority of glycerol from crude biodiesel. Afterwards, the crude biodiesel was sent for characterization analysis of DSC and Gas chromatography Mass spectrometry (GCMS) before proceeding with the biodiesel purification process via SAC.

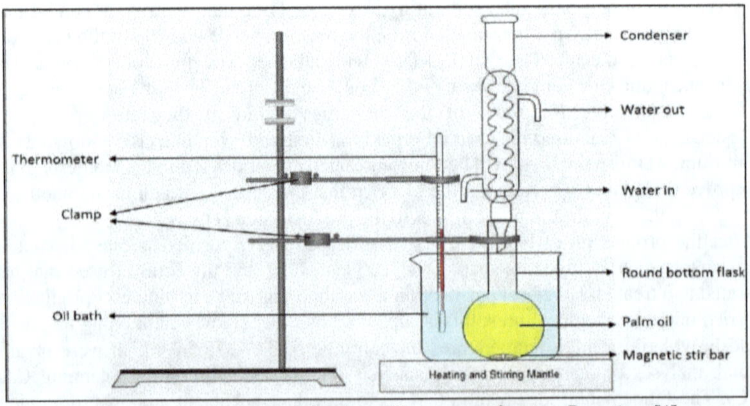

Fig. 1. Experimental Setup of Catalyzed Transesterification Reaction [4]

Purification of biodiesel via stirring solvent aided crystallization. The fabricated setup and major apparatus employed for this study are shown in Figure 2. A volume of 500mL of crude coconut oil, derived from the preceding transesterification reaction, was introduced into the cylindrical vessel along with C4H10O as solvent. Subsequently, the vessel was immersed in a coolant comprising C2H6O2 within the refrigerated bath, following the attainment of the desired temperature set by the chiller (630D, PROTECH, Selangor, Malaysia). The temperature also varied according to the operational conditions. The rationale behind altering the crystallizer temperature was to identify the optimal temperature that would result in the highest purity of biodiesel. Upon completing the SAC process, the biodiesel which is the liquid phase was extracted from the cylindrical vessel and its volume was measured. In addition, the extracted liquid was also placed into a glass vial and sent for various characterization analysis such as GCMS, pour point, viscosity, cloud point, density, TAN and oxidative stability. The biodiesel obtained following gravitational settling was characterized by using two complementary techniques: gas chromatography- mass spectrometry (GC-MS) (Agilent, Santa Clara, CA, USA) and DSC (TA Instruments-Waters, LLC, New Castle, DE, USA). This combined approach provided comprehensive information on the biodiesel's composition (GC-MS) and thermal properties (DSC). biodiesel. Upon completing the SAC process, the biodiesel which is the liquid phase was extracted from the cylindrical vessel and its volume was measured. In addition, the extracted liquid was also placed into a glass vial and sent for various characterization analysis such as GCMS, pour point, viscosity, cloud point, density, TAN and oxidative stability. The biodiesel obtained following gravitational settling was characterized by using two complementary techniques: gas chromatography- mass spectrometry (GC-MS) (Agilent, Santa Clara, CA, USA) and DSC (TA Instruments-Waters, LLC, New Castle, DE, USA). This combined approach provided comprehensive information on the biodiesel's composition (GC-MS) and thermal properties (DSC).

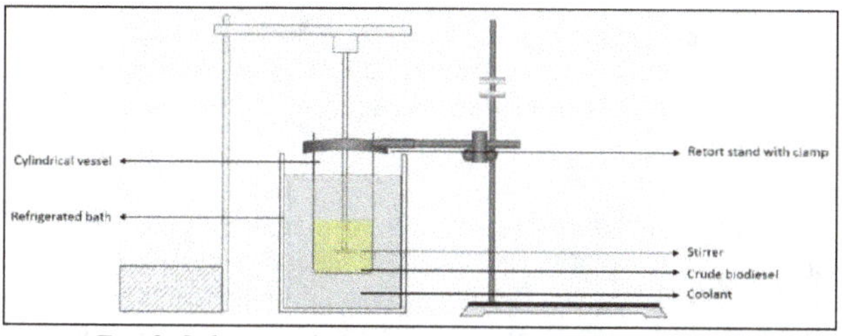

Fig. 2. Solvent-aided crystallization method setup [4]

Parameter Screening. Parameter screening experiments were performed using crude biodiesel from Coconut Oil as feed. The aim of these parameters screening experiments was to determine the range of working conditions for the equipment and to obtain information on how best to improve the equipment in order to attain a higher purity of biodiesel. During the parameters screening experiments, the coolant temperature, crystallization time and shaking speed were tested on the range between -3 to 9°C, 15 to 35 minutes and 181 rpm to 261 rpm respectively. Therefore, the minimum range for coolant temperature, crystallization time and shaking speed are -3°C, 15 minutes and 181 rpm, respectively. Meanwhile, the maximum range for coolant temperature, crystallization time and shaking speed are 9°C, 35 minutes and 261 rpm, respectively. The parameters along with their range of investigation applied in the experiments are presented in Table 1.

Table 1 Ranges of Parameters for Stirring SAC

Parameter	Range
Coolant Temperature (°C)	-3 to 9
Crystallization Time (minutes)	15 to 35
Shaking Speed (rpm)	201 rpm to 261 rpm

Investigations on the Effect of Operating Parameter. The focus of stirring SAC is to produce high purity, yield and excellent quality of biodiesel that adhere to biodiesel fuels standards of ASTM D6751 and EN 14214. The purity of biodiesel was determined by using GCMS analysis. The yield of biodiesel was measured based on the volume of liquid biodiesel obtained after SAC process while the quality of biodiesel was characterized based on its pour point, viscosity value, cloud point, density value and TAN. Based on screening experiment explained earlier, the constant value of 3°C for the coolant temperature, 25 minutes for crystallization time and 221 rpm for stirring speed were chosen.

Temperature is a key factor in the crystallization process of biodiesel components in Solvent Aided Crystallization (SAC). The optimal coolant temperature range is -3°C to 9°C, which maximizes the biodiesel crystallization process and prevents FAMEs from becoming less soluble in the solvent. The upper limit of 9°C balances effective crystallization with manageable handling conditions. Stirring rate is crucial for homogenous mixtures, promoting consistent nucleation and crystal development. Research shows that reasonable stirring rates balance crystal growth and mixing efficiency. The shaking speed range of 201 to 261 revolutions per minute (rpm) is chosen for efficient mixing and consistent crystallization. The amount of time needed to crystallize biodiesel impacts its yield and purity. Longer crystallization durations can enhance separation and purity by allowing contaminants to stay in the solution. However, prolonged crystallization durations can lead to secondary crystallization effects or crystal agglomeration, negatively impacting yield and making the separation process more difficult.

Efficiency Assessment. The efficiency of a stirring SAC system was assessed from the biodiesel purity (%) and biodiesel yield (ml). In addition, these parameters were studied to allow the tracking of the stirring SAC system.

Results and Discussions

Investigations on the Effect of Operating Parameter: As previously mentioned, the focus of stirring SAC is to produce high purity, yield and quality of biodiesel that meet the specification of biodiesel fuels standards of ASTM D6751 and EN 14214. The purity and yield were evaluated based on three different parameters which were coolant temperature, crystallization time and stirring speed. And the effect of each parameter was discussed individually.

Effect of Coolant Temperature. This subsection investigated the impact of coolant temperature on the SAC process. The coolant temperature varied from -3 °C to 9 °C in 3 °C increments, while the crystallization time and stirring rate were held constant at 25 minutes and 221 rpm, respectively. Coolant temperature is a critical factor influencing the rate and extent of solid glycerol crystallization, ultimately affecting biodiesel purity and yield [8, 9]. The relationship between biodiesel purity and yield at different coolant temperatures is depicted in Figure 3 (insert reference for the figure). Biodiesel yield exhibited a generally increasing trend with rising coolant temperature. The yield started at 426 mL at -3 °C and gradually increased to reach a maximum of 490 mL at 9 °C. In contrast, the trend observed for biodiesel purity was less consistent. A slight decrease in purity was observed from 99.39% at -3 °C to 99.24% at 0 °C. However, purity increased to 99.68% at 3 °C. This was followed by another decrease to 99.36% at 6 °C, and a subsequent sharp increase to 100% at the highest coolant temperature of 9 °C. The observed inconsistency in the purity trend warrants further investigation. Experiment repetition under identical conditions is necessary to confirm these findings and obtain statistically significant data.

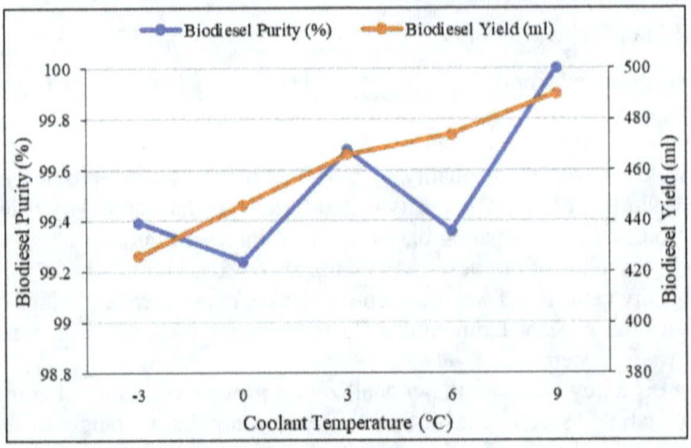

Fig. 3. Effect of Coolant Temperature on Biodiesel Purity and Yield

Effect of Crystallization Time. This subsection examined the influence of crystallization time on the SAC process. Crystallization time was varied from 15 to 35 minutes in 5-minute increments, while the coolant temperature and stirring rate were maintained at 3 °C and 221 rpm, respectively. Crystallization time is a critical parameter that governs the growth and properties of solid glycerol crystals, ultimately impacting biodiesel purity and yield. The trends observed in biodiesel purity and yield with varying crystallization time exhibited some inconsistencies (Figure X, insert reference for the figure). Biodiesel purity initially increased from 99.60% at 15 minutes to 100% at 20 minutes. However, a gradual decrease followed, with purity reaching 99.04% at 35 minutes. Biodiesel yield also displayed a fluctuating trend. It decreased from 458 mL at 15 minutes to 443 mL at 20 minutes. Interestingly, a significant increase was observed at 25 minutes, reaching 466 m. This was followed by a sharp decline to 432 mL at 30 minutes, with a slight recovery to 438 mL at 35 minutes. These somewhat inconsistent results suggest the need for further investigation to fully understand the interplay between crystallization time and SAC performance. Future studies could explore a wider range of crystallization times or employ more controlled experimental conditions.

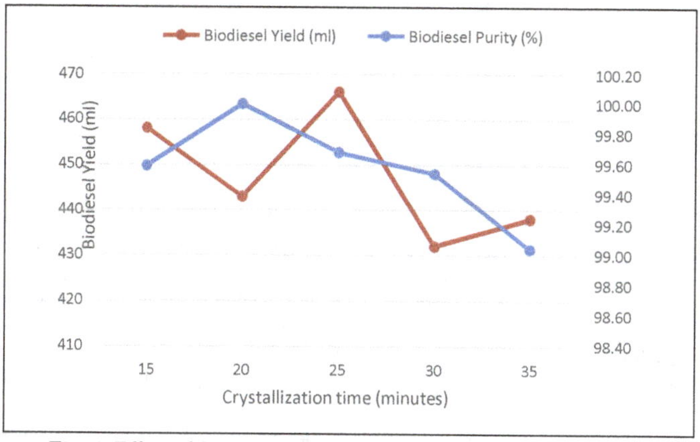

Fig. 4. Effect of Crystallization time on Biodiesel Purity and Yield

Effect of Stirring rate. This section investigated the impact of stirring rate on the SAC process. The stirring rate was varied from 181 to 261 rpm at increments of 20 rpm, while the coolant temperature and crystallization time were held constant at 3 °C and 25 minutes, respectively. Stirring rate is a crucial factor influencing the formation and morphology of solid glycerol crystals, which can ultimately affect biodiesel purity and yield. The relationship between biodiesel purity and yield at different stirring rates is depicted in Figure 3. The biodiesel yields initially exhibited a slight decrease from 432 mL at 181 rpm to 429 mL at 201 rpm. However, a significant increase in yield was observed at 221 rpm, reaching 466 ml. This was followed by a sharp decrease to 448 mL at 241 rpm, with a final minor increase to 453 mL at 261 rpm. Biodiesel purity displayed a less defined trend across the stirring rate range. A slight reduction in purity was observed from 98% to 100% as the stirring rate increased from 181 rpm to 241 rpm. However, a more substantial decrease occurred between 241 rpm and 261 rpm, with purity dropping from 100% to 90%. It is important to note that the observed inconsistencies in biodiesel purity warrant further investigation. To confirm these findings and obtain more statistically significant data, experiment repetition is necessary under identical conditions.

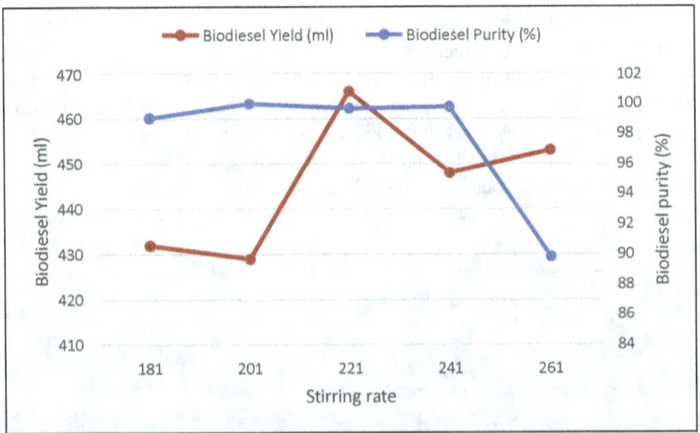

Fig. 5. Effect of Stirring rate on Biodiesel Purity and Yield

Conclusion

This study addressed a critical knowledge gap in the biodiesel production industry by investigating the application of SAC for purifying coconut oil biodiesel. Through a series of experiments and analyses, we demonstrated the effectiveness of SAC in achieving high biodiesel purities. This approach has the potential to significantly enhance the quality and efficiency of coconut oil-derived biodiesel production.

These findings contribute to the development of sustainable and environmentally friendly alternatives to traditional diesel fuels, aligning with the ongoing pursuit of renewable energy solutions. This research not only expands the scientific understanding of biodiesel purification using SAC but also provides valuable insights for future industrial applications, ultimately promoting a greener and more sustainable future.

The study investigated the impact of three key operating parameters on SAC effectiveness: stirring rate, crystallization time, and coolant temperature. The results revealed that all tested conditions yielded high biodiesel purities ranging from 89.83% to 100%. Coolant temperature variation exhibited the most significant effect on purity, followed by crystallization time. Stirring rate variation showed the least impact on achieving high purity.

Limitations and Future Work

The study encountered limitations due to the use of partially malfunctioning equipment. These included an unstable temperature controller and a variable-speed stirring motor, both of which contributed to data inconsistencies. Additionally, time constraints limit our ability to repeat experiments for complete result verification. Future research efforts will benefit from employing well-maintained equipment and allocating sufficient time for experiment repetition to confirm findings and obtain even more accurate results.

Acknowledgements

The authors would like to acknowledge the financial assistance from PETRONAS via YUTP-FRG (Cost Centre: 015LC0-378), and the facilities support from HICoE – Centre for Biofuel and Biochemical Research (CBBR), and Chemical Engineering Department, Universiti Teknologi PETRONAS, support from the Ministry of Education Malaysia through the HICoE award to CBBR is duly acknowledged.

References

[1] M. Berrios and R. L. Skelton, "Comparison of purification methods for biodiesel," *Chemical Engineering Journal,* vol. 144, no. 3, pp. 459-465, 2008/11/01/ 2008. https://doi.org/https://doi.org/10.1016/j.cej.2008.07.019

[2] L. C. Meher, D. Vidya Sagar, and S. N. Naik, "Technical aspects of biodiesel production by transesterification—a review," *Renewable and Sustainable Energy Reviews,* vol. 10, no. 3, pp. 248- 268, 2006/06/01/ 2006. https://doi.org/https://doi.org/10.1016/j.rser.2004.09.002

[3] M. A. Ahmad and S. Samsuri, "Biodiesel Purification via Ultrasonic-Assisted Solvent-Aided Crystallization," *Crystals,* vol. 11, no. 2, p. 212, 2021. [Online]. Available: https://www.mdpi.com/2073-4352/11/2/212

[4] W. N. A. Wan Osman, N. A. I. Badrol, and S. Samsuri, "Biodiesel Purification by Solvent-Aided Crystallization Using 2-Methyltetrahydrofuran," *Molecules,* vol. 28, no. 3, p. 1512, 2023. [Online]. Available: https://www.mdpi.com/1420-3049/28/3/1512

[5] M. Lapuerta, J. Rodríguez-Fernández, D. Fernández-Rodríguez, and R. Patiño-Camino, "Cold flow and filterability properties of n-butanol and ethanol blends with diesel and biodiesel fuels," *Fuel,* vol. 224, pp. 552-559, 2018/07/15/ 2018. https://doi.org/https://doi.org/10.1016/j.fuel.2018.03.083

[6] E. Alptekin and M. Canakci, "Determination of the density and the viscosities of biodiesel–diesel fuel blends," *Renewable Energy,* vol. 33, pp. 2623-2630, 12/01 2008. https://doi.org/10.1016/j.renene.2008.02.020

[7] H. Siva Subramanian, "Examination of Thermo-physical properties of Ethanol and Gasoline numerically and its behaviour at different temperatures," *International Journal of Advanced Research,* vol. 4, 10/31 2016. https://doi.org/10.21474/IJAR01/2006

[8] W. N. M. Wan Ghazali, R. Mamat, H. H. Masjuki, and G. Najafi, "Effects of biodiesel from different feedstocks on engine performance and emissions: A review," *Renewable and Sustainable Energy Reviews,* vol. 51, no. C, pp. 585-602, 2015. [Online]. Available: https://EconPapers.repec.org/RePEc:eee:rensus:v:51:y:2015:i:c:p:585-602

Preliminary synthesis of the TiO_2 photocatalyst for hydrogen production

NUR AIN ATIQAH Mohd Amin[1,2], HAYYIRATUL FATIMAH Mohd Zaid[1,2,*]

[1]Department of Chemical Engineering, Universiti Teknologi PETRONAS, Perak, Malaysia

[2]Centre of Innovative Nanostructure and Nanodevoces (COINN), Universiti Teknologi PETRONAS, Perak, Malaysia

*hayyiratul.mzaid@utp.edu.my

Keywords: Photoelectrochemical Cell, Titanium Dioxide, Calcination Temperature, Photocatalytic Activity

Abstract. The preliminary analysis of the proposed work, specifically the development of the titanium dioxide (TiO_2) photocatalyst, was described in this chapter. The results of the analysis are useful for comparing the composite structures of TiO_2 photocatalysts with different calcination temperatures. Various crystalline and amorphous structures of TiO_2, such as anatase, rutile, and brookite, have garnered significant attention for their diverse applications. This study investigates the synthesis, characterization, and performance evaluation of TiO_2 photocatalysts for hydrogen production applications. TiO_2, known for its stability, non-toxicity, and low cost, has garnered significant interest due to its unique properties. Utilizing various experimental methods including sol-gel hydrothermal synthesis and calcination at different temperatures, the researchers synthesized anatase and rutile TiO_2 photocatalysts. Characterization techniques such as thermogravimetric analysis (TGA), X-ray diffraction (XRD), field emission scanning electron microscopy (FESEM), surface area analysis, electrochemical impedance spectroscopy (EIS), and linear sweep voltammetry (LSV) were employed to analyze the structural and performance properties of the photocatalysts. Results revealed that calcination temperature significantly influenced the crystallinity, particle size, surface area, and pore structure of TiO_2, with an optimal temperature of 500°C yielding superior photocatalytic activity. Electrochemical analysis indicated smoother photocharge carrier transfer kinetics and enhanced photo electrocatalytic activity for TiO_2 calcined at 500°C, highlighting its potential for efficient hydrogen production. This comprehensive investigation contributes valuable insights into the development of sustainable TiO_2 photocatalysts for clean energy applications.

Introduction

In various experimental studies in recent years, titanium dioxide (TiO_2) has emerged as a crucial experimental material, primarily due to its remarkable properties, such as unique stability, non-toxicity, recyclability, low cost, and unique photochemical and photophysical benefits, particularly in hydrogen production. It is extensively used in waste treatment applications, including water purification, biological sensors, and solar cells [1], [2], [3]. Inorganic materials of different sizes and specific structural properties, such as nanotubes, nanorods, flower-like structures, fan-shaped structures, and microspheres (whether hollow, solid, or porous), have gained significant attention for their innovative characteristics [4], [5].

TiO_2 has emerged as a promising catalyst for hydrogen production through various methods, including photocatalysis, photoelectrochemical (PEC) water splitting, and thermochemical processes [6]. These approaches leverage the unique properties of TiO_2 to drive the decomposition of water into hydrogen and oxygen, offering a sustainable and renewable source of clean energy.

Materials Research Proceedings 53 (2025) 584-595 https://doi.org/10.21741/9781644903575-59

Photocatalytic water-splitting using TiO_2 involves harnessing solar energy to initiate the reaction, where TiO_2 nanoparticles act as catalysts to facilitate the conversion of water into hydrogen and oxygen [7]. This process holds immense potential for addressing the global energy demand while mitigating environmental concerns associated with conventional fossil fuel-based energy sources. Researchers are actively exploring strategies to enhance the efficiency and stability of TiO_2 photocatalysts by modifying their structure, composition, and surface properties, thereby improving their performance for large-scale hydrogen production applications [8].

TiO_2 exhibits various crystalline forms, including anatase, rutile, and brookite, each characterized by distinct structural properties. Anatase and rutile are tetragonal, while brookite is orthorhombic, with all featuring the TiO_6 octahedral configuration but differing in bonding arrangements. The unit cell constants for rutile (a=4.59, c=2.96) and anatase (a=3.79, c=9.51) highlight their structural differences. Rutile comprises linear chains with oxygen atoms at chain corners, while anatase lacks these corner oxygen atoms, resulting in unique tetragonal configurations [9]. As shown in Fig. 1, each Ti^{4+} ion in the crystal is surrounded by six O^{2-} ions. In rutile crystals, the octahedral structure is slightly orthorhombic, while in anatase TiO_2, the structure is significantly distorted to a symmetry less than orthorhombic. The bond lengths between Ti-Ti in anatase (3.79 Å and 3.04 Å) are longer than in rutile (3.57 Å and 2.96 Å), whereas the Ti-O bond lengths in anatase (1.934 Å and 1.980 Å) are shorter than in rutile (1.949 Å and 1.980 Å) [10]. These structural differences are the main reason anatase TiO_2 is more reactive than rutile TiO_2, leading to variations in electronic band structures and bulk densities. Rutile TiO_2, with the highest refractive index among the three phases, is commonly used in paint raw materials [10].

The anatase crystal form exhibits the highest photocatalytic activity compared to other crystal forms [11]. Consequently, anatase TiO_2 is the focus of a significant portion of research studies. Calcination temperature plays a crucial role in determining the properties and performance of TiO_2 photocatalysts for hydrogen production. TiO_2 is one of the most widely studied photocatalysts due to its high stability, low cost, and non-toxicity. In addition, studies have shown how calcination temperature affects structural morphology and photocatalytic activity of various semiconductors [12], [13]. Calcination temperature influences the particle size and morphology of TiO_2 nanoparticles. Higher calcination temperatures tend to result in larger particle sizes and changes in morphology. The size and morphology of nanoparticles significantly impact their surface area, light absorption, and charge transfer properties, all of which affect the photocatalytic activity. Lower calcination temperatures usually result in higher surface areas of TiO_2 nanoparticles due to less agglomeration. A higher surface area provides more active sites for photocatalytic reactions, potentially enhancing hydrogen production rates. Calcination temperature also influences the crystallinity of TiO_2. Higher calcination temperatures typically lead to increased crystallinity, which can affect photocatalytic activity. An optimal crystalline structure is essential for efficient charge carrier generation and separation. TiO_2 exists in different crystal phases, such as anatase, rutile, and brookite. Calcination temperature can influence the phase composition of TiO_2. The anatase phase is usually more favorable for photocatalytic activity due to its higher surface energy and better charge separation properties [14]. In this paper, we present a sol-gel hydrothermal approach with varying calcination temperatures for synthesizing anatase and rutile TiO_2. This study aims to advance fundamental research on various synthesis methods and to investigate and detail the photocatalytic properties of these two forms of TiO_2.

Materials Research Proceedings 53 (2025) 584-595 https://doi.org/10.21741/9781644903575-59

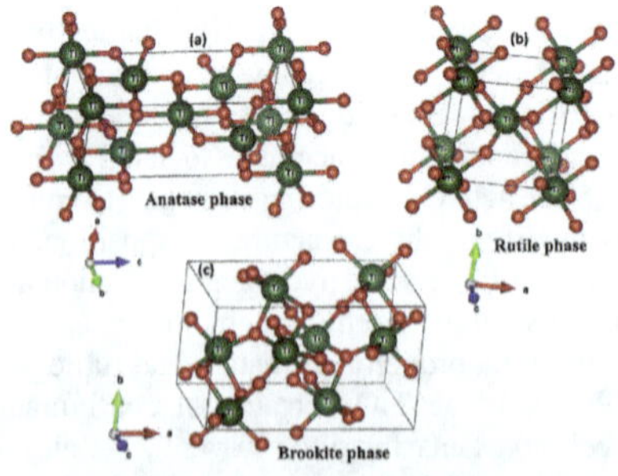

Figure 1 Three basic polymeric crystal structures

Experimental

Synthesizing the TiO_2 photocatalyst. Monometallic and bimetallic photocatalysts were synthesized using titanium isopropoxide (TTIP) through a hydrothermal process followed by calcination. The typical synthesis procedure involved adding 27 mL of TTIP mixed with 100 mL of n-heptane (Fig. 2). The mixture was stirred at 400 rpm for 3 hours, after which 7.98 mL of distilled water was added dropwise to achieve a TTIP:H_2O molar ratio of 1:5. The mixture was then left to age for 72 hours at room temperature. The resulting sol suspension was transferred to a 100 mL Teflon-lined autoclave and heated to 240°C for 12 hours. The synthesized photocatalyst particles were separated by centrifugation at 4,000 rpm for 15 minutes and rinsed three times with a mixture of deionized water and ethanol. The precipitate was dried in an oven at 70°C for 12 hours. Post drying, the TiO_2 was calcined in a furnace at 400°C, 500°C, and 600°C in ambient air for 12 hours, with a temperature ramp of 4°C/min. The TGA profile in Fig. 3 was used to determine the calcination temperatures. This calcination process was designed to eliminate remaining organic materials and facilitate the conversion of hydroxides into anatase TiO_2. The optimal calcination temperature was chosen based on the XRD patterns of the photocatalyst presented in Fig 4. The photocatalysts are labeled as TiO_2_T, with 'T' indicating the calcination temperature in °C. For instance, TiO_2_500 denotes TiO_2 calcined at 500°C.

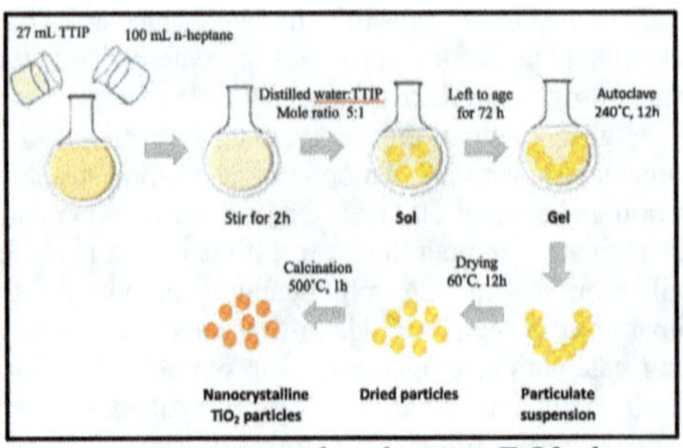

Fig. 2 Schematic representation of synthesizing TiO2 photocatalyst

Fabrication of TiO_2 Photoanode. The FTO glass substrate thin film (15 Ω/square) was cleaned using detergent, acetone, alcohol, and distilled water. This was followed by sintering at 400°C to eliminate organic compounds and improve adhesion for the printed films. The application of the first paste layer using the Doctor Blade method and subsequent drying in a belt furnace at sequential belt temperatures of 75, 85, 95, and 85°C with a speed of 125 mm/min to the glass paste printing in a belt furnace (Hengli Eletek Co. Ltd.) at sequential belt temperatures of 75, 85, 95, and 85°C with a speed of 125 mm/min. Additionally, the sintering process in a high-temperature belt furnace under airflow, with defined temperature zones: zone 1 at 300°C for 5 minutes, zone 2 at 400°C for 10 minutes, zone 3 at 350°C for 5 minutes, zone 4 at 270°C, and cooling steps at 180°C for 5 minutes, mirrors the sintering studies conducted with copper inks on FTO coated glass substrates. Lastly, the use of a conductive silver ink strip before the final sintering process is reminiscent of the conductive circuit preparation process on glass-based circuit boards.

Characterization of TiO_2 Photocatalyst. For the pre-treatment of the photocatalysts, thermogravimetric analysis (TGA) was performed using the Perkin Elmer TG (Pyris 1) system to determine the approximate decomposition temperature of the photocatalysts and to estimate the appropriate calcination temperature and duration. The structural properties of the synthesized nanomaterials were analyzed using X-ray diffraction (XRD, Cu-Kα radiation, Bruker AXS D8) with a scan range (2θ) from 10° to 80°, a step size of 0.02°, and a scan speed of 4°C/min. This analysis aimed to assess the dispersion of the dopant into TiO_2. The photocatalysts' morphologies, including their crystallite shape, size, and metal dispersion, were investigated using field emission scanning electron microscopy (FESEM, Carl Zeiss, Supra 55 VP) along with energy-dispersive X-ray analysis (EDX). The prepared photocatalysts' morphologies were observed with an electron high tension (EHT) of 10 kV, a working distance of 2-4 mm, and a magnification of 100 kX. The specific surface area, pore volume, and pore size distribution were determined using a surface area analyzer and porosity analysis (SAP) based on the nitrogen adsorption-desorption method (Micromeritics Tristar II). The effective area of the photoelectrode was 1.0 cm^2. Electrochemical impedance spectroscopy (EIS) measurements were conducted on the photoelectrochemical (PEC) cell using a Metrohm Autolab potentiostat (PGSTAT 128N) under open-circuit conditions, with a frequency range from 0.1 to 100 kHz, utilizing a frequency response analyzer (FRA 32M).

Results and Discussion

Characterization of TiO_2 Photocatalyst. Determining the optimal calcination temperature and duration is essential for transforming the as-synthesized amorphous TiO_2 into the active anatase phase while preventing the formation of the inactive rutile phase. Figure 3 presents the thermogram of TiO_2 prepared via the sol-gel hydrothermal method. The first weight loss of 11%, occurring between 50 and 155°C, signifies the elimination of physically adsorbed water and the organic solvent, n-propanol. The second weight loss of 10%, between 155 and 240°C, corresponds to the combustion of alkoxide groups attached to the Ti atom. A third, slower weight loss of 5% occurs from 240 to 580°C, indicating the conversion of titanium hydroxide ($Ti(OH)_4$) into anatase TiO_2. The total observed weight loss is 26%, with a plateau indicating no further weight loss. This thermogram curve helped estimate a minimum calcination temperature of 350°C for the as-synthesized TiO_2, leading to the selection of calcination temperatures between 400 and 600°C.

The TGA thermogram for TiO_2 calcined at 500°C for 1 hour is shown in Fig. 3(b). It indicates a weight loss of only 5.14% between 50 and 400°C, which is attributed to the physically absorbed moisture and the removal of various chemisorbed hydroxyl groups present in the photocatalyst [16]. The anatase phase begins to appear around 400°C and fully transforms by 500°C. When the calcination temperature exceeds 500°C, the anatase phase converts to the rutile phase (TiO_2 (anatase) → TiO_2 (rutile)), as confirmed by the XRD results in Fig. 4. Therefore, heat treatment of the amorphous powder at 500°C for 1 hour is sufficient to remove any residual entities in the photocatalyst.

Materials Research Forum LLC
https://doi.org/10.21741/9781644903575-59

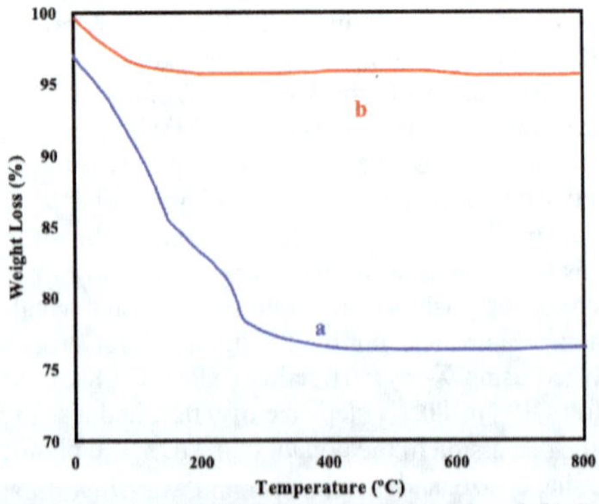

Fig. 3 TGA curves of (a) TiO2_as synthesized and (b) TiO2_500

The XRD patterns of as-synthesized and calcined TiO_2 are shown in Fig. 4, compared with commercial TiO_2 (Degussa P25). The as-synthesized TiO_2 exhibited a very weak anatase phase. However, as the calcination temperature rose, the intensity of the anatase peaks increased accordingly. Major peaks were observed at 24.43°, 36.89°, 49.21°, 53.29°, 61.84°, 68.21°, and 74.31°, corresponding to the reflection planes (101), (004), (200), (105), (211), (204), and (215), respectively, confirming the presence of a well-crystallized pure anatase TiO_2 phase. Peaks at 30.56° and 37.04°, corresponding to the reflection planes (110) and (111), indicate a minor rutile TiO_2 phase. All photocatalysts displayed a preferred orientation with higher peak intensity corresponding to the (101) plane. It is well known that XRD peaks broaden as grain size decreases. Therefore, the broad diffraction peaks for photocatalysts calcined at 400°C indicate a smaller crystallite size compared to those calcined at 600°C, where narrower peaks indicate larger particle sizes due to the sintering effect. The average crystallite size was calculated using the Scherrer Equation (eq 1).

$$d = k\lambda/(FWHM \cos\theta) \quad (1)$$

In eq 1, where d represents the crystallite size, k is a dimensionless shape factor typically around 0.89, λ denotes the X-ray wavelength of Cu (0.5406 nm), θ stands for the Bragg angle corresponding to the anatase (101) peak, and B represents the line broadening at half the maximum intensity (FWHM). Upon calcination of the photocatalyst at 600°C, a rutile phase (2θ = 27.56° (110)) emerged, indicating complete crystallization of the anatase phase. This suggests that the phase transformation from anatase to rutile in the as-synthesized TiO_2 occurred above 500°C, consistent with the reported range of 500 to 600°C. The conversion from anatase to rutile phase is generally unfavorable as it diminishes the photocatalytic activity. Watthanaatun et al. reported pure anatase formation at 500°C [17], while Lopez et al. indicated that the transition from anatase to rutile began at a calcination temperature of 550°C, with complete conversion to the rutile phase achieved at 900°C [18]. Considering these findings, 500°C is chosen as the optimal calcination temperature for further investigations.

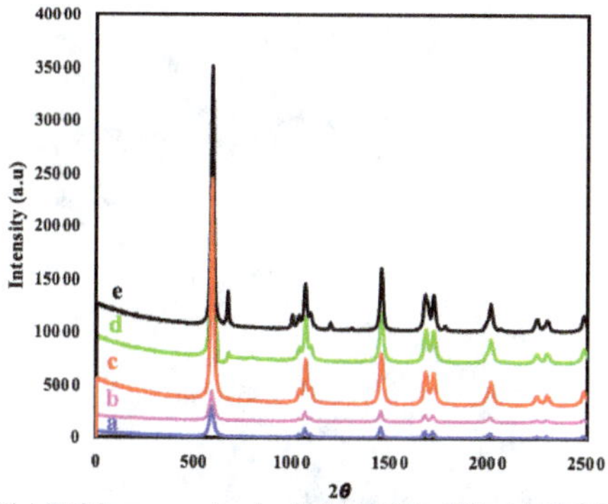

Fig. 4 XRD patterns of (a) TiO2_as synthesized, (b) TiO2_400, (c) TiO2_500, (d) TiO2_600, and (e) P25 Degussa

Table 1 presents the calculated crystallite sizes for all photocatalysts before and after undergoing calcination at various temperatures. The crystallite size exhibited a rise with increasing calcination temperature. However, at 600°C, the appearance of a rutile phase suggested that the chosen temperature was excessively high.

Table 1 Average crystallite (XRD) and particle sizes (FESEM) of photocatalyst photocatalysts prepared

Photocatalyst	Average crystallite size (nm)		Average Particle size (nm)
	Anatase	Rutile	
P25	33.10	37.85	33.14
TiO_2_as-synt	11.27	-	9.11
TiO_2_400	17.91	-	18.20
TiO_2_500	21.22	-	21.26
TiO_2_600	32.38	4.09	33.27

FESEM images of TiO_2 subjected to varying calcination temperatures are presented in Fig. 5 (a-d). The particle size of P25 ranged from approximately 33.10 to 37.85 nm. The FESEM micrographs (Fig. 4 (b-d)) illustrate relatively uniform structures for TiO_2_400 (b), TiO_2_500 (c), and TiO_2_600 (d). The particle size increased with increasing calcination temperature, as observed for TiO_2_600, due to the sintering of small particles into larger ones. These findings align with the values reported in Table 1.

FESEM combined with EDX (FESEM-EDX) allows for the determination of the elemental composition of the photocatalyst. Fig. 5 (e) displays the EDX spectrum of the photocatalyst TiO_2_500. In the spectrum, each component may exhibit multiple peaks, each represented by the $K_{\alpha 1}$ transition. However, only transitions with the highest signal are considered when calculating the composition for each element.

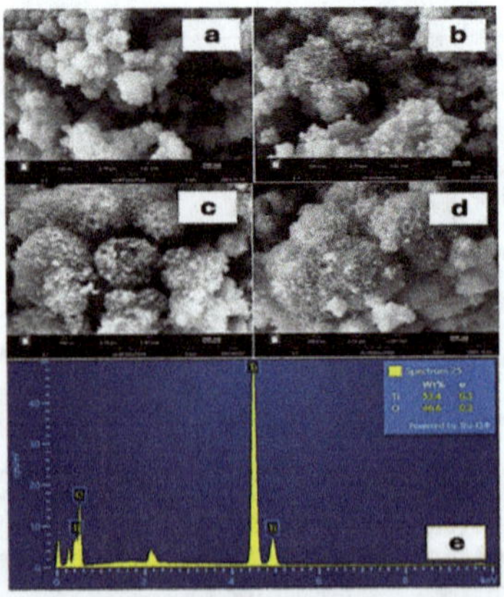

Fig. 5 FESEM micrographs of (a) TiO2_as synthesized, (b) TiO2_400, (c) TiO2_500 and (d) TiO2_600, and (e) EDX spectrum of TiO2_500

Table 2 Element composition of TiO2_500 photocatalyst

Element composition	Weight %	Atomic %
O K	46.6	72.31
Ti K	53.4	27.69
Total	100	100

Table 2 presents the elemental composition of the photocatalyst. To provide representative data, the EDX spectrum was obtained from three different locations. The data acquired from the instrument is presented in terms of weight % and atomic %. These terms, weight %, atomic %, and mol%, are commonly used in catalysis studies and are interrelated. Overall, the elemental composition determined by EDX mapping is slightly lower than the theoretical values for titanium. It's important to note that EDX mapping analyzes only a very small portion of the photocatalyst, and the results are significantly influenced by the homogeneity of the sample.

Fig. 6 illustrates the hysteresis loops of the synthesized TiO_2 materials before and after calcination at 400, 500, and 600°C. The loops exhibit the classical type IV N hysteresis, indicating the presence of mesoporous structures as per the IUPAC classification of porous materials. The inflection points shift to a higher P/Po range of 0.8-0.9 for TiO_2 calcined at 400°C, resulting in a relatively large hysteresis loop due to capillary condensation in the mesoporous channels. At higher calcination temperatures of 500°C and 600°C, the N adsorption and desorption curves substantially increase to the P/P_o range of 0.8-1.0. The volume adsorbed at calcination temperatures of 400°C and 500°C is similar but decreases for TiO_2_600. These results suggest a decrease in adsorbed volume with increasing calcination temperature, leading to a reduction in surface area. The pore distribution of TiO_2 was determined using the Barrett-Joyner-Halenda (BJH) method. From Table 3, it is evident that all synthesized TiO_2 materials were mesoporous. According to the IUPAC definition, porous materials are categorized into three classes:

microporous (pore size <2 nm), mesoporous (pore size 2-50 nm), and macroporous (pore size >50 nm) materials.

The BET surface areas of all photocatalysts are detailed in Table 3. Increasing the calcination temperature led to a decrease in both surface area and pore volume due to particle size enlargement and enhanced agglomeration caused by sintering within the material. Conversely, pore diameter expanded with rising calcination temperature. Mesoporous TiO_2 spheres consist of nanocrystalline TiO_2 aggregates with mesopores situated intercrystallite. Elevated calcination temperatures prompted crystallite size enlargement, thereby increasing pore diameter as small pores collapsed and larger ones expanded. Crystallite sizes were measured at 18.13 nm, 21.22 nm, and 32.38 nm for respective temperatures. The composite morphology featured agglomerated granular particles, with agglomeration intensifying as temperature increased [19].

Table 1 SAP analysis of the TiO2 photocatalyst

Photocatalyst	BET Surface Area (m^2/g)	Pore Volume (cm^3/g)	Pore Diameter (nm)
TiO_2_as-synt	112.2354	0.7811	12.3329
TiO_2_400	90.5691	0.4303	18.9865
TiO_2_500	58.5629	0.2802	19.1139
TiO_2_500	56.6881	0.2718	25.7267

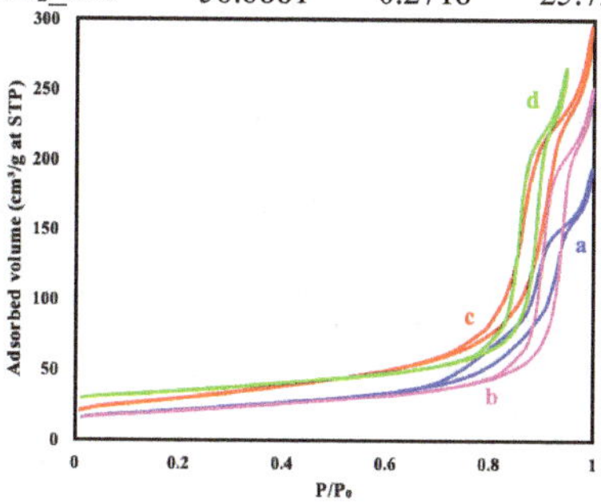

Figure 6 N2 adsorption-desorption isotherm of (a) TiO2_as synthesized, (b) TiO2_400, (c) TiO2_500, (d) TiO2_600

Electrochemical Impedance Analysis. Electrochemical impedance spectroscopy (EIS) allows for the investigation of interfacial charge transport phenomena in the PEC cell. Fig. 7 depicts the Nyquist plot of the TiO2 photocatalysts synthesized at various calcination temperatures assessed at 0.7 V vs. Ag/AgCl, with a bias applied voltage of 10 mV [20]. The small semicircle observed in the high-frequency range represents the charge transfer resistance, which is associated with charge transfer at the interfaces of electrolyte/Pt counter electrode and FTO/TiO2.

The arc radius observed on the Nyquist plot can indeed serve as an indicator of interface layer resistance and reflect the reaction rate at the surface of the electrode. A smaller arc radius typically indicates lower photo charge transfer resistance, suggesting a more efficient charge transfer process. This phenomenon is closely related to the detailed relationship between arc duration, contact resistance, and electrode surface profile [21]. It was noted that the arc radius was the smallest for TiO2_500, while TiO2_as-synthesised exhibited the largest arc radius compared to other photocatalysts. Consequently, the measured charge transfer resistance (Rct) values for TiO2_as synthesized, TiO2_400, TiO2_500, and TiO2_600 photocatalysts were 21.6, 11.3, 7.2, and 9.4 Ω, respectively. A higher Rct value suggests slower charge transfer kinetics, often associated with sluggish reaction rates or increased impedance at the electrode-electrolyte interface. Hence, these findings imply that the photo charge carriers' transfer kinetics in TiO2_500 photocatalyst was smoother and more efficient, resulting in improved photo electrocatalytic activity. The presence of higher oxygen vacancy sites on TiO2_500 photocatalyst, indeed contributes to additional chemisorption reactions between water molecules and the photoelectrode, consequently reducing the resistance to photo charge carrier transfer [22]. This phenomenon enhances the photocatalytic performance by facilitating efficient charge separation and transfer processes [23]. Moreover, The Nyquist plot displaying only one semicircle suggests that Faradaic charge transfer at the photoanode interface is the primary limiting factor for the photoelectrochemical (PEC) performance, emphasizing the significance of efficient charge transfer mechanisms in achieving enhanced PEC efficiency [24]. The synergy between oxygen vacancies and the photoanode interface plays a crucial role in optimizing the photocatalytic activity of TiO2 materials for various applications, including water splitting and pollutant degradation.

Linear sweep voltammetry (LSV) is a crucial method for assessing electrode performance in photoelectrochemical cells designed for hydrogen production. Photocurrent density analysis was conducted via LSV in a three-electrode system, covering a potential scan range of -0.2 to 1.0V vs RHE under AM 1.5 light illumination, as shown in Fig. 7. In line with the EIS Nyquist analysis, the TiO_2_500 photocatalyst achieved the highest photocurrent density of 1.88 mA/cm^2 at 1.23 V vs RHE. This indicates a significant enhancement in photocurrent density with the optimal calcination temperature of TiO_2, suggesting a notable reduction in the rate of photocharge carrier recombination. In comparison, TiO_2_as-synthesized, TiO_2_400, and TiO_2_600 photocatalysts exhibited photocurrent density performances of 0.23, 0.41, and 1.71 mA/cm^2, respectively. The photocurrent density for all photocatalysts increased toward the water-splitting potential, indicating an expansion in the space charge layer at the semiconductor/electrolyte interface and a faster electron transfer rate. The outstanding performance observed in the TiO_2_500 photocatalyst can be attributed to enhanced photocharge carrier transfer kinetics due to minimal photocharge carrier resistance. Additionally, the large pore diameter and volume, as identified in the SAP analysis, facilitate high accessibility for the migration of photocharge carriers through porous channels in the TiO_2_500 photocatalyst.

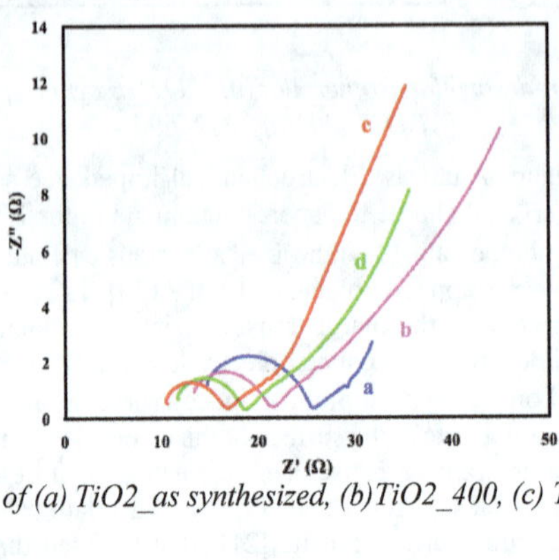

Fig. 7 EIS Nyquist plot of (a) TiO2_as synthesized, (b)TiO2_400, (c) TiO2_500, (d) TiO2_600

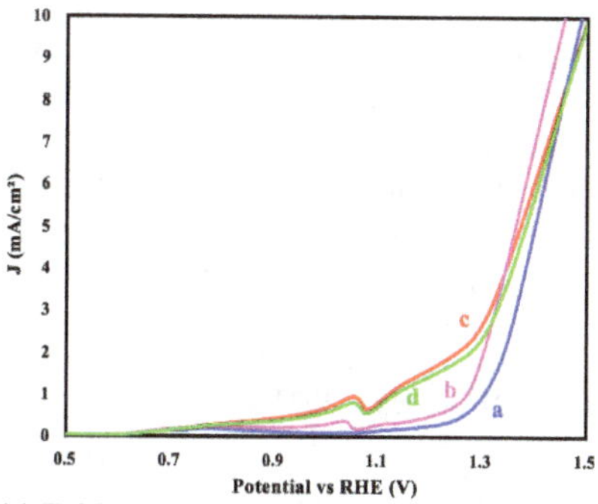

Fig. 8 LSV curves of (a) TiO2_as synthesized, (b)TiO2_400, (c) TiO2_500, (d) TiO2_600

Conclusion

In conclusion, the XRD, BET, and FESEM analyses indicate that 500°C is the optimal calcination temperature, yielding a pure anatase phase with high crystallinity, moderate particle size, and adequate surface area. This structural optimization leads to enhanced photocatalytic performance as evidenced by the electrochemical analyses. The pure anatase phase at 500°C is crucial for high photocatalytic activity due to its better charge separation properties than the rutile phase. The BET and FESEM analyses show that the specific surface area and particle morphology significantly impact the photocatalytic efficiency. Higher surface area at lower calcination temperatures provides more active sites, while appropriate particle sizes facilitate better light absorption and charge transfer.

Acknowledgment

The authors gratefully acknowledge Universiti Teknologi PETRONAS (UTP) for providing financial support through the Graduate Research Assistant (GRA) Scheme. This research was funded by Yayasan Universiti Teknologi PETRONAS (YUTP) - 015LC0-501.Literature

References

[1] Ebrahimi, M., Zakery, A., Karimipour, M., & Molaei, M. (2016). Nonlinear optical properties and optical limiting measurements of graphene oxide-Ag@TiO2 compounds. Optical materials, 57, 146-152. https://doi.org/10.1016/j.optmat.2016.04.039

[2] Le, L., Xu, J., Zhou, Z., Wang, H., Xiong, R., & Shi, J. (2018). Effect of oxygen vacancies and Ag deposition on the magnetic properties of Ag/N co-doped TiO2 single-crystal films. Materials Research Bulletin, 102, 337-341. https://doi.org/10.1016/j.materresbull.2018.01.045

[3] Wang, T., Wei, J., Shi, H., Zhou, M., Zhang, Y., Chen, Q., & Zhang, Z. (2017). Preparation of electrospun Ag/TiO2 nanotubes with enhanced photocatalytic activity based on water/oil phase separation. Physica E: Low-dimensional Systems and Nanostructures, 86, 103-110. https://doi.org/10.1016/j.physe.2016.10.016

[4] Ye, M. M., Chen, Z. L., Wang, W. S., Zhen, L., & Shen, J. M. (2008). Template-free hydrothermal preparation of mesoporous TiO2 microspheres on a large scale. Chemistry letters, 37(9), 938-939. https://doi.org/10.1246/cl.2008.938

[5] Zhang, Q., Ge, J., Goebl, J., Hu, Y., Lu, Z., & Yin, Y. (2009). Rattle-type silica colloidal particles prepared by a surface-protected etching process. Nano Research, 2, 583-591. https://doi.org/10.1007/s12274-009-9060-5

[6] Yu, M., Liang, H., Zhan, R., Xu, L., & Niu, J. (2021). Sm-doped g-C3N4/Ti3C2 MXene heterojunction for visible-light photocatalytic degradation of ciprofloxacin. Chinese Chemical Letters, 32(7), 2155-2158. https://doi.org/10.1016/j.cclet.2020.11.069

[7] Liu, W., Sun, M., Ding, Z., Gao, B., & Ding, W. (2021). Ti3C2 MXene embellished g- C3N4 nanosheets for improving photocatalytic redox capacity. Journal of Alloys and Compounds, 877, 160223. https://doi.org/10.1016/j.jallcom.2021.160223

[8] Zhou, Y., Yu, M., Zhan, R., Wang, X., Peng, G., & Niu, J. (2021). Ti3C2 MXene-induced interface electron separation in g-C3N4/Ti3C2 MXene/MoSe2 Z-scheme heterojunction for enhancing visible light-irradiated enoxacin degradation. Separation and Purification Technology, 275, 119194. https://doi.org/10.1016/j.seppur.2021.119194

[9] Fujishima, A. K. I. R. A., Rao, T. N., & Tryk, D. A. (2000). TiO2 photocatalysts and diamond electrodes. Electrochimica acta, 45(28), 4683-4690. https://doi.org/10.1016/S0013-4686(00)00620-4

[10] Zhao, J., Bowman, L., Zhang, X., Vallyathan, V., Young, S. H., Castranova, V., & Ding, M. (2009). Titanium dioxide (TiO2) nanoparticles induce JB6 cell apoptosis through activation of the caspase-8/Bid and mitochondrial pathways. Journal of toxicology and environmental health, part A, 72(19), 1141-1149. https://doi.org/10.1080/15287390903091764

[11] Liu, X., Zhou, K., Wang, L., Wang, B., & Li, Y. (2009). Oxygen vacancy clusters promoting reducibility and activity of ceria nanorods. Journal of the American Chemical Society, 131(9), 3140-3141. https://doi.org/10.1021/ja808433d

[12] Fuziki, M. E. K., Brackmann, R., Dias, D. T., Tusset, A. M., Specchia, S., & Lenzi, G. G. (2021). Effects of synthesis parameters on the properties and photocatalytic activity of the magnetic catalyst TiO2/CoFe2O4 applied to selenium photoreduction. Journal of Water Process Engineering, 42, 102163. https://doi.org/10.1016/j.jwpe.2021.102163

[13] Ali, W., Ullah, H., Zada, A., Alamgir, M. K., Muhammad, W., Ahmad, M. J., & Nadhman, A. (2018). Effect of calcination temperature on the photoactivities of ZnO/SnO2 nanocomposites for the degradation of methyl orange. Materials Chemistry and Physics, 213, 259-266. https://doi.org/10.1016/j.matchemphys.2018.04.015

[14] Peng, T., Zhao, D., Dai, K., Shi, W., & Hirao, K. (2005). Synthesis of titanium dioxide nanoparticles with mesoporous anatase wall and high photocatalytic activity. The journal of physical chemistry B, 109(11), 4947-4952. https://doi.org/10.1021/jp044771r

[15] Li, J., Zhu, L., Wu, Y., Harima, Y., Zhang, A., & Tang, H. (2006). Hybrid composites of conductive polyaniline and nanocrystalline titanium oxide prepared via self-assembling and graft polymerization. Polymer, 47(21), 7361-7367. https://doi.org/10.1016/j.polymer.2006.08.059

[16] Funda, A. K., Birlik, I., Keskin, Ö. Y., & Delice, T. K. Improvement of Photocatalytic Degradation of Titanium Dioxide Nanomaterials by Non-metal Doping. Afyon Kocatepe Üniversitesi Fen Ve Mühendislik Bilimleri Dergisi, 23(4), 874-882. https://doi.org/10.35414/akufemubid.1256778

[17] Watthanaarun, J., Pavarajarn, V., & Supaphol, P. (2005). Titanium (IV) oxide nanofibers by combined sol-gel and electrospinning techniques: preliminary report on effects of preparation conditions and secondary metal dopant. Science and Technology of Advanced Materials, 6(3-4), 240-245. https://doi.org/10.1016/j.stam.2005.02.002

[18] Castro-López, C. A., Centeno, A., & Giraldo, S. A. (2010). Fe-modified TiO2 photocatalysts for the oxidative degradation of recalcitrant water contaminants. Catalysis Today, 157(1-4), 119-124. https://doi.org/10.1016/j.cattod.2010.04.050

[19] Usman, A., Aris, A., Labaran, B., Darwish, M., & Jagaba, A. (2022). Effect of calcination temperature on the morphology, crystallinity, and photocatalytic activity of ZnO/TiO2 in selenite photoreduction from aqueous phase. J. New Mater. Electrochem. Syst, 25(4), 251- 258. https://doi.org/10.14447/jnmes.v25i4.a05

[20] Ullah, H., Tahir, A. A., Bibi, S., Mallick, T. K., & Karazhanov, S. Z. (2018). Electronic properties of β-TaON and its surfaces for solar water splitting. Applied Catalysis B: Environmental, 229, 24-31. https://doi.org/10.1016/j.apcatb.2018.02.001

[21] Mou, H., Song, C., Zhou, Y., Zhang, B., & Wang, D. (2018). Design and synthesis of porous Ag/ZnO nanosheets assemblies as super photocatalysts for enhanced visible-light degradation of 4-nitrophenol and hydrogen evolution. Applied Catalysis B: Environmental, 221, 565-573. https://doi.org/10.1016/j.apcatb.2017.09.061

[22] Manik photocatalytic dye degradation and hydrogen evolution performance of Cu encapsulated BiVO4 under visible light irradiation. Chemical Physics Impact, 6, 100178. https://doi.org/10.1016/j.chphi.2023.100178

[23] Safaei, J., Ullah, H., Mohamed, N. A., Noh, M. F. M., Soh, M. F., Tahir, A. A., ... & Teridi, M. A. M. (2018). Enhanced photoelectrochemical performance of Z-scheme g- C3N4/BiVO4 photocatalyst. Applied Catalysis B: Environmental, 234, 296-310. https://doi.org/10.1016/j.apcatb.2018.04.056

[24] Liu, C., Yang, Y., Li, J., Chen, S., Li, W., & Tang, X. (2017). An in-situ transformation approach for fabrication of BiVO4/WO3 heterojunction photoanode with high photoelectrochemical activity. Chemical Engineering Journal, 326, 603-611. https://doi.org/10.1016/j.cej.2017.05.179

Multiphase flow regimes study in microchannel reactor for CO_2 absorption application

ALAM Mohammad Fahmi Al[1,a]*, AZHAR Amiratul Husna[1,b] and LAZIZ Afiq Mohd[1,c]

[1]Universiti Teknologi Petronas, 32610 Seri Iskandar, Malaysia

[a]fahmialalam@gmail.com, [b]husnaazhar136@gmail.com, [c]afiq.laziz@utp.edu.my

Keywords: CO_2 Absorption, Microchannel Reactor, Multiphase Flow

Abstract. The use of sodium hydroxide (NaOH) as the absorbent for carbon dioxide (CO_2) capture is considered interesting in commercial point of view due to the obtained products, sodium carbonate (Na_2CO_3) and sodium bicarbonate ($NaHCO_3$). Conventional absorbers have been used to remove the impurity in the natural gas but the low interfacial area due to their large volume has resulted in poor interfacial mass transport. Microchannel reactor which believed able to improve the mixing and mass transfer has been used to solve the problems. Hence, the purpose of this work is to study the characteristics of multiphase flow regimes in the CO_2-NaOH system and to determine the optimum condition for the CO_2 absorption in the microchannel reactor by varying the gas-to-liquid, G/L ratio and the NaOH concentration. The G/L ratio varied from 2.33 to 9.00 and the NaOH concentration varied from 0.05 M to 2.00 M. The characteristics of the multiphase flow regimes were observed through digital camera by measuring the slug length, distance of consecutive slug and the interfacial area (S/V). The amounts of CO_2 absorbed by the system were estimated based on the quantities of products that existed in the sample to determine the effect of parameters variation. From the experiment, it was found that at highest G/L ratio, the slug length and distance produced was the longest that contributed to the highest S/V due to increase of the gas flowrate. Different from G/L ratio, the increased NaOH concentration would decrease the slug length, distance of consecutive slug, and the interfacial area produced by the slug due to increase of the rate of reaction. As for the effect of G/L ratio and NaOH concentration towards the CO_2 absorption, it was observed that the G/L ratio is inversely proportional to the amount of CO_2 absorbs while the increase of NaOH concentration would increase amount of CO_2 absorbed by the system due to the availability of Na+ to react in the system, therefore the rate of reaction increase.

Introduction

The high concentration of carbon dioxide in the natural gas and fossil fuels product is one of the major issues associated with environmental repercussions such as increased the earth surface temperature and increased the sea level besides affecting the fuel quality [1]. According to Beben and Steliga [2], the presence of CO_2 in natural gas can lead to significant issues such as pressure drops during transmission and the formation of acidic solutions that cause corrosion in gas transmission lines and related equipment which accordingly increase the cost of operation and utilization. Besides, the presence of significant amount of CO_2 will reduce the thermal value of the natural gas. Hence, CO_2 removal is an important step in upstream process for natural gas.

CO_2 absorption is one of the most effective for reducing CO_2 emissions into the atmosphere and have relatively lower operating cost [3]. CO_2 absorption is preferred by using chemical absorption as the process can be done at atmospheric pressure [4]. However, this process is highly dependent on the surface area of the two phases. Therefore, microchannel reactor will be used to solve the problem. Microchannel reactor has a large surface area between phases caused by the capillary force and surface tension, these devices improve mixing, mass transfer, and heat transfer [4].

Absorbent that commonly used for CO_2 absorption is alkanolamine solutions. However, amine solution known to be a toxic solvent that can cause high level of corrosion, high volatility and required high level of energy for the regeneration process. Therefore, sodium hydroxide (NaOH) solution will be used instead. Beside the byproduct of NaOH and CO_2 that is useful for multiple cleaning ingredients, not many research regarding the use of NaOH as CO_2 absorbent has been conducted. NaOH is also a waste product in some chemicals production such as chlorine production which can lower the cost of the process [5]. This research aims to characterize the multiphase flow regime between CO_2 and NaOH inside microchannel reactor in terms of total interfacial area. The optimum condition for CO_2 absorption during the process is also investigated.

Methodology

The experiments were conducted at room temperature, 26°C and atmospheric pressure, 0.1 MPa. The total flowrate of gas and liquid is kept constant at 10 mL/min. The CO_2 flowrate used for the experiment is in the range of 1.0 – 10.0 mL/min and NaOH flowrate = 1.0 - 3.0 mL/min. This experiment was conducted to determine the droplet characteristics and the optimum flow conditions for CO_2 absorption by manipulating the G/L ratio and NaOH concentration. G/L used for this experiment are 9.00 (G = 9 mL/min; L = 1 mL/min), 4.00 (G = 8 mL/min; L = 2 mL/min), and 2.33 (G = 7 mL/min; L = 3 mL/min) while the NaOH concentration used are 0.05 M, 1.00 M, and 2.00 M.

Experiment Setup. The experimental setup that has been used to study the characteristics of the flow regimes in the multiphase flow system is shown in Fig. 3. CO_2 gas flowed from the cylinder and the sodium hydroxide, NaOH solution flowed from the syringe pump. CO_2 and NaOH came in contact in the microchannel reactor for the gas absorption process. The images of the flow formed during the two-phase flow in the microchannel reactor were observed by using the digital camera to determine the droplet length, L and droplet distance, d as shown in Fig. 1. Fig. 2 demonstrated ways to estimate the droplet length and distance by using graph paper. From the information, the total surface area, and the interfacial area which from the ratio of surface-area to volume (S/V) can be calculated. The products were collected to determine the amount of CO_2 absorbed by the system.

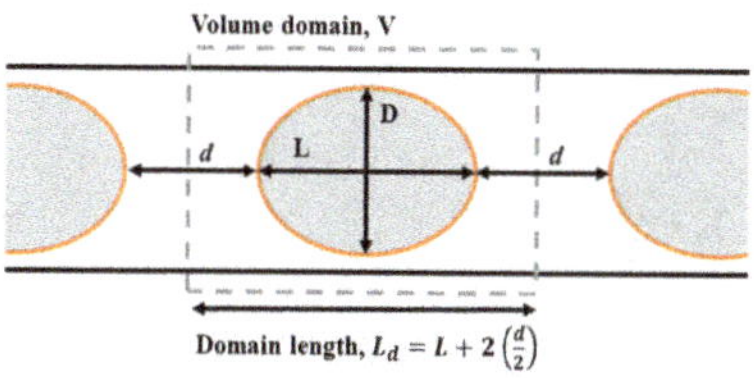

Figure 1 Slug size measurement inside microchannel

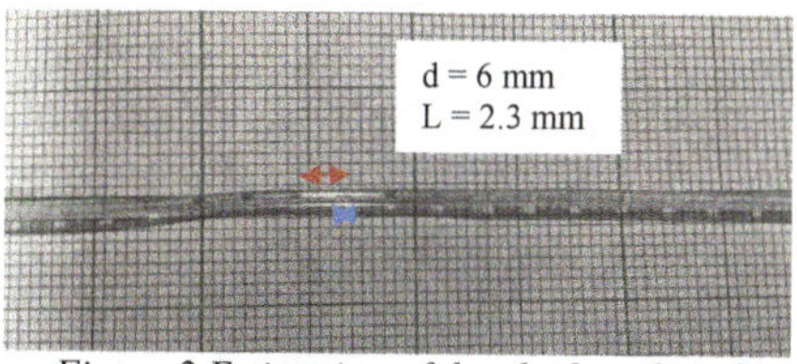

Figure 2 Estimation of droplet length and distance

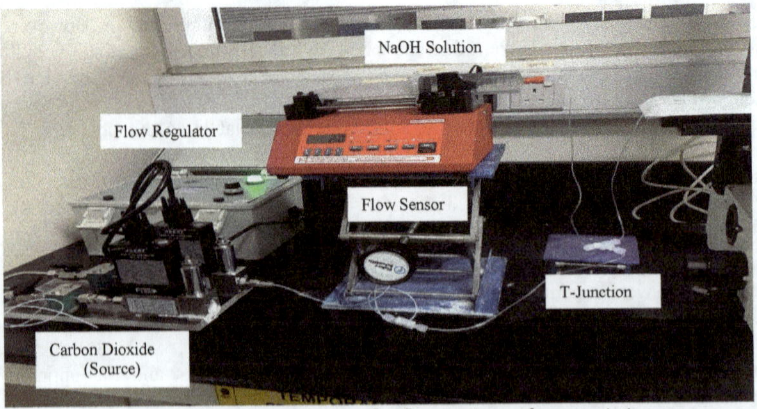

Figure 3 Experimental setup for flow regimes characteristics

The number of droplets produced in the microchannel can be determined by using Eq. 2 [6]. Eq. 4 and Eq. 5 are used to calculate the total surface area and the interfacial area. The constant volume of the microchannel was computed for a total length of 1 m and an interior diameter of 1 mm by using the formula for a cylindrical shape, $V = \pi D^2 L/4$. As for the surface area, it can be calculated based on approximation of capsule-like shape since length, L >>> width, W.

$$Domain\ length, L_d = L + d. \tag{1}$$

$$number\ of\ droplets, N = \frac{Distance\ occupied\ by\ 1\ droplet, L_d}{Total\ length\ of\ microchannel, L_T}. \tag{2}$$

$$Surface\ area\ , s = \pi D^2 + \pi D(D - L). \tag{3}$$

$$Total\ Surface\ Area, A_s = Surface\ Area\ \times number\ of\ droplets. \tag{4}$$

$$Interfacial\ area, S/_V = \frac{Surface\ Area}{Volume}. \tag{5}$$

Determination Amount of Carbon Dioxide Absorb. To determine the amount of CO_2 absorbed by the system, the double titration method also known as Warder's method have been used [7]. The exiting products from the microchannel are expected to contain a mixture of sodium carbonate and sodium bicarbonate. For this titration, hydrochloric acid (HCl) and two indicators which are phenolphthalein and methyl orange have been used. The two indicators were used to indicate the products existing in the sample. According to Nicolas [8], when all NaOH has already reacted, the species presence is carbonate and bicarbonate. Na_2CO_3 reaction with HCl is indicated by using the phenolphthalein while the reaction of $NaHCO_3$ with HCl will be indicated by using the methyl orange indicators. Fig. 4 shows the steps of double titration that been done to determine the amounts of CO_2 absorbed by the CO_2-NaOH system while Table 1 shows the reactions involved for the titration method.

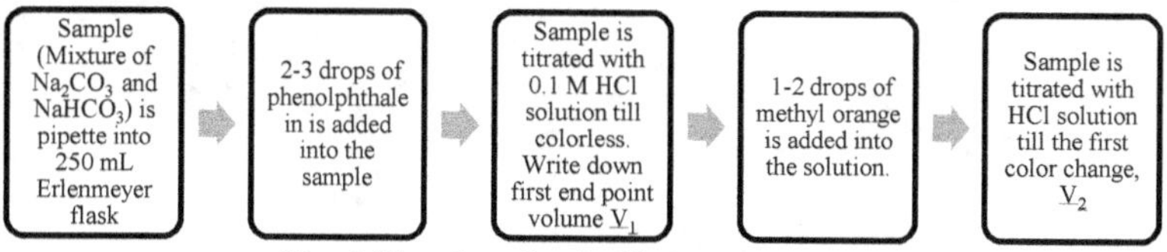

Fig. 4 Flowchart for double Titration method

Table 1 Indicators used to indicate the product in the sample.

Indicators	Reaction
Phenolphthalein	$Na_2CO_3 + HCl \rightarrow NaHCO_3 + NaCl$
Methyl Orange	$NaHCO_3 + HCl \rightarrow NaCl + H_2O + CO_2$

The concentration of the sodium hydroxide and sodium carbonate ion and bicarbonate ion can be calculated by using the Eq. 6, Eq. 7, Eq. 8, and Eq. 9.

$$V_{CO_3^{2-}} = 2 \, . \, V_1. \tag{6}$$

$$V_{HCO_3^-} = V_2 - V_1. \tag{7}$$

$$[CO_3^{2-}] = \frac{[HCl] \, . \, V_{CO_3^{2-}}}{V_s}. \tag{8}$$

$$[HCO_3^-] = \frac{[HCl] \, . \, V_{HCO_3^-}}{V_s}. \tag{9}$$

Droplet Characteristics. To observe the droplets characteristics, the slugs were characterized by the length and distance between consecutive slugs. The experiment was carried out by manipulating the G/L ratio and the concentration of sodium hydroxide solution. Fig. 5 shows the slug length and distance by varying the G/L ratio at different NaOH concentrations. It can be observed that the slug length and slug distance are directly proportional to the G/L ratio. The typical process of slug formation normally starts when the NaOH solution starts absorbing the CO_2 gases at the junction of inlet and main channel. Afiq, et al [6] stated that the increase of the slug length is caused by the higher flow rate of CO_2 that elongates during the slug formation at the junction. It can be observed from the figure, the higher slug length produce was at G/L ratio = 9 when the CO_2 flowrate was at 9 mL/min, the highest compared to others G/L ratio.

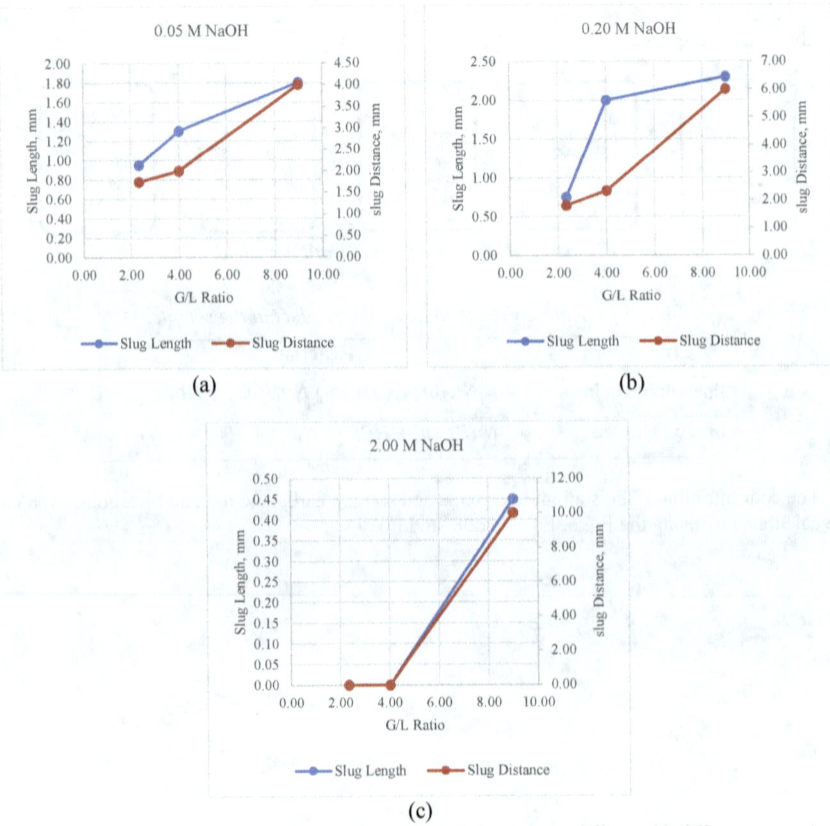

Fig. 5 Effect of G/L ratio to the slug length and slug distance at different NaOH concentration (a) 0.05 M (b) 0.20 M and (c) 2.00 M

Besides the G/L ratio, concentration of the NaOH solution also shows significant effect on the slug length and slug distance. As shown in Fig. 5, the slug length and slug distance are decreasing as the concentration of NaOH increases. The slug length is longer at a fixed axial distance for lower concentrations of NaOH than for higher concentrations. The reason for this is that when the concentration of NaOH increases, the rate of reaction also increases, therefore the rate of absorption increases [7]. When the NaOH concentration is raised to a fixed mixture of gas and liquid flow rates, the slug length at any axial position along the channel is reduced as a result.

The slug length and distance between consecutive slugs are relatively important parameters as it helps in determining the number of slugs formed in a 1 m long microchannel. Another important droplet characteristic that can be observed from the experiment is the total surface area, As. Fig. 6 shows the relation of the number of droplets produced and total surface area of slugs at varied G/L ratio and different NaOH concentration. It can be deduced that the number of droplets and total surface are inversely proportional to the G/L ratio. As the G/L ratio increased from 2.33 to 9.00, the number of droplets produced, and the total surface area of the slug decreased. The highest total surface area was 1451.27 mm2 at G/L ratio = 4.00 and 0.20 M NaOH and 135.30 mm2 is the lowest surface area produced at G/L ratio = 9.00 and 2.00 M NaOH. The higher total surface area indicated that more CO_2 had been exposed to the NaOH solution.

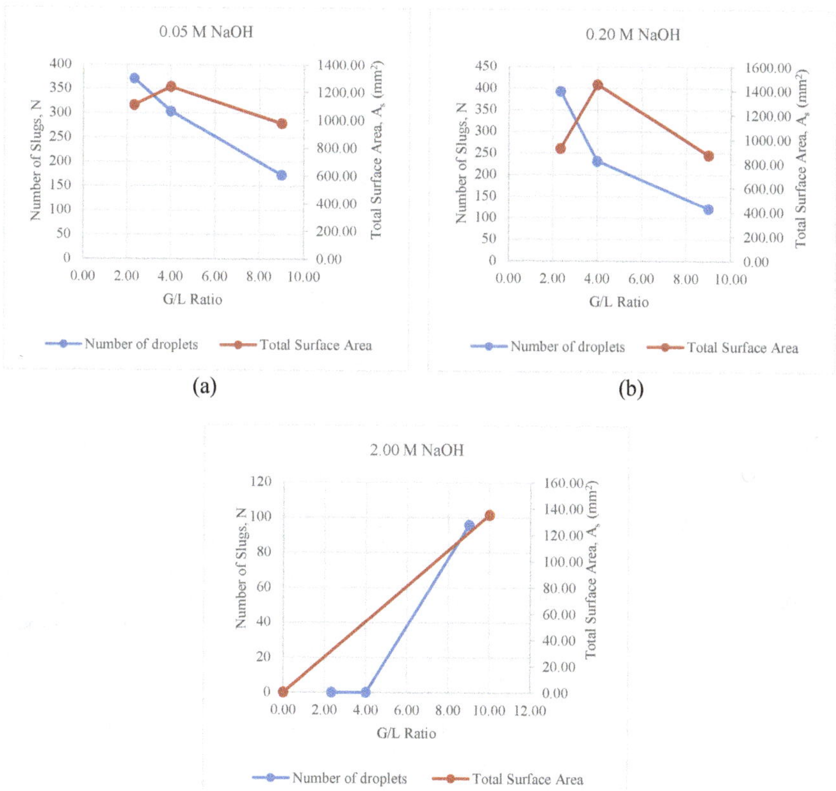

(a) (b)

(c)

Fig. 6 Effect of G/L ratio to the number of slugs produced and total interfacial area of slug at different NaOH concentration (a) 0.05 M (b) 0.20 M and (c) 2.00 M

Besides the results above, the interfacial area was also determined from the surface area-to-volume ratio, S/V value. Fig. 7 shows the S/V comparison for all flow regimes which include different sizes of droplet created by T-junction microchannel. The S/V ratio has a linear relationship with the G/L ratio and is inversely proportional to the NaOH concentration. As shown, the S/V ratio gradually increases when the G/L ratio increases. The S/V values increase with the G/L ratio due to the larger surface area with longer droplet length [9]. Besides G/L ratio, concentration of NaOH also affects the S/V ratio values. As the concentration of the NaOH solution increased, the ratio S/V value decreased. At G/L ratio = 9.00, the highest ratio S/V is 1.8476 mm2/mm3 at 0.20 M NaOH while the lowest S/V value is 0.1722 mm2/mm3 at 2.00 M NaOH.

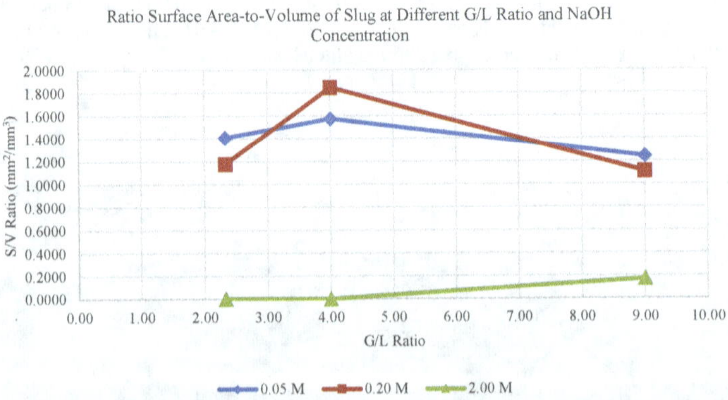

Fig. 7 ratio surface area-to-total volume of slug produced at different G/L ratio

Based on the results, it shows that the solution concentration and G/L ratio significantly affects the droplets characteristics. It can be concluded that the highest G/L ratio produces the longest slug length and distance that contribute to the highest S/V value same as founded by Wang [10]. Different from G/L ratio, the increased NaOH concentration would decrease the slug length, distance of consecutive slug, and ratio S/V produced by the slug due to the rate of reaction have increase.

Carbon Dioxide Absorption by Sodium Hydroxide. To determine the amount of carbonate and bicarbonate produced from the reaction, double titration using phenolphthalein and methyl orange were carried out. Fig. 8 shows the amount of carbonate and bicarbonate ions produced from the reaction of sodium hydroxide with CO_2. Based on the amount of carbonate ions produced, the amount of CO_2 absorbed by the NaOH solution at different concentration and G/L ratio can be estimated. The amount of carbonate and bicarbonate ions produced increase relatively when the concentration of NaOH solution increased and decreased as the G/L ratio increased. As the G/L ratio increased from 2.33 to 9.00, the amounts of carbonate and bicarbonate ions produced decreased at constant NaOH concentration. The increased G/L ratio will reduce the NaOH flowrate and will cause less CO_2 to be absorbed by the system. Different with the NaOH concentration, as the NaOH concentration been increased, more Na+ ions available to reacts with CO_2 gases, hence more CO_2 been absorbed by the system. Durgadevi & Pushpavanam [7] stated that the absorption increases as the concentration increase due to the increase of the driving force. The increase of carbonate products produce indicated more CO_2 been absorbed by the NaOH during the reaction. From Figure 10 above, it can be concluded that the CO_2 been absorbed the most at G/L ratio = 2.33 and 2.00 M by NaOH solution.

Materials Research Forum LLC
https://doi.org/10.21741/9781644903575-60

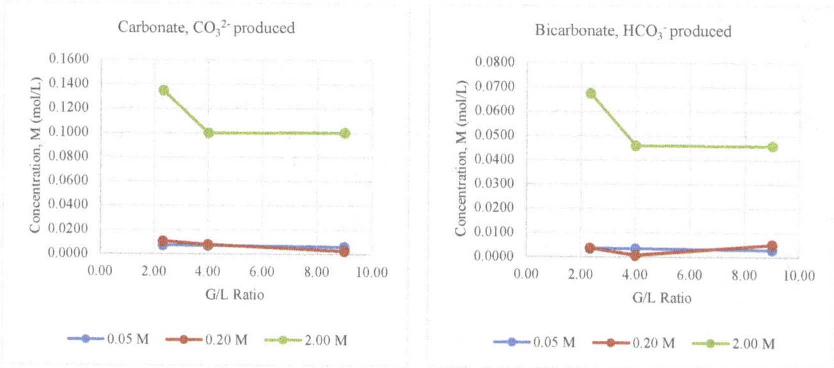

(a) (b)

Fig. 8 Amount of products exist after NaOH absorbed CO_2 gases (a) carbonate and (b) bicarbonate

Conclusion

In conclusion, the changes of gas-to-liquid ratio and sodium hydroxide concentration changed the hydrodynamic flow characteristics in term of interfacial area (S/V) and the amount of CO_2 able to absorb by the system. The interfacial area of the slugs is highest at the highest gas-to-liquid ratio and decrease when the concentration of NaOH solution increase. The highest interfacial area was found at G/L ratio = 9.00 and 0.20 M NaOH which was 1.8476 mm^2/mm^3. As for the amount of CO_2 absorbed by the system based on the amount of product exist in the sample, it was found that more CO_2 been absorbed by the system when G/L ratio is at the lowest and at the concentration of the NaOH the highest due to more Na+ ions been supplied to the system. By knowing the effect of the G/L ratio and NaOH concentration towards the hydrodynamic flow and amounts of CO_2 been absorbed by the system, the optimum condition for the CO_2 absorption in the microchannel can be determine. Based on the experiment, the optimum flow condition to absorb CO_2 in the microchannel by using NaOH as absorbent is at G/L ratio = 2.33 and 2.00 M NaOH.

Reference

[1] N. Akkarawatkhoosith, W. Nopcharoenkul, A. Kaewchada, and A. Jaree, "Mass Transfer Correlation and Optimization of Carbon Dioxide Capture in a Microchannel Contactor: A Case of CO2-Rich Gas," *Energies*, vol. 13, no. 20, p. 5465, Oct. 2020. https://doi.org/10.3390/en13205465

[2] D. Bęben and T. Steliga, "Monitoring and Preventing Failures of Transmission Pipelines at Oil and Natural Gas Plants," *Energies*, vol. 16, no. 18, p. 6640, Sep. 2023. https://doi.org/10.3390/en16186640

[3] P. Chalermthai, N. Akkarawatkhoosith, A. Kaewchada, and A. Jaree, "Carbon dioxide Removal via Absorption using Artificial Seawater in a Microchannel for the Case of CO2-Rich Gas," *Chem. Eng. Process. - Process Intensif.*, vol. 175, p. 108928, May 2022. https://doi.org/10.1016/j.cep.2022.108928

[4] M. Pasha, S. Liu, J. Zhang, M. Qiu, and Y. Su, "Recent Advancements on Hydrodynamics and Mass Transfer Characteristics for CO 2 Absorption in Microreactors," *Ind. Eng. Chem. Res.*, vol. 61, no. 34, pp. 12249–12268, Aug. 2022. https://doi.org/10.1021/acs.iecr.2c01982

[5] Z. Liang *et al.*, "Chlor-alkali membrane cell process for industrial waste salt utilization: Fundamentals and challenges," *Desalination*, vol. 587, p. 117921, Oct. 2024. https://doi.org/10.1016/j.desal.2024.117921

[6] A. Mohd Laziz, K. KuShaari, J. Chin, and J. Denecke, "Quantitative analysis of hydrodynamic effect on transesterification process in T-junction microchannel reactor system," *Chem. Eng. Process. - Process Intensif.*, vol. 140, pp. 91–99, Jun. 2019. https://doi.org/10.1016/j.cep.2019.04.019

[7] D. A and P. S, "An experimental and theoretical investigation of pure carbon dioxide absorption in aqueous sodium hydroxide in glass millichannels," *J. CO2 Util.*, vol. 26, pp. 133–142, Jul. 2018. https://doi.org/10.1016/j.jcou.2018.05.002

[8] N. Cambier, "Carbon dioxide capture using sodium hydroxide solution : comparison between an absorption column and a membrane contactor," Université Catholique de Louvain, 2017.

[9] K. Kushaari, A. M. Laziz, and N. H. Hamid, "Quantitative Analysis of Interfacial Area on Liquid-liquid Multiphase Flow of Transesterification Process in Cross-junction Microchannel Reactor," *J. Eng. Technol. Sci.*, vol. 51, no. 4, pp. 585–596, Aug. 2019. https://doi.org/10.5614/j.eng.technol.sci.2019.51.4.9

[10] L. lin Wang, "Study on effect of gas-liquid two phase physical feature on slug flow in microchannels," *Front. Phys.*, vol. 11, Mar. 2023. https://doi.org/10.3389/fphy.2023.1125220

Decarbonization Technology: ICDT2024 Materials Research Forum LLC
Materials Research Proceedings 53 (2025) 605-615 https://doi.org/10.21741/9781644903575-61

Investigating the role of ionic liquids in CO_2 electrochemical reduction: AI meets quantum chemistry

FASEE Ullah[1,a *], SULAFA Abdalmageed Saadaldeen Mohammed[2,b], RABEEA Jaffari[3,c], WAN ZAIREEN NISA Yahya [4,d] and MANZOOR Ahmed Hashmani[5,e]

[1]Computer and Information Sciences Department, Universiti Teknologi Petronas, Seri Iskander, 32610, Perak, Malaysia

[2]Department of Energy Technology, Aalborg University, 9220, Aalborg, Denmark

[3]Software Engineering Department, Mehran University of Engineering and Technology, Jamshoro, 76260, Sindh, Pakistan

[4]Chemical Engineering Department, Universiti Teknologi Petronas, Seri Iskander, 32610, Perak, Malaysia

[5]Computer and Information Sciences Department, Universiti Teknologi Petronas, Seri Iskander, 32610, Perak, Malaysia

[a]fasee.ullah@utp.edu.my, [b]sasm@energy.aau.dk, [c]rabeea.jaffari@faculty.muet.edu.pk, [d]zaireen.yahya@utp.edu.my, [e]manzoor.hashmani@utp.edu.my

Keywords: CO_2 Electrochemical Reduction, ILs, COSMO-RS, AI, Feature Selection

Abstract. Recently, ionic liquids (ILs) have garnered remarkable attention as electrolytes for CO_2 electrochemical reduction (CO_2ER) due to their unique properties viz. thermal and chemical stability, good CO_2 solubility, and their potential to reduce overpotential. While many researchers have explored the catalytic performance of ILs in CO_2ER, a comprehensive understanding of the parameters affecting the catalytic performance is still absent. Experimental methods for evaluating the catalytic performance have limitations, given the unclear understanding of the reaction mechanism. Recently, Artificial Intelligence (AI) methods have gained increased attention across diverse applications including chemical engineering. These methods play a pivotal role in extracting insights, understanding patterns, and mitigating uncertainty within datasets. In this study, we leverage AI for investigating the critical factors affecting the CO_2ER catalytic performance via Gibbs free energy and capacity. We formulate a novel dataset of 30 electronic and geometric properties of 90 ILs using the Conductor-like Screening Model for Realistic Solvents (COSMO-RS) and TURBOMOLE. Despite the conventional literature that emphasizes the impact of anions on CO_2 solubility and the conventional association of cations as co-catalysts proximal to the negatively charged electrode, our findings underscore the indispensable role of anions, specifically those featuring Sulphur (S), Fluorine (F), and Oxygen (O), in influencing both CO_2 solubility and catalytic processes, despite their relatively far distance from the electrode. Moreover, our study provides an explanation for the difference between solubility trends and catalytic activity, focusing on interaction types. These outcomes will contribute to the more effective selection of ionic liquids for CO_2ER.

Introduction and Related Works

The CO_2 electrochemical reduction (CO_2ER) is a viable solution for the direct conversion of CO_2 to valuable fuels and chemicals at standard conditions [1, 2], leading to potential integration with renewable energy such as wind and solar energy [3] as depicted in Fig. 1.

Ionic liquids (ILs) have received attention as co-catalysts for CO_2ER because of their unique properties, such as low vapor pressures, low heat capacity, liquid state over a wide temperature range, good CO_2 solubility, and high thermal and chemical stability [4]. Much research has been reported on the catalytic performance of ILs in CO_2ER using computational and experimental methods [5-9].

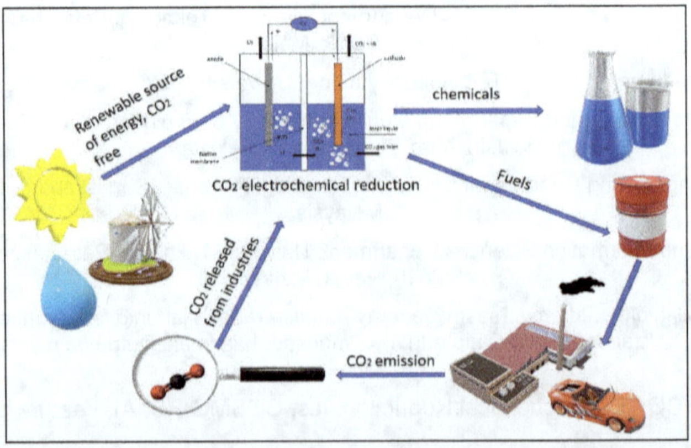

Fig 1. CO_2 Electrochemical Reduction (CO_2ER) Process. Adapted from [3]

On the other hand, several mechanisms have also been suggested to explain the role of ILs in CO_2ER [3, 10]. Lately, computational methods have contributed to elucidating the role of ILs on CO_2ER [9]. This is primarily due to the possibility of studying various ILs at low cost, as well as the ability to detect and study different parameters changes. However, the reaction mechanism and the catalytic roles remain mysterious.

AI is playing a pivotal role in a variety of fields [11, 12] including chemical engineering for optimizing processes through predictive modeling, enhancing material design by analyzing complex molecular structures, and expediting drug discovery through intelligent data analysis [13, 14]. A recent research [15] offers promising results for predicting catalytic activity in CO_2ER using AI machine learning (ML) models. However, the studies [15-17] discussed in this work focus on the adsorption or binding energies of certain intermediates in CO_2ER which could only determine the activity of CO_2ER, but the selectivity or dominating reaction pathways are not distinguishable. Moreover, these works [15-17] do not focus on the ILs but rather on other types of CO_2ER electrocatalysts and are more concerned on the development of the ML models for predicting the adsorption energy without paying any heed to the feature selection process [18] which is essential for determining the effectiveness of the ML model. Hence, in this research we propose to utilize the Gibbs free energy (ΔG) and CO_2 capacity of ILs as catalytic performance indicators and observe the effect of various electronic and geometric properties of ILs on these indicators via AI feature selection.

The direction of a chemical reaction is determined by the change in Gibbs free energy (ΔG), and thus the equilibrium constants [9, 19], the knowledge of its magnitude, and sign can aid in a better understanding of reactions in general, and in the analysis of reaction mechanisms and operation optimization. The activity coefficients ($1/CO_2$ capacity) provide insights into the non-ideality of a solution and serve as a precise indicator of CO_2 solubility [20]. With the presence of high-performance computers and well-established computational methods, the chemical equilibria can be calculated on a regular basis. Thermodynamics can be used to describe reactions in a solution as it calculates the free energy contributions of every reactive species in the gas phase and as well as in the solution separately [19]. CO_2ER occurs in the presence of electrolyte which is considered as co-catalyst in terms of reducing the cell potential (E_{cell}) which is related to the ΔG by (1) [21].

$$\Delta G = -nFE_{cell} \tag{1}$$

where n is the number of electrons involved in the redox reaction and F is the Faraday constant (96,485 coulombs/mole).

The Gibbs free energy for a reaction in solution (ΔG_{reac}) can be calculated through (2) [19].

$$\Delta G_{reac} = \Delta E_{gas} + \Delta G_{therm} + \Delta G_{solv} \tag{2}$$

where the compounds free energy of solvation (G_{solv}), reaction energy in the gas phase (ΔE_{gas}), and the thermodynamics (finite temperature) contribution (ΔG_{therm}) [19] defines the link between gaseous and liquid states in the thermodynamical cycle. ΔE_{gas} can be calculated through the energy difference between the gas phase electronic energies of the product and the reactant using ab initio quantum mechanics (QM) for isolated molecules at zero Kelvin [22]. Ab initio QM can also be used to determine ΔG_{therm}, to obtain the molar thermodynamic functions where the harmonic oscillator and ideal gas approximations are commonly used [22]. TURBOMOLE software [23] can be used for the calculation of Ab initio QM while ΔG_{solv} can be calculated using COSMO-RS [24]. COSMO-RS software is a quantum chemical calculation tool to compute the chemical potential differences of molecules in liquids. It has gained attention because of its ability to predict the thermodynamic properties and behavior of ILs in a variety of applications [24-26].

In this study, the G_{solv} for 90 ILs was obtained through computational modeling using COSMO-RS to have insight about the parameters that enhance the catalytic performance of ILs in CO_2ER using AI feature selection. This is done primarily to investigate the reason why ILs that have good CO_2 solubility usually do not have good catalytic performance. Hence, the contributions of this study are two-fold namely:

- A novel dataset (CO_2ERIL) comprising of electronic and geometric properties of 90 ILs developed via COSMO-RS [24] and TmoleX [27] software.
- Application of AI feature selection on the formulated dataset to determine the parameters affecting the CO_2ER catalytic performance via G_{solv} and CO_2 capacity.

The rest of the paper is organized as follows: section II discusses the research methods employed for this study, section III discusses the results of AI feature selection on the developed dataset, followed by conclusion and directions for future work at the end in section IV.

Research Method

An overview of the methodology is depicted in Fig. 2, followed by a detailed description of all the relevant materials and methods.

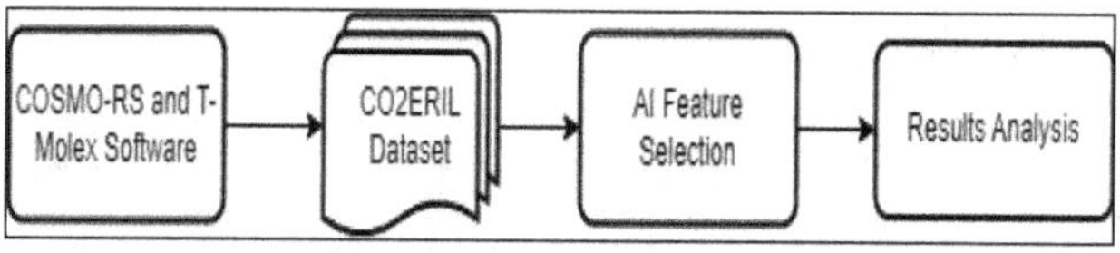

Fig 2. Research Methodology for the Investigation of Factors Affecting the Catalytic Performance of CO_2ER

CO2ERIL Dataset. In this work, we create the novel CO_2ER IL dataset by studying the molecular mechanism of ILs as co-catalyst in CO_2ER using quantum chemical calculation software: COSMO-RS and TmoleX. TmoleX, the GUI for TURBOMOLE, is used to optimize the selected structures and to extract the LUMO and HOMO values of the anions and cations of the ILs. TmoleX is employed based on the density functional theory (DFT) using triple-zeta valence with polarization (TZVP) parametrization. COSMO-RS is used for the evaluation of hydrogen bond interaction energies, free energy of solvation (G_{solv}), and sigma surface. BP-TZVP is selected as the basis set, since it is considered as one of the two "high quality" levels available in COSMO-RS. It uses a COSMO charge surface calculation on a geometry optimized on BP DFT with def-TZVP basis set which is available in TmoleX [19, 22, 23]. Along with the analysis, the COSMO-RS related theory and the corresponding importance of the simulated properties are already discussed by the developer [24-26]. The Sigma surface is employed to visually illustrate the charge distribution on the structure surface. Color coding is utilized to discern the molecule's charge distribution: green denoting a neutral nature, yellow indicating a partial negative charge, blue representing the positive charge region, and red portraying the negative charge region as depicted in Fig. 5 ahead. The activity coefficient at infinite dilution quantified the extent of non-ideal behaviour in the solvents, as expressed in (3).

$$\ln(\gamma i) = (\mu S, \infty i - \mu pi)/\mathrm{RT} \qquad (3)$$

where $\mu S, \infty i$, and μpi are chemical potentials of both solvent and pure compounds, respectively. The inverse activity coefficient of a solute i, obtained at infinite dilution of the solute in the IL solvent ($\gamma IL, \infty i$), is used to describe a solute's capacity. Solubility capacity $CILi$ in an IL solvent is described in (4).

$$CILi = 1/\gamma IL, \infty i \qquad (4)$$

The IL capacity $CILi$ thus corresponds to a non-iterative and unnormalized liquid solubility of solute i in the IL in mole fraction units [mol$i$/molIL]. The IL stoichiometry scales the IL capacity calculated by COSMO-RS. A snapshot of the formulated CO_2ER IL dataset is depicted in Fig. 3. The formulated dataset is a supervised/labelled dataset and comprises of a total of 30 electronic and geometric properties of 90 ILs to gauge their effect on the target variables of catalytic performance: G_{solv} and CO_2 capacity (highlighted by a green border in Fig. 3). The properties include the types of atoms (14 atoms) of cations and anions, the LUMO and HOMO values for the structures, band gaps, total interaction energy, molar volume of cation and anion, density, viscosity, conductivity, misfit interaction, hydrogen interaction, and van der Waals.

	IL	Cation (lumo)	Cation (homo)	Anion (lumo)	Anion (homo)	Band_gap_IL	C (cat)	N (cat)	H (cat)	O (cat)	B (cat)	F (cat)	S (cat)	C (ani)	N (ani)	H (ani)	O (ani)	B (ani)	Capacity	Gsolv (kcal/mol)
2	[AEAMIM][ALA]	-1.8233	5.9202	0.2244	-5.2	3.3787	6	4	18	0	0	0	0	2	1	6	2	0	2.40788957	-607180.2434
3	[ACHMIM][ALA]	-1.9814	-5.9375	0.2244	-5.2	3.2186	6	2	12	1	0	0	0	2	1	6	2	0	1.73304783	-619662.9744
4	[BBT][ALA]	2.3675	7.7733	0.2244	-5.2	2.8325	10	8	20	0	0	0	0	2	1	6	2	0	2.98579107	-671185.3179
5	[DMIM][ALA]	-1.7779	-6.9106	0.2244	-5.2	3.4221	8	2	15	0	0	0	0	2	1	6	2	0	2.95189419	-587038.0011
6	[BMTr8][ALA]	2.9958	7.0079	0.2244	-5.2	2.2042	7	3	14	0	0	0	0	2	1	6	2	0	2.35896844	-597080.2035
7	[DMTr4][ALA]	-2.2733	-7.6751	0.2244	-5.2	2.9267	7	3	14	0	0	0	0	2	1	6	2	0	2.3371492	-597101.3839
8	[EAMIM][ALA]	1.7809	6.5018	0.2244	-5.2	3.4191	6	3	12	0	0	0	0	2	1	6	2	0	2.86656129	-572487.816
9	[DHMIM][ALA]	-1.5022	-6.9741	0.2244	-5.2	3.2178	6	2	11	1	0	0	0	2	1	6	2	0	1.58041782	-584913.8741
10	[EMIM][ALA]	1.7643	6.8949	0.2244	-5.2	3.4357	6	2	11	0	0	0	0	2	1	6	2	0	3.14724897	-587082.5873
11	[EMPe][ALA]	-3.9149	-5.1607	0.2244	-5.2	1.2851	3	5	8	0	0	0	0	2	1	6	2	0	3.09163035	-567838.2383
12	[EMPy][ALA]	2.2015	7.1957	0.2244	-5.2	2.9885	6	2	11	0	0	0	0	2	1	6	2	0	3.05199046	-587002.944
13	[EMTe][ALA]	-3.3932	-8.6286	0.2244	-5.2	1.8068	4	4	9	0	0	0	0	2	1	6	2	0	2.93249966	-537793.8608
14	[EMTr4][ALA]	2.2456	7.9167	0.2244	-5.2	2.9544	5	3	10	0	0	0	0	2	1	6	2	0	2.42300511	-597101.8899
15	[HEAMIM][ALA]	-1.0715	-6.5154	0.2244	-5.2	3.3285	6	3	12	1	0	0	0	2	1	6	2	0	1.68284896	-619662.8928
16	[HEHMIM][ALA]	1.9515	6.0148	0.2244	-5.2	3.2485	6	2	11	2	0	0	0	2	1	6	2	0	1.18635335	-682140.0286
17	[HMIM][ALA]	-1.8352	-6.8818	0.2244	-5.2	3.3648	10	2	19	0	0	0	0	2	1	6	2	0	2.93570057	-636391.8304
18	[METr3][ALA]	2.9928	7.0998	0.2244	-5.2	2.2072	5	3	10	0	0	0	0	2	1	6	2	0	2.5466942	-597080.2035
19	[PMIM][ALA]	-1.7702	-6.8891	0.2244	-5.2	3.4298	7	2	13	0	0	0	0	2	1	6	2	0	2.98805404	-562360.4035
20	[AEAMIM][BF4]	1.8233	5.9202	1.3069	8.028	6.2047	6	4	13	0	0	0	0	0	0	0	0	1	1.14547639	670799.73
21	[ACHMIM][BF4]	-1.9814	-5.9375	1.3065	-8.028	6.0466	6	2	12	1	0	0	0	0	0	0	0	1	0.66959879	-683272.9219
22	[BBT][BF4]	2.3675	7.7755	1.3069	8.028	5.6605	10	8	20	0	0	0	0	0	0	0	0	1	2.550945	714747.8544
23	[illegible]	[illegible]	[illegible]	[illegible]	[illegible]	[illegible]	[illegible]	[illegible]	[illegible]	[illegible]	[illegible]	[illegible]	[illegible]	[illegible]	[illegible]	[illegible]	[illegible]	[illegible]	[illegible]	[illegible]

Fig 3. Sample Screeshot of the Formulated CO_2ERIL dataset

Materials Research Proceedings 53 (2025) 605-615 https://doi.org/10.21741/9781644903575-61

AI Feature Selection. Feature selection is an important concept in ML and statistics as it identifies the most relevant input features which contribute towards achieving the target/output variable [28]. The supervised feature selection techniques can be divided into three broad categories: (1) Filter methods, (2) Wrapper methods, and (3) Embedded methods. Filter methods are the simplest, fastest, and less prone to over-fitting than all the other methods. This is because the filter methods are based on statistical measures and hence, do not require to train any ML model beforehand to identify the most significant features contributing towards the target variable [29]. Since the filter methods are based on statistical measures, these are highly dependent on the data type of the input variables or features. A comprehensive summary of various filter methods and their applicability is illustrated in Fig. 4.

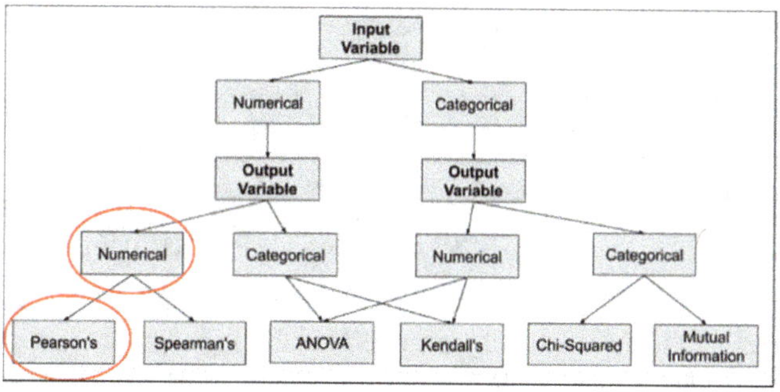

Fig 4. Comprehensive Summary of Various Filter Feature Selection Methods. Adapted from [30]

It should be noted that all the filter methods depicted in Fig. 4 are univariate in nature meaning that the statistical measures employed in these methods are always calculated between one input feature and one target variable at a particular time [29]. Since, our target variables (G_{solv} and CO_2 capacity) are numeric and continuous in nature (as depicted in Fig. 3), hence, we employ the Pearson's correlation coefficient [31] filter method (highlighted by red in Fig. 4) for finding their correlation with the 30 input features one at a time. The Pearson's correlation coefficient finds linear correlation between two sets of data with values ranging from -1 to 1 where -1 signifies a strong negative correlation, 1 signifies a strong positive correlation, and 0 signifies no correlation at all. We calculate the Pearson's correlation coefficient for our CO_2ERIL dataset using sklearn Python library [32]. The results are discussed in Section 3 ahead. The threshold for the results is set to 0.3 and the magnitude is ignored because we are just concerned about finding the existing correlation.

Results and Discussion

The correlation results for G_{solv} are depicted in Table 1.

Materials Research Proceedings 53 (2025) 605-615 https://doi.org/10.21741/9781644903575-61

Table 1: Correlation Coefficients for G_{solv} on CO_2ERIL Dataset

S.#	Input Feature	Correlation Coefficient
1	Band gap (IL)	0.377876
2	Anion (homo)	0.417626
3	S (anion)	0.987037
4	F (anion)	0.802249
5	O (anion)	0.758393
6	H (anion)	0.421032
7	Anion (lumo)	0.832966
8	Molar volume of anion	0.765284
9	Density	0.810226
10	Viscosity	0.537069
11	Conductivity	0.349779
12	Misfit interaction energy	0.517422
13	Hydrogen bond energy	0.436116

The data presented in Table 1 highlights the considerable impact of anions in comparison to cations. This observation can be attributed to the acid/base interaction between CO_2 and the anion. It was reported that CO_2 exhibits dual properties, acting as both an acid and a base depending on the surrounding conditions [33]. This behaviour arises from the electron-accepting properties of the carbon atom and the electron-donating tendencies of the oxygen atom. However, the stronger electron-accepting ability renders it relatively more acidic as a gas. Given that CO_2 is recognized as an acidic gas [21], and the anion in ILs is considered a base according to the Lewis acid–base theory (where an acid is an electron pair acceptor and a base is an electron pair donor), elucidates the significant impact of anions in this context. From Table 1, the following observations can be made:

- HOMO and LUMO have significant effects on G_{solv}. Based on the frontier molecular orbital theory [34], the overlap between the donor's HOMO and the acceptor's LUMO can be used to gauge how intense the donor–acceptor interactions are, which also correlated with the acid/base interactions.
- Misfit and Hydrogen bond interactions significantly affect the G_{solv}. This observation further supports the influence of acid/base interactions on catalytic performance.
- The molar volume of the anion exhibits significant effects. This aligns with the existing literature [35], suggesting that CO_2 tends to occupy the available space between ions and thus increase the interaction between CO_2 and the anion of the ILs. From Table 1, it can also be noted, that the density and viscosity of the ILs, highly influence the G_{solv}, since higher density is often associated with higher viscosity in liquids. And since the viscosity is a measure of a liquid's resistance to flow. The higher viscosity can impede the movement of gas molecules through the liquid and reduce the interaction energy, which is also supported by the available literature [36].
- The G_{solv} is notably influenced by the types of atoms. The fact that the distribution of polarity can be significantly influenced by the types of atoms is also depicted in Fig. 5.

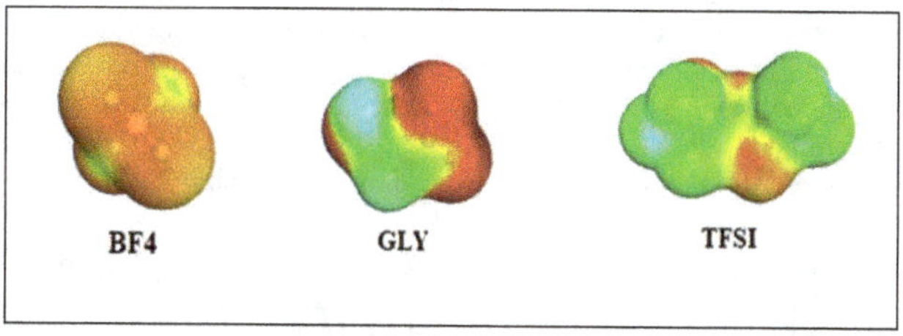

Fig 5. Sigma Surface for the Chosen Cations and Anions by Employing COSMO-RS

Table 1 also shows that the sulphur highly influences the G_{solv} which enhances the catalytic performance, which can explain the result that showed that TFSI-based ILs showed better catalytic performance than BF4-based. The The experimental results from [21] were used to further validate the use of G_{solv} as a good indication for ILs' catalytic performance as described in Table 2 and illustrated in Fig. 6 collectively.

Table 2: The Gibss Free Energy And Co2 Reduction Potential

S.#	ILs	Gibbs Free Energy (G_{solv})	CO_2 Reduction Potential
1	py12	-82.535	-2.0
2	py124	-81.086	-2.025
3	py123	-80.55	-2.05
4	py1234	-79.42	-2.075
5	py1235	-78.89	-2.1

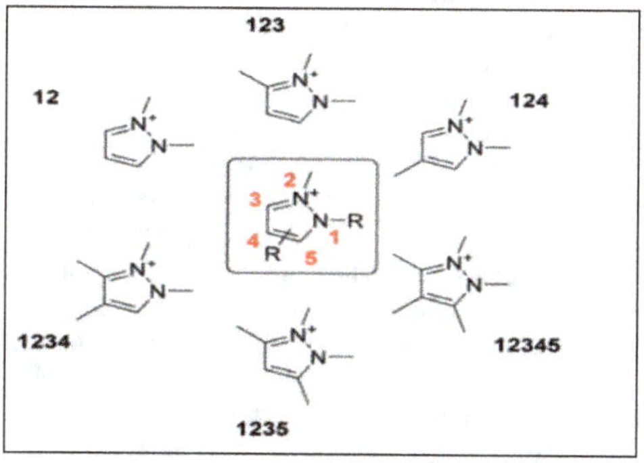

Fig 6. The Structures Used in the Study [21]. Also described in Table 2

The results showed a linear relationship between the reduction potential and Gibbs free energy with (R^2=0.95). It was reported [37] that the anions' CO_2ER activity results could not demonstrate the correlation with the increasing solubility of CO_2 in different anions. In our previous study [38], it can be observed that the type of CO_2 absorption in ILs is highly dependent on molecular interactions. In physical absorption, non-polar/non-polar interactions dominate, whereas in chemical absorption, acid/base interactions are predominant. It can be concluded that the trend of CO_2 capacity is not same as the activity when the nonpolar/nonpolar interaction between the ILs and CO_2 is dominant.

The correlation results for CO_2 capacity are depicted in Table 3.

Table 3 Correlation Coefficients for CO_2 capacity on CO_2ERIL Dataset

S.#	Input Feature	Correlation Coefficient
1	Band gap IL	0.500392
2	Anion (homo)	0.499378
3	O (ani)	0.515251
4	C (cat)	0.30
5	Anion (lumo)	0.461186
6	C (ani)	0.426023
7	N (ani)	0.532049
8	H (ani)	0.320938
9	B (ani)	0.640810
10	Total interaction energy	0.329089
11	Molar volume of anion	0.376808
12	Misfit interaction energy	0.574340
13	Vander Waal's	0.554530

From Table 3, it can be observed that the alkyl chain length of the cations and the Vander Waal's force have special effects on CO_2 capacity. This because CO_2 can be considered as non-polar because the two polar bonds in carbon dioxide (C=O) are symmetrically aligned in opposite orientations due to its linear geometry. As a result, the bond dipoles have equal magnitudes and oriented in the opposite directions, creating a net dipole moment of zero and enhancing London dispersion force.

It can be concluded that, in nonpolar molecules, the primary force at play is the London dispersion force, which is a Vander Waal's force arising from temporary fluctuations in electron distribution creating temporary dipoles. This force is weaker than other intermolecular forces, allowing CO_2 molecules to weakly attach to ILs and enhance CO_2 absorption. However, it does not lead to strong chemical interactions, which is crucial for reducing CO_2 stability and, consequently, enhancing catalytic performance and reduce the overpotential.

Moreover, it was reported that the activity coefficients provide insights into the non-ideality of a solution, considering interactions between solute and solvent molecules [20]. It can be noted that the activity coefficient ($1/CO_2$ capacity) is more accurate as an indication for CO_2 capacity compared to G_{solv}, since the activity coefficient considers all the types of molecular interactions.

Conclusion and Future Work

In summary, while much research has traditionally emphasized the impact of anions on CO_2 solubility, with cations often viewed as co-catalysts due to their proximity to the negatively charged electrode, our study challenges this norm. We demonstrated that anions play a crucial role in both CO_2 solubility and catalysis. Specifically, anions containing S, F, and O exhibit robust catalytic effects for Gibbs free energy. Additionally, we offered an explanation for the observed disparity between solubility trends and catalytic activity, attributing it to the influence of interaction types. Notably, we introduced the use of COSMO-RS for the first time to predict the catalytic performance of CO_2 electroreduction. Our findings indicate that Gibbs free energy serves as a highly effective parameter for predicting catalytic performance. A robust linear regression (R^2=0.95) was achieved when comparing predicted Gibbs free energy data to experimental results for CO_2 reduction potential.

As far as future work is concerned, we intend to explore various other types of AI feature selection methods on our CO_2ERIL dataset and compare the results with our current findings. We also intend to train various ML models on our CO_2ERIL dataset using the significant features highlighted in this study for predicting the catalytic performance.

Acknowledgment
The authors acknowledge the Center of Research in Ionic Liquis (CoRIL) and High-Performance Cloud Computing Centre (HPC3) at Universiti Teknologi PETRONAS (UTP), Malaysia for their support to complete this research study. The authors would also like to express their gratitude to International Centre of Theoretical Physics (ICTP), Italy for providing the required scientific research training: High Performance Computing for Sustainable Development (smr 3828). This research work is a part of an ongoing research project jointly funded by UTP, Malaysia and Kyushu Institute of Technology (Kyutech), Japan.

References

[1] B. M. Tackett, E. Gomez, and J. Chen, "Net reduction of CO2 via its thermocatalytic and electrocatalytic transformation reactions in standard and hybrid processes," J Nature Catalysis, vol. 2, no. 5, pp. 381-386, 2019. https://doi.org/10.1038/s41929-019-0266-y

[2] M. Aresta, A. Dibenedetto, and A. Angelini, "Catalysis for the valorization of exhaust carbon: from CO2 to chemicals, materials, and fuels. Tec[1hnological use of CO2," Chemical reviews, vol. 114, no. 3, pp. 1709-1742, 2014. https://doi.org/10.1021/cr4002758

[3] S. A. S. Mohammed, W. Z. N. Yahya, M. A. Bustam, and M. G. Kibria, "Elucidation of the Roles of Ionic Liquid in CO2 Electrochemical Reduction to Value-Added Chemicals and Fuels," molecules, vol. 26, no. 22, p. 6962, 2021. [Online]. Available: https://www.mdpi.com/1420-3049/26/22/6962. https://doi.org/10.3390/molecules26226962

[4] K. Ghandi, "A review of ionic liquids, their limits and applications," Green sustainable chemistry vol. 2014, 2014. https://doi.org/10.4236/gsc.2014.41008

[5] Y. Zhao, D. Cao, X. Xu, X. Zhang, and Y. Hu, "Ionic Liquid [C3mim] OTf aqueous solution: Green High Efficiency Electroreduction for Carbon Dioxide at Room-Temperature," Microchemical Journal p. 106559, 2021. https://doi.org/10.1016/j.microc.2021.106559

[6] Y. Yang et al., "Aromatic Ester-Functionalized Ionic Liquid for Highly Efficient CO2 Electrochemical Reduction to Oxalic Acid," ChemSusChem, vol. 13, no. 18, pp. 4900-4905, 2020. https://doi.org/10.1002/cssc.202001194

[7] Y. Oh and X. Hu, "Ionic liquids enhance the electrochemical CO2 reduction catalyzed by MoO2," Chemical Communications, vol. 51, pp. 13698-13701, 2015, doi: 10.1039/c5cc05263g. https://doi.org/10.1039/C5CC05263G

[8] H. Liu, Q. Huang, W. An, Y. Wang, Y. Men, and S. Liu, "Dual-atom active sites embedded in two-dimensional C2N for efficient CO2 electroreduction: A computational study," Journal of Energy Chemistry, vol. 61, pp. 507-516, 2021. https://doi.org/10.1016/j.jechem.2021.02.007

[9] A. Rendon-Calle, S. Builes, and F. Calle-Vallejo, "A brief review of the computational modeling of CO2 electroreduction on Cu electrodes," Current Opinion in Electrochemistry, vol. 9, pp. 158-165, 2018. https://doi.org/10.1016/j.coelec.2018.03.012

[10] D. V. Vasilyev, S. Shyshkanov, E. Shirzadi, S. A. Katsyuba, M. K. Nazeeruddin, and P. J. Dyson, "Principal descriptors of ionic liquid co-catalysts for the electrochemical reduction of CO2," ACS Applied Energy Materials, vol. 3, no. 5, pp. 4690-4698, 2020. https://doi.org/10.1021/acsaem.0c00330

[11] R. Jaffari, M. A. Hashmani, H. Taib, N. Abdullah, and S. S. H. Rizvi, "A Novel Image-Based Framework for Process Monitoring and Fault Diagnosis of Mooring Lines," Journal of Hunan University Natural Sciences, vol. 47, no. 10, 2020.

[12] N. A. i. A. Rahman, Z. May, R. Jaffari, and M. Hanif, "Failure Severity Prediction for Protective-Coating Disbondment via the Classification of Acoustic Emission Signals," Sensors, vol. 23, no. 15, p. 6833, 2023. https://doi.org/10.3390/s23156833

[13] L. K. Vora, A. D. Gholap, K. Jetha, R. R. S. Thakur, H. K. Solanki, and V. P. Chavda, "Artificial intelligence in pharmaceutical technology and drug delivery design," Pharmaceutics, vol. 15, no. 7, p. 1916, 2023. https://doi.org/10.3390/pharmaceutics15071916

[14] G. Schneider, "Automating drug discovery," Nature reviews drug discovery, vol. 17, no. 2, pp. 97-113, 2018. https://doi.org/10.1038/nrd.2017.232

[15] E. Hu, C. Liu, W. Zhang, and Q. Yan, "Machine Learning Assisted Understanding and Discovery of CO2 Reduction Reaction Electrocatalyst," The Journal of Physical Chemistry C, vol. 127, no. 2, pp. 882-893, 2023. https://doi.org/10.1021/acs.jpcc.2c08343

[16] Z. Li, S. Wang, W. S. Chin, L. E. Achenie, and H. Xin, "High-throughput screening of bimetallic catalysts enabled by machine learning," Journal of Materials Chemistry A, vol. 5, no. 46, pp. 24131-24138, 2017. https://doi.org/10.1039/C7TA01812F

[17] S. Back, J. Yoon, N. Tian, W. Zhong, K. Tran, and Z. W. Ulissi, "Convolutional neural network of atomic surface structures to predict binding energies for high-throughput screening of catalysts," The journal of physical chemistry letters, vol. 10, no. 15, pp. 4401-4408, 2019. https://doi.org/10.1021/acs.jpclett.9b01428

[18] J. Li et al., "Feature selection: A data perspective," ACM computing surveys (CSUR), vol. 50, no. 6, pp. 1-45, 2017. https://doi.org/10.1145/3136625

[19] A. Hellweg and F. Eckert, "Brick by brick computation of the gibbs free energy of reaction in solution using quantum chemistry and COSMO-RS," AIChE Journal, vol. 63, no. 9, pp. 3944-3954, 2017. https://doi.org/10.1002/aic.15716

[20] S. D. Bachchhav, S. Roy, and M. Mukhopadhyay, "Effect of solution nonideality on cholesterol supersaturation for liquid antisolvent crystallization," Chemical Engineering Communications, vol. 209, no. 9, pp. 1191-1202, 2022. https://doi.org/10.1080/00986445.2021.1951717

[21] C. Chen, K. Xu, X. Ji, L. Miao, and J. Jiang, "Enhanced adsorption of acidic gases (CO 2, NO 2 and SO 2) on light metal decorated graphene oxide," Physical Chemistry Chemical Physics, vol. 16, no. 22, pp. 11031-11036, 2014. https://doi.org/10.1039/c4cp00702f

[22] F. Eckert and A. Klamt, "Fast solvent screening via quantum chemistry: COSMO-RS approach," AIChE Journal, vol. 48, no. 2, pp. 369-385, 2002. https://doi.org/10.1002/aic.690480220

[23] F. Furche, R. Ahlrichs, C. Hättig, W. Klopper, M. Sierka, and F. Weigend, "Turbomole," Wiley Interdisciplinary Reviews: Computational Molecular Science, vol. 4, no. 2, pp. 91-100, 2014. https://doi.org/10.1002/wcms.1162

[24] A. Klamt, COSMO-RS: from quantum chemistry to fluid phase thermodynamics and drug design. Germany: Elsevier, 2005.

[25] A. Klamt, "Conductor-like screening model for real solvents: a new approach to the quantitative calculation of solvation phenomena," J. Phys. Chem., vol. 99, no. 7, pp. 2224-2235, 1995. https://doi.org/10.1021/j100007a062

[26] A. Klamt, V. Jonas, T. Bürger, and J. C. Lohrenz, "Refinement and parametrization of COSMO-RS," J. Phys. Chem A, vol. 102, no. 26, pp. 5074-5085, 1998. https://doi.org/10.1021/jp980017s

[27] C. Steffen, K. Thomas, U. Huniar, A. Hellweg, O. Rubner, and A. Schroer, "TmoleX-a graphical user interface for TURBOMOLE," Journal of computational chemistry, vol. 31, no. 16, pp. 2967-2970, 2010. https://doi.org/10.1002/jcc.21576

[28] V. Kumar and S. Minz, "Feature selection: a literature review," SmartCR, vol. 4, no. 3, pp. 211-229, 2014. https://doi.org/10.6029/smartcr.2014.03.007

[29] M. Kuhn and K. Johnson, Applied predictive modeling. Springer, 2013. https://doi.org/10.1007/978-1-4614-6849-3

[30] J. Brownlee. "How to Choose a Feature Selection Method For Machine Learning." Machine Learning Mastery. https://machinelearningmastery.com/feature-selection-with-real-and-categorical-data/ (accessed 02 June 2023).

[31] S. M. Stigler, "Francis Galton's account of the invention of correlation," Statistical Science, pp. 73-79, 1989. https://doi.org/10.1214/ss/1177012580

Materials Research Proceedings 53 (2025) 616-625 https://doi.org/10.21741/9781644903575-62

CO_2 Channelling in porous media: A three-phase displacement study using the lab-on-a-chip for CO_2 sequestration in depleted oil reservoirs

RAJAT Dehury[1], BALAKANNAN Murali[1] and JITENDRA S. Sangwai[1,*]

[1]Department of Chemical Engineering, Indian Institute of Technology Madras, Chennai, Tamil Nadu 600 036, India

*jitendrasangwai@iitm.ac.in

Keywords: Carbon Capture and Storage (CCS), Lab-On-A-Chip, Net-Zero Emissions, Porous Media, Snap-Offs, Viscous Fingering

Abstract. Carbon dioxide (CO_2) sequestration in depleted hydrocarbon reservoirs emerged as a promising strategy for mitigating global climate change. Along with large and long-term CO_2 storage opportunities, the pre-existing installations of hydrocarbon fields can be used for CO_2 injection and monitoring processes, making depleted oil and gas reservoirs an encouraging carbon sink with minimal environmental influences. However, the complex multiphase flow observed in underground geo-formations needs further investigation through the flow dynamics of CO_2 injection and displacement of native fluids, which are crucial for efficient and safe CO_2 storage. This study investigates CO_2/brine/oil three-phase displacement dynamics in a porous media following the lab-on-a-chip approach. A highly heterogeneous porous microchannel is used to mimic the underground porous media, which is placed under a high-resolution digital microscope for visualization and quantification of CO_2 trapping. It was observed that initially, CO_2 prefers high permeable flow paths, which results in channelling due to high viscosity contrast (viscosity ratio, $M = 0.103$). Subsequently, CO_2 invades oil from dead-end portions of the microfluidic device by elongating the oil droplets, owing to a reduction in surrounding phase viscosity and interfacial tension (IFT). The elongation of oil increases the snap-off of oil at the small throat radius, which increases oil recovery. CO_2 snap-offs were also observed due to high-pressure drop during the invasion of flow paths with smaller throat radii. Unlike viscous fingering, snap-off events increase the trapping of CO_2 in the porous media. Surfactant additives form emulsions and amplify the immiscible multiphase interaction, optimizing CO_2 trapping through structural and capillary trapping mechanisms. The findings can be valuable while strategizing efficient, enhanced, and secured subsurface CO_2 sequestration.

Introduction

The upstream petroleum industry is one of the important industries to cater the energy needs of mankind. Fossil fuels will remain important energy sources in the foreseeable future, even though the contribution of renewable energy is expected to increase in the energy basket. There is a gradual increase in energy consumption from 564.01×10^{18} J in 2020 to 595.15×10^{18} J by 2021 [1]. Fossil fuels, viz., coal, oil, and gas, have contributed about 80% of this demand between 2020 and 2021. In this, oil and gas contributed about 55% of the world's energy consumption. In addition, petroleum and petrochemical products have been widely used in our daily lives. This indicates that the petroleum industry will continue to dominate our lives in the near future. In this scenario, one of the important issues is the changing climate due to anthropogenic activities resulting in significant CO_2 emissions. Climate change is mainly due to the exploitation of fossil fuels for meeting energy demands. A significant increase in greenhouse gas (GHG) emissions, primarily CO_2, has been observed in the past few decades, increasing the atmospheric CO_2 concentration to about 425 ppm.

Hence, currently, climate change due to CO_2 emissions remains one of the important issues mankind is facing. In recent times, how to address climate change and CO_2 emissions has gained major attention from researchers worldwide. Researchers are making efforts to find new ways to capture, store, utilise, and sequestrate CO_2. In 2023, the total CO_2 emissions stood at about 37 gigatons (Gt) [2]. This CO_2 increase has resulted in an increase in global temperature, affecting the ice poles, the rainy seasons, draughts, and many more. Almost half of these emissions are absorbed by the ocean and terrestrial atmosphere, leaving 50% in the atmosphere, affecting climate change. In order to mitigate the negative impact of CO_2 emissions, the Paris Agreement recommends limiting the global average temperature rise to less than 2 °C [2]. This necessitates immediate actions to capture and sequester CO_2 safely. Various methods have been evolved to capture and store CO_2 safely. Some of these methods are being used at field scale, which include CO_2 capture from flue gases, direct air capture, and CO_2 sequestration in oil and gas reservoirs for enhanced oil recovery [3]. While more technologies are necessary to capture CO_2 from atmosphere, its safe storage remains the challenge. The oil and gas industry is the pioneer in utilizing CO_2 for enhance oil recovery purposes while increasing the oil and gas production and storing CO_2 safely underground [3]. Saline aquifers are one of the potential CO_2 storage sites. In addition, depleted oil and gas reservoirs also offer great potential to store large amounts of CO_2 in subsurface formations. While doing so, the existing infrastructure of the oil and gas industry can be utilized to sequester CO_2, and hence, it has been found to be one of the cost-effective methods rather than converting CO_2 to any value-added products.

Studies involving the CO_2 fluid flow under subsurface reservoirs will add value to improve the CO_2 storage capacity of saline aquifers and depleted oil and gas reservoirs. Often depleted oil and gas reservoirs have some oil and gas remaining in situ. These oil and gas reservoirs also have high temperature and high-pressure conditions. The reservoir rock properties are also varied due to the precipitation of heavier hydrocarbons, such as waxes and asphaltenes onto the rock surface. It is expected that while the CO_2 is injected into the reservoir's pore matrix, there will be some interaction of CO_2 with the remaining reservoir fluids, such as oil and gas, thus affecting the storage potential. Often, field studies make it difficult to analyse the fundamental insights into the fluid flow behaviours, and hence laboratory-based studies are a vital component to get more insights into the overall governing process of CO_2 sequestration at the pore level. In these directions, various experimental methods are being explored at the lab scale. This includes core flood, Amott cell, sand-pack reactor, micromodel, etc. However, there still remains a challenge to visualize the fluid flow behaviours at small pore networks. Recently, microfluidics has gained interest in studying various fluid flow behaviour problems, such as microbiology, microelectronics, pharmacy, and chemical industries [4]. Its use in understanding the oil and gas behaviour at reservoir conditions has also shown promising potential. Microfluidic apparatus can be used as a tool for oil and gas industries to understand the CO_2 sequestration potential by means of fluid propagation, emulsification, foam generation, droplet formation, wettability alteration, and flooding process behaviour. The microfluidic device was also found to replace the expensive pressure–volume–temperature (PVT) cells [5]. It is observed that the microfluidic apparatus is a faster and more economical method than conventional methods due to the requirement of fewer samples and more straightforward operation. Also, this method helps to visualize the fluid flow behaviour at the pore level and thus helps to understand many underlying phenomena [6, 7]. A recently developed microfluidic set-up has been used to study the oil recovery phenomena using various chemical agents [8]. In this work, the microfluidics set-up was used to study the CO_2 sequestration process at the pore level and understand how the fluid flow behaviour takes place so as to improve the CO_2 storage capacity of a reservoir.

Materials and Methods

Materials. The detailed list of all the chemicals used in this study, along with their purities and sources, is given below in Table 1. In this study, two non-ionic surfactants, namely Tween 80 and TX-100, were used due to their low costs, low toxicity, greater inactivity, and high solubilization. Non-ionic surfactants are considered instead of ionic surfactants to avoid any chemical interactions while only focusing on physical fluid displacement behavior. The deionized water (DIW) used in all experiments has a resistivity of 18.2 MΩ cm at 25 °C.

A light crude oil with an API gravity of 43.3 was used for the microfluidic experimental procedure. The source of this oil was the Narimanam oilfield in Tamil Nadu, India, which was provided by Chennai Petroleum Corporation Limited (CPCL). The oil properties such as saturates, aromatics, resins, and asphaltene were 43.2, 35, 21, and 0.8%, respectively.

Table 1 List of all chemicals used in this study

Chemicals	Purity	Supplier
Ethanol	Analytical grade	SRL Pvt. Ltd., India
Methanol	Analytical grade	FINAR Chemicals
Toluene	Analytical grade	SRL Pvt. Ltd., India
Tween 80 (Polysorbate 80)	≥ 99%	SRL Pvt. Ltd., India
TX-100 (Octyl phenol ethoxylate)	≥ 98%	SRL Pvt. Ltd., India
Sodium hydroxide	Analytical grade	RANKEM Chemicals

The seawater (brine) was obtained from Thiruvanmiyur Beach in Chennai, Tamil Nadu, India. The seawater was filtered to remove impurities using Whatman filter paper (grade 1, diameter 125 mm). X-ray diffraction (XRD) was performed on the sea salts obtained by evaporating the filtered seawater in a hot air oven, and spectra were obtained for further analysis. The prominent diffraction peaks were compared with the corresponding peaks in the reported data from the International Centre for Diffraction Data (ICDD). The diffraction peaks at 2θ of 32.17, 66.65, 75.65, and 84.30° correspond to the crystalline NaCl salt were observed. Additionally, the prominent peaks observed at 2θ of 27.80, 45.90, and 56.83° correspond to calcium, magnesium, and potassium salts, respectively.

The microfluidic chip utilized for this experiment was a physical rock network with one inlet and outlet. The properties of this chip are showcased in the table 2. The porous media micromodel employed in this study for displacement experiments is shown in Figure 1a. The experimental set-up, as shown in Figure 1b, consists of a syringe pump, microfluidic chip, high-power microscope, pump controller, gas cylinder, camera, computer, and waste collector. The syringe pump utilized is a Prosense B.V. (USA), which floods the microfluidic chip with high precision and controllable flow rates needed for the pore-scale three-phase displacement. It is also used in the chip-cleaning process to inject various cleaning agents. The microfluidic chip is held in the chip holder, which has an interface for connecting the inlet and outlet tubes, along with a pressure detector that is attached to the tubing. The chip was selected in this study due to its realistic representation of underground rock pore geometry.

Table 2 Properties of the Microfluidic chip used in this three-phase displacement experiment

Properties	Values
Chip material	Borosilicate glass
Permeability	2.5 D
Porosity	57%
Network dimensions	20 mm x 10 mm x 0.02 mm
Average pore-radius	56.21 μm
Average throat-radius	63.21 μm
Chip pore volume	5.7 μL
Wettability	water-wet

Experimental Procedure. Low-salinity seawater preparation. The low-salinity seawater (LSW) was prepared by diluting the seawater in 2:1 (DIW to seawater) ratio. The optimum salinity of 10,000 ppm seawater was selected due to the minimum contact angle achieved, leading to a change from an oil-wet to a water-wet state. The contact angles were found to be affected by salinity due to the presence of salt ions within the adsorbed layer between crude oil droplets and the sandstone substrate. The result was an altered sandstone substrate having a state of intermediate wettability.

Surfactant solution preparation. The surfactant solutions employed were TX-100 and Tween 80 which were prepared in the LSW that was filtered and diluted. Initially, the critical micelle concentration (CMC) of both surfactants was found using the conductometric method, which relies on finding a breaking point on curves that describe the concentration dependence of conductivity. The point at which micelle formation starts is indicated by the concentration dependence of specific conductivity (κ) as a breaking point (a change in slope), caused by the less mobile micelles that don't carry charge as well as monomers. For each surfactant, various concentrations (in ppm) were prepared by dissolving the surfactant in 35 mL batches, with the conductivity being measured using PC2700 Eutech instrumentation. The conductivity rod was dipped in the solution until the sensor was fully covered, and the conductivity values were noted upon the stoppage of reading fluctuation. Between each trial, the rod was cleaned using DIW, and a new sample at a different concentration was prepared. Following this, a graph was plotted between concentration (ppm) and conductivity (μS). The concentrations were chosen with the knowledge of CMC values for both surfactants in DIW, which would provide an estimated range for CMC determination in LSW. The CMC values of TX-100 and Tween 80 were found to be 150 ppm and 40 ppm, respectively. After the determination of CMC, a fresh solution was prepared at a concentration slightly above CMC for the pore-scale displacement investigation.

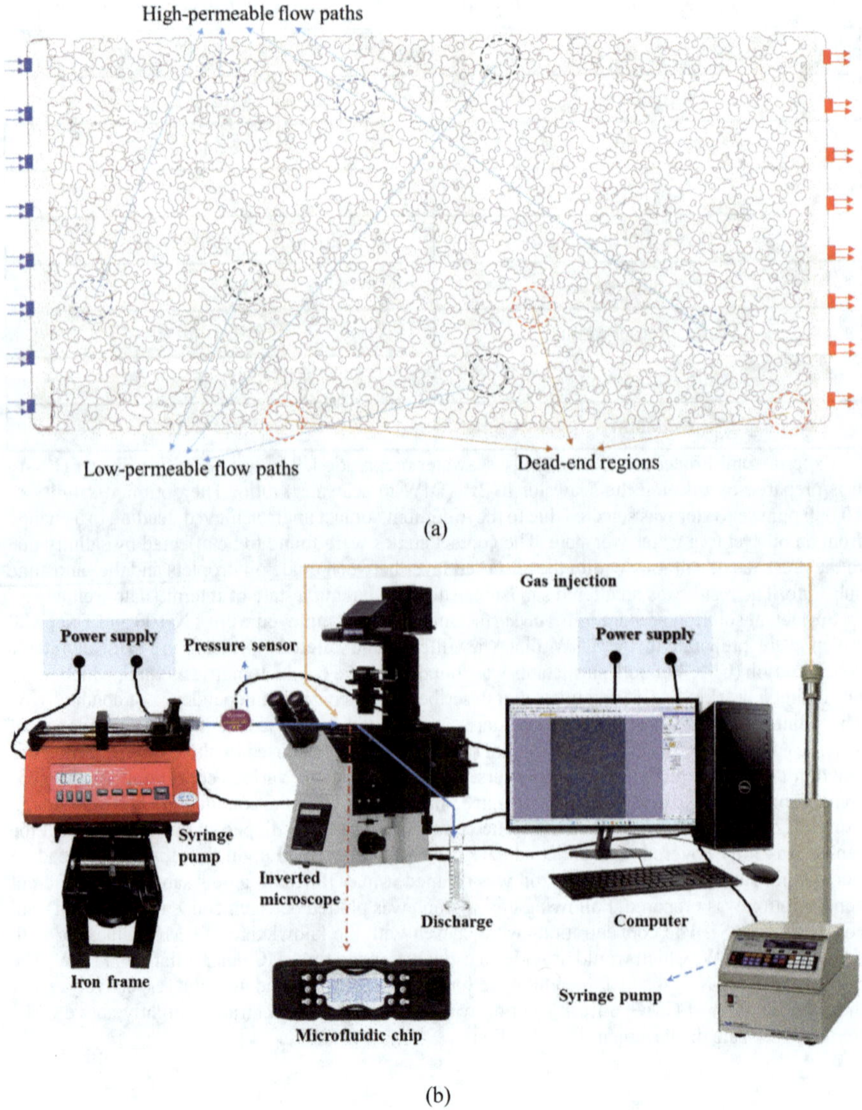

Fig. 1 (a) Microfluidic chip used in the study showing different permeable flow paths, (b) experimental set-up for three-phase microfluidic displacement study.

Surface tension measurement. The surface tension (SFT) of the surfactants at CMC was measured using a tensiometer (DataPhysics DCAT-11EC, Germany). The instrument works on the force-based measurement principle and has a high-precision electrodynamically compensated weighing system. The IFT was measured using a Wilhelmy plate (rectangular-shaped, thickness 0.2 mm, and length 19.9 mm). The accuracy of the instrument is ± 0.03 mN/m. The instrument was calibrated, and the plate was cleaned with acetone, followed by torching until red hot prior to each experiment. The SCAT software was used to take readings (around 30 seconds) and control the height of the sample holder. Each experiment was performed iteratively around ten times in order to obtain an accurate and precise SFT reading.

Flooding procedure. The clean microfluidic chip was initially saturated with crude oil at a flow rate of 0.12 mL/min, which was controlled using the syringe pump. The tubing was used to connect the syringe on the syringe pump to the inlet port on the chip holder. Following saturation, in order to account for the adsorption time, the crude oil-saturated microfluidic chip was aged under ambient pressure and temperature conditions for about 12 hours, which also prevented the escape of lighter hydrocarbons and any air bubbles. The oil displacement experiments were performed under ambient pressure and temperature conditions using the prepared surfactant solutions in brine. The course of each displacement process was visually observed, and images were captured using a microscopic camera attached to the inverted microscope. The process of displacement involved injecting brine with the surfactant to displace the saturated oil in the chip, followed by injection of CO_2 using a syringe pump controller that was attached to the CO_2 cylinder, as illustrated in Figure 1b. Following this, images were captured to observe the displacement patterns in various regions of the chip with differing permeabilities, along with a live-action video. The CO_2 was pulled into the syringe at 1 bar pressure using a tubing connection to the high-pressure syringe pump controller, ensuring the absence of foreign gasses and impurities. After a sufficient time, the CO_2 was injected at the same flow rate to observe the displacement and storage patterns.

Microfluidic chip cleaning. A borosilicate microfluidic chip was used in this investigation. It is essential to clean the microfluidic chip efficiently, as the same chip was used for all the displacement studies to maintain uniformity of the pore network and avoid any errors. Prior to each experiment, the chip was cleaned to ensure unbiased and unaffected results. Initially, 1 M NaOH (prepared by dissolving 3.2 g of NaOH pellets in 80 mL of deionized water) was flooded at various flow rates to rinse the chip. Dry air was injected at the same flow rate to remove any water droplets. This was performed periodically to ensure the removal of a maximum amount of impurities. After this, toluene was injected to remove any crude oil that was present after each microfluidic experiment. Finally, the microfluidic chip was rinsed with ethanol to remove any organic residue. All the effluents were disposed of in the waste collector and carefully discarded.

Results and discussion

A highly heterogeneous porous microchannel with intricate structure was used for Brine/oil/CO_2 three-phase flow investigation. This section investigates the unstable flow displacement pattern observed in the porous media along different permeable regions. Oil displacement due to the injection of brine with varying values of salinity and non-ionic surfactants was discussed. CO_2 injection resulted in viscous fingering and channeling through high-permeable flow paths, which limits storage in the porous domain.

Oil/Brine/CO_2 displacement study. Initially, the microfluidic chip is saturated with light crude oil. Three different sections representing high, mid, and low-permeable flow paths of the microfluidic chip are captured for further investigations. The injection of less viscous CO_2 gas created channeling, as shown in Figure 2a-d. A typical flooding pattern is shown in Figure 2a. The CO_2 channeling can be observed after CO_2 flooding. The channeling is due to the unstable viscous fingering between the oil/brine and CO_2 phase. Viscous fingering occurs when there is a huge contrast between the displacing and displaced phases. For this case, the viscosity ratio (M) between displacing and displaced phase is 0.103.

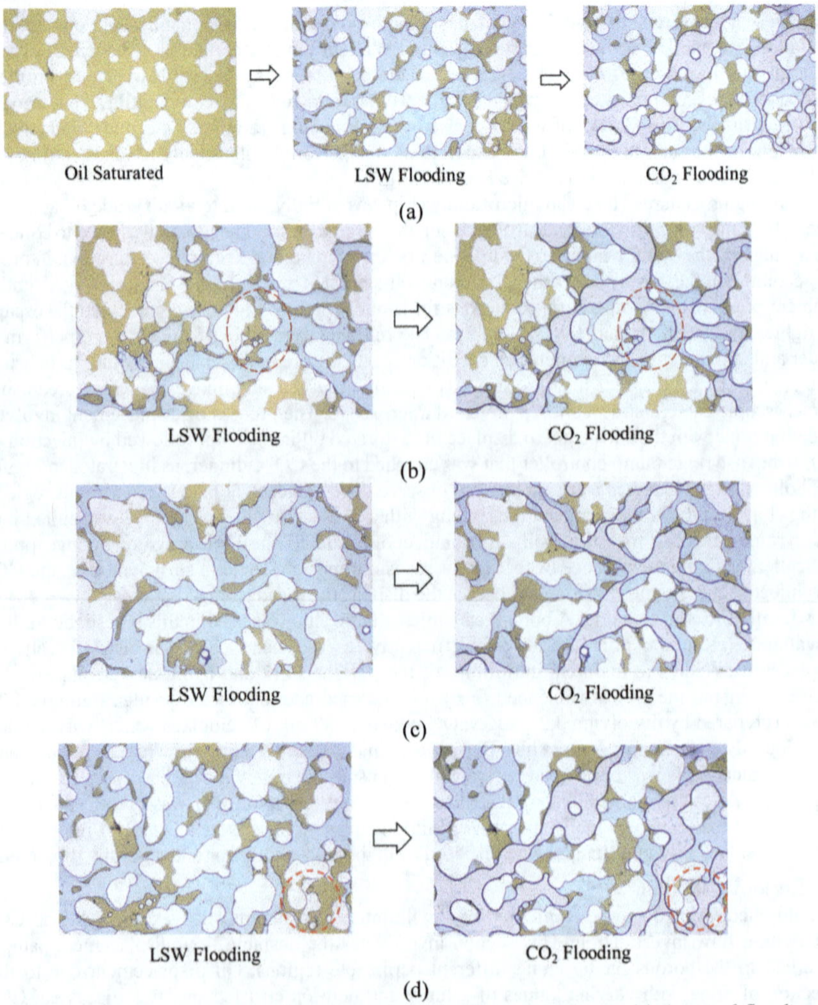

Fig. 2: Microscopic images of the three-phase displacement pattern. (a) A typical flooding pattern, flooding patterns for (b) low-permeable zone, (c) mid-permeable zone, and (d) high permeable zone. (Yellow: Oil, Blue: Brine, Grey: CO_2 gas)

Figure 2b-d illustrates the flooding patterns for low, mid, and high-permeable zones of the porous microdomain. As expected, more oil remains in the low-permeable section of the chip (Figure 2b) compared to high-permeable regions. CO_2 flooding recovers smaller size oil ganglia and improves oil recovery. CO_2 channeling is prominent in the high-preamble sections due to a positive feedback mechanism between pressure and fluid velocity [9]. These channeling patterns reduce the CO_2 trapping inside the porous domain. Undermining these patterns can lead to leakage, compromising the integrity of the storage reservoir. Injection of CO_2 also improves the oil recovery from the porous domain and reduces the residual oil ganglia sizes. The reduced oil saturation in the microchannel can be observed in Figures 2b and d, highlighted in red circles. CO_2 penetrates the oil in the dead-end sections of the microfluidic device, causing the oil droplets to elongate due to a decrease in the viscosity of the surrounding phase and the interfacial tension (IFT). The elongation of the oil droplets promotes the breaking off of oil at the narrow throat radius, leading to increased oil recovery.

Moreover, CO_2 snap-off events were observed during the displacement experiments, where CO_2 bubbles were disconnected from the continuous gas phase and moved independently. Figure 3 shows a snap-off event observed during CO_2 flooding. The constricted CO_2 flow paths get disconnected at the smaller pore throats during displacement dynamics, resulting in a snap-off event. This snapped bubble can increase the solubility trapping of CO_2 as the contact area between CO_2 and the brine phase increases in the manifold. As a result, snap-off events can be favourable for CO_2 storage in underground porous formations.

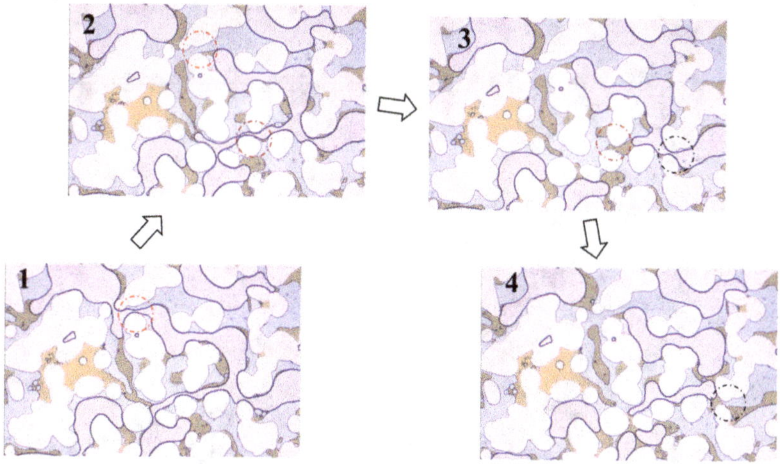

Fig. 3: CO_2 snap-off event during three-phase flow dynamics in a porous microchannel.

Oil/Surfactant/CO_2 Displacement study. Various research studies have used surfactants for chemically enhanced oil recovery. Here, CO_2 is injected, following surfactant injection, into the porous micro-structure to comprehend the oil recovery and CO_2 sequestration. Non-ionic surfactants significantly reduce the interfacial tension between oil and water, as well as oil and gas. This reduction in interfacial tension promotes better oil mobilization and displacement, allowing for increased oil recovery. The impact of IFT reduction has yet to be studied in depth in CO_2 geo-sequestration. Following a similar set of experiments as discussed above, the influence of CO_2 injection into the oil reservoir after tertiary recovery with oil/surfactant phase present in porous media is investigated here.

Figures 4a and b show the displacement patterns with the use of surfactant instead of brine. Both surfactants form emulsions with air bubbles, which helps in additional oil recovery. The combined effect of surfactant and CO_2 creates a more favorable environment for oil displacement, driving it towards production wells. The emulsion resulted in CO_2 bubbles, which increased the quantity of CO_2 trapped in the porous domain. However, the observed patterns may change with time, as observed with hydrogen in core flooding experiments [10, 11]. Miscibility studies of CO_2

Materials Research Proceedings 53 (2025) 616-625 https://doi.org/10.21741/9781644903575-62

and oil need to be carried out at high pressure and high temperature conditions to predict storage capacity in depleted oil reservoirs accurately.

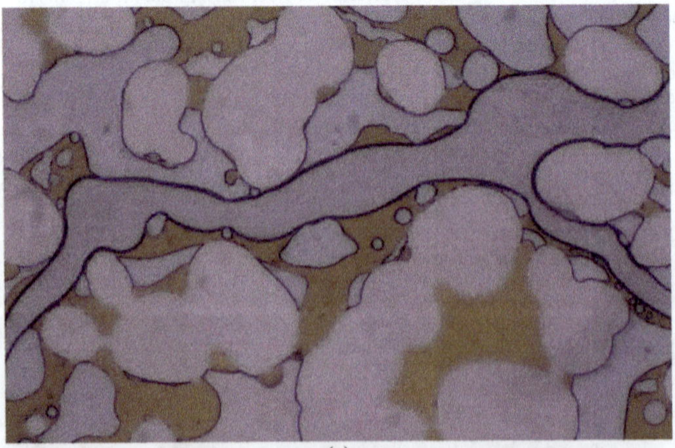

(a)

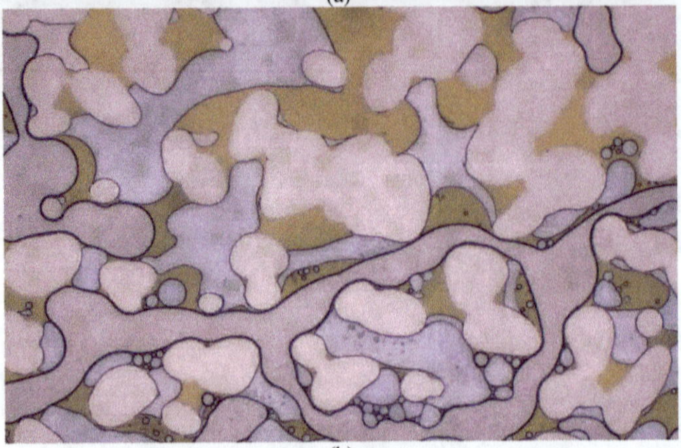

(b)

Fig. 4: Microscopic images of a three-phase displacement study with the injection of non-ionic surfactants (a) TX-100 and (b) Tween-80.

Conclusion

The lab-on-a-chip technique was used to analyse the three-phase displacement dynamics of CO_2/brine/oil in porous media. The study employed a highly heterogeneous porous microchannel with a 56 μm pore radius and 63 μm throat radius, mimicking subsurface porous media. The porous microchannel is examined under a microscope to see and quantify CO_2 trapping. CO_2 channelling can be seen following CO_2 flooding, which reduces CO_2 sequestration capability. CO_2 injection results in enhanced oil elongation, which is responsible for an increase in the snap-off of oil at the narrow throat radius, leading to higher oil recovery. In addition, CO_2 snap-offs occur due to the significant pressure drop caused by the invasion of flow channels with smaller throat radii. Unlike viscous fingering, snap-off events lead to more CO_2 being trapped in the porous material. With surfactant, lower IFT produced more snap-off CO_2 bubbles, improving CO_2 trapping capacity. The findings can be useful for developing strategies for efficient, increased, and secure subsurface CO_2 sequestration.

References

[1] S. K. Prasad, J. S. Sangwai, and H. S. Byun, "A review of the supercritical CO2 fluid applications for improved oil and gas production and associated carbon storage," Jun. 01, 2023, *Elsevier Ltd.* https://doi.org/10.1016/j.jcou.2023.102479

[2] Internet. International Energy Agency, "CO_2 Emissions in 2023," 2023. [Online]. Available: www.iea.org

[3] X. Zhou *et al.*, "Review of Carbon dioxide utilization and sequestration in depleted oil reservoirs," *Renewable and Sustainable Energy Reviews*, vol. 202, p. 114646, Sep. 2024. https://doi.org/10.1016/j.rser.2024.114646

[4] A. D. Stroock, S. K. W. Dertinger, A. Ajdari, I. M. Mezic′, H. A. Stone, and G. M. Whitesides, "Chaotic Mixer for Microchannels." [Online]. Available: https://www.science.org

[5] S. Molla, F. Mostowfi, Microfluidic PVT-Saturation Pressure and Phase-Volume Measurement of Black Oils. SPE Reserv. Eval. Eng. 2017, 20, 233– 239.

[6] P. Mehrabianfar, H. Bahraminejad, and A. K. Manshad, "An introductory investigation of a polymeric surfactant from a new natural source in chemical enhanced oil recovery (CEOR)," *J Pet Sci Eng*, vol. 198, Mar. 2021. https://doi.org/10.1016/j.petrol.2020.108172

[7] R. Farajzadeh, S. Kahrobaei, A. A. Eftekhari, R. A. Mjeni, D. Boersma, J. Bruining, Chemical Enhanced Oil Recovery and the Dilemma of More and Cleaner Energy. Sci. Rep. 2021, 11, 1– 14.

[8] G. Kumar, E. Mani, and J. S. Sangwai, "Microfluidic Investigation of Surfactant-Assisted Functional Silica Nanofluids in Low-Salinity Seawater for Enhanced Oil Recovery Using Reservoir-on-a-Chip," *Energy & Fuels*, Jun. 2023. https://doi.org/10.1021/acs.energyfuels.3c00563

[9] R. Dehury, S. Chowdhury, and J. S. Sangwai, "Dynamics of hydrogen storage in subsurface saline aquifers: A computational and experimental pore-scale displacement study," *Int J Hydrogen Energy*, vol. 69, pp. 817–836, Jun. 2024. https://doi.org/10.1016/j.ijhydene.2024.05.009

[10] S. Goodarzi, Y. Zhang, S. Foroughi, B. Bijeljic, and M. J. Blunt, "Trapping, hysteresis and Ostwald ripening in hydrogen storage: A pore-scale imaging study," *Int J Hydrogen Energy*, vol. 56, pp. 1139–1151, Feb. 2024. https://doi.org/10.1016/j.ijhydene.2023.12.029

[11] Y. Zhang, B. Bijeljic, Y. Gao, S. Goodarzi, S. Foroughi, and M. J. Blunt, "Pore-Scale Observations of Hydrogen Trapping and Migration in Porous Rock: Demonstrating the Effect of Ostwald Ripening," *Geophys Res Lett*, vol. 50, no. 7, Apr. 2023. https://doi.org/10.1029/2022GL102383

Decarbonization Technology: ICDT2024 | Materials Research Forum LLC
Materials Research Proceedings 53 (2025) 626-636 | https://doi.org/10.21741/9781644903575-63

Graphene-manganese dioxide based aptasensor for highly sensitive detection of Ochratoxin A

NABIHAH Mohammad Illias[1], VEERADASAN Perumal[2,3], MOHAMED SHUAIB Mohamed Saheed[2,3*]

[1]Department of Fundamental & Applied Sciences, Universiti Teknologi PETRONAS, 32610 Seri Iskandar, Perak, Malaysia

[2]Department of Mechanical Engineering, Universiti Teknologi PETRONAS, 32610 Seri Iskandar, Perak.] , Malaysia

[3]Center of Innovative Nanostructures & Nanodevices (COINN), Universiti Teknologi PETRONAS, 32610 Seri Iskandar, Perak, Malaysia

*shuaib.saheed@utp.edu.my

Keywords: Ochratoxin A, Biosensor, Graphene, Manganese Dioxide Nanoparticles, Aptasensor

Abstract. Biosensors have recently emerged as a simpler and cheaper alternative to traditional methods like HPLC for detecting ochratoxin A (OTA). This study aims to design a highly sensitive and selective aptasensor using graphene oxide (GO) and manganese dioxide nanoparticles (MnO_2NPs). The aptasensor, made with screen-printed carbon electrode (SPCE), was fabricated using hydrothermally synthesized MnO_2NPs at concentrations of 0.2M, 0.3M, and 0.4M. electrochemical impedance spectroscopy (EIS) assessed the biosensor's sensitivity, specificity, and selectivity, revealing a detection limit of 3.33 ng/mL for OTA. The aptasensor showed high selectivity for OTA even in the presence of other mycotoxins like aflatoxin B1 (AFB1), deoxynivalenol (DON), and zearalenone (ZEA) and effective detected OTA in spiked cocoa and oats. The fabrication of GO-MnO_2NPs helps in enhancing the aptasensor's stability and performance due to their high surface area, biocompatibility, and excellent electrical conductivity. This study establishes the foundation for creating a functional, economical, and efficient biosensing system to detect OTA in food. The extensive implementation of such technology can significantly help improve food safety and safeguard consumer interests.

Introduction

Ochratoxin A (OTA) is considered one of the most toxic among mycotoxins. Belongs to fungus species of *Aspergillus* and *Penicillium* such as *Aspergillus ochraceus, Penicillium verrucosum, Aspergillus carbonarius,* and *Penicillium viride.* It has been classified as possibly carcinogen (class 2B) according to the *International Agency for Research on Cancer* (IARC) and owing to its susceptibility for food stocks resulting in European Union (EU) determined a permissible limits for OTA in food and feed, allowing for lower concentrations [1]. A proper detection method for the presence of OTA in the commodities is becoming necessaries due to the remarkable stability of OTA throughout the process of pre-planting, planting, harvesting, post-harvest, processing, storage, as well as transportation of the commodities in various locations as in Figure 1.1. Moreover, OTA has a high capacity for bioaccumulation and can travel up the food chain [2]. Therefore, convenient and precise quantitative detection of OTA is of extreme relevance in minimizing the risks.

Recently, graphene has become a common name of materials used in biosensor due to its outstanding properties. Its remarkable properties of having high charge-carrier mobility, large surface area, high thermal conductivity, low cost, superior electronic properties and electrical properties, as well as chemical stability would contribute in developing the aptasensor. Furthermore, it also has a higher adsorption rate to firmly hold the biomolecules [3]. Despite the excellent physical and chemical properties of GO, agglomeration tend to occurs between GO sheets due to the π-π interactions, reducing their surface area and overall performance as an electrode material. To minimize the agglomeration, incorporating metal dioxide for enhancing the specific surface area was necessary [4].

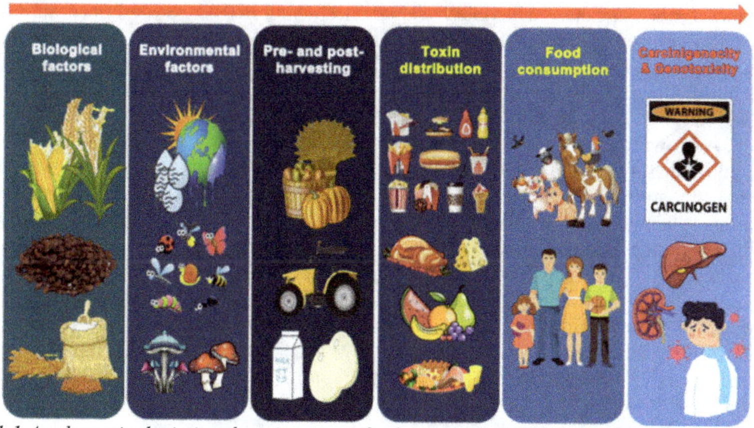

Fig. 1.1 A schematic depicting the migration of ochratoxin A (OTA) from plantation to pre- and post-harvest to human

MnO_2 nanoparticles possessed a high surface-to-volume ratio and were environmentally friendly as well as inexpensive. However, MnO_2 has disadvantage of having poor electrical conductivity due to the presence of non-conductive oxide. Hence, by combining GO with MnO_2 can address the issue and significantly enhance the conductivity of GO. This can be seen as MnO_2 has been widely used with carbon-based materials in improving its electrical conductivity, increase the electrochemical performance and mass loading of ultrathin MnO_2 films [5].

Biosensor has been widely studied due to its excellent capability in having a detection through sensitivity and selectivity. However, it also able to suffers limitations such as having poor detections and having high production costs. To ensure the developed biosensor achieving its capability, nanomaterials of GO and MnO_2 nanoparticles were employ in constructing the biosensor in this research. Besides, the development of biosensors often involves complex techniques, including the use of noble metals, which are time-consuming, require trained personnel, and are expensive due to the harsh chemicals used. By excluding noble metals, the overall cost of material preparation can be significantly reduced. Therefore, this research develops an aptasensor comprising the materials of GO and MnO_2 nanoparticles fabricated on a screen-printed carbon electrode (SPCE).

Experimental

Materials and Chemicals. The biosensor utilized commercially available graphene from Ugent Tech. Sdn. Bhd. Manganese dioxide nanoparticles (MnO_2NPs) were synthesized utilizing manganese acetate tetrahydrate ($Mn(CH_3COO)_2{\cdot}4H_2O$) from Sigma Aldrich (USA), potassium permanganate ($KMnO_4$) obtained from Bendosen Laboratory Chemicals, and sodium hydroxide (NaOH). Deionized water (DI) served as the solvent for the entire duration of the experiment. The sensing study employed a Metrohm DropSens screen-printed carbon electrode (SPCE). The

Materials Research Proceedings 53 (2025) 626-636 https://doi.org/10.21741/9781644903575-63

utilization of 95% and 70% ethanol solutions ensured the sterility of the material and chemical processing. Sigma Aldrich (USA) provided (3-aminopropyl) triethoxysilane (APTES), glutaraldehyde, phosphate buffer saline solution (PBS), ethanolamine, and ochratoxin A (OTA), while Apical Scientific Sdn. Bhd. synthesized OTA aptamer using published procedure [6]. The sequence of the aptamer is as follows; 5'-NH2-C6-SH-(CH2)6-GAT CGG GTG TGG GTG GCG TAA AGG GAG CAT CGG ACA-3'. All chemicals were stored and utilized without purification, as per the supplier's instructions.

Synthesis of Manganese Dioxide Nanoparticles (MnO_2NPs). As described in [7], this investigation utilized a modified hydrothermal technique to produce MnO_2 nanoparticles. 0.2M MnO_2 nanoparticles were synthesized through the dissolution of manganese acetate tetrahydrate in a 50 ml of DI water. On its own, a $KMnO_4$ solution was made by stirring it for 5 minutes in 50 ml of DI water. The two solutions were gently mixed under a fume hood to commence the redox reaction. A brownish colloidal Mn precursor was produced by gradually adding NaOH solution (prepared by dissolving 2 g NaOH granules in 50 ml of DI water) to the resulting Mn solution precursor after stirring it for 30 minutes at room temperature. After transferring the precursor to a Teflon-lined stainless steel autoclave, it was heated for 12 hours at 120ºC. After the autoclave had cooled, it was depressurized and unsealed in a secure manner. Following centrifugation, the precipitate of manganese was purified with ethanol and DI water to eliminate impurities. The MnO_2NPs sediment was subsequently dessicated at 60ºC overnight. Nanoparticles of MnO_2 were synthesized at concentrations of 0.2M, 0.3M, and 0.4M to assess their biosensing characteristics. Therefore, the handling of each concentration of 0.3M and 0.4M was identical. The overall process of synthesis has been recorded in the flow chart at Figure 2.1.

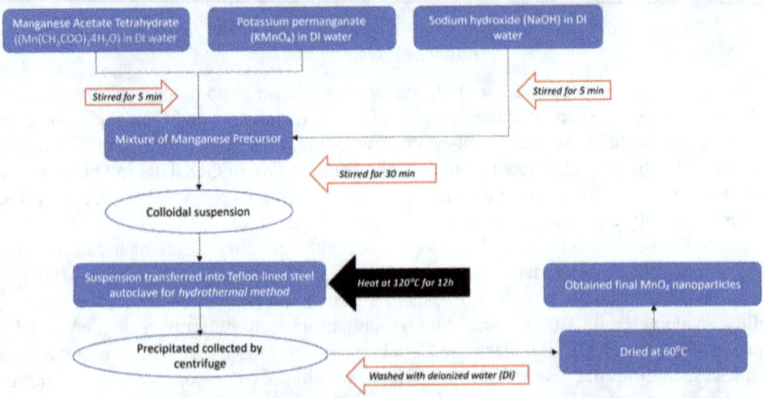

Fig. 2.2 Flow chart on the synthesis of Mn_2 nanoparticles.

Surface Characterizations. Specific characterization technologies enabled the examination of the materials' morphological structure, chemical composition, and instrinsic qualities. These comprehensive characterizations allow research of MnO_2 nanoparticles and graphene. The material was examined using FESEM (Zeiss Supra55 VP) and energy dispersive x-ray (EDX) analysis. XPS (Thermo Scientific, K-alpha) confirms the chemical composition of graphene and MnO_2NPs, whereas Raman spectroscopy detects and confirms their structural fingerprint. XRD analysis equipment is designed to determine the crystallinity of treated materials. Fourier-transform infrared spectroscopy (FTIR, model; PerkinElmer, Spectrum One) was used to verify biomolecular immobilization and surface modification.

Materials Research Proceedings 53 (2025) 626-636 https://doi.org/10.21741/9781644903575-63

Fabrication of Screen-printed electrode (SPCE)/ Graphene (GO) - Manganese dioxide nanoparticles (MnO_2NPs). Graphene and MnO_2 nanoparticles were synthesized and utilized to form an apta-sensor, with SPCE as its foundation. MnO_2 nanoparticles were characterized to determine the right concentration for fabrication. All of the supplies had been prepared earlier. For performance testing, graphene was diluted with DI water. MnO_2 nanoparticles were dispersed in ethanol and Nafion, which served as a binding solution. This solution was sonicated for 30 minutes. Clean SPCE electrodes underwent layer-by-layer deposition. First, graphene was added, then MnO_2 nanoparticles. The surface was dried before the final pipetting of nanomaterials onto the SPCE. Graphene and MnO_2NPs were applied in an identical ratio.

Performance of Biosensing Application For Ochratoxin A (OTA) Detection. Biosensing involved surface chemical functionalization and biomolecular immobilization. The biosensor is built layer-by-layer based on the described articles, with slight modifications [7]. Begin by casting graphene with MnO_2 nanoparticles on SPCE (SPCE/ GO- MnO_2NPs) substrate. Next, the linkers, 2.5% APTES, and 3% glutaraldehyde were drop-casted onto the substrate and incubated for one hour each layer. The addition of APTES to the biosensor changes the amine group surfaces, allowing for aptamer immobilization while glutaraldehyde completes the substrate's aldehyde group. MnO_2 nanoparticles with graphene were employed to immobilize the pre-heated OTA aptamer. This step takes 30 minutes of substrate incubation. Add ethanolamine as a blocking agent and leave for 15 minutes. During this time, the ethanolamine blocked the aldehyde group that was not attached to the aptamer. Finally, OTA was serially diluted at 1 fg/mL, 10 fg/mL, 100 fg/mL, 1 pg/mL, 10 pg/mL, 100 pg/mL, and 1 ng/mL levels. The biosensor's sensitivity was determined by immobilizing OTA from low to higher concetrations. Washing with 30-fold PBS solutions was performed between the functionalization and immobilization layers.

This research used electrochemical impedance spectroscopy (EIS) and cyclic voltammetry (CV). Charge transfer resistance (Rct) measurements monitored electrochemical changes that occurred in response to surface modifications. Rct was determined for each phase between 1E + 06 and 0.1 Hz. To evaluate the OTA sensitivity of the biosensor, CV was conducted at 0.05 V/s on potentials ranging from -1 to 1 V. PBS applied to SPCE/GO- MnO_2NPs prior to measurement to keep them hydrated. Electrochemical circle fitting is frequently used to validate Nyquist charts in EIS. Electrochemical circle fittings guarantee circuit measurements that are equivalent to Randles.

Surface Characterizations. Achieving consistent readings across all manufactured biosensors requires that the biosensor be reproducible. By means of EIS, the performance of a number of SPCE/GO- MnO_2NPs devices that were farbricated was assessed.

Sensitivity Test. A biosensor sensitivity test ascertains the LOD of the analyte. OTA concentrations of 1 fg/mL, 10 fg/mL, 100 fg/mL, 1 pg/mL, 10 pg/mL, 100 pg/mL, and 1 ng/mL were utilized to evaluate the biosensor. LOD was determined by applying a specific formula. The value is interpolated onto the graph of the calibration curve [8].

Limit of detection (LOD):

$$LOD = 3\sigma + Rct\ of\ Aptamer$$

σ = standard error of aptamer Rct

Selectivity Test. Given the vast number of mycotoxins produced by fungi today, it is crucial to evaluate the biosensor for its specificity towards the analyte. Mycotoxins such as aflatoxin B1 (AFB1), ochratoxin A (OTA), deoxynivalenol (DON), and zearalenone (ZEA) often coexist in contaminated crops. The fabricated biosensor developed with an OTA-specific aptamer to enhance its selectivity for OTA. Thus, the binding of the specific analyte to the aptamer is determined using the EIS measurements.

Spiking Test of a Real-life Sample. Finally, the constructed SPCE/GO-MnO_2NPs biosensor is evaluated for its efficacy in detecting OTA in real-life samples. This assessment utilizes rolled oats and cocoa powder for the spiking test. Prior to analysis, the rolled oats are ground into powder. These samples are selected due to their frequent contamination by OTA [9]. Both samples were procured from a local grocery market and diluted in PBS solutions (1 mg/mL) before evalution with the aptamer surface. The diluted samples, considered raw, were further diluted with PBS solution at ratios of 1:9, 1:99, and 1:999.

During evalution, the Rct values of these concentrations were recorded using EIS, and the samples with the lowest resistance values were selected for OTA spiking. OTA was spiked into the rolled oat and cocoa powder samples at concentrations of 1 fg/mL, 1 pg/mL, and 1 ng/mL. Figure 2.2 provides an overview of the spiking test process conducted throughout the experiment.

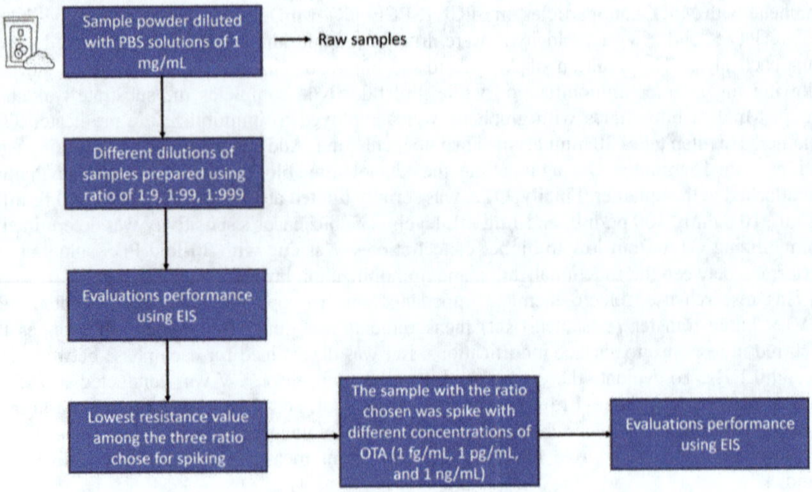

Fig. 2.3 Schematic diagram on the process of spiking test.

Results and Discussion

Physicochemical Characterization of Graphene and MnO_2 Nanoparticles: Surface Morphology and Elemental Composition. The field emission scanning electron microscopy (FESEM) was employed for the examination of graphene oxide (GO) nanostructures and manganese dioxide nanoparticles (MnO_2NPs). The fesem imaging of graphene shows aggregated, multi-layered, sheet-like structures as in Figure 1B. EDX analysis indicates that the GO consists mainly carbon of 85.98%, and oxygen (4.49%), with minor contaminants of 9.53% like aluminium, sulphur, and copper. The results has been reported in the cited article [10].

As in the reported articles [10], the prepared MnO_2 nanoparticles presented a flower-like morphology across three concentrations of 0.2 M, 0.3 M, and 0.4 M. Across the three concentrations of MnO_2NPs, 0.3 M exhibits the highest weight percentage of Mn. The elemental analysis revealed that the MnO_2NPs mostly consist of oxygen (O) and manganese (Mn), with negligible impurities, potentially originating from contamination during the analysis process. The inclusion of oxygen-based functional groups in MnO_2 nanoparticles is crucial for the coupling of biomolecules, resulting in enhanced electrochemical activities, particularly in sensing applications.

Physicochemical Characterization of Graphene and MnO_2 nanoparticles. The XPS spectrum for graphene nanomaterials displays distinct peaks of C1s and O1s. As shown in Figure 3.1, the

Materials Research Proceedings 53 (2025) 626-636 https://doi.org/10.21741/9781644903575-63

element peak of C1s was deconvoluted to four components at 285.69 eV, 287.36 eV, 289.76 eV, and 285.21 eV which correspond to C-C, single-bonded carbon, C=O, and O=C-OH. of MnO_2 nanoparticles displays peak. The figure also shows the XPS spectra of O1s deconvoluted into two components with peaks at 533.18 eV and 534.51 eV representing C-O-C and H_2O. The XPS spectra of graphene aligns with the EDX analysis presented. The atomic concentration of oxygen-containing moieties, such as carbonyl and carboxyl groups, indicates that the carbon-based materials are hydrophilic, which is essential for biosensing application [11].

As supported from EDX analysis, the MnO_2 nanoparticles displays the XPS spectra of Mn2p and O1s as in Figure 3.2. The XPS spectra presents an element peak of Mn2p deconvoluted to two components of Mn $2p_{1/2}$ at 642.3 eV and Mn $2p_{3/2}$ at 654.3 eV with binding energy of 12.0 eV, indicating a predominant Mn^{4+} oxidation state as reported in the literature [12], [13]. The deconvoluted Mn $2p_{3/2}$ peaks illustrated in Figure 3.2.2 indicated a various Mn^{x+} ions ascribed to Mn^{4+}, Mn^{3+}, and Mn^{2+} ions. The presence of MnO_2 was confirmend by Mn in its +4 valence state [14], [15]. The O1s peak measured at 530.0 eV deconvoluted into four components representing interaction of oxygen with different elements at 530.0 eV, 531.3 eV, 529.2 eV, and 532.7 eV. These peaks correspond to lattice oxygen, oxygen vacancies, and residual water (Mn-O-Mn, Mn-O-H, K-O-H, and H_2O). the number of oxygen vacancies is directly related the material's bandgap and enhances electrical conductivity [16].

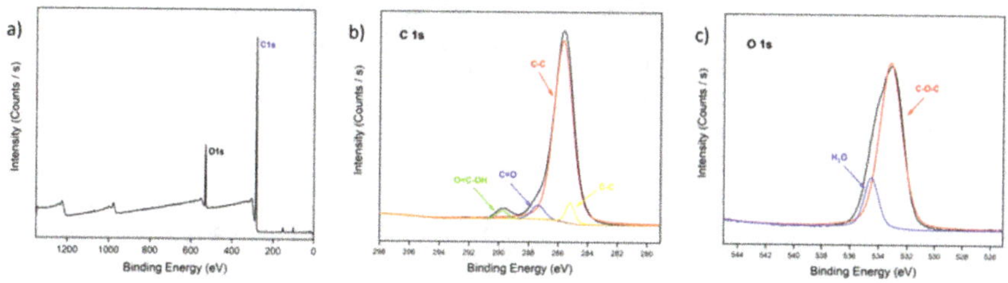

Fig. 3.4 XPS of graphene, (a) the overall XPS analysis of the materials, (b) the carbon, C, elemental composition analysis, (c) the oxygen, O, elemental composition analysis.

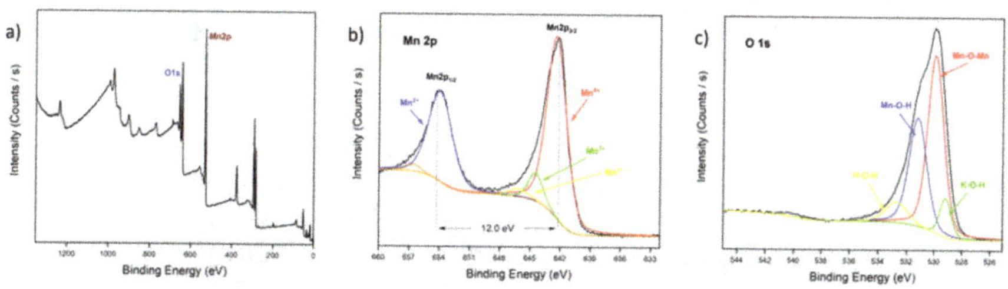

Fig. 3.2: XPS of MnO_2NPs, (a) the overall XPS analysis of the materials, (b) the manganese, Mn, elemental composition analysis, (c) the oxygen, O, elemental composition analysis.

Reproducibility of SPCE/GO-MnO_2NPs. Before testing the sensing capabilities of the fabricated SPCE/GO-MnO_2NPs biosensor, its performance and reproducibility were assessed. Figure 3.3 presents the reproducibility of three fabricated aptasensors, highlighting their distinct Rct values. The initial Rct values for the SPCE/GO electrodes were 363.65 Ω, 321.66 Ω, and 318.95 Ω. After adding MnO_2 nanoparticles, these value decreased to 288.39 Ω, 257.95 Ω, and 245.89 Ω indicating an improvement in conductivity. The relative standard deviations (RSD) were

7.49% for SPCE/GO and 8.29% for SPCE/GO-MnO_2NPs, suggesting a uniform distribution of the nanoparticles across the electrodes. The addition of MnO_2 nanoparticles contribute in enhancing the aptasensor electrodes' conductivity. Furthermore, fabricating graphene with MnO_2 nanoparticles, increased the surface area and electrocatalytic properties [17], as demonstrated by the impedance spectroscopy (EIS) measurements (Figure 3.3).

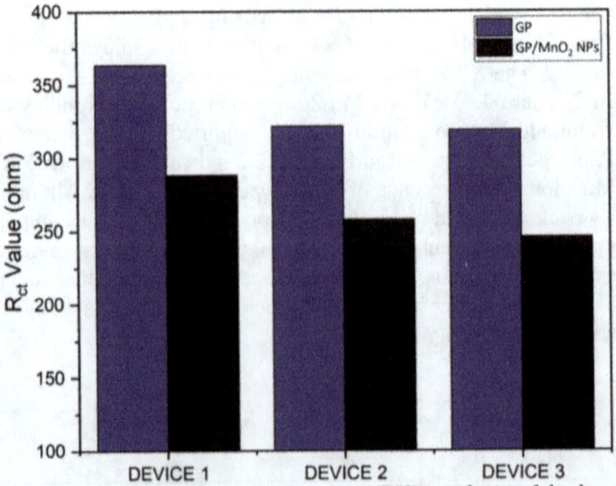

Fig. 3.5 The electrochemical impedance spectroscopy (EIS) analysis of the bare SPCE-modified with graphene and MnO_2 nanoparticles.

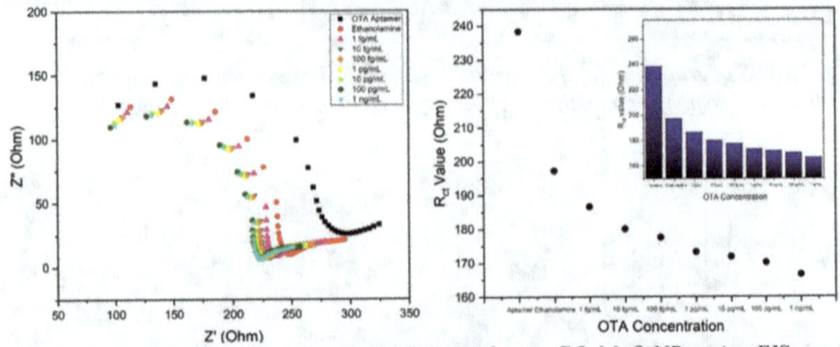

Fig. 3.6 The analysis of sensitivity test employing GO-MnO_2NPs using EIS

Sensitivity Analysis and Limit of Detection (LOD). The performance of this biosensor was assessed by analyzing the impedance changes caused by various concentrations of OTA ranging from 1 fg/mL to 1 ng/mL (Figure 3.4). The charge transfer resistance (Rct) in EIS relates to energy dissipation due to activation energy rather than electron transfer [18]. The results recorded presents that the immobilization of OTA aptamer on the SPCE/GO-MnO_2NPs electrode reduced the Rct to 244 Ω. Ethanolamine used in the research used to block the non-specific binding sites, significantly decreasing the resistance to 184 Ω proving its excellent conductivity. As OTA was added to the aptasensor, the Rct value decreased with increase in OTA concentrations from 1 fg/mL to 1 ng/mL. Specifically, 1 fg/mL of OTA showed Rct value of 174 Ω, while 1 ng/mL has

Materials Research Proceedings 53 (2025) 626-636 https://doi.org/10.21741/9781644903575-63

Rct value of 153 Ω. The aptasensor displays a decreasing resistance trend likely due to the high conductivity of OTA and the electrostatic repulsion between the electrolyte and OTA's negatively charged carboxyl and hydroxyl groups [19]. Consequently, the OTA concentration-dependent Rct can be used to measure OTA.

Using the LOD equations, Figure 3.5 reveals a correlation value 0.96 that indicates a linear trend between 1 fg/mL and 1 ng/mL of OTA. The LOD value for this aptasensor recorded at 3.33 ng/mL. While the SPCE/GO-MnO_2NPS biosensor shows a lower LOD, the synthesis of gold nanostars is complex and time-consuming, making them less common [9]. Moreover, most developed sensors were using gold (Au), which is an expensive material for an electrode.

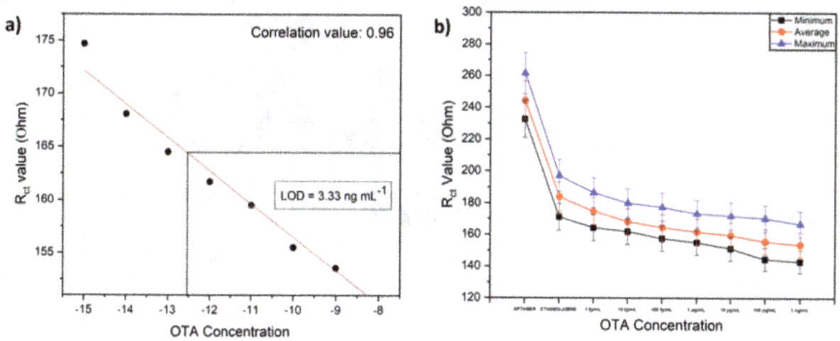

Figure 3.7 a) The limit of detection (LOD) of the biosensor. b) The overall data recorded across three devices for detecting OTA.

Selectivity Analysis. A biosensor must be both sensitive and specific, making the biorecognition element essential for selectivity. The selectivity of the developed aptasensor was evaluated by disputing the functionalized platform using three different mycotoxins: aflatoxin (AFB1), deoxynivalenone (DON), and zearalenone (ZEA). These mycotoxins were applied on the fabricated aptasensor at similar concentrations. These mycotoxins chosen due to their common presence in commodities and AFB1 is a toxin that has similar genus as OTA [20]. The signal responses recorded in Figure 3.7 revealing that the SPCE/GO-MnO_2NPs aptasensor has a high selectivity for OTA in comparison to the interfering compounds of AFB1, DON, and ZEA. By referring to the sensitivity analysis, the OTA readings for Rct were the lowest from its aptamer. The recorded Rct values at 1ng/mL were; 201.21 Ω (OTA), 215.45 Ω (AFB1), 228.58 Ω (DON), and 222.39 Ω (ZEA). The aptasensor detected OTA with a significantly lower Rct compared to DON, validating the "lock and key" interaction between the OTA aptamer and its target.

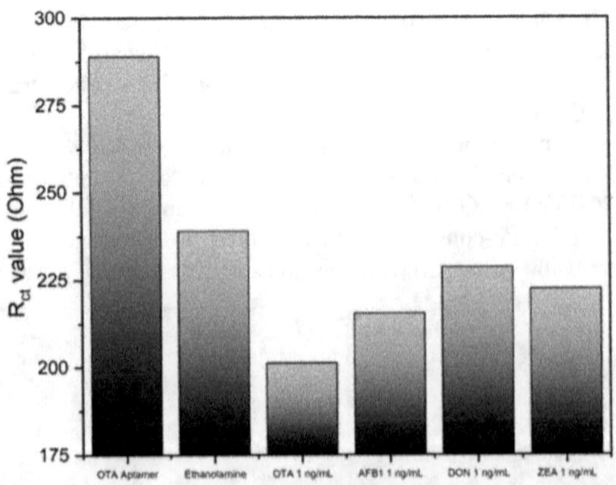

Fig. 3.8 Selectivity test using OTA (as bio-target) with three counter mycotoxins such as AFB1, DON, and ZEA.

Spiking Test: A Real-life Sample Analysis. After verifying the selective detection of OTA, we employed the aptasensor to detect OTA in cocoa powder and rolled oats. OTA is a prevalent and toxic contaminant found in various food commodities, particularly in cocoa powder and oats, which among highly infected agricultural products [4], [21]. Both samples were prepared as powders, diluted with PBS to concentrations of 1, 0.01, and 0.001 mg/mL, and tested on the OTA aptamer surface. The optimal dilution to spike with OTA was determined to 1 mg/mL, which produced the slightest background noise, resulting in Rct values of 208.0 Ω for cocoa powder and 157.0 Ω for oats.

As in Figure 3.8, the OTA was spiked into 1 mg/mL of cocoa powder and oats at concentrations of 1 fg/mL, 1 pg/mL, and 1 ng/mL. the Rct values for cocoa powder were 217.0 Ω , 213.0 Ω , and 212.0 Ω , respectively, while for oats, they were 180.0 Ω , 175.0 Ω, and 169.0 Ω. The recovery rates were 102% to 105% for cocoa samples and 107% and 115% for oat samples. The developed SPCE/GO-MnO_2NPs biosensor successfully detected OTA in cocoa powder and oats at quantities below the maximum levels sets by the European Union (EU), demonstrating its effectivess in real-life sample analysis.

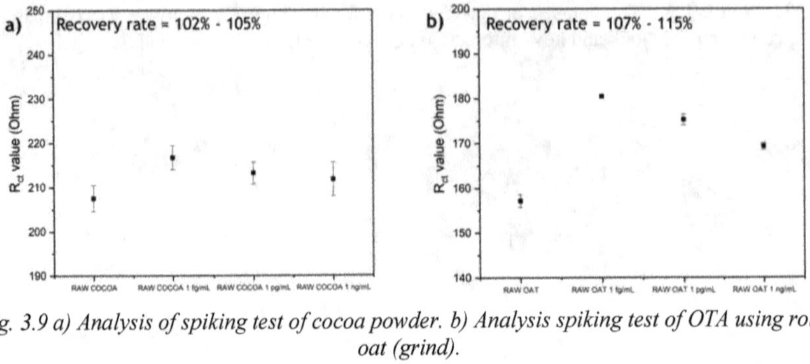

Fig. 3.9 a) Analysis of spiking test of cocoa powder. b) Analysis spiking test of OTA using rolled oat (grind).

Conclusion

In summary, the constructed electrochemical biosensor based on GO-MnO_2NPs modified SPCEs was successfully developed. Notably, the aptasensor analyzed OTA with a low detection limit of 3.33 ng/mL, and was successfully applied to detect OTA through selectivity test by using interference compounds of AFB1, DON, and ZEA. Moreover, the aptasensor also effectively applied to detect OTA on a real-life sample of cocoa powder and rolled oat. However, as the materials development specifically MnO_2 nanoparticles has only one-way of synthesis according to its flow-chart, a better approach of synthesis as well as the application of binding compounds use need to be explore in the near future.

Acknowledgement

This work was supported by University Research Internal Fund (URIF) with Grant No.: 015LB0-023, Universiti Teknologi PETRONAS (UTP)

References

[1] M. Bryła *et al.*, "Toxico-pathological effects of ochratoxin A and its diastereoisomer under in ovo conditions and in vitro evaluation of the toxicity of these toxins against the embryo Gallus gallus fibroblast cell line," *Poult. Sci.*, vol. 102, no. 2, pp. 1–11, 2023. https://doi.org/10.1016/j.psj.2022.102413

[2] W. C. Liu *et al.*, "Ochratoxin A as an alarming health threat for livestock and human: A review on molecular interactions, mechanism of toxicity, detection, detoxification, and dietary prophylaxis," *Toxicon*, vol. 213, no. April, pp. 59–75, 2022. https://doi.org/10.1016/j.toxicon.2022.04.012

[3] V. Vukojevi *et al.*, "Enzymatic glucose biosensor based on manganese dioxide nanoparticles decorated on graphene nanoribbons," *J. Electroanal. Chem.*, vol. 823, pp. 610–616, 2018. https://doi.org/10.1016/j.jelechem.2018.07.013

[4] H. Wang *et al.*, "A fluorescence and surface-enhanced Raman scattering dual-mode aptasensor for rapid and sensitive detection of ochratoxin A," *Biosens. Bioelectron.*, vol. 207, no. March, p. 114164, 2022. https://doi.org/10.1016/j.bios.2022.114164

[5] D. A. Oliveira, J. L. Lutkenhaus, and J. R. Siqueira, "Building up nanostructured layer-by-layer films combining reduced graphene oxide-manganese dioxide nanocomposite in supercapacitor electrodes," *Thin Solid Films*, vol. 718, no. December 2020, 2021. https://doi.org/10.1016/j.tsf.2020.138483

[6] R. K. Mishra, A. Hayat, G. Catanante, C. Ocaña, and J.-L. Marty, "A label free aptasensor for Ochratoxin A detection in cocoa beans: An application to chocolate industries," *Anal. Chim. Acta*, vol. 889, pp. 106–112, 2015. https://doi.org/10.1016/j.aca.2015.06.052

[7] K. Vijayalakshmi, A. Renitta, K. Alagusundaram, and A. Monamary, "Novel two-step process for the fabrication of MnO2 nanostructures on tantalum for enhanced electrochemical H2O2 detection," *Mater. Chem. Phys.*, vol. 214, pp. 431–439, 2018. https://doi.org/10.1016/j.matchemphys.2018.04.108

[8] S. Centane and T. Nyokong, "Aptamer versus antibody as probes for the impedimetric biosensor for human epidermal growth factor receptor," *J. Inorg. Biochem.*, vol. 230, p. 111764, 2022. https://doi.org/10.1016/j.jinorgbio.2022.111764

[9] Y. Wang *et al.*, "Ascorbic acid-mediated in situ growth of gold nanostars for photothermal immunoassay of ochratoxin A," *Food Chem.*, vol. 419, 2023. https://doi.org/10.1016/j.foodchem.2023.136049

[10] N. M. Illias, V. Perumal, and M. S. Mohamed, "Highly Crystalline Manganese Dioxide Nanoparticles via One-Pot- Hydrothermal Technique with Graphene Oxide Paste," *Proc. Int. Conf. Glob. Optim. Its Appl. 2022 (ICoGOIA 2022)*, pp. 57–65, 2022, [Online]. Available: https://apc01.safelinks.protection.outlook.com/?url=https%3A%2F%2Ficogoia.utem.edu.my%2Fproceedings.html&data=05%7C01%7Cnabihah_18003516%40utp.edu.my%7C8b1f798250b042cc2fa308dbf52be0fb%7C84187be3037e41ec889ca150fe476432%7C0%7C0%7C638373341587651387%7CUnkn

[11] R. Al-Gaashani, A. Najjar, Y. Zakaria, S. Mansour, and M. A. Atieh, "XPS and structural studies of high quality graphene oxide and reduced graphene oxide prepared by different chemical oxidation methods," *Ceram. Int.*, vol. 45, no. 11, pp. 14439–14448, 2019. https://doi.org/10.1016/j.ceramint.2019.04.165

[12] M. Dinesh, C. Revathi, Y. Haldorai, and R. T. Rajendra Kumar, "Birnessite MnO2 decorated MWCNTs composite as a nonenzymatic hydrogen peroxide sensor," *Chem. Phys. Lett.*, vol. 731, no. May, p. 136612, 2019. https://doi.org/10.1016/j.cplett.2019.136612

[13] L. V. F. Oliveira, S. Bennici, L. Josien, L. Limousy, M. A. Bizeto, and F. F. Camilo, "Free-standing cellulose film containing manganese dioxide nanoparticles and its use in discoloration of indigo carmine dye," *Carbohydr. Polym.*, vol. 230, no. August 2019, p. 115621, 2020. https://doi.org/10.1016/j.carbpol.2019.115621

[14] Y. Qiu *et al.*, "Immobilization of manganese dioxide nanoparticles on modified poly 2,4-dichlorostyrene microspheres: A highly efficient and recyclable catalyst for borrowing hydrogen reactions," *Org. Chem. Front.*, vol. 6, no. 19, pp. 3420–3427, 2019. https://doi.org/10.1039/c9qo00892f

[15] T. Pimklang, A. Watthanaphanit, and P. Pakawatpanurut, "Novel green synthesis of graphene oxide-manganese dioxide using solution plasma process for energy storage," *Chem. Eng. J.*, vol. 442, no. P2, p. 136244, 2022. https://doi.org/10.1016/j.cej.2022.136244

[16] Z. Ni, X. Liang, L. Zhao, H. Zhao, B. Ge, and W. Li, "Tin doping manganese dioxide cathode materials with the improved stability for aqueous zinc-ion batteries," *Mater. Chem. Phys.*, vol. 287, no. May, p. 126238, 2022. https://doi.org/10.1016/j.matchemphys.2022.126238

[17] P. K. Kalambate *et al.*, "Progress, challenges, and opportunities of two-dimensional layered materials based electrochemical sensors and biosensors," *Mater. Today Chem.*, vol. 26, p. 101235, 2022. https://doi.org/10.1016/j.mtchem.2022.101235

[18] T. A. Burinaru *et al.*, "Electrochemical impedance spectroscopy based microfluidic biosensor for the detection of circulating tumor cells," *Mater. Today Commun.*, vol. 32, no. June, 2022. https://doi.org/10.1016/j.mtcomm.2022.104016

[19] Y. Hou *et al.*, "Development of a Nafion-MWCNTs and in-situ generated Au nanopopcorns dual-amplification electrochemical aptasensor for ultrasensitive detection of OTA," *Food Chem.*, vol. 403, no. May 2022, p. 134375, 2023. https://doi.org/10.1016/j.foodchem.2022.134375

[20] I. Beukes, G. J. Marais, and K. Jacobs, "The occurrence of mycotoxigenic fungi in abalone feed in South Africa," vol. 2338, 2018. https://doi.org/10.2989/1814232X.2018.1534699

[21] P. Kumar *et al.*, "Ochratoxins in food and feed: Occurrence and its impact on human health and management strategies," *Toxicon*, vol. 187, no. July, pp. 151–162, 2020. https://doi.org/10.1016/j.toxicon.2020.08.031

Materials Research Proceedings 53 (2025) 637-641 https://doi.org/10.21741/9781644903575-64

Carbon quantum dots enhanced high valence matel oxidation for decontamination

YUNPENG Wang[1,a], SHAOBIN Wang[1,b]

[1]School of Chemical Engineering, The University of Adelaide, Adelaide, SA 5005, Australia

[a]yunpeng.wang@adelaide.edu.au, [b]shaobin.wang@adelaide.edu.au

Keywords: Oxidation, Ferrate, Carbon Quantum Dots, Decontamination

Abstract. Fe(VI) demonstrates remarkable versatility in water purification, with ongoing efforts to enhance its efficacy in removing pollutants. Recent investigations have revealed that, alongside intermediate iron species [Fe(IV)/Fe(V)], direct electron transfer (DET) plays a pivotal role in organic pollutant degradation within the Fe(VI)/carbon quantum dots (CQDs) system. Initially, 30 μM of Fe(VI) oxidized approximately 52% of phenol. Upon introduction of CQDs (4 mg/L), the oxidation rates increased significantly to 87%. CQDs not only mitigated the adverse effects of solution pH on Fe(VI) pollutant abatement but also acted as catalysts, facilitating DET from organic pollutants to Fe(VI). Notably, DET mediated by CQDs proved to be a more efficient means of utilizing Fe(VI)'s oxidation potential compared to iron species oxidation alone. This DET mechanism offers a promising avenue for enhancing pollution control capabilities of Fe(VI).

Introduction

With escalating environmental challenges, decarbonization technologies have emerged as imperative tools for combatting pollution and reducing carbon emissions.[1] In the field of water treatment, advance oxidation processes (AOPs) are considered as promising approach for decontamination across various technologies. AOPs harness the power of highly reactive oxygen species to degrade and eliminate organic pollutants, offering efficient and sustainable means to mitigate environmental degradation [2].

Ferrate [Fe(VI)] is effective for the abatement of pollutants due to the oxidation capacity of high valence iron species and the adsorption capacity of in situ formed ferric(hydro) oxides [3]. Previous investigations have shown that the intermediate iron species [Fe(IV)/Fe(V)] formed during the reduction of Fe(VI) exhibit higher reactivity towards certain organic pollutants than Fe(VI).[4] However, despite their high reactivity towards pollutants, intermediate iron species are also reactive with water and background constituents such as dissolved organic matter, $H_2PO_4^-$, HCO_3^-.[5]. The decomposition or competitive consumption of intermediate iron species in aqueous environment negatively impacts their potential for pollutant abatement. By comparison, DET process has advantage of resisting negative impact of background constituents, resilience over a wide range of pH and maintained the oxidation efficiency, high selectivity toward electron- rich organic pollutants and avoiding radical self-quenching or autodecomposition of intermediate metal species during the reaction. Therefore, developing DET process in AOPs is in need. Carbonaceous materials have conjugated carbon network, which can facilitate electron transfer in the carbon configuration.[6] Carbon quantum dots (CQDs) are a new-fangled type of nanomaterial. They are made up of unique, quasi- spherical carbon nanoparticles with dimensions less than 10 nm [7]. In this study, the catalytic efficiency of CQDs on Fe(VI) oxidation are explored and their potential DET mechanism was investigated, which pave the way for decarbonization in water treatment.

Materials and Methods

Phenol, and 2,2′-azino-bis(3-ethylbenzothiazoline-6-sulfonate) (ABTS) were purchased from Sigma- Aldrich (Germany). All other chemicals were at least of analytical grade and used without further purification. All experimental solutions were prepared with ultrapure water produced by a Millipore system. K_2FeO_4 [Fe(VI)] was synthesized in the laboratory. To prevent its oxidation by air, Fe(VI) solution was freshly prepared for each experiment by dissolving solid crystals of Fe(VI) in Milli-Q water. The solution was filtered through a hydrophilic acetate fiber membrane ($\varphi = 0.22$ μm) and the concentration of Fe(VI) was determined using a UV-Visible spectrophotometry at 510 nm. The concentration of phenol was analyzed with high-performance liquid chromatography (HPLC, Waters 2995) equipped with a Symmetry C18 column (4.6 × 150 mm, 5 μM of particle size, Waters) and a photodiode array detector (PDAD, Waters 2998). The mobile phase consisted of methanol (phase A) and 0.1% acetic acid (phase B) with a ratio of 40:60 (v:v) at a flow rate of 1 mL/min. CQDs was synthesized by an electrochemical exfoliation method. Two commercial graphite rods (99.99%) were inserted into 500 mL of the electrolyte (pH = 5.0, 30 V, direct current power) at a distance of 6 cm. After electrolysis for 72 h, the transparent solution turned into a homogenous dark-yellow solution. Then, the obtained solution was filtered with a PTFE syringe filter ($\varphi = 0.22$ μm) and centrifuged at 15000 rpm for 40 min to remove the large particles.

Results and Discussion

Effect of CQDs on Fe(VI) Oxidation. The pollution control ability of Fe(VI) and Fe(VI)/CQDs were comparatively compared by oxidation of phenol. In the reaction with Fe(VI) (30 μM, 5.94 mg/L), around 52% of phenol was oxidized. When 4 mg/L of CQDs were added into the polluted water, less than 2% of phenol was abated. The effect of CQDs on the oxidation or adsorption was inferior. For the combination of Fe(VI) (30 μM, 5.94 mg/L) with CQDs (4 mg/L), the determined oxidation ratios of phenol increased to 87% (Figure 1).

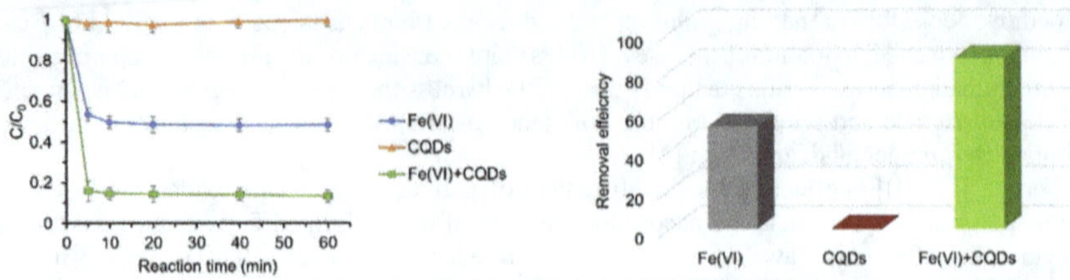

Fig. 1. The abatement of phenol in Fe(VI), CQDs, and Fe(VI)/CQDs system, respectively.

Effect of pH. The oxidizing ability of Fe(VI) is greatly influenced by pH. We examined the impact of pH on phenol oxidation within the Fe(VI)/CQDs system (Figure 2). Our findings demonstrate that the Fe(VI)/CQDs combination exhibits notably higher oxidation efficiency across all pH ranges compared to Fe(VI) alone. Specifically, at pH 5.0 and 6.0, the Fe(VI)/CQDs system achieved oxidation efficiencies of 73% and 81%, respectively, representing a significant enhancement of 15% and 19% over Fe(VI) alone, which oxidized 58% and 64% of phenol within 5 minutes under the same conditions. This heightened reactivity of both systems can be attributed to the potent oxidative capacity of Fe(VI) under acidic conditions, where direct Fe(VI) oxidation predominates the reaction. At pH 7.0, the Fe(VI)/CQDs system exhibited a 36% increase in oxidation efficiency compared to Fe(VI). Despite displaying moderate intermediate Fe oxidative capacity under neutral conditions, this system yielded the highest oxidation efficiency. Although alkaline conditions promote the stability of Fe(VI), its reactivity is compromised. At pH 8.0, Fe(VI) oxidized 46% of phenol within 10 minutes, while the Fe(VI)/CQDs system achieved a significantly higher oxidation rate, oxidizing 78% of phenol within the same timeframe. Similarly, although Fe(VI) oxidation of phenol was hindered at pH 9.0, the Fe(VI)/CQDs system still exhibited a slight enhancement compared to Fe(VI). Overall, the Fe(VI)/CQDs system demonstrates promising pH-resistant potential for phenol oxidation.

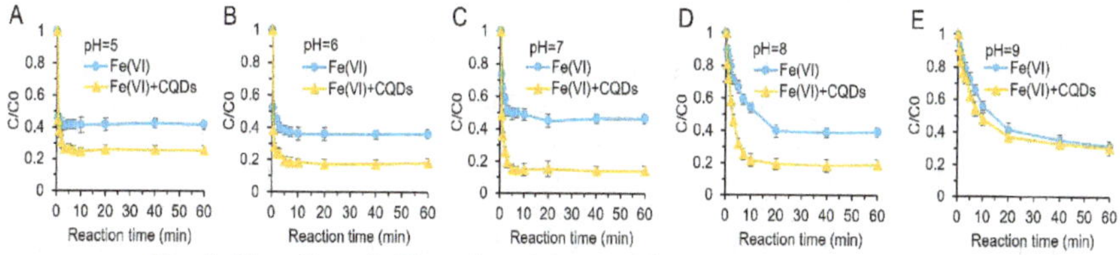

Fig. 2. The effect of pH on phenol removal by Fe(VI) and Fe(VI)/CQDs.

Properties of CQDs Before and After the Reaction with Fe(VI). The physicochemical characteristics of CQDs were examined both prior to and following their interaction with Fe(VI). The synthesized CQDs exhibited high dispersion, uniformity in size, and a predominantly spherical morphology (Figure 3A, B). Their average diameter measured at 6.67 nm aligned with size distributions determined by Malvern analysis (Figure 3C). FTIR analysis indicated the presence of characteristic stretching vibrations at 1647 cm^{-1} (C=O), 1540 cm^{-1} [v(C=C), indicative of skeletal polycyclic aromatic hydrocarbons], and 1337 cm^{-1} (C-O-C, suggesting the presence of CQDs in a complex structure). Remarkably, there was minimal alteration in the intensity of these peaks observed pre- and post-reaction with Fe(VI) (Figure 3B). XPS analysis revealed that the C 1s photoelectron spectrum of CQDs could be deconvoluted into distinct peaks representing C=C, C-C, C-H (at 284.3 and 285 eV), C-OH (at 286 eV), C=O (at 287.6 eV), and COOH (at 288.6 eV) (Figure 3C). Additionally, examination of the O 1s XPS spectrum corroborated the presence of oxygen functional groups, notably divided into C-O (at 533.2 eV) and C=O (at 531.6 eV) signals (Figure 3D). These XPS findings indicated that both the quantity and types of functional groups within the CQDs remained relatively unchanged before and after exposure to Fe(VI), a consistency mirrored by the FTIR results. This suggests that CQDs may indeed serve as catalysts, facilitating the Fe(VI) oxidation of organic pollutants.

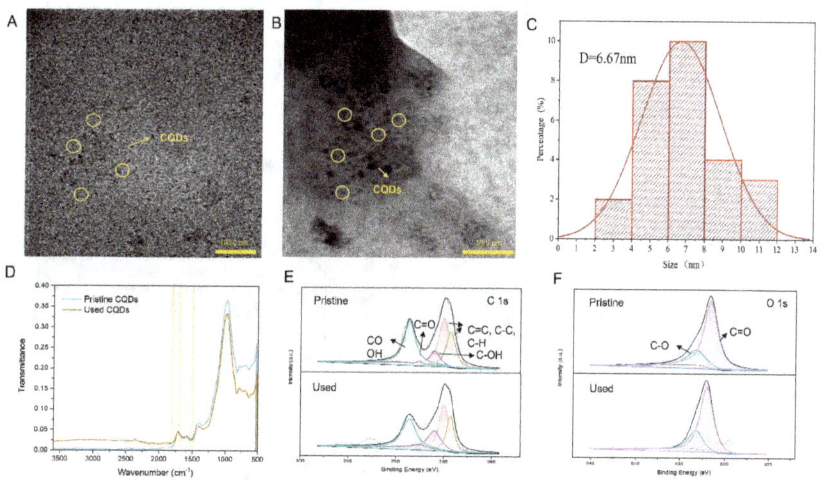

Fig. 3. (A, B) TEM image of CQDs; (C) the particle size distribution of CQDs; (D) FTIR results about CQDs before and after reaction with Fe(VI); (E) High resolution XPS of C1s before and after the reaction with Fe(VI); (F) High resolution XPS of O before and after the reaction with Fe(VI).

Reaction Mechanism. Intermediate iron species are known to play a critical role in the abatement of many pollutants when using Fe(VI).[8, 9] Carbonaceous materials, including CNTs, graphene oxide, and biochar had been demonstrated to promote the formation of intermediate iron species in Fe(VI) oxidation process.[10, 11] In this study, the possible contribution of intermediate iron species to the abatement of pollutants in Fe(VI)/CQDs system was investigated by PMSO probe. When 30 μM of Fe(VI) was reacted with 10 μM of PMSO at pH 7.0, 9.4 μM of PMSO was oxidized with the formation of 9.4 μM of $PMSO_2$ by intermediate Fe (Figure 4A). However, when Fe(VI)/CQDs was reacted with 10 μM of PMSO, only 6.0 μM of PMSO was oxidized with the formation of 6.0 μM of $PMSO_2$ (Figure 4B). The oxidation ratio of PMSO in the Fe(VI)/CQDs system was 34% lower than that in the Fe(VI) system, which indicated that intermediate iron species were not responsible for the enhanced oxidation.

CQDs have excellent ability of electron transfer. Therefore, the possibility of direct electron transfer was examined by chronoamperometric experiment (Figure 4C). The sharply changed current output with phenol addition revealed that CQDs could mediate the electron transfer process from pollutants to the CQDs-Fe(VI) in a wide range. The decay of Fe(VI) in different systems were further carried out to identify the electron transfer regime. The addition of CQDs cannot accelerate the decomposition of Fe(VI), while with further addition of phenol, the content of Fe(VI) was almost depleted in 10 min (Figure 4D). This evidenced that electrons can be transferred from phenol to Fe(VI) though CQDs.

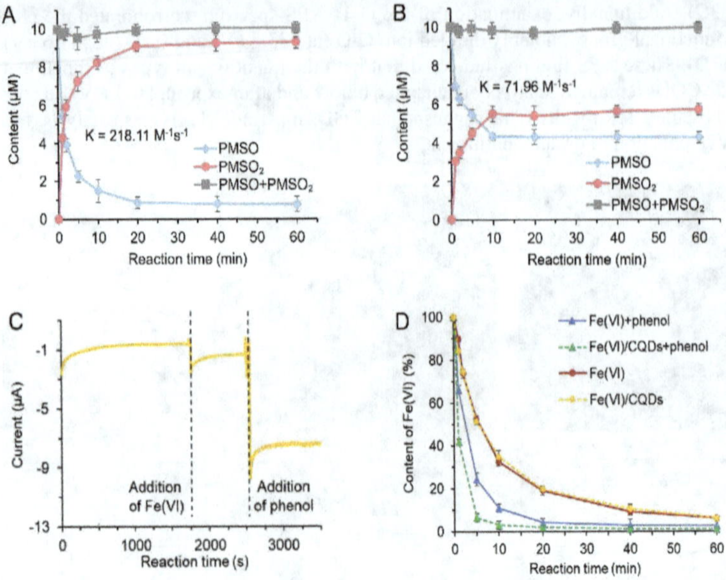

Fig. 4. (A) The transformation of PMSO by Fe(VI); (B) the transformation of PMSO by Fe(VI)/CQDs; (C) Amperometric i-t curve measurements using CQDs-GCE as the working electrode; (D) the decay of Fe(VI) species in relevant systems.

Conclusion

Herein, it was found that besides stimulating Fe(IV)/Fe(V) formation, utilization of CQDs as electron shuttles to facilitate direct electron transfer from pollutants to Fe(VI) is also an effective method for enhancing the abatement of pollutants. The CQDs mediated Fe(VI) oxidation process showed a remarkable 35% improvement in oxidation. CQDs are capable of complexing with Fe(VI) rather than engaging in redox reactions similar to other carbonaceous materials in Fe(VI) system. Moreover, the Fe(VI)/CQDs system had excellent efficiency over a wide range of pH values. Considering that the preparation of CQDs is easy and cost-effective, the integration of Fe(VI) and CQDs is promising for the elimination of contaminants from polluted water.

References

[1] Ekins, P.; Zenghelis, D., (2021). The costs and benefits of environmental sustainability, Sustain Sci, 16, 3, 949-965. https://doi.org/10.1007/s11625-021-00910-5

[2] Deng, Y.; Zhao, R., (2015). Advanced Oxidation Processes (AOPs) in Wastewater Treatment, Current Pollution Reports, 1, 3, 167-176. https://doi.org/10.1007/s40726-015-0015-z

[3] Liu, Y.; Wang, L.; Wang, X.; Huang, Z.; Xu, C.; Yang, T.; Zhao, X.; Qi, J.; Ma, J., (2017). Highly efficient removal of trace thallium from contaminated source waters with ferrate: Role of in situ formed ferric nanoparticle, Water Res, 124, 149-157. https://doi.org/10.1016/j.watres.2017.07.051

[4] Sharma, V. K.; Feng, M.; Dionysiou, D. D.; Zhou, H. C.; Jinadatha, C.; Manoli, K.; Smith, M. F.; Luque, R.; Ma, X.; Huang, C. H., (2022). Reactive High-Valent Iron Intermediates in Enhancing Treatment of Water by Ferrate, Environ Sci Technol, 56, 1, 30-47. https://doi.org/10.1021/acs.est.1c04616

[5] Wang, Y. P.; Liu, Y. L.; Tian, S. Q.; Yang, J. J.; Wang, L.; Ma, J., (2021). Straw biochar enhanced removal of heavy metal by ferrate, J Hazard Mater, 416, 126-128. https://doi.org/10.1016/j.jhazmat.2021.126128

[6] Liu, Y.; Chen, C.; Duan, X.; Wang, S.; Wang, Y., (2021). Carbocatalytic ozonation toward advanced water purification, Journal of Materials Chemistry A, 9, 35, 18994-19024. https://doi.org/10.1039/D1TA02953C

[7] Khan, M. E.; Mohammad, A.; Yoon, T., (2022). State-of-the-art developments in carbon quantum dots (CQDs): Photo-catalysis, bio-imaging, and bio-sensing applications. Chemosphere, 302, 134815. https://doi.org/10.1016/j.chemosphere.2022.134815

[8] Shao, B.; Dong, H.; Sun, B.; Guan, X., (2019). Role of Ferrate(IV) and Ferrate(V) in Activating Ferrate(VI) by Calcium Sulfite for Enhanced Oxidation of Organic Contaminants, Environ Sci Technol, 53, 2, 894-902. https://doi.org/10.1021/acs.est.8b04990

[9] Yang, T.; Mai, J.; Cheng, H.; Zhu, M.; Wu, S.; Tang, L.; Liang, P.; Jia, J.; Ma, J., (2022). UVA-LED-Assisted Activation of the Ferrate(VI) Process for Enhanced Micropollutant Degradation: Important Role of Ferrate(IV) and Ferrate(V), Environ Sci Technol, 56, 2, 1221-1232. https://doi.org/10.1021/acs.est.1c03725

[10] Pan, B.; Feng, M.; McDonald, T. J.; Manoli, K.; Wang, C.; Huang, C.-H.; Sharma, V. K., (2020). Enhanced ferrate(VI) oxidation of micropollutants in water by carbonaceous materials: Elucidating surface functionality. Chemical Engineering Journal, 398, 125607. https://doi.org/10.1016/j.cej.2020.125607

[11] Sun, S.; Jiang, J.; Qiu, L.; Pang, S.; Li, J.; Liu, C.; Wang, L.; Xue, M.; Ma, J., (2019). Activation of ferrate by carbon nanotube for enhanced degradation of bromophenols: Kinetics, products, and involvement of Fe(V)/Fe(IV). Water Res, 156, 1-8. https://doi.org/10.1016/j.watres.2019.02.057

Decarbonization Technology: ICDT2024 Materials Research Forum LLC
Materials Research Proceedings 53 (2025) 642-650 https://doi.org/10.21741/9781644903575-65

Kinetic study of biomass activated carbon impregnated with copper (AC/Cu) as a photocatalyst in degradation of ammoniacal nitrogen

SELVARAJ Hemaraj[1,a], YAKUB Ibrahim[1,b *], MOHAMAD SAID Khairul Anwar[1,c], MOHAMED AMIN Mohamed Afizal[1,d], AHMAD ZAUZI Nur Syuhada[1,e], and BAINI Rubiyah[1,f]

[1]Department of Chemical Engineering & Energy Sustainability, Faculty of Engineering, Universiti Malaysia Sarawak, Jalan Dato Mohd Musa, 94300 Kota Samarahan, Sarawak, Malaysia

[a]hemarajselvaraj@gmail.com, [b]yibrahim@unimas.my, [c]mskanwar@unimas.my, [d]mamafizal@unimas.my, [e]aznsyuhada@unimas.my, [f]ruby@unimas.my

Keywords: Biomass, Activated Carbon, Catalyst, Environment, Ammoniacal Nitrogen

Abstract. Generation of wastewater including leachate that contains ammoniacal nitrogen (AN) is increasing and requires urgent solution. The current status of the photocatalyst development for the degradation of this pollutant is promising but further improvement is desired. This study explored the possibility of using biomass activated carbon as a catalyst support to provide more surface area. Activated carbon from coconut shell was used to support 30 wt% copper catalyst in a wet impregnation procedure for the catalyst synthesis. Characterization of the photocatalyst (AC/Cu) was performed using scanning electron microscopy and energy dispersive X-ray (SEM-EDX), BET surface area analysis, thermogravimetric and differential thermal analysis (TG/DTA), and Fourier-Transform Infra-Red spectroscopy (FTIR). It was revealed that the catalyst has BET surface area of 430 m^2/g, 0.254 cc/g total pore volume, 60% carbon content, and surface functional groups that include nitriles, alkenes and aromatics at low intensity. The batch experiment on AN removal showed the best catalyst dosage in this system is 0.5 g while 90 spm shaker speed resulted in the highest AN removal. The initial AN concentration adversely affected the performance of the catalyst where the higher the concentration, the lower the removal. Between pseudo-first and pseudo-second order kinetic models, the latter exhibited the best fit to the experimental data explaining the mechanism of AN removal is attributable to chemisorption. The photocatalyst supported over biomass carbon has larger BET surface area compared to the conventional titania-based photocatalyst and the system could be further improved to cater higher concentration of AN for practical applications.

Introduction

Malaysia is a growing nation that is quickly becoming more urbanized and seeing a rise in its overall population. It is anticipated that the population would reach 37.4 million by the year 2030. This state has driven the country to an increasing demand for the usage of water, which has led to an increase in the level of water pollution correspondingly [1]. The surge in population has also resulted in a considerable increase in the volume of solid waste. When precipitation or rainfall runs through the garbage in a landfill, the waste that was collected there produces a liquid waste that is known as leachate. This waste product is comprised of a complex combination of chemicals. The seepage of leachates into soil, surface water, and subsurface is a common problem associated with landfills. This problem creates a number of dangers that have a negative effect on the ecosystem. This pollution has an effect on water sources, and as a result, it may present a significant risk to both human and animal life, as well as marine life, and potentially lead to the downfall of the ecology overall [2].

Ammoniacal nitrogen (AN) is one of the most important characteristics and indicators that determines the quality of water. The inorganic carbon species that have the most reduced forms are those that include AN, which is the sum of unionized ammonia and ammonium. In the majority of natural fluids, AN presents itself largely as ammonium in the pH spectrum; however, the proportion of unionized ammonia increases with increasing pH levels as well as temperature [3]. The presence of AN has an impact on marine habitats, which may have subsequent negative repercussions for both economic development and the health of humans.

A material known as photocatalyst is one that takes in light, raises the light's overall energy level by absorbing it, and then transfers that energy to a reacting agent so that a chemical reaction may take place. Fujishima and Honda are credited for being the pioneers in the development of the photocatalyst in 1972. Thereafter, the use of a photocatalyst in many areas has attracted vast research and development initiatives, with the primary focus being placed on the implementation to pollution control [4]. However, this material, usually titania-based, suffers from several challenges especially low surface area for pollutant adsorption [5]. Hypothetically, supporting a photocatalyst over porous material like activated carbon could provide more sites for adsorption and thus photodegradation. Activated carbon derived from biomass waste such as coconut shell and palm kernel shell are known to offer benefits due to its cost-effectiveness [5]. Copper (Cu) is one of the best photocatalysts besides titania, because of their availability, cost, and ability to provide high photoexcited reducing power, making it an attractive photo redox catalysts [6]. Therefore, this study explored the possibility of utilizing biomass waste and Cu in photodegradation of AN.

Materials and Method

Catalyst synthesis. The AC was ground to obtain size less than 600 nm. The copper impregnated with activated carbon (AC/Cu) was prepared via wet impregnation method. Copper sulfate pentahydrate 99.0 % (Merck, USA) was dissolved in acetone 99.8 % (Merck, USA) before stirring for 1 hour. After that, the mixture was heated to 60 °C for 1 day until complete dryness. The resulting solid mixture will be calcined in a nitrogen flow at 600 °C for 1 hour using a tube furnace.

Characterization. Scanning electron microscopy (SEM) and energy dispersive X-ray (EDX) were used to analyze the surface morphology of the AC and AC/Cu by using TM4000 Plus SEM-EDX (Hitachi, Germany). Besides, surface area was studied using Brunauer-Emmett-Teller (BET) method in Autosorb iQ (Quantachrome, USA). Meanwhile, thermal behavior was studied using thermogravimetric analysis (TGA) and differential thermal analysis (DTA) in DTG-60H TG/DTA (Shimadzu, Japan). In addition, surface functional groups were observed of their presence using IRAffinity-1 FTIR spectrometer (Shimadzu, Japan).

Experiment. The AN removal experiment was conducted in a customized UV-irradiated batch reactor. A diluted ammonia solution was prepared in a beaker (150 mL working volume) to simulate the synthetic landfill leachate condition of certain AN concentration (5, 10, 15 or 20 ppm). Then, a particular amount of AC/Cu (0.25, 0.50, 0.75 or 1.00 g) was added in. The reactor was closed to isolate the system from external light sources and only be irradiated under UV light (UV-C lamp at 254 nm). The shaker speed was set to a desired value (35, 65 or 90 strokes per minute) and maintained at constant room temperature using a water bath. In an interval time of 20 minutes, by using the DR900 Calorimeter (Hanna, USA), an aliquot was extracted from the system to determine the concentration of the AN.

Materials Research Proceedings 53 (2025) 642-650 https://doi.org/10.21741/9781644903575-65

The AN removal efficiency is defined as the percentage of AN removed and illustrated in equation 1:

$$R = \frac{(C_0 - C_t)}{C_0}\% \quad (1)$$

where, R is the AN removal efficiency (%); C_0 is the initial AN concentration (ppm); and C_t is the AN concentration at time t.

Two kinetics models are selected – pseudo-first order (equation 2) and pseudo-second order (equation 3) kinetics models.

$$log(q_e - q_t) = log(q_e) - k_1 t \quad (2)$$

$$\frac{t}{q_t} = \frac{1}{k_2 q_e^2} - \frac{1}{q_e} \quad (3)$$

where, q_e is the adsorption capacity at equilibrium (mg/g); q_t is the adsorption capacity at time t; k_1 and k_2 are kinetic constants for first and second order model, respectively; and t is time (s).

Results and Discussion

Surface morphology. The images obtained from SEM analysis are shown in Fig. 1 to observe the surface characteristics of the parent carbon (AC) and the resulted catalyst (AC/Cu).

a) b)

Fig. 1. Scanning electron micrographs for a) AC, and b) AC/Cu

It can be seen from the figure that the AC surface has a relatively smooth and flat appearance, and it is dispersed with micropores. Following the calcination process, some minute copper granules are observable at the exterior surface of the AC, displaying an uneven and spiky structure. This made the AC/Cu to exhibit the appearance of a powdery agglomeration which was also reported by other study [4].

Fig. 2a shows the energy dispersive X-ray spectrum for AC/Cu where it shows presence of copper (Cu), sulphur (S), silicon (Si), carbon (C), oxygen (O) and nitrogen (N). Cu can be seen dissipated over the carbon surface after the impregnation process as also shown in other study [4]. The elemental composition as obtained from the EDX is shown in Fig. 2b. The deviation from the copper design loading composition of 30% is about 2% which could be due to the limitation of the synthesis procedure.

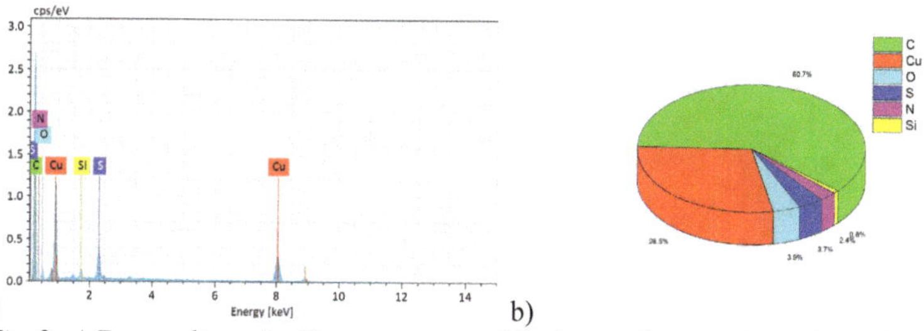

a) b)

Fig. 2. a) Energy dispersive X-ray pattern, and b) elemental composition, for AC/Cu

Surface area. Brunauer-Emmett-Teller (BET) analysis was used to analyse and compare the surface area, average pore diameter, and total volume of AC and AC/Cu. The results are tabulated in Table 1. Even though both materials are still considered as microporous, the synthesis process had caused some wall structures to collapse due to the high calcination temperature and the impregnated copper filled in the pores, reducing the numbers of micropores on the carbon surface [7]. Besides,. Therefore, the total pore volume and BET surface area can be seen to decrease. However, the BET surface area of AC/Cu is considerably higher than the conventional photocatalyst titania nanoparticles which has a surface area of around 200 m^2/g [8].

Table 1. Comparison of surface area between AC and AC/Cu

Sample	BET surface area (m^2/g)	Average pore diameter (nm)	Total pore volume (cc/g)
AC	501.376	0.122	0.306
AC/Cu	431.609	0.118	0.254

Thermal behaviour. Fig. 3 shows the thermogravimetric (TGA) and differential thermal (DTA) graph for AC and AC/Cu. Both analysis showed the presence of adsorbed water at the first stage between 50 to 200 °C, carboneceous material at the second stage between 400 to 600 °C and ash at the final stage above 600 °C [9]. After the impregnation process, the carbonaceous content decreased tremendously from 67 wt% to 40 wt% while the ash content increased from 19 wt% to 40 wt% due to the inclusion of copper which decomposed at higher temperatures.

Surface functional groups. Fig. 4 shows the Fourier-Transform Infra-Red (FTIR) peaks for both materials. Most of the surface functional groups presence on AC, including nitriles (around 2400 cm^{-1}), unsaturated alkenes (around 1500 cm^{-1}) and aromatics (around 600 cm^{-1}) [10] were highly reduced, after the synthesis.

Effect of catalyst dosage. Fig. 5 shows the effect of increasing catalyst dosage on the ammoniacal nitrogen (AN) removal. It can be observed that AC/Cu dosage of 0.5 g has the highest removal efficiency at the end of 120 minutes with 37%, followed by 1.0 g dosage with 36%, 0.75 g dosage with 35% and the least efficient was 0.25 g of AC/Cu dosage with just 31%. According to Jiang et. al. (2018) [4], higher catalyst dosage could provide more active sites for the reaction to take place, while lower dosage of photocatalyst increased the competitive molecules for the limited number of sites. However from Fig. 5, this explanation is valid only until 0.5 g of dosage whereas higher catalyst dosage showed lower removal efficiency. This is highly due to the agglomeration effects at higher dosage that reduced the number of available active sites.

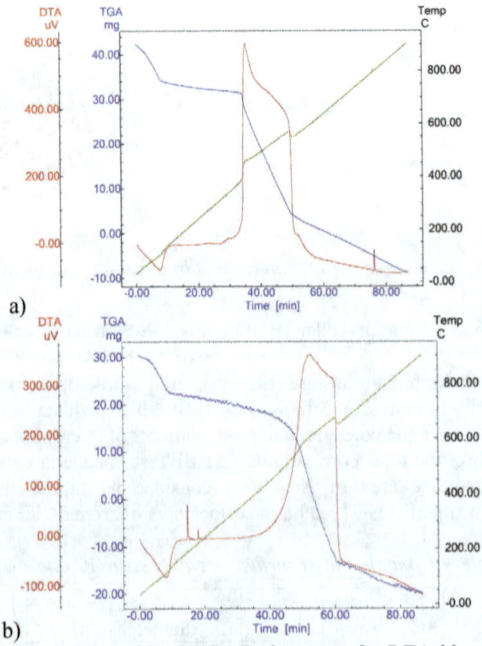

Fig. 3. TGA and DTA for a) AC, and b) AC/Cu. Red curve is for DTA, blue for TGA and green is temperature profile.

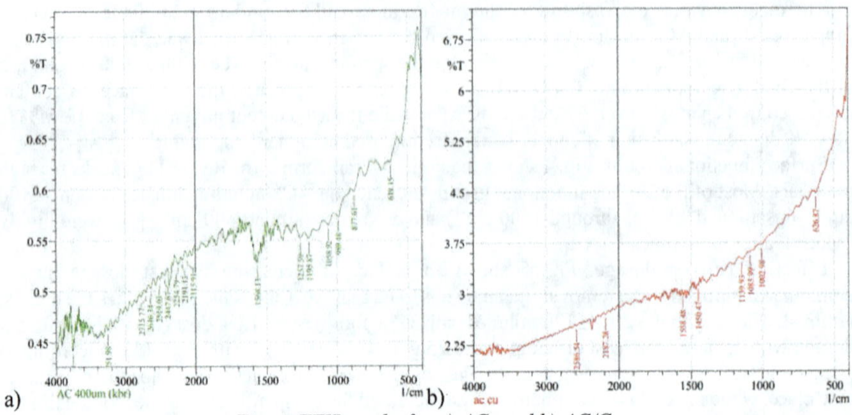

Fig. 4. FTIR peaks for a) AC, and b) AC/Cu

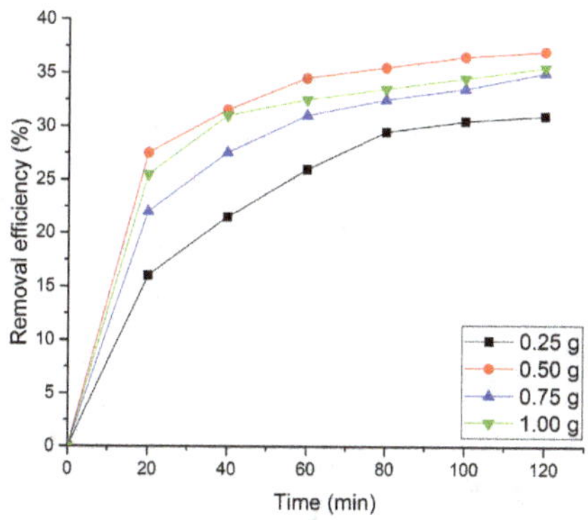

Fig. 5. Effect of AC/Cu dosage on AN removal at 20 mg/L initial AN concentration, 90 spm and 25 °C

Effect of shaker speed. Fig. 6 shows the effect of increasing the shaker speed on the AN removal. As the speed increased, the removal efficiency increased where the highest final removal at 120 min was 37% when the speed was at maximum. At high shaker speed, the mixing between the catalyst particles and the solution is more vigorous allowing the contact between the active sites and the reactant. The effective contact is also contributed by the thinner interfacial film formed between the bulk solution and the particle surface.

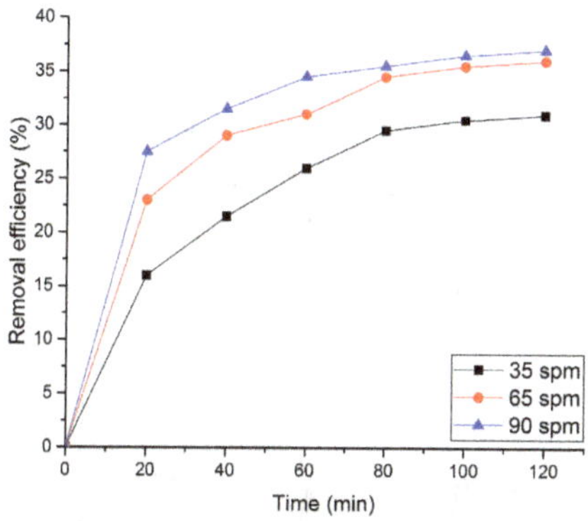

Fig. 6. Effect of shaker speed on AN removal at 20 mg/L initial AN concentration, 0.5 g catalyst dosage and 25 °C

Effect of initial concentration. Fig. 7 shows the effect of increasing the initial AN concentration on the AN removal. According to the figure, 5 ppm of initial AN has the highest removal efficiency of 78%, followed by 10 ppm with 73%, 15 ppm with 47% and 20 ppm with the lowest removal efficiency of 37%. Thus, the lower the concentration, the higher the removal efficiency. This is because less amount of molecules competed against the limited number of sites available in the system [11].

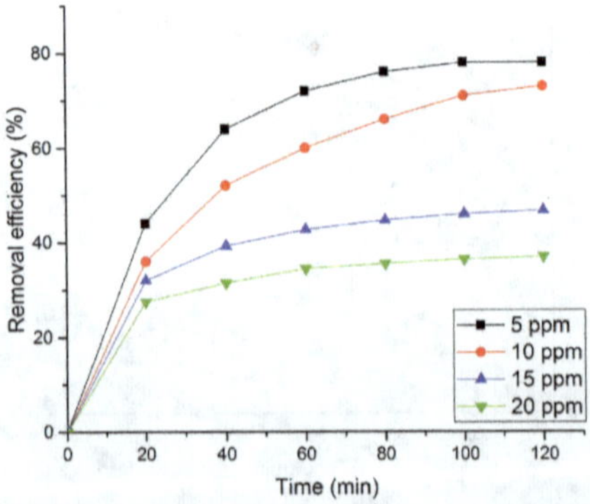

Fig. 7. Effect of initial AN concentration on AN removal at 0.5 g catalyst dosage, 90 spm and 25 °C

Kinetic analysis. The plots for the two selected kinetic models are shown in Fig 8. The linearized plot assuming pseudo-first order kinetic exhibited low correlation with the experimental data where the R^2-values varied as low as 0.67 to only 0.96. Meanwhile, pseudo-second order kinetic model showed high correlation with the experimental data with R^2-values distributed between 0.98 to 0.99. Therefore, the removal of AN when using AC/Cu is said to follow the assumptions in pseudo-second order kinetic model which implies that chemisorption is the limiting step and adsorption is highly depending on the adsorption capacity. The parameters for this model is tabulated in Table 2 for all AN concentrations used in this study.

Table 2. Kinetic model parameters for removal of AN using AC/Cu

Pseudo-first order			
Initial concentration (ppm)	q_e	K_1	R^2
5	0.660	1.88E-04	0.916
10	0.235	2.83E-04	0.968
15	2.654	-1.21E-04	0.825
20	3.782	-6.56E-05	0.671
Pseudo-second order			
Initial concentration (ppm)	q_e	k_2	R^2
5	4.367	0.052	0.981
10	15.640	0.017	0.959
15	13.141	0.045	0.992
20	14.438	0.053	0.995

Materials Research Proceedings 53 (2025) 642-650 https://doi.org/10.21741/9781644903575-65

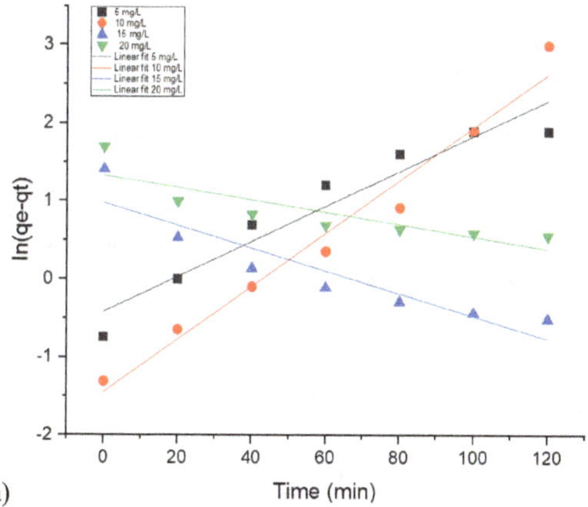

a)

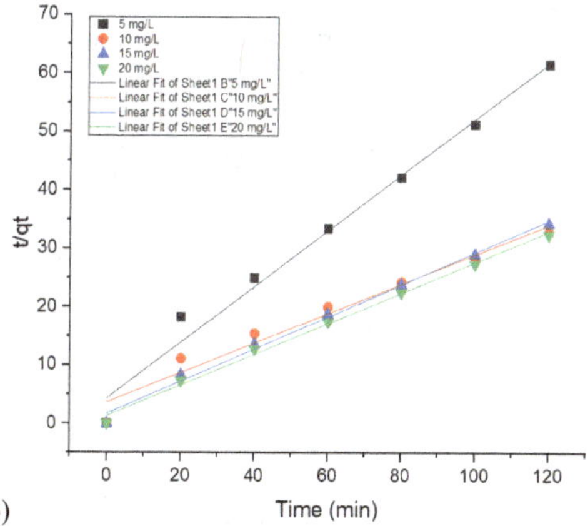

b)

Fig. 8. Kinetic plots of a) pseudo-first order kinetic, and b) pseudo-second order kinetic models

Conclusion

This study aimed to utilize biomass activated carbon and copper as a photocatalyst in degrading ammoniacal nitrogen (AN). The photocatalyst was synthesized using wet impregnation method with acetone as the solvent. Characterization was performed to investigate the surface morphology, surface area, thermal behavior, and surface functional groups. It was revealed that after the synthesis process, the carbon surface became rougher, BET surface area and pore volume decreased due to the collapse of micropores, carbonaceous content reduced, and surface functional groups were reduced in terms of nitriles, alkenes and aromatics. However, AC/Cu has higher surface area compared to the conventional photocatalyst which is titania-based. The photocatalyst was also tested in a batch isothermal reactor with varying catalyst dosage, shaker speed, and initial AN concentration. It showed that increasing the catalyst dosage increased the AN removal but only until 0.50 g (removal of 37% after 120 mins), whereas higher dosage would only agglomerated and reduced the removal rate. Besides, higher shaker speed gave the best performance of the catalyst. As for the AN initial concentration, the lowest value gave the highest

AN removal due to the lower amount of competitive molecules (78% removal at 5 ppm AN). Additionally, kinetic analysis was performed to fit the experimental data. Pseudo-first and pseudo-second models were compared and R^2-value suggested the pseudo-second order gave the best fit to describe the kinetics of AN removal. Further improvement is required to ensure higher AN concentration can be treated using this catalyst in real application.

Acknowledgement
The authors would like to acknowledge the Ministry of Higher Education Malaysia and University Malaysia Sarawak for the support of this work under the Fundamental Research Grant Scheme FRGS/1/2020/WAB02/UNIMAS/03/1.

References

[1] C. P. C. Bong, W. S. Ho, H. Hashim, et al., Review on the renewable energy and solid waste management policies towards biogas development in Malaysia. Renewable and Sustainable Energy Reviews 70 (2017) 988-998.. https://doi.org/s://doi.org/10.1016/j.rser.2016.12.004

[2] H. A. Aziz & S. F. Ramli, Recent development in sanitary landfilling and landfill leachate treatment in Malaysia, Int. J. Environ. Eng. 9 (2018), 201-229.. https://doi.org/s://doi.org/10.1504/IJEE.2018.097517

[3] K. Lin, Y. Zhu, Y. Zhang & H. Lin, Determination of ammonia nitrogen in natural waters: Recent advances and applications, Trends in Environ. Anal. Chem. 24 (2019) 1-13.. https://doi.org/s://doi.org/10.1016/j.teac.2019.e00073

[4] X. Jiang, H. Xia, L. Zhang, et al., Synthesis of copper- loaded activated carbon for enhancing the photocatalytic removal of methylene blue, J. Mol. Liq. (2018) 1-27.. https://doi.org/s://doi.org/10.1016/j.molliq.2018.09.087

[5] J. Moma & J. Baloyi, Modified titanium dioxide for photocatalytic applications, in K. Akhtar & S. Bahadar, Photocatalysts - Applications and Attributes, Intechopen Limited, London, 2018, pp. 37-56.. https://doi.org/s://doi.org/10.5772/intechopen.79374

[6] Y. Zhang, Q. Wang, Y. Zongsheng, G. Ma & Y. Zheng, Visible-light-mediated copper photocatalysis for organic syntheses, Beilstein J. Org. Chem. 17 (2021) 2520-2542.. https://doi.org/s://doi.org/10.3762/bjoc.17.169

[7] S. B. Barrabas, R. Baini, N. M. Sutan, et al., HC-SCR: NOx reduction using Mn and Cu catalysts impregnated in coconut and palm kernel shell activated carbon, MATEC Web Conf. 87 (2017) 1-6.. https://doi.org/s://doi.org/10.1051/matecconf/20178703004

[8] C. P. M. de Oliveira, A. L. A. Lage, D. C. S. Martins, et al., High surface area TiO2 nanoparticles: impact of carboxylporphyrin sensitizers in the photocatalytic activity, Surfaces & Interfaces 21 (2020) 1-8.. https://doi.org/s://doi.org/10.1016/j.surfin.2020.100774

[9] C. C. de Souza, L. Z. M. de Souza, M. Yilmaz, et al., Activated carbon of Coriandrum sativum for adsorption of methylene blue: Equilibrium and kinetic modeling, J. Cleaner Mat. 3 (2022) 1-10.. https://doi.org/s://doi.org/10.1016/j.clema.2022.100052

[10] I. Yakub & J. McGregor, Catalytic reduction of nitric oxide with hydrogen using carbon-supported d-metal catalysts, Waste & Biomass Valor. 13 (2022) 1665-1680.. https://doi.org/s://doi.org/10.1007/s12649-021-01623-7

[11] M. A. El-Nemr, A. El-Nemr, M. A. Hassaan, et al., Microporous activated carbon from Pisum sativum Pods using various activation methods and tested for adsorption of acid orange 7 dye from water, Mol. 27 (2022) 1-37.. https://doi.org/s://doi.org/10.3390/molecules27154840

Keyword Index

About the Editors

Dr Nik Abdul Hadi bin Sapiaa@Md Nordin (N.A.H.M. Nordin) graduated with chemical engineering degree with honors from Universiti Teknologi Malaysia (UTM) on 2011. Right after, he continues his doctorate study in UTM under fast-track program. In 2015, he graduated his PhD focused on the development of zeolitic imidazole framework 8 (ZIF-8) mixed matrix membrane for CO2 separation. He then served as postdoctoral fellow where he was involved in developing composite ZIF-8 with graphene oxide for gas separation and antibacterial UF membrane. He is currently serving in Universiti Teknologi PETRONAS (UTP) as a senior lecturer in Chemical Engineering Department and researcher in Centre of Research in Ionic Liquid (CORIL) UTP. His current research interest focuses on the ionic liquid modified fillers for gas separation, membrane integration, process simulations, and technoeconomics analysis. He is actively involved with industrial projects focuses on material development for hydrogen production via electrolysis and CO2 direct air capture via adsorption. Aside from research activity, he is also responsible for undergraduates Process Simulation SYMMETRY® Module Development with collaboration with SLB Digital. Since 2014, he has published 185 research articles, cited 3558 times with h-index of 33.

Ts. Dr. Lam Man Kee is an Associate Professor in the Chemical Engineering Department at Universiti Teknologi PETRONAS and serves as the Deputy Head of the HICoE—Centre for Biofuel and Biochemical Research at the same university. His research interests include microalgae cultivation for biomass and biofuel production, CO_2 capture, wastewater treatment, and nanoparticle production. To date, Dr. Lam has published over 100 Q1 and Q2 journal papers, accumulating 11,000 citations and achieving an H-index of 58, as recorded in the SCOPUS database. From 2021 to 2024, he was recognized as one of the global Top 2% Scientists in a study conducted by Stanford University. His research on microalgae cultivation for biofuel production has also attracted significant industrial attention, leading to an industrial grant and consultancy project with PETRONAS to scale up the process. Dr. Lam has successfully secured 13 research grants as Principal Investigator, with a total funding amount of RM 1.4 million. He currently serves as an Editor for *Energy Conversion and Management: X*, published by Elsevier.

ChM Ts Dr Khairulazhar Jumbri is an Associate Professor with a Professional Technologist title at the Department of Fundamental and Applied Sciences, Universiti Teknologi PETRONAS. He received his PhD degree in Physical Chemistry (including Computational Chemistry) from Universiti Putra Malaysia in 2015. His PhD study was funded by National Science Fellowship (NSF), from Ministry of Science, Technology and Innovation (MOSTI) Malaysia. In terms of research, his areas of specializations are related to oil & gas such as carbon capture and separation (CCS), material development, contaminant management and ionic liquids. ChM Ts Dr Khairulazhar received numerous research product awards obtained through various competitions at international and national exhibitions. His products have been submitted for intellectual properties submissions through patent. For his professional memberships, he is currently the associate member of the Royal Society of Chemistry (RSC), UK and American Chemical Society (ACS), USA. Accordingly, he has published many research papers in recognized publishers such as RSC, Elsevier and MDPI including book chapters. Previously he has been selected by the Council for the Lindau Nobel Laureate Meetings to attend the 67th Lindau Nobel Laureate Meeting (dedicated to Chemistry) in 2017.

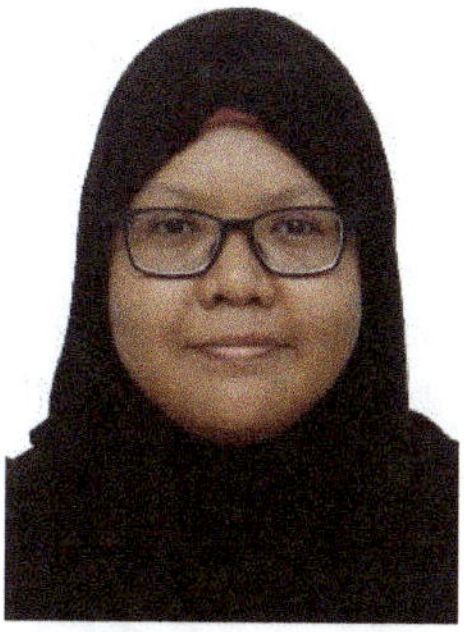

Dr Syaza Izyanni Ahmad is a lecturer in Chemical Engineering Department, UTP, an accredited trainer under HRDCorp, and currently a member of Centre for Advanced Process Safety (CAPS). She is an Associate Member of Institution of Chemical Engineers (IChemE), Graduate Technologists of Malaysia Board of Technologists (MBOT) and Graduate Engineer of Board of Engineer Malaysia (BEM). She has PhD in inherent safety assessment and since then, she has actively embarked in research related to safety considerations of green technologies, risk assessment on hydrogen production and distribution process, safety considerations on fusion processes, GHG emission, GHG accounting, and GHG emission reduction initiatives as well as calculation of fugitive emissions. Recently, she was certified as the ISO 14064 GHG Accounting Lead Verifier.

Ts Dr Bamidele Victor Ayodele is a Senior Lecturer at the Department of Chemical Engineering, Universiti Teknologi PETRONAS (UTP), Malaysia and actively involved in research at the Centre of Carbon Capture Utilization, and Storage, Institute of Sustainable Energy and Research, UTP. His research focus on energy stainability and decarbonization, CO2 utilization, process modelling, optimization using machine learning, and catalysis for energy applications. He has published over 100 articles indexed in Web of Science with H-index of 28 and citation of 2440 (Google Scholar). He serves on the Editorial Board of Frontiers in Energy Research.

www.ingramcontent.com/pod-product-compliance
Lightning Source LLC
LaVergne TN
LVHW052206200926
848481LV00001B/45